—

The Rock Physics Handbook, Third Edition

Responding to the latest developments in rock physics research, this popular reference book has been thoroughly updated while retaining its comprehensive coverage of the fundamental theory, concepts, and laboratory results. It brings together the vast literature from the field to address the relationships between geophysical observations and the underlying physical properties of Earth materials – including water, hydrocarbons, gases, minerals, rocks, ice, magma, and methane hydrates. This third edition includes expanded coverage of topics such as effective medium models, viscoelasticity, attenuation, anisotropy, electrical-elastic cross relations, and highlights applications in unconventional reservoirs. Appendices have been enhanced with new materials and properties, while worked examples (supplemented by online datasets and MATLAB codes) enable readers to implement the workflows and models in practice. This significantly revised edition will continue to be the go-to reference for students and researchers interested in rock physics, near-surface geophysics, seismology, and for professionals in the oil and gas industries.

Gary Mavko is Professor Emeritus at the Department of Geophysics, Stanford University. He was awarded Honorary Membership by the Society of Exploration Geophysicists (SEG) in 2001, was named the spring SEG Distinguished Lecturer in 2006, and shared the ENI medal for New Frontiers in Hydrocarbons in 2014. He is also coauthor of *Quantitative Seismic Interpretation* (Cambridge University Press, 2005).

Tapan Mukerji is Professor at the Department of Energy Resources Engineering, Stanford University. He received the SEG's Karcher Award in 2000, and shared the 2014 ENI award for pioneering innovations in theoretical and practical rock physics for seismic reservoir characterization. He is also coauthor of *Quantitative Seismic Interpretation* (Cambridge University Press, 2005) and *Value of Information in the Earth Sciences* (Cambridge University Press, 2015).

Jack Dvorkin is Research Fellow and Program Leader at King Fahd University of Petroleum and Minerals, the Kingdom of Saudi Arabia. He was awarded Honorary Membership by the SEG in 2014, and shared the 2014 ENI medal for New Frontiers in Hydrocarbons. He is also coauthor of *Seismic Reflections of Rock Properties* (Cambridge University Press, 2014).

"Rock physics finds order in complexity. In this book, the authors' diverse and deep experience across this subject allows them to link theory to measurements to practical models, providing an orderly and useful toolbox for even the seasoned practitioner. Just opening it up to a random page will yield some nugget of interest."

Joe Stefani, Chevron

"This third edition provides a significant update with an expanded coverage of topics and discussion. Geophysicists and students will value this concisely written book as an excellent and handy reference with well-summarized theoretical and empirical relations of rock physics, and clearly presented assumptions and limitations. I would highly recommend keeping it on your shelf."

Chen Guo, Chang'an University, China

"*The Rock Physics Handbook* offers a condensed synopsis of the most relevant theoretical concepts related to elasticity, effective medium models, fluid flow, electrical properties, and seismic wave propagation, accompanied with a comprehensive chapter on rock physics empirical relationships. It provides a balanced combination of theory and practice and is a must-have reference for academics and practitioners in petrophysics, geomechanics, and quantitative seismic interpretation."

Juan-Mauricio Florez, BHP

The Rock Physics
Handbook, Third Edition

Gary Mavko
Stanford University, California

Tapan Mukerji
Stanford University, California

Jack Dvorkin
King Fahd University of Petroleum and Minerals, Saudi Arabia

CAMBRIDGE
UNIVERSITY PRESS

CAMBRIDGE
UNIVERSITY PRESS

University Printing House, Cambridge CB2 8BS, United Kingdom

One Liberty Plaza, 20th Floor, New York, NY 10006, USA

477 Williamstown Road, Port Melbourne, VIC 3207, Australia

314–321, 3rd Floor, Plot 3, Splendor Forum, Jasola District Centre, New Delhi – 110025, India

79 Anson Road, #06–04/06, Singapore 079906

Cambridge University Press is part of the University of Cambridge.

It furthers the University's mission by disseminating knowledge in the pursuit of education, learning, and research at the highest international levels of excellence.

www.cambridge.org
Information on this title: www.cambridge.org/9781108420266
DOI: 10.1017/9781108333016

First edition published 1998
Second edition published 2009
Third edition published 2020

A catalogue record for this publication is available from the British Library.

Library of Congress Cataloging-in-Publication Data
Names: Mavko, Gary, 1949– author. | Mukerji, Tapan, 1965– author. | Dvorkin, Jack, 1953– author.
Title: The rock physics handbook / Gary Mavko (Stanford University, California), Tapan Mukerji (,Stanford University, California), Jack Dvorkin (King Fahd University of Petroleum and Minerals, Saudi Arabia).
Description: Third edition | New York : 2019. | Includes bibliographical references and index.
Identifiers: LCCN 2019019465 | ISBN 9781108420266
Subjects: LCSH: Rocks. | Geophysics.
Classification: LCC QE431.6.P5 M38 2019 | DDC 552/.06–dc23
LC record available at https://lccn.loc.gov/2019019465

ISBN 978-1-108-42026-6 Hardback

Additional resources for this publication at [www.cambridge.org/9781108420266].

Contents

Preface to the Third Edition   *page* xi

1   Basic Tools   1

1.1   The Fourier Transform   1
1.2   The Hilbert Transform and Analytic Signal   7
1.3   The Laplace Transform   10
1.4   Statistics and Probability   12
1.5   Coordinate Transformations   21
1.6   Tensor Properties and Operations   25

2   Elasticity and Hooke's Law   37

2.1   Elastic Moduli: Isotropic Form of Hooke's Law   37
2.2   Anisotropic Form of Hooke's Law   44
2.3   Thomsen's Notation for Weak Elastic Anisotropy   57
2.4   Sayers' Simplified Notation for Weak VTI Anisotropy   61
2.5   Tsvankin's Extended Thomsen Parameters for Orthorhombic Media   62
2.6   Third-Order Nonlinear Elasticity   64
2.7   Effective Stress Properties of Rocks   66
2.8   Stress-Induced Anisotropy in Rocks   70
2.9   Strain Components and Equations of Motion in Cylindrical
       and Spherical Coordinate Systems   80
2.10   Deformation of Inclusions and Cavities in Elastic Solids   81
2.11   Deformation of a Circular Hole: Borehole Stresses   96
2.12   Eshelby's General Solution for Ellipsoidal Inclusions   102
2.13   Mohr's Circles   109
2.14   Static and Dynamic Moduli   112
2.15   Stress Intensity Factors   115

3   Seismic Wave Propagation   121

3.1   Seismic Velocities   121
3.2   Phase, Group, and Energy Velocities   123
3.3   NMO in Isotropic and Anisotropic Media   126
3.4   Impedance, Reflectivity, and Transmissivity   133
3.5   Reflectivity and Amplitude Variations with Offset (AVO) in Isotropic
       Media   136
3.6   Plane-Wave Reflectivity in Anisotropic Media   144
3.7   Elastic Impedance   154
3.8   Viscoelasticity and Q 160
3.9   Kramers–Kronig Relations between Velocity Dispersion and Q 179
3.10   Waves in Layered Media: Full-Waveform Synthetic Seismograms   181
3.11   Waves in Layered Media: Stratigraphic Filtering and Velocity Dispersion   185
3.12   Waves in Layered Media: Frequency-Dependent Anisotropy, Dispersion,
       and Attenuation   189
3.13   Scale-Dependent Seismic Velocities in Heterogeneous Media   197
3.14   Scattering Attenuation   201
3.15   Waves in Cylindrical Rods: the Resonant Bar   206
3.16   Waves in Boreholes   211

4   Effective Elastic Media: Bounds and Mixing Laws   220

4.1   Voigt and Reuss Bounds   220
4.2   Hashin–Shtrikman–Walpole Bounds   222
4.3   Improvements on the Hashin–Shtrikman–Walpole Bounds   229
4.4   Wood's Formula   234
4.5   Voigt–Reuss–Hill Average Moduli Estimate   235
4.6   Composite with Uniform Shear Modulus   236
4.7   Rock and Pore Compressibilities and Some Pitfalls   238
4.8   General Comments on Inclusion-Based Estimation Models   241
4.9   Mori–Tanaka Formulation for Effective Moduli   248
4.10   Kuster and Toksöz Formulation for Effective Moduli   249
4.11   Self-Consistent Approximations of Effective Moduli   251
4.12   Differential Effective Medium Model   255
4.13   Hudson's Model for Cracked Media   258
4.14   Eshelby-Cheng Model for Cracked Anisotropic Media   267
4.15 T-Matrix Inclusion Models for Effective Moduli   269
4.16   Elastic Constants in Finely Layered Media: Backus Average   275
4.17   Elastic Constants in Finely Layered Media: General Layer Anisotropy   279

4.18   Poroelastic and Viscoelastic Backus Average   280
4.19   Seismic Response to Fractures   285
4.20   Bound-Filling Models   290
4.21   Effective Moduli of Polycrystalline Aggregates   295
4.22   Comments on the Representative Volume Element   304
4.23   A Few Theorems on Strain in an Effective Medium   306

5   Granular Media   309

5.1   Packing and Sorting of Spheres and Irregular Particles   309
5.2   Percolation of Random Ellipsoidal Packs   325
5.3   Thomas–Stieber–Yin-Marion Model for Sand–Shale Systems   329
5.4   Particle Size and Sorting   335
5.5   Random Spherical Grain Packings: Contact Models and Effective Moduli   337
5.6   Ordered Spherical Grain Packings: Effective Moduli   364

6   Fluid Effects on Wave Propagation   367

6.1   Biot's Velocity Relations   367
6.2   Geertsma–Smit Approximations of Biot's Relations   372
6.3   Gassmann's Relations: Isotropic Form   374
6.4   Bounds on Fluid Substitution   384
6.5   Brown and Korringa's Generalized Gassmann Equations for Mixed
       Mineralogy   385
6.6   Fluid Substitution in Anisotropic Rocks   387
6.7   Generalized Gassmann's Equations for Composite Porous Media   390
6.8   Solid Substitution of Frame or Pore-Filling Phases   393
6.9   Fluid Substitution in Thinly Laminated Reservoirs   407
6.10   BAM: Marion's Bounding Average Method   412
6.11   Mavko–Jizba Squirt Relations   413
6.12   Extension of Mavko–Jizba Squirt Relations for All Frequencies   416
6.13   Biot–Squirt Model   419
6.14   Chapman *et al.* Squirt Model   421
6.15   Anisotropic Squirt   423
6.16   Common Features of Fluid-Related Velocity Dispersion Mechanisms   427
6.17   Dvorkin-Mavko Attenuation Model   433
6.18   Partial and Multiphase Saturations   438
6.19   Partial Saturation: White and Dutta-Odé Model for Velocity
       Dispersion and Attenuation   444

6.20   Velocity Dispersion, Attenuation, and Dynamic Permeability
        in Heterogeneous Poroelastic Media   449
6.21   Waves in a Pure Viscous Fluid   455
6.22   Physical Properties of Gases and Fluids   457

7   Empirical Relations   474

7.1   Velocity–Porosity Models: Critical Porosity and Modified
        Upper and Lower Bounds   474
7.2   Velocity–Porosity Models: Wyllie's Time Average and Geertsma's
        Empirical Relations for Compressibility   478
7.3   Vernik–Kachanov Clastics Models   480
7.4   Velocity–Porosity Models: Raymer–Hunt–Gardner Relations   482
7.5   Velocity–Porosity–Clay Models: Han's Empirical Relations for Shaly
        Sandstones   485
7.6   Velocity–Porosity–Clay Models: Tosaya's Empirical Relations for Shaly
        Sandstones   486
7.7   Velocity–Porosity–Clay Models: Castagna's Empirical Relations for
        Velocities   487
7.8 V_P–V_S–Density Models: Brocher's Compilation   488
7.9 V_P–V_S Relations   492
7.10   Velocity–Density Relations   508
7.11   Eaton and Bowers Pore-Pressure Relations   511
7.12   Kan and Swan Pore-Pressure Relations   512
7.13   Attenuation and Quality Factor Relations   513
7.14   Velocity–Porosity–Strength Relations   515
7.15   Birch's Law   517
7.16   Kerogen Properties   519

8   Flow and Diffusion   525

8.1   Darcy's Law   525
8.2   Viscous Flow   533
8.3   Capillary Forces   538
8.4   Kozeny–Carman Relation for Flow   542
8.5   Permeability Relations with S_{wi} 557
8.6   Permeability of Fractured Formations   560
8.7   Diffusion and Filtration: Special Cases   562
8.8   Heavy Oil Viscosity and Shear Modulus   564

8.9   Particles and Bubbles in a Viscoelastic Background   567
8.10   Viscosity of Silicate Melts and Magma   571

9   Electrical Properties   577

9.1   Bounds and Effective Medium Models   577
9.2   Velocity Dispersion and Attenuation   582
9.3   Empirical Relations for Composites   585
9.4   Electrical Conductivity in Porous Rocks   588
9.5   Cross-Property Bounds and Relations between Elastic and Electrical
       Parameters   596
9.6   Brine Resistivity   608
9.7   Dielectric Constants   611

Appendices   613

A.1   Typical Rock Properties   613
A.2   Conversions   634
A.3   Physical Constants   638
A.4   Moduli and Density of Common Minerals   641
A.5   Properties of Mantle Minerals   650
A.6   Properties of Melts, Magma, and Igneous Rocks   653
A.7   Velocities and Moduli of Ice, Methane Hydrate, and Sea Water   659
A.8   Physical Properties of Common Gases   663
A.9   Velocity, Moduli, and Density of Carbon Dioxide   669
A.10   Standard Temperature and Pressure   672

       References   673
       Index   716

Preface to the Third Edition

In the decade since publication of the Second Edition of the *Rock Physics Handbook*, research and application of rock physics have continued to thrive and mature. Today, areas of study where rock physics plays a role include oil and gas exploration and production, geothermal resources, ground water management, near-surface geophysics, earthquake seismology, geodynamics, geotechnical engineering, rock mechanics, glaciology, and even lunar and planetary sciences.

While preparing the Third Edition, our objective was still to summarize in a convenient form many of the commonly needed theoretical and empirical relations of rock physics. Our approach was to present results, with a few of the key assumptions and limitations, and almost never any derivations. Our intention was to create a quick reference and not a textbook. Hence, we chose to encapsulate a broad range of topics rather than to give in-depth coverage of a few. Even so, there are many topics that we have not addressed. While we have summarized the assumptions and limitations of each result, we hope that the brevity of our discussions does not give the impression that application of any rock physics result to real rocks is free of pitfalls. We assume that the reader will generally be aware of the various topics, and, if not, we provide a few references to the more complete descriptions in books and journals.

The Third Edition contains 121 sections on basic mathematical tools, elasticity theory, wave propagation, effective media, elasticity and poroelasticity, granular media, and pore-fluid flow and diffusion, plus overviews of dispersion mechanisms, fluid substitution, and V_P–V_S relations. The book also presents empirical results derived from reservoir rocks, sediments, and granular media, as well as tables of data on minerals, magmas and melts, gases, ice, hydrates, and an atlas of reservoir rock properties. The emphasis is still on elastic and seismic topics, though the discussion of electrical and cross seismic–electrical relations has grown. An associated website (http://srb.stanford.edu/books) offers MATLAB codes for many of the models and results described in the Third Edition.

In this Third Edition, Chapter 1 has been expanded to include the Laplace Transform and basic tensor operations. Chapter 2 includes new discussions on two-dimensional elasticity, anisotropic Poisson's ratio and Young's modulus, Eshelby's theory, and stress intensity factors. Chapter 3 has a more extensive discussion of viscoelasticity, relevant to both wave dispersion and attenuation, and time–domain creep. Chapter 4 has gained a more comprehensive discussion of elastic bounds, an expanded discussion of effective medium inclusion models, and estimates and bounds on the effective moduli of polycrystalline rock aggregates. Chapter 5, on granular media, has an expanded discussion of sand–clay mixing models, ideal and non-ideal particle mixture properties, and percolation in particle packs. Chapter 6 has an expanded discussion of petroleum oil, gas, and condensates, and new elastic discussions on fluid and solid substitution, rigorous bounds on fluid substitution, and elastic mineral substitution. Chapter 7 has gained new elastic models of clastic sediments, including shales. Chapter 8 has a new discussion on extensions to the Kozeny–Carman relations, diffusion of pressure, flow properties of viscous fluids containing particles and bubbles, and viscosity of silicate melts and magmas. The Appendix has been expanded to include new plots of representative rock properties, and extensive new tables of properties of crustal and mantle rocks and minerals, melts and magmas, fresh-water and sea-water ices, polycrystalline salt, and commonly used dimensionless numbers.

This *Handbook* is complementary to a number of other excellent books. For in-depth discussions of specific rock physics topics, we recommend *Fundamentals of Rock Mechanics*, 4th edition, by Jaeger, Cook, and Zimmerman; *Compressibility of Sandstones* by Zimmerman; *Physical Properties of Rocks: Fundamentals and Principles of Petrophysics* by Schön; *Acoustics of Porous Media* by Bourbié, Coussy, and Zinszner; *Introduction to the Physics of Rocks* by Guéguen and Palciauskas; *A Geoscientist's Guide to Petrophysics* by Zinszner and Pellerin; *Theory of Linear Poroelasticity* by Wang; *Underground Sound* by White; *Mechanics of Composite Materials* by Christensen; *Viscoelastic Materials* by Lakes; *The Theory of Composites* by Milton; *Random Heterogeneous Materials* by Torquato; *Rock Physics and Phase Relations* edited by Ahrens; *Offset Dependent Reflectivity: Theory and Practice of AVO Analysis* edited by Castagna and Backus; *Seismic Petrophysics in Quantitative Interpretation* by Vernik; *Seismic Amplitude: An Interpreter's Handbook* by Simm and Bacon; and *Seismic Reflections of Rock Properties* by Dvorkin, Gutierrez, and Grana.

We wish to thank the students, scientific staff, and industrial affiliates of the Stanford Rock Physics and Borehole Geophysics (SRB) project for many valuable comments and insights. Mustafa Al Ibrahim contributed heavily to the section on kerogen properties; Li Teng contributed to the chapter on anisotropic AVOZ; and Ran Bachrach contributed to the chapter on dielectric properties. We benefited extensively from discussions with Amos Nur, Nishank Saxena, Vishal Das, Sabrina Aliyeva, Yu Xia, Priyanka Dutta, Nattavadee Srisutthiyakorn, Abdulla Kerimov, Uri Wollner, Iris (Yunfei) Yang, Wei Chu, Rayan Kanfar, Obai Shaikh, Salma Alsinan, Abrar AlAbbad, Salman Alkhater, Chen Guo, and Juan Pablo Daza. Juan Pablo also created and graciously shared the digital rock image shown on the cover. We thank the readers who pointed out errors in the previous editions.

We hope you find this updated edition useful.

Gary Mavko, Tapan Mukerji, and Jack Dvorkin

1 Basic Tools

1.1 The Fourier Transform

Synopsis

The **Fourier transform** of $f(x)$ is defined as

$$F(s) = \int_{-\infty}^{\infty} f(x)e^{-i2\pi xs}dx \tag{1.1.1}$$

The inverse Fourier transform is given by

$$f(x) = \int_{-\infty}^{\infty} F(s)e^{+i2\pi xs}ds \tag{1.1.2}$$

> **Caution:**
>
> In this section, we use the symbol s to represent the frequency, or Fourier transform variable, following the terminology of Bracewell (1965). When x refers to time, the symbol f is often used for frequency instead of s; when x refers to space, the symbol k is often used for spatial frequency instead of s. Do not confuse this with the use of s when defining the Laplace transform (Section 1.3) or S when defining the analytic signal (Section 1.2).

Evenness and Oddness

A function $E(x)$ is *even* if $E(x) = E(-x)$. A function $O(x)$ is *odd* if $O(x) = -O(-x)$.

The Fourier transform has the following properties for even and odd functions:

- *Even functions.* The Fourier transform of an even function is even. A *real even* function transforms to a *real even* function. An *imaginary even* function transforms to an *imaginary even* function.
- *Odd functions.* The Fourier transform of an odd function is odd. A *real odd* function transforms to an *imaginary odd* function. An *imaginary odd* function transforms to

a *real odd* function (i.e., the "realness" flips when the Fourier transform of an odd function is taken).

Any function can be expressed in terms of its even and odd parts:

$$f(x) = E(x) + O(x) \tag{1.1.3}$$

where

$$E(x) = \frac{1}{2}[f(x) + f(-x)] \tag{1.1.4}$$

$$O(x) = \frac{1}{2}[f(x) - f(-x)] \tag{1.1.5}$$

As a consequence, a real function $f(x)$ has a Fourier transform that is *Hermitian*, $F(s) = F^*(-s)$, where $*$ refers to the complex conjugate.

For a more general complex function, $f(x)$, we can tabulate some additional properties (Bracewell, 1965):

$$f(x) \Leftrightarrow F(s)$$
$$f^*(x) \Leftrightarrow F^*(-s)$$
$$f^*(-x) \Leftrightarrow F^*(s)$$
$$f(-x) \Leftrightarrow F(-s)$$
$$2 \operatorname{Re} f(x) \Leftrightarrow F(s) + F^*(-s)$$
$$2 \operatorname{Im} f(x) \Leftrightarrow F(s) - F^*(-s)$$
$$f(x) + f^*(-x) \Leftrightarrow 2 \operatorname{Re} F(s)$$
$$f(x) - f^*(-x) \Leftrightarrow 2 \operatorname{Im} F(s)$$

The **convolution** of two functions $f(x)$ and $g(x)$ is

$$f(x) * g(x) = \int_{-\infty}^{+\infty} f(z)g(x-z)dz = \int_{-\infty}^{+\infty} f(x-z)g(z)dz \tag{1.1.6}$$

Convolution Theorem

If $f(x)$ has the Fourier transform $F(s)$, and $g(x)$ has the Fourier transform $G(s)$, then the Fourier transform of the convolution $f(x)*g(x)$ is the product $F(s)\ G(s)$.

The **cross-correlation** of two functions $f(x)$ and $g(x)$ is

$$f^*(x) \star g(x) = \int_{-\infty}^{+\infty} f^*(z-x)g(z)\,dz = \int_{-\infty}^{+\infty} f^*(z)g(z+x)\,dz \tag{1.1.7}$$

where f^* refers to the complex conjugate of f. When the two functions are the same, $f^*(x) \star f(x)$ is called the **autocorrelation** of $f(x)$.

Energy Spectrum

The modulus squared of the Fourier transform $|F(s)|^2 = F(s) F^*(s)$ is sometimes called the **energy spectrum** or simply the **spectrum**.

If $f(x)$ has the Fourier transform $F(s)$, then the autocorrelation of $f(x)$ has the Fourier transform $|F(s)|^2$.

Phase Spectrum

The Fourier transform $F(s)$ is most generally a complex function, which can be written as

$$F(s) = |F(s)|e^{i\varphi(s)} = \text{Re } F(s) + i\text{Im } F(s) \tag{1.1.8}$$

where $|F|$ is the modulus and φ is the **phase**, given by

$$\varphi(s) = \tan^{-1}[\text{Im } F(s)/\text{Re } F(s)] \tag{1.1.9}$$

The function $\varphi(s)$ is sometimes also called the **phase spectrum**.

Obviously, both the modulus and phase must be known to completely specify the Fourier transform $F(s)$ or its transform pair in the other domain, $f(x)$. Consequently, an infinite number of functions $f(x) \Leftrightarrow F(s)$ are consistent with a given spectrum $|F(s)|^2$.

The zero-phase equivalent function (or zero-phase equivalent wavelet) corresponding to a given spectrum is

$$F(s) = |F(s)| \tag{1.1.10}$$

$$f(x) = \int_{-\infty}^{\infty} |F(s)|e^{+i2\pi xs} \, ds \tag{1.1.11}$$

which implies that $F(s)$ is real and $f(x)$ is Hermitian. In the case of zero-phase *real* wavelets, both $F(s)$ and $f(x)$ are real even functions.

The **minimum-phase** equivalent function or wavelet corresponding to a spectrum is the unique one that is both *causal* and *invertible*. A simple way to compute the minimum-phase equivalent of a spectrum $|F(s)|^2$ is to perform the following steps (Claerbout, 1992):
(1) Take the logarithm, $B(s) = \ln |F(s)|$.
(2) Take the Fourier transform, $B(s) \Rightarrow b(x)$.
(3) Multiply $b(x)$ by zero for $x < 0$ and by 2 for $x > 0$. If done numerically, leave the values of b at zero and the Nyquist frequency unchanged.
(4) Transform back, giving $B(s) + i\varphi(s)$, where φ is the desired phase spectrum.
(5) Take the complex exponential to yield the minimum-phase function:
$F_{\text{mp}}(s) = \exp[B(s) + i\varphi(s)] = |F(s)|e^{i\varphi(s)}$.
(6) The causal minimum-phase wavelet is the Fourier transform of $F_{\text{mp}}(s) \Rightarrow f_{\text{mp}}(x)$.
Another way of saying this is that the phase spectrum of the minimum-phase equivalent function is the Hilbert transform (see Section 1.2 on the Hilbert transform) of the log of the energy spectrum.

Sampling Theorem

A function $f(x)$ is said to be *band limited* if its Fourier transform is nonzero only within a finite range of frequencies, $|s| < s_c$, where s_c is sometimes called the *cut-off frequency*. The function $f(x)$ is fully specified if sampled at equal spacing not exceeding $\Delta x = 1/(2s_c)$. Equivalently, a time series sampled at interval Δt adequately describes the frequency components out to the *Nyquist frequency* $f_N = 1/(2\Delta t)$.

The numerical process to recover the intermediate points between samples is to convolve with the *sinc function*:

$$2s_c \, \mathrm{sinc}(2s_c x) = 2s_c \sin(\pi 2s_c x)/\pi 2s_c x \tag{1.1.12}$$

where

$$\mathrm{sinc}(x) \equiv \frac{\sin(\pi x)}{\pi x} \tag{1.1.13}$$

which has the properties:

$$\left.\begin{array}{l} \mathrm{sinc}(0) = 1 \\ \mathrm{sinc}(n) = 0 \end{array}\right\} n = \text{nonzero integer} \tag{1.1.14}$$

The Fourier transform of $\mathrm{sinc}(x)$ is the boxcar function $\Pi(s)$:

$$\Pi(s) = \begin{cases} 0 & |s| > \dfrac{1}{2} \\[2mm] 1/2 & |s| = \dfrac{1}{2} \\[2mm] 1 & |s| < \dfrac{1}{2} \end{cases} \tag{1.1.15}$$

Plots of the function $\mathrm{sinc}(x)$ and its Fourier transform $\Pi(s)$ are shown in Figure 1.1.1. One can see from the convolution and similarity theorems below that convolving with $2s_c \, \mathrm{sinc}(2s_c x)$ is equivalent to multiplying by $\Pi(s/2s_c)$ in the

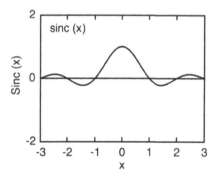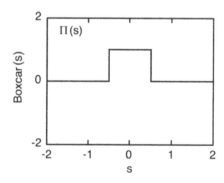

Figure 1.1.1 Plots of the function $\mathrm{sinc}(x)$ and its Fourier transform $\Pi(s)$

frequency domain (i.e., zeroing out all frequencies $|s| > s_c$ and passing all frequencies $|s| < s_c$).

Spectral Estimation and Windowing

It is often desirable in rock physics and seismic analysis to estimate the spectrum of a wavelet or seismic trace. The most common, easiest, and, in some ways, the worst way is simply to chop out a piece of the data, take the Fourier transform, and find its magnitude. The problem is related to sample length. If the true data function is $f(t)$, a small sample of the data can be thought of as

$$f_{\text{sample}}(t) = \begin{cases} f(t), & a \leq t \leq b \\ 0, & \text{elsewhere} \end{cases} \tag{1.1.16}$$

or

$$f_{\text{sample}}(t) = f(t)\,\Pi\left(\frac{t - \frac{1}{2}(a+b)}{b-a}\right) \tag{1.1.17}$$

where $\Pi(t)$ is the boxcar function previously discussed. Taking the Fourier transform of the data sample gives

$$F_{\text{sample}}(s) = F(s) * [|b-a|\,\text{sinc}\left((b\text{–}a)s\right) e^{-i\pi(a+b)s}] \tag{1.1.18}$$

More generally, we can "window" the sample with some other function $\omega(t)$:

$$f_{\text{sample}}(t) = f(t)\omega(t) \tag{1.1.19}$$

yielding

$$F_{\text{sample}}(s) = F(s) * W(s) \tag{1.1.20}$$

Thus, the estimated spectrum can be highly contaminated by the Fourier transform of the window, often with the effect of smoothing and distorting the spectrum due to the convolution with the window spectrum $W(s)$. This can be particularly severe in the analysis of ultrasonic waveforms in the laboratory, where often only the first 1 to $1\frac{1}{2}$ cycles are included in the window. The solution to the problem is not easy, and there is extensive literature (e.g., Jenkins and Watts, 1968; Marple, 1987) on spectral estimation. Our advice is to be aware of the artifacts of windowing and to experiment to determine the sensitivity of the results, such as the spectral ratio or the phase velocity, to the choice of window size and shape.

Table 1.1.1 *Fourier transform theorems*

Theorem	x-domain		s-domain
Similarity	$f(ax)$	$\Leftrightarrow$	$\dfrac{1}{\|a\|}F\left(\dfrac{s}{a}\right)$
Addition	$f(x) + g(x)$	$\Leftrightarrow$	$F(s) + G(s)$
Shift	$f(x - a)$	$\Leftrightarrow$	$e^{-i2\pi as}F(s)$
Modulation	$f(x)\cos\omega x$	$\Leftrightarrow$	$\dfrac{1}{2}F\left(s - \dfrac{\omega}{2\pi}\right) + \dfrac{1}{2}F\left(s + \dfrac{\omega}{2\pi}\right)$
Convolution	$f(x)*g(x)$	$\Leftrightarrow$	$F(s)\,G(s)$
Autocorrelation	$f(x)*f^*(-x)$	$\Leftrightarrow$	$\|F(s)\|^2$
Derivative	$f'(x)$	$\Leftrightarrow$	$i2\pi s F(s)$

Table 1.1.2 *Some additional theorems*

Derivative of convolution	$\dfrac{d}{dx}[f(x)*g(x)] = f'(x)*g(x) = f(x)*g'(x)$
Rayleigh	$\displaystyle\int_{-\infty}^{\infty}\|f(x)\|^2dx = \int_{-\infty}^{\infty}\|F(s)\|^2ds$
Power	$\displaystyle\int_{-\infty}^{\infty}f(x)g^*(x)dx = \int_{-\infty}^{\infty}F(s)G^*(s)ds$
(f and g real)	$\displaystyle\int_{-\infty}^{\infty}f(x)g(-x)dx = \int_{-\infty}^{\infty}F(s)G(s)ds$

Fourier Transform Theorems

Tables 1.1.1 and 1.1.2 summarize some useful theorems (Bracewell, 1965). If $f(x)$ has the Fourier transform $F(s)$, and $g(x)$ has the Fourier transform $G(s)$, then the Fourier transform pairs in the x-domain and the s-domain are as shown in the tables. Table 1.1.3 lists some useful Fourier transform pairs. The delta function $\delta(x)$ is defined as:

$$\delta(x) = 0,\ x\neq0\ ;\quad \int_{-\infty}^{\infty}\delta(x)dx = 1 \tag{1.1.21}$$

The Hartley Transform

The Hartley transform, $F_H(\omega)$, of a function $f(t)$ is defined as

$$F_H(\omega) = \frac{1}{\sqrt{2\pi}}\int_{-\infty}^{\infty}f(t)\,\text{cas}(\omega t)dt, \tag{1.1.22}$$

where

$$\text{cas}(t) = \cos(t) + \sin(t) = \sqrt{2}\,\sin(t + \pi/4) = \sqrt{2}\,\cos(t - \pi/4) \tag{1.1.23}$$

Table 1.1.3 *Some Fourier transform pairs*

x-domain	Transform domain	
$\sin \pi x$	$\dfrac{i}{2}\left[\delta\left(s+\dfrac{1}{2}\right) - \delta\left(s-\dfrac{1}{2}\right)\right]$	
$\cos \pi x$	$\dfrac{1}{2}\left[\delta\left(s+\dfrac{1}{2}\right) + \delta\left(s-\dfrac{1}{2}\right)\right]$	
$\delta(x)$	1	
$\operatorname{sinc}(x)$	$\Pi(s)$	
$\operatorname{sinc}^2(x)$	$\Lambda(s)$	
$e^{-\pi x^2}$	$e^{-\pi s^2}$	
$-1/\pi x$	$i\operatorname{sgn}(s)$	
$\dfrac{x_0}{x_0^2 + x^2}$	$\pi \exp(-2\pi x_0|s|)$	
$e^{-|x|}$	$\dfrac{2}{1 + (2\pi s)^2}$	
$|x|^{-1/2}$	$|s|^{-1/2}$	

The Hartley transform is its own inverse. The Hartley transform is related to the Fourier transform $F(\omega)$ by

$$F(\omega) = \frac{F_H(\omega) + F_H(-\omega)}{2} - i\frac{F_H(\omega) - F_H(-\omega)}{2} \tag{1.1.24}$$

In other words, the real and imaginary parts of the Fourier transform are the even and odd parts of the Hartley transform.

1.2   The Hilbert Transform and Analytic Signal

Synopsis

The **Hilbert transform** of $f(x)$ is defined as

$$F_{\mathrm{Hi}}(x) = \frac{1}{\pi}\int_{-\infty}^{\infty} \frac{f(z)\,dz}{z - x} \tag{1.2.1}$$

which can be expressed as a convolution of $f(x)$ with $(-1/\pi x)$ by

$$F_{\text{Hi}} = -\frac{1}{\pi x} * f(x)$$

The Fourier transform of $(-1/\pi x)$ is $(i \, \text{sgn}(s))$, that is, $+i$ for positive s and $-i$ for negative s. Hence, applying the Hilbert transform keeps the Fourier amplitudes or spectrum the same but changes the phase. Under the Hilbert transform, $\sin(kx)$ is converted to $\cos(kx)$, and $\cos(kx)$ is converted to $-\sin(kx)$. Similarly, the Hilbert transforms of even functions are odd functions and vice versa.

Caution:

In this section, we use the symbol S to represent the analytic signal. Do not confuse this with the use of s when defining the Fourier transform (Section 1.1) or the Laplace transform (Section 1.3).

The inverse of the Hilbert transform is itself the Hilbert transform with a change of sign:

$$f(x) = -\frac{1}{\pi} \int_{-\infty}^{\infty} \frac{F_{\text{Hi}}(z)dz}{z - x} \tag{1.2.2}$$

or

$$f(x) = -\left(-\frac{1}{\pi x}\right) * F_{\text{Hi}} \tag{1.2.3}$$

The **analytic signal** associated with a real function, $f(t)$, is the complex function

$$S(t) = f(t) - iF_{\text{Hi}}(t) \tag{1.2.4}$$

As discussed in the following, the Fourier transform of $S(t)$ is zero for negative frequencies.

The **instantaneous envelope** of the analytic signal is

$$E(t) = \sqrt{f^2(t) + F_{\text{Hi}}^2(t)} \tag{1.2.5}$$

The **instantaneous phase** of the analytic signal is

$$\begin{aligned}\varphi(t) &= \tan^{-1}[-F_{\text{Hi}}(t)/f(t)] \\ &= \text{Im}\left[\ln\left(S(t)\right)\right]\end{aligned} \tag{1.2.6}$$

The **instantaneous frequency** of the analytic signal is

$$\omega = \frac{d\varphi}{dt} = \text{Im}\left[\frac{d}{dt}\ln(S)\right] = \text{Im}\left(\frac{1}{S}\frac{dS}{dt}\right) \tag{1.2.7}$$

Claerbout (1992) has suggested that ω can be numerically more stable if the denominator is rationalized and the functions are locally smoothed, as in the following equation:

$$\bar{\omega} = \mathrm{Im} \left[\frac{\left\langle S^*(t) \frac{dS(t)}{dt} \right\rangle}{\left\langle S^*(t)S(t) \right\rangle} \right] \tag{1.2.8}$$

where $\langle \cdot \rangle$ indicates some form of running average or smoothing.

Causality

The **impulse response, $I(t)$**, of a real physical system must be causal, that is,

$$I(t) = 0, \text{ for } t < 0 \tag{1.2.9}$$

The Fourier transform $T(f)$ of the impulse response of a causal system is sometimes called the **transfer function**:

$$T(f) = \int_{-\infty}^{\infty} I(t)e^{-i2\pi ft} dt \tag{1.2.10}$$

$T(f)$ must have the property that the real and imaginary parts are Hilbert transform pairs, that is, $T(f)$ will have the form

$$T(f) = G(f) + iB(f) \tag{1.2.11}$$

where $B(f)$ is the Hilbert transform of $G(f)$:

$$B(f) = \frac{1}{\pi} \int_{-\infty}^{\infty} \frac{G(z)dz}{z - f} \tag{1.2.12}$$

$$G(f) = -\frac{1}{\pi} \int_{-\infty}^{\infty} \frac{B(z)dz}{z - f} \tag{1.2.13}$$

Similarly, if we reverse the domains, an analytic signal of the form

$$S(t) = f(t) - iF_{\mathrm{Hi}}(t) \tag{1.2.14}$$

must have a Fourier transform that is zero for negative frequencies. In fact, one convenient way to implement the Hilbert transform of a real function is by performing the following steps:
(1) Take the Fourier transform.
(2) Multiply the Fourier transform by zero for $f < 0$.
(3) Multiply the Fourier transform by 2 for $f > 0$.
(4) If done numerically, leave the samples at $f = 0$ and the Nyquist frequency unchanged.
(5) Take the inverse Fourier transform.
The imaginary part of the result will be the negative Hilbert transform of the real part.

1.3   The Laplace Transform

The **Laplace transform** of a function $f(t)$ is defined as

$$\mathcal{L}[f(t)] \equiv F(s) \equiv \int_0^\infty f(t)\, e^{-st} dt \tag{1.3.1}$$

where the transform variable s may be complex. If $F(s)$ is an analytic function of s except at isolated singular points, then the inverse transform is given by

$$f(t) = \mathcal{L}^{-1}[F(s)] = \frac{1}{2\pi i} \int_{\gamma-i\infty}^{\gamma+i\infty} F(s)\, e^{st} ds \tag{1.3.2}$$

where $\mathrm{Re}(s) = \gamma$ is to the right of all singularities of $F(s)$. In geophysics, the Laplace transform is often used to model transient behavior of materials with viscoelastic creep.

Laplace Transform Theorems

Table 1.3.1 shows some useful Laplace transform theorems (Bracewell, 1965). Table 1.3.2 lists some Laplace transform pairs in the t-domain and the s-domain, given that $f(t)$ has the Laplace transform $F(s)$, and $g(t)$ has the Laplace transform $G(s)$.

Table 1.3.1 *Some Laplace transform theorems*

Theorem	
	$\mathcal{L}\big(f(t)\big) = F(s)$
Derivative	$\mathcal{L}\left(\dfrac{df(t)}{dt}\right) = sF(s) - f(0)$
Integral	$\mathcal{L}\left(\displaystyle\int_0^t f(\tau)d\tau\right) = \dfrac{F(s)}{s}$
Multiplication by time	$\mathcal{L}\big(tf(t)\big) = -\dfrac{d}{ds}F(s)$
Division by time	$\mathcal{L}\left(\dfrac{f(t)}{t}\right) = \displaystyle\int_s^\infty F(u)\, du$
Multiplication by exponential	$\mathcal{L}\big(e^{-at}f(t)\big) = F(s+a)$
Time shift	$\mathcal{L}\big(f(t-\tau)\mathcal{H}(t-\tau)\big) = e^{-\tau s}F(s)$
Scale change	$\mathcal{L}\big(f(at)\big) = \dfrac{1}{a}F(s/a)$
Convolution	$\mathcal{L}\left(\displaystyle\int_0^t f(\tau)g(t-\tau)d\tau\right) = F(s) \cdot G(s)$

Table 1.3.2 *Some Laplace transform pairs*

Time domain function $f(t)$, $t \geq 0$	Laplace transform $F(s)$
$\delta(t)$	1
$\delta(t-a)$	e^{-as}
$\mathcal{H}(t)^1$	$\dfrac{1}{s}$
$\mathcal{H}(t-a)^1$	$\dfrac{1}{s}e^{-as}$
$t\,\mathcal{H}(t)^1$	$\dfrac{1}{s^2}$
te^{-at}	$\dfrac{1}{(s+a)^2}$
e^{-at}	$\dfrac{1}{s+a}$
$\dfrac{t^{n-1}e^{-at}}{(n-1)!}$	$\dfrac{1}{(s+a)^n}$
$\dfrac{1}{a}(1-e^{-at})$	$\dfrac{1}{s(s+a)}$
$\dfrac{e^{bt}-e^{at}}{b-a}$	$\dfrac{1}{(s-a)(s-b)}$, for $a \neq b$
$\dfrac{be^{bt}-ae^{at}}{b-a}$	$\dfrac{s}{(s-a)(s-b)}$, for $a \neq b$
$\cosh(at)$	$\dfrac{s}{s^2-a^2}$
$\sinh(at)$	$\dfrac{a}{s^2-a^2}$
$\cos(at)$	$\dfrac{s}{s^2+a^2}$
$\sin(at)$	$\dfrac{a}{s^2+a^2}$
$\dfrac{\sin(at)}{t}$	$\tan^{-1}(a/s)$
$t\cos(at)$	$\dfrac{s^2-a^2}{(s^2+a^2)^2}$
$t\sin(at)$	$\dfrac{2as}{(s^2+a^2)^2}$
$e^{-at}\cos(\omega t)$	$\dfrac{s+a}{(s+a)^2+\omega^2}$
$e^{-at}\sin(\omega t)$	$\dfrac{\omega}{(s+a)^2+\omega^2}$
t^n	$\dfrac{\Gamma(n+1)}{s^{n+1}}$
$\dfrac{t^{n-1}}{(n-1)!}$	$\dfrac{1}{s^n}$
$t^n e^{-at}$	$\dfrac{\Gamma(n+1)}{(s+a)^{n+1}}$

^{1} $\mathcal{H}(t)$ is the Heaviside step function.

1.4   Statistics and Probability

Synopsis

The **sample mean**, m, of a set of n data points, x_i, is the arithmetic average of the data values:

$$m = \frac{1}{n} \sum_{i=1}^{n} x_i \tag{1.4.1}$$

The **median** is the midpoint of the observed values if they are arranged in increasing order. The sample variance, σ^2, is the average squared difference of the observed values from the mean:

$$\sigma^2 = \frac{1}{n} \sum_{i=1}^{n} (x_i - m)^2 \tag{1.4.2}$$

(An unbiased estimate of the **population variance** is often found by dividing the sum given above by $(n-1)$ instead of by n.)

The **standard deviation**, σ, is the square root of the variance, while the **coefficient of variation** is σ/m. The **mean deviation**, α, is

$$\alpha = \frac{1}{n} \sum_{i=1}^{n} |x_i - m| \tag{1.4.3}$$

Regression

When trying to determine whether two different data variables, x and y, are related, we often estimate the **correlation coefficient**, ρ, given by (e.g., Young, 1962)

$$\rho = \frac{\frac{1}{n} \sum_{i=1}^{n} (x_i - m_x)(y_i - m_y)}{\sigma_x \sigma_y}, \quad \text{where } |\rho| \leq 1 \tag{1.4.4}$$

where σ_x and σ_y are the standard deviations of the two distributions and m_x and m_y are their means. The correlation coefficient gives a measure of how close the points come to falling along a straight line in a scatter plot of x versus y. $|\rho| = 1$ if the points lie perfectly along a line, and $|\rho| < 1$ if there is scatter about the line. The numerator of this expression is the **sample covariance**, C_{xy}, which is defined as

$$C_{xy} = \frac{1}{n} \sum_{i=1}^{n} (x_i - m_x)(y_i - m_y) \tag{1.4.5}$$

It is important to remember that the correlation coefficient is a measure of the *linear* relation between x and y. If they are related in a nonlinear way, the correlation coefficient will be misleadingly small.

The simplest recipe for estimating the linear relation between two variables, x and y, is **linear regression**, in which we assume a relation of the form:

$$y = ax + b \tag{1.4.6}$$

The coefficients that provide the best fit to the measured values of y, in the least-squares sense, are

$$a = \rho \frac{\sigma_y}{\sigma_x}, \quad b = m_y - am_x \tag{1.4.7}$$

More explicitly,

$$a = \frac{n \sum x_i y_i - \left(\sum x_i\right)\left(\sum y_i\right)}{n \sum x_i^2 - \left(\sum x_i\right)^2}, \qquad \text{slope} \tag{1.4.8}$$

$$b = \frac{\left(\sum y_i\right)\left(\sum x_i^2\right) - \left(\sum x_i y_i\right)\left(\sum x_i\right)}{n \sum x_i^2 - \left(\sum x_i\right)^2}, \quad \text{intercept} \tag{1.4.9}$$

The scatter or variation of y-values around the regression line can be described by the sum of the squared errors as

$$E^2 = \sum_{i=1}^{n} (y_i - \hat{y}_i)^2 \tag{1.4.10}$$

where $\hat{y}_i$ is the value predicted from the regression line. This can be expressed as a variance around the regression line as

$$\hat{\sigma}_y^2 = \frac{1}{n} \sum_{i=1}^{n} (y_i - \hat{y}_i)^2 \tag{1.4.11}$$

The square of the correlation coefficient ρ is the coefficient of determination, often denoted by r^2, which is a measure of the regression variance relative to the total variance in the variable y, expressed as

$$r^2 = \rho^2 = 1 - \frac{\text{variance of } y \text{ around the linear regression}}{\text{total variance of } y}$$
$$= 1 - \frac{\sum_{i=1}^{n} (y_i - \hat{y}_i)^2}{\sum_{i=1}^{n} (y_i - m_y)^2} = 1 - \frac{\hat{\sigma}_y^2}{\sigma_y^2} \tag{1.4.12}$$

The inverse relation is

$$\hat{\sigma}_y^2 = \sigma_y^2 (1 - r^2) \tag{1.4.13}$$

Often, when doing a linear regression the choice of dependent and independent variables is arbitrary. The form above treats x as independent and exact and assigns errors to y. It often makes just as much sense to reverse their roles, and we can find a regression of the form

$$x = a'y + b' \tag{1.4.14}$$

Generally $a \neq 1/a'$ unless the data are perfectly correlated. In fact, the correlation coefficient, ρ, can be written as $\rho = \sqrt{aa'}$.

The coefficients of the linear regression among three variables of the form

$$z = a + bx + cy \tag{1.4.15}$$

are given by

$$
\begin{aligned}
b &= \frac{C_{xz}C_{yy} - C_{xy}C_{yz}}{C_{xx}C_{yy} - C_{xy}^2} \\
c &= \frac{C_{xx}C_{yz} - C_{xy}C_{xz}}{C_{xx}C_{yy} - C_{xy}^2} \\
a &= m_z - m_x b - m_y c
\end{aligned}
\tag{1.4.16}
$$

The coefficients of the n-dimensional linear regression of the form

$$z = c_0 + c_1 x_1 + c_2 x_2 + \cdots + c_n x_n \tag{1.4.17}$$

are given by

$$
\begin{bmatrix} c_0 \\ c_1 \\ c_2 \\ \vdots \\ c_n \end{bmatrix}
= (M^T M)^{-1} M^T
\begin{bmatrix} z^{(1)} \\ z^{(2)} \\ \vdots \\ z^{(k)} \end{bmatrix}
\tag{1.4.18}
$$

where the k sets of independent variables form columns 2:$(n + 1)$ in the matrix M:

$$
M =
\begin{bmatrix}
1 & x_1^{(1)} & x_2^{(1)} & \cdots & x_n^{(1)} \\
1 & x_1^{(2)} & x_2^{(2)} & \cdots & x_n^{(2)} \\
\vdots & \vdots & \vdots & & \vdots \\
1 & x_1^{(k)} & x_2^{(k)} & \cdots & x_n^{(k)}
\end{bmatrix}
\tag{1.4.19}
$$

Variogram and Covariance Function

In geostatistics, variables are modeled as random fields, $X(\mathbf{u})$, where $\mathbf{u}$ is the spatial position vector. Spatial correlation between two random fields $X(\mathbf{u})$ and $Y(\mathbf{u})$ is described by the cross-covariance function $C_{XY}(\mathbf{h})$, defined by

$$C_{XY}(\mathbf{h}) = E\{[X(\mathbf{u}) - m_X(\mathbf{u})][Y(\mathbf{u} + \mathbf{h}) - m_Y(\mathbf{u} + \mathbf{h})]\} \tag{1.4.20}$$

where $E\{\}$ denotes the expectation operator, m_X and m_Y are the means of X and Y, and **h** is called the lag vector. For stationary fields, m_X and m_Y are independent of position. When X and Y are the same function, the equation represents the auto-covariance function $C_{XX}(\mathbf{h})$. A closely related measure of two-point spatial variability is the semivariogram, $\gamma(\mathbf{h})$. For stationary random fields $X(\mathbf{u})$ and $Y(\mathbf{u})$, the cross-variogram $2\gamma_{XY}(\mathbf{h})$ is defined as

$$2\gamma_{XY}(\mathbf{h}) = E\{[X(\mathbf{u}+\mathbf{h}) - X(\mathbf{u})][Y(\mathbf{u}+\mathbf{h}) - Y(\mathbf{u})]\} \qquad (1.4.21)$$

When X and Y are the same, the equation represents the variogram of $X(\mathbf{h})$. For a stationary random field, the variogram and covariance function are related by

$$\gamma_{XX}(\mathbf{h}) = C_{XX}(0) - C_{XX}(\mathbf{h}) \qquad (1.4.22)$$

where $C_{XX}(0)$ is the stationary variance of X.

Distributions

A population of n elements possesses $\binom{n}{r}$ (pronounced "n choose r") different sub-populations of size $r \leq n$, where

$$\binom{n}{r} = \frac{(n)_r}{r!} = \frac{n(n-1)\cdots(n-r+1)}{1\cdot 2 \cdots (r-1)r} = \frac{n!}{r!(n-r)!} \qquad (1.4.23)$$

Expressions of this kind are called **binomial coefficients**. Another way to say this is that a subset of r elements can be chosen in $\binom{n}{r}$ different ways from the original set.

The **binomial distribution** gives the probability of n successes in N independent trials, if p is the probability of success in any one trial. The binomial distribution is given by

$$f_{N,p}(n) = \binom{N}{n} p^n (1-p)^{N-n} \qquad (1.4.24)$$

The mean of the binomial distribution is given by

$$m_{\mathrm{b}} = Np \qquad (1.4.25)$$

and the variance of the binomial distribution is given by

$$\sigma_{\mathrm{b}}^2 = Np(1-p) \qquad (1.4.26)$$

The **Poisson distribution** is the limit of the binomial distribution as $N \to \infty$ and $p \to 0$ so that $\lambda = Np$ remains finite. The Poisson distribution is given by

$$f_\lambda(n) = \frac{\lambda^n e^{-\lambda}}{n!} \qquad (1.4.27)$$

The Poisson distribution is a discrete probability distribution and expresses the probability of n events occurring during a given interval of time if the events have an average (positive real) rate λ, and the events are independent of the time since the previous event. n is a non-negative integer.

The mean of the Poisson distribution is given by

$$m_P = \lambda \tag{1.4.28}$$

and the variance of the Poisson distribution is given by

$$\sigma_P^2 = \lambda \tag{1.4.29}$$

The **uniform** distribution is given by

$$f(x) = \begin{cases} \dfrac{1}{b-a}, & a \leq x \leq b \\ 0, & \text{elsewhere} \end{cases} \tag{1.4.30}$$

The mean of the uniform distribution is

$$m = \frac{(a+b)}{2} \tag{1.4.31}$$

and the standard deviation of the uniform distribution is

$$\sigma = \frac{|b-a|}{\sqrt{12}} \tag{1.4.32}$$

The **Gaussian** or **normal** distribution is given by

$$f(x) = \frac{1}{\sigma\sqrt{2\pi}} e^{-(x-m)^2/2\sigma^2} \tag{1.4.33}$$

where σ is the standard deviation and m is the mean. The mean deviation for the Gaussian distribution is

$$\alpha = \sigma\sqrt{\frac{2}{\pi}} \tag{1.4.34}$$

When m measurements are made of n quantities, the situation is described by the n-dimensional **multivariate Gaussian** probability density function (pdf):

$$f_n(x) = \frac{1}{(2\pi)^{n/2}|\mathbf{C}|^{1/2}} \exp\left[-\frac{1}{2}(x-m)^T \mathbf{C}^{-1}(x-m)\right] \tag{1.4.35}$$

where $x^{\mathrm{T}} = (x_1, x_2, \ldots, x_n)$ is the vector of observations, $m^{\mathrm{T}} = (m_1, m_2, \ldots, m_n)$ is the vector of means of the individual distributions, and $\mathbf{C}$ is the covariance matrix:

$$\mathbf{C} = [C_{ij}] \tag{1.4.36}$$

where the individual covariances, C_{ij}, are as defined in equation 1.4.20. Notice that this reduces to the single variable normal distribution when $n = 1$.

When the natural logarithm of a variable, $x = \ln(y)$, is normally distributed, it belongs to a **lognormal distribution** expressed as

$$f(y) = \frac{1}{\sqrt{2\pi}\beta y} \exp\left[-\frac{1}{2}\left(\frac{\ln(y) - \alpha}{\beta}\right)^2\right] \tag{1.4.37}$$

where α is the mean and β^2 is the variance. The relations among the arithmetic and logarithmic parameters are

$$m = e^{\alpha + \beta^2/2}, \quad \alpha = \ln(m) - \beta^2/2 \tag{1.4.38}$$

$$\sigma^2 = m^2(e^{\beta^2} - 1), \quad \beta^2 = \ln(1 + \sigma^2/m^2) \tag{1.4.39}$$

The **truncated exponential distribution** is given by

$$P(x) = \begin{cases} \frac{1}{X}\exp(-x/X) & x \geq 0 \\ 0, & x < 0 \end{cases} \tag{1.4.40}$$

The mean m and variance σ^2 of the exponential distribution are given by

$$m = X \tag{1.4.41}$$

$$\sigma^2 = X^2 \tag{1.4.42}$$

The **truncated exponential distribution** and the **Poisson distribution** are closely related. Many random events in nature are described by the Poisson distribution. For example, in a well log, consider that the occurrence of a flooding surface is a random event. Then in every interval of log of length D, the probability of finding exactly n flooding surfaces is

$$P_{\lambda D}(n) = \frac{(\lambda D)^n}{n!}\exp(-\lambda D) \tag{1.4.43}$$

The mean number of occurrences is λD, where λ is the mean number of occurrences per unit length. The mean thickness between events is $D/(\lambda D) = 1/\lambda$. The interval thicknesses d between flooding events are governed by the truncated exponential

$$P(d) = \begin{cases} \lambda\exp(-\lambda d), & x \geq 0 \\ 0, & x < 0 \end{cases} \tag{1.4.44}$$

the mean of which is $1/\lambda$.

The **logistic distribution** is a continuous distribution with probability density function given by

$$P(x) = \frac{e^{-(x-\mu)/s}}{s[1 + e^{-(x-\mu)/s}]^2} = \frac{1}{4s} \text{sech}^2\left(\frac{x-\mu}{2s}\right) \tag{1.4.45}$$

The mean m and variance σ^2 of the logistic distribution are given by

$$m = \mu$$
$$\sigma^2 = \frac{\pi^2 s^2}{3} \tag{1.4.46}$$

The **Weibull distribution** is a continuous distribution with probability density function

$$P(x) = \begin{cases} \frac{k}{\lambda}\left(\frac{x}{\lambda}\right)^{k-1} e^{-(x/\lambda)^k}, & x \geq 0 \\ 0, & x < 0 \end{cases} \tag{1.4.47}$$

where $k > 0$ is the *shape* parameter and $\lambda > 0$ is the *scale* parameter for the distribution. The mean m and variance σ^2 of the logistic distribution are given by

$$m = \lambda\Gamma\left(1 + \frac{1}{k}\right)$$
$$\sigma^2 = \lambda^2\Gamma\left(1 + \frac{2}{k}\right) - m^2 \tag{1.4.48}$$

where Γ is the gamma function. The Weibull distribution is often used to describe failure rates. If the failure rate decreases over time, $k < 1$; if the failure rate is constant, $k = 1$; and if the failure rate increases over time, $k > 1$. When $k = 3$, the Weibull distribution is a good approximation to the normal distribution, and when $k = 1$, the Weibull distribution reduces to the exponential distribution.

Monte Carlo Simulations

Statistical simulation is a powerful numerical method for tackling many probabilistic problems. One of the steps is to draw samples X_i from a desired probability distribution function $F(x)$. This procedure is often called *Monte Carlo simulation*, a term made popular by physicists working on the bomb during the Second World War. In general, Monte Carlo simulation can be a very difficult problem, especially when X is multivariate with correlated components, and $F(x)$ is a complicated function. For the simple case of a univariate X and a completely known $F(x)$ (either analytically or numerically), drawing X_i amounts to first drawing *uniform* random variates U_i between 0 and 1, and then evaluating the inverse of the desired cumulative distribution function (CDF) at these U_i: $X_i = F^{-1}(U_i)$. The inverse of the CDF is called the

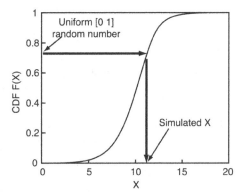

Figure 1.4.1 Schematic of a univariate Monte Carlo simulation

quantile function. When $F^{-1}(X)$ is not known analytically, the inversion can be easily done by table-lookup and interpolation from the numerically evaluated or nonparametric CDF derived from data. A graphical description of univariate Monte Carlo simulation is shown in Figure 1.4.1.

Many modern computer packages have random number generators not only for uniform and normal (Gaussian) distributions but also for a large number of well-known, analytically defined statistical distributions.

Often Monte Carlo simulations require simulating correlated random variables (e.g., V_P, V_S). Correlated random variables may be simulated sequentially, making use of the chain rule of probability, which expresses the joint probability density in terms of the conditional and marginal densities: $P(V_P, V_S) = P(V_S|V_P) P(V_P)$.

A simple procedure for correlated Monte Carlo draws is as follows:
- draw a V_P sample from the V_P distribution;
- compute a V_S from the drawn V_P and the V_P–V_S regression; and
- add to the computed V_S a random Gaussian error with zero mean and variance equal to the variance of the residuals from the V_P–V_S regression.

This gives a random, correlated (V_P, V_S) sample. A better approach is to draw V_S from the conditional distributions of V_S for each given V_P value, instead of using a simple V_P–V_S regression. Given sufficient V_P–V_S training data, the conditional distributions of V_S for different V_P can be computed.

Bootstrap

"Bootstrap" is a very powerful computational statistical method for assigning measures of accuracy to statistical estimates (e.g., Efron and Tibshirani, 1993). The general idea is to make multiple replicates of the data by drawing from the original data *with replacement*. Each of the bootstrap data replicates has the *same number* of samples as the original data set, but since they are drawn with replacement, some of the data may be represented more than once in the replicate data sets, while others might be missing. Drawing with replacement from the data is equivalent to Monte Carlo realizations from the empirical CDF. The statistic of interest is computed on all of the replicate bootstrap data sets. The distribution of the bootstrap replicates of the statistic is a measure of uncertainty of the statistic.

Drawing bootstrap replicates from the empirical CDF in this way is sometimes termed **nonparametric bootstrap**. In **parametric bootstrap** the data are first modeled by a parametric CDF (e.g., a multivariate Gaussian), and then bootstrap data replicates are drawn from the modeled CDF. Both simple bootstrap techniques previously described assume the data are independent and identically distributed. More sophisticated bootstrap techniques exist that can account for data dependence.

Statistical Classification

The goal in statistical classification problems is to predict the class of an unknown sample based on observed attributes or features of the sample. For example, the observed attributes could be P and S impedances, and the classes could be lithofacies, such as sand and shale. The classes are sometimes also called states, outcomes, or responses, while the observed features are called the predictors. Discussions concerning many statistical classification methods may be found in Fukunaga (1990), Duda *et al.* (2000), Hastie *et al.* (2001), and Bishop (2006).

There are two general types of statistical classification: **supervised classification**, which uses a training data set of samples for which both the attributes and classes have been observed; and **unsupervised learning**, for which only the observed attributes are included in the data. Supervised classification uses the training data to devise a classification rule, which is then used to predict the classes for new data, where the attributes are observed but the outcomes are unknown. Unsupervised learning tries to cluster the data into groups that are statistically different from each other based on the observed attributes.

A fundamental approach to the supervised classification problem is provided by Bayesian decision theory. Let x denote the univariate or multivariate input attributes, and let c_j, $j = 1, \ldots, N$ denote the N different states or classes. The Bayes formula expresses the probability of a particular class given an observed x as

$$P(c_j|x) = \frac{P(x, c_j)}{P(x)} = \frac{P(x|c_j)P(c_j)}{P(x)} \qquad (1.4.49)$$

where $P(x, c_j)$ denotes the joint probability of x and c_j; $P(x|c_j)$ denotes the conditional probability of x given c_j; and $P(c_j)$ is the prior probability of a particular class. Finally, $P(x)$ is the marginal or unconditional pdf of the attribute values across all N states. It can be written as

$$P(X) = \sum_{j=1}^{N} P(X|c_j)P(c_j) \qquad (1.4.50)$$

and serves as a normalization constant. The class-conditional pdf, $P(x \mid c_j)$, is estimated from the training data or from a combination of training data and forward models.

The Bayes classification rule says:

classify as class c_k if $P(c_k \mid x) > P(c_j \mid x)$ for all $j \neq k$.

This is equivalent to choosing c_k when $P(x \mid c_k)P(c_k) > P(x \mid c_j)P(c_j)$ for all $j \neq k$.

The Bayes classification rule is the optimal one that minimizes the misclassification error and maximizes the posterior probability. Bayes classification requires estimating the complete set of class-conditional pdfs $P(x|c_j)$. With a large number of attributes, getting a good estimate of the highly multivariate pdf becomes difficult.

Classification based on traditional **discriminant analysis** uses only the means and covariances of the training data, which are easier to estimate than the complete pdfs. When the input features follow a multivariate Gaussian distribution, discriminant classification is equivalent to Bayes classification, but with other data distribution patterns, the discriminant classification is not guaranteed to maximize the posterior probability. Discriminant analysis classifies new samples according to the minimum Mahalanobis distance to each class cluster in the training data. The Mahalanobis distance is defined as follows:

$$M^2 = (\mathbf{x} - \boldsymbol{\mu_j})^{\mathrm{T}} \Sigma^{-1} (\mathbf{x} - \boldsymbol{\mu_j}) \tag{1.4.51}$$

where $\mathbf{x}$ is the sample feature vector (measured attribute), $\boldsymbol{\mu_j}$ are the vectors of the attribute means for the different categories or classes, and Σ is the training data covariance matrix. The Mahalanobis distance can be interpreted as the usual Euclidean distance scaled by the covariance, which decorrelates and normalizes the components of the feature vector. When the covariance matrices for all the classes are taken to be identical, the classification gives rise to linear discriminant surfaces in the feature space. More generally, with different covariance matrices for each category, the discriminant surfaces are quadratic. If the classes have unequal prior probabilities, the term $\ln[P(\text{class})_j]$ is added to the right-hand side of the equation for the Mahalanobis distance, where $P(\text{class})_j$ is the prior probability for the jth class. Linear and quadratic discriminant classifiers are simple, robust classifiers and often produce good results, performing among the top few classifier algorithms.

1.5   Coordinate Transformations

Synopsis

It is often necessary to transform vector and tensor quantities in one coordinate system to another more suited to a particular problem. Consider two right-hand rectangular Cartesian coordinates (x, y, z) and (x', y', z') with the same origin, but with their axes rotated arbitrarily with respect to each other. The relative orientation of the two sets of axes is given by the direction cosines β_{ij}, where each element is defined as the cosine of the angle between the new i'-axis and the original j-axis. The variables β_{ij} constitute the elements of a 3×3 rotation matrix $[\beta]$. Thus, β_{23} is the cosine of the angle between the 2-axis of the primed coordinate system and the 3-axis of the unprimed coordinate system.

The general transformation law for tensors is

$$M'_{ABCD\ldots} = \beta_{Aa}\beta_{Bb}\beta_{Cc}\beta_{Dd}\cdots M_{abcd\ldots} \qquad (1.5.1)$$

where summation over repeated indices is implied. The left-hand subscripts (A, B, C, D, ...) on the βs match the subscripts of the transformed tensor $\mathbf{M'}$ on the left, and the right-hand subscripts (a, b, c, d, ...) match the subscripts of $\mathbf{M}$ on the right. Thus, vectors, which are first-rank tensors, transform as

$$v'_i = \beta_{ij}v_j \qquad (1.5.2)$$

or, in matrix notation, as

$$\begin{pmatrix} v'_1 \\ v'_2 \\ v'_3 \end{pmatrix} = \begin{pmatrix} \beta_{11} & \beta_{12} & \beta_{13} \\ \beta_{21} & \beta_{22} & \beta_{23} \\ \beta_{31} & \beta_{32} & \beta_{33} \end{pmatrix} \begin{pmatrix} v_1 \\ v_2 \\ v_3 \end{pmatrix} \qquad (1.5.3)$$

whereas second-rank tensors, such as stresses and strains, obey

$$\sigma'_{ij} = \beta_{ik}\beta_{jl}\sigma_{kl} \qquad (1.5.4)$$

or

$$[\sigma'] = [\beta]\,[\sigma]\,[\beta]^{\mathrm{T}} \qquad (1.5.5)$$

in matrix notation. Elastic stiffnesses and compliances are, in general, fourth-order tensors and hence transform according to

$$c'_{ijkl} = \beta_{ip}\beta_{jq}\beta_{kr}\beta_{ls}c_{pqrs} \qquad (1.5.6)$$

Often c_{ijkl} and s_{ijkl} are expressed as the 6×6 matrices using either the abbreviated 2-index **Voigt** or **Kelvin** notations, as defined in Section 2.2 on anisotropic elasticity. When using the 6×6 *Voigt* notation for elastic constants, C_{ij} and S_{ij}, the usual tensor transformation law is no longer valid, and the change of coordinates is more efficiently performed with the 6×6 **Bond transformation matrices M** and **N**, as explained in this section (Auld, 1990):

$$[C'] = [M]\,[C]\,[M]^{\mathrm{T}} \qquad (1.5.7)$$

$$[S'] = [N]\,[S]\,[N]^{\mathrm{T}} \qquad (1.5.8)$$

The elements of the 6×6 **M** and **N** matrices are given in terms of the direction cosines as follows:

$$\mathbf{M} = \begin{bmatrix} \beta_{11}^2 & \beta_{12}^2 & \beta_{13}^2 & 2\beta_{12}\beta_{13} & 2\beta_{13}\beta_{11} & 2\beta_{11}\beta_{12} \\ \beta_{21}^2 & \beta_{22}^2 & \beta_{23}^2 & 2\beta_{22}\beta_{23} & 2\beta_{23}\beta_{21} & 2\beta_{21}\beta_{22} \\ \beta_{31}^2 & \beta_{32}^2 & \beta_{33}^2 & 2\beta_{32}\beta_{33} & 2\beta_{33}\beta_{31} & 2\beta_{31}\beta_{32} \\ \beta_{21}\beta_{31} & \beta_{22}\beta_{32} & \beta_{23}\beta_{33} & \beta_{22}\beta_{33}+\beta_{23}\beta_{32} & \beta_{21}\beta_{33}+\beta_{23}\beta_{31} & \beta_{22}\beta_{31}+\beta_{21}\beta_{32} \\ \beta_{31}\beta_{11} & \beta_{32}\beta_{12} & \beta_{33}\beta_{13} & \beta_{12}\beta_{33}+\beta_{13}\beta_{32} & \beta_{11}\beta_{33}+\beta_{13}\beta_{31} & \beta_{11}\beta_{32}+\beta_{12}\beta_{31} \\ \beta_{11}\beta_{21} & \beta_{12}\beta_{22} & \beta_{13}\beta_{23} & \beta_{22}\beta_{13}+\beta_{12}\beta_{23} & \beta_{11}\beta_{23}+\beta_{13}\beta_{21} & \beta_{22}\beta_{11}+\beta_{12}\beta_{21} \end{bmatrix}$$

$$(1.5.9)$$

and

$$\mathbf{N} = \begin{bmatrix} \beta_{11}^2 & \beta_{12}^2 & \beta_{13}^2 & \beta_{12}\beta_{13} & \beta_{13}\beta_{11} & \beta_{11}\beta_{12} \\ \beta_{21}^2 & \beta_{22}^2 & \beta_{23}^2 & \beta_{22}\beta_{23} & \beta_{23}\beta_{21} & \beta_{21}\beta_{22} \\ \beta_{31}^2 & \beta_{32}^2 & \beta_{33}^2 & \beta_{32}\beta_{33} & \beta_{33}\beta_{31} & \beta_{31}\beta_{32} \\ 2\beta_{21}\beta_{31} & 2\beta_{22}\beta_{32} & 2\beta_{23}\beta_{33} & \beta_{22}\beta_{33}+\beta_{23}\beta_{32} & \beta_{21}\beta_{33}+\beta_{23}\beta_{31} & \beta_{22}\beta_{31}+\beta_{21}\beta_{32} \\ 2\beta_{31}\beta_{11} & 2\beta_{32}\beta_{12} & 2\beta_{33}\beta_{13} & \beta_{12}\beta_{33}+\beta_{13}\beta_{32} & \beta_{11}\beta_{33}+\beta_{13}\beta_{31} & \beta_{11}\beta_{32}+\beta_{12}\beta_{31} \\ 2\beta_{11}\beta_{21} & 2\beta_{12}\beta_{22} & 2\beta_{13}\beta_{23} & \beta_{22}\beta_{13}+\beta_{12}\beta_{23} & \beta_{11}\beta_{23}+\beta_{13}\beta_{21} & \beta_{22}\beta_{11}+\beta_{12}\beta_{21} \end{bmatrix}$$

$$(1.5.10)$$

The advantage of the Bond method for transforming stiffnesses and compliances is that it can be applied directly to the elastic constants given in 2-index notation, as they almost always are in handbooks and tables.

When using the 6×6 *Kelvin* notation (see Section 2.2) for elastic constants, $\hat{C}_{ij}$ and $\hat{S}_{ij}$, the Bond transformation can be replaced by the more symmetric form (Mehrabadi and Cowin, 1990):

$$\hat{\mathbf{C}}' = \hat{\mathbf{Q}}\hat{\mathbf{C}}\hat{\mathbf{Q}}^T \tag{1.5.11}$$

$$\hat{\mathbf{S}}' = \hat{\mathbf{Q}}\hat{\mathbf{S}}\hat{\mathbf{Q}}^T \tag{1.5.12}$$

$$\hat{\boldsymbol{\sigma}}' = \hat{\mathbf{Q}}\hat{\boldsymbol{\sigma}} \tag{1.5.13}$$

$$\hat{\boldsymbol{\varepsilon}}' = \hat{\mathbf{Q}}\hat{\boldsymbol{\varepsilon}} \tag{1.5.14}$$

$$\hat{Q} = \begin{bmatrix} \beta_{11}^2 & \beta_{12}^2 & \beta_{13}^2 & \sqrt{2}\beta_{12}\beta_{13} & \sqrt{2}\beta_{11}\beta_{13} & \sqrt{2}\beta_{11}\beta_{12} \\ \beta_{21}^2 & \beta_{22}^2 & \beta_{23}^2 & \sqrt{2}\beta_{22}\beta_{23} & \sqrt{2}\beta_{21}\beta_{23} & \sqrt{2}\beta_{22}\beta_{21} \\ \beta_{31}^2 & \beta_{32}^2 & \beta_{33}^2 & \sqrt{2}\beta_{33}\beta_{32} & \sqrt{2}\beta_{33}\beta_{31} & \sqrt{2}\beta_{31}\beta_{32} \\ \sqrt{2}\beta_{21}\beta_{31} & \sqrt{2}\beta_{22}\beta_{32} & \sqrt{2}\beta_{23}\beta_{33} & \beta_{22}\beta_{33}+\beta_{23}\beta_{32} & \beta_{21}\beta_{33}+\beta_{31}\beta_{23} & \beta_{21}\beta_{32}+\beta_{31}\beta_{22} \\ \sqrt{2}\beta_{11}\beta_{31} & \sqrt{2}\beta_{12}\beta_{32} & \sqrt{2}\beta_{13}\beta_{33} & \beta_{12}\beta_{33}+\beta_{32}\beta_{13} & \beta_{11}\beta_{33}+\beta_{13}\beta_{31} & \beta_{11}\beta_{32}+\beta_{31}\beta_{12} \\ \sqrt{2}\beta_{11}\beta_{21} & \sqrt{2}\beta_{12}\beta_{22} & \sqrt{2}\beta_{13}\beta_{23} & \beta_{12}\beta_{23}+\beta_{22}\beta_{13} & \beta_{11}\beta_{23}+\beta_{21}\beta_{13} & \beta_{11}\beta_{22}+\beta_{21}\beta_{12} \end{bmatrix}$$

$$(1.5.15)$$

Note that $\hat{Q}^{-1} = \hat{Q}^T$.

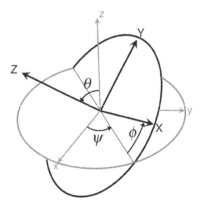

Figure 1.5.1. Euler angles. Original coordinate axes in gray. Rotated coordinate axes in black.

Euler Angles

Euler angles, θ, ψ, and ϕ, describe an arbitrary coordinate rotation in three dimensions (see Figure 1.5.1). Let x, y, and z be the original Cartesian coordinate axes, and X, Y, and Z be the axes after rotation. The new orientation can be accomplished by first applying a rotation about the z-axis by angle ψ, next applying a rotation about the X-axis by an angle θ, and finally applying a rotation about the Z-axis by an angle ϕ.

The coordinate rotation matrices corresponding to this sequence are (Helbig, 1994)

$$\mathbf{A}_\psi^{(3)} = \begin{bmatrix} \cos\psi & \sin\psi & 0 \\ -\sin\psi & \cos\psi & 0 \\ 0 & 0 & 1 \end{bmatrix} \tag{1.5.16}$$

$$\mathbf{A}_\theta^{(1)} = \begin{bmatrix} 1 & 0 & 0 \\ 0 & \cos\theta & \sin\theta \\ 0 & -\sin\theta & \cos\theta \end{bmatrix} \tag{1.5.17}$$

$$\mathbf{A}_\phi^{(3)} = \begin{bmatrix} \cos\phi & \sin\phi & 0 \\ -\sin\phi & \cos\phi & 0 \\ 0 & 0 & 1 \end{bmatrix} \tag{1.5.18}$$

The matrix of the compound set of rotations is

$$\mathbf{A} = \mathbf{A}_\phi^{(3)} \mathbf{A}_\theta^{(1)} \mathbf{A}_\psi^{(3)}$$
$$= \begin{bmatrix} \cos\psi\,\cos\phi - \sin\psi\,\cos\theta\,\sin\phi & \sin\psi\,\cos\phi + \cos\psi\,\cos\theta\,\sin\phi & \sin\theta\,\sin\phi \\ -\cos\psi\,\sin\phi - \sin\psi\,\cos\theta\,\cos\phi & -\sin\psi\,\sin\phi + \cos\psi\,\cos\theta\,\cos\phi & \sin\theta\,\cos\phi \\ \sin\psi\,\sin\theta & -\cos\psi\,\sin\theta & \cos\theta \end{bmatrix} \tag{1.5.19}$$

It can also be shown that (Helbig, 1994)

$$\tan\psi = -\frac{\beta_{31}}{\beta_{32}} \; ; \quad \cos\theta = \beta_{33} \; ; \quad \tan\phi = \frac{\beta_{13}}{\beta_{23}} \tag{1.5.20}$$

where β_{ij} are the coordinates of the three unit vectors defining the new coordinate system with respect to the old coordinate system – that is, the direction cosine of the new ith axis and the old jth axis.

Assumptions and Limitations

Coordinate transformations presuppose right-handed rectangular coordinate systems.

1.6   Tensor Properties and Operations

Synopsis

A tensor is a mathematical object that can be used to describe physical properties or the linear relation between two tensor physical properties. For example, the linear elastic stiffness tensor relates the stress tensor to the infinitesimal strain tensor (Hooke's law); the electrical conductivity tensor relates gradient of electrical potential to the vector of current density (ohm's law). Because they represent physical objects or physical properties, tensors have an existence independent of any coordinate system, yet are most conveniently described when quantified with respect to a particular coordinate system.

Tensors are classified by rank or order, according to the form of the coordinate transformation law they obey. In N-dimensional space, a tensor of order p has N^p components. In three dimensions, a tensor of order 0 has one component and is a scalar; a tensor of order 1 has three components and is a vector; a tensor of order 2 has nine components, and so on.

A tensor is positive definite if it is symmetric and all of its eigenvalues are positive.

Tensor Operations

The **Dyadic product** of symmetric second-order tensors $a_{ij} = a_{ji}$ and $b_{ij} = b_{ji}$ is the symmetric fourth-order tensor $C_{ijkl} = a_{ij} b_{kl}$ and is written as $\mathbf{C} = \mathbf{a} \otimes \mathbf{b}$.

Simple tensor contraction refers to the product of tensors with repeated summation over one shared index. For example, the contraction of second-order tensor $\mathbf{T}$ and vector $\mathbf{a}$ is the vector

$$\mathbf{Ta} = T_{ij} a_j \mathbf{e}_i, \tag{1.6.1}$$

where $\mathbf{e}_i$ are unit vectors along the Cartesian axes. The contraction of two second-order tensors is given by

$$\mathbf{TS} = T_{ij} S_{jk} (\mathbf{e}_i \otimes \mathbf{e}_k). \tag{1.6.2}$$

The result of a simple tensor contraction is a tensor whose order is *two* less than the sum of the original orders.

Double tensor contraction is a product that sums over two indices, so that the order of the result is *four* less than the sum of the original orders. For example, the double contraction of two second-order tensors is a scalar:

$$\mathbf{T} : \mathbf{S} = T_{ij}S_{ij}. \tag{1.6.3}$$

The double contraction over two fourth-order tensors is another fourth-order tensor:

$$\mathbf{M} : \mathbf{N} = M_{ijpq}N_{pqkl}(\mathbf{e}_i\otimes\mathbf{e}_j\otimes\mathbf{e}_k\otimes\mathbf{e}_l) \tag{1.6.4}$$

Functions of Tensors

$\mathbf{P}$ is a positive definite $n \times n$ symmetric, real matrix. The **spectral decomposition** of $\mathbf{P}$ is given by

$$\mathbf{P} = \sum_{i=1}^{n} \lambda_i v_i v_i^T \tag{1.6.5}$$

where λ_i are the eigenvalues and v_i are the eigenvectors of $\mathbf{P}$, satisfying $(\lambda_i > 0)$ and $(v_i^T v_j = \delta_{ij})$. Functions $f(\mathbf{P})$ can be evaluated using *Sylvester's formula* (Norris, 2006; Itskov, 2013):

$$f(\mathbf{P}) = \sum_{i=1}^{n} f(\lambda_i) v_i v_i^T. \tag{1.6.6}$$

For example

$$Log(\mathbf{P}) = \sum_{i=1}^{n} Log(\lambda_i) v_i v_i^T \tag{1.6.7}$$

Closest Tensor of Greater Symmetry

Moakher and Norris (2006) describe three different measures of the distance between any pair of tensors $\mathbf{C}_1$ and $\mathbf{C}_2$: the **Euclidean or Frobenius metric** d_F, the **log-Euclidean metric** d_L, and the **Riemannian metric** d_R:

$$d_F(\mathbf{C}_1, \mathbf{C}_2) = \|\mathbf{C}_1 - \mathbf{C}_2\| \tag{1.6.8}$$

$$d_L(\mathbf{C}_1, \mathbf{C}_2) = \|\ln(\mathbf{C}_1) - \ln(\mathbf{C}_2)\| \tag{1.6.9}$$

$$d_R(\mathbf{C}_1, \mathbf{C}_2) = \|\ln(\mathbf{C}_1^{-1/2}\mathbf{C}_2\mathbf{C}_1^{-1/2})\| \tag{1.6.10}$$

The Euclidean metric $\|\mathbf{D}\| = \sqrt{D_{ijkl}D_{ijkl}}$ is most commonly used and is the easiest to implement, though it is not unique under inversion; i.e., the Euclidean distance between two tensors is generally not equal to the distance between their inverses. The log-Euclidean and Riemannian distances are each invariant under inversion.

The problem of finding the closest elastic stiffness tensor of a specified symmetry to another tensor with lesser symmetry has been treated by a number of authors (e.g., Fedorov, 1968; Helbig, 1996; Gazis *et al.*, 2004; Browaeys and Chevrot, 2004; Dellinger, 2005; Moakher and Norris, 2006; Norris, 2006; Caro, 2014). The concept of one tensor being closest to another translates into minimizing the chosen metric. Reasons for looking for the closest tensor with greater symmetry include looking for a simpler approximation to the elasticity tensor or filtering out lower symmetry features introduced by measurement errors or from measurements made on a sample that is smaller than the representative elementary volume. Fedorov (1968) considered the problem of the closest isotropic elastic tensor to one of lesser symmetry using the criterion of best matching the acoustic wave velocities. Norris (2006) showed that Fedorov's criterion is satisfied by the Euclidean metric.

When looking for the closest tensor of greater symmetry, it is important that the lower symmetry tensor is expressed in a canonical coordinate system that is natural for its material symmetry. The Cartesian coordinate axis $\mathbf{e}_3$ is placed along the normal to the monoclinic symmetry plane, the normal to the plane spanning normals to the planes of reflection symmetry in trigonal and tetragonal symmetry, the axis of transversely isotropic symmetry, or any cube axis (Moakher and Norris, 2006). The Cartesian $\mathbf{e}_2$ is placed along the normal to the (010) plane for triclinic or cubic symmetry. For hexagonal or transversely isotropic media the orientation for $\mathbf{e}_2$ does not matter. Coordinate axis $\mathbf{e}_1$ is determined by the vector cross product, $\mathbf{e}_1 = \mathbf{e}_2 \times \mathbf{e}_3$. Caro (2014) gives a procedure for determining the optimum orientation by optimizing over all rotations for the smallest distance between the tensors of higher and lower symmetry.

For a given orientation, the closest higher symmetry tensor using the Euclidean distance can be found as a projection. For example, the elastic tensor can be expressed as a vector of length 21, defined as

$$
\begin{aligned}
\mathbf{X} &= (x_1, x_2, x_3, x_4, x_5, x_6, x_7, x_8, x_9, x_{10}, x_{11}, x_{12}, x_{13}, x_{14}, x_{15}, x_{16}, x_{17}, x_{18}, x_{19}, x_{20}, x_{21})^T \\
&= (c_{11}, c_{22}, c_{33}, \sqrt{2}c_{23}, \sqrt{2}c_{13}, \sqrt{2}c_{12}, 2c_{44}, 2c_{55}, 2c_{66}, 2c_{14}, 2c_{25}, 2c_{36}, 2c_{34}, 2c_{15}, 2c_{26}, \\
&\quad\; 2c_{24}, 2c_{35}, 2c_{16}, 2\sqrt{2}c_{56}, 2\sqrt{2}c_{46}, 2\sqrt{2}c_{45})^T \\
&= (\hat{c}_{11}, \hat{c}_{22}, \hat{c}_{33}, \sqrt{2}\hat{c}_{23}, \sqrt{2}\hat{c}_{13}, \sqrt{2}\hat{c}_{12}, \hat{c}_{44}, \hat{c}_{55}, \hat{c}_{66}, \sqrt{2}\hat{c}_{14}, \sqrt{2}\hat{c}_{25}, \sqrt{2}\hat{c}_{36}, \sqrt{2}\hat{c}_{34}, \sqrt{2}\hat{c}_{15}, \\
&\quad\; \sqrt{2}\hat{c}_{26}, \sqrt{2}\hat{c}_{24}, \sqrt{2}\hat{c}_{35}, \sqrt{2}\hat{c}_{16}, \sqrt{2}\hat{c}_{56}, \sqrt{2}\hat{c}_{46}, \sqrt{2}\hat{c}_{45})^T
\end{aligned}
$$

$$(1.6.11)$$

where c_{ij} are the elastic constants in the 6x6 Voigt notation, and $\hat{c}_{ij}$ are the elastic constants in the Kelvin notation. The factors of 2 and $\sqrt{2}$ ensure that the vector has the same Euclidean norm as the second-order, 6-dimensional Kelvin tensor or the fourth rank, 3-dimensional tensor. The projections of vector $\mathbf{X}$ onto the vector representing higher symmetry $\mathbf{X}_{sym}$ can be expressed as

$$\mathbf{X}_{sym} = \mathbf{P}\mathbf{X} \tag{1.6.12}$$

where **P** is a 21x21 matrix projector given as follows (Browaeys and Chevrot, 2004):

The **monoclinic** projector, matrix $\mathbf{P}_{Mon}$, is the unit (21x21) square matrix, but with diagonal coefficients $P_{i,i}$ with i= 10, 11, 13, 14, 16, 17, 19, 20 equal to zero.

The **orthorhombic** projector, matrix $\mathbf{P}_{Ort}$, is the unit (21x21) square matrix, but with diagonal coefficients $P_{i,i}$ with i= 10, ..., 21 equal zero.

The **tetragonal** projector, matrix $\mathbf{P}_{Tet}$, is

$$\mathbf{P}_{Tet} = \begin{pmatrix} \mathbf{M}_{Tet} & 0_{9x12} \\ 0_{12x9} & 0_{12x12} \end{pmatrix} \tag{1.6.13}$$

where

$$\mathbf{M}_{Tet} = \begin{bmatrix} 1/2 & 1/2 & 0 & 0 & 0 & 0 & 0 & 0 & 0 \\ 1/2 & 1/2 & 0 & 0 & 0 & 0 & 0 & 0 & 0 \\ 0 & 0 & 1 & 0 & 0 & 0 & 0 & 0 & 0 \\ 0 & 0 & 0 & 1/2 & 1/2 & 0 & 0 & 0 & 0 \\ 0 & 0 & 0 & 1/2 & 1/2 & 0 & 0 & 0 & 0 \\ 0 & 0 & 0 & 0 & 0 & 1 & 0 & 0 & 0 \\ 0 & 0 & 0 & 0 & 0 & 0 & 1/2 & 1/2 & 0 \\ 0 & 0 & 0 & 0 & 0 & 0 & 1/2 & 1/2 & 0 \\ 0 & 0 & 0 & 0 & 0 & 0 & 0 & 0 & 1 \end{bmatrix} \tag{1.6.14}$$

The **transverse isotropy** projector, matrix $\mathbf{P}_{Hex}$, is

$$\mathbf{P}_{Hex} = \begin{pmatrix} \mathbf{M}_{Hex} & 0_{9x12} \\ 0_{12x9} & 0_{12x12} \end{pmatrix} \tag{1.6.15}$$

where

$$\mathbf{M}_{Hex} = \begin{bmatrix} 3/8 & 3/8 & 0 & 0 & 0 & 1/4\sqrt{2} & 0 & 0 & 1/4 \\ 3/8 & 3/8 & 0 & 0 & 0 & 1/4\sqrt{2} & 0 & 0 & 1/4 \\ 0 & 0 & 1 & 0 & 0 & 0 & 0 & 0 & 0 \\ 0 & 0 & 0 & 1/2 & 1/2 & 0 & 0 & 0 & 0 \\ 0 & 0 & 0 & 1/2 & 1/2 & 0 & 0 & 0 & 0 \\ 1/4\sqrt{2} & 1/4\sqrt{2} & 0 & 0 & 0 & 3/4 & 0 & 0 & -1/2\sqrt{2} \\ 0 & 0 & 0 & 0 & 0 & 0 & 1/2 & 1/2 & 0 \\ 0 & 0 & 0 & 0 & 0 & 0 & 1/2 & 1/2 & 0 \\ 1/4 & 1/4 & 0 & 0 & 0 & -1/2\sqrt{2} & 0 & 0 & 1/2 \end{bmatrix} \tag{1.6.16}$$

The **cubic** projector, matrix $\mathbf{P}_{Cub}$, is

$$\mathbf{P}_{Cub} = \begin{pmatrix} \mathbf{M}_{Cub} & 0_{9x12} \\ 0_{12x9} & 0_{12x12} \end{pmatrix} \tag{1.6.17}$$

where

$$
\mathbf{M}_{Cub} = \begin{bmatrix}
1/3 & 1/3 & 1/3 & 0 & 0 & 0 & 0 & 0 & 0 \\
1/3 & 1/3 & 1/3 & 0 & 0 & 0 & 0 & 0 & 0 \\
1/3 & 1/3 & 1/3 & 0 & 0 & 0 & 0 & 0 & 0 \\
0 & 0 & 0 & 1/3 & 1/3 & 1/3 & 0 & 0 & 0 \\
0 & 0 & 0 & 1/3 & 1/3 & 1/3 & 0 & 0 & 0 \\
0 & 0 & 0 & 1/3 & 1/3 & 1/3 & 0 & 0 & 0 \\
0 & 0 & 0 & 0 & 0 & 0 & 1/3 & 1/3 & 1/3 \\
0 & 0 & 0 & 0 & 0 & 0 & 1/3 & 1/3 & 1/3 \\
0 & 0 & 0 & 0 & 0 & 0 & 1/3 & 1/3 & 1/3
\end{bmatrix} \tag{1.6.18}
$$

The **isotropy** projector, matrix $\mathbf{P}_{Iso}$, is

$$
\mathbf{P}_{Iso} = \begin{pmatrix} \mathbf{M}_{Iso} & 0_{9x12} \\ 0_{12x9} & 0_{12x12} \end{pmatrix} \tag{1.6.19}
$$

where

$$
\mathbf{M}_{Iso} = \begin{bmatrix}
3/15 & 3/15 & 3/15 & \sqrt{2}/15 & \sqrt{2}/15 & \sqrt{2}/15 & 2/15 & 2/15 & 2/15 \\
3/15 & 3/15 & 3/15 & \sqrt{2}/15 & \sqrt{2}/15 & \sqrt{2}/15 & 2/15 & 2/15 & 2/15 \\
3/15 & 3/15 & 3/15 & \sqrt{2}/15 & \sqrt{2}/15 & \sqrt{2}/15 & 2/15 & 2/15 & 2/15 \\
\sqrt{2}/15 & \sqrt{2}/15 & \sqrt{2}/15 & 4/15 & 4/15 & 4/15 & -\sqrt{2}/15 & -\sqrt{2}/15 & -\sqrt{2}/15 \\
\sqrt{2}/15 & \sqrt{2}/15 & \sqrt{2}/15 & 4/15 & 4/15 & 4/15 & -\sqrt{2}/15 & -\sqrt{2}/15 & -\sqrt{2}/15 \\
\sqrt{2}/15 & \sqrt{2}/15 & \sqrt{2}/15 & 4/15 & 4/15 & 4/15 & -\sqrt{2}/15 & -\sqrt{2}/15 & -\sqrt{2}/15 \\
2/15 & 2/15 & 2/15 & -\sqrt{2}/15 & -\sqrt{2}/15 & -\sqrt{2}/15 & 1/5 & 1/5 & 1/5 \\
2/15 & 2/15 & 2/15 & -\sqrt{2}/15 & -\sqrt{2}/15 & -\sqrt{2}/15 & 1/5 & 1/5 & 1/5 \\
2/15 & 2/15 & 2/15 & -\sqrt{2}/15 & -\sqrt{2}/15 & -\sqrt{2}/15 & 1/5 & 1/5 & 1/5
\end{bmatrix} \tag{1.6.20}
$$

The resulting projections are given in Kelvin notation as follows:

Monoclinic Symmetry

$$
\mathbf{C}_{Mon} = \begin{pmatrix}
\hat{c}_{11} & \hat{c}_{12} & \hat{c}_{13} & 0 & 0 & \hat{c}_{16} \\
\hat{c}_{12} & \hat{c}_{22} & \hat{c}_{23} & 0 & 0 & \hat{c}_{26} \\
\hat{c}_{13} & \hat{c}_{23} & \hat{c}_{33} & 0 & 0 & \hat{c}_{36} \\
0 & 0 & 0 & \hat{c}_{44} & \hat{c}_{45} & 0 \\
0 & 0 & 0 & \hat{c}_{45} & \hat{c}_{55} & 0 \\
\hat{c}_{16} & \hat{c}_{26} & \hat{c}_{36} & 0 & 0 & \hat{c}_{66}
\end{pmatrix} \tag{1.6.21}
$$

$$
\text{norm}: \|C_{Mon}\|^2 = \hat{c}_{11}^2 + \hat{c}_{22}^2 + \hat{c}_{33}^2 + \hat{c}_{44}^2 + \hat{c}_{55}^2 + \hat{c}_{66}^2
$$
$$
+ 2\left(\hat{c}_{12}^2 + \hat{c}_{13}^2 + \hat{c}_{23}^2 + \hat{c}_{45}^2 + \hat{c}_{16}^2 + \hat{c}_{26}^2 + \hat{c}_{36}^2 \right) \tag{1.6.22}
$$

Orthorhombic Symmetry

$$\mathbf{C}_{Ort} = \begin{pmatrix} \hat{c}_{11} & \hat{c}_{12} & \hat{c}_{13} & 0 & 0 & 0 \\ \hat{c}_{12} & \hat{c}_{22} & \hat{c}_{23} & 0 & 0 & 0 \\ \hat{c}_{13} & \hat{c}_{23} & \hat{c}_{33} & 0 & 0 & 0 \\ 0 & 0 & 0 & \hat{c}_{44} & 0 & 0 \\ 0 & 0 & 0 & 0 & \hat{c}_{55} & 0 \\ 0 & 0 & 0 & 0 & 0 & \hat{c}_{66} \end{pmatrix} \tag{1.6.23}$$

$$\text{norm}: \|C_{Ort}\|^2 = \hat{c}_{11}^2 + \hat{c}_{22}^2 + \hat{c}_{33}^2 + \hat{c}_{44}^2 + \hat{c}_{55}^2 + \hat{c}_{66}^2 + 2(\hat{c}_{12}^2 + \hat{c}_{13}^2 + \hat{c}_{23}^2) \tag{1.6.24}$$

Tetragonal Symmetry

$$\mathbf{C}_{Tet} = \begin{pmatrix} \frac{1}{2}(b+q) & \frac{1}{2}(b-q) & \frac{c}{\sqrt{2}} & 0 & 0 & \frac{r}{\sqrt{2}} \\ \frac{1}{2}(b-q) & \frac{1}{2}(b+q) & \frac{c}{\sqrt{2}} & 0 & 0 & -\frac{r}{\sqrt{2}} \\ \frac{c}{\sqrt{2}} & \frac{c}{\sqrt{2}} & a & 0 & 0 & 0 \\ 0 & 0 & 0 & g & 0 & 0 \\ 0 & 0 & 0 & 0 & g & 0 \\ \frac{r}{\sqrt{2}} & -\frac{r}{\sqrt{2}} & 0 & 0 & 0 & p \end{pmatrix} \tag{1.6.25}$$

$$\text{norm}: \|C_{tet}\|^2 = a^2 + b^2 + 2c^2 + p^2 + q^2 + 2r^2 + 2g^2 \tag{1.6.26}$$

where

$$a = \hat{c}_{33}, \quad b = \frac{1}{2}(\hat{c}_{11} + \hat{c}_{22} + 2\hat{c}_{12}), \quad c = \frac{1}{\sqrt{2}}(\hat{c}_{13} + \hat{c}_{23}), \quad g = \frac{1}{2}(\hat{c}_{44} + \hat{c}_{55}) \tag{1.6.27}$$

$$p = \hat{c}_{66}, \quad q = \frac{1}{2}(\hat{c}_{11} + \hat{c}_{22} - 2\hat{c}_{12}), \quad r = \frac{1}{\sqrt{2}}(\hat{c}_{16} - \hat{c}_{26}), \tag{1.6.28}$$

Trigonal Symmetry

$$\mathbf{C}_{Tri} = \begin{pmatrix} \frac{1}{2}(b+p) & \frac{1}{2}(b-p) & \frac{1}{\sqrt{2}}c & \frac{1}{\sqrt{2}}v & \frac{1}{\sqrt{2}}s & 0 \\ \frac{1}{2}(b-p) & \frac{1}{2}(b+p) & \frac{1}{\sqrt{2}}c & -\frac{1}{\sqrt{2}}v & -\frac{1}{\sqrt{2}}s & 0 \\ \frac{1}{\sqrt{2}}c & \frac{1}{\sqrt{2}}c & a & 0 & 0 & 0 \\ \frac{1}{\sqrt{2}}v & -\frac{1}{\sqrt{2}}v & 0 & t & 0 & -s \\ \frac{1}{\sqrt{2}}s & -\frac{1}{\sqrt{2}}s & 0 & 0 & t & v \\ 0 & 0 & 0 & -s & v & p \end{pmatrix} \tag{1.6.29}$$

norm : $\|C_{tet}\|^2 = a^2 + b^2 + 2c^2 + 2p^2 + 2t^2 + 4v^2 + 4s^2$ (1.6.30)

$a = \hat{c}_{33}, \quad b = \frac{1}{2}(\hat{c}_{11} + \hat{c}_{22} + 2\hat{c}_{12}), \quad c = \frac{1}{\sqrt{2}}(\hat{c}_{13} + \hat{c}_{23}), \quad p = \hat{c}_{66},$ (1.6.31)

$t = \frac{1}{2}(\hat{c}_{44} + \hat{c}_{55}), \quad v = \frac{1}{2\sqrt{2}}(\hat{c}_{14} - \hat{c}_{24}) + \frac{1}{2}\hat{c}_{56}, \quad s = \frac{1}{2\sqrt{2}}(\hat{c}_{15} - \hat{c}_{25}) - \frac{1}{2}\hat{c}_{46}$ (1.6.32)

Transverse Isotropy

$$\mathbf{C}_{Hex} = \begin{pmatrix} \frac{1}{2}(b+f) & \frac{1}{2}(b-f) & \frac{c}{\sqrt{2}} & 0 & 0 & 0 \\ \frac{1}{2}(b-f) & \frac{1}{2}(b+f) & \frac{c}{\sqrt{2}} & 0 & 0 & 0 \\ \frac{c}{\sqrt{2}} & \frac{c}{\sqrt{2}} & a & 0 & 0 & 0 \\ 0 & 0 & 0 & g & 0 & 0 \\ 0 & 0 & 0 & 0 & g & 0 \\ 0 & 0 & 0 & 0 & 0 & f \end{pmatrix}$$ (1.6.33)

norm : $\|C_{Hex}\|^2 = a^2 + b^2 + 2c^2 + 2f^2 + 2g^2$ (1.6.34)

where

$a = \hat{c}_{33}, \quad b = \frac{1}{2}(\hat{c}_{11} + \hat{c}_{22} + 2\hat{c}_{12}), \quad c = \frac{1}{\sqrt{2}}(\hat{c}_{13} + \hat{c}_{23}), \quad g = \frac{1}{2}(\hat{c}_{44} + \hat{c}_{55})$ (1.6.35)

$f = \frac{1}{2}(p+q) = \frac{1}{4}(\hat{c}_{11} + \hat{c}_{22} + 2\hat{c}_{66} - 2\hat{c}_{12})$ (1.6.36)

Cubic Symmetry

$$\mathbf{C}_{Cub} = \begin{pmatrix} \frac{1}{3}(a'+2c') & \frac{1}{3}(a'-c') & \frac{1}{3}(a'-c') & 0 & 0 & 0 \\ \frac{1}{3}(a'-c') & \frac{1}{3}(a'+2c') & \frac{1}{3}(a'-c') & 0 & 0 & 0 \\ \frac{1}{3}(a'-c') & \frac{1}{3}(a'-c') & \frac{1}{3}(a'+2c') & 0 & 0 & 0 \\ 0 & 0 & 0 & b' & 0 & 0 \\ 0 & 0 & 0 & 0 & b' & 0 \\ 0 & 0 & 0 & 0 & 0 & b' \end{pmatrix}$$ (1.6.37)

norm : $\|C_{Cub}\|^2 = a'^2 + 3b'^2 + 2c'^2$ (1.6.38)

where

$$a' = \frac{1}{3}(\hat{c}_{11} + \hat{c}_{22} + \hat{c}_{33} + 2\hat{c}_{12} + 2\hat{c}_{13} + 2\hat{c}_{23}), \tag{1.6.39}$$

$$b' = \frac{1}{3}(\hat{c}_{44} + \hat{c}_{55} + \hat{c}_{66}), \tag{1.6.40}$$

$$c' = \frac{1}{3}(\hat{c}_{11} + \hat{c}_{22} + \hat{c}_{33} - \hat{c}_{12} - \hat{c}_{13} - \hat{c}_{23}), \tag{1.6.41}$$

Isotropy

$$\mathbf{C}_{Iso} = \begin{pmatrix} k + \frac{4}{3}\mu & k - \frac{2}{3}\mu & k - \frac{2}{3}\mu & 0 & 0 & 0 \\ k - \frac{2}{3}\mu & k + \frac{4}{3}\mu & k - \frac{2}{3}\mu & 0 & 0 & 0 \\ k - \frac{2}{3}\mu & k - \frac{2}{3}\mu & k + \frac{4}{3}\mu & 0 & 0 & 0 \\ 0 & 0 & 0 & 2\mu & 0 & 0 \\ 0 & 0 & 0 & 0 & 2\mu & 0 \\ 0 & 0 & 0 & 0 & 0 & 2\mu \end{pmatrix} \tag{1.6.42}$$

$$9k = (\hat{c}_{11} + \hat{c}_{22} + \hat{c}_{33} + 2\hat{c}_{12} + 2\hat{c}_{13} + 2\hat{c}_{23}) \tag{1.6.43}$$

$$30\mu = 2(\hat{c}_{11} + \hat{c}_{22} + \hat{c}_{33} - \hat{c}_{12} - \hat{c}_{23} - \hat{c}_{31}) + 3(\hat{c}_{44} + \hat{c}_{55} + \hat{c}_{66}) \tag{1.6.44}$$

$$\text{norm}: \|C_{Iso}\|^2 = 9k^2 + 20\mu^2 \tag{1.6.45}$$

Torquato (2002) defines the following two projection tensors onto isotropic symmetry:

$$(\mathbf{\Lambda}_h)_{ijkl} = \frac{1}{3}\delta_{ij}\delta_{kl} \tag{1.6.46}$$

$$(\mathbf{\Lambda}_s)_{ijkl} = \frac{1}{2}[\delta_{ik}\delta_{jl} + \delta_{il}\delta_{jk}] - \frac{1}{3}\delta_{ij}\delta_{kl} \tag{1.6.47}$$

In terms of these projection tensors, an isotropic stiffness tensor can be written as

$$\mathbf{C}^{iso} = 3K\mathbf{\Lambda}_h + 2\mu\mathbf{\Lambda}_s \tag{1.6.48}$$

The projection tensors can be written in Kelvin notation (see Section 2.2) as

$$\hat{\mathbf{\Lambda}}_h = \frac{1}{3}\begin{pmatrix} 1 & 1 & 1 & 0 & 0 & 0 \\ 1 & 1 & 1 & 0 & 0 & 0 \\ 1 & 1 & 1 & 0 & 0 & 0 \\ 0 & 0 & 0 & 0 & 0 & 0 \\ 0 & 0 & 0 & 0 & 0 & 0 \\ 0 & 0 & 0 & 0 & 0 & 0 \end{pmatrix} \tag{1.6.49}$$

$$
\hat{\mathbf{\Lambda}}_s = \frac{1}{3}
\begin{pmatrix}
2 & -1 & -1 & 0 & 0 & 0 \\
-1 & 2 & -1 & 0 & 0 & 0 \\
-1 & -1 & 2 & 0 & 0 & 0 \\
0 & 0 & 0 & 3 & 0 & 0 \\
0 & 0 & 0 & 0 & 3 & 0 \\
0 & 0 & 0 & 0 & 0 & 3
\end{pmatrix}
\tag{1.6.50}
$$

Quantities in Kelvin notation are indicated by $\hat{\ }$ over symbols. Note that the Euclidean norm of $\mathbf{\Lambda}_h$ in either 4-subscript or Kelvin notations is $\|\hat{\mathbf{\Lambda}}_h\| = 1$, and $\|\mathbf{\Lambda}_h\| = 1$. (*If using the 6x6 notation, it is important to do this in Kelvin notation to preserve the norms.*) The Euclidean norm of $\mathbf{\Lambda}_s$ in either format is $\|\hat{\mathbf{\Lambda}}_s\| = \sqrt{5}$, and $\|\mathbf{\Lambda}_s\| = \sqrt{5}$. Therefore, $\|\mathbf{C}_{iso}\|^2 = 9K^2 + 20\mu^2$ (Moakher and Norris, 2006).

The bulk and shear moduli are found by taking the inner product of the stiffness tensor with the projections:

$$
3K = \hat{\mathbf{C}} : \hat{\mathbf{\Lambda}}_h / \|\hat{\mathbf{\Lambda}}_h\|^2 = \frac{1}{3} \mathbf{C}_{iijj}
\tag{1.6.51}
$$

$$
2\mu = \hat{\mathbf{C}} : \hat{\mathbf{\Lambda}}_s / \|\hat{\mathbf{\Lambda}}_s\|^2 = \mathbf{C}_{ijij} - \frac{2}{3}\mathbf{C}_{iijj}
\tag{1.6.52}
$$

Or, more explicitly:

$$
3K = \frac{1}{3}
\begin{pmatrix}
\hat{c}_{11} & \hat{c}_{12} & \hat{c}_{13} & \hat{c}_{14} & \hat{c}_{15} & \hat{c}_{16} \\
\hat{c}_{21} & \hat{c}_{22} & \hat{c}_{23} & \hat{c}_{24} & \hat{c}_{25} & \hat{c}_{26} \\
\hat{c}_{31} & \hat{c}_{32} & \hat{c}_{33} & \hat{c}_{34} & \hat{c}_{35} & \hat{c}_{36} \\
\hat{c}_{41} & \hat{c}_{42} & \hat{c}_{43} & \hat{c}_{44} & \hat{c}_{45} & \hat{c}_{46} \\
\hat{c}_{51} & \hat{c}_{52} & \hat{c}_{53} & \hat{c}_{54} & \hat{c}_{55} & \hat{c}_{56} \\
\hat{c}_{61} & \hat{c}_{62} & \hat{c}_{63} & \hat{c}_{64} & \hat{c}_{65} & \hat{c}_{66}
\end{pmatrix}
:
\begin{pmatrix}
1 & 1 & 1 & 0 & 0 & 0 \\
1 & 1 & 1 & 0 & 0 & 0 \\
1 & 1 & 1 & 0 & 0 & 0 \\
0 & 0 & 0 & 0 & 0 & 0 \\
0 & 0 & 0 & 0 & 0 & 0 \\
0 & 0 & 0 & 0 & 0 & 0
\end{pmatrix}
\tag{1.6.53}
$$

$$
2\mu = \frac{1}{15}
\begin{pmatrix}
\hat{c}_{11} & \hat{c}_{12} & \hat{c}_{13} & \hat{c}_{14} & \hat{c}_{15} & \hat{c}_{16} \\
\hat{c}_{21} & \hat{c}_{22} & \hat{c}_{23} & \hat{c}_{24} & \hat{c}_{25} & \hat{c}_{26} \\
\hat{c}_{31} & \hat{c}_{32} & \hat{c}_{33} & \hat{c}_{34} & \hat{c}_{35} & \hat{c}_{36} \\
\hat{c}_{41} & \hat{c}_{42} & \hat{c}_{43} & \hat{c}_{44} & \hat{c}_{45} & \hat{c}_{46} \\
\hat{c}_{51} & \hat{c}_{52} & \hat{c}_{53} & \hat{c}_{54} & \hat{c}_{55} & \hat{c}_{56} \\
\hat{c}_{61} & \hat{c}_{62} & \hat{c}_{63} & \hat{c}_{64} & \hat{c}_{65} & \hat{c}_{66}
\end{pmatrix}
:
\begin{pmatrix}
2 & -1 & -1 & 0 & 0 & 0 \\
-1 & 2 & -1 & 0 & 0 & 0 \\
-1 & -1 & 2 & 0 & 0 & 0 \\
0 & 0 & 0 & 3 & 0 & 0 \\
0 & 0 & 0 & 0 & 3 & 0 \\
0 & 0 & 0 & 0 & 0 & 3
\end{pmatrix}
\tag{1.6.54}
$$

Curie Theorem on the Symmetry of Physical Property Tensors

Several important theorems relate the symmetry of tensor properties of solids to the symmetries of the underlying material structure and external influences (Nye, 1957; Wadhawan, 1987; Tinder, 2008).

Curie's (1894) **principle** is composed of several parts (Nakamura and Nagahama, 2000): (1) the symmetry group of the causes is a sub-group of the symmetry group of the effects. An example is stressed-induced seismic anisotropy: *if an initially isotropic material develops anisotropy due to applied stress, the anisotropy must have at least orthorhombic symmetry.* (2) If the known effect lacks certain symmetries, then they are also lacking in the cause. (3) A crystal under an external influence will exhibit only those symmetry elements that are common to both the crystal and the external influences. Curie's principle has also been applied at the geologic scale (Nur, 1971; Rasolofosaon, 1998; Nakamura and Nagahama, 2000). Whatever the contributing factors, the symmetry that is common to them cannot exceed the symmetry of the resulting geologic texture (Sanders, 1930). See, for example, the table by Nur (1971) in Section 2.8.

Neumann's (1885) **principle** states that the symmetry elements of a physical property tensor must include the symmetry elements of the point group of the crystal (Nye, 1957; Tinder, 2008). The physical property tensor may be more symmetric than the point group, but the point group cannot be more symmetric than the physical property tensor. This resembles Curie's principle in the sense that the crystal point group is a cause of the resulting physical property symmetry. One consequence of Neumann's principle is that *any cubic crystal system is isotropic with respect to any symmetric, second rank property tensor* (electrical conductivity, thermal expansion coefficient, thermal conductivity). Permeability is a symmetric second-rank tensor, but not a crystal property, per se. Whatever the contributing causes, the symmetry that is common to them cannot be higher than the symmetry of the rock property; similarly, any symmetry element missing from the physical property must also be missing from at least one of the causes (Nakamura and Nagahama, 2000).

The **Hermann** (1934) **theorem** states that if a rank-r physical property tensor has an N-fold symmetry axis and $N > r$, the property tensor has an axis of infinite fold symmetry axis. An example is that the optical indicatrix, which has rank r=2, is isotropic in cubic crystals since cubic crystals have more than one symmetry axis of fold $N>2$. Electrical conductivity is also isotropic in cubic crystals. The elastic properties (tensor rank 4) are rotationally invariant with respect to any n-fold axis with $n>4$. Thus, hexagonal symmetry is transversely isotropic with respect to the six-fold axis.

Tensor Invariants

Certain scalar properties of tensors remain invariant under coordinate rotation. The **principal invariants** I_1, I_2, and I_3 of a 3-dimensional second-order tensor **D** are determined from the coefficients of the characteristic equation of **D** (Ghosh *et al.*, 2012):

$$I_1 = tr(\mathbf{D}) = \lambda_1 + \lambda_2 + \lambda_3 \tag{1.6.55}$$

$$I_2 = \frac{1}{2}[tr(\mathbf{D})^2 - tr(\mathbf{D}^2)] = \lambda_1\lambda_2 + \lambda_2\lambda_3 + \lambda_3\lambda_1 \tag{1.6.56}$$

$$I_3 = \det(\mathbf{D}) = \lambda_1\lambda_2\lambda_3 \tag{1.6.57}$$

where λ_i are the eigenvalues of D.

The basic invariants J_1, J_2, and J_3 of a second-order tensor **D** are defined by

$$J_1 = tr(\mathbf{D}) = \lambda_1 + \lambda_2 + \lambda_3 \tag{1.6.58}$$

$$J_2 = tr(\mathbf{D}^2) = I_1^2 - 2I_2 = \lambda_1^2 + \lambda_2^2 + \lambda_3^2 \tag{1.6.59}$$

$$J_3 = tr(\mathbf{D}^3) = \lambda_1^3 + \lambda_2^3 + \lambda_3^3 \tag{1.6.60}$$

It can be show algebraically that:

$$I_3 = (J_1^3 - 3J_1J_2 + 2J_3)/6 \tag{1.6.61}$$

Functions of invariants are also invariant. Hence, the set of principal and basic invariants can be solved from one another.

3D Fourth-Order Tensor Invariants of Totally Symmetric (a_{ijkl})

The *basic invariants* J_i of a 3-dimensional fourth-order tensor are given by (Betten, 1987):

$$J_1 = tr(\mathbf{A}) \tag{1.6.62}$$

$$J_2 = tr(\mathbf{A}^2) \tag{1.6.63}$$

$$J_3 = tr(\mathbf{A}^3) \tag{1.6.64}$$

$$J_4 = tr(\mathbf{A}^4) \tag{1.6.65}$$

$$J_5 = tr(\mathbf{A}^5) \tag{1.6.66}$$

$$J_6 = tr(\mathbf{A}^6) \tag{1.6.67}$$

where **A** is the 6-D second-order matrix mapped from (a_{ijkl}). **For elasticity, A** in Kelvin's notation for the elastic stiffness tensor is (Moakher, 2008):

$$\mathbf{A} = \begin{bmatrix} a_{1111} & a_{1122} & a_{1133} & \sqrt{2}a_{1123} & \sqrt{2}a_{1113} & \sqrt{2}a_{1112} \\ a_{1122} & a_{2222} & a_{2233} & \sqrt{2}a_{2223} & \sqrt{2}a_{2213} & \sqrt{2}a_{2212} \\ a_{1133} & a_{2233} & a_{3333} & \sqrt{2}a_{3323} & \sqrt{2}a_{3313} & \sqrt{2}a_{3312} \\ \sqrt{2}a_{1123} & \sqrt{2}a_{2223} & \sqrt{2}a_{3323} & 2a_{2323} & 2a_{2313} & 2a_{2312} \\ \sqrt{2}a_{1113} & \sqrt{2}a_{2213} & \sqrt{2}a_{3313} & 2a_{2313} & 2a_{1313} & 2a_{1312} \\ \sqrt{2}a_{1112} & \sqrt{2}a_{2212} & \sqrt{2}a_{3312} & 2a_{2312} & 2a_{1312} & 2a_{1212} \end{bmatrix} \tag{1.6.68}$$

Eigenvalues and Eigentensors

In linear algebra, an eigenvector of a linear transformation is a vector whose direction does not change when the transformation is applied to it. For example, stress at a point in 3-dimensional space is a second-order tensor, usually written as

$$\tilde{\sigma} = \begin{pmatrix} \sigma_{11} & \sigma_{12} & \sigma_{13} \\ \sigma_{21} & \sigma_{22} & \sigma_{23} \\ \sigma_{31} & \sigma_{32} & \sigma_{33} \end{pmatrix}. \tag{1.6.69}$$

Cauchy's formula, $\vec{\mathbf{T}} = \tilde{\sigma}\hat{\mathbf{v}}$, yields the traction vector $\vec{\mathbf{T}}$ acting at the point on a surface with orientation defined by its unit normal vector $\hat{\mathbf{v}}$. We can always find three mutually orthogonal surfaces at which the traction vectors are parallel to the unit normal vectors. These normals, $\hat{\mathbf{v}}_i$, are the principal directions (eigenvectors), of the tensor $\tilde{\sigma}$ and satisfy the relations

$$\tilde{\sigma}\hat{\mathbf{v}}_i = \sigma_i\hat{\mathbf{v}}_i = \vec{\mathbf{T}}_i, \; i = 1, 2, 3. \tag{1.6.70}$$

The scalar constants, σ_i, are the three **principal stresses** (eigenvalues of stress) at the point. If we rotate our chosen coordinate system so that the axes align with the eigenvectors, then the stress tensor is diagonalized with the principal stresses along the diagonal.

Similarly the elastic stiffness tensor $\mathbf{C}$ is a linear transformation that, when applied to the strain tensor $\tilde{\varepsilon}$, yields the stress tensor $\tilde{\sigma}$:

$$\sigma_{ij} = C_{ijkl}\varepsilon_{kl} \tag{1.6.71}$$

The elastic stiffness tensor is fourth order in 3-dimensional space. If the elastic material is isotropic, then the total stress and total strain tensors can each be decomposed into the sum of two eigentensors, the deviatoric and hydrostatic parts, which have identical form (Mehrabadi and Cowin, 1990) such that

$$\tilde{\sigma}_0 = \frac{K}{3}\tilde{\varepsilon}_0 \text{ and } \tilde{\sigma}_d = 2\mu\tilde{\varepsilon}_d \tag{1.6.72}$$

For elastic stiffness tensors of any symmetry the stress and strain tensors can always be decomposed into a sum of six or fewer eigentensors of identical form, i.e., each stress eigentensor will be proportional to the corresponding strain tensor. Additional discussion on eigentensors can be found in Section 2.2.

2 Elasticity and Hooke's Law

2.1 Elastic Moduli: Isotropic Form of Hooke's Law

Synopsis

In an isotropic, linear elastic material, the stress and strain are related by **Hooke's law** as follows (e.g., Timoshenko and Goodier, 1934):

$$\sigma_{ij} = \lambda \delta_{ij} \varepsilon_{\alpha\alpha} + 2\mu \varepsilon_{ij} \tag{2.1.1}$$

or

$$\varepsilon_{ij} = \frac{1}{E} \left[(1 + v)\sigma_{ij} - v\delta_{ij}\sigma_{\alpha\alpha} \right] \tag{2.1.2}$$

where

ε_{ij} = elements of the strain tensor

σ_{ij} = elements of the stress tensor

$\varepsilon_{\alpha\alpha}$ = volumetric strain (sum over repeated index)

$\sigma_{\alpha\alpha}$ = mean stress times 3 (sum over repeated index)

δ_{ij} = 0 if $i \neq j$ and $\delta_{ij} = 1$ if $i = j$

In an isotropic, linear elastic medium, only two constants are needed to specify the stress–strain relation completely (for example, Lame's coefficient and shear modulus $[\lambda, \mu]$ in the first equation or Young's modulus and Poisson's ratio $[E, v]$, which can be derived from $[\lambda, \mu]$, in the second equation). Other useful and convenient moduli can be defined, but they are always relatable to just two constants. The three moduli that follow are examples.

The bulk modulus, K, is defined as the ratio of the hydrostatic stress, σ_0, to the volumetric strain:

$$\sigma_0 = \frac{1}{3}\sigma_{\alpha\alpha} = K\varepsilon_{\alpha\alpha} \tag{2.1.3}$$

The bulk modulus is the reciprocal of the **compressibility**, β, which is widely used to describe the volumetric compliance of a liquid, solid, or gas:

$$\beta = \frac{1}{K} \tag{2.1.4}$$

Caution:

Occasionally in the literature, authors have used the term *incompressibility* as an alternate name for Lame's constant, λ, even though λ *is not the reciprocal of the compressibility.*

The shear modulus, μ, is defined as the ratio of the shear stress to the shear strain:

$$\sigma_{ij} = 2\mu\varepsilon_{ij}, \quad i \neq j \tag{2.1.5}$$

Young's modulus, E, is defined as the ratio of the extensional stress to the extensional strain in a *uniaxial stress state:*

$$\sigma_{zz} = E\varepsilon_{zz}, \quad \sigma_{xx} = \sigma_{yy} = \sigma_{xy} = \sigma_{xz} = \sigma_{yz} = 0 \tag{2.1.6}$$

Poisson's ratio, which is defined as minus the ratio of the lateral strain to the axial strain in a *uniaxial stress* state:

$$\nu = -\frac{\varepsilon_{xx}}{\varepsilon_{zz}}, \quad \sigma_{xx} = \sigma_{yy} = \sigma_{xy} = \sigma_{xz} = \sigma_{yz} = 0 \tag{2.1.7}$$

P-wave modulus, $M = \rho V_P^2$, defined as the ratio of the axial stress to the axial strain in a *uniaxial strain* state:

$$\sigma_{zz} = M\varepsilon_{zz}, \quad \varepsilon_{xx} = \varepsilon_{yy} = \varepsilon_{xy} = \varepsilon_{xz} = \varepsilon_{yz} = 0 \tag{2.1.8}$$

Note that the moduli (λ, μ, K, E, M) all have the same units as stress (force/area), whereas Poisson's ratio is dimensionless.

Energy considerations require that the following relations always hold. If they do not, one should suspect experimental errors or that the material is not isotropic:

$$K = \lambda + \frac{2\mu}{3} \geq 0; \quad \mu \geq 0 \tag{2.1.9}$$

or

$$-1 < \nu \leq \frac{1}{2}; \quad E \geq 0 \tag{2.1.10}$$

In rocks, we seldom, if ever, observe a Poisson's ratio of less than 0. Although permitted by equation 2.1.10, a negative measured value is usually treated with suspicion. A Poisson's ratio of 0.5 can mean an infinitely incompressible rock (not possible) or a liquid. A suspension of particles in fluid, or extremely soft, water-saturated sediments under essentially zero effective stress, such as pelagic ooze, can have a Poisson's ratio approaching 0.5. Table 2.1.1 summarizes useful relations among the constants of linear isotropic elastic media.

Table 2.1.1 *Relationships among elastic constants in an isotropic material (after Birch, 1961)*

K	E	λ	v	M	μ
$\lambda + 2\mu/3$	$\mu\dfrac{3\lambda + 2\mu}{\lambda + \mu}$	—	$\dfrac{\lambda}{2(\lambda + \mu)}$	$\lambda + 2\mu$	—
—	$9K\dfrac{K - \lambda}{3K - \lambda}$	—	$\dfrac{\lambda}{3K - \lambda}$	$3\,K - 2\lambda$	$3(K - \lambda)/2$
—	$\dfrac{9K\mu}{3K + \mu}$	$K - 2\mu/3$	$\dfrac{3K - 2\mu}{2(3K + \mu)}$	$K + 4\mu/3$	—
$\dfrac{E\mu}{3(3\mu - E)}$	—	$\mu\dfrac{E - 2\mu}{(3\mu - E)}$	$E/(2\mu) - 1$	$\mu\dfrac{4\mu - E}{3\mu - E}$	—
—	—	$3K\dfrac{3K - E}{9K - E}$	$\dfrac{3K - E}{6K}$	$3K\dfrac{3K + E}{9K - E}$	$\dfrac{3KE}{9K - E}$
$\lambda\dfrac{1 + v}{3v}$	$\lambda\dfrac{(1 + v)(1 - 2v)}{v}$	—	—	$\lambda\dfrac{1 - v}{v}$	$\lambda\dfrac{1 - 2v}{2v}$
$\mu\dfrac{2(1 + v)}{3(1 - 2v)}$	$2\mu(1 + v)$	$\mu\dfrac{2v}{1 - 2v}$	—	$\mu\dfrac{2 - 2v}{1 - 2v}$	—
—	$3K(1 - 2\,v)$	$3K\dfrac{v}{1 + v}$	—	$3K\dfrac{1 - v}{1 + v}$	$3K\dfrac{1 - 2v}{2 + 2v}$
$\dfrac{E}{3(1 - 2v)}$	—	$\dfrac{Ev}{(1 + v)(1 - 2v)}$	—	$\dfrac{E(1 - v)}{(1 + v)(1 - 2v)}$	$\dfrac{E}{2 + 2v}$
$M - \dfrac{4}{3}\mu$	$\dfrac{\mu(3M - 4\mu)}{M - \mu}$	$M - 2\mu$	$\dfrac{M - 2\mu}{2(M - \mu)}$	—	—

Although any one of the isotropic constants (λ, μ, K, M, E, and v) can be derived in terms of the others, μ and K have a special significance as *eigenelastic* constants (Mehrabadi and Cowin, 1990) or *principal elasticities* of the material (Kelvin, 1856). The stress and strain eigentensors associated with μ and K are orthogonal, as discussed in Section 2.2. Such an orthogonal significance does not hold for the pair λ and μ.

The second-order stress and strain tensors can always be decomposed into the sum of the **hydrostatic** part (sometimes called the volumetric part) and the **deviatoric** part:

$$\widetilde{\sigma}^{vol} = \frac{1}{3} tr(\widetilde{\sigma})\widetilde{\mathbf{I}} \quad \text{and} \quad \widetilde{\sigma}^{dev} = \widetilde{\sigma} - \widetilde{\sigma}^{vol} \tag{2.1.11}$$

$$\widetilde{\varepsilon}^{vol} = \frac{1}{3} tr(\widetilde{\varepsilon})\widetilde{\mathbf{I}} \quad \text{and} \quad \widetilde{\varepsilon}^{dev} = \widetilde{\varepsilon} - \widetilde{\varepsilon}^{vol} \tag{2.1.12}$$

Isotropic Hooke's law can now be expressed in a decoupled form as follows:

$$\widetilde{\sigma}^{vol} = 3K\widetilde{\varepsilon}^{vol} \text{ and } \widetilde{\sigma}^{dev} = 2\mu\widetilde{\varepsilon}^{dev} \tag{2.1.13}$$

The isotropic elastic stiffness tensor that is nearest to any other stiffness tensor of lesser symmetry can be found via a projections process as described in Section 1.6.

The von Mises stress is a measure of the deviatoric stress at a point. It can be written in terms of the second invariant of the deviatoric stress tensor, $\widetilde{\sigma}^{dev}$:

$$\sigma_{vM} = \sqrt{\frac{3}{2}\sigma_{ij}^{dev}\sigma_{ij}^{dev}} \tag{2.1.14}$$

In terms of the total stress tensor:

$$\sigma_{vM} = \sqrt{\frac{(\sigma_{11} - \sigma_{22})^2 + (\sigma_{22} - \sigma_{33})^2 + (\sigma_{33} - \sigma_{11})^2 + 6(\sigma_{12}^2 + \sigma_{23}^2 + \sigma_{31}^2)}{2}} \tag{2.1.15}$$

In terms of the principal stresses:

$$\sigma_{vM} = \sqrt{\frac{(\sigma_1 - \sigma_2)^2 + (\sigma_2 - \sigma_3)^2 + (\sigma_3 - \sigma_1)^2}{2}} \tag{2.1.16}$$

Two-Dimensional Elasticity

An object is considered to be under a state of "plane stress" in the XY plane when the traction vectors acting on all surfaces normal to the z-axis are zero (i.e., stress components $\sigma_{zx} = \sigma_{zy} = \sigma_{zz} = 0$). This is equivalent to a state where, at every point within the object, one of the principal stresses is parallel to the z-axis and has a value of zero. This is often a suitable description for thin plates under tractions in the XY plane only. A state of "plane strain" exists when all displacements in the z-direction are zero (i.e., strain components $\varepsilon_{zx} = \varepsilon_{zy} = \varepsilon_{zz} = 0$) at every point throughout the object. This is often a suitable description for thick slabs with transversely isotropic symmetry. In both plane stress and plane strain, the material properties and the material geometry are assumed to be independent of the z-direction, and $\partial\varepsilon_{ij}/\partial z = \partial\sigma_{ij}/\partial z = 0$. In either plane stress or plane strain problems, the in-plane stresses and strains are functions of x and y only.

The stress–strain relation of a linear elastic three-dimensional (3D) material is given by Hooke's law, using the elastic stiffnesses c_{ijkl}, or in the case of an isotropic material, the bulk modulus, K, and shear modulus, μ. However, since variations in the z-direction are zero, and either the displacements or tractions in the z-direction are zero, such problems are often studied using the theory of *two-dimensional (2D) elasticity*. In 2D elasticity, subscripts of stress, σ_{ij}, strain, ε_{ij}, and stiffness, c_{ijkl}, vary over only x and y ($i,j,k,l = 1,2$), and isotropic elastic moduli are *redefined* to describe in-plane stress and strain.

We can generalize Hooke's law for isotropic materials to (Eischen and Torquato, 1993)

$$\widetilde{\sigma} = 2\mu^{(d)}\widetilde{\varepsilon} + (K^{(d)} - 2\mu^{(d)}/d)Tr(\widetilde{\varepsilon})\mathbf{I} \tag{2.1.17}$$

or

$$\sigma_{ij} = 2\mu^{(d)}\varepsilon_{ij} + (K^{(d)} - 2\mu^{(d)}/d)\delta_{ij}\varepsilon_{\alpha\alpha}, \quad i,j,\alpha = 1,\ldots d, \tag{2.1.18}$$

where $K^{(d)}$ is the bulk modulus, $\mu^{(d)}$ is the shear modulus, and d is the spatial dimension. Note that the stress $\widetilde{\sigma}$ and strain $\widetilde{\varepsilon}$ in the previous expressions of Hooke's law are $d \times d$ tensors.

The 2D moduli are sometimes called the "area" moduli, since they only describe the relations among stresses and strains in the two planar directions x and y:

$$K^{(2)} \equiv \frac{1}{2} \frac{(\sigma_{xx} + \sigma_{yy})}{(\varepsilon_{xx} + \varepsilon_{yy})} \tag{2.1.19}$$

$$\mu^{(2)} \equiv \frac{1}{2} \frac{\sigma_{xy}}{\varepsilon_{xy}} \tag{2.1.20}$$

$$E^{(2)} \equiv \frac{\sigma_{xx}}{\varepsilon_{xx}} \quad \text{and} \quad v^{(2)} \equiv -\frac{\varepsilon_{yy}}{\varepsilon_{xx}}, \quad \text{when } \sigma_{yy} = 0 \tag{2.1.21}$$

The planar shear modulus $\mu^{(2)}$ (either plane stress or plane strain) is equal to the 3D shear modulus $\mu^{(3)}$. The relations among the other 2D and 3D moduli can be more complicated, as shown in Table 2.1.2 (Torquato, 2002). *The relations given in Table 2.1.2 between 2D and 3D moduli only apply to homogeneous isotropic materials and not to composites or porous materials.*

Caution:

While we define 2D elastic constants to have the same familiar symbols as used in 3D (e.g., $E^{(2)}$, $K^{(2)}$, $v^{(2)}$), their numerical values differ between 2D and 3D, and their occurrences in Hooke's law and equilibrium equations are different. Some authors use 3D elastic constants to represent 2D problems, while other authors use 2D elastic constants. It is important to distinguish whether 2D or 3D moduli are being used.

Table 2.1.2 *The relations between 2D and 3D elastic moduli vary, depending on whether plane strain or plane stress elasticity is being considered.*

2D plane strain	2D plane stress
$\mu^{(2)} = \mu^{(3)}$	$\mu^{(2)} = \mu^{(3)}$
$K^{(2)} = K^{(3)} + \mu^{(3)}/3$	$K^{(2)} = \dfrac{9K^{(3)}\mu^{(3)}}{3K^{(3)} + 4\mu^{(3)}}$
$K^{(3)} = K^{(2)} - \mu^{(2)}/3$	$K^{(3)} = \dfrac{4K^{(2)}\mu^{(2)}}{9\mu^{(2)} - 3K^{(2)}}$
$v^{(2)} = \dfrac{v^{(3)}}{\left(1 - v^{(3)}\right)}$	$v^{(2)} = v^{(3)}$
$v^{(3)} = \dfrac{v^{(2)}}{\left(1 + v^{(2)}\right)}$	
$E^{(2)} = \dfrac{E^{(3)}}{\left(1 - v^{(3)}\right)\left(1 + v^{(3)}\right)}$	$E^{(2)} = E^{(3)}$

Table 2.1.3 *The table shows relations among various isotropic elastic moduli in each space. The 2D expressions apply among plane strain moduli, and among plane stress moduli (Torquato, 2002).*

3D	2D
$\mu^{(3)} = \dfrac{E^{(3)}}{2(1 + \nu^{(3)})}$	$\mu^{(2)} = \dfrac{E^{(2)}}{2(1 + \nu^{(2)})}$
$K^{(3)} = \dfrac{E^{(3)}}{3\left(1 - 2\nu^{(3)}\right)}$	$K^{(2)} = \dfrac{E^{(2)}}{2\left(1 - \nu^{(2)}\right)}$
$\dfrac{9}{E^{(3)}} = \dfrac{1}{K^{(3)}} + \dfrac{3}{\mu^{(3)}}$	$\dfrac{4}{E^{(2)}} = \dfrac{1}{K^{(2)}} + \dfrac{1}{\mu^{(2)}}$
$\nu^{(3)} = \dfrac{3K^{(3)} - 2\mu^{(3)}}{6K^{(3)} + 2\mu^{(3)}}$	$\nu^{(2)} = \dfrac{K^{(2)} - \mu^{(2)}}{K^{(2)} + \mu^{(2)}}$
$-1 \le \nu^{(3)} \le \dfrac{1}{2}$	$-1 \le \nu^{(2)} \le 1$
$\lambda^{(3)} = K^{(3)} - \dfrac{2\mu^{(3)}}{3}$	$\lambda^{(2)} = K^{(2)} - \mu^{(2)}$

Table 2.1.3 shows how the various elastic moduli are related to each other within each of the dimensional domains. Note that in three dimensions, the Poisson's ratio must fall within the range $-1 \le \nu^{(3)} \le 1/2$, while in two dimensions, the Poisson's ratio must fall in the range $-1 \le \nu^{(2)} \le 1$. The interrelations among the 2D moduli in Table 2.1.3 apply to both plane stress and plane strain problems.

A homogeneous isotropic elastic plate has 2D Young's modulus, $E_1^{(2)}$, and Poisson's ratio, $\nu_1^{(2)}$. If a statistically isotropic set of arbitrarily shaped 2D holes are created, the normalized 2D effective Young's modulus $E_e^{(2)}/E_1^{(2)}$ depends on the porosity and hole geometry but is independent of $\nu_1^{(2)}$. That is, $E_e^{(2)}/E_1^{(2)}$ is independent of the material making up the plate (Milton, 2002). When there are so many holes that the plate is about to fall apart (the critical porosity), the Poisson's ratio goes to a universal value that depends on geometry but is independent of either Young's modulus or Poisson's ratio of the plate (Milton, 2002; Thorpe and Jasiuk, 1992).

The values of effective plane strain moduli of a composite are generally not the same as the corresponding effective plane stress moduli (Eischen and Torquato, 1993). However, the functional form of the expressions for effective moduli will be the same in both domains. For example, the effective *plain strain* bulk modulus of a two-phase composite will be a function of the microgeometry, the volume fractions, and the individual phase moduli:

$$K_e^{pstrain} = f(K_1^{pstrain}, K_2^{pstrain}, \mu_1^{pstrain}, \mu_2^{pstrain}), \tag{2.1.22}$$

and the effective *plane stress* bulk modulus will be

$$K_e^{pstress} = f(K_1^{pstress}, K_2^{pstress}, \mu_1^{pstress}, \mu_2^{pstress}). \tag{2.1.23}$$

Note, however, that the plane strain effective moduli depend on the plane strain phase moduli, while the plane-stress effective moduli depend on the plane stress phase moduli.

A specific example is the Hashin–Shtrikman bounds for two dimensions. For a two-phase composite with 2D phase moduli $(K_1^{(2)}, \mu_1^{(2)})$ and $(K_2^{(2)}, \mu_2^{(2)})$ and volume fractions $f_1 + f_2 = 1$, the 2D Hashin–Shtrikman bounds are (Milton, 2002; Hashin, 1965)

$$K_{2D}^{HS+/-} = f_1 K_1^{(2)} + f_2 K_2^{(2)} - \frac{f_1 f_2 \left(K_2^{(2)} - K_1^{(2)}\right)^2}{f_2 K_1^{(2)} + f_1 K_2^{(2)} + \mu_{\min/\max}^{(2)}}. \tag{2.1.24}$$

The upper bound on planar bulk modulus is found when $\mu_{\min/\max}^{(2)} = \max\left(\mu_1^{(2)}, \mu_2^{(2)}\right)$ and the lower bound when $\mu_{\min/\max}^{(2)} = \min\left(\mu_1^{(2)}, \mu_2^{(2)}\right)$. The Hashin–Shtrikman bounds on planar shear modulus are

$$\mu_{2D}^{HS+/-} = f_1 \mu_1^{(2)} + f_2 \mu_2^{(2)} - \frac{f_1 f_2 \left(\mu_2^{(2)} - \mu_1^{(2)}\right)^2}{f_2 \mu_1^{(2)} + f_1 \mu_2^{(2)} + H_{\min/\max}^{(2)}}, \tag{2.1.25}$$

where

$$H_{\min/\max}^{(2)} = \mu_{\min/\max}^{(2)} \left[\frac{K_{\min/\max}^{(2)} + 3\mu_{\min/\max}^{(2)}/2}{K_{\min/\max}^{(2)} + 2\mu_{\min/\max}^{(2)}}\right], \tag{2.1.26}$$

This expression for $\mu_{2D}^{HS+/-}$ is for the well-ordered case $(K_2^{(2)} - K_1^{(2)})(\mu_2^{(2)} - \mu_1^{(2)}) \geq 0$. In this case the upper bound is obtained when $K_{\min/\max}^{(2)} = \max\left(K_1^{(2)}, K_2^{(2)}\right)$ and $\mu_{\min/\max}^{(2)} = \max\left(\mu_1^{(2)}, \mu_2^{(2)}\right)$, and the lower bound is obtained when $K_{\min/\max}^{(2)} = \min\left(K_1^{(2)}, K_2^{(2)}\right)$ and $\mu_{\min/\max}^{(2)} = \min\left(\mu_1^{(2)}, \mu_2^{(2)}\right)$. These expressions yield the bounds on *plane strain* effective moduli if the inputs are *plane strain* phase moduli. Similarly, they are bounds on *plane stress* effective moduli if the inputs are *plane–stress* phase moduli.

Caution:

The functional forms of the Hashin–Shtrikman bounds are different in 2D than in 3D.

Assumptions and Limitations

The preceding equations assume isotropic, linear elastic media.

2.2   Anisotropic Form of Hooke's Law

Synopsis

Hooke's law for a general anisotropic, linear, elastic solid states that the stress σ_{ij} is linearly proportional to the strain ε_{ij}, as expressed by

$$\sigma_{ij} = c_{ijkl}\varepsilon_{kl} \tag{2.2.1}$$

in which summation (over 1, 2, 3) is implied over the repeated subscripts k and l. The **elastic stiffness tensor**, with elements c_{ijkl}, is a fourth-rank tensor obeying the laws of tensor transformation and has a total of 81 components. However, not all 81 components are independent. The symmetry of stresses and strains implies that

$$c_{ijkl} = c_{jikl} = c_{ijlk} = c_{jilk} \tag{2.2.2}$$

reducing the number of independent constants to 36. In addition, the existence of a unique strain energy potential requires that

$$c_{ijkl} = c_{klij} \tag{2.2.3}$$

further reducing the number of independent constants to 21. This is the maximum number of independent elastic constants that any homogeneous linear elastic medium can have. Additional restrictions imposed by symmetry considerations reduce the number much further. **Isotropic**, linear elastic materials, which have maximum symmetry, are completely characterized by two independent constants, whereas materials with **triclinic** symmetry (the minimum symmetry) require all 21 constants.

Alternatively, the strains may be expressed as a linear combination of the stresses by the following expression:

$$\varepsilon_{ij} = s_{ijkl}\sigma_{kl} \tag{2.2.4}$$

In this case, s_{ijkl} are elements of the **elastic compliance tensor**, which has the same symmetry as the corresponding stiffness tensor. The compliance and stiffness are tensor inverses, denoted by

$$c_{ijkl}s_{klmn} = I_{ijmn} = \frac{1}{2}\left(\delta_{im}\delta_{jn} + \delta_{in}\delta_{jm}\right) \tag{2.2.5}$$

The stiffness and compliance tensors must always be positive definite. One way to express this requirement is that all of the eigenvalues of the elasticity tensor (described later in this section) must be positive.

Voigt Notation

It is a standard practice in elasticity to use an abbreviated *Voigt* notation for the stresses, strains, and stiffness and compliance tensors, for doing so simplifies some of the key equations (Auld, 1990). In this abbreviated notation, the stresses and strains are written as six-element column vectors rather than as nine-element square matrices:

$$
T = \begin{bmatrix} \sigma_1 = \sigma_{11} \\ \sigma_2 = \sigma_{22} \\ \sigma_3 = \sigma_{33} \\ \sigma_4 = \sigma_{23} \\ \sigma_5 = \sigma_{13} \\ \sigma_6 = \sigma_{12} \end{bmatrix} \quad E = \begin{bmatrix} \varepsilon_1 = \varepsilon_{11} \\ \varepsilon_2 = \varepsilon_{22} \\ \varepsilon_3 = \varepsilon_{33} \\ \varepsilon_4 = 2\varepsilon_{23} \\ \varepsilon_5 = 2\varepsilon_{13} \\ \varepsilon_6 = 2\varepsilon_{12} \end{bmatrix} \tag{2.2.6}
$$

Note the factor of 2 in the definitions of strains, but not in the definition of stresses.

With the Voigt notation, four subscripts of the stiffness and compliance tensors are reduced to two. Each pair of indices $ij(kl)$ is replaced by one index $I(J)$ using the following convention:

$ij(kl)$	$I(J)$
11	1
22	2
33	3
23; 32	4
13; 31	5
12; 21	6

The relation, therefore, is $c_{IJ} = c_{ijkl}$ and $s_{IJ} = s_{ijkl}$ **N**, where

$$
N = \begin{cases} 1 & \text{for } I \text{ and } J = 1, 2, 3 \\ 2 & \text{for } I \text{ or } J = 4, 5, 6 \\ 4 & \text{for } I \text{ and } J = 4, 5, 6 \end{cases}
$$

Note how the definition of s_{IJ} differs from that of c_{IJ}. This results from the factors of 2 introduced in the definition of strains in the abbreviated notation. Hence the Voigt matrix representation of the elastic stiffness is

$$
\begin{pmatrix} c_{11} & c_{12} & c_{13} & c_{14} & c_{15} & c_{16} \\ c_{12} & c_{22} & c_{23} & c_{24} & c_{25} & c_{26} \\ c_{13} & c_{23} & c_{33} & c_{34} & c_{35} & c_{36} \\ c_{14} & c_{24} & c_{34} & c_{44} & c_{45} & c_{46} \\ c_{15} & c_{25} & c_{35} & c_{45} & c_{55} & c_{56} \\ c_{16} & c_{26} & c_{36} & c_{46} & c_{56} & c_{66} \end{pmatrix} = \begin{pmatrix} c_{1111} & c_{1122} & c_{1133} & c_{1123} & c_{1113} & c_{1112} \\ c_{1122} & c_{2222} & c_{2233} & c_{2223} & c_{2213} & c_{2212} \\ c_{1133} & c_{2233} & c_{3333} & c_{3323} & c_{3313} & c_{3312} \\ c_{1123} & c_{2223} & c_{3323} & c_{2323} & c_{2313} & c_{2312} \\ c_{1113} & c_{2213} & c_{3313} & c_{2313} & c_{1313} & c_{1312} \\ c_{1112} & c_{2212} & c_{3312} & c_{2312} & c_{1312} & c_{1212} \end{pmatrix} \tag{2.2.7}
$$

and similarly, the Voigt matrix representation of the elastic compliance is

$$
\begin{pmatrix}
s_{11} & s_{12} & s_{13} & s_{14} & s_{15} & s_{16} \\
s_{12} & s_{22} & s_{23} & s_{24} & s_{25} & s_{26} \\
s_{13} & s_{23} & s_{33} & s_{34} & s_{35} & s_{36} \\
s_{14} & s_{24} & s_{34} & s_{44} & s_{45} & s_{46} \\
s_{15} & s_{25} & s_{35} & s_{45} & s_{55} & s_{56} \\
s_{16} & s_{26} & s_{36} & s_{46} & s_{56} & s_{66}
\end{pmatrix}
=
\begin{pmatrix}
s_{1111} & s_{1122} & s_{1133} & 2s_{1123} & 2s_{1113} & 2s_{1112} \\
s_{1122} & s_{2222} & s_{2233} & 2s_{2223} & 2s_{2213} & 2s_{2212} \\
s_{1133} & s_{2233} & s_{3333} & 2s_{3323} & 2s_{3313} & 2s_{3312} \\
2s_{1123} & 2s_{2223} & 2s_{3323} & 4s_{2323} & 4s_{2313} & 4s_{2312} \\
2s_{1113} & 2s_{2213} & 2s_{3313} & 4s_{2313} & 4s_{1313} & 4s_{1312} \\
2s_{1112} & 2s_{2212} & 2s_{3312} & 4s_{2312} & 4s_{1312} & 4s_{1212}
\end{pmatrix}
\tag{2.2.8}
$$

The Voigt stiffness and compliance matrices are symmetric. The upper triangle contains 21 constants, enough to contain the maximum number of independent constants that would be required for the least symmetric linear elastic material.

Using the Voigt notation, we can write Hooke's law as

$$
\begin{pmatrix}
\sigma_1 \\
\sigma_2 \\
\sigma_3 \\
\sigma_4 \\
\sigma_5 \\
\sigma_6
\end{pmatrix}
=
\begin{pmatrix}
c_{11} & c_{12} & c_{13} & c_{14} & c_{15} & c_{16} \\
c_{12} & c_{22} & c_{23} & c_{24} & c_{25} & c_{26} \\
c_{13} & c_{23} & c_{33} & c_{34} & c_{35} & c_{36} \\
c_{14} & c_{24} & c_{34} & c_{44} & c_{45} & c_{46} \\
c_{15} & c_{25} & c_{35} & c_{45} & c_{55} & c_{56} \\
c_{16} & c_{26} & c_{36} & c_{46} & c_{56} & c_{66}
\end{pmatrix}
\begin{pmatrix}
\varepsilon_1 \\
\varepsilon_2 \\
\varepsilon_3 \\
\varepsilon_4 \\
\varepsilon_5 \\
\varepsilon_6
\end{pmatrix}
\tag{2.2.9}
$$

It is very important to note that the stress (strain) vector and stiffness (compliance) matrix in Voigt notation are not tensors. We recommend that the reader compare the Voigt notation with the Kelvin notation, which is described later in this section.

Caution:

Some forms of the abbreviated notation adopt different definitions of strains, moving the factors of 2 and 4 from the compliances to the stiffnesses. However, the form previously given is the more common convention. In the two-index notation, c_{IJ} and s_{IJ} can conveniently be represented as 6×6 matrices. However, these matrices no longer follow the laws of tensor transformation. Care must be taken when transforming from one coordinate system to another. One way is to go back to the four-index notation and then use the ordinary laws of coordinate transformation. A more efficient method is to use the Bond transformation matrices, which are explained in Section 1.5 on coordinate transformations.

Voigt Stiffness Matrix Structure for Common Anisotropy Classes

The nonzero components of the more symmetric anisotropy classes commonly used in modeling rock properties are given in this chapter in Voigt notation.

Isotropic: Two Independent Constants

The structure of the Voigt elastic stiffness matrix for an isotropic linear elastic material has the following form:

$$
\begin{bmatrix}
c_{11} & c_{12} & c_{12} & 0 & 0 & 0 \\
c_{12} & c_{11} & c_{12} & 0 & 0 & 0 \\
c_{12} & c_{12} & c_{11} & 0 & 0 & 0 \\
0 & 0 & 0 & c_{44} & 0 & 0 \\
0 & 0 & 0 & 0 & c_{44} & 0 \\
0 & 0 & 0 & 0 & 0 & c_{44}
\end{bmatrix}, \quad c_{12} = c_{11} - 2c_{44}
\tag{2.2.10}
$$

The relations between the elements c and Lamé's parameters λ and μ of isotropic linear elasticity are

$$
c_{11} = \lambda + 2\mu, \qquad c_{12} = \lambda, \qquad c_{44} = \mu
\tag{2.2.11}
$$

The corresponding isotropic compliance tensor elements can be written in terms of the stiffnesses:

$$
s_{11} = \frac{c_{11} + c_{12}}{(c_{11} - c_{12})(c_{11} + 2c_{12})}, \quad s_{44} = \frac{1}{c_{44}}
\tag{2.2.12}
$$

Energy considerations require that for an isotropic linear elastic material, the following conditions must hold:

$$
K = c_{11} - \frac{4}{3}c_{44} > 0, \quad \mu = c_{44} > 0
\tag{2.2.13}
$$

Cubic: Three Independent Constants

When each Cartesian coordinate plane is aligned with a symmetry plane of a material with cubic symmetry, the Voigt elastic stiffness matrix has the following form:

$$
\begin{bmatrix}
c_{11} & c_{12} & c_{12} & 0 & 0 & 0 \\
c_{12} & c_{11} & c_{12} & 0 & 0 & 0 \\
c_{12} & c_{12} & c_{11} & 0 & 0 & 0 \\
0 & 0 & 0 & c_{44} & 0 & 0 \\
0 & 0 & 0 & 0 & c_{44} & 0 \\
0 & 0 & 0 & 0 & 0 & c_{44}
\end{bmatrix}
\tag{2.2.14}
$$

The corresponding nonzero cubic compliance tensor elements can be written in terms of the stiffnesses:

$$
s_{11} = \frac{c_{11} + c_{12}}{(c_{11} - c_{12})(c_{11} + 2c_{12})}, \quad s_{12} = \frac{-c_{12}}{(c_{11} - c_{12})(c_{11} + 2c_{12})}, \quad s_{44} - \frac{1}{c_{44}}
\tag{2.2.15}
$$

Energy considerations require that for a linear elastic material with cubic symmetry, the following conditions must hold:

$$c_{44} \geq 0, \quad c_{11} > |c_{12}|, \quad c_{11} + 2c_{12} > 0 \tag{2.2.16}$$

Hexagonal or Transversely Isotropic: Five Independent Constants

When the axis of symmetry of a transversely isotropic material lies along the x3-axis, the Voigt stiffness matrix has the form:

$$\begin{bmatrix} c_{11} & c_{12} & c_{13} & 0 & 0 & 0 \\ c_{12} & c_{11} & c_{13} & 0 & 0 & 0 \\ c_{13} & c_{13} & c_{33} & 0 & 0 & 0 \\ 0 & 0 & 0 & c_{44} & 0 & 0 \\ 0 & 0 & 0 & 0 & c_{44} & 0 \\ 0 & 0 & 0 & 0 & 0 & c_{66} \end{bmatrix}; \quad c_{66} = \frac{1}{2}(c_{11} - c_{12}) \tag{2.2.17}$$

The corresponding nonzero hexagonal compliance tensor elements can be written in terms of the stiffnesses:

$$s_{11} + s_{12} = \frac{c_{33}}{c_{33}(c_{11} + c_{12}) - 2c_{13}^2}, \quad s_{11} - s_{12} = \frac{1}{c_{11} - c_{12}} \tag{2.2.18}$$

$$s_{13} = -\frac{c_{13}}{c_{33}(c_{11} + c_{12}) - 2c_{13}^2}, \quad s_{33} = \frac{c_{11} + c_{12}}{c_{33}(c_{11} + c_{12}) - 2c_{13}^2}, \quad s_{44} = \frac{1}{c_{44}} \tag{2.2.19}$$

Energy considerations require that for a linear elastic material with transversely isotropic symmetry the following conditions must hold:

$$c_{44} \geq 0, \quad c_{11} > |c_{12}|, \quad (c_{11} + c_{12})c_{33} > 2c_{13}^2 \tag{2.2.20}$$

Orthorhombic: Nine Independent Constants

When each Cartesian coordinate plane is aligned with a symmetry plane of a material with orthorhombic symmetry, the Voigt elastic stiffness matrix has the following form:

$$\begin{bmatrix} c_{11} & c_{12} & c_{13} & 0 & 0 & 0 \\ c_{12} & c_{22} & c_{23} & 0 & 0 & 0 \\ c_{13} & c_{23} & c_{33} & 0 & 0 & 0 \\ 0 & 0 & 0 & c_{44} & 0 & 0 \\ 0 & 0 & 0 & 0 & c_{55} & 0 \\ 0 & 0 & 0 & 0 & 0 & c_{66} \end{bmatrix} \tag{2.2.21}$$

Monoclinic: 13 Independent Constants

When the symmetry plane of a monoclinic medium is orthogonal to the x3-axis, the Voigt elastic stiffness matrix has the following form:

$$\begin{bmatrix} c_{11} & c_{12} & c_{13} & 0 & c_{15} & 0 \\ c_{12} & c_{22} & c_{23} & 0 & c_{25} & 0 \\ c_{13} & c_{23} & c_{33} & 0 & c_{35} & 0 \\ 0 & 0 & 0 & c_{44} & 0 & c_{46} \\ c_{15} & c_{25} & c_{35} & 0 & c_{55} & 0 \\ 0 & 0 & 0 & c_{46} & 0 & c_{66} \end{bmatrix} \tag{2.2.22}$$

Phase Velocities for Several Elastic Symmetry Classes

For *isotropic* symmetry, the phase velocities of propagating elastic body waves are given by

$$V_P = \sqrt{\frac{c_{11}}{\rho}}, \qquad V_S = \sqrt{\frac{c_{44}}{\rho}} \tag{2.2.23}$$

where ρ is the density.

In *anisotropic* media there are, in general, three modes of propagation (quasi-longitudinal, quasi-shear, and pure shear) with mutually orthogonal polarizations.

For a medium with **transversely isotropic** (hexagonal) symmetry, the wave slowness surface is always rotationally symmetric about the axis of symmetry. The phase velocities of the three modes in any plane containing the symmetry axis are given as

quasi-longitudinal mode (transversely isotropic):

$$V_P = (c_{11}\sin^2\theta + c_{33}\cos^2\theta + c_{44} + \sqrt{M})^{1/2}(2\rho)^{-1/2} \tag{2.2.24}$$

quasi-shear mode (transversely isotropic):

$$V_{SV} = (c_{11}\sin^2\theta + c_{33}\cos^2\theta + c_{44} - \sqrt{M})^{1/2}(2\rho)^{-1/2} \tag{2.2.25}$$

pure shear mode (transversely isotropic):

$$V_{SH} = \left(\frac{c_{66}\sin^2\theta + c_{44}\cos^2\theta}{\rho}\right)^{1/2} \tag{2.2.26}$$

where

$$M = [(c_{11} - c_{44})\sin^2\theta - (c_{33} - c_{44})\cos^2\theta]^2 + (c_{13} + c_{44})^2\sin^2 2\theta \tag{2.2.27}$$

and θ is the angle between the wave vector and the x_3-axis of symmetry ($\theta = 0$ for propagation along the x_3-axis). The five components of the stiffness tensor for a transversely isotropic material are obtained from five velocity measurements: $V_P(0°)$, $V_P(90°)$, $V_P(45°)$, $V_{SH}(90°)$, and $V_{SH}(0°) = V_{SV}(0°)$:

$$c_{11} = \rho V_P^2(90°) \tag{2.2.28}$$

$$c_{12} = c_{11} - 2\rho V_{SH}^2(90°) \tag{2.2.29}$$

$$c_{33} = \rho V_P^2(0°) \tag{2.2.30}$$

$$c_{44} = \rho V_{SH}^2(0°) \tag{2.2.31}$$

and

$$c_{13} = -c_{44} + \sqrt{4\rho^2 V_{P(45°)}^4 - 2\rho V_{P(45°)}^2(c_{11} + c_{33} + 2c_{44}) + (c_{11} + c_{44})(c_{33} + c_{44})} \tag{2.2.32}$$

For the more general **orthorhombic** symmetry, the phase velocities of the three modes propagating in the three symmetry planes (*XZ*, *YZ*, and *XY*) are given as follows:

quasi-longitudinal mode (orthorhombic, *XZ* plane):

$$V_P = \left(c_{55} + c_{11}\sin^2\theta + c_{33}\cos^2\theta + \sqrt{(c_{55} + c_{11}\sin^2\theta + c_{33}\cos^2\theta)^2 - 4A}\right)^{1/2}(2\rho)^{-1/2} \tag{2.2.33}$$

quasi-shear mode (orthorhombic, *XZ* plane):

$$V_{SV} = \left(c_{55} + c_{11}\sin^2\theta + c_{33}\cos^2\theta - \sqrt{(c_{55} + c_{11}\sin^2\theta + c_{33}\cos^2\theta)^2 - 4A}\right)^{1/2}(2\rho)^{-1/2} \tag{2.2.34}$$

pure shear mode (orthorhombic, *XZ* plane):

$$V_{SH} = \left(\frac{c_{66}\sin^2\theta + c_{44}\cos^2\theta}{\rho}\right)^{1/2} \tag{2.2.35}$$

where

$$A = (c_{11}\sin^2\theta + c_{55}\cos^2\theta)(c_{55}\sin^2\theta + c_{33}\cos^2\theta) - (c_{13} + c_{55})^2\sin^2\theta\cos^2\theta \tag{2.2.36}$$

quasi-longitudinal mode (orthorhombic, *YZ* plane):

$$V_P = \left(c_{44} + c_{22}\sin^2\theta + c_{33}\cos^2\theta + \sqrt{(c_{44} + c_{22}\sin^2\theta + c_{33}\cos^2\theta)^2 - 4B}\right)^{1/2}(2\rho)^{-1/2} \tag{2.2.37}$$

quasi-shear mode (orthorhombic, *YZ* plane):

$$V_{SV} = \left(c_{44} + c_{22}\sin^2\theta + c_{33}\cos^2\theta - \sqrt{(c_{44} + c_{22}\sin^2\theta + c_{33}\cos^2\theta)^2 - 4B}\right)^{1/2}(2\rho)^{-1/2} \tag{2.2.38}$$

pure shear mode (orthorhombic, *YZ* plane):

$$V_{\text{SH}} = \left(\frac{c_{66} \sin^2\theta + c_{55} \cos^2\theta}{\rho} \right)^{1/2} \tag{2.2.39}$$

where

$$B = (c_{22} \sin^2\theta + c_{44} \cos^2\theta)(c_{44} \sin^2\theta + c_{33} \cos^2\theta) - (c_{23} + c_{44})^2 \sin^2\theta \cos^2\theta \tag{2.2.40}$$

quasi-longitudinal mode (orthorhombic, *XY* plane):

$$V_{\text{P}} = \left(c_{66} + c_{22} \sin^2\varphi + c_{11} \cos^2\varphi + \sqrt{(c_{66} + c_{22} \sin^2\varphi + c_{11} \cos^2\varphi)^2 - 4C} \right)^{1/2} (2\rho)^{-1/2} \tag{2.2.41}$$

quasi-shear mode (orthorhombic, *XY* plane):

$$V_{\text{SH}} = \left(c_{66} + c_{22} \sin^2\varphi + c_{11} \cos^2\varphi - \sqrt{(c_{66} + c_{22} \sin^2\varphi + c_{11} \cos^2\varphi)^2 - 4C} \right)^{1/2} (2\rho)^{-1/2} \tag{2.2.42}$$

pure shear mode (orthorhombic, *XY* plane):

$$V_{\text{SV}} = \left(\frac{c_{55} \cos^2\varphi + c_{44} \sin^2\varphi}{\rho} \right)^{1/2} \tag{2.2.43}$$

where

$$C = (c_{66} \sin^2\varphi + c_{11} \cos^2\varphi)(c_{22} \sin^2\varphi + c_{66} \cos^2\varphi) - (c_{12} + c_{66})^2 \sin^2\varphi \cos^2\varphi \tag{2.2.44}$$

and θ and φ are the angles of the wave vector relative to the x_3 and x_1 axes, respectively.

Kelvin Notation

In spite of its almost exclusive use in the geophysical literature, the abbreviated Voigt notation has several mathematical disadvantages. For example, certain norms of the fourth-rank stiffness tensor are not equal to the corresponding norms of the Voigt stiffness matrix (Thomson, 1878; Helbig, 1994; Dellinger *et al.*, 1998), and the eigenvalues of the Voigt stiffness matrix are not the eigenvalues of the stiffness tensor.

The lesser-known **Kelvin notation** is very similar to the Voigt notation, except that each element of the 6×6 matrix is weighted according to how many elements of the actual stiffness tensor it represents. Kelvin matrix elements having indices 1, 2, or 3 each map to only a single index pair in the fourth-rank notation, 11, 22, and 33, respectively, so any matrix stiffness element with index 1, 2, or 3 is given a weight of 1. Kelvin elements with indices 4, 5, or 6 each represent two index pairs (23, 32), (13, 31), and (12, 21), respectively, so each element containing 4, 5, or 6 receives a weight of $\sqrt{2}$. The weighting must be applied for each Kelvin index. For example, the Kelvin notation would map $\hat{c}_{11} = c_{1111}$, $\hat{c}_{14} = \sqrt{2}c_{1123}$, $\hat{c}_{66} = 2c_{1212}$. One way to convert a Voigt stiffness matrix into a Kelvin stiffness matrix is to pre- and post-multiply by the following weighting matrix (Dellinger et $al.$, 1998):

$$
\begin{bmatrix}
1 & 0 & 0 & 0 & 0 & 0 \\
0 & 1 & 0 & 0 & 0 & 0 \\
0 & 0 & 1 & 0 & 0 & 0 \\
0 & 0 & 0 & \sqrt{2} & 0 & 0 \\
0 & 0 & 0 & 0 & \sqrt{2} & 0 \\
0 & 0 & 0 & 0 & 0 & \sqrt{2}
\end{bmatrix}
\tag{2.2.45}
$$

<u>yielding</u>

$$
\begin{pmatrix}
\hat{c}_{11} & \hat{c}_{12} & \hat{c}_{13} & \hat{c}_{14} & \hat{c}_{15} & \hat{c}_{16} \\
\hat{c}_{12} & \hat{c}_{22} & \hat{c}_{23} & \hat{c}_{24} & \hat{c}_{25} & \hat{c}_{26} \\
\hat{c}_{13} & \hat{c}_{23} & \hat{c}_{33} & \hat{c}_{34} & \hat{c}_{35} & \hat{c}_{36} \\
\hat{c}_{14} & \hat{c}_{24} & \hat{c}_{34} & \hat{c}_{44} & \hat{c}_{45} & \hat{c}_{46} \\
\hat{c}_{15} & \hat{c}_{25} & \hat{c}_{35} & \hat{c}_{45} & \hat{c}_{55} & \hat{c}_{56} \\
\hat{c}_{16} & \hat{c}_{26} & \hat{c}_{36} & \hat{c}_{46} & \hat{c}_{56} & \hat{c}_{66}
\end{pmatrix}
\tag{2.2.46}
$$

$$
=
\begin{pmatrix}
c_{11} & c_{12} & c_{13} & \sqrt{2}c_{14} & \sqrt{2}c_{15} & \sqrt{2}c_{16} \\
c_{12} & c_{22} & c_{23} & \sqrt{2}c_{24} & \sqrt{2}c_{25} & \sqrt{2}c_{26} \\
c_{13} & c_{23} & c_{33} & \sqrt{2}c_{34} & \sqrt{2}c_{35} & \sqrt{2}c_{36} \\
\sqrt{2}c_{14} & \sqrt{2}c_{24} & \sqrt{2}c_{34} & 2c_{44} & 2c_{45} & 2c_{46} \\
\sqrt{2}c_{15} & \sqrt{2}c_{25} & \sqrt{2}c_{35} & 2c_{45} & 2c_{55} & 2c_{56} \\
\sqrt{2}c_{16} & \sqrt{2}c_{26} & \sqrt{2}c_{36} & 2c_{46} & 2c_{56} & 2c_{66}
\end{pmatrix}
\tag{2.2.47}
$$

where the $\hat{c}_{ij}$ are the Kelvin elastic elements and c_{ij} are the Voigt elements.

Similarly, the stress and strain elements take on the following weights in the Kelvin notation:

$$
\hat{T} =
\begin{bmatrix}
\hat{\sigma}_1 & = & \sigma_{11} \\
\hat{\sigma}_2 & = & \sigma_{22} \\
\hat{\sigma}_3 & = & \sigma_{33} \\
\hat{\sigma}_4 & = & \sqrt{2}\sigma_{23} \\
\hat{\sigma}_5 & = & \sqrt{2}\sigma_{13} \\
\hat{\sigma}_6 & = & \sqrt{2}\sigma_{12}
\end{bmatrix}
;\quad
\hat{E} =
\begin{bmatrix}
\hat{\varepsilon}_1 & = & \varepsilon_{11} \\
\hat{\varepsilon}_2 & = & \varepsilon_{22} \\
\hat{\varepsilon}_3 & = & \varepsilon_{33} \\
\hat{\varepsilon}_4 & = & \sqrt{2}\varepsilon_{23} \\
\hat{\varepsilon}_5 & = & \sqrt{2}\varepsilon_{13} \\
\hat{\varepsilon}_6 & = & \sqrt{2}\varepsilon_{12}
\end{bmatrix}
\tag{2.2.48}
$$

Hooke's law again takes on the familiar form in the Kelvin notation:

$$
\begin{pmatrix} \hat{\sigma}_1 \\ \hat{\sigma}_2 \\ \hat{\sigma}_3 \\ \hat{\sigma}_4 \\ \hat{\sigma}_5 \\ \hat{\sigma}_6 \end{pmatrix} = \begin{pmatrix} \hat{c}_{11} & \hat{c}_{12} & \hat{c}_{13} & \hat{c}_{14} & \hat{c}_{15} & \hat{c}_{16} \\ \hat{c}_{12} & \hat{c}_{22} & \hat{c}_{23} & \hat{c}_{24} & \hat{c}_{25} & \hat{c}_{26} \\ \hat{c}_{13} & \hat{c}_{23} & \hat{c}_{33} & \hat{c}_{34} & \hat{c}_{35} & \hat{c}_{36} \\ \hat{c}_{14} & \hat{c}_{24} & \hat{c}_{34} & \hat{c}_{44} & \hat{c}_{45} & \hat{c}_{46} \\ \hat{c}_{15} & \hat{c}_{25} & \hat{c}_{35} & \hat{c}_{45} & \hat{c}_{55} & \hat{c}_{56} \\ \hat{c}_{16} & \hat{c}_{26} & \hat{c}_{36} & \hat{c}_{46} & \hat{c}_{56} & \hat{c}_{66} \end{pmatrix} \begin{pmatrix} \hat{\varepsilon}_1 \\ \hat{\varepsilon}_2 \\ \hat{\varepsilon}_3 \\ \hat{\varepsilon}_4 \\ \hat{\varepsilon}_5 \\ \hat{\varepsilon}_6 \end{pmatrix}
\tag{2.2.49}
$$

The **Kelvin notation** for the stiffness matrix has a number of advantages. The Kelvin elastic matrix is a tensor. It preserves the norm of the $3 \times 3 \times 3 \times 3$ notation:

$$
\sum_{i,j=1,\dots,6} \hat{c}_{ij}^2 = \sum_{i,j,k,l=1,\dots,3} c_{ijkl}^2
\tag{2.2.50}
$$

The eigenvalues and eigenvectors (or eigentensors) of the Kelvin 6×6 C-matrix are geometrically meaningful (Kelvin, 1856; Mehrabadi and Cowin, 1990; Dellinger *et al.*, 1998). Each Kelvin eigentensor corresponds to a state of the medium where the stress and strain ellipsoids are aligned and have the same aspect ratios. Furthermore, a stiffness matrix is physically realizable if and only if all of the Kelvin eigenvalues are non-negative, which is useful for inferring elastic constants from laboratory data.

Elastic Eigentensors and Eigenvalues

The eigenvectors of the 3D fourth-rank anisotropic elasticity tensor are called *eigentensors* when projected into 3D space. The maximum number of eigentensors for any elastic symmetry is six (Kelvin, 1856; Mehrabadi and Cowin, 1990). In the case of isotropic linear elasticity, there are two unique eigentensors. The total stress tensor $\widetilde{\sigma}$ and the total strain tensor $\widetilde{\varepsilon}$ for the isotropic case can be decomposed in terms of the deviatoric second-rank tensors ($\widetilde{\sigma}^{dev}$ and $\widetilde{\varepsilon}^{dev}$) and scaled unit tensors $[\frac{1}{3} tr(\widetilde{\sigma})\widetilde{I}$ and $\frac{1}{3} tr(\widetilde{\varepsilon})\widetilde{I}]$:

$$
\widetilde{\sigma} = \widetilde{\sigma}^{dev} + \frac{1}{3} tr(\widetilde{\sigma})\,\widetilde{I}
\tag{2.2.51}
$$

$$
\widetilde{\varepsilon} = \widetilde{\varepsilon}^{dev} + \frac{1}{3} tr(\widetilde{\varepsilon})\,\widetilde{I}
\tag{2.2.52}
$$

For an isotropic material, Hooke's law can be written as

$$
\widetilde{\sigma} = \lambda tr(\widetilde{\varepsilon})\widetilde{I} + 2\mu\widetilde{\varepsilon}
\tag{2.2.53}
$$

where λ is Lamé's constant and μ is the shear modulus. However, if expressed in terms of the stress and strain eigentensors, Hooke's law becomes two uncoupled equations:

$$
tr(\widetilde{\sigma})\widetilde{I} = 3K\, tr(\widetilde{\varepsilon})\widetilde{I}, \quad \widetilde{\sigma}^{dev} = 2\mu\widetilde{\varepsilon}^{dev}
\tag{2.2.54}
$$

Table 2.2.1 *Eigenelastic constants for several different symmetries, expressed in Voigt notation*

Symmetry	Eigenvalue	Multiplicity of eigenvalue
Isotropic	$c_{11} + 2c_{12} = 3K$	1
	$2c_{44} = 2\mu$	5
Cubic	$c_{11} + 2c_{12}$	1
	$c_{11} - c_{12}$	2
	$2c_{44}$	3
Transverse Isotropy	$c_{33} + \sqrt{2}c_{13}(\beta + \sqrt{\beta^2 + 1}), \beta = \dfrac{\sqrt{2}}{4c_{13}}(c_{11} + c_{12} - c_{33})$	1
	$c_{33} + \sqrt{2}c_{13}(\beta - \sqrt{\beta^2 + 1})$	1
	$c_{11} - c_{12}$	2
	$2c_{44}$	2

where K is the bulk modulus. Similarly, the strain energy U for an isotropic material can be written as

$$2U = 2\mu\, tr[(\widetilde{\boldsymbol{\varepsilon}}^{dev})^2] + K[tr(\widetilde{\boldsymbol{\varepsilon}})]^2 \tag{2.2.55}$$

where μ and K scale two energy modes that do not interact.

Hence we can see that, although any two of the isotropic constants (λ, μ, K, Young's modulus, and Poisson's ratio) can be derived in terms of the others, μ and K have a special significance as *eigenelastic constants* (Mehrabadi and Cowin, 1990) or *principal elasticities* of the material (Kelvin, 1856). The stress and strain eigen-tensors associated with μ and K are orthogonal, as shown above. Such an orthogonal significance does not hold for the pair λ and μ. Eigenelastic constants for a few other symmetries are shown in Table 2.2.1, expressed in terms of the Voigt constants (Mehrabadi and Cowin, 1990).

Poisson's Ratio Defined for Anisotropic Elastic Materials

Familiar isotropic definitions for elastic constants are sometimes extended to anisotropic materials. For example, consider a transversely isotropic (TI) material with uniaxial stress, applied along the symmetry (x_3-) axis, such that

$$\sigma_{33} \neq 0 \quad \sigma_{11} = \sigma_{12} = \sigma_{13} = \sigma_{22} = \sigma_{23} = 0 \tag{2.2.56}$$

One can define a transversely isotropic Young's modulus associated with this experiment as

$$E_{33} = \frac{\sigma_{33}}{\varepsilon_{33}} = c_{33} - \frac{2c_{31}^2}{c_{11} + c_{12}} \tag{2.2.57}$$

where the c_{ij} are the elastic stiffnesses in Voigt notation. A pair of TI Poisson's ratios can similarly be defined in terms of the same experiment:

$$v_{31} = v_{32} = -\frac{\varepsilon_{11}}{\varepsilon_{33}} = \frac{c_{31}}{c_{11} + c_{12}} \tag{2.2.58}$$

If the uniaxial stress field is rotated normal to the symmetry axis, such that

$$\sigma_{11} \neq 0 \quad \sigma_{12} = \sigma_{13} = \sigma_{22} = \sigma_{23} = \sigma_{33} = 0 \tag{2.2.59}$$

then one can define another TI Young's modulus and pair of Poisson's ratios as

$$E_{11} = \frac{\sigma_{11}}{\varepsilon_{11}} = c_{11} + \frac{c_{31}^2(-c_{11} + c_{12}) + c_{12}(-c_{33}c_{12} + c_{31}^2)}{c_{33}c_{11} - c_{31}^2} \tag{2.2.60}$$

$$v_{13} = v_{23} = -\frac{\varepsilon_{33}}{\varepsilon_{11}} = \frac{c_{31}(c_{11} - c_{12})}{c_{33}c_{11} - c_{31}^2} \tag{2.2.61}$$

$$v_{12} = v_{21} = -\frac{\varepsilon_{22}}{\varepsilon_{11}} = \frac{c_{33} c_{12} - c_{31}^2}{c_{33}c_{11} - c_{31}^2} \tag{2.2.62}$$

Caution:

Just as with the isotropic case, definitions of Young's modulus and Poisson's ratio in terms of stresses and strains are only true for the uniaxial stress state. Definitions in terms of elastic stiffnesses are independent of stress state.

In spite of their similarity to their isotropic analogs, these TI Poisson's ratios have several distinct differences. For example, the bounds on the isotropic Poisson's ratio require that

$$-1 \leq v_{\text{isotropic}} \leq \frac{1}{2} \tag{2.2.63}$$

In contrast, the bounds on the TI Poisson's ratios defined here are (Christensen, 2005)

$$|v_{12}| \leq 1; \quad |v_{31}| \leq \left(\frac{E_{33}}{E_{11}}\right)^{1/2}; \quad |v_{13}| \leq \left(\frac{E_{11}}{E_{33}}\right)^{1/2} \tag{2.2.64}$$

Another distinct difference is the relation of Poisson's ratio to seismic velocities. In an isotropic material, Poisson's ratio is directly related to the V_P/V_S ratio as follows:

$$(V_P/V_S)^2 = \frac{2(1-v)}{(1-2v)} \tag{2.2.65}$$

or, equivalently,

$$v = \frac{(V_P/V_S)^2 - 2}{2[(V_P/V_S)^2 - 1]} \tag{2.2.66}$$

However, in the TI material, the ratio of velocities propagating along the symmetry (x_3-) axis is

$$\left(\frac{V_P}{V_S}\right)^2 = \frac{c_{33}}{c_{44}} \tag{2.2.67}$$

and is not simply related to the v_{31} Poisson's ratio.

Caution:

The definition of Young's modulus and Poisson's ratio for anisotropic materials, while possible, can be misinterpreted, especially when compared with isotropic formulas

For *orthorhombic* and higher symmetries, the "engineering" parameters – Poisson's ratios, v_{ij}, shear moduli, G_{ij}, and Young's moduli, E_{ij} – can be conveniently defined in terms of elastic *compliances* (Jaeger *et al.*, 2007):

$$\begin{array}{lll}
E_{11} = 1/s_{11} & E_{22} = 1/s_{22} & E_{33} = 1/s_{33} \\
v_{21} = -s_{12}E_{22} & v_{31} = -s_{13}E_{33} & v_{32} = -s_{23}E_{33} \\
G_{12} = 1/s_{66} & G_{13} = 1/s_{55} & G_{23} = 1/s_{44}
\end{array} \tag{2.2.68}$$

Existence of a strain energy leads to the following constraints:

$$G_{12}, \ G_{13}, \ G_{23}, \ E_{11}, \ E_{22}, \ E_{33} > 0 \tag{2.2.69}$$

$$\frac{E_{11}}{E_{22}}(v_{21})^2 < 1, \quad \frac{E_{11}}{E_{33}}(v_{31})^2 < 1, \quad \frac{E_{22}}{E_{33}}(v_{32})^2 < 1 \tag{2.2.70}$$

Bulk Modulus Defined for Anisotropic Elastic Materials

For a linear elastic material with isotropic or cubic elastic symmetry, the elastic bulk modulus is unambiguously defined as the ratio of mean stress to volumetric strain:

$$K = \frac{1}{3}\frac{(\sigma_{11} + \sigma_{22} + \sigma_{33})}{(\varepsilon_{11} + \varepsilon_{22} + \varepsilon_{33})}, \tag{2.2.71}$$

where σ_{ii} and ε_{ii} are the traces of the stress and strain tensors, respectively. For the six other commonly considered elastic symmetries (hexagonal, tetragonal, trigonal, orthorhombic, monoclinic, and triclinic), the definition of bulk modulus is less clear (Berryman, 2011). For example, if a uniform volume strain $\varepsilon_{ij} = \varepsilon_0\delta_{ij}$ is imposed on the sample, the normal stresses are:

$$\begin{bmatrix} \sigma_1 \\ \sigma_2 \\ \sigma_3 \end{bmatrix} = \begin{bmatrix} c_{11} & c_{12} & c_{13} \\ c_{12} & c_{22} & c_{23} \\ c_{13} & c_{23} & c_{33} \end{bmatrix} \begin{bmatrix} \varepsilon_0 \\ \varepsilon_0 \\ \varepsilon_0 \end{bmatrix},$$ (2.2.72)

and the inferred bulk modulus (as the ratio in equation 2.2.71) in terms of the Voigt notation stiffness components is

$$K_1 = \frac{1}{9}(c_{11} + c_{22} + c_{33} + 2c_{12} + 2c_{13} + 2c_{23}),$$ (2.2.73)

which is the most commonly quoted definition. If, instead, a uniform normal stress $\sigma_{ij} = \sigma_0 \delta_{ij}$ is imposed, the normal strains are:

$$\begin{bmatrix} \varepsilon_1 \\ \varepsilon_2 \\ \varepsilon_3 \end{bmatrix} = \begin{bmatrix} s_{11} & s_{12} & s_{13} \\ s_{12} & s_{22} & s_{23} \\ s_{13} & s_{23} & s_{33} \end{bmatrix} \begin{bmatrix} \sigma_0 \\ \sigma_0 \\ \sigma_0 \end{bmatrix},$$ (2.2.74)

and the inferred bulk modulus in terms of Voigt notation compliance components is

$$K_2 = (s_{11} + s_{22} + s_{33} + 2s_{12} + 2s_{13} + 2s_{23})^{-1}.$$ (2.2.75)

Note that $K_2 \leq K_1$. Equality comes when the material has isotropic or cubic elastic symmetry (Avellaneda and Milton, 1989). For less symmetric materials, the ratio of mean stress to volumetric strain depends on the boundary conditions.

Assumptions and Limitations

The preceding equations assume anisotropic, linear elastic media.

2.3 Thomsen's Notation for Weak Elastic Anisotropy

Synopsis

A transversely isotropic elastic material is completely specified by five independent constants. In terms of the abbreviated Voigt notation (see Section 2.2 on elastic anisotropy), the elastic constants can be represented as

$$\begin{bmatrix} c_{11} & c_{12} & c_{13} & 0 & 0 & 0 \\ c_{12} & c_{11} & c_{13} & 0 & 0 & 0 \\ c_{13} & c_{13} & c_{33} & 0 & 0 & 0 \\ 0 & 0 & 0 & c_{44} & 0 & 0 \\ 0 & 0 & 0 & 0 & c_{44} & 0 \\ 0 & 0 & 0 & 0 & 0 & c_{66} \end{bmatrix}, \text{ where } c_{66} = \frac{1}{2}(c_{11} - c_{12})$$ (2.3.1)

and where the axis of symmetry lies along the x_3-axis.

Thomsen (1986) suggested the following convenient notation for the elasticity of a VTI (transversely isotropic with vertical symmetry axis) material. His notation uses the P- and S-wave velocities (denoted by α and β, respectively) propagating along the symmetry axis, plus three additional constants, ε, γ, and δ:

$$\alpha = \sqrt{c_{33}/\rho}, \quad \beta = \sqrt{c_{44}/\rho}, \quad \varepsilon = \frac{c_{11} - c_{33}}{2c_{33}} \tag{2.3.2}$$

$$\gamma = \frac{c_{66} - c_{44}}{2c_{44}}, \quad \delta = \frac{(c_{13} + c_{44})^2 - (c_{33} - c_{44})^2}{2c_{33}(c_{33} - c_{44})} \tag{2.3.3}$$

> The five **Thomsen parameters** in equations 2.3.2 and 2.3.3 are valid for any strength of VTI anisotropy, since they are just definitions. However the expressions for phase velocities given in this section in terms of ε, γ, and δ are approximations valid only for weak anisotropy.

In terms of the Thomsen parameters, the three phase velocities for weak anisotropy can be approximated as

$$V_P(\theta) \approx \alpha(1 + \delta \sin^2\theta \cos^2\theta + \varepsilon \sin^4\theta) \tag{2.3.4}$$

$$V_{SV}(\theta) \approx \beta \left[1 + \frac{\alpha^2}{\beta^2}(\varepsilon - \delta) \sin^2\theta \cos^2\theta\right] \tag{2.3.5}$$

$$V_{SH}(\theta) \approx \beta(1 + \gamma \sin^2\theta) \tag{2.3.6}$$

where θ is the angle of the wave vector relative to the x_3-axis; V_{SH} is the wavefront velocity of the pure shear wave, which has no component of polarization in the symmetry axis direction; V_{SV} is the pseudo-shear wave polarized normal to the pure shear wave; and V_P is the pseudo-longitudinal wave.

Berryman (2008) extends the validity of Thomsen's expressions for P- and quasi-SV-wave velocities to wider ranges of angles and stronger anisotropy with the following expressions:

$$V_P(\theta) \approx \alpha \left[1 + \varepsilon \sin^2\theta - (\varepsilon - \delta)\frac{2 \sin^2\theta_m \sin^2\theta \cos^2\theta}{1 - \cos 2\theta_m \cos 2\theta}\right] \tag{2.3.7}$$

$$V_{SV}(\theta) \approx \beta \left[1 + \left(\frac{\alpha^2}{\beta^2}\right)(\varepsilon - \delta)\frac{2 \sin^2\theta_m \sin^2\theta \cos^2\theta}{1 - \cos 2\theta_m \cos 2\theta}\right] \tag{2.3.8}$$

where

$$\tan^2\theta_m = \frac{c_{33} - c_{44}}{c_{11} - c_{44}} \tag{2.3.9}$$

Berryman's formulas give more accurate velocities at larger angles. They are designed to give the angular location of the peak (or trough) of the quasi-SV-velocity closer to the correct location; hence, the quasi-SV-velocities are more accurate than those from Thomsen's equations. Thomsen's P-wave velocities will sometimes be more accurate than Berryman's at small θ and sometimes worse.

For weak anisotropy, the constant ε can be seen to describe the fractional difference between the P-wave velocities parallel and orthogonal to the symmetry axis (in the weak anisotropy approximation):

$$\varepsilon \approx \frac{V_P(90°) - V_P(0°)}{V_P(0°)} \tag{2.3.10}$$

Therefore, ε best describes what is usually called the "P-wave anisotropy."

Similarly, for weak anisotropy, the constant γ can be seen to describe the fractional difference between the SH-wave velocities parallel and orthogonal to the symmetry axis, which is equivalent to the difference between the velocities of S-waves polarized parallel and normal to the symmetry axis, both propagating normal to the symmetry axis:

$$\gamma \approx \frac{V_{SH}(90°) - V_{SV}(90°)}{V_{SV}(90°)} = \frac{V_{SH}(90°) - V_{SH}(0°)}{V_{SH}(0°)} \tag{2.3.11}$$

The small-offset normal moveout (NMO) velocity is affected by VTI anisotropy. In terms of the Thomsen parameters, NMO velocities, $V_{NMO,P}$, $V_{NMO,SV}$, and $V_{NMO,SH}$ for P-, SV-, and SH-modes are (Tsvankin, 2001):

$$V_{NMO,P} = \alpha\sqrt{1 + 2\delta} \tag{2.3.12}$$

$$V_{NMO,SV} = \beta\sqrt{1 + 2\sigma}, \quad \sigma = \left(\frac{\alpha}{\beta}\right)^2 (\varepsilon - \delta) \tag{2.3.13}$$

$$V_{NMO,SH} = \beta\sqrt{1 + 2\gamma} \tag{2.3.14}$$

An additional *anellipticity* parameter, η, was introduced by Alkhalifah and Tsvankin (1995):

$$\eta = \frac{\varepsilon - \delta}{1 + 2\delta} \tag{2.3.15}$$

η is important in quantifying the effects of anisotropy on nonhyperbolic moveout and P-wave time-processing steps (Tsvankin, 2001), including NMO, DMO (dip moveout), and time migration.

The Thomsen parameters can be inverted for the elastic constants as follows:

$$c_{33} = \rho\alpha^2, \quad c_{44} = \rho\beta^2 \tag{2.3.16}$$

$$c_{11} = c_{33}(1 + 2\varepsilon), \quad c_{66} = c_{44}(1 + 2\gamma) \tag{2.3.17}$$

$$c_{13} = \pm\sqrt{2c_{33}(c_{33} - c_{44})\delta + (c_{33} - c_{44})^2} - c_{44} \tag{2.3.18}$$

Table 2.3.1 *Ranges of Thomsen parameters expected for thin laminations of isotropic materials (Berryman* et al., *1999)*

	$\mu = $ const	$\lambda = $ const $\mu \neq $ const	$\lambda + 2\mu = $ const $\mu \neq $ const	$\lambda + \mu = $ const $\mu \neq $ const	$v = $ const $\lambda, \mu \neq $ const
ε	0	≥ 0	0	≤ 0	≥ 0
δ	0	≤ 0	≤ 0	≤ 0	0
γ	0	≥ 0	≥ 0	≥ 0	≥ 0
$\varepsilon - \delta$	0	≥ 0	≥ 0	≥ 0	≥ 0

Note the nonuniqueness in c_{13} that results from uncertainty in the sign of $(c_{13} + c_{44})$. Tsvankin (2001) argues that for most cases, it can be assumed that $(c_{13} + c_{44}) > 0$, and therefore, the +sign in the equation for c_{13} is usually appropriate.

Tsvankin (2001) summarizes some bounds on the values of the Thomsen parameters:

- the lower bound for δ: $\delta \geq -(1 - \beta^2/\alpha^2)/2$;
- an *approximate* upper bound for δ: $\delta \leq 2/(\alpha^2/\beta^2 - 1)$; and
- in VTI materials resulting from thin layering, $\varepsilon > \delta$, and $\gamma > 0$.

Transversely isotropic media consisting of thin isotropic layers always have Thomsen (1986) parameters, such that $\varepsilon - \delta \geq 0$ and $\gamma \geq 0$ (Tsvankin, 2001). Berryman *et al.* (1999) find the additional conditions summarized in Table 2.3.1, based on Backus (1962) average analysis and Monte Carlo simulations of thinly laminated media. Although all of the cases in the table have $\delta \leq 0$, Berryman *et al.* find that δ can be positive if the layers have significantly varying and positively correlated shear modulus, μ, and Poisson's ratio, v.

Bakulin *et al.* (2000) studied the Thomsen parameters for anisotropic HTI (transversely isotropic with horizontal symmetry axis) media resulting from aligned vertical fractures with crack normals along the horizontal x_1-axis in an isotropic background. For example, when the Hudson (1980) penny-shaped crack model (Section 4.13) is used to estimate weak anisotropy resulting from crack density, e, the dry-rock Thomsen parameters in the vertical plane containing the x_1-axis can be approximated as

$$\varepsilon_{\text{dry}}^{(V)} = \frac{c_{11} - c_{33}}{2c_{33}} = -\frac{8}{3}e \leq 0; \tag{2.3.19}$$

$$\delta_{\text{dry}}^{(V)} = \frac{(c_{13} + c_{55})^2 - (c_{33} - c_{55})^2}{2c_{33}(c_{33} - c_{55})} = -\frac{8}{3}e\left[1 + \frac{g(1 - 2g)}{(3 - 2g)(1 - g)}\right] \leq 0; \tag{2.3.20}$$

$$\gamma_{\text{dry}}^{(V)} = \frac{c_{66} - c_{44}}{2c_{44}} = -\frac{8e}{3(3 - 2g)} \leq 0; \tag{2.3.21}$$

$$\eta_{\text{dry}}^{(V)} = \frac{\varepsilon_{\text{dry}}^{(V)} - \delta_{\text{dry}}^{(V)}}{1 + 2\delta_{\text{dry}}^{(V)}} = \frac{8}{3}e\left[\frac{g(1 - 2g)}{(3 - 2g)(1 - g)}\right] \geq 0 \tag{2.3.22}$$

where $g = V_S^2 / V_P^2$ is a property of the unfractured background rock. In the case of fluid-saturated penny-shaped cracks, such that the crack aspect, α, is much less than the ratio of the fluid bulk modulus to the mineral bulk modulus, $K_{fluid}/K_{mineral}$, then the weak-anisotropy Thomsen parameters can be approximated as

$$\varepsilon_{\text{saturated}}^{(V)} = \frac{c_{11} - c_{33}}{2c_{33}} = 0 \tag{2.3.23}$$

$$\delta_{\text{saturated}}^{(V)} = \frac{(c_{13} + c_{55})^2 - (c_{33} - c_{55})^2}{2c_{33}(c_{33} - c_{55})} = -\frac{32ge}{3(3 - 2g)} \leq 0 \tag{2.3.24}$$

$$\gamma_{\text{saturated}}^{(V)} = \frac{c_{66} - c_{44}}{2c_{44}} = -\frac{8e}{3(3 - 2g)} \leq 0 \tag{2.3.25}$$

$$\eta_{\text{saturated}}^{(V)} = -\delta_{\text{saturated}}^{(V)} = \frac{32ge}{3(3 - 2g)} \geq 0 \tag{2.3.26}$$

In intrinsically anisotropic shales, Sayers (2004) finds that δ can be positive or negative, depending on the contact stiffness between microscopic clay domains and on the distribution of clay domain orientations. He shows, via modeling, that distributions of clays domains, each having $\delta \leq 0$, can yield a composite with $\delta > 0$ overall if the domain orientations vary significantly from parallel.

Rasolofosaon (1998) shows that under the assumptions of third-order elasticity, an isotropic rock obtains ellipsoidal symmetry with respect to the propagation of qP waves. Hence $\varepsilon = \delta$ in the symmetry planes. (See Section 2.5 on Tsvankin's extended Thomsen parameters.)

Uses

Thomsen's notation for weak elastic anisotropy is useful for conveniently characterizing the elastic constants of a transversely isotropic linear elastic medium.

Assumptions and Limitations

The preceding equations are based on the following assumptions:
- material is linear, elastic, and transversely isotropic;
- expressions for anisotropic velocities assume that anisotropy is weak, $\varepsilon, \gamma, \delta \ll 1$, however the definitions of the Thomsen parameters in terms of elastic constants are not limited to weak anisotropy.

2.4   Sayers' Simplified Notation for Weak VTI Anisotropy

Sayers (1995c) finds simplified anisotropy parameters for P-wave NMO, SV-wave NMO, and anellipticity that are valid when the difference between P-wave NMO velocity and vertical P-wave velocity differ by less than 25%. P-wave NMO parameter δ can be written as

$$\delta = \chi + \frac{\chi^2}{2(1 - c_{44}/c_{33})},$$ (2.4.1)

where

$$\chi = \frac{(c_{13} + 2c_{44})}{c_{33}} - 1.$$ (2.4.2)

Similarly, the SV-wave NMO parameter σ can be approximated as

$$\sigma \approx \frac{c_{11} + c_{33} - 2(c_{13} + 2c_{44})}{2c_{44}}.$$ (2.4.3)

Anellipticity can be written as

$$\eta = \frac{a}{2(1 - a)},$$ (2.4.4)

where

$$a \approx \frac{c_{11} + c_{33} - 2(c_{13} + 2c_{44})}{c_{11}}$$ (2.4.5)

Thus P-wave NMO, S-wave NMO, and anellipticity depend only on c_{11}, c_{33}, c_{44}, and the combination $c_{13} + 2c_{44}$.

2.5   Tsvankin's Extended Thomsen Parameters for Orthorhombic Media

The well-known Thomsen (1986) parameters for weak anisotropy are well suited for TI media (see Section 2.3). They allow the five independent elastic constants c_{11}, c_{33}, c_{12}, c_{13}, and c_{44} to be expressed in terms of the more intuitive P-wave velocity, α, and S-wave velocity, β, along the symmetry axis, plus additional constants ε, γ, and δ. For orthorhombic media, requiring nine independent elastic constants, the conventional Thomsen parameters are insufficient.

Analogs of the Thomsen parameters suitable for orthorhombic media can be defined (Tsvankin, 1997), recognizing that wave propagation in the x_1-x_3 and x_2-x_3 symmetry planes (pseudo-P and pseudo-S polarized in each plane and SH polarized normal to each plane) is analogous to propagation in two different VTI media. We once again define vertically propagating (along the x3-axis) P- and S-wave velocities, α and β, respectively:

$$\alpha = \sqrt{c_{33}/\rho}, \quad \beta = \sqrt{c_{55}/\rho}$$ (2.5.1)

Unlike in a VTI medium, S-waves propagating along the x_3-axis in an orthorhombic medium can have two different velocities, $\beta_{x_2} = \sqrt{c_{44}/\rho}$ and $\beta_{x_1} = \sqrt{c_{55}/\rho}$, for waves polarized in the x_2 and x_1 directions, respectively. Either polarization can be chosen as a reference, though here we take $\beta = \sqrt{c_{55}/\rho}$ following the definitions of Tsvankin (1997). Some results shown in later sections will use redefined polarizations in the definition of β.

For the seven constants, we can write

$$\varepsilon^{(1)} = \frac{c_{22} - c_{33}}{2c_{33}} \tag{2.5.2}$$

$$\delta^{(1)} = \frac{(c_{23} + c_{44})^2 - (c_{33} - c_{44})^2}{2c_{33}(c_{33} - c_{44})} \tag{2.5.3}$$

$$\gamma^{(1)} = \frac{c_{66} - c_{55}}{2c_{55}} \tag{2.5.4}$$

$$\varepsilon^{(2)} = \frac{c_{11} - c_{33}}{2c_{33}} \tag{2.5.5}$$

$$\delta^{(2)} = \frac{(c_{13} + c_{55})^2 - (c_{33} - c_{55})^2}{2c_{33}(c_{33} - c_{55})} \tag{2.5.6}$$

$$\gamma^{(2)} = \frac{c_{66} - c_{44}}{2c_{44}} \tag{2.5.7}$$

$$\delta^{(3)} = \frac{(c_{12} + c_{66})^2 - (c_{11} - c_{66})^2}{2c_{11}(c_{11} - c_{66})} \tag{2.5.8}$$

Here, the superscripts (1), (2), and (3) refer to the TI-analog parameters in the symmetry planes normal to x_1, x_2, and x_3, respectively. These definitions assume that one of the symmetry planes of the orthorhombic medium is horizontal and that the vertical symmetry axis is along the x_3 direction.

These Thomsen–Tsvankin parameters play a useful role in modeling wave propagation and reflectivity in anisotropic media.

Uses

Tsvankin's notation for weak elastic anisotropy is useful for conveniently characterizing the elastic constants of an orthorhombic elastic medium.

Assumptions and Limitations

The preceding equations are based on the following assumptions:
- material is linear, elastic, and has orthorhombic or higher symmetry;
- the constants are definitions. They sometimes appear in expressions for anisotropy of arbitrary strength, but at other times the applications assume that the anisotropy is weak, so that $\varepsilon, \gamma, \delta \ll 1$.

2.6   Third-Order Nonlinear Elasticity

Synopsis

Seismic velocities in crustal rocks are almost always sensitive to stress. Since so much of geophysics is based on *linear* elasticity, it is common to extend the familiar linear elastic terminology and refer to the "stress-dependent linear elastic moduli" – which can have meaning for the local slope of the strain-curves at a given static state of stress. If the relation between stress and strain has no hysteresis and no dependence on rate, it is more accurate to say that the rocks are **nonlinearly elastic** (e.g., Truesdell, 1965; Helbig, 1998; Rasolofosaon, 1998). Nonlinear elasticity (i.e., stress-dependent velocities) in rocks is due to the presence of compliant mechanical defects, such as cracks and grain contacts (e.g., Walsh, 1965; Jaeger and Cook, 1969; Bourbié *et al.*, 1987).

In a material with **third-order nonlinear elasticity**, the strain energy function E (for arbitrary anisotropy) can be expressed as (Helbig, 1998)

$$E = \frac{1}{2} c_{ijkl} \varepsilon_{ij} \varepsilon_{kl} + \frac{1}{6} c_{ijklmn} \varepsilon_{ij} \varepsilon_{kl} \varepsilon_{mn} \tag{2.6.1}$$

where c_{ijkl} and c_{ijklmn} designate the components of the second- and third-order elastic tensors, respectively, and repeated indices in a term imply summation from 1 to 3. The components c_{ijkl} are the usual elastic constants in Hooke's law, discussed earlier. Hence, linear elasticity is often referred to as **second-order elasticity** because the strain energy in a linear elastic material is second order in strain. The linear elastic tensor (c_{ijkl}) is fourth rank, having a minimum of two independent constants for a material with the highest symmetry (isotropic) and a maximum of 21 independent constants for a material with the lowest symmetry (triclinic). The additional tensor of third-order elastic coefficients (c_{ijklmn}) is rank six, having a minimum of three independent constants (isotropic) and a maximum of 56 independent constants (triclinic) (Rasolofosaon, 1998).

Third-order elasticity is sometimes used to describe the stress-sensitivity of seismic velocities and apparent elastic constants in rocks. The apparent fourth-rank stiffness tensor, $\tilde{c}^{\text{eff}}$, which determines the speeds of infinitesimal-amplitude waves in a rock under applied static stress can be written as

$$\tilde{c}_{ijkl}^{\text{eff}} = c_{ijkl} + c_{ijklmn} \varepsilon_{mn} \tag{2.6.2}$$

where ε_{mn} are the principal strains associated with the applied static stress.

Approximate expressions, in Voigt notation, for the effective elastic constants of a stressed VTI solid can be written as (Rasolofosaon, 1998; Sarkar *et al.*, 2003; Prioul *et al.*, 2004)

$$c_{11}^{\text{eff}} \approx c_{11}^0 + c_{111} \varepsilon_{11} + c_{112}(\varepsilon_{22} + \varepsilon_{33}) \tag{2.6.3}$$

$$c_{22}^{\text{eff}} \approx c_{11}^0 + c_{111} \varepsilon_{22} + c_{112}(\varepsilon_{11} + \varepsilon_{33}) \tag{2.6.4}$$

$$c_{33}^{\text{eff}} \approx c_{33}^0 + c_{111} \varepsilon_{33} + c_{112}(\varepsilon_{11} + \varepsilon_{22}) \tag{2.6.5}$$

$$c_{12}^{\text{eff}} \approx c_{12}^{0} + c_{112}(\varepsilon_{11} + \varepsilon_{22}) + c_{123}\varepsilon_{33} \tag{2.6.6}$$

$$c_{13}^{\text{eff}} \approx c_{13}^{0} + c_{112}(\varepsilon_{11} + \varepsilon_{33}) + c_{123}\varepsilon_{22} \tag{2.6.7}$$

$$c_{23}^{\text{eff}} \approx c_{13}^{0} + c_{112}(\varepsilon_{22} + \varepsilon_{33}) + c_{123}\varepsilon_{11} \tag{2.6.8}$$

$$c_{66}^{\text{eff}} \approx c_{66}^{0} + c_{144}\varepsilon_{33} + c_{155}(\varepsilon_{11} + \varepsilon_{22}) \tag{2.6.9}$$

$$c_{55}^{\text{eff}} \approx c_{44}^{0} + c_{144}\varepsilon_{22} + c_{155}(\varepsilon_{11} + \varepsilon_{33}) \tag{2.6.10}$$

$$c_{44}^{\text{eff}} \approx c_{44}^{0} + c_{144}\varepsilon_{11} + c_{155}(\varepsilon_{22} + \varepsilon_{33}) \tag{2.6.11}$$

where the constants c_{11}^{0}, c_{33}^{0}, c_{13}^{0}, c_{44}^{0}, c_{66}^{0} are the VTI elastic constants at the unstressed reference state, with $c_{12}^{0} = c_{11}^{0} - 2c_{66}^{0}$. ε_{11}, ε_{22}, and ε_{33} are the principal strains, computed from the applied stress using the conventional Hooke's law, $\varepsilon_{ij} = s_{ijkl}\,\sigma_{kl}$. For these expressions, it is assumed that the direction of the applied principal stress is aligned with the VTI symmetry (x_3-) axis. Furthermore, for these expressions, it is assumed that the stress-sensitive third-order tensor is isotropic, defined by the three independent constants, c_{111}, c_{112}, and c_{123} with $c_{144} = (c_{112} - c_{123})/2$, $c_{155} = (c_{111} - c_{112})/4$, and $c_{456} = (c_{111} - 3c_{112} + 2c_{123})/8$. It is generally observed (Prioul et al., 2004) that $c_{111} < c_{112} < c_{123}, c_{155} < c_{144}, c_{155} < 0$, and $c_{456} < 0$. A sample of experimentally determined values from Prioul and Lebrat (2004) using laboratory data from Wang (2002) are shown in the Table 2.6.1.

The third-order elasticity as used in most of geophysics and rock physics (Bakulin and Bakulin, 1999; Prioul et al., 2004) is called the **Murnaghan** (1951) formulation of finite deformations, and the third-order constants are also called the **Murnaghan constants**.

Various representations of the third-order constants that can be found in the literature (Rasolofosaon, 1998) include the **crystallographic set** ($c_{111}, c_{112}, c_{123}$) presented here, the Murnaghan (1951) constants (l, m, n), and **Landau's set** (A, B, C) (Landau and Lifschitz, 1959). The relations among these are (Rasolofosaon, 1998)

$$c_{111} = 2A + 6B + 2C = 2l + 4m \tag{2.6.12}$$

$$c_{112} = 2B + 2C = 2l \tag{2.6.13}$$

$$c_{123} = 2C = 2l - 2m + n \tag{2.6.14}$$

Chelam (1961) and Krishnamurty (1963) looked at fourth-order elastic coefficients, based on an extension of Murnaghan's theory. It turns out from group theory that there will only be n independent nth-order coefficients for isotropic solids, and $n^2 - 2n + 3$ independent nth-order constants for cubic systems. Triclinic solids have 126 fourth-order elastic constants.

Uses

Third order elasticity provides a way to parameterize the stress dependence of seismic velocities. It also allows for a compact description of stress-induced anisotropy, which is discussed later.

Table 2.6.1 *Experimentally determined third-order elastic constants c_{111}, c_{112}, and c_{123} and derived constants c_{144}, c_{155}, and c_{456}, determined by Prioul and Lebrat (2004), using laboratory data from Wang (2002). Six different sandstone and six different shale samples are shown.*

	c_{111} (GPa)	c_{112} (GPa)	c_{123} (GPa)	c_{144} (GPa)	c_{155} (GPa)	c_{456} (GPa)
Sandstones						
	−10245	−966	−966	0	−2320	−1160
	−9482	−1745	−1745	0	−1934	−967
	−6288	−1744	−1744	0	−1136	−568
	−8580	−527	−527	0	−2013	−1006
	−8460	−1162	−1162	0	−1825	−912
	−12440	−3469	−3094	−188	−2243	−1027
Shales						
	−6903	−976	−976	0	−1482	−741
	−4329	−2122	−1019	−552	−552	0
	−7034	−2147	296	−1222	−1222	0
	−4160	−2013	−940	−536	−536	0
	1294	−510	−119	−196	−196	0
	−1203	−637	−354	−141	−141	0

Assumptions and Limitations

- The preceding equations assume that the material is **hyperelastic**, i.e., there is no hysteresis or rate dependence in the relation between stress and strain, and there exists a unique strain energy function.
- This formalism assumes that strains are infinitesimal. When strains become finite, an additional source of nonlinearity, called **geometrical** or **kinetic** nonlinearity, appears, related to the difference between Lagrangian and Eulerian descriptions of motion (Zarembo and Krasil'nikov, 1971; Johnson and Rasolofosaon, 1996).
- Third-order elasticity is often not general enough to describe the shapes of real stress–strain curves over large ranges of stress and strain. Third-order elasticity is most useful when describing stress–strain within a small range around a reference state of stress and strain.

2.7   Effective Stress Properties of Rocks

Synopsis

Because rocks are deformable, many rock properties are sensitive to applied stresses and pore pressure. Stress-sensitive properties include porosity, permeability, electrical resistivity, sample volume, pore-space volume, and elastic moduli. Empirical evidence (Hicks and Berry, 1956; Wyllie *et al.*, 1958; Todd and Simmons, 1972; Christensen and Wang, 1985; Prasad and Manghnani, 1997; Siggins and Dewhurst, 2003; Hoffman *et al.*, 2005) and theory (Brandt, 1955; Nur and Byerlee, 1971; Zimmerman, 1991a; Gangi and Carlson, 1996; Berryman, 1992a, 1993; Gurevich, 2004) suggest that the pressure dependence of each of these rock properties, X, can be represented as a function, f_X, of a linear combination of the hydrostatic confining stress, P_C, and the pore pressure, P_P:

$$X = f_X(P_C - n_X P_P); \quad n_X \leq 1 \tag{2.7.1}$$

The combination $P_{\text{eff}} = P_C - n_X P_P$ is called the **effective pressure**, or more generally, the tensor $\sigma_{ij}^{\text{eff}} = \sigma_{ij}^C - n_X P_P \delta_{ij}$ is the **effective stress**. The parameter n_X is called the "effective-stress coefficient," which can itself be a function of stress. The negative sign on the pore pressure indicates that the pore pressure approximately counteracts the effect of the confining pressure. An expression such as $X = f_X(P_C - n_X P_P)$ is sometimes called the **effective-stress law** for the property X. It is important to point out that each rock property might have a different function f_i and a different value of n_i (Zimmerman, 1991a; Berryman, 1992a, 1993; Gurevich, 2004). Extensive discussions on the effective-stress behavior of elastic moduli, permeability, resistivity, and thermoelastic properties can be found in Berryman (1992a, 1993). Zimmerman (1991a) gives a comprehensive discussion of effective-stress behavior for strain and elastic constants.

Zimmerman (1991a) points out the distinction between the effective-stress behavior for finite pressure changes versus the effective-stress behavior for infinitesimal increments of pressure. For example, increments of the bulk-volume strain, ε_b, and the pore-volume strain, ε_p, can be written as

$$\varepsilon_b(P_C, P_P) = -C_{bc}(P_C - m_b P_P)(\delta P_C - n_b \delta P_P); \quad C_{bc} = -\frac{1}{V_T}\left(\frac{\delta V_T}{\delta P_C}\right)_{P_P} \tag{2.7.2}$$

$$\varepsilon_p(P_C, P_P) = -C_{pc}(P_C - m_p P_P)(\delta P_C - n_p \delta P_P); \quad C_{pc} = -\frac{1}{V_P}\left(\frac{\delta V_P}{\delta P_C}\right)_{P_P} \tag{2.7.3}$$

where the compressibilities C_{bc} and C_{pc} are functions of $P_C - m P_P$. The coefficients m_b and m_P govern the way the compressibilities vary with P_C and P_P. In contrast, the coefficients n_b and n_P describe the relative increments of additional strain resulting from pressure *increments* δP_C and δP_P. For example, in a laboratory experiment, ultrasonic velocities will depend on the values of C_{pc}, the local slope of the stress–strain curve at the static values of P_C and P_P. On the other hand, the sample length change monitored within the pressure vessel is the total strain, obtained by integrating the strain over the entire stress path.

The existence of an effective-stress law, i.e., that a rock property depends only on the state of stress, requires that the rock be elastic – possibly nonlinearly elastic. The deformation of an elastic material depends only on the state of stress, and is independent of the stress history and the rate of loading. Furthermore, the existence of an effective-stress law requires that there is no hysteresis in stress–strain cycles. Since no rock is perfectly elastic, all effective-stress laws for rocks are approximations. In fact, deviation from elasticity makes estimating the effective-stress coefficient from laboratory data sometimes ambiguous. Another condition required for the existence of an effective-pressure law is that the pore pressure is well defined and uniform throughout the pore space. Todd and Simmons (1972) show that the effect of pore pressure on velocities varies with the *rate* of pore-pressure change and whether the pore pressure has enough time to equilibrate in thin cracks and poorly connected pores. Slow changes in pore pressure yield more stable results, describable with an effective-stress law and with a larger value of *n* for velocity.

Much discussion focuses on the value of the effective-stress constants, *n* (and *m*). Biot and Willis (1957) predicted theoretically that the pressure-induced *volume* increment, δV_T, of a sample of linear poroelastic material depends on pressure increments $(\delta P_C - n_X \delta P_P)$. For this special case, $n_X = \alpha = 1 - K/K_S$, where α is known as the Biot coefficient or Biot-Willis coefficient. K is the dry-rock (drained) bulk modulus and K_S is the mineral bulk modulus (or some appropriate average of the moduli if there is mixed mineralogy), defined below. Explicitly,

$$\frac{\delta V_T}{V_T} = -\frac{1}{K}\left[\delta P_C - \left(1 - \frac{K}{K_S}\right)\delta P_P\right] \tag{2.7.4}$$

where δV_T, δP_C, and δP_P signify increments relative to a reference state.

Pitfall:

A common error is to assume that the Biot-Willis effective-stress coefficient α for volume change also applies to other deformation-related rock properties. For example, although rock elastic moduli vary with crack and pore deformation, there is no theoretical justification for extrapolating α to elastic moduli and seismic velocities.

Other factors determining the apparent effective-stress coefficient observed in the laboratory include the rate of change of pore pressure, the connectivity of the pore space, the presence or absence of hysteresis, heterogeneity of the rock mineralogy, and variation of pore-fluid compressibility with pore pressure.

There is still a need to reconcile theoretical predictions of effective stress with certain laboratory data. For example, simple, yet rigorous, theoretical considerations (Zimmerman, 1991a; Berryman, 1992a; Gurevich, 2004) predict that $n_{\text{velocity}} = 1$ for monomineralic, elastic rocks. Experimentally observed values for n_{velocity} are

Table 2.7.1 *Four defined rock moduli that appear in the effective stress laws shown in Table 2.7.2, where $P_d = P_C - P_P$ is the differential pressure, V_T is the sample bulk (i.e., total) volume, and V_ϕ is the pore volume. The negative sign for each of these rock properties follows from defining pressures as being positive in compression and volumes positive in expansion*

$\dfrac{1}{K} = -\dfrac{1}{V_T}\left(\dfrac{\partial V_T}{\partial P_d}\right)_{P_P};$	K = modulus of the drained porous frame,
$\dfrac{1}{K_S} = -\dfrac{1}{V_T}\left(\dfrac{\partial V_T}{\partial P_P}\right)_{P_d};$	K_S = unjacketed modulus; if monomineralic,
	$K_S = K_{\text{mineral}}$, otherwise K_S is a poorly understood average of the mixed mineral moduli
$\dfrac{1}{K_{\phi\phi}} = -\dfrac{1}{V_\phi}\left(\dfrac{\partial V_\phi}{\partial P_P}\right)_{P_d};$	if monomineralic, $K_S = K_{\phi\phi} = K_{\text{mineral}}$
$\dfrac{1}{K_P} = -\dfrac{1}{V_\phi}\left(\dfrac{\partial V_\phi}{\partial P_d}\right)_{P_P} = \dfrac{1}{\phi}\left(\dfrac{1}{K} - \dfrac{1}{K_S}\right)$	

Table 2.7.2 *Theoretically predicted effective-stress laws for incremental changes in confining and pore pressures, compiled from Zimmerman (1984a), Berryman (1992a, 1993), and H. F. Wang (2000)*

Property	General mineralogy
Sample volume[a]	$\dfrac{\delta V_T}{V_T} = -\dfrac{1}{K}(\delta P_C - \alpha \delta P_P)$
Pore volume[b]	$\dfrac{\delta V_\phi}{V_\phi} = -\dfrac{1}{K_P}(\delta P_C - \beta \delta P_P)$
Porosity[c]	$\dfrac{\delta \phi}{\phi} = -\left(\dfrac{\alpha - \phi}{\phi K}\right)(\delta P_C - \chi \delta P_P)$
Solid volume[d]	$\dfrac{\delta V_S}{V_S} = -\dfrac{1}{(1-\phi)K_S}(\delta P_C - \sigma \delta P_P)$
Permeability[e]	$\dfrac{\delta k}{k} = -\left[h\left(\dfrac{\alpha - \phi}{\phi K}\right) + \dfrac{2}{3K}\right](\delta P_C - \kappa \delta P_P)$
Velocity/elastic moduli[f]	$\dfrac{\delta V_P}{V_P} = f(\delta P_C - \theta P_P)$

Notes:

[a] V_T is the total volume. $\alpha = 1 - K/K_S$, Biot coefficient; usually in the range $\phi \leq \alpha \leq 1$; if monomineralic, $\alpha = 1 - K/K_{\text{mineral}}$.

[b] $V_\phi = \phi V_T$, pore volume. $\beta = 1 - K_P/K_{\phi\phi}$, usually, $\phi \leq \beta \leq 1$, but it is possible that $\beta > 1$.

[c] $\chi = \left(\dfrac{\beta - \phi}{\alpha - \phi}\right)\alpha$; if monomineralic, $\chi = 1$.

[d] $\sigma = \alpha - \left(\dfrac{\chi - \alpha}{1 - \alpha}\right)(\alpha - \phi)$; if monomineralic, $\sigma = \phi$. $\sigma \leq \alpha \leq \beta$.

[e] $\kappa = 1 - \dfrac{2\phi(1-\alpha)}{3h(\alpha - \phi) + 2\phi} \leq 1$. $h \approx 2 + m \approx 4$, where m is Archie's cementation exponent.

[f] $\theta = 1 - \dfrac{\partial(1/K_S)/\partial P_C}{\partial(1/K)/\partial P_C}$; if monomineralic, $\theta = 1$.

sometimes close to 1, and sometimes less than one. Speculations for the variations have included mineral heterogeneity, poorly connected pore space, pressure-related changes in pore-fluid properties, incomplete correction for fluid-related velocity dispersion in ultrasonic measurements, poorly equilibrated or characterized pore pressure, and inelastic deformation.

Uses

Characterization of the stress sensitivity of rock properties makes it possible to invert for rock-property changes from changes in seismic or electrical measurements. It also provides a means of understanding how rock properties might change in response to tectonic stresses or pressure changes resulting from reservoir or aquifer production.

Assumptions and Limitations

- The existence of effective-pressure laws assumes that the rocks are *hyperelastic*, i.e., there is no hysteresis or rate dependence in the relation between stress and strain.
- Rocks are extremely variable, so effective-pressure behavior can likewise be variable.

2.8   Stress-Induced Anisotropy in Rocks

Synopsis

The closing of cracks under compressive stress (or, equivalently, the stiffening of compliant grain contacts) tends to increase the effective elastic moduli of rocks (see also Section 2.6 on third-order elasticity).

When the crack population is anisotropic, either in the original unstressed condition or as a result of the stress field, this condition can impact the overall elastic anisotropy of the rock. Laboratory demonstrations of stress-induced anisotropy have been reported by numerous authors (Nur and Simmons, 1969a; Lockner *et al.*, 1977; Zamora and Poirier, 1990; Sayers *et al.*, 1990; Yin, 1992; Cruts *et al.*, 1995).

The simplest case to understand is a rock with a random (isotropic) distribution of cracks embedded in an isotropic mineral matrix. In the initial unstressed state, the rock is elastically isotropic. If a *hydrostatic* compressive stress is applied, cracks in all directions respond similarly, and the rock remains isotropic but becomes stiffer. However, if a *uniaxial* compressive stress is applied, cracks with normals parallel or nearly parallel to the applied-stress axis will tend to close preferentially, and the rock will take on an axial or transversely isotropic symmetry.

An initially isotropic rock with arbitrary stress applied will have at least orthorhombic symmetry (Nur, 1971; Rasolofosaon, 1998), provided that the stress-induced changes in moduli are small relative to the absolute moduli.

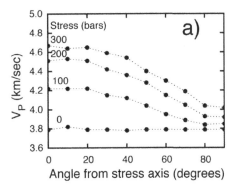

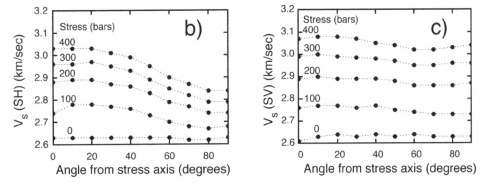

Figure 2.8.1 The effects of stress-induced crack alignment on seismic velocity anisotropy measured in the laboratory (Nur and Simmons, 1969a)

Figure 2.8.1 illustrates the effects of stress-induced crack alignment on seismic-velocity anisotropy discovered in the laboratory by Nur and Simmons (1969a). The crack porosity of the dry granite sample is essentially isotropic at low stress.

As uniaxial stress is applied, crack anisotropy is induced. The velocities (compressional and two polarizations of shear) clearly vary with direction relative to the stress-induced crack alignment. Table 2.8.1 summarizes the elastic symmetries that result when various applied-stress fields interact with various initial crack symmetries (Paterson and Weiss, 1961; Nur, 1971).

A rule of thumb is that a wave is most sensitive to cracks when its direction of propagation or direction of polarization is perpendicular (or nearly so) to the crack faces.

The most common approach to modeling the stress-induced anisotropy is to assume angular distributions of idealized penny-shaped cracks (Nur, 1971; Sayers, 1988a, 1998b; Gibson and Toksöz, 1990). The stress dependence is introduced by assuming or inferring distributions or spectra of crack aspect ratios with various orientations.

The assumption is that a crack will close when the component of applied compressive stress normal to the crack faces causes a normal displacement of the crack faces equal to the original half-width of the crack. This allows us to estimate the crack **closing stress** as follows:

Table 2.8.1 *Dependence of symmetry of induced velocity anisotropy on initial crack distribution and applied stress and its orientation*

Symmetry of initial crack distribution	Applied stress	Orientation of applied stress	Symmetry of induced velocity anisotropy	Number of elastic constants
Random	Hydrostatic		Isotropic	2
	Uniaxial		Axial	5
	Triaxial[a]		Orthorhombic	9
Axial	Hydrostatic		Axial	5
	Uniaxial	Parallel to axis of symmetry	Axial	5
	Uniaxial	Normal to axis of symmetry	Orthorhombic	9
	Uniaxial	Inclined	Monoclinic	13
	Triaxial[a]	Parallel to axis of symmetry	Orthorhombic	9
	Triaxial[a]	Inclined	Monoclinic	13
Orthorhombic	Hydrostatic		Orthorhombic	9
	Uniaxial	Parallel to axis of symmetry	Orthorhombic	9
	Uniaxial	Inclined in plane of symmetry	Monoclinic	13
	Uniaxial	Inclined	Triclinic	21
	Triaxial[a]	Parallel to axis of symmetry	Orthorhombic	9
	Triaxial[a]	Inclined in plane of symmetry	Monoclinic	13
	Triaxial[a]	Inclined	Triclinic	21

Note:
[a] Three generally unequal principal stresses.

$$\sigma_{\text{close}} = \frac{3\pi(1-2\nu)}{4(1-\nu^2)}\alpha K_0 = \frac{\pi}{2(1-\nu)}\alpha\mu_0 \tag{2.8.1}$$

where α is the aspect ratio of the crack, and ν, μ_0, and K_0 are the Poisson's ratio, shear modulus, and bulk modulus of the mineral, respectively (see Section 2.10 on the deformation of inclusions and cavities in elastic solids). Hence, the thinnest cracks will close first, followed by thicker ones. This allows one to estimate, for a given aspect ratio distribution, how many cracks remain open in each direction for any applied stress field. These inferred crack distributions and their orientations can be put into one of the popular crack models (e.g., Hudson, 1981) to estimate the resulting effective elastic moduli of the rock. Although these penny-shaped crack models have been relatively successful and provide a useful physical interpretation, they are limited to low crack concentrations and might not effectively represent a broad range of crack geometries (see Section 4.13 on Hudson's model for cracked media).

As an alternative, Mavko *et al.* (1995) presented a simple recipe for estimating stress-induced velocity anisotropy directly from measured values of isotropic V_P and V_S versus hydrostatic pressure. This method differs from the inclusion models, because it is relatively independent of any assumed crack geometry and is not limited to small crack densities. To invert for a particular crack distribution, one needs to assume crack shapes and aspect-ratio spectra. However, if rather than inverting for a crack distribution, we instead directly transform hydrostatic velocity-pressure data to stress-induced velocity anisotropy, we can avoid the need for parameterization in terms of ellipsoidal cracks and the resulting limitations to low crack densities. In this sense, the method of Mavko *et al.* (1995) provides not only a simpler but also a more general solution to this problem, for ellipsoidal cracks are just one particular case of the general formulation.

The procedure is to estimate the generalized pore-space compliance from the measurements of isotropic V_P and V_S. The physical assumption that the compliant part of the pore space is crack-like means that the pressure dependence of the generalized compliances is governed primarily by *normal* tractions resolved across cracks and defects. These defects can include grain boundaries and contact regions between clay platelets (Sayers, 1995c). This assumption allows the measured pressure dependence to be mapped from the hydrostatic stress state to any applied nonhydrostatic stress.

The method applies to rocks that are approximately isotropic under hydrostatic stress and in which the anisotropy is caused by *crack closure* under stress. Sayers (1988b) also found evidence for some stress-induced *opening* of cracks, which is ignored in this method. The potentially important problem of stress–strain hysteresis is also ignored.

The anisotropic elastic compliance tensor $S_{ijkl}(\boldsymbol{\sigma})$ at any given stress state $\boldsymbol{\sigma}$ may be expressed as

$$
\begin{aligned}
\Delta S_{ijkl}(\boldsymbol{\sigma}) &= S_{ijkl}(\boldsymbol{\sigma}) - S^0_{ijkl} \\
&= \int_{\theta=0}^{\pi/2} \int_{\phi=0}^{2\pi} \left[W'_{3333}(\hat{\mathbf{m}}^{\mathrm{T}} \boldsymbol{\sigma} \hat{\mathbf{m}}) - 4W'_{2323}(\hat{\mathbf{m}}^{\mathrm{T}} \boldsymbol{\sigma} \hat{\mathbf{m}}) \right] m_i m_j m_k m_l \, \sin\theta d\theta d\phi \\
&\quad + \int_{\theta=0}^{\pi/2} \int_{\phi=0}^{2\pi} W'_{2323}(\hat{\mathbf{m}}^{\mathrm{T}} \boldsymbol{\sigma} \hat{\mathbf{m}}) [\delta_{ik} m_j m_l + \delta_{il} m_j m_k + \delta_{jk} m_i m_l \\
&\quad\quad + \delta_{jl} m_i m_k] \sin\theta d\theta d\phi
\end{aligned}
\tag{2.8.2}
$$

where

$$
W'_{3333}(p) = \frac{1}{2\pi} \Delta S^{\mathrm{iso}}_{jjkk}(p)
\tag{2.8.3}
$$

$$
W'_{2323}(p) = \frac{1}{8\pi} \left[\Delta S^{\mathrm{iso}}_{\alpha\beta\alpha\beta}(p) - \Delta S^{\mathrm{iso}}_{\alpha\alpha\beta\beta}(p) \right]
\tag{2.8.4}
$$

The tensor S^0_{ijkl} denotes the reference compliance at some very large confining hydrostatic pressure when all of the compliant parts of the pore space are closed. The expression $\Delta S_{ijkl}^{iso}(p) = S_{ijkl}^{iso}(p) - S_{ijkl}^{0}$ describes the difference between the compliance under a hydrostatic effective pressure p and the reference compliance at high pressure. These are determined from measured P- and S-wave velocities versus the hydrostatic pressure. The tensor elements W'_{3333} and W'_{2323} are the *measured* normal and shear crack compliances and include all interactions with neighboring cracks and pores. These could be approximated, for example, by the compliances of idealized ellipsoidal cracks, interacting or not, but this would immediately reduce the generality. The expression $\hat{\mathbf{m}} \equiv (\sin\theta \cos\phi, \sin\theta \sin\phi, \cos\theta)^{T}$ denotes the unit normal to the crack face, where θ and ϕ are the polar and azimuthal angles in a spherical coordinate system, respectively.

An important physical assumption in the preceding equations is that, for thin cracks, the *crack compliance tensor* W'_{ijkl} is *sparse*, and thus only W'_{3333}, W'_{1313}, and W'_{2323} are nonzero. This is a general property of planar crack formulations and reflects an approximate decoupling of normal and shear deformation of the crack and decoupling of the in-plane and out-of-plane deformations. This allows us to write $W'_{jjkk} \approx W'_{3333}$. Furthermore, it is assumed that the two unknown shear compliances are approximately equal: $W'_{1313} \approx W'_{2323}$. A second important physical assumption is that for a thin crack under any stress field, it is primarily the *normal* component of stress, $\sigma_n = \hat{\mathbf{m}}^{T} \boldsymbol{\sigma} \hat{\mathbf{m}}$, resolved on the faces of a crack, that causes it to close and to have a stress-dependent compliance. Any open crack will have both normal and shear deformation under normal and shear loading, but it is only the normal stress that determines crack closure.

For the case of uniaxial stress σ_0 applied along the 3-axis to an initially isotropic rock, the normal stress in any direction is $\sigma_n = \sigma_0 \cos^2\theta$. The rock takes on a transversely isotropic symmetry, with five independent elastic constants. The five independent components of ΔS_{ijkl} become

$$\Delta S_{3333}^{uni} = 2\pi \int_{0}^{\pi/2} [W'_{3333}(\sigma_0 \cos^2\theta) - 4W'_{2323}(\sigma_0 \cos^2\theta)] \cos^4\theta \sin\theta \, d\theta$$

$$+ 2\pi \int_{0}^{\pi/2} 4W'_{2323}(\sigma_0 \cos^2\theta) \cos^2\theta \sin\theta \, d\theta \tag{2.8.5}$$

$$\Delta S_{1111}^{uni} = 2\pi \int_{0}^{\pi/2} \frac{3}{8} [W'_{3333}(\sigma_0 \cos^2\theta) - 4W'_{2323}(\sigma_0 \cos^2\theta)] \sin^4\theta \sin\theta \, d\theta$$

$$+ 2\pi \int_{0}^{\pi/2} 2W'_{2323}(\sigma_0 \cos^2\theta) \sin^2\theta \sin\theta \, d\theta \tag{2.8.6}$$

$$\Delta S_{1122}^{uni} = 2\pi \int_{0}^{\pi/2} \frac{1}{8} [W'_{3333}(\sigma_0 \cos^2\theta) - 4W'_{2323}(\sigma_0 \cos^2\theta)] \sin^4\theta \sin\theta \, d\theta \tag{2.8.7}$$

$$\Delta S_{1133}^{\text{uni}} = 2\pi \int_0^{\pi/2} \frac{1}{2} \left[W'_{3333} \left(\sigma_0 \cos^2\theta \right) - 4W'_{2323} \left(\sigma_0 \cos^2\theta \right) \right] \sin^2\theta \cos^2\theta \sin\theta \, d\theta$$

(2.8.8)

$$\Delta S_{2323}^{\text{uni}} = 2\pi \int_0^{\pi/2} \frac{1}{2} \left[W'_{3333} \left(\sigma_0 \cos^2\theta \right) - 4W'_{2323} \left(\sigma_0 \cos^2\theta \right) \right] \sin^2\theta \cos^2\theta \sin\theta \, d\theta$$

$$+ 2\pi \int_0^{\pi/2} \frac{1}{2} W'_{2323} \left(\sigma_0 \cos^2\theta \right) \sin^2\theta \sin\theta \, d\theta$$

$$+ 2\pi \int_0^{\pi/2} W'_{2323} \left(\sigma_0 \cos^2\theta \right) \cos^2\theta \sin\theta \, d\theta$$

(2.8.9)

Note that in the preceding equations, the terms in parentheses with W'_{2323} () and W'_{3333} () are arguments to the W'_{2323} and W'_{3333} pressure functions, not multiplicative factors.

Sayers and Kachanov (1991, 1995) have presented an equivalent formalism for stress-induced anisotropy. The elastic compliance S_{ijkl} is once again written in the form

$$\Delta S_{ijkl} = S_{ijkl}(\boldsymbol{\sigma}) - S_{ijkl}^0$$

(2.8.10)

where S_{ijkl}^0 is the compliance in the absence of compliant cracks and grain boundaries and ΔS_{ijkl} is the excess compliance due to the cracks. ΔS_{ijkl} can be written as

$$\Delta S_{ijkl} = \frac{1}{4} \left(\delta_{ik}\alpha_{jl} + \delta_{il}\alpha_{jk} + \delta_{jk}\alpha_{il} + \delta_{jl}\alpha_{ik} \right) + \beta_{ijkl}$$

(2.8.11)

where α_{ij} is a second-rank tensor and β_{ijkl} is a fourth-rank tensor defined by

$$\alpha_{ij} = \frac{1}{V} \sum_r B_T^{(r)} n_i^{(r)} n_j^{(r)} A^{(r)}$$

(2.8.12)

$$\beta_{ijkl} = \frac{1}{V} \sum_r \left(B_N^{(r)} - B_T^{(r)} \right) n_i^{(r)} n_j^{(r)} n_k^{(r)} n_l^{(r)} A^{(r)}$$

(2.8.13)

In these expressions, the summation is over all grain contacts and microcracks within the rock volume V. $B_N^{(r)}$ and $B_T^{(r)}$ are the normal and shear compliances of the rth discontinuity, which relate the displacement discontinuity across the crack to the applied traction across the crack faces; $n_i^{(r)}$ is the ith component of the normal to the discontinuity, and $A^{(r)}$ is the area of the discontinuity.

A completely different strategy for quantifying stress-induced anisotropy is to use the formalism of third-order elasticity, described in Section 2.6 (e.g., Helbig, 1994; Johnson and Rasolofosaon, 1996; Prioul *et al.*, 2004). The third-order elasticity approach is phenomenological, avoiding the physical mechanisms of stress sensitivity, but providing a compact notation. For example, Prioul *et al.* (2004) found that for a stressed VTI (transversely isotropic with vertical symmetry axis) material, the effective elastic constants can be approximated in Voigt notation as

$$c_{11}^{\text{eff}} \approx c_{11}^0 + c_{111}\varepsilon_{11} + c_{112}(\varepsilon_{22} + \varepsilon_{33}) \tag{2.8.14}$$

$$c_{22}^{\text{eff}} \approx c_{11}^0 + c_{111}\varepsilon_{22} + c_{112}(\varepsilon_{11} + \varepsilon_{33}) \tag{2.8.15}$$

$$c_{33}^{\text{eff}} \approx c_{33}^0 + c_{111}\varepsilon_{33} + c_{112}(\varepsilon_{11} + \varepsilon_{22}) \tag{2.8.16}$$

$$c_{12}^{\text{eff}} \approx c_{12}^0 + c_{112}(\varepsilon_{11} + \varepsilon_{22}) + c_{123}\varepsilon_{33} \tag{2.8.17}$$

$$c_{13}^{\text{eff}} \approx c_{13}^0 + c_{112}(\varepsilon_{11} + \varepsilon_{33}) + c_{123}\varepsilon_{22} \tag{2.8.18}$$

$$c_{23}^{\text{eff}} \approx c_{13}^0 + c_{112}(\varepsilon_{22} + \varepsilon_{33}) + c_{123}\varepsilon_{11} \tag{2.8.19}$$

$$c_{66}^{\text{eff}} \approx c_{66}^0 + c_{144}\varepsilon_{33} + c_{155}(\varepsilon_{11} + \varepsilon_{22}) \tag{2.8.20}$$

$$c_{55}^{\text{eff}} \approx c_{44}^0 + c_{144}\varepsilon_{22} + c_{155}(\varepsilon_{11} + \varepsilon_{33}) \tag{2.8.21}$$

$$c_{44}^{\text{eff}} \approx c_{44}^0 + c_{144}\varepsilon_{11} + c_{155}(\varepsilon_{22} + \varepsilon_{33}) \tag{2.8.22}$$

where the constants $c_{11}^0, c_{33}^0, c_{13}^0, c_{44}^0, c_{66}^0$ are the VTI elastic constants at the unstressed reference state, with $c_{12}^0 = c_{11}^0 - 2c_{66}^0$. The quantities ε_{11}, ε_{22}, and ε_{33} are the principal strains, computed from the applied stress using the conventional Hooke's law, $\varepsilon_{ij} = S_{ijkl}\sigma_{kl}$. For these expressions, it is assumed that the direction of the applied principal stress is aligned with the VTI symmetry (x_3-) axis. Furthermore, for these expressions it is assumed that the stress-sensitive third-order tensor is isotropic, defined by the three independent constants, c_{111}, c_{112} and c_{123} with $c_{144} = (c_{112} - c_{123})/2$ and $c_{155} = (c_{111} - c_{112})/4$.

In practice, the elastic constants (five for the unstressed VTI rock and three for the third-order elasticity) can be estimated from laboratory measurements, as illustrated by Sarkar *et al.* (2003) and Prioul *et al.* (2004). For an intrinsically VTI rock, hydrostatic loading experiments provide sufficient information to invert for all three of the third-order constants, provided that the anisotropy is not too weak. However, for an intrinsically isotropic rock, nonhydrostatic loading is required in order to provide enough independent information to determine the constants. Once the constants are determined, the stress-induced elastic constants corresponding to any (aligned) stress field can be determined. The full expressions for stress-induced anisotropy of an originally VTI rock are given by Sarkar *et al.* (2003) and Prioul *et al.* (2004).

Sarkar *et al.* (2003) give expressions for stress-induced changes in Thomsen parameters of an originally VTI rock, provided that both the original and stress-induced anisotropies are weak:

$$\varepsilon^{(1)} = \varepsilon^0 + \frac{c_{155}}{c_{33}^0 c_{55}^0}(\sigma_{22} - \sigma_{33}); \qquad \varepsilon^{(2)} = \varepsilon^0 + \frac{c_{155}}{c_{33}^0 c_{55}^0}(\sigma_{11} - \sigma_{33})$$

$$\delta^{(1)} = \delta^0 + \frac{c_{155}}{c_{33}^0 c_{55}^0}(\sigma_{22} - \sigma_{33}); \qquad \delta^{(2)} = \delta^0 + \frac{c_{155}}{c_{33}^0 c_{55}^0}(\sigma_{11} - \sigma_{33})$$

$$\gamma^{(1)} = \gamma^0 + \frac{c_{456}}{2c_{55}^0 c_{55}^0}(\sigma_{22} - \sigma_{33}); \qquad \gamma^{(2)} = \gamma^0 + \frac{c_{456}}{2c_{55}^0 c_{55}^0}(\sigma_{11} - \sigma_{33})$$

$$\delta^{(3)} = \frac{c_{155}}{c_{33}^0 c_{55}^0}(\sigma_{22} - \sigma_{11})$$

$$c_{155} = \frac{1}{4}(c_{111} - c_{112}); \qquad c_{456} = \frac{1}{8}(c_{111} - 3c_{112} + 2c_{123})$$

The results are shown in terms of Tsvankin's extended Thomsen parameters (Section 2.5). Orthorhombic symmetry occurs when $\sigma_{11} \neq \sigma_{22}$. (Strictly speaking, three different principal stresses applied to a VTI rock do not result in perfect orthorhombic symmetry. However, orthorhombic parameters are adequate in the case of weak anisotropy.)

Shapiro and Kaselow (2005) and Shapiro (2017) have presented a formulation for stress effects on elastic anisotropy in dry, porous, fractured rocks. This porosity deformation approach (also termed piezosensitivity approach) models the pore space as consisting of a stiff component and a compliant crack-like part. The stress-dependence of the effective elastic compliance is then derived by considering stress-dependent contribution of both the stiff and compliant parts of the pore space. There are no assumptions about the shapes of the stiff and compliant pores, and the general theory (Shapiro, 2017) is valid for arbitrary pore geometries, and all anisotropic symmetries, including isotropy as the limiting case. The principal axes of the applied stresses do not have to coincide with the symmetry axes of anisotropic medium. The piezosensitivity tensor, describing the stress dependence is itself assumed symmetric and isotropic. It is also assumed that the stress-induced changes in the stiff and compliant pore spaces are independent of each other. In the principal coordinate axes of the applied stress (with principal stresses σ_1, σ_2, and σ_3), a convenient parametrization of the dry compliance tensor and its stress sensitivity is given as (Shapiro, 2017):

$$S_{ijkl}^{dry} = S_{ijkl}^{dry0} + K_{ijkl}^{(1)}\sigma_1 + K_{ijkl}^{(2)}\sigma_2 + K_{ijkl}^{(3)}\sigma_3$$

$$+ \delta_{ik}B_{jl}e^{F_c(\sigma_j + \sigma_l)/2} + \delta_{il}B_{jk}e^{F_c(\sigma_j + \sigma_k)/2} + \delta_{jk}B_{il}e^{F_c(\sigma_i + \sigma_l)/2} + \delta_{jl}B_{ik}e^{F_c(\sigma_i + \sigma_k)/2} \qquad (2.8.23)$$

Stress is taken to be negative in compression. The coefficients $K_{ijkl}^{(\alpha)}$ (18 independent coefficients), relate to the stiff pore space deformation, while B_{ij} (6 independent coefficients) and F_c describe the deformation of the soft, compliant parts of the pore space. These coefficients can be obtained by fit to experimental data (Sviridov *et al.*,

2019). For low to moderate applied stresses, the contributions from $K_{ijkl}^{(\alpha)}$ can be neglected. The reference dry compliance S_{ijkl}^{dry0} is the compliance at a reference state where the compliant crack-like part of the pore space is closed (absent), and the stiff pores are in an undeformed state (under no load). This reference state is an artificial geometric configuration, and S_{ijkl}^{dry0} can also be estimated by fitting to data. The stress-dependent porosity is given by:

$$
\begin{aligned}
\phi = \phi^{s0} &+ (\delta S_{1111} + \delta S_{1122} + \delta S_{1133})\sigma_1 \\
&+ (\delta S_{1122} + \delta S_{2222} + \delta S_{2233})\sigma_2 + (\delta S_{1133} + \delta S_{2233} + \delta S_{3333})\sigma_3 \\
&+ \frac{4}{F_c}(B_{11}e^{F_c\sigma_1} + B_{22}e^{F_c\sigma_2} + B_{33}e^{F_c\sigma_3})
\end{aligned}
\tag{2.8.24}
$$

where ϕ^{s0} is the stiff porosity in the reference state, and δ denotes the difference from the reference state. For an orthorhombic medium, under uniaxial stress σ_1 along coordinate axis x_1, the general equations reduce to (Sviridov et al., 2019):

$$S_{1111}^{dry} = S_{1111}^{dry0} + K_{1111}^{(1)}\sigma_1 + 4B_{11}e^{F_c\sigma_1} \tag{2.8.25}$$

$$S_{2222}^{dry} = S_{2222}^{dry0} + K_{2222}^{(1)}\sigma_1 + 4B_{22} \tag{2.8.26}$$

$$S_{3333}^{dry} = S_{3333}^{dry0} + K_{3333}^{(1)}\sigma_1 + 4B_{33} \tag{2.8.27}$$

$$S_{2323}^{dry} = S_{2323}^{dry0} + K_{2323}^{(1)}\sigma_1 + B_{33} + B_{22} \tag{2.8.28}$$

$$S_{1313}^{dry} = S_{1313}^{dry0} + K_{1313}^{(1)}\sigma_1 + B_{33} + B_{11}e^{F_c\sigma_1} \tag{2.8.29}$$

$$S_{1212}^{dry} = S_{1212}^{dry0} + K_{1212}^{(1)}\sigma_1 + B_{22} + B_{11}e^{F_c\sigma_1} \tag{2.8.30}$$

$$S_{1133}^{dry} = S_{1133}^{dry0} + K_{1133}^{(1)}\sigma_1; \ S_{2233}^{dry} = S_{2233}^{dry0} + K_{2233}^{(1)}\sigma_1; \ S_{1122}^{dry} = S_{1122}^{dry0} + K_{1122}^{(1)}\sigma_1 \tag{2.8.31}$$

Similarly, for a transversely isotropic medium under uniaxial stress σ_3 along the x_3 symmetry axis, the equations are (Sviridov et al., 2019):

$$S_{2222}^{dry} = S_{1111}^{dry} = S_{1111}^{dry0} + K_{1111}^{(3)}\sigma_3 + 4B_{11} \tag{2.8.32}$$

$$S_{3333}^{dry} = S_{3333}^{dry0} + K_{3333}^{(3)}\sigma_3 + 4B_{33}e^{F_c\sigma_3} \tag{2.8.33}$$

$$S_{2323}^{dry} = S_{1313}^{dry} = S_{1313}^{dry0} + K_{1313}^{(3)}\sigma_3 + B_{33}e^{F_c\sigma_3} + B_{11} \tag{2.8.34}$$

$$S_{1212}^{dry} = S_{1212}^{dry0} + K_{1212}^{(3)}\sigma_3 + B_{22} + B_{11} \tag{2.8.35}$$

$$S_{2233}^{dry} = S_{1133}^{dry} = S_{1133}^{dry0} + K_{1133}^{(3)}\sigma_3 \tag{2.8.36}$$

The theoretically predicted stress-dependence consisting of a linear term and an exponential term is similar in form to the empirical relations (for the isotropic case) obtained earlier by Eberhart-Phillips (1989) using laboratory ultrasonic data. Shapiro (2017) relates the piezosensitivity coefficients from the porosity deformation theory to the third-order elastic coefficients C_{111}, C_{112}, and C_{113}, from third-order elasticity theory. These predictions are in rough agreement with data. For an isotropic rock with nonlinearity (stress-dependence) only due to compliant pores, the theory predicts the following relations (Shapiro, 2017):

$$\frac{C_{111}}{C_{112}} = \frac{(\lambda + 2\mu)^3 + 2\lambda^3}{2\lambda(\lambda + 2\mu)(\lambda + \mu) + \lambda^3} \tag{2.8.37}$$

$$\frac{C_{111}}{C_{123}} = \frac{(\lambda + 2\mu)^3 + 2\lambda^3}{3\lambda^2(\lambda + 2\mu)} \tag{2.8.38}$$

where λ and μ are the usual second-order isotropic elastic constants (Lame's constants).

Uses

Understanding or, at least, empirically describing the stress dependence of velocities is useful for quantifying the change of velocities in seismic time-lapse data due to changes in reservoir pressure, as well as in certain types of naturally occurring overpressure. Since the state of stress *in situ* is seldom hydrostatic, quantifying the impact of stress on anisotropy can often improve on the usual isotropic analysis.

Assumptions and Limitations

- Most models for predicting or describing stress-induced anisotropy that are based on cracks and crack-like flaws assume an isotropic, linear, elastic solid mineral material.
- Methods based on ellipsoidal cracks or spherical contacts are limited to idealized geometries and low crack densities.
- The method of Mavko *et al.* (1995) has been shown to sometimes under-predict stress-induced anisotropy when the maximum stress difference is comparable to the mean stress magnitude. More generally, there is evidence (Johnson and Rasolofosaon, 1996) that classical elastic formulations (nonlinear elasticity) can fail to describe the behavior of rocks at low stresses.
- The methods presented here assume that the strains are infinitesimal. When strains become finite, an additional source of nonlinearity, called *geometrical* or *kinetic* non-linearity, appears, related to the difference between Lagrangian and Eulerian descriptions of motion (Zarembo and Krasil'nikov, 1971; Johnson and Rasolofosaon, 1996).

2.9 Strain Components and Equations of Motion in Cylindrical and Spherical Coordinate Systems

Synopsis

The equations of motion and the expressions for small-strain components in cylindrical and spherical coordinate systems differ from those in a rectangular coordinate system. Figure 2.9.1 shows the variables used in the equations that follow.

In the **cylindrical** coordinate system (r, ϕ, z), the coordinates are related to those in the rectangular coordinate system (x, y, z) as

$$r = \sqrt{x^2 + y^2}, \quad \tan(\phi) = \frac{y}{x} \tag{2.9.1}$$

$$x = r\cos(\phi), \quad y = r\sin(\phi), \quad z = z \tag{2.9.2}$$

The small-strain components can be expressed through the displacements u_r, u_ϕ, and u_z (which are in the directions r, ϕ, and z, respectively) as

$$e_{rr} = \frac{\partial u_r}{\partial r}, \quad e_{\phi\phi} = \frac{1}{r}\frac{\partial u_\phi}{\partial \phi} + \frac{u_r}{r}, \quad e_{zz} = \frac{\partial u_z}{\partial z} \tag{2.9.3}$$

$$e_{r\phi} = \frac{1}{2}\left(\frac{1}{r}\frac{\partial u_r}{\partial \phi} + \frac{\partial u_\phi}{\partial r} - \frac{u_\phi}{r}\right) \tag{2.9.4}$$

$$e_{\phi z} = \frac{1}{2}\left(\frac{\partial u_\phi}{\partial z} + \frac{1}{r}\frac{\partial u_z}{\partial \phi}\right), \quad e_{zr} = \frac{1}{2}\left(\frac{\partial u_z}{\partial r} + \frac{\partial u_r}{\partial z}\right) \tag{2.9.5}$$

The equations of motion are

$$\frac{\partial \sigma_{rr}}{\partial r} + \frac{1}{r}\frac{\partial \sigma_{r\phi}}{\partial \phi} + \frac{\partial \sigma_{zr}}{\partial z} + \frac{\sigma_{rr} - \sigma_{\phi\phi}}{r} = \rho\frac{\partial^2 u_r}{\partial t^2} \tag{2.9.6}$$

$$\frac{\partial \sigma_{r\phi}}{\partial r} + \frac{1}{r}\frac{\partial \sigma_{\phi\phi}}{\partial \phi} + \frac{\partial \sigma_{\phi z}}{\partial z} + \frac{2\sigma_{r\phi}}{r} = \rho\frac{\partial^2 u_\phi}{\partial t^2} \tag{2.9.7}$$

$$\frac{\partial \sigma_{rz}}{\partial r} + \frac{1}{r}\frac{\partial \sigma_{\phi z}}{\partial \phi} + \frac{\partial \sigma_{zz}}{\partial z} + \frac{\sigma_{rz}}{r} = \rho\frac{\partial^2 u_z}{\partial t^2} \tag{2.9.8}$$

where ρ denotes density and t time.

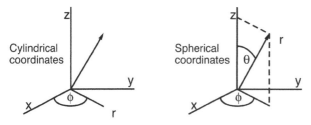

Figure 2.9.1 The variables used for converting between Cartesian, spherical, and cylindrical coordinates

In the **spherical** coordinate system (r, ϕ, θ) the coordinates are related to those in the rectangular coordinate system (x, y, z) as

$$r = \sqrt{x^2 + y^2 + z^2}, \; \tan(\phi) = \frac{y}{x}, \; \cos(\theta) = \frac{z}{\sqrt{x^2 + y^2 + z^2}} \tag{2.9.9}$$

$$x = r\sin(\theta)\cos(\phi) \quad y = r\sin(\theta)\sin(\phi) \quad z = r\cos(\theta)$$

The small-strain components can be expressed through the displacements u_r, u_ϕ, and u_θ (which are in the directions r, ϕ, and θ, respectively) as

$$e_{rr} = \frac{\partial u_r}{\partial r}, \; e_{\phi\phi} = \frac{1}{r\sin(\theta)}\frac{\partial u_\phi}{\partial \phi} + \frac{u_r}{r} + \frac{u_\theta}{r\tan(\theta)}, \; e_{\theta\theta} = \frac{1}{r}\frac{\partial u_\theta}{\partial \theta} + \frac{u_r}{r} \tag{2.9.10}$$

$$e_{r\phi} = \frac{1}{2}\left[\frac{1}{r\sin(\theta)}\frac{\partial u_r}{\partial \phi} + \frac{\partial u_\phi}{\partial r} - \frac{u_\phi}{r}\right] \tag{2.9.11}$$

$$e_{\phi\theta} = \frac{1}{2}\left[\frac{1}{r}\frac{\partial u_\phi}{\partial \theta} - \frac{u_\phi}{r\tan(\theta)} + \frac{1}{r\sin(\theta)}\frac{\partial u_\theta}{\partial \phi}\right] \tag{2.9.12}$$

$$e_{r\theta} = \frac{1}{2}\left(\frac{\partial u_\theta}{\partial r} - \frac{u_\theta}{r} + \frac{1}{r}\frac{\partial u_r}{\partial \theta}\right) \tag{2.9.13}$$

The equations of motion are

$$\frac{\partial \sigma_{rr}}{\partial r} + \frac{1}{r\sin(\theta)}\frac{\partial \sigma_{r\phi}}{\partial \phi} + \frac{1}{r}\frac{\partial \sigma_{r\theta}}{\partial \theta} + \frac{2\sigma_{rr} + \sigma_{r\theta}\cot(\theta) - \sigma_{\phi\phi} - \sigma_{\theta\theta}}{r} = \rho\frac{\partial^2 u_r}{\partial t^2} \tag{2.9.14}$$

$$\frac{\partial \sigma_{r\phi}}{\partial r} + \frac{1}{r\sin(\theta)}\frac{\partial \sigma_{\phi\phi}}{\partial \phi} + \frac{1}{r}\frac{\partial \sigma_{\phi\theta}}{\partial \theta} + \frac{3\sigma_{r\phi} + \sigma_{\phi\theta}\cot(\theta)}{r} = \rho\frac{\partial^2 u_\phi}{\partial t^2} \tag{2.9.15}$$

$$\frac{\partial \sigma_{r\theta}}{\partial r} + \frac{1}{r\sin(\theta)}\frac{\partial \sigma_{\phi\theta}}{\partial \phi} + \frac{1}{r}\frac{\partial \sigma_{\phi\theta}}{\partial \theta} + \frac{3\sigma_{r\theta} + (\sigma_{\theta\theta} - \sigma_{\phi\phi})\cot(\theta)}{r} = \rho\frac{\partial^2 u_\theta}{\partial t^2} \tag{2.9.16}$$

Uses

The foregoing equations are used to solve elasticity problems where cylindrical or spherical geometries are most natural.

Assumptions and Limitations

The equations presented assume that the strains are small.

2.10   Deformation of Inclusions and Cavities in Elastic Solids

Synopsis

Many problems in effective-medium theory and poroelasticity can be solved or estimated in terms of the elastic behavior of cavities and inclusions. Some **static** and

quasistatic results for cavities are presented here. It should be remembered that often these are also valid for certain limiting cases of dynamic problems. Excellent treatments of cavity deformation and pore compressibility are given by Jaeger and Cook (1969) and by Zimmerman (1991a).

General Pore Deformation

Effective Dry Compressibility

Consider a homogeneous linear elastic solid that has an arbitrarily shaped pore space – either a single cavity or a collection of pores. The effective dry compressibility (i.e., the reciprocal of the dry bulk modulus) of the porous solid can be written as

$$\frac{1}{K_{dry}} = \frac{1}{K_0} + \frac{\phi}{v_p} \frac{\partial v_p}{\partial \sigma}\Big|_{dry} \tag{2.10.1}$$

where K_{dry} is the effective bulk modulus of the dry porous solid, K_0 is the bulk modulus of the solid mineral material, ϕ is the porosity, v_p is the pore volume, $\partial v_p / \partial \sigma|_{dry}$ is the derivative of the pore volume with respect to the externally applied hydrostatic stress. We also assume that no inelastic effects such as friction or viscosity are present. This expression is strictly true, regardless of the pore geometry and the pore concentration. The preceding equation can be rewritten slightly as

$$\frac{1}{K_{dry}} = \frac{1}{K_0} + \frac{\phi}{K_\phi} \tag{2.10.2}$$

where $K_\phi = v_p / (\partial v_p / \partial \sigma)|_{dry}$ is defined as the dry pore-space stiffness. These equations state simply that the porous rock compressibility is equal to the intrinsic mineral compressibility plus an additional compressibility caused by the pore space.

Caution:

"Dry rock" is not the same as gas-saturated rock. The dry-frame modulus refers to the incremental bulk deformation resulting from an increment of applied confining pressure with pore pressure held constant. This corresponds to a "drained" experiment in which pore fluids can flow freely in or out of the sample to ensure constant pore pressure. Alternatively, the dry-frame modulus can correspond to an undrained experiment in which the pore fluid has zero bulk modulus and in which pore compressions, therefore, do not induce changes in pore pressure. This is approximately the case for an air-filled sample at standard temperature and pressure. However, at reservoir conditions, the gas takes on a non-negligible bulk modulus and should be treated as a saturating fluid.

An equivalent expression for the dry rock compressibility or bulk modulus is

$$K_{dry} = K_0(1 - \beta) \tag{2.10.3}$$

where β is sometimes called the **Biot coefficient**, which describes the ratio of pore-volume change Δv_p to total bulk-volume change ΔV under dry or drained conditions:

$$\beta = \frac{\Delta v_p}{\Delta V}\bigg|_{\text{dry}} = \frac{\phi K_{\text{dry}}}{K_\phi} \tag{2.10.4}$$

Stress-Induced Pore Pressure: Skempton's Coefficient

If this arbitrary pore space is filled with a pore fluid with bulk modulus, K_{fl}, the saturated solid is stiffer under compression than the dry solid, because an increment of pore-fluid pressure is induced that resists the volumetric strain. The ratio of the induced pore pressure, dP, to the applied compressive stress, $d\sigma$, is sometimes called **Skempton's coefficient** and can be written as

$$B \equiv \frac{dP}{d\sigma} = \frac{1}{1 + K_\phi(1/K_{\text{fl}} - 1/K_0)}$$
$$= \frac{1}{1 + \phi(1/K_{\text{fl}} - 1/K_0)(1/K_{\text{dry}} - 1/K_0)^{-1}} \tag{2.10.5}$$

where K_ϕ is the dry pore-space stiffness defined earlier in this section. For this definition to be true, the pore pressure must be uniform throughout the pore space, as will be the case if:
(1) there is only one pore,
(2) all pores are well connected and the frequency and viscosity are low enough for any pressure differences to equilibrate, or
(3) all pores have the same dry pore stiffness.
    Given these conditions, there is no additional limitation on pore geometry or concentration. All of the necessary information concerning pore stiffness and geometry is contained in the parameter K_ϕ.

Saturated Stress-Induced Pore-Volume Change

The corresponding change in fluid-saturated pore volume, v_p, caused by the remote stress is

$$\frac{1}{v_p}\frac{dv_p}{d\sigma}\bigg|_{\text{sat}} = \frac{1}{K_{\text{fl}}}\frac{dP}{d\sigma} = \frac{1/K_{\text{fl}}}{1 + K_\phi(1/K_{\text{fl}} - 1/K_0)} \tag{2.10.6}$$

Low-Frequency Saturated Compressibility

The low-frequency saturated bulk modulus, K_{sat}, can be derived from Gassmann's equation (see Section 6.3 on Gassmann). One equivalent form is

$$\frac{1}{K_{\text{sat}}} = \frac{1}{K_0} + \frac{\phi}{K_\phi + \dfrac{K_0 K_{\text{fl}}}{K_0 - K_{\text{fl}}}} \approx \frac{1}{K_0} + \frac{\phi}{K_\phi + K_{\text{fl}}} \tag{2.10.7}$$

where, again, all of the necessary information concerning pore stiffness and geometry is contained in the dry pore stiffness K_ϕ, and we must ensure that the stress-induced pore pressure is uniform throughout the pore space.

Three-Dimensional Ellipsoidal Cavities

Many effective media models are based on ellipsoidal inclusions or cavities. These are mathematically convenient shapes and allow quantitative estimates of, for example, K_ϕ, which was defined earlier in this section. Eshelby (1957) discovered that the strain, ε_{ij}, inside an isolated ellipsoidal inclusion is homogeneous when a homogeneous strain, ε_{ij}^0, (or stress) is applied at infinity and the background material is homogeneous. Because the inclusion strain is homogeneous, operations such as determining the inclusion stress or integrating to obtain the displacement field are straightforward. Eshelby's solution is summarized in Section 2.12.

It is very important to remember that the following results assume a single isolated cavity in an infinite medium. Therefore, substituting them directly into the preceding formulas for dry and saturated moduli gives estimates that are strictly valid only for low concentrations of pores (see also Sections 4.8–4.15 on effective medium theories).

Spherical Cavity under Remote Hydrostatic Stress

For a single spherical cavity with volume $v_\mathrm{p} = \dfrac{4}{3}\pi R^3$ and a hydrostatic stress, $d\sigma$, applied at infinity, the radial strain of the cavity is

$$\frac{dR}{R} = \frac{1}{K_0}\frac{(1-v)}{2(1-2v)}d\sigma \tag{2.10.8}$$

where v and K_0 are the Poisson's ratio and bulk modulus of the solid material, respectively. The change of pore volume is

$$dv_\mathrm{p} = \frac{1}{K_0}\frac{3(1-v)}{2(1-2v)}v_\mathrm{p}d\sigma \tag{2.10.9}$$

Then, the volumetric strain of the sphere is

$$\varepsilon_{ii} = \frac{dv_\mathrm{p}}{v_\mathrm{p}} = \frac{1}{K_0}\frac{3(1-v)}{2(1-2v)}d\sigma \tag{2.10.10}$$

and the single-pore stiffness is

$$\frac{1}{K_\phi} = \frac{1}{v_\mathrm{p}}\frac{dv_\mathrm{p}}{d\sigma} = \frac{1}{K_0}\frac{3(1-v)}{2(1-2v)} \tag{2.10.11}$$

Remember that this estimate of K_ϕ assumes a single isolated spherical cavity in an infinite medium.

Under a remotely applied homogeneous *shear stress*, τ_0, corresponding to remote shear strain $\varepsilon_0 = \tau_0/2\mu_0$, the effective shear strain in the spherical cavity is

$$\varepsilon = \frac{15(1-v)}{2\mu_0(7-5v)}\tau_0 \tag{2.10.12}$$

where μ_0 is the shear modulus of the solid. Note that this results in approximately twice the strain that would occur without the cavity.

Spherical Cavity under Remote Triaxial Loading

A linear elastic solid with shear modulus μ and Poisson's ratio v contains a spherical cavity (radius R) and is loaded at infinity with principal stresses σ_1, σ_2, σ_3 and pressurized inside by $-p_0$. The displacements within the solid are given by (Warren, 1983)

$$\begin{aligned}
2\mu\frac{u_r}{R} = & \frac{1}{2}p_0\frac{R^2}{r^2} + [\sigma_1\sin^2\theta\cos^2\gamma + \sigma_2\sin^2\theta\sin^2\gamma + \sigma_3\cos^2\theta] \\
& \times \left[\frac{r}{R} + \frac{5(5-4v)}{2(7-5v)}\frac{R^2}{r^2} - \frac{9}{2(7-5v)}\frac{R^4}{r^4}\right] \\
& -(\sigma_1+\sigma_2+\sigma_3)\left[\frac{v}{1+v}\frac{r}{R} + \frac{(6-5v)}{2(7-5v)}\frac{R^2}{r^2} - \frac{3}{2(7-5v)}\frac{R^4}{r^4}\right]
\end{aligned} \tag{2.10.13}$$

$$\begin{aligned}
2\mu\frac{u_\theta}{R} = & \sin\theta\cos\theta[\sigma_1\cos^2\gamma + \sigma_2\sin^2\gamma - \sigma_3] \\
& \times \left[\frac{r}{R} + \frac{5(1-2v)}{(7-5v)}\frac{R^2}{r^2} + \frac{3}{(7-5v)}\frac{R^4}{r^4}\right],
\end{aligned} \tag{2.10.14}$$

$$2\mu\frac{u_\gamma}{R} = (\sigma_2-\sigma_1)\sin\theta\cos\gamma\sin\gamma\left[\frac{r}{R} + \frac{5(1-2v)}{(7-5v)}\frac{R^2}{r^2} + \frac{3}{(7-5v)}\frac{R^4}{r^4}\right], \tag{2.10.15}$$

where angles θ and γ are defined in Figure 2.10.1. Stresses at the cavity surface $r = R$ are

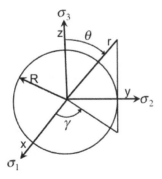

Figure 2.10.1 Coordinates used to describe the deformation of a spherical cavity

$$\sigma_{\theta\theta}|_{r=R} = \frac{1}{2}p_0 + \frac{3}{2(7-5v)}\{3\sigma_3(3-5v) - (1-5v)(\sigma_1+\sigma_2)$$
$$+ 10[\sigma_1\cos^2\gamma + \sigma_2\sin^2\gamma - \sigma_3](\cos^2\theta - v)\} \tag{2.10.16}$$

$$\sigma_{\gamma\gamma}|_{r=R} = \frac{1}{2}p_0 + \frac{3}{2(7-5v)}\{-(11-5v)\sigma_3 + (9-5v)(\sigma_1+\sigma_2)$$
$$+ 10[\sigma_1\cos^2\gamma + \sigma_2\sin^2\gamma - \sigma_3](v\cos^2\theta - 1)\} \tag{2.10.17}$$

$$\sigma_{\theta\gamma}|_{r=R} = \frac{3}{2(7-5v)}[10(1-v)(\sigma_2-\sigma_1)\cos\theta\cos\gamma\sin\gamma] \tag{2.10.18}$$

$$\sigma_{rr}|_{r=R} = -p_0 \tag{2.10.19}$$

Expressions for the complete stress field can be found in Brethauer (1974). The hoop stress concentration at $\theta = 0$ and $\theta = 90°$ under uniaxial stress σ_{33} is

$$\sigma_{\theta\theta}|_{r=R} = \begin{cases} \sigma_3 \dfrac{-3(1+5v)}{2(7-5v)}; & \theta = 0° \\[3mm] \sigma_3 \dfrac{3(9-5v)}{2(7-5v)}; & \theta = 90° \end{cases} \tag{2.10.20}$$

Similar to the stress concentration around a circular (2D) borehole, the stress is negative at the top of the sphere and magnified positive at the sides. Unlike the 2D problem, the stress concentrations depend on the Poisson's ratio.

When the spherical cavity is in a transversely isotropic background, the stress concentrations deviate somewhat from the isotropic case (Chen, 1968). If the remote principal stress along the symmetry axis of the TI medium is σ_3^0 and the remote horizontal principal stresses are σ_{rr}^0, the problem is rotationally symmetric. The hoop stress at the equator is given by

$$\sigma_{\theta\theta} = \sigma_3^0 + A_1(1+k_1)c_{44}\psi_{11} + A_2(1+k_2)c_{44}\psi_{12} \tag{2.10.21}$$

and the hoop stress at the poles is

$$\sigma_{\theta\theta} = \sigma_{rr}^0 + A_1c_{44}\left[\frac{2(1+k_1)}{s_1} - \frac{(c_{11}-c_{12})}{c_{44}}\right]\psi_{21} + A_2c_{44}\left[\frac{2(1+k_2)}{s_2} - \frac{(c_{11}-c_{12})}{c_{44}}\right]\psi_{22} \tag{2.10.22}$$

Dimensionless parameters s_1 and s_2 are the roots of the following quadratic equation in terms of the background elastic constants c_{ij}:

$$c_{11}c_{44}s^2 + [c_{13}(c_{13}+2c_{44}) - c_{11}c_{33}]s + c_{33}c_{44} = 0. \tag{2.10.23}$$

The constants A_1 and A_2 are determined by solving the two linear equations

$$2A_1(1+k_1)\psi_{21} + 2A_2(1+k_2)\psi_{22} = \frac{\sigma_{33}^0}{c_{44}} \tag{2.10.24}$$

$$A_1\left\{\frac{(1+k_1)}{s_1}\psi_{11} + \frac{(c_{11}-c_{12})}{c_{44}}\psi_{21}\right\} + A_2\left\{\frac{(1+k_2)}{s_2}\psi_{12} + \frac{(c_{11}-c_{12})}{c_{44}}\psi_{22}\right\} = \frac{\sigma_{rr}}{c_{44}} \tag{2.10.25}$$

where

$$k_\alpha = (c_{11}s_\alpha - c_{44})/(c_{13} + c_{44}), \quad \alpha = 1, 2 \tag{2.10.26}$$

$$\psi_{0\alpha} = \ln\left(s_\alpha^{-1/2} + \sqrt{s_\alpha^{-1} - 1}\right) \tag{2.10.27}$$

$$\psi_{1\alpha} = \psi_{0\alpha} - \sqrt{1 - s_\alpha} \tag{2.10.28}$$

$$\psi_{2\alpha} = -[\psi_{0\alpha} - \sqrt{1 - s_\alpha}/s_\alpha]/2 \tag{2.10.29}$$

As an example, for an isotropic medium loaded remotely by σ_3^0, the hoop stresses at the poles and equator are $-0.6818\sigma_3^0$ and $2.0455\sigma_3^0$. If instead, the medium is weakly anisotropic, with Thomsen's parameters $\varepsilon = \gamma = 0.05$ and $\delta = 0$, the hoop stresses are $-0.6992\sigma_3^0$ and $2.0508\sigma_3^0$, respectively.

Deformation of Spherical Shells

The radial displacement u_r of a spherical elastic shell (Figure 2.10.2) with inner pore radius R_p and outer radius R_c under applied confining pressure P_c and pore pressure P_p is given by (Ciz *et al.*, 2008)

$$u_r = \frac{r}{3K(1 - \phi)}\left(\phi P_p - P_c\right) + \frac{R_p^3}{4r^2\mu(1 - \phi)}\left(P_p - P_c\right) \tag{2.10.30}$$

where r is the radial distance from the center of the sphere, K is the bulk modulus of the solid elastic material, μ is the shear modulus of the solid elastic material, and ϕ is the porosity. The associated volume change of the inner spherical pore is

$$dV_\phi = \frac{3V_\phi}{1 - \phi}\left[\left(\frac{\phi}{3K} + \frac{1}{4\mu}\right)dP_p - \left(\frac{1}{3K} + \frac{1}{4\mu}\right)dP_c\right]. \tag{2.10.31}$$

The volume change of the total sphere (shell plus pore) is

$$dV = \frac{3V}{1 - \phi}\left[\left(\frac{1}{3K} + \frac{1}{4\mu}\right)dP_p - \left(\frac{1}{3K} + \frac{\phi}{4\mu}\right)dP_c\right]. \tag{2.10.32}$$

Similarly, the radial displacement of a double-layer sphere (Figure 2.10.2) is given by (Ciz *et al.*, 2008)

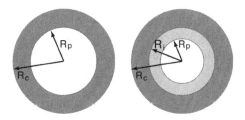

Figure 2.10.2 Deformation of a single- and double-layer spherical shell

$$u_r = \frac{A_n r}{3} + \frac{B_n}{r^2} \tag{2.10.33}$$

where R_p is the pore radius, R_c is the outer sphere radius, and R_i is the radius at the boundary between the inner and outer elastic shells. The elastic moduli of the inner solid shell are K_1 and μ_1, and the moduli of the outer shell are K_2 and μ_2. Displacements in the inner shell are found using n=1, and in the outer shell using n=2.

$$A_1 = \frac{1}{\Delta}\left\{ \left[(4\mu_1\mu_2 + 3K_2\mu_2)R_i^3 + (3K_2\mu_1 - 3K_2\mu_2)R_c^3\right]R_p^3 P_p - (3K_2\mu_1 + 4\mu_1\mu_2)R_c^3 R_i^3 P_c \right\} \tag{2.10.34}$$

$$B_1 = \frac{1}{\Delta}\left\{ \left[\left(\frac{3}{4}K_1K_2 + K_2\mu_2\right)R_c^3 + (K_1\mu_2 + K_2\mu_2)R_i^3\right]R_p^3 R_i^3 P_p - \left(\frac{3}{4}K_1K_2 + K_1\mu_2\right)R_c^3 R_i^3 R_p^3 P_c \right\} \tag{2.10.35}$$

$$A_2 = \frac{1}{\Delta}\left\{ \left[(3K_1\mu_1 - 3K_1\mu_2)R_p^3 - (3K_1\mu_2 + 4\mu_1\mu_2)R_i^3\right]R_c^3 P_c + (3K_1\mu_2 - 4\mu_1\mu_2)R_i^3 R_p^3 P_p \right\} \tag{2.10.36}$$

$$B_2 = \frac{1}{\Delta}\left\{ \left[(K_1\mu_1 - K_2\mu_1)R_i^3 - \left(\frac{3}{4}K_1K_2 + K_2\mu_1\right)R_p^3\right]R_c^3 R_i^3 P_c + \left(\frac{3}{4}K_1K_2 + K_1\mu_1\right)R_c^3 R_i^3 R_p^3 P_p \right\} \tag{2.10.37}$$

$$\Delta = K_2\mu_1(3K_1 + 4\mu_2)R_c^3 R_i^3 - K_1\mu_2(3K_2 + 4\mu_1)R_i^3 R_p^3 + 4\mu_1\mu_2(K_1 - K_2)R_i^6$$
$$+ 3K_1K_2(\mu_2 - \mu_1)R_c^3 R_p^3 \tag{2.10.38}$$

Penny-Shaped Crack: Oblate Spheroid

Consider a dry penny-shaped ellipsoidal cavity with semiaxes $a \ll b = c$. When a remote uniform tensional stress, $d\sigma$, is applied normal to the plane of the crack, each crack face undergoes an outward displacement, U, normal to the plane of the crack, given by the radially symmetric distribution

$$U(r) = \frac{4(1 - v^2)c\sqrt{1 - (r/c)^2}}{3\pi K_0(1 - 2v)} d\sigma \tag{2.10.39}$$

where r is the radial distance from the axis of the crack. For any arbitrary homogeneous remote stress, $d\sigma$ is the component of stress *normal* to the plane of the crack. Thus, $d\sigma$ can also be thought of as a remote hydrostatic stress field.

This displacement function is also an ellipsoid with semiminor axis, $U(r = 0)$, and semimajor axis, c. Therefore, the volume change, dv, which is simply the integral of $U(r)$ over the faces of the crack, is just the volume of the displacement ellipsoid:

$$dv_p = \frac{4}{3}\pi U(0)c^2 = \frac{16c^3}{9K_0}\frac{(1 - v^2)}{(1 - 2v)} d\sigma \tag{2.10.40}$$

Then the volumetric strain of the cavity is

$$\varepsilon_{ii} = \frac{dv_\mathrm{p}}{v_\mathrm{p}} = \frac{4(c/a)}{3\pi K_0} \frac{(1-v^2)}{(1-2v)} do \tag{2.10.41}$$

and the pore stiffness is

$$\frac{1}{K_\phi} = \frac{1}{v_\mathrm{p}} \frac{dv_\mathrm{p}}{d\sigma} = \frac{4(c/a)}{3\pi K_0} \frac{(1-v^2)}{(1-2v)} \tag{2.10.42}$$

An interesting case is an applied compressive stress causing a displacement, $U(0)$, equal to the original half-width of the crack, thus closing the crack. Setting $U(0) = a$ allows one to compute the **closing stress**:

$$\sigma_\mathrm{close} = \frac{3\pi(1-2v)}{4(1-v^2)} \alpha K_0 = \frac{\pi}{2(1-v)} \alpha \mu_0 \tag{2.10.43}$$

where $\alpha = (a/c)$ is the aspect ratio. Note that for a thin, penny-shaped crack, $\sigma_\mathrm{close} = K_\phi$.

Under a remotely applied homogeneous *shear stress, τ_0* (traction parallel to the plane of the crack), corresponding to remote shear strain $\varepsilon_0 = \tau_0/2\mu_0$, the effective **shear strain** in the cavity is

$$\varepsilon = \tau_0 \frac{2(c/a)}{\pi\mu_0} \frac{(1-v)}{(2-v)} \tag{2.10.44}$$

Next, consider the case where the penny-shaped cavity is lying in an elastic background with VTI symmetry, with Voigt notation elastic constants c_{ij}. The symmetry axes of the cavity and the VTI medium are aligned. The normal displacement $U(r)$ of the cavity surface under uniform internal pressure p_0 is (Fabrikant, 1989; Kachanov *et al.,* 2003)

$$U(r) = \frac{2p_0}{\pi} \left[\frac{(\gamma_1+\gamma_2)c_{11}}{(c_{11}c_{33}-c_{13}^2)} \right] \left[\frac{m_1}{m_1-1} + \frac{m_2}{m_2-1} \right] \sqrt{c^2-r^2} \tag{2.10.45}$$

where r is the radial distance from the center, and

$$m_i = \frac{-(2c_{13}c_{44} - c_{11}c_{33} + c_{13}^2 + 2c_{44}^2) \pm \sqrt{-(-c_{13}^2 + c_{11}c_{33})(c_{13}^2 + 4c_{13}c_{44} + 4c_{44}^2 - c_{11}c_{33}))}}{2c_{44}(c_{13}+c_{44})},$$

$$i = 1,2 \tag{2.10.46}$$

$$\gamma_i = \sqrt{(c_{44} + m_i(c_{13}+c_{44}))/c_{11}}. \tag{2.10.47}$$

Needle-Shaped Pore: Prolate Spheroid

Consider a dry needle-shaped ellipsoidal cavity with semiaxes $a \gg b = c$ and with pore volume $v = \frac{4}{3}\pi ac^2$. When a remote hydrostatic stress, do, is applied, the pore volume change is

$$dv_\mathrm{p} = \frac{(5-4v)}{3K_0(1-2v)} v_\mathrm{p} do \tag{2.10.48}$$

The volumetric strain of the cavity is then

$$\varepsilon_{ii} = \frac{dv_p}{v_p} = \frac{(5 - 4v)}{3K_0(1 - 2v)} d\sigma \tag{2.10.49}$$

and the pore stiffness is

$$\frac{1}{K_\phi} = \frac{1}{v_p} \frac{dv_p}{d\sigma} = \frac{(5 - 4v)}{3K_0(1 - 2v)} \tag{2.10.50}$$

Note that in the limit of very large a/c, these results are exactly the same as for a 2D circular cylinder.

Example: Estimate the increment of pore pressure induced in a water-saturated rock when a 1 bar increment of hydrostatic confining pressure is applied. Assume that the rock consists of stiff, spherical pores in a quartz matrix. Compare this with a rock with thin, penny-shaped cracks (aspect ratio $\alpha = 0.001$) in a quartz matrix. The elastic constants of the individual constituents are $K_{quartz} = 36$ GPa, $K_{water} = 2.2$ GPa, and $v_{quartz} = 0.07$.

The pore-space stiffnesses are given by

$$K_{\phi-sphere} = K_{quartz} \frac{2(1 - 2v_{quartz})}{3(1 - v_{quartz})} = 22.2 \text{ GPa}$$

$$K_{\phi-crack} = \frac{3\pi\alpha K_{quartz}}{4} \frac{(1 - 2v_{quartz})}{(1 - v_{quartz}^2)} = 0.0733 \text{ GPa}$$

The pore-pressure increment is computed from Skempton's coefficient:

$$\frac{\Delta P_{pore}}{\Delta P_{confining}} = B = \frac{1}{1 + K_\phi(K_{water}^{-1} - K_{quartz}^{-1})}$$

$$B_{sphere} = \frac{1}{1 + 22.2[(1/2.2) - (1/36)]} = 0.095$$

$$B_{crack} = \frac{1}{1 + 0.0733[(1/2.2) - (1/36)]} = 0.970$$

Therefore, the pore pressure induced in the spherical pores is 0.095 bar and the pore pressure induced in the cracks is 0.98 bar.

Two-Dimensional Tubes

A special 2D case of long, tubular pores was treated by Mavko (1980) to describe melt or fluids arranged along the edges of grains. The cross-sectional shape is described by the equations

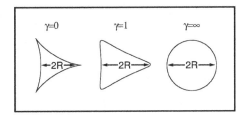

Figure 2.10.3 The cross-sectional shapes of various two-dimensional tubes described by equations 2.10.51 and 2.10.52

$$x = R\left(\cos\theta + \frac{1}{2+\gamma}\cos2\theta\right) \tag{2.10.51}$$

$$y = R\left(-\sin\theta + \frac{1}{2+\gamma}\sin2\theta\right) \tag{2.10.52}$$

where γ is a parameter describing the roundness (Figure 2.10.3).

Consider, in particular, the case on the left, $\gamma = 0$. The pore volume is $\frac{1}{2}\pi aR^2$, where $a \gg R$ is the length of the tube. When a remote hydrostatic stress, $d\sigma$, is applied, the pore volume change is

$$d\upsilon_p = \frac{(13 - 4v - 8v^2)}{3K_0(1-2v)}\upsilon_p d\sigma \tag{2.10.53}$$

The volumetric strain of the cavity is then

$$\varepsilon_{ii} = \frac{d\upsilon_p}{\upsilon_p} = \frac{(13 - 4v - 8v^2)}{3K_0(1-2v)}d\sigma \tag{2.10.54}$$

and the pore stiffness is

$$\frac{1}{K_\phi} = \frac{1}{\upsilon_p}\frac{d\upsilon_p}{d\sigma} = \frac{(13 - 4v - 8v^2)}{3K_0(1-2v)} \tag{2.10.55}$$

In the extreme, $\gamma \to \infty$, the shape becomes a circular cylinder, and the expression for pore stiffness, K_ϕ, is exactly the same as that previously derived for the needle-shaped pores. Note that the triangular cavity ($\gamma = 0$) has about half the pore stiffness of the circular one; that is, the triangular tube can give approximately the same effective modulus as the circular tube with about half the porosity.

Caution:

These expressions for K_ϕ, $d\upsilon_p$, and ε_{ii} include an estimate of tube shortening as well as a reduction in pore cross-sectional area under hydrostatic stress. Hence, the deformation is neither plane stress nor plane strain.

Plane Strain

The **plane strain** compressibility in terms of the reduction in cross-sectional area A is given by

$$\frac{1}{K'_\phi} = \frac{1}{A}\frac{dA}{d\sigma} = \begin{cases} \dfrac{6(1-v)}{\mu_0}, & \gamma \to 0 \\[2ex] \dfrac{2(1-v)}{\mu_0}, & \gamma \to \infty \end{cases} \tag{2.10.56}$$

The latter case ($\gamma \to \infty$) corresponds to a tube with a circular cross-section and agrees (as it should) with the expression given in this chapter for the limiting case of a tube with an elliptical cross-section with aspect ratio unity.

A general method of determining K'_ϕ for nearly arbitrarily shaped 2D cavities under *plane-strain* deformation was developed by Zimmerman (1986, 1991a) and involves conformal mapping of the tube shape into circular pores. For example, pores with cross-sectional shapes that are n-sided hypotrochoids given by

$$x = \cos(\theta) + \frac{1}{(n-1)}\cos(n-1)\theta \tag{2.10.57}$$

$$y = -\sin(\theta) + \frac{1}{(n-1)}\sin(n-1)\theta \tag{2.10.58}$$

(see the examples labeled (1) in Table 2.10.1) have plane-strain compressibilities

$$\frac{1}{K'_\phi} = \frac{1}{A}\frac{dA}{d\sigma} = \frac{1}{K'^{\text{cir}}_\phi}\frac{1+(n-1)^{-1}}{1-(n-1)^{-1}} \tag{2.10.59}$$

where $1/K'^{\text{cir}}_\phi$ is the plane-strain compressibility of a circular tube given by

$$\frac{1}{K'^{\text{cir}}_\phi} = \frac{2(1-v)}{\mu_0} \tag{2.10.60}$$

Table 2.10.1 summarizes a few plane-strain pore compressibilities.

Two-Dimensional Thin Cracks

A convenient description of very thin 2D cracks is in terms of elastic line dislocations. Consider a crack lying along $-c < x < c$ in the $y = 0$ plane and very long in the z direction. The **total relative displacement** of the crack faces $u(x)$, defined as the displacement of the negative face ($y = 0^-$) relative to the positive face ($y = 0^+$), is related to the dislocation density function by

$$B(x) = -\frac{\partial u}{\partial x} \tag{2.10.61}$$

Table 2.10.1 *Plane-strain compressibility normalized by the compressibility of a circular tube. The numbers below each shape are the corresponding normalized quantity* $\mu_0/(2(1-\nu)K'_\phi)$.

(1)	(1)	(1)	(1)	(1)	(2)	(2)	(2)
◯ circle	△ triangle	(curved triangle)	◇ diamond	✦ four-pointed star	▭ slit	⬭ ellipse	◊ lens (tapered)
1	1.581	3	1.188	2	$\pi/8a$	$1/2a$	$2/3a$

Notes:

(1) $x = \cos(\theta) + \dfrac{1}{(n-1)}\cos(n-1)\theta$, where n = number of sides; $y = -\sin(\theta) + \dfrac{1}{(n-1)}\sin(n-1)\theta$.

(2) $y = 2b\sqrt{1 - (x/c)^2}$ (ellipse). (3) $y = 2b[1 - (x/c)^2]^{3/2}$ (nonelliptical, "tapered" crack).

where $B(x)\,dx$ represents the total length of the Burger vectors of the dislocations lying between x and $x + dx$. The stress change in the plane of the crack that results from introduction of a dislocation line with unit Burger vector at the origin is

$$\sigma = \frac{\mu_0}{2\pi Dx} \tag{2.10.62}$$

where $D = 1$ for screw dislocations and $D = (1 - v)$ for edge dislocations (v is Poisson's ratio, and μ_0 is the shear modulus). Edge dislocations can be used to describe mode I and mode II cracks; screw dislocations can be used to describe mode III cracks.

The stress here is the component of traction in the crack plane parallel to the displacement: normal stress for mode I deformation, in-plane shear for mode II deformation, and out-of-plane shear for mode III deformation.

Then the stress resulting from the distribution $B(x)$ is given by the convolution

$$\sigma(x) = \frac{\mu_0}{2\pi D} \int_{-c}^{c} \frac{B(x')dx'}{x - x'} \tag{2.10.63}$$

The special case of interest for nonfrictional cavities is the deformation for stress-free crack faces under a remote uniform tensional stress, $d\sigma$, acting normal to the plane of the crack. The outward displacement distribution of *each* crack face is given by (using edge dislocations)

$$U(x) = \frac{c\sigma(1 - v)\sqrt{1 - (x/c)^2}}{\mu_0} = \frac{2c\sigma(1 - v^2)\sqrt{1 - (x/c)^2}}{3K_0(1 - 2v)} \tag{2.10.64}$$

The cross-sectional area change is then given by

$$\frac{dA}{d\sigma} = \pi U(0)c = \frac{2\pi c^2 (1 - v^2)}{3K_0 (1 - 2v)} \tag{2.10.65}$$

It is important to note that *these results for displacement and volume change apply to any 2D crack of arbitrary cross-section* as long as it is very thin and approximately planar. They are not limited to cracks of elliptical cross-section.

For the special case in which the very thin crack is elliptical in cross-section with half-width b in the thin direction, the cross-sectional area is $A = \pi bc$, and the pore stiffness under plane-strain deformation is given by

$$\frac{1}{K'_\phi} = \frac{1}{A}\frac{dA}{d\sigma} = \frac{(c/b)(1 - v)}{\mu_0} = \frac{2(c/b)(1 - v^2)}{3K_0(1 - 2v)} \tag{2.10.66}$$

Another special case is a crack of *nonelliptical* form (Mavko and Nur, 1978) with initial shape given by

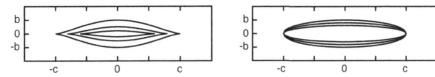

Figure 2.10.4 A nonelliptical crack shortens as well as narrows under compression. An elliptical crack only narrows.

$$U_0(x) = 2b\left[1 - \left(\frac{x}{c_0}\right)^2\right]^{3/2} \tag{2.10.67}$$

where c_0 is the crack half-length and $2b$ is the maximum crack width. This crack is plotted in Figure 2.10.4. Note that unlike elliptical cracks that have rounded or blunted ends, this crack has tapered tips where faces make a smooth, tangent contact. If we apply a pressure, P, the crack shortens as well as thins, and the pressure-dependent length is given by

$$c = c_0\left[1 - \frac{2(1-v)}{3\mu_0(b/c_0)}P\right]^{1/2} \tag{2.10.68}$$

Then the deformed shape is

$$U(x, P) = 2b\left(\frac{c}{c_0}\right)^3\left[1 - \left(\frac{x}{c}\right)^2\right]^{3/2}, \quad |x| \leq c \tag{2.10.69}$$

An important consequence of the smoothly tapered crack tips and the gradual crack shortening is that there is no stress singularity at the crack tips. In this case, crack closure occurs (i.e., $U \to 0$) as the crack length goes to zero ($c \to 0$). The closing stress is

$$\sigma_{\text{close}} = \frac{3}{2(1-v)}\alpha_0\mu_0 = \frac{3}{4(1-v^2)}\alpha_0 E_0 \tag{2.10.70}$$

where $\alpha_0 = b/c_0$ is the original crack aspect ratio, and μ_0 and E_0 are the shear and Young's moduli of the solid material, respectively. This expression is consistent with the usual rule of thumb that the crack-closing stress is numerically $\sim\alpha_0 E_0$. The exact factor depends on the details of the original crack shape. In comparison, the pressure required to close a 2D elliptical crack of aspect ratio α_0 is

$$\sigma_{\text{close}} = \frac{1}{2(1-v^2)}\alpha_0 E_0 \tag{2.10.71}$$

Ellipsoidal Cracks of Finite Thickness

The pore compressibility under plane-strain deformation of a *2D* elliptical cavity of arbitrary aspect ratio α is given by (Zimmerman, 1991a)

$$\frac{1}{K'_\phi} = \frac{1}{A}\frac{dA}{d\sigma} = \frac{1-v}{\mu_0}\left(\alpha + \frac{1}{\alpha}\right) = \frac{2(1-v^2)}{3K_0(1-2v)}\left(\alpha + \frac{1}{\alpha}\right) \tag{2.10.72}$$

where μ_0, K_0, and v are the shear modulus, bulk modulus, and Poisson's ratio of the mineral material, respectively. Circular pores (tubes) correspond to aspect ratio $\alpha = 1$ and the pore compressibility is given as

$$\frac{1}{K'_\phi} = \frac{1}{A}\frac{dA}{d\sigma} = \frac{2(1-v)}{\mu_0} = \frac{4(1-v^2)}{3K_0(1-2v)} \tag{2.10.73}$$

Uses

The equations presented in this section are useful for computing deformation of cavities in elastic solids and estimating effective moduli of porous solids.

Assumptions and Limitations

The equations presented in this section are based on the following assumptions.
- Solid material must be homogeneous, isotropic, linear, and elastic.
- Results for specific geometries, such as spheres and ellipsoids, are derived for single isolated cavities. Therefore, estimates of effective moduli based on these are limited to relatively low pore concentrations where pore elastic interaction is small.
- Pore-pressure computations assume that the induced pore pressure is uniform throughout the pore space, which will be the case if (i) there is only one pore, (2) all pores are well connected and the frequency and viscosity are low enough for any pressure differences to equilibrate, or (3) all pores have the same dry pore stiffness.

2.11   Deformation of a Circular Hole: Borehole Stresses

Synopsis

Presented here are some solutions related to a circular hole in a stressed, linear, elastic, and poroelastic isotropic medium.

Hollow Cylinder with Internal and External Pressures

The cylinder's internal radius is R_1 and the external radius is R_2. Hydrostatic stress p_1 is applied at the interior surface at R_1 and hydrostatic stress p_2 is applied at the exterior surface at R_2. The resulting (plane-strain) outward displacement U and radial and tangential stresses are

$$U = \frac{(p_2 R_2^2 - p_1 R_1^2)}{2(\lambda + \mu)(R_2^2 - R_1^2)}r + \frac{(p_2 - p_1)R_1^2 R_2^2}{2\mu(R_2^2 - R_1^2)}\frac{1}{r} \tag{2.11.1}$$

$$\sigma_{rr} = \frac{(p_2 R_2^2 - p_1 R_1^2)}{(R_2^2 - R_1^2)} - \frac{(p_2 - p_1) R_1^2 R_2^2}{(R_2^2 - R_1^2)} \frac{1}{r^2} \tag{2.11.2}$$

$$\sigma_{\theta\theta} = \frac{(p_2 R_2^2 - p_1 R_1^2)}{(R_2^2 - R_1^2)} + \frac{(p_2 - p_1) R_1^2 R_2^2}{(R_2^2 - R_1^2)} \frac{1}{r^2} \tag{2.11.3}$$

where λ and μ are the Lamé coefficient and shear modulus, respectively. If $R_1 = 0$, we have the case of a **solid cylinder** under external pressure, with displacement and stress denoted by the following:

$$U = \frac{p_2 r}{2(\lambda + \mu)} \tag{2.11.4}$$

$$\sigma_{rr} = \sigma_{\theta\theta} = p_2 \tag{2.11.5}$$

If, instead, $R_2 \rightarrow \infty$, then

$$U = \frac{p_2 r}{2(\lambda + \mu)} + \frac{(p_2 - p_1) R_1^2}{2\mu r} \tag{2.11.6}$$

$$\sigma_{rr} = p_2 \left(1 - \frac{R_1^2}{r^2}\right) + \frac{p_1 R_1^2}{r^2} \tag{2.11.7}$$

$$\sigma_{\theta\theta} = p_2 \left(1 + \frac{R_1^2}{r^2}\right) - \frac{p_1 R_1^2}{r^2} \tag{2.11.8}$$

These results for **plane strain** can be converted to **plane stress** by replacing v by $v/(1 + v)$, where v is the Poisson's ratio.

Circular Hole with Principal Stresses at Infinity

The circular hole with radius R lies along the z-axis. A principal stress, σ_{xx}, is applied at infinity. The stress solution is then

$$\sigma_{rr} = \frac{\sigma_{xx}}{2}\left(1 - \frac{R^2}{r^2}\right) + \frac{\sigma_{xx}}{2}\left(1 - \frac{4R^2}{r^2} + \frac{3R^4}{r^4}\right)\cos 2\theta \tag{2.11.9}$$

$$\sigma_{\theta\theta} = \frac{\sigma_{xx}}{2}\left(1 + \frac{R^2}{r^2}\right) - \frac{\sigma_{xx}}{2}\left(1 + \frac{3R^4}{r^4}\right)\cos 2\theta \tag{2.11.10}$$

$$\sigma_{r\theta} = -\frac{\sigma_{xx}}{2}\left(1 + \frac{2R^2}{r^2} - \frac{3R^4}{r^4}\right)\sin 2\theta \tag{2.11.11}$$

$$\frac{8\mu U_r}{R\sigma_{xx}} = (\chi - 1 + 2\cos 2\theta)\frac{r}{R} + \frac{2R}{r}\left[1 + \left(\chi + 1 - \frac{R^2}{r^2}\right)\cos 2\theta\right] \tag{2.11.12}$$

$$\frac{8\mu U_\theta}{R\sigma_{xx}} = \left[-\frac{2r}{R} + \frac{2R}{r}\left(1 - \chi - \frac{R^2}{r^2}\right)\right]\sin 2\theta \tag{2.11.13}$$

where θ is measured from the x-axis, and

$\chi = 3 - 4v$, for plane strain　(2.11.14)

$\chi = \dfrac{3 - v}{1 + v}$, for plane stress　(2.11.15)

At the cavity surface, $r = R$,

$\sigma_{rr} = \sigma_{r\theta} = 0$　(2.11.16)

$\sigma_{\theta\theta} = \sigma_{xx}(1 - 2\cos2\theta)$　(2.11.17)

Thus, we see the well-known result that the borehole creates a stress concentration of

$\sigma_{\theta\theta} = 3\sigma_{xx}$ at $\theta = 90°$　(2.11.18)

Stress Concentration around an Elliptical Hole

If, instead, the borehole is elliptical in cross-section with a shape denoted by (Lawn and Wilshaw, 1975)

$$\dfrac{x^2}{b^2} + \dfrac{y^2}{c^2} = 1$$　(2.11.19)

where b is the semiminor axis, c is the semimajor axis, and the principal stress σ_{xx} is applied at infinity, the largest stress concentration occurs at the tip of the long axis ($y = c$; $x = 0$). This is the same location at $\theta = 90°$ as for the circular hole. The stress concentration is

$$\sigma_{\theta\theta} = \sigma_{xx}\left[1 + 2(c/\rho)^{1/2}\right]$$　(2.11.20)

where ρ is the radius of curvature at the tip given by

$$\rho = \dfrac{b^2}{c}$$　(2.11.21)

When $b \ll c$, the stress concentration is approximately

$$\dfrac{\sigma_{\theta\theta}}{\sigma_{xx}} \approx \dfrac{2c}{b} = 2\sqrt{\dfrac{c}{\rho}}$$　(2.11.22)

Stress around an Inclined Cylindrical Hole

We now consider the case of a cylindrical borehole of radius R inclined at an angle i to the vertical axis, in a linear, isotropic elastic medium with Poisson's ratio v in a nonhydrostatic remote stress field (Jaeger and Cook, 1969; Bradley, 1979; Fjaer et al., 2008). The borehole coordinate system is denoted by (x, y, z) with the z-axis along the axis of the borehole. The remote principal stresses are denoted by

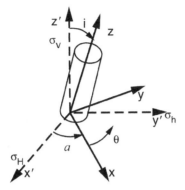

Figure 2.11.1 The coordinate system for the remote stress field around an inclined, cylindrical borehole.

σ_v, the vertical stress,
σ_H, the maximum horizontal stress, and
σ_h, the minimum horizontal stress

The coordinate system for the remote stress field (shown in Figure 2.11.1) is denoted by (x', y', z') with x' along the direction of the maximum horizontal stress, and z' along the vertical. The angle a represents the azimuth of the borehole x-axis with respect to the x'-axis, while θ is the azimuthal angle around the borehole measured from the x-axis. Assuming plane-strain conditions with no displacements along the z-axis, the stresses as a function of radial distance r and azimuthal angle θ are given by (Fjaer *et al.*, 2008):

$$\sigma_{rr} = \left(\frac{\sigma_{xx}^0 + \sigma_{yy}^0}{2}\right)\left(1 - \frac{R^2}{r^2}\right) + \left(\frac{\sigma_{xx}^0 - \sigma_{yy}^0}{2}\right)\left(1 + 3\frac{R^4}{r^4} - 4\frac{R^2}{r^2}\right)\cos2\theta$$

$$+ \sigma_{xy}^0\left(1 + 3\frac{R^4}{r^4} - 4\frac{R^2}{r^2}\right)\sin2\theta + p_w\frac{R^2}{r^2} \qquad (2.11.23)$$

$$\sigma_{\theta\theta} = \left(\frac{\sigma_{xx}^0 + \sigma_{yy}^0}{2}\right)\left(1 + \frac{R^2}{r^2}\right) - \left(\frac{\sigma_{xx}^0 - \sigma_{yy}^0}{2}\right)\left(1 + 3\frac{R^4}{r^4}\right)\cos2\theta$$

$$- \sigma_{xy}^0\left(1 + 3\frac{R^4}{r^4}\right)\sin2\theta - p_w\frac{R^2}{r^2} \qquad (2.11.24)$$

$$\sigma_{zz} = \sigma_{zz}^0 - v\left[2(\sigma_{xx}^0 - \sigma_{yy}^0)\frac{R^2}{r^2}\cos2\theta + 4\sigma_{xy}^0\frac{R^2}{r^2}\sin2\theta\right] \qquad (2.11.25)$$

$$\sigma_{r\theta} = \left(\frac{\sigma_{yy}^0 - \sigma_{xx}^0}{2}\right)\left(1 - 3\frac{R^4}{r^4} + 2\frac{R^2}{r^2}\right)\sin2\theta + \sigma_{xy}^0\left(1 - 3\frac{R^4}{r^4} + 2\frac{R^2}{r^2}\right)\cos2\theta \qquad (2.11.26)$$

$$\sigma_{\theta z} = (-\sigma_{xz}^0\sin\theta + \sigma_{yz}^0\cos\theta)\left(1 + \frac{R^2}{r^2}\right) \qquad (2.11.27)$$

$$\sigma_{rz} = (\sigma_{xz}^0 \cos\theta + \sigma_{yz}^0 \sin\theta)\left(1 - \frac{R^2}{r^2}\right) \tag{2.11.28}$$

In the previous equations, p_w represents the well-bore pressure and σ_{ij}^0 is the remote stress tensor expressed in the borehole coordinate system through the usual coordinate transformation (see Section 1.5 on coordinate transformations) involving the direction cosines of the angles between the (x', y', z') axes and the (x, y, z) axes as follows:

$$\sigma_{xx}^0 = \beta_{xx'}^2 \sigma_H + \beta_{xy'}^2 \sigma_h + \beta_{xz'}^2 \sigma_v \tag{2.11.29}$$

$$\sigma_{yy}^0 = \beta_{yx'}^2 \sigma_H + \beta_{yy'}^2 \sigma_h + \beta_{yz'}^2 \sigma_v \tag{2.11.30}$$

$$\sigma_{zz}^0 = \beta_{zx'}^2 \sigma_H + \beta_{zy'}^2 \sigma_h + \beta_{zz'}^2 \sigma_v \tag{2.11.31}$$

$$\sigma_{xy}^0 = \beta_{xx'}\beta_{yx'} \sigma_H + \beta_{xy'}\beta_{yy'} \sigma_h + \beta_{xz'}\beta_{yz'} \sigma_v \tag{2.11.32}$$

$$\sigma_{yz}^0 = \beta_{yx'}\beta_{zx'} \sigma_H + \beta_{yy'}\beta_{zy'} \sigma_h + \beta_{yz'}\beta_{zz'} \sigma_v \tag{2.11.33}$$

$$\sigma_{zx}^0 = \beta_{zx'}\beta_{xx'} \sigma_H + \beta_{zy'}\beta_{xy'} \sigma_h + \beta_{zz'}\beta_{xz'} \sigma_v \tag{2.11.34}$$

and

$$\beta_{xx'} = \cos a \cos i, \quad \beta_{xy'} = \sin a \cos i, \quad \beta_{xz'} = -\sin i, \tag{2.11.35}$$

$$\beta_{yx'} = -\sin a, \quad \beta_{yy'} = \cos a, \quad \beta_{yz'} = 0 \tag{2.11.36}$$

$$\beta_{zx'} = \cos a \sin i, \quad \beta_{zy'} = \sin a \sin i, \quad \beta_{zz'} = \cos i \tag{2.11.37}$$

Stress around a Vertical Cylindrical Hole in a Poroelastic Medium

The circular hole with radius R_i lies along the z-axis. Pore pressure, p_f, at the permeable borehole wall equals the pressure in the well bore, p_w. At a remote boundary $R_0 \gg R_i$ the stresses and pore pressure are

$$\sigma_{zz}(R_0) = \sigma_v \tag{2.11.38}$$

$$\sigma_{rr}(R_0) = \sigma_{\theta\theta}(R_0) = \sigma_h \tag{2.11.39}$$

$$p_f(R_0) = p_{f0} \tag{2.11.40}$$

The stresses as a function of radial distance r from the hole are given by (Risnes et al., 1982; Bratli et al., 1983; Fjaer et al., 2008)

$$\sigma_{rr} = \sigma_h + (\sigma_h - p_w)\frac{R_i^2}{R_0^2 - R_i^2}\left[1 - \left(\frac{R_0}{r}\right)^2\right] - (p_{f0} - p_w)\frac{1 - 2v}{2(1-v)}$$

$$\times \alpha\left\{\frac{R_i^2}{R_0^2 - R_i^2}\left[1 - \left(\frac{R_0}{r}\right)^2\right] + \frac{\ln(R_0/r)}{\ln(R_0/R_i)}\right\} \tag{2.11.41}$$

$$\sigma_{\theta\theta} = \sigma_h + (\sigma_h - p_w)\frac{R_i^2}{R_0^2 - R_i^2}\left[1 + \left(\frac{R_0}{r}\right)^2\right] - (p_{f0} - p_w)\frac{1 - 2v}{2(1 - v)}$$

$$\times \alpha\left\{\frac{R_i^2}{R_0^2 - R_i^2}\left[1 + \left(\frac{R_0}{r}\right)^2\right] + \frac{\ln(R_0/r) - 1}{\ln(R_0/R_i)}\right\} \tag{2.11.42}$$

$$\sigma_{zz} = \sigma_v + 2v(\sigma_h - p_w)\frac{R_i^2}{R_0^2 - R_i^2} - (p_{f0} - p_w)\frac{1 - 2v}{2(1 - v)}$$

$$\times \alpha\left[v\frac{2R_i^2}{R_0^2 - R_i^2} + \frac{2\ln(R_0/r) - v}{\ln(R_0/R_i)}\right] \tag{2.11.43}$$

where v is the dry (drained) Poisson's ratio of the poroelastic medium, $\alpha = 1 - K_{dry}/K_0$ is the Biot coefficient, K_{dry} is effective bulk modulus of dry porous solid, and K_0 is the bulk modulus of solid mineral material. In the limit $R_0/R_i \to \infty$ the expressions simplify to (Fjaer *et al.*, 2008)

$$\sigma_{rr} = \sigma_h - (\sigma_h - p_w)\left(\frac{R_i}{r}\right)^2 + (p_{f0} - p_w)\frac{1 - 2v}{2(1 - v)}\alpha\left[\left(\frac{R_i}{r}\right)^2 - \frac{\ln(R_0/r)}{\ln(R_0/R_i)}\right] \tag{2.11.44}$$

$$\sigma_{\theta\theta} = \sigma_h + (\sigma_h - p_w)\left(\frac{R_i}{r}\right)^2 - (p_{f0} - p_w)\frac{1 - 2v}{2(1 - v)}\alpha\left[\left(\frac{R_i}{r}\right)^2 + \frac{\ln(R_0/r)}{\ln(R_0/R_i)}\right] \tag{2.11.45}$$

$$\sigma_{zz} = \sigma_v - (p_{f0} - p_w)\frac{1 - 2v}{2(1 - v)}\alpha\frac{2\ln(R_0/r) - v}{\ln(R_0/R_i)} \tag{2.11.46}$$

Uses

The equations presented in this section can be used for the following:
- estimating the stresses around a borehole resulting from tectonic stresses;
- estimating the stresses and deformation of a borehole caused by changes in borehole fluid pressure.

Assumptions and Limitations

The equations presented in this section are based on the following assumptions: the material is linear, isotropic, and elastic or poroelastic.

Extensions

More complicated remote stress fields can be constructed by superimposing the solutions for the corresponding principal stresses.

2.12   Eshelby's General Solution for Ellipsoidal Inclusions

Eshelby (1957) solved the theoretical problem of determining the deformation of a single ellipsoidal elastic inclusion embedded in a uniform elastic background material subject to uniform strains at infinity.

Ellipsoidal Inclusion Problem

Eshelby's method began with the following thought experiment, which he called the *inclusion problem*: (1) Carve out an ellipsoid from the originally unstrained uniform background material. (2) Apply a uniform *stress-free strain*, ε_{kl}^*, to the removed piece, which causes its size and shape to change. (The stress-free strain, ε_{kl}^*, could be realized, for example, via thermal expansion or a phase transformation of crystalline material.) Mura (1986) refers to the stress-free strain as the *eigenstrain*. (3) Apply surface forces to the extracted piece that will induce strains $-\varepsilon_{kl}^*$, restoring it to its original size and shape. (4) Insert the piece into the hole, and relax the surface forces. The strain change that occurs as the surfaces forces relax, ε_{kl}^C, is called the *constrained strain*. It is the perturbation in the strain field, relative to the strain in the initially uniform material, caused by the presence of the transformed inclusion.

It is convenient to write the relation between eigenstrain and constrained strain as

$$\varepsilon_{ij}^C = S_{ijpq}\varepsilon_{pq}^*, \tag{2.12.1}$$

where S_{ijpq} is known as Eshelby's tensor (also known as a depolarization tensor). For isotropic materials, Eshelby's tensor depends only on the shape of the inclusion and the Poisson's ratio of the background material. Because the tensor S_{ijkl} relates two symmetric strain tensors, it has certain symmetries:

$$S_{ijkl} = S_{jikl} = S_{ijlk}, \tag{2.12.2}$$

but it does not have major symmetry, i.e., $S_{1122} \neq S_{2211}$. Elements relating extension to shear ($S_{1112}, S_{1123}, S_{2311}, \ldots$) or relating one shear to another (S_{1223}) are zero (Eshelby, 1957).

If we superimpose a uniform strain field, ε_{ij}^0, onto the material with its inclusion, the total strain at the surface of the inclusion is $\varepsilon_{ij}^C + \varepsilon_{ij}^0$, and the total strain field at infinity is ε_{ij}^0. The stresses inside the inclusion are given by Hooke's law:

$$\boldsymbol{\sigma} = \mathbf{C}_0(\boldsymbol{\varepsilon}^C + \boldsymbol{\varepsilon}^0 - \boldsymbol{\varepsilon}^*), \tag{2.12.3}$$

where $\mathbf{C}_0$ is the elastic stiffness tensor. (Note the extra stress contributions $-\mathbf{C}_0\boldsymbol{\varepsilon}^*$ arising from the stress required to undo the eigenstrain.)

Eshelby (1957) showed the very important result that the strains within an ellipsoidal inclusion are uniform if the strain applied at infinity is uniform. In fact the strain within the inclusion is uniform if and only if the single inclusion is an ellipsoid.

Ellipsoidal Heterogeneity Problem

Eshelby showed how to adapt his method to the case when the inclusion has a different set of elastic constants than those of the background material. In this case, there is no explicit eigenstrain imposed. When the uniform strains, ε_{ij}^0, are imposed at infinity, the stresses inside the heterogeneity can be written as

$$\boldsymbol{\sigma} = \mathbf{C}_1(\boldsymbol{\varepsilon}^C + \boldsymbol{\varepsilon}^0), \tag{2.12.4}$$

where $\mathbf{C}_1$ is the elastic stiffness tensor of the inclusion material. In this case, the strain perturbations, ε_{kl}^C, are caused by the elastic stiffness contrast between the inclusion and background materials. Eshelby noted that the strain perturbation caused by the modulus contrasts could be mimicked by an "equivalent" eigenstrain with no modulus contrast. Stresses within the equivalent inclusion and heterogeneity will be equal if

$$\mathbf{C}_1(\boldsymbol{\varepsilon}^C + \boldsymbol{\varepsilon}^0) = \mathbf{C}_0(\boldsymbol{\varepsilon}^C + \boldsymbol{\varepsilon}^0 - \boldsymbol{\varepsilon}^*). \tag{2.12.5}$$

In practice, we often specify ε_{ij}^0, $\mathbf{C}_0$, and $\mathbf{C}_1$ and wish to solve for the deformation of the heterogeneity. By inverting the previous equation, we obtain

$$\mathbf{T} = [\mathbf{I} + \mathbf{S} : \mathbf{C}_0^{-1} : (\mathbf{C}_1 - \mathbf{C}_0)]^{-1}, \tag{2.12.6}$$

where $\mathbf{T}$ is the *strain concentration* tensor (Torquato, 2002). Then the total uniform strain field, ε_{ij}, inside an ellipsoidal heterogeneity can be predicted from the applied strain using

$$\boldsymbol{\varepsilon} = \mathbf{T} : \boldsymbol{\varepsilon}^0. \tag{2.12.7}$$

Values of $\mathbf{T}$ in terms of $\mathbf{S}$ are given by (correcting an error in Wu, 1966):

$$T_{1212} = T_{2121} = \frac{1}{(1 + 2AS_{1212})}, \tag{2.12.8}$$

$$T_{2323} = T_{3232} = \frac{1}{(1 + 2AS_{2323})}, \tag{2.12.9}$$

$$T_{3131} = T_{1313} = \frac{1}{(1 + 2AS_{3131})}, \tag{2.12.10}$$

and

$$\begin{bmatrix} T_{1111} & T_{1122} & T_{1133} \\ T_{2211} & T_{2222} & T_{2233} \\ T_{3311} & T_{3322} & T_{3333} \end{bmatrix} = \begin{bmatrix} 1 + AS_{1111} + BS_1 & AS_{1122} + BS_1 & AS_{1133} + BS_1 \\ AS_{2211} + BS_2 & 1 + AS_{2222} + BS_2 & AS_{2233} + BS_2 \\ AS_{3311} + BS_3 & AS_{3322} + BS_3 & 1 + AS_{3333} + BS_3 \end{bmatrix}^{-1},$$ (2.12.11)

where

$$A = \frac{\mu_1}{\mu_0} - 1, \ B = \frac{1}{3}\left(\frac{K_1}{K_0} - \frac{\mu_1}{G_0}\right), \ 0=\text{matrix; } 1=\text{inclusion}$$

$$S_1 = S_{1111} + S_{1122} + S_{1133},$$ (2.12.12)

$$S_2 = S_{2211} + S_{2222} + S_{2233},$$ (2.12.13)

$$S_3 = S_{3311} + S_{3322} + S_{3333}.$$ (2.12.14)

All other

$$T_{ijkl} = 0.$$ (2.12.15)

Values of Eshelby's Tensor

For the **general ellipsoid** with semi-axes $a \geq b \geq c$ along the x_1, x_2, and x_3 directions, and for isotropic materials, the elements of Eshelby's tensor are given by the following:

$$S_{1111} = \frac{3}{8\pi(1 - v)}a^2 I_{11} + \frac{1 - 2v}{8\pi(1 - v)}I_1$$ (2.12.16)

$$S_{1122} = \frac{1}{8\pi(1 - v)}b^2 I_{12} - \frac{1 - 2v}{8\pi(1 - v)}I_1$$ (2.12.17)

$$S_{1133} = \frac{1}{8\pi(1 - v)}c^2 I_{13} - \frac{1 - 2v}{8\pi(1 - v)}I_1$$ (2.12.18)

$$S_{1212} = \frac{a^2 + b^2}{16\pi(1 - v)}I_{12} + \frac{1 - 2v}{16\pi(1 - v)}(I_1 + I_2)$$ (2.12.19)

$$S_{1112} = S_{1223} = S_{1232} = 0$$ (2.12.20)

$$S_{2222} = \frac{3}{8\pi(1 - v)}b^2 I_{22} + \frac{1 - 2v}{8\pi(1 - v)}I_2$$ (2.12.21)

$$S_{2233} = \frac{1}{8\pi(1 - v)}c^2 I_{23} - \frac{1 - 2v}{8\pi(1 - v)}I_2$$ (2.12.22)

$$S_{2211} = \frac{1}{8\pi(1 - v)}a^2 I_{21} - \frac{1 - 2v}{8\pi(1 - v)}I_2$$ (2.12.23)

$$S_{2323} = \frac{b^2 + c^2}{16\pi(1 - v)}I_{23} + \frac{1 - 2v}{16\pi(1 - v)}(I_2 + I_3)$$ (2.12.24)

$$S_{2223} = S_{2331} = S_{2313} = 0$$ (2.12.25)

$$S_{3333} = \frac{3}{8\pi(1 - v)}c^2 I_{33} + \frac{1 - 2v}{8\pi(1 - v)}I_3$$ (2.12.26)

$$S_{3311} = \frac{1}{8\pi(1-v)}a^2 I_{31} - \frac{1-2v}{8\pi(1-v)} I_3 \tag{2.12.27}$$

$$S_{3322} = \frac{1}{8\pi(1-v)}b^2 I_{32} - \frac{1-2v}{8\pi(1-v)} I_3 \tag{2.12.28}$$

$$S_{3131} = \frac{c^2+a^2}{16\pi(1-v)}I_{31} + \frac{1-2v}{16\pi(1-v)}(I_3 + I_1) \tag{2.12.29}$$

$$S_{3331} = S_{3112} = S_{3121} = 0 \tag{2.12.30}$$

All other S_{ijkl} are zero.

The quantities I_k in equations 2.12.16–2.12.30 are given by

$$I_1 = \frac{4\pi abc}{(a^2-b^2)(a^2-c^2)^{1/2}}[F(\theta,k) - E(\theta,k)] , \tag{2.12.31}$$

$$I_3 = \frac{4\pi abc}{(b^2-c^2)(a^2-c^2)^{1/2}}\left[\frac{b(a^2-c^2)^{1/2}}{ac} - E(\theta,k)\right] , \tag{2.12.32}$$

$$I_1 + I_2 + I_3 = 4\pi . \tag{2.12.33}$$

$E(\theta,k)$ and $F(\theta,k)$ are standard elliptic integrals:

$$F(\theta,k) = \int_0^\theta \frac{dw}{(1-k^2\sin^2 w)^{1/2}} , \tag{2.12.34}$$

$$E(\theta,k) = \int_0^\theta (1-k^2\sin^2 w)^{1/2}dw , \tag{2.12.35}$$

with

$$\theta = \arcsin\sqrt{\frac{a^2-c^2}{a^2}} , \tag{2.12.36}$$

$$k = \sqrt{\frac{a^2-b^2}{a^2-c^2}} . \tag{2.12.37}$$

The quantities I_{ij} are given by

$$I_{12} = \frac{I_2-I_1}{a^2-b^2} \tag{2.12.38}$$

$$I_{23} = \frac{I_3-I_2}{b^2-c^2} \tag{2.12.39}$$

$$I_{31} = \frac{I_1-I_3}{c^2-a^2} \tag{2.12.40}$$

$$3I_{11} + I_{12} + I_{13} = \frac{4\pi}{a^2} \tag{2.12.41}$$

$$3I_{22} + I_{23} + I_{21} = \frac{4\pi}{b^2} \tag{2.12.42}$$

$$3I_{33} + I_{31} + I_{32} = \frac{4\pi}{c^2} \tag{2.12.43}$$

$$3a^2 I_{11} + b^2 I_{12} + c^2 I_{13} = 3I_1 \tag{2.12.44}$$

$$3b^2 I_{22} + c^2 I_{23} + a^2 I_{21} = 3I_2 \tag{2.12.45}$$

$$3c^2 I_{33} + a^2 I_{31} + b^2 I_{32} = 3I_3. \tag{2.12.46}$$

For the case of a **general spheroid** with semi-axes $a = b$, c in the x_1, x_2, and x_3 directions, the elliptic integrals can be written explicitly. When the spheroid symmetry axis is aligned along the x_3 direction, the Eshelby tensor elements are:

$$S_{3333} = \frac{1}{2(1-v)} \left[2(1-v)(1-g) + g - Ar^2 \frac{3g-2}{Ar^2-1} \right] \tag{2.12.47}$$

$$S_{1111} = S_{2222} = \frac{1}{4(1-v)} \left[2(2-v)g - \frac{1}{2} - \left(Ar^2 - \frac{1}{4} \right) \frac{3g-2}{Ar^2-1} \right] \tag{2.12.48}$$

$$S_{3311} = S_{3322} = \frac{1}{4(1-v)} \left[4v(1-g) - g + Ar^2 \frac{3g-2}{Ar^2-1} \right] \tag{2.12.49}$$

$$S_{1122} = S_{2211} = \frac{1}{4(1-v)} \left[-(1-2v)g + \frac{1}{2} - \frac{1}{4} \frac{3g-2}{Ar^2-1} \right] \tag{2.12.50}$$

$$S_{1133} = S_{2233} = \frac{1}{4(1-v)} \left[-(1-2v)g + Ar^2 \frac{3g-2}{Ar^2-1} \right] \tag{2.12.51}$$

$$S_{3131} = S_{3232} = \frac{1}{4(1-v)} \left[(1-v)(2-g) - g + Ar^2 \frac{3g-2}{Ar^2-1} \right] \tag{2.12.52}$$

$$S_{1212} = \frac{S_{1111} - S_{3311}}{2} \tag{2.12.53}$$

where $Ar = c/a$ is the aspect ratio, and

$$g = \frac{Ar}{(1-Ar^2)^{3/2}} [\cos^{-1} Ar - Ar(1-Ar^2)^{1/2}]; \quad 0 < Ar < 1 \tag{2.12.54}$$

$$g = \frac{Ar}{(Ar^2-1)^{3/2}} [Ar(Ar^2-1)^{1/2} - \cosh^{-1} Ar]; \quad 1 < Ar < \infty \tag{2.12.55}$$

$$S_{ijkl} = S_{jikl} = S_{ijlk} = S_{jilk}. \tag{2.12.56}$$

(Note that these equations for the general spheroid do not yield the correct answer for a sphere with $Ar = 1$. The asymptotic behavior of the expressions near $Ar = 1$ has to be considered more carefully.)

For a **spherical inclusion** ($a=b=c$):

$$S_{ijkl} = \frac{5v - 1}{15(1 - v)} \delta_{ij}\delta_{kl} + \frac{4 - 5v}{15(1 - v)} (\delta_{ik}\delta_{jl} + \delta_{il}\delta_{jk}) \tag{2.12.57}$$

$$I_1 = I_2 = I_3 = 4\pi/3 \tag{2.12.58}$$

$$I_{11} = I_{22} = I_{33} = I_{12} = I_{23} = I_{31} = 4\pi/5a^2 \tag{2.12.59}$$

$$S_{1111} = S_{2222} = S_{3333} = \frac{7 - 5v}{15(1 - v)} \tag{2.12.60}$$

$$S_{1122} = S_{2233} = S_{3311} = S_{1133} = S_{2211} = S_{3322} = \frac{5v - 1}{15(1 - v)} \tag{2.12.61}$$

$$S_{1212} = S_{2323} = S_{3131} = \frac{4 - 5v}{15(1 - v)} \tag{2.12.62}$$

For a **penny-shaped ellipsoid** (a=b≫c) with x3 symmetry axis:

$$S_{1111} = S_{2222} = \frac{\pi(13 - 8v)}{32(1 - v)} \frac{c}{a} \tag{2.12.63}$$

$$S_{3333} = 1 - \frac{\pi(1 - 2v)}{4(1 - v)} \frac{c}{a} \tag{2.12.64}$$

$$S_{1122} = S_{2211} = \frac{\pi(8v - 1)}{32(1 - v)} \frac{c}{a} \tag{2.12.65}$$

$$S_{1133} = S_{2233} = \frac{\pi(2v - 1)}{8(1 - v)} \frac{c}{a} \tag{2.12.66}$$

$$S_{3311} = S_{3322} = \frac{v}{1 - v} \left(1 - \frac{\pi(4v + 1)}{8v} \frac{c}{a} \right) \tag{2.12.67}$$

$$S_{1212} = \frac{\pi(7 - 8v)}{32(1 - v)} \frac{c}{a} \tag{2.12.68}$$

$$S_{1313} = S_{2323} = \frac{1}{2} \left(1 + \frac{\pi(v - 2)}{4(1 - v)} \frac{c}{a} \right) \tag{2.12.69}$$

$$S_{kk11} = S_{kk22} = \frac{\pi(1 - 2v)}{4(1 - v)} \frac{c}{a} + \frac{v}{(1 - v)} \tag{2.12.70}$$

$$S_{kk33} = 1 - \frac{\pi(1 - 2v)}{2(1 - v)} \frac{c}{a} \tag{2.12.71}$$

where

$$I_1 = I_2 = \pi^2(c/a) \tag{2.12.72}$$

$$I_3 = 4\pi - 2\pi^2(c/a) \tag{2.12.73}$$

$$I_{12} = I_{21} = 3\pi^2 c/4a^3 \tag{2.12.74}$$

$$I_{13} = I_{23} = I_{31} = I_{32} = 3\left(\frac{4}{3}\pi - \pi^2(c/a)\right)/a^2 \tag{2.12.75}$$

$$I_{11} = I_{22} = 3\pi^2 c/4a^3 \tag{2.12.76}$$

$$I_{33} = \frac{4}{3}\pi/c^2 \tag{2.12.77}$$

For an **oblate spheroid** $(a = b > c)$ with x3 symmetry axis:

$$I_1 = I_2 = \frac{2\pi a^2 c}{(a^2 - c^2)^{3/2}}\left\{\cos^{-1}\frac{c}{a} - \frac{c}{a}\left(1 - \frac{c^2}{a^2}\right)^{1/2}\right\} \tag{2.12.78}$$

$$I_3 = 4\pi - 2I_1 \tag{2.12.79}$$

$$I_{11} = I_{22} = I_{12} \tag{2.12.80}$$

$$I_{12} = \frac{\pi}{a^2} - \frac{1}{4}I_{13} = \frac{\pi}{a^2} - \frac{(I_1 - I_3)}{4(c^2 - a^2)} \tag{2.12.81}$$

$$I_{13} = I_{23} = \frac{(I_1 - I_3)}{(c^2 - a^2)} \tag{2.12.82}$$

$$3I_{33} = \frac{4\pi}{c^2} - 2I_{13} \tag{2.12.83}$$

For a prolate spheroid $(a = b < c)$ with symmetry axis in the x_1 direction:

$$I_1 = I_2 = \frac{2\pi c a^2}{(c^2 - a^2)^{3/2}}\left\{\frac{c}{a}\left(\frac{c^2}{a^2} - 1\right)^{1/2} - \cosh^{-1}\frac{c}{a}\right\} \tag{2.12.84}$$

$$I_3 = 4\pi - 2I_1 \tag{2.12.85}$$

$$I_{31} = \frac{(I_1 - I_3)}{(c^2 - a^2)} \tag{2.12.86}$$

$$3I_{33} = \frac{4\pi}{c^2} - 2I_{31} \tag{2.12.87}$$

$$I_{11} = I_{22} = I_{12} \tag{2.12.88}$$

$$3I_{11} = \frac{4\pi}{a^2} - I_{12} - \frac{(I_1 - I_3)}{(c^2 - a^2)} \tag{2.12.89}$$

$$I_{12} = \frac{\pi}{a^2} - \frac{(I_1 - I_3)}{4(c^2 - a^2)} \tag{2.12.90}$$

The Eshelby tensor can also be written in 6x6 Kelvin notation as

$$
\mathbf{S} = \begin{pmatrix}
\hat{s}_{11} & \hat{s}_{12} & \hat{s}_{13} & \hat{s}_{14} & \hat{s}_{15} & \hat{s}_{16} \\
\hat{s}_{21} & \hat{s}_{22} & \hat{s}_{23} & \hat{s}_{24} & \hat{s}_{25} & \hat{s}_{26} \\
\hat{s}_{31} & \hat{s}_{32} & \hat{s}_{33} & \hat{s}_{34} & \hat{s}_{35} & \hat{s}_{36} \\
\hat{s}_{41} & \hat{s}_{42} & \hat{s}_{43} & \hat{s}_{44} & \hat{s}_{45} & \hat{s}_{46} \\
\hat{s}_{51} & \hat{s}_{52} & \hat{s}_{53} & \hat{s}_{54} & \hat{s}_{55} & \hat{s}_{56} \\
\hat{s}_{61} & \hat{s}_{62} & \hat{s}_{63} & \hat{s}_{64} & \hat{s}_{65} & \hat{s}_{66}
\end{pmatrix}
$$

$$
= \begin{pmatrix}
s_{1111} & s_{1122} & s_{1133} & \sqrt{2}s_{1123} & \sqrt{2}s_{1113} & \sqrt{2}s_{1112} \\
s_{2211} & s_{2222} & s_{2233} & \sqrt{2}s_{2223} & \sqrt{2}s_{2213} & \sqrt{2}s_{2212} \\
s_{3311} & s_{3322} & s_{3333} & \sqrt{2}s_{3323} & \sqrt{2}s_{3313} & \sqrt{2}s_{3312} \\
\sqrt{2}s_{2311} & \sqrt{2}s_{2322} & \sqrt{2}s_{2333} & 2s_{2323} & 2s_{2313} & 2s_{2312} \\
\sqrt{2}s_{1311} & \sqrt{2}s_{1322} & \sqrt{2}s_{1333} & 2s_{1323} & 2s_{1313} & 2s_{1312} \\
\sqrt{2}s_{1211} & \sqrt{2}s_{1222} & \sqrt{2}s_{1233} & 2s_{1223} & 2s_{1213} & 2s_{1212}
\end{pmatrix} \tag{2.12.91}
$$

which relates Eshelby's eigenstrain to constrained strain, $\varepsilon^C = \mathbf{S}\varepsilon^*$:

$$
\begin{pmatrix}
\varepsilon_{11} \\
\varepsilon_{22} \\
\varepsilon_{33} \\
\sqrt{2}\varepsilon_{23} \\
\sqrt{2}\varepsilon_{13} \\
\sqrt{2}\varepsilon_{12}
\end{pmatrix}_C
= \begin{pmatrix}
s_{1111} & s_{1122} & s_{1133} & \sqrt{2}s_{1123} & \sqrt{2}s_{1113} & \sqrt{2}s_{1112} \\
s_{2211} & s_{2222} & s_{2233} & \sqrt{2}s_{2223} & \sqrt{2}s_{2213} & \sqrt{2}s_{2212} \\
s_{3311} & s_{3322} & s_{3333} & \sqrt{2}s_{3323} & \sqrt{2}s_{3313} & \sqrt{2}s_{3312} \\
\sqrt{2}s_{2311} & \sqrt{2}s_{2322} & \sqrt{2}s_{2333} & 2s_{2323} & 2s_{2313} & 2s_{2312} \\
\sqrt{2}s_{1311} & \sqrt{2}s_{1322} & \sqrt{2}s_{1333} & 2s_{1323} & 2s_{1313} & 2s_{1312} \\
\sqrt{2}s_{1211} & \sqrt{2}s_{1222} & \sqrt{2}s_{1233} & 2s_{1223} & 2s_{1213} & 2s_{1212}
\end{pmatrix}
\begin{pmatrix}
\varepsilon_{11} \\
\varepsilon_{22} \\
\varepsilon_{33} \\
\sqrt{2}\varepsilon_{23} \\
\sqrt{2}\varepsilon_{13} \\
\sqrt{2}\varepsilon_{12}
\end{pmatrix}_*
\tag{2.12.92}
$$

The Eshelby tensor in Voigt notation is expressed as

$$
\mathbf{S} = \begin{pmatrix}
S_{11} & S_{12} & S_{13} & S_{14} & S_{15} & S_{16} \\
S_{21} & S_{22} & S_{23} & S_{24} & S_{25} & S_{26} \\
S_{31} & S_{32} & S_{33} & S_{34} & S_{35} & S_{36} \\
S_{41} & S_{42} & S_{43} & S_{44} & S_{45} & S_{46} \\
S_{51} & S_{52} & S_{53} & S_{54} & S_{55} & S_{56} \\
S_{61} & S_{62} & S_{63} & S_{64} & S_{65} & S_{66}
\end{pmatrix}
= \begin{pmatrix}
S_{1111} & S_{1122} & S_{1133} & S_{1123} & S_{1113} & S_{1112} \\
S_{2211} & S_{2222} & S_{2233} & S_{2223} & S_{2213} & S_{2212} \\
S_{3311} & S_{3322} & S_{3333} & S_{3323} & S_{3313} & S_{3312} \\
2s_{2311} & 2s_{2322} & 2s_{2333} & 2s_{2323} & 2s_{2313} & 2s_{2312} \\
2s_{1311} & 2s_{1322} & 2s_{1333} & 2s_{1323} & 2s_{1313} & 2s_{1312} \\
2s_{1211} & 2s_{1222} & 2s_{1233} & 2s_{1223} & 2s_{1213} & 2s_{1212}
\end{pmatrix}
\tag{2.12.93}
$$

$$
\begin{pmatrix}
\varepsilon_{11} \\
\varepsilon_{22} \\
\varepsilon_{33} \\
2\varepsilon_{23} \\
2\varepsilon_{13} \\
2\varepsilon_{12}
\end{pmatrix}_C
= \begin{pmatrix}
S_{1111} & S_{1122} & S_{1133} & S_{1123} & S_{1113} & S_{1112} \\
S_{2211} & S_{2222} & S_{2233} & S_{2223} & S_{2213} & S_{2212} \\
S_{3311} & S_{3322} & S_{3333} & S_{3323} & S_{3313} & S_{3312} \\
2s_{2311} & 2s_{2322} & 2s_{2333} & 2s_{2323} & 2s_{2313} & 2s_{2312} \\
2s_{1311} & 2s_{1322} & 2s_{1333} & 2s_{1323} & 2s_{1313} & 2s_{1312} \\
2s_{1211} & 2s_{1222} & 2s_{1233} & 2s_{1223} & 2s_{1213} & 2s_{1212}
\end{pmatrix}
\begin{pmatrix}
\varepsilon_{11} \\
\varepsilon_{22} \\
\varepsilon_{33} \\
2\varepsilon_{23} \\
2\varepsilon_{13} \\
2\varepsilon_{12}
\end{pmatrix}_*
\tag{2.12.94}
$$

2.13 Mohr's Circles

Synopsis

Mohr's circles provide a graphical representation of how the tractions on a plane depend on the angular orientation of the plane within a given stress field.

Consider a stress state with principal stresses $\sigma_1 \geq \sigma_2 \geq \sigma_3$ and coordinate axes defined along the corresponding principal directions x_1, x_2, x_3. The traction vector, $\mathbf{T}$, acting on a plane with outward unit normal vector, $\mathbf{n} = (n_1, n_2\ n_3)$, is given by Cauchy's formula as

$$\mathbf{T} = \boldsymbol{\sigma}\mathbf{n} \tag{2.13.1}$$

where $\boldsymbol{\sigma}$ is the stress tensor. The components of $\mathbf{n}$ are the direction cosines of $\mathbf{n}$ relative to the coordinate axes and are denoted by

$$n_1 = \cos\phi, \quad n_2 = \cos\gamma, \quad n_3 = \cos\theta \tag{2.13.2}$$

and

$$n_1^2 + n_2^2 + n_3^2 = 1 \tag{2.13.3}$$

where ϕ, γ, and θ are the angles between $\mathbf{n}$ and the axes x_1, x_2, x_3 (Figure 2.13.1).

The normal component of traction, σ, and the *magnitude* of the shear component, τ, acting on the plane are given by

$$\sigma = n_1^2\sigma_1 + n_2^2\sigma_2 + n_3^2\sigma_3 \tag{2.13.4}$$

$$\tau^2 = n_1^2\sigma_1^2 + n_2^2\sigma_2^2 + n_3^2\sigma_3^2 - \sigma^2 \tag{2.13.5}$$

Three-Dimensional Mohr's Circle

The numerical values of σ and τ can be read graphically from the 3D Mohr's circle shown in Figure 2.13.2. All permissible values of σ and τ must lie in the shaded area.

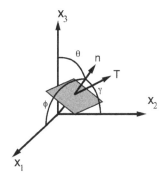

Figure 2.13.1 Angle and vector conventions for Mohr's circles

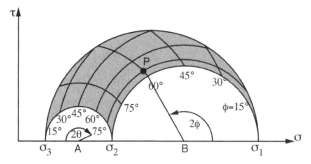

Figure 2.13.2 Three-dimensional Mohr's circle

To determine σ and τ from the orientation of the plane (ϕ, γ, and θ), perform the following procedure.

(1) Plot $\sigma_1 \geq \sigma_2 \geq \sigma_3$ on the horizontal axis and construct the three circles, as shown. The outer circle is centered at $(\sigma_1 + \sigma_3)/2$ and has radius $(\sigma_1 - \sigma_3)/2$. The left-hand inner circle is centered at $(\sigma_2 + \sigma_3)/2$ and has radius $(\sigma_2 - \sigma_3)/2$. The right-hand inner circle is centered at $(\sigma_1 + \sigma_2)/2$ and has radius $(\sigma_1 - \sigma_2)/2$.

(2) Mark angles 2θ and 2ϕ on the small circles centered at **A** and **B**. For example, $\phi = 60°$ plots at $2\phi = 120°$ from the horizontal, and $\theta = 75°$ plots at $2\theta = 150°$ from the horizontal, as shown. Be certain to include the factor of 2, and note the different directions defined for the positive angles.

(3) Draw a circle centered at point **A** that intersects the right-hand small circle at the mark for ϕ.

(4) Draw another circle centered at point **B** that intersects the left-hand small circle at the point for θ.

(5) The intersection of the two constructed circles at point **P** gives the values of σ and τ.

Reverse the procedure to determine the orientation of the plane having particular values of σ and τ.

Two-Dimensional Mohr's Circle

When the plane of interest contains one of the principal axes, the tractions on the plane depend only on the two remaining principal stresses; therefore, using Mohr's circle is simplified. For example, when $\theta = 90°$ (i.e., the x_3-axis lies in the plane of interest), all stress states lie on the circle centered at **B** in Figure 2.13.2. The stresses then depend only on σ_1 and σ_2 and on the angle ϕ, and we need only draw the single circle, as shown in Figure 2.13.3.

Uses

Mohr's circle is used for graphical determination of normal and shear tractions acting on a plane of arbitrary orientation relative to the principal stresses.

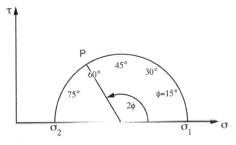

Figure 2.13.3 Two-dimensional Mohr's circle

2.14   Static and Dynamic Moduli

In a uniaxial stress experiment (Figure 2.14.1), Young's modulus E is defined as the ratio of the axial stress σ to the axial strain ε_a, while Poisson's ratio v is defined as the (negative) ratio of the radial strain ε_r to the axial strain:

$$E = \frac{\sigma}{\varepsilon_a}, \quad v = -\frac{\varepsilon_r}{\varepsilon_a} \tag{2.14.1}$$

It follows from these definitions that Poisson's ratio is zero if the sample does not expand radially during axial loading and Poisson's ratio is 0.5 if the radial strain is half the axial strain, which is the case for fluids and incompressible solids. Poisson's ratio must lie within the range $-1 < v \le 0.5$.

The speeds of elastic waves in the solid are linked to the elastic moduli and the bulk density ρ by the wave equation. The corresponding expressions for Poisson's ratio and Young's modulus are:

$$v = \frac{1}{2} \frac{(V_P/V_S)^2 - 2}{(V_P/V_S)^2 - 1}, \quad E = 2\rho V_s^2 (1 + v) \tag{2.14.2}$$

where V_P and V_S are the P- and S-wave velocities, respectively. The elastic moduli calculated from the elastic-wave velocities and density are the **dynamic moduli**. In contrast, the elastic moduli calculated from deformational experiments, such as the one shown in Figure 2.14.1, are the **static moduli**.

In most cases, the static moduli are different from the dynamic moduli for the same sample of rock. There are several reasons for this. One is that stress–strain relations for rocks are often nonlinear. As a result, the ratio of stress to strain over a large-strain

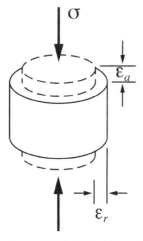

Figure 2.14.1 Uniaxial loading experiment. Dashed lines show undeformed sample, and solid lines show deformed sample.

measurement is different from the ratio of stress to strain over a very small-strain measurement. Another reason is that rocks are often inelastic, affected, for example, by frictional sliding across microcracks and grain boundaries. More internal deformation can occur over a large-strain experiment than over very small-strain cycles. The strain magnitude relevant to geomechanical processes, such as hydrofracturing, is of the order of 10^{-2}, while the strain magnitude due to elastic-wave propagation is of the order of 10^{-7} or less. This large strain difference affects the difference between the static and dynamic moduli.

Relations between the dynamic and static moduli are not simple and universal because:

(a) the elastic wave velocity in a sample and the resulting dynamic elastic moduli depend on the conditions of the measurement, specifically on the effective pressure and pore fluid; and

(b) the static moduli depend on details of the loading experiment. Even for the same type of experiment – axial loading – the static Young's modulus may be strongly affected by the overall pressure applied to the sample, as well as by the axial deformation magnitude.

Some results have been reviewed by Schön (1996) and Wang and Nur (2000). Presented in this section and in Figure 2.14.2 are some of their equations, where the moduli are in GPa and the impedance is in km/s g/cc. In all the examples, E_{stat} is the static Young's modulus and E_{dyn} is the dynamic Young's modulus.

Data on microcline-granite, by Belikov *et al.* (1970):

$$E_{stat} = 1.137E_{dyn} - 9.685 \tag{2.14.3}$$

Igneous and metamorphic rocks from the Canadian Shield, by King (1983):

$$E_{stat} = 1.263E_{dyn} - 29.5 \tag{2.14.4}$$

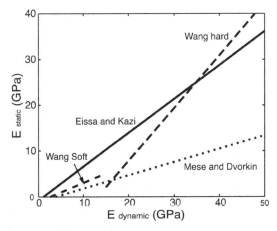

Figure 2.14.2 Comparison of selected relations between dynamic and static Young's moduli

Granites and Jurassic sediments in the UK, by McCann and Entwisle (1992):

$$E_{\text{stat}} = 0.69E_{\text{dyn}} + 6.4 \tag{2.14.5}$$

A wide range of rock types, by Eissa and Kazi (1988):

$$E_{\text{stat}} = 0.74E_{\text{dyn}} - 0.82 \tag{2.14.6}$$

Shallow soil samples, by Gorjainov and Ljachowickij (1979) for clay:

$$E_{\text{stat}} = 0.033E_{\text{dyn}} + 0.0065 \tag{2.14.7}$$

and for sandy, wet soil:

$$E_{\text{stat}} = 0.061E_{\text{dyn}} + 0.00285 \tag{2.14.8}$$

Wang and Nur (2000) for soft rocks (defined as rocks with the static Young's modulus < 15 GPa):

$$E_{\text{stat}} = 0.41E_{\text{dyn}} - 1.06 \tag{2.14.9}$$

Wang and Nur (2000) for hard rocks (defined as rocks with the static Young's modulus > 15 GPa):

$$E_{\text{stat}} = 1.153E_{\text{dyn}} - 15.2 \tag{2.14.10}$$

Mese and Dvorkin (2000) related the static Young's modulus and static Poisson's ratio (v_{stat}) to the dynamic shear modulus calculated from the shear-wave velocity in shales and shaly sands:

$$E_{\text{stat}} = 0.59\mu_{\text{dyn}} - 0.34, \quad v_{\text{stat}} = -0.0208\mu_{\text{dyn}} + 0.37 \tag{2.14.11}$$

where μ_{dyn} is the dynamic shear modulus $\mu_{dyn} = \rho V_S^2$.
The same data were used to obtain relations between the static moduli and the dynamic S-wave impedance, $I_{\text{S\_dyn}}$:

$$E_{\text{stat}} = 1.99I_{\text{S\_dyn}} - 3.84, \quad v_{\text{stat}} = -0.07I_{\text{S\_dyn}} + 0.5 \tag{2.14.12}$$

as well as between the static moduli and the dynamic Young's modulus:

$$E_{\text{stat}} = 0.29E_{\text{dyn}} - 1.1, \quad v_{\text{stat}} = -0.00743E_{\text{dyn}} + 0.34 \tag{2.14.13}$$

Jizba (1991) compared the static bulk modulus, K_{stat} and the dynamic bulk modulus, K_{dyn}, and found the following empirical relations as a function of confining pressure in dry tight sandstones from the Travis Peak formation (Table 2.14.1).

Assumptions and Limitations

There are only a handful of well-documented experimental data where large-strain deformational experiments have been conducted with simultaneous measurement of the dynamic P- and S-wave velocity. One of the main uncertainties in applying laboratory static moduli data *in situ* comes from the fact that the *in-situ* loading conditions are often unknown. Moreover, in most cases, the *in-situ* conditions are so complex that they are

Table 2.14.1 *Empirical relations between static and dynamic bulk moduli in dry tight sandstones from the Travis Peak formation*

Pressure (MPa)	$K_{stat} = a + bK_{dyn}$ (GPa)	
	a (GPa)	b
5	0.98	0.490
20	3.16	0.567
40	1.85	0.822
125	−1.85	1.13

virtually impossible to reproduce in the laboratory. In most cases, static data exhibit a *strongly nonlinear stress dependence*, so that it is never clear which data point to use as the static modulus at *in-situ* conditions. Because of the strongly nonlinear dependence of static moduli on the strain magnitude, the isotropic linear elasticity equations that relate various elastic moduli to each other might not be applicable to static moduli.

Finally, this section refers only to isotropic descriptions of static and dynamic moduli.

2.15   Stress Intensity Factors

A crack with zero radius of curvature at its tip has a $r^{-1/2}$ stress singularity when loaded, where r is the distance from the tip. The singularity is weighted by the stress intensity factor, K, which depends on the loading, the crack dimensions, and the overall geometry of the crack and the object containing the crack.

In **modes I and II** deformation (Figure 2.15.1), both stresses and displacements are independent of the 3-direction ($\partial/\partial z = 0$). The mode I displacements near the crack front are given by (François *et al.*, 2013)

$$\begin{Bmatrix} u_1 \\ u_2 \end{Bmatrix} = \frac{K_I}{2\mu}\sqrt{\frac{r}{2\pi}} \begin{Bmatrix} \cos(\theta/2)[\kappa - 1 + 2\sin^2(\theta/2)] \\ \sin(\theta/2)[\kappa + 1 - 2\cos^2(\theta/2)] \end{Bmatrix} + O(r) \tag{2.15.1}$$

(a)   (b)   (c)

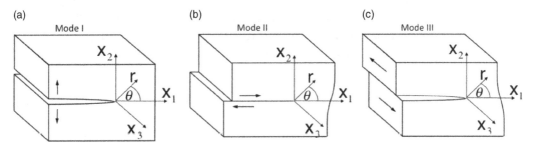

Figure 2.15.1 Schematic displacements of mode I, II, and III deformation

where μ is the shear modulus, $\kappa = 3 - 4v$ in plane strain, and $\kappa = (3 - v)/(1 + v)$ in plane stress, v being the Poisson's ratio. Variable r is the distance from the crack tip, and θ is the angle from the plane of the crack, as shown in Figure 1.15.1. Mode I stresses are given by (François *et al.*, 2013)

$$\begin{Bmatrix} \sigma_{11} \\ \sigma_{22} \\ \sigma_{12} \end{Bmatrix} = \frac{K_I}{\sqrt{2\pi r}} \cos(\theta/2) \begin{Bmatrix} 1 - \sin(\theta/2)\sin(3\theta/2) \\ 1 + \sin(\theta/2)\sin(3\theta/2) \\ \sin(\theta/2)\cos(3\theta/2) \end{Bmatrix} + O(r). \tag{2.15.2}$$

Stress $\sigma_{33} = v(\sigma_{11} + \sigma_{22}) = (K_I/\sqrt{2\pi r})\cos(\theta/2)$ for plane strain and $\sigma_{33} = 0$ for plane stress.

Mode II displacements near the crack front are given by (François *et al.*, 2013)

$$\begin{Bmatrix} u_1 \\ u_2 \end{Bmatrix} = \frac{K_{II}}{2\mu} \sqrt{\frac{r}{2\pi}} \begin{Bmatrix} \sin(\theta/2)[\kappa + 1 + 2\cos^2(\theta/2)] \\ -\cos(\theta/2)[\kappa - 1 - 2\sin^2(\theta/2)] \end{Bmatrix} + O(r). \tag{2.15.3}$$

Mode II stresses are given by

$$\begin{Bmatrix} \sigma_{11} \\ \sigma_{22} \\ \sigma_{12} \end{Bmatrix} = \frac{K_{II}}{\sqrt{2\pi r}} \begin{Bmatrix} -\sin(\theta/2)[2 + \cos(\theta/2)\cos(3\theta/2)] \\ \sin(\theta/2)\cos(\theta/2)\cos(3\theta/2) \\ \cos(\theta/2)[1 - \sin(\theta/2)\sin(3\theta/2)] \end{Bmatrix} + O(r), \tag{2.15.4}$$

and

$$\sigma_{\theta\theta} = \frac{K_{II}}{\sqrt{2\pi r}} \frac{3}{2} \sin(\theta) \cos(\theta/2). \tag{2.15.5}$$

In mode III deformation (also known as anti-plane strain), the in-plane displacements are zero ($u_1 = u_2 = 0$) and both stresses and displacements are independent of the z-direction ($\partial/\partial z = 0$). The displacements and stresses near the crack front are given by

$$u_3 = \frac{2K_{III}}{\mu} \sqrt{\frac{r}{2\pi}} \sin(\theta/2) + O(r), \tag{2.15.6}$$

$$\begin{Bmatrix} \sigma_{13} \\ \sigma_{23} \end{Bmatrix} = \frac{K_{III}}{\sqrt{2\pi r}} \begin{Bmatrix} -\sin(\theta/2) \\ \cos(\theta/2) \end{Bmatrix} + O(r). \tag{2.15.7}$$

Stress intensity factors for several simple crack geometries are shown in this section.

A **2D crack in an infinite plate** (Figure 2.15.2), with tensional stress σ normal to crack faces has stress intensity factor (François *et al.*, 2013)

$$K_I = \sigma\sqrt{\pi a}, \tag{2.15.8}$$

where a is the crack half-width. The same crack with zero remote stress but internal pressure p has stress intensity factor

$$K_I = p\sqrt{\pi a}. \tag{2.15.9}$$

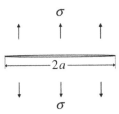

Figure 2.15.2 Tensional loading normal to a 2D planar crack with width $2a$

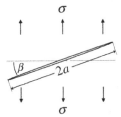

Figure 2.15.3 A 2D planar crack inclined by angle β relative to the tension axis

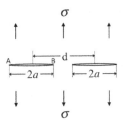

Figure 2.15.4 Two co-planar 2D cracks, each with width $2a$ and with centers separated by distance d.

A **2D crack in an infinite plate**, tilted by angle β relative to direction of tension σ (Figure 2.15.3) has stress intensities (Murakami, 1987)

$$K_I = \sigma\sqrt{\pi a}\,\cos^2\beta; \quad K_{II} = \sigma\sqrt{\pi a}\,\cos\beta\,\sin\beta \tag{2.15.10}$$

Two coplanar 2D cracks in an infinite plate (Figure 2.15.4), each with width $2a$ and centers separated by d, with tension σ normal to crack faces has stress intensity factor (Murakami, 1987)

$$K_I = F_I\sigma\sqrt{\pi a} \tag{2.15.11}$$

$$F_{I,A} = \frac{d+2a}{4a}\left(\frac{d+2a}{d}\right)^{1/2}\left(1 - \frac{E(k)}{K(k)}\right) \text{ (outer crack tips)} \tag{2.15.12}$$

$$F_{I,B} = \frac{d-2a}{4a}\left(\frac{d-2a}{d}\right)^{1/2}\left[\left(\frac{d+2a}{d-2a}\right)^2\frac{E(k)}{K(k)} - 1\right] \text{ (inner crack tips)} \tag{2.15.13}$$

$$k = \left[1 - \left(\frac{d - 2a}{d + 2a} \right)^2 \right]^{1/2} \tag{2.15.14}$$

$$K(k) = \int_0^{\pi/2} (1 - k^2 \sin^2 \theta)^{-1/2} d\theta, \ E(k) = \int_0^{\pi/2} (1 - k^2 \sin^2 \theta)^{1/2} d\theta \tag{2.15.15}$$

Approximate expressions with accuracy 5% are

$$F_{I,A} \approx 1 - 0.0037\lambda + 0.1613\lambda^2 - 0.1628\lambda^3 + 0.1560\lambda^4; \tag{2.15.16}$$

$$F_{I,B} \approx 1 - 0.0426\lambda + 0.5461\lambda^2 - 1.1654\lambda^3 + 1.2368\lambda^4; \tag{2.15.17}$$

$$\lambda = 2a/d; \ 0 \le \lambda \le 0.8 \tag{2.15.18}$$

Two parallel 2D cracks in an infinite plate (Figure 2.15.5), each with width $2a$ and separated by d, with tension σ normal to crack faces (Murakami, 1987) have stress intensity factor (accuracy 5%)

$$K_I = F_I \sigma \sqrt{\pi a} \tag{2.15.19}$$

$$F_I \approx 1 - 0.0007\lambda - 0.4130\lambda^2 + 0.2687\lambda^3; \tag{2.15.20}$$

$$\lambda = 2a/d; \ 0 \le \lambda \le 0.8 \tag{2.15.21}$$

A 2D crack extending from a circular hole with radius R (Figure 2.15.6), subject to uniaxial stress σ normal to crack faces (Liu, 1996) has stress intensity factor

$$K_I = \sigma \sqrt{\pi c} \ \varphi_n \tag{2.15.22}$$

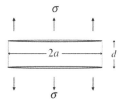

Figure 2.15.5 Two parallel 2D cracks, each with width $2a$ and with centers separated by distance d.

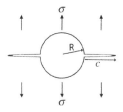

Figure 2.15.6 Two-dimensional cracks (length c) extending from a circular hole (radius R)

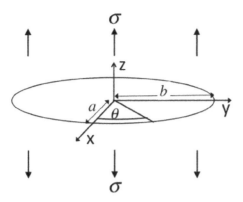

Figure 2.15.7 A 3D crack, elliptical in the plane of the crack, subject to tension σ

$$\varphi_1 \approx 0.707 - 0.18\lambda + 6.55\lambda^2 - 10.54\lambda^3 + 6.85\lambda^4. \quad \text{single crack} \tag{2.15.23}$$

$$\varphi_2 \approx 1 - 0.15\lambda + 3.46\lambda^2 - 4.47\lambda^3 + 3.52\lambda^4. \quad \text{two symmetric cracks} \tag{2.15.24}$$

$$\lambda = \frac{1}{1 + c/R} \tag{2.15.25}$$

Note that the radius of the crack c is measured from the surface of the circular hole, not its center.

A 3D penny crack of radius a in an infinite medium with tension σ normal to crack faces (François *et al.*, 2013) has stress intensity factor

$$K_I = \frac{2}{\pi}\sigma\sqrt{\pi a}. \tag{2.15.26}$$

The same crack with zero remote stress but internal pressure p has stress intensity factor

$$K_I = \frac{2}{\pi}p\sqrt{\pi a}. \tag{2.15.27}$$

A **3D planar crack** with tension σ normal to crack faces has the shape of an ellipse with semi-axes a and b in the plane of the crack (Figure 2.15.7). The stress intensity factor varies with position around the tip (François *et al.*, 2013):

$$K_I = \frac{\sigma\sqrt{\pi a}}{E(k)}\left(\sin^2\theta + \frac{a^2}{b^2}\cos^2\theta\right)^{1/4}, \tag{2.15.28}$$

$$E(k) = \int_0^{\pi/2}(1 - k^2\cos^2\theta)^{1/2}\,d\theta, \tag{2.15.29}$$

$$k^2 = 1 - a^2/b^2. \tag{2.15.30}$$

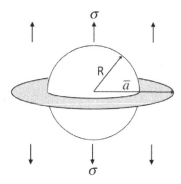

Figure 2.15.8 A circular crack (radius $\bar{a}$) extending from a spherical hole (radius R)

A **3D circular crack extending from a spherical cavity** (Figure 2.15.8), with tension σ normal to crack faces (Figure 2.15.8) has approximate stress intensity factor (Paris *et al.*, 2009)

$$K_I = \sigma\sqrt{\pi\bar{a}}\left[\sqrt{\frac{1}{2}\left(1-\frac{R}{\bar{a}}\right)}\right]F_\sigma(R/\bar{a}) \tag{2.15.31}$$

$$F_\sigma(R/\bar{a}) \approx 0.900 + 0.085(R/\bar{a}) + 0.015(R/\bar{a})^2 \tag{2.15.32}$$

Crack radius $\bar{a}$ is measured from the center of the circular hole.

3 Seismic Wave Propagation

3.1 Seismic Velocities

Synopsis

The velocities of various types of seismic waves in homogeneous, isotropic, elastic media are given by

$$V_P = \sqrt{\frac{K + \frac{4}{3}\mu}{\rho}} = \sqrt{\frac{\lambda + 2\mu}{\rho}} \tag{3.1.1}$$

$$V_S = \sqrt{\frac{\mu}{\rho}} \tag{3.1.2}$$

$$V_E = \sqrt{\frac{E}{\rho}} \tag{3.1.3}$$

where V_P is the P-wave velocity, V_S is the S-wave velocity, and V_E is the extensional wave velocity in a narrow bar.

In addition, ρ is the density, K is the bulk modulus, μ is the shear modulus, λ is Lamé's coefficient, E is Young's modulus, and v is Poisson's ratio.

In terms of Poisson's ratio one can also write

$$\frac{V_P^2}{V_S^2} = \frac{2(1-v)}{(1-2v)} \tag{3.1.4}$$

$$\frac{V_E^2}{V_P^2} = \frac{(1+v)(1-2v)}{(1-v)} \tag{3.1.5}$$

$$\frac{V_E^2}{V_S^2} = 2(1+v) \tag{3.1.6}$$

$$v = \frac{V_P^2 - 2V_S^2}{2(V_P^2 - V_S^2)} = \frac{V_E^2 - 2V_S^2}{2V_S^2} \tag{3.1.7}$$

The various wave velocities are related by

$$\frac{V_P^2}{V_S^2} = \frac{4 - V_E^2/V_S^2}{3 - V_E^2/V_S^2} \tag{3.1.8}$$

$$\frac{V_E^2}{V_S^2} = \frac{3V_P^2/V_S^2 - 4}{V_P^2/V_S^2 - 1} \tag{3.1.9}$$

The elastic moduli can be extracted from measurements of density and any two wave velocities. For example,

$$\mu = \rho V_S^2 \tag{3.1.10}$$

$$K = \rho\left(V_P^2 - \frac{4}{3}V_S^2\right) \tag{3.1.11}$$

$$E = \rho V_E^2 \tag{3.1.12}$$

$$v = \frac{V_P^2 - 2V_S^2}{2(V_P^2 - V_S^2)} \tag{3.1.13}$$

The **Rayleigh wave** phase velocity V_R (White, 1983) at the surface of an isotropic homogeneous elastic half-space is given by the solution to the equation

$$\left(2 - \frac{V_R^2}{V_S^2}\right)^2 - 4\left(1 - \frac{V_R^2}{V_P^2}\right)^{1/2}\left(1 - \frac{V_R^2}{V_S^2}\right)^{1/2} = 0 \tag{3.1.14}$$

(Note that the equivalent equation given in Bourbié *et al.* [1987] is in error.) The wave speed is plotted in Figure 3.1.1 and is given by

$$\frac{V_R^2}{V_S^2} = \left[-\frac{q}{2} + \left(\frac{q^2}{4} + \frac{p^3}{27}\right)^{1/2}\right]^{1/3} + \left[-\frac{q}{2} - \left(\frac{q^2}{4} + \frac{p^3}{27}\right)^{1/2}\right]^{1/3} + \frac{8}{3}$$

$$\text{for } \left(\frac{q^2}{4} + \frac{p^3}{27}\right) > 0 \tag{3.1.15}$$

$$\frac{V_R^2}{V_S^2} = -2\left(\frac{-p}{3}\right)^{1/2}\cos\left[\frac{\pi - \cos^{-1}(-27q^2/4p^3)^{1/2}}{3}\right] + \frac{8}{3}$$

$$\text{for } \left(\frac{q^2}{4} + \frac{p^3}{27}\right) < 0 \tag{3.1.16}$$

$$p = \frac{8}{3} - \frac{16V_S^2}{V_P^2}; \quad q = \frac{272}{27} - \frac{80V_S^2}{3V_P^2} \tag{3.1.17}$$

The Rayleigh velocity at the surface of a homogeneous elastic half-space is non-dispersive (i.e., independent of frequency).

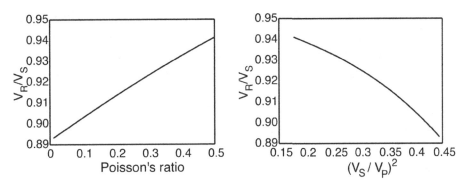

Figure 3.1.1 Rayleigh wave phase velocity normalized by shear velocity

Assumptions and Limitations

These equations assume isotropic, linear, elastic media.

3.2   Phase, Group, and Energy Velocities

Synopsis

In the physics of wave propagation we often talk about different velocities (the **phase**, **group**, and **energy** velocities: V_p, V_g, and V_e, respectively) associated with the wave phenomenon. In laboratory measurements of core sample velocities using finite bandwidth signals and finite-sized transducers, the velocity given by the first arrival does not always correspond to an easily identified velocity.

A general time-harmonic wave may be defined as

$$U(\mathbf{x}, t) = U_0(\mathbf{x})\cos[\omega t - p(\mathbf{x})] \tag{3.2.1}$$

where ω is the angular frequency and U_0 and p are functions of position x; U can be any field of interest such as pressure, stress, or electromagnetic fields. The surfaces given by $p(\mathbf{x})$ = constant are called cophasal or wave surfaces. In particular, for plane waves, $p(\mathbf{x})$ = $\mathbf{k} \cdot \mathbf{x}$, where $\mathbf{k}$ is the wave vector, or the propagation vector, and is in the direction of propagation. For the phase to be the same at $(\mathbf{x}, t)$ and $(\mathbf{x} + d\mathbf{x}, t + dt)$, we must have

$$\omega dt - (\text{grad } p) \cdot dx = 0 \tag{3.2.2}$$

from which the **phase velocity** is defined as

$$V_p = \frac{\omega}{|\text{grad } p|} \tag{3.2.3}$$

For *plane waves* grad p = $\mathbf{k}$, and hence $V_p = \omega/k$. The reciprocal of the phase velocity is often called the **slowness**, and a polar plot of slowness versus the direction of propagation is termed the slowness surface. Phase velocity is the speed of advance of the cophasal surfaces. Born and Wolf (1980) consider the phase velocity to be devoid of any physical significance because it does not correspond to the velocity of propagation of any signal and cannot be directly determined experimentally.

Waves encountered in rock physics are rarely perfectly monochromatic but instead have a finite bandwidth, $\Delta\omega$, centered around some mean frequency $\overline{\omega}$. The wave may be regarded as a superposition of monochromatic waves of different frequencies, which then gives rise to the concept of wave packets or wave groups. Wave packets, or modulation on a wave containing a finite band of frequencies, propagate with the **group velocity** defined as

$$V_g = \frac{1}{|\mathrm{grad}(\partial p/\partial\omega)_{\overline{\omega}}|} \tag{3.2.4}$$

which for plane waves becomes

$$V_g = \left(\frac{\partial\omega}{\partial k}\right)_{\overline{\omega}} \tag{3.2.5}$$

The group velocity may be considered to be the velocity of propagation of the envelope of a modulated carrier wave. The group velocity can also be expressed in various equivalent ways as

$$V_g = V_p - \lambda\frac{dV_p}{d\lambda} \tag{3.2.6}$$

$$V_g = V_p + k\frac{dV_p}{dk} \tag{3.2.7}$$

or

$$\frac{1}{V_g} = \frac{1}{V_p} - \frac{\omega}{V_p^2}\frac{dV_p}{d\omega} \tag{3.2.8}$$

where λ is the wavelength. These equations show that the group velocity is different from the phase velocity when the phase velocity is *frequency dependent, direction dependent*, or both. When the phase velocity is frequency dependent (and hence different from the group velocity), the medium is said to be **dispersive**. Dispersion is termed *normal* if the group velocity decreases with frequency and *anomalous* or *inverse* if it increases with frequency (Elmore and Heald, 1985; Bourbié *et al.*, 1987). In elastic, isotropic media, dispersion can arise as a result of geometric effects such as propagation along waveguides. As a rule, such geometric dispersion (Rayleigh waves, waveguides) is *normal* (i.e., the group velocity decreases with frequency). In a homogeneous viscoelastic medium, on the other hand, dispersion is *anomalous* or *inverse* and arises owing to intrinsic dissipation.

The **energy velocity** V_e represents the velocity at which energy propagates and may be defined as

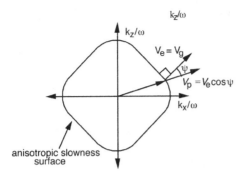

Figure 3.2.1 In anisotropic media, energy propagates along V_e, which is always normal to the slowness surface and in general is deflected by the angle ψ away from V_p and the wave vector **k**.

$$V_e = \frac{P_{av}}{E_{av}} \qquad (3.2.9)$$

where P_{av} is the average power flow density and E_{av} is the average total energy density.

In *isotropic, homogeneous, elastic media*, all three velocities are the same. In a *lossless homogeneous* medium (of arbitrary symmetry), V_g and V_e are identical, and energy propagates with the group velocity. In this case, the energy velocity may be obtained from the group velocity, which is usually somewhat easier to compute. If the medium is not strongly dispersive and a wave group can travel a measurable distance without appreciable "smearing" out, the group velocity *may* be considered to represent the velocity at which the energy is propagated (though this is not strictly true in general).

In *anisotropic, homogeneous, elastic* media, the phase velocity, in general, differs from the group velocity (which is equal to the energy velocity because the medium is elastic) except along certain symmetry directions, where they coincide. The direction in which V_e is deflected away from **k** (which is also the direction of V_p) is obtained from the slowness surface (shown in Figure 3.2.1), for V_e (= V_g in elastic media) must always be normal to the slowness surface (Auld, 1990).

The group velocity in anisotropic media may be calculated by differentiation of the dispersion relation obtained in an implicit form from the Christoffel equation given by

$$\left| \mathbf{k}^2 c_{ijkl} n_j n_l - \rho\omega^2 \delta_{ik} \right| = \Phi(\omega, k_x, k_y, k_z) = 0 \qquad (3.2.10)$$

where c_{ijkl} is the stiffness tensor, n_i are the direction cosines of **k**, ρ is the density, and δ_{ij} is the Kronecker delta function. The group velocity is then evaluated as

$$V_g = -\frac{\nabla_k \Phi}{\partial\Phi/\partial\omega} \qquad (3.2.11)$$

where the gradient is with respect to k_x, k_y, and k_z.

The concept of group velocity is not strictly applicable to attenuating viscoelastic media, but the energy velocity is still well defined (White, 1983). The energy propagation velocity in a dissipative medium is neither the group velocity nor the phase velocity *except* when

(1) the medium is infinite, homogeneous, linear, and viscoelastic, and

(2) the wave is monochromatic and homogeneous, i.e., planes of equal phase are parallel to planes of equal amplitude, or, in other words, the real and imaginary parts of the complex wave vector point in the same direction (in general they do not), in which case the energy velocity is equal to the phase velocity (Ben-Menahem and Singh, 1981; Bourbié *et al.*, 1987).

For the special case of a Voigt solid (see Section 3.8 on viscoelasticity) the energy transport velocity is equal to the phase velocity at all frequencies. For wave propagation in dispersive, viscoelastic media, one sometimes defines the limit

$$V_\infty = \lim_{\omega \to \infty} V_p(\omega) \tag{3.2.12}$$

which describes the propagation of a well-defined wavefront and is referred to as the **signal velocity** (Beltzer, 1988).

Sometimes it is not clear which velocities are represented by the recorded travel times in laboratory ultrasonic core sample measurements, especially when the sample is anisotropic. For elastic materials, there is no ambiguity for propagation along symmetry directions because the phase and group velocities are identical. For non-symmetry directions, the energy does not necessarily propagate straight up the axis of the core from the transducer to the receiver. Numerical modeling of laboratory experiments (Dellinger and Vernik, 1992) indicates that, for typical transducer widths (10 mm), the recorded travel times correspond closely to the *phase velocity*. Accurate measurement of group velocity along nonsymmetry directions would essentially require point transducers of less than 2 mm width.

According to Bourbié *et al.* (1987), the velocity measured by a resonant-bar standing-wave technique corresponds to the phase velocity.

Assumptions and Limitations

In general, phase, group, and energy velocities may differ from each other in both magnitude and direction. Under certain conditions two or more of them may become identical. For homogeneous, linear, isotropic, elastic media all three are the same.

3.3   NMO in Isotropic and Anisotropic Media

Synopsis

The two-way seismic travel time, t, of a pure (nonconverted) mode from the surface, through a homogeneous, isotropic, elastic layer, to a horizontal reflector is hyperbolic

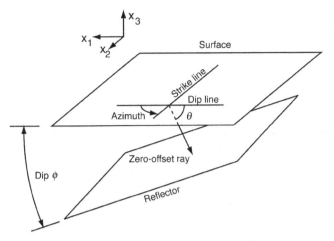

Figure 3.3.1 Schematic showing the geometry for NMO with a dipping reflector

$$t^2 = t_0^2 + \frac{x^2}{V^2} \tag{3.3.1}$$

where x is the *offset* between the source and the receiver, t_0 is the zero-offset, two-way travel time, and V is the velocity of the layer. The increase in travel time with distance, specifically the extra travel time relative to t_0, is called **normal moveout** (or NMO), and the velocity that determines the NMO, is called the **NMO velocity**, V_{NMO}. In the case of a homogeneous isotropic layer above a horizontal reflector, $V_{\mathrm{NMO}} = V$.

When the reflector is dipping with angle ϕ from the horizontal (Figure 3.3.1), the travel time equation in the vertical plane along dip becomes (Levin, 1971)

$$t^2 = t_0^2 + \frac{x^2}{(V/\cos\phi)^2} \tag{3.3.2}$$

or $V_{\mathrm{NMO}}(\phi) = V/\cos\phi$. Again, t_0 is the zero-offset, two-way travel time. More generally, the azimuthally varying NMO velocity is (Levin, 1971)

$$V_{\mathrm{NMO}}^2(\zeta, \phi) = \frac{V^2}{1 - \cos^2\zeta \, \sin^2\phi} \tag{3.3.3}$$

where ϕ is the reflector dip and ζ is the azimuth relative to a horizontal axis in the dip direction. Note that in the strike direction $V_{\mathrm{NMO}} = V$.

When the Earth consists of horizontal, homogeneous, isotropic layers down to the reflector, the two-way travel time equation is approximately hyperbolic, with the approximation being best for small offsets:

$$t^2 \approx t_0^2 + \frac{x^2}{V_{\mathrm{NMO}}^2} \tag{3.3.4}$$

At zero dip, the NMO velocity is often approximated as the **root mean squared (RMS) velocity**, $V_{NMO} \approx V_{RMS}$, where

$$V_{RMS}^2 = \sum_{i=1}^{N} V_i^2 t_i \Big/ \sum_{i=1}^{N} t_i \tag{3.3.5}$$

where V_i is the velocity of the ith layer and t_i, is the two-way, zero-offset travel time through the ith layer. The summations are over all layers from the surface to the reflector. The approximation $V_{NMO} \approx V_{RMS}$ is best at offsets that are small relative to the reflector depth.

If the NMO velocity is estimated at two horizontal reflectors – e.g., $V_{NMO-(n)}$ at the base of layer n and $V_{NMO-(n-1)}$ at the base of layer $(n-1)$ – then the RMS equation can be inverted to yield the **Dix equation** (Dix, 1955) for the **interval velocity**, V_n, of the nth layer:

$$V_n^2 \approx \frac{1}{t_n} \left(V_{NMO-(n)}^2 \sum_{i=1}^{n} t_i - V_{NMO-(n-1)}^2 \sum_{i=1}^{n-1} t_i \right) \tag{3.3.6}$$

It is important to remember that this model assumes: (1) flat, homogeneous, isotropic layers; (2) the offsets are small enough for the RMS velocity to be a reasonable estimate of the moveout velocity; and (3) the estimated interval velocities are themselves averages over velocities of thinner layers lying below the resolution of the seismic data.

The equivalent of the Dix equation for anisotropic media is shown in the following section.

NMO in an Anisotropic Earth

For a heterogeneous elastic Earth with arbitrary anisotropy and arbitrary dip, the NMO velocity of pure modes (at offsets generally less than the depth) can be written as (Grechka and Tsvankin, 1998; Tsvankin, 2001)

$$\frac{1}{V_{NMO}^2(\zeta)} = W_{11} \cos^2\zeta + 2W_{12} \sin\zeta \cos\zeta + W_{22} \sin^2\zeta \tag{3.3.7}$$

where ζ is the azimuth relative to the x_1-axis. The components W_{ij} are elements of a symmetric matrix $\widetilde{\mathbf{W}}$ and are defined as $W_{ij} = \tau_0(\partial p_i/\partial x_j)$, where $p_i = \partial\tau/\partial x_i$ are horizontal components of the slowness vector for rays between the zero-offset reflection point and the surface location $[x_i, x_j]$, $\tau(x_1, x_2)$ is the one-way travel time from the zero-offset reflection point, and τ_0 is the one-way travel time to the CMP (common midpoint) location $x_1 = x_2 = 0$. The derivatives are evaluated at the CMP location. This result assumes a sufficiently smooth reflector and sufficiently smooth lateral velocity heterogeneity, such that the travel time field exists (e.g., no shadow zones) at the CMP point and the derivatives can be evaluated. This can be rewritten as

$$\frac{1}{V_{\mathrm{NMO}}^2(\zeta)} = \lambda_1 \cos^2(\zeta - \psi) + \lambda_2 \sin^2(\zeta - \psi) \tag{3.3.8}$$

where λ_1 and λ_2 are the eigenvalues of $\widetilde{\mathbf{W}}$, and ψ is the rotation of the eigenvectors of $\widetilde{\mathbf{W}}$ relative to the coordinate system. *λ_1 and λ_2 are typically positive, in which case V_{NMO} (ζ) is an ellipse in the horizontal plane.* The elliptical form allows the exact expression for V_{NMO} (ζ) to be determined by only three parameters: the values of the NMO velocity at the axes of the ellipse, and the orientation of the ellipse relative to the coordinate axes. The elliptical form simplifies modeling of azimuthally varying NMO for various geometries and anisotropies, since only V_{NMO} (ζ) along the ellipse axes needs to be derived. For example, we can write

$$\frac{1}{V_{\mathrm{NMO}}^2(\zeta)} = \frac{\cos^2\zeta}{V_{\mathrm{NMO}}^2(\zeta = 0)} + \frac{\sin^2\zeta}{V_{\mathrm{NMO}}^2(\zeta = \pi/2)} \tag{3.3.9}$$

where the axes of the ellipse are at $\zeta = 0$ and $\zeta = \pi/2$.

A simple example of the elliptical form is the azimuthally varying moveout velocity for a homogeneous, isotropic layer above a dipping reflector, previously given, which can be written as (Grechka and Tsvankin, 1998)

$$\frac{1}{V_{\mathrm{NMO}}^2(\zeta, \phi)} = \frac{\sin^2\zeta}{V^2} + \frac{\cos^2\zeta}{(V/\cos\phi)^2} \tag{3.3.10}$$

where ϕ is the dip and ζ is the azimuth relative to a horizontal axis in the dip direction.

VTI Symmetry with Horizontal Reflector

NMO velocities in a **VTI medium** (vertical symmetry axis) overlying a horizontal reflector are given for P, SV, and SH modes (Tsvankin, 2001) by

$$V_{\mathrm{NMO,P}}(0) = V_{\mathrm{P0}}\sqrt{1 + 2\delta} \tag{3.3.11}$$

$$V_{\mathrm{NMO,SV}}(0) = V_{\mathrm{S0}}\sqrt{1 + 2\sigma}; \quad \sigma = \left(\frac{V_{\mathrm{P0}}}{V_{\mathrm{S0}}}\right)^2 (\varepsilon - \delta) \tag{3.3.12}$$

$$V_{\mathrm{NMO,SH}}(0) = V_{\mathrm{S0}}\sqrt{1 + 2\gamma} \tag{3.3.13}$$

where V_{P0} and V_{S0} are the vertical P- and S-wave velocities, and ε, δ, and γ are the Thomsen parameters for VTI media. In this case, the NMO ellipse is a circle, giving an azimuthally independent NMO velocity. These expressions for NMO velocity hold for anisotropy of arbitrary strength (Tsvankin, 2001).

In the special case of **elliptical anisotropy**, with vertical symmetry axis, $\varepsilon = \delta$, the expressions for NMO velocity become (Tsvankin, 2001):

$$V_{\mathrm{NMO,P}}(0) = V_{\mathrm{P0}}\sqrt{1 + 2\varepsilon} \tag{3.3.14}$$

$$V_{\mathrm{NMO,SV}}(0) = V_{\mathrm{S0}} \tag{3.3.15}$$

$$V_{\mathrm{NMO,SH}}(0) = V_{\mathrm{S0}}\sqrt{1 + 2\gamma} \tag{3.3.16}$$

Vertical Symmetry Axis with Dipping Reflector

A dipping reflector creates an azimuthal dependence for the NMO velocity, even with a VTI medium. For general VTI symmetry (vertical symmetry axis) and assuming weak anisotropy, the NMO velocities overlying a dipping reflector are given for P and SV modes by Tsvankin (2001). In the dip direction,

$$V_{\mathrm{NMO,P}}(0,\phi) = \frac{V_{\mathrm{P0}}(1 + \delta)}{\cos\phi}\left[1 + \delta\sin^2\phi + 3(\varepsilon - \delta)\sin^2\phi(2 - \sin^2\phi)\right] \tag{3.3.17}$$

$$V_{\mathrm{NMO,SV}}(0,\phi) = \frac{V_{\mathrm{S0}}(1 + \sigma)}{\cos\phi}\left[1 - 5\sigma\sin^2\phi + 3\sigma\sin^4\phi\right]; \quad \sigma = \left(\frac{V_{\mathrm{P0}}}{V_{\mathrm{S0}}}\right)^2(\varepsilon - \delta) \tag{3.3.18}$$

where ϕ is the dip, V_{P0} and V_{S0} are the vertical P- and S-wave velocities, and ε, δ, and γ are the Thomsen parameters for VTI media. In the strike direction with VTI symmetry:

$$V_{\mathrm{NMO,P}}\left(\frac{\pi}{2},\phi\right) = V_{\mathrm{P0}}(1 + \delta)[1 + (\varepsilon - \delta)\sin^2\phi(2 - \sin^2\phi)] \tag{3.3.19}$$

$$V_{\mathrm{NMO,SV}}\left(\frac{\pi}{2},\phi\right) = V_{\mathrm{S0}}(1 + \sigma)\left[1 - \sigma\sin^2\phi(2 - \sin^2\phi)\right]; \quad \sigma = \left(\frac{V_{\mathrm{P0}}}{V_{\mathrm{S0}}}\right)^2(\varepsilon - \delta) \tag{3.3.20}$$

The complete azimuthal dependence can be easily written as an ellipse, where the strike and dip NMO velocities are the semiaxes.

For the special case of **elliptical symmetry** ($\varepsilon = \delta$) of arbitrary strength anisotropy, the dip-direction NMO velocities are (Tsvankin, 2001):

$$V_{\mathrm{NMO,P}}(0,\phi) = \frac{V_{\mathrm{P0}}\sqrt{1 + 2\varepsilon}}{\cos\phi}\sqrt{1 + 2\varepsilon\sin^2\phi} \tag{3.3.21}$$

$$V_{\mathrm{NMO,SV}}(0,\phi) = \frac{V_{\mathrm{S0}}}{\cos\phi} \tag{3.1.22}$$

$$V_{\mathrm{NMO,SH}}(0,\phi) = \frac{V_{\mathrm{S0}}\sqrt{1 + 2\gamma}}{\cos\phi}\sqrt{1 + 2\gamma\sin^2\phi} \tag{3.3.23}$$

where ϕ is the layer dip angle. In the strike direction,

$$V_{\text{NMO,P}}\left(\frac{\pi}{2}, \phi\right) = V_{P0}\sqrt{1 + 2\varepsilon} \tag{3.3.24}$$

$$V_{\text{NMO,SV}}\left(\frac{\pi}{2}, \phi\right) = V_{S0} \tag{3.3.25}$$

$$V_{\text{NMO,SH}}\left(\frac{\pi}{2}, \phi\right) = V_{S0}\sqrt{1 + 2\gamma} \tag{3.3.26}$$

Tilted TI Symmetry with Dipping Reflector

For TI symmetry with symmetry axis tilting an angle v from the vertical and assuming weak anisotropy, the NMO velocities are given by Tsvankin (2001). In the dip direction,

$$
\begin{aligned}
V_{\text{NMO,P}}(0, \phi) = \frac{V_{P0}}{\cos\phi} & \left\{ 1 + \delta + \delta \sin^2(\phi - v) + 3(\varepsilon - \delta) \sin^2(\phi - v)\left[2 - \sin^2(\phi - v)\right] \right. \\
& \left. + \frac{2 \sin v \sin(\phi - v)}{\cos\phi} \left[\delta + 2(\varepsilon - \delta) \sin^2(\phi - v)\right] \right\}
\end{aligned}
\tag{3.1.27}
$$

$$
\begin{aligned}
V_{\text{NMO,SV}}(0, \phi) = \frac{V_{S0}}{\cos\phi} & \left\{ 1 + \sigma + \sigma \sin^2(\phi - v) - 3\sigma \sin^2(\phi - v)\left[2 - \sin^2(\phi - v)\right] \right. \\
& \left. + 2\sigma \frac{\sin v \sin(\phi - v)\cos 2(\phi - v)}{\cos\phi} \right\}
\end{aligned}
\tag{3.3.28}
$$

where ϕ is the reflector dip, V_{P0} and V_{S0} are the symmetry axis P- and S-wave velocities, and ε, δ, and γ are the Thomsen parameters for TI media. It is assumed that the azimuth of the symmetry axis tilt is the same as the azimuth of the reflector dip. In the strike direction, for weak tilted TI symmetry

$$V_{\text{NMO,P}}\left(\frac{\pi}{2}, \phi\right) = V_{P0}(1 + \delta)\left\{1 + (\varepsilon - \delta) \sin^2(\phi - v)\left[2 - \sin^2(\phi - v)\right]\right\} \tag{3.3.29}$$

$$V_{\text{NMO,SV}}\left(\frac{\pi}{2}, \phi\right) = V_{S0}(1 + \sigma)\left\{1 - \sigma \sin^2(\phi - v)\left[2 - \sin^2(\phi - v)\right]\right\} \tag{3.3.30}$$

$$\sigma = \left(\frac{V_{P0}}{V_{S0}}\right)^2 (\varepsilon - \delta) \tag{3.3.31}$$

In the special case of **tilted elliptical symmetry** with a tilted symmetry axis, the NMO velocity in the dip direction is

$$V_{\text{NMO,P}}(0, \phi) = \frac{V_{P0}}{\cos\phi}\sqrt{1 + 2\delta}\sqrt{1 + 2\delta \sin^2(\phi - v)}\left[1 - 2\delta\frac{\sin v \sin(\phi - v)}{\cos\phi}\right]^{-1} \tag{3.3.32}$$

$$V_{\text{NMO,SV}}(\phi) = \frac{V_{S0}}{\cos\phi} \tag{3.3.33}$$

$$V_{\text{NMO,SH}}(0,\phi) = \frac{V_{S0}}{\cos\phi}\sqrt{1+2\gamma}\sqrt{1+2\gamma\sin^2(\phi-v)}\left[1-2\gamma\frac{\sin v\,\sin(\phi-v)}{\cos\phi}\right]^{-1} \tag{3.3.34}$$

In the strike direction for **tilted elliptical symmetry**,

$$V_{\text{NMO,P}}\left(\frac{\pi}{2},\phi\right) = V_{P0}(1+\delta) \tag{3.3.35}$$

$$V_{\text{NMO,SV}}\left(\frac{\pi}{2},\phi\right) = V_{S0} \tag{3.3.36}$$

Orthorhombic Symmetry with Horizontal Reflector

We now consider an orthorhombic layer of arbitrary strength anisotropy over a horizontal reflector. A symmetry plane of the orthorhombic medium is also horizontal. The vertically propagating (along the x_3-axis) P-wave velocity (V_{P0}) and vertically propagating S-wave velocities, polarized in the x_1 (V_{S0}) and x_2 (V_{S1}) directions are

$$V_{P0} = \sqrt{c_{33}/\rho}; \quad V_{S0} = \sqrt{c_{55}/\rho}; \quad V_{S1} = \sqrt{c_{44}/\rho} \tag{3.3.37}$$

where ρ is the density. Additional constants necessary for this discussion are

$$\varepsilon^{(2)} = \frac{c_{11}-c_{33}}{2c_{33}} \qquad\qquad \varepsilon^{(1)} = \frac{c_{22}-c_{33}}{2c_{33}}$$

$$\delta^{(2)} = \frac{(c_{13}+c_{55})^2-(c_{33}-c_{55})^2}{2c_{33}(c_{33}-c_{55})} \qquad \delta^{(1)} = \frac{(c_{23}+c_{44})^2-(c_{33}-c_{44})^2}{2c_{33}(c_{33}-c_{44})}$$

$$\gamma^{(2)} = \frac{c_{66}-c_{44}}{2c_{44}} \qquad\qquad \gamma^{(1)} = \frac{c_{66}-c_{55}}{2c_{55}}$$

$$\sigma^{(2)} = \left(\frac{V_{P0}}{V_{S0}}\right)^2[\varepsilon^{(2)}-\delta^{(2)}] \qquad \sigma^{(1)} = \left(\frac{V_{P0}}{V_{S1}}\right)^2[\varepsilon^{(1)}-\delta^{(1)}]$$

For a CMP line in the x_1 direction, the NMO velocities are

$$V_{\text{NMO,P}}(\zeta=0,\phi=0) = V_{P0}\sqrt{1+2\delta^{(2)}} \tag{3.3.38}$$

$$V_{\text{NMO,SV}}(\zeta=0,\phi=0) = V_{S0}\sqrt{1+2\sigma^{(2)}} \tag{3.3.39}$$

$$V_{\text{NMO,SH}}(\zeta=0,\phi=0) = V_{S1}\sqrt{1+2\gamma^{(2)}} \tag{3.3.40}$$

where ϕ is the dip and ζ is the azimuth measured from the x_1-axis. For a line in the x_2 direction,

$$V_{\text{NMO,P}}\left(\zeta=\frac{\pi}{2},\phi=0\right) = V_{P0}\sqrt{1+2\delta^{(1)}} \tag{3.3.41}$$

$$V_{\text{NMO,SV}}\left(\zeta=\frac{\pi}{2},\phi=0\right) = V_{S1}\sqrt{1+2\sigma^{(1)}} \tag{3.3.42}$$

$$V_{\text{NMO,SH}}\left(\zeta = \frac{\pi}{2}, \phi = 0\right) = V_{S0}\sqrt{1 + 2\gamma^{(1)}} \tag{3.3.43}$$

NMO in a Horizontally Layered Anisotropic Earth

The effective NMO for a stack of horizontal homogeneous layers above a dipping reflector can be written as (Tsvankin, 2001)

$$V_{\text{NMO}}^2 = \frac{1}{t_0}\sum_{i=1}^{N}[V_{\text{NMO}}^{(i)}(p)]^2 t_0^i(p) \tag{3.3.44}$$

where $t_0^i(p)$ is the interval travel time in layer i computed along the zero-offset ray, $t_0 = \sum_{i=1}^{N} t_0^i$ is the total zero-offset time, and $V_{\text{NMO}}^{(i)}(p)$ is the interval NMO velocity for the ray parameter p of the zero-offset ray. The ray parameter for the zero-offset ray is

$$p = \frac{\sin\phi}{V_N(\phi)} \tag{3.3.45}$$

where ϕ is the reflector dip, and $V_N(\phi)$ is the velocity at angle ϕ from the vertical in the Nth layer at the reflector.

Assumptions and Limitations

The expressions for NMO in a layered and/or anisotropic elastic medium generally work best for small offsets.

3.4　Impedance, Reflectivity, and Transmissivity

Synopsis

The impedance, I, of an elastic medium is the ratio of the stress to the particle velocity (Aki and Richards, 1980) and is given by ρV, where ρ is the density and V is the wave propagation velocity. At a plane interface between two thick, homogeneous, isotropic, elastic layers, the normal incidence reflectivity for waves traveling from medium 1 to medium 2 is the ratio of the displacement amplitude, A_r, of the reflected wave to that of the incident wave, A_i, and is given by

$$\begin{aligned} R_{12} &= \frac{A_r}{A_i} = \frac{I_2 - I_1}{I_2 + I_1} = \frac{\rho_2 V_2 - \rho_1 V_1}{\rho_2 V_2 + \rho_1 V_1} \\ &\approx \frac{1}{2}\ln(I_2/I_1) \end{aligned} \tag{3.4.1}$$

The logarithmic approximation is reasonable for $|R| < 0.5$ (Castagna, 1993). A normally incident P-wave generates only reflected and transmitted P-waves. A normally incident S-wave generates only reflected and transmitted S-waves. There is no mode conversion.

The previous expression for the reflection coefficient is obtained when the *particle displacements are measured with respect to the direction of the wave vector* (equivalent to the slowness vector or the direction of propagation). A displacement is taken to be positive when its component along the interface has the same phase (or the same direction) as the component of the wave vector along the interface. For P-waves, this means that positive displacement is along the direction of propagation. Thus, a positive reflection coefficient implies that a compression is reflected as a compression, whereas a negative reflection coefficient implies a phase inversion (Sheriff, 1991). When the *displacements are measured with respect to a space-fixed coordinate system*, and not with respect to the wave vector, the reflection coefficient is given by

$$R_{12} = \frac{A_r}{A_i} = \frac{I_1 - I_2}{I_2 + I_1} = \frac{\rho_1 V_1 - \rho_2 V_2}{\rho_2 V_2 + \rho_1 V_1} \tag{3.4.2}$$

The normal incidence transmissivity in both coordinate systems is

$$T_{12} = \frac{A_t}{A_i} = \frac{2I_1}{I_2 + I_1} = \frac{2\rho_1 V_1}{\rho_2 V_2 + \rho_1 V_1} \tag{3.4.3}$$

where A_t is the displacement amplitude of the transmitted wave. Continuity at the interface requires

$$A_i + A_r = A_t \tag{3.4.4}$$

$$1 + R = T \tag{3.4.5}$$

This choice of signs for A_i and A_r is for a space-fixed coordinate system. Note that the transmission coefficient for wave amplitudes can be greater than 1. Sometimes the reflection and transmission coefficients are defined in terms of scaled displacements A', which are proportional to the square root of energy flux (Aki and Richards, 1980; Kennett, 1983). The scaled displacements are given by

$$A' = A\sqrt{\rho V \cos\theta} \tag{3.4.6}$$

where θ is the angle between the wave vector and the normal to the interface. The normal incidence reflection and transmission coefficients in terms of these scaled displacements are

$$R'_{12} = \frac{A'_r}{A'_i} = R_{12} \tag{3.4.7}$$

$$T'_{12} = \frac{A'_t}{A'_i} = T_{12} \frac{\sqrt{\rho_2 V_2}}{\sqrt{\rho_1 V_1}} = \frac{2\sqrt{\rho_1 V_1 \rho_2 V_2}}{\rho_2 V_2 + \rho_1 V_1} \tag{3.4.8}$$

Reflectivity and transmissivity for energy fluxes, R^e and T^e, respectively, are given by the squares of the reflection and transmission coefficients for scaled displacements. For normal incidence they are

$$R^e_{12} = \frac{E_r}{E_i} = (R'_{12})^2 = \frac{(\rho_1 V_1 - \rho_2 V_2)^2}{(\rho_2 V_2 + \rho_1 V_1)^2} \tag{3.4.9}$$

$$T^e_{12} = \frac{E_t}{E_i} = (T'_{12})^2 = \frac{4\rho_1 V_1 \rho_2 V_2}{(\rho_2 V_2 + \rho_1 V_1)^2} \tag{3.4.10}$$

where E_i, E_r, and E_t are the incident, reflected, and transmitted energy fluxes, respectively. Energy reflection coefficients were first given by Knott (1899). Conservation of energy at an interface where no trapping of energy occurs requires that

$$E_i = E_r + E_t \tag{3.4.11}$$

$$1 = R^e + T^e \tag{3.4.12}$$

The reflection and transmission coefficients for energy fluxes can never be greater than 1.

Simple Band-Limited Inverse of Reflectivity Time Series

Consider a flat-layered Earth, with an impedance time series $I_n = \rho\,(n\Delta t)\,V\,(n\Delta t)$, where ρ is density, V is velocity (either P or S), and $t = n\Delta t$, where $n = 0, 1, 2, \ldots, N$ is the normal incidence two-way travel time, equally sampled in intervals Δt. In these expressions, $n\Delta t$ is the argument of $\rho\,(\cdot)$ and $V(\cdot)$, not a factor multiplying ρ and V.

The reflectivity time series can be approximated as

$$R_{n+1} = \frac{I_{n+1} - I_n}{I_{n+1} + I_n} \tag{3.4.13}$$

Solving for I_{n+1} yields a simple recursive algorithm

$$I_{n+1} = I_n \frac{(1 + R_{n+1})}{(1 - R_{n+1})} \quad ; \quad n = 0 : N - 1 \tag{3.4.14}$$

where I_0, the impedance at time $t = 0$, must be supplied. For small Δt and small R_n, this equation can be written as

$$I_n \approx I_0 \exp\left[\sum_{i=1}^{n}\left(2R_i + \frac{2}{3}R_i^3 + \frac{2}{5}R_i^5 + \cdots\right)\right] \tag{3.4.15}$$

In contrast, a simple running summation of R_n leads to

$$I_n \approx I_0 \exp\left[\sum_{i=1}^{n}(2R_i)\right] \tag{3.4.16}$$

which is a low-order approximation of the first algorithm.

Rough Surfaces

Random interface roughness at scales smaller than the wavelength causes incoherent scattering and a decrease in amplitude of the coherent reflected and transmitted waves. This could be one of the explanations for the observation that amplitudes of multiples in synthetic seismograms are often larger than the amplitudes of corresponding multiples in the data (Frazer, 1994). Kuperman (1975) gives results that modify the reflectivity and transmissivity to include scattering losses at the interface. With the mean-squared departure from planarity of the rough interface denoted by σ^2, the modified coefficients are

$$\widetilde{R}_{12} = R_{12}(1 - 2k_1^2\sigma^2) = R_{12}[1 - 8\pi^2(\sigma/\lambda_1)^2] \tag{3.4.17}$$

$$\widetilde{T}_{12} = T_{12}\left[1 - \frac{1}{2}(k_1 - k_2)^2\sigma^2\right] = T_{12}\left[1 - 2\pi^2(\lambda_2 - \lambda_1)^2\left(\frac{\sigma}{\lambda_1\lambda_2}\right)^2\right] \tag{3.4.18}$$

where $k_1 = \omega/V_1$, $k_2 = \omega/V_2$ are the wavenumbers, and λ_1 and λ_2 are the wavelengths in media 1 and 2, respectively.

Uses

The equations presented in this section can be used for the following purposes:
- to calculate amplitudes and energy fluxes of reflected and transmitted waves at interfaces in elastic media; and
- to estimate the decrease in wave amplitude caused by scattering losses during reflection and transmission at rough interfaces.

Assumptions and Limitations

The equations presented in this section apply only under the following conditions:
- normal incidence, plane-wave, time-harmonic propagation in isotropic, linear, elastic media with a single interface, or multiple interfaces well separated by thick layers with thickness much greater than the wavelength;
- no energy losses or trapping at the interface; and
- rough surface results are valid for small deviations from planarity (small σ).

3.5   Reflectivity and Amplitude Variations with Offset (AVO) in Isotropic Media

Synopsis

The **seismic impedance** is the product of velocity and density (see Section 3.4), as expressed by

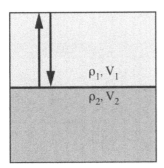

Figure 3.5.1 Reflection of a normal-incidence wave at an interface between two thick homogeneous, isotropic, elastic layers

$$I_P = \rho V_P \quad I_S = \rho V_S \tag{3.5.1}$$

where I_P, I_S are P- and S-wave impedances, V_P, V_S are P- and S-wave velocities, and ρ is density.

At an interface between two thick homogeneous, isotropic, elastic layers, the **normal incidence reflectivity**, defined as the ratio of the reflected wave amplitude to the incident wave amplitude, is

$$R_{PP} = \frac{\rho_2 V_{P2} - \rho_1 V_{P1}}{\rho_2 V_{P2} + \rho_1 V_{P1}} = \frac{I_{P2} - I_{P1}}{I_{P2} + I_{P1}}$$

$$\approx \frac{1}{2} \ln(I_{P2}/I_{P1}) \tag{3.5.2}$$

$$R_{SS} = \frac{\rho_2 V_{S2} - \rho_1 V_{S1}}{\rho_2 V_{S2} + \rho_1 V_{S1}} = \frac{I_{S2} - I_{S1}}{I_{S2} + I_{S1}}$$

$$\approx \frac{1}{2} \ln(I_{S2}/I_{S1}) \tag{3.5.3}$$

where R_{PP} is the normal incidence P-to-P reflectivity, R_{SS} is the S-to-S reflectivity, and the subscripts 1 and 2 refer to the first and second media, respectively (Figure 3.5.1). The logarithmic approximation is reasonable for $|R| < 0.5$ (Castagna, 1993). A normally incident P-wave generates only reflected and transmitted P-waves. A normally incident S-wave generates only reflected and transmitted S-waves. There is no mode conversion.

AVO: Amplitude Variations with Offset

For non-normal incidence, the situation is more complicated. An incident P-wave generates reflected P- and S-waves and transmitted P- and S-waves. The reflection and transmission coefficients depend on the angle of incidence as well as on the material properties of the two layers. An excellent review is given by Castagna (1993).

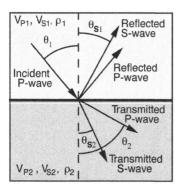

Figure 3.5.2 The angles of the incident, reflected, and transmitted rays of a P-wave with non-normal incidence

The angles of the incident, reflected, and transmitted rays (Figure 3.5.2) are related by Snell's law as follows:

$$p = \frac{\sin\theta_1}{V_{P1}} = \frac{\sin\theta_2}{V_{P2}} = \frac{\sin\theta_{S1}}{V_{S1}} = \frac{\sin\theta_{S2}}{V_{S2}} \tag{3.5.4}$$

where p is the **ray parameter**. θ and θ_S are the angles of P- and S-wave propagation, respectively, relative to the reflector normal. Subscripts 1 and 2 indicate angles or material properties of layers 1 and 2, respectively.

The complete solution for the amplitudes of transmitted and reflected P- and S-waves for both incident P- and S-waves is given by the Knott–Zoeppritz equations (Knott, 1899; Zoeppritz, 1919; Aki and Richards, 1980; Castagna, 1993).

Aki and Richards (1980) give the results in the following convenient matrix form:

$$\begin{pmatrix} \overset{\downarrow\uparrow}{PP} & \overset{\downarrow\uparrow}{SP} & \overset{\uparrow\uparrow}{PP} & \overset{\uparrow\uparrow}{SP} \\[6pt] \overset{\downarrow\uparrow}{PS} & \overset{\downarrow\uparrow}{SS} & \overset{\uparrow\uparrow}{PS} & \overset{\uparrow\uparrow}{SS} \\[6pt] \overset{\downarrow\downarrow}{PP} & \overset{\downarrow\downarrow}{SP} & \overset{\uparrow\downarrow}{PP} & \overset{\uparrow\downarrow}{SP} \\[6pt] \overset{\downarrow\downarrow}{PS} & \overset{\downarrow\downarrow}{SS} & \overset{\uparrow\downarrow}{PS} & \overset{\uparrow\downarrow}{SS} \end{pmatrix} = \mathbf{M}^{-1}\mathbf{N} \tag{3.5.5}$$

where each matrix element is a reflection or transmission coefficient for displacement amplitudes. The first letter designates the type of incident wave, and the second letter designates the type of reflected or transmitted wave. The arrows indicate downward ↓ and upward ↑ propagation, so that a combination ↑↓ indicates a reflection coefficient, while a combination ↓↓ indicates a transmission coefficient. The matrices **M** and **N** are given by

$$M = \begin{pmatrix} -\sin\theta_1 & -\cos\theta_{S1} & \sin\theta_2 & \cos\theta_{S2} \\ \cos\theta_1 & -\sin\theta_{S1} & \cos\theta_2 & -\sin\theta_{S2} \\ 2\rho_1 V_{S1}\sin\theta_{S1}\cos\theta_1 & \rho_1 V_{S1}(1-2\sin^2\theta_{S1}) & 2\rho_2 V_{S2}\sin\theta_{S2}\cos\theta_2 & \rho_2 V_{S2}(1-2\sin^2\theta_{S2}) \\ -\rho_1 V_{P1}(1-2\sin^2\theta_{S1}) & \rho_1 V_{S1}\sin2\theta_{S1} & \rho_2 V_{P2}(1-2\sin^2\theta_{S2}) & -\rho_2 V_{S2}\sin2\theta_{S2} \end{pmatrix}$$

(3.5.6)

$$N = \begin{pmatrix} \sin\theta_1 & \cos\theta_{S1} & -\sin\theta_2 & -\cos\theta_{S2} \\ \cos\theta_1 & -\sin\theta_{S1} & \cos\theta_2 & -\sin\theta_{S2} \\ 2\rho_1 V_{S1}\sin\theta_{S1}\cos\theta_1 & \rho_1 V_{S1}(1-2\sin^2\theta_{S1}) & 2\rho_2 V_{S2}\sin\theta_{S2}\cos\theta_2 & \rho_2 V_{S2}(1-2\sin^2\theta_{S2}) \\ \rho_1 V_{P1}(1-2\sin^2\theta_{S1}) & -\rho_1 V_{S1}\sin2\theta_{S1} & -\rho_2 V_{P2}(1-2\sin^2\theta_{S2}) & \rho_2 V_{S2}\sin2\theta_{S2} \end{pmatrix}$$

(3.5.7)

Results for incident P and incident S, given explicitly by Aki and Richards (1980), are as follows, where $R_{PP} = \overset{\downarrow\uparrow}{PP}$, $R_{PS} = \overset{\downarrow\uparrow}{PS}$, $T_{PP} = \overset{\downarrow\downarrow}{PP}$, $T_{PS} = \overset{\downarrow\downarrow}{PS}$, $R_{SS} = \overset{\downarrow\uparrow}{SS}$, $R_{SP} = \overset{\downarrow\uparrow}{SP}$, $T_{SP} = \overset{\downarrow\downarrow}{SP}$, $T_{SS} = \overset{\downarrow\downarrow}{SS}$:

$$R_{PP} = \left[\left(b\frac{\cos\theta_1}{V_{P1}} - c\frac{\cos\theta_2}{V_{P2}}\right)F - \left(a + d\frac{\cos\theta_1}{V_{P1}}\frac{\cos\theta_{S2}}{V_{S2}}\right)Hp^2\right]/D$$

(3.5.8)

$$R_{PS} = \left[-2\frac{\cos\theta_1}{V_{P1}}\left(ab + cd\frac{\cos\theta_2}{V_{P2}}\frac{\cos\theta_{S2}}{V_{S2}}\right)pV_{P1}\right]/(V_{S1}D)$$

(3.5.9)

$$T_{PP} = 2\rho_1\frac{\cos\theta_1}{V_{P1}}FV_{P1}/(V_{P2}D)$$

(3.5.10)

$$T_{PS} = 2\rho_1\frac{\cos\theta_1}{V_{P1}}HpV_{P1}/(V_{S2}D)$$

(3.5.11)

$$R_{SS} = -\left[\left(b\frac{\cos\theta_{S1}}{V_{S1}} - c\frac{\cos\theta_{S2}}{V_{S2}}\right)E - \left(a + d\frac{\cos\theta_2}{V_{P2}}\frac{\cos\theta_{S1}}{V_{S1}}\right)Gp^2\right]/D$$

(3.5.12)

$$R_{SP} = -2\frac{\cos\theta_{S1}}{V_{S1}}\left(ab + cd\frac{\cos\theta_2}{V_{P2}}\frac{\cos\theta_{S2}}{V_{S2}}\right)pV_{S1}/(V_{P1}D)$$

(3.5.13)

$$T_{SP} = -2\rho_1\frac{\cos\theta_{S1}}{V_{S1}}GpV_{S1}/(V_{P2}D)$$

(3.5.14)

$$T_{SS} = 2\rho_1\frac{\cos\theta_{S1}}{V_{S1}}EV_{S1}/(V_{S2}D)$$

(3.5.15)

where

$$a = \rho_2(1 - 2\sin^2\theta_{S2}) - \rho_1(1 - 2\sin^2\theta_{S1})$$

(3.5.16)

$$b = \rho_2(1 - 2\sin^2\theta_{S2}) + 2\rho_1\sin^2\theta_{S1}$$

(3.5.17)

$$c = \rho_1(1 - 2\sin^2\theta_{S1}) + 2\rho_2\sin^2\theta_{S2}$$

(3.5.18)

$$d = 2(\rho_2 V_{S2}^2 - \rho_1 V_{S1}^2)$$

(3.5.19)

$$D = EF + GHp^2 = (\det \mathbf{M})/(V_{P1}V_{P2}V_{S1}V_{S2}) \tag{3.5.20}$$

$$E = b\frac{\cos\theta_1}{V_{P1}} + c\frac{\cos\theta_2}{V_{P2}} \tag{3.5.21}$$

$$F = b\frac{\cos\theta_{S1}}{V_{S1}} + c\frac{\cos\theta_{S2}}{V_{S2}} \tag{3.5.22}$$

$$G = a - d\frac{\cos\theta_1}{V_{P1}}\frac{\cos\theta_{S2}}{V_{S2}} \tag{3.5.23}$$

$$H = a - d\frac{\cos\theta_2}{V_{P2}}\frac{\cos\theta_{S1}}{V_{S1}} \tag{3.5.24}$$

and p is the ray parameter.

Approximate Forms

Although the complete Zoeppritz equations can be evaluated numerically, it is often useful and more insightful to use one of the simpler approximations.

Bortfeld (1961) linearized the Zoeppritz equations by assuming small contrasts between layer properties as follows:

$$R_{PP}(\theta_1) \approx \frac{1}{2}\ln\left(\frac{V_{P2}\rho_2\,\cos\theta_1}{V_{P1}\rho_1\,\cos\theta_2}\right) + \left(\frac{\sin\theta_1}{V_{P1}}\right)^2(V_{S1}^2 - V_{S2}^2)\left[2 + \frac{\ln(\rho_2/\rho_1)}{\ln(V_{S2}/V_{S1})}\right] \tag{3.5.25}$$

Aki and Richards (1980) also derived a simplified form by assuming small layer contrasts. The results are conveniently expressed in terms of contrasts in V_P, V_S, and ρ as follows:

$$R_{PP}(\theta) \approx \frac{1}{2}(1 - 4p^2\overline{V}_S^2)\frac{\Delta\rho}{\overline{\rho}} + \frac{1}{2\cos^2\theta}\frac{\Delta V_P}{\overline{V}_P} - 4p^2\overline{V}_S^2\frac{\Delta V_S}{\overline{V}_S} \tag{3.5.26}$$

$$R_{PS}(\theta) \approx \frac{-p\overline{V}_P}{2\cos\theta_S}\left[\left(1 - 2\overline{V}_S^2 p^2 + 2\overline{V}_S^2\frac{\cos\theta}{\overline{V}_P}\frac{\cos\theta_S}{\overline{V}_S}\right)\frac{\Delta\rho}{\overline{\rho}}\right.$$

$$\left. - \left(4p^2\overline{V}_S^2 - 4\overline{V}_S^2\frac{\cos\theta}{\overline{V}_P}\frac{\cos\theta_S}{\overline{V}_S}\right)\frac{\Delta V_S}{\overline{V}_S}\right] \tag{3.5.27}$$

$$T_{PP}(\theta) \approx 1 - \frac{1}{2}\frac{\Delta\rho}{\overline{\rho}} + \left(\frac{1}{2\cos^2\theta} - 1\right)\frac{\Delta V_P}{\overline{V}_P} \tag{3.5.28}$$

$$T_{PS}(\theta) \approx \frac{p\overline{V}_P}{2\cos\theta_S}\left[\left(1 - 2\overline{V}_S^2 p^2 - 2\overline{V}_S^2\frac{\cos\theta}{\overline{V}_P}\frac{\cos\theta_S}{\overline{V}_S}\right)\frac{\Delta\rho}{\overline{\rho}}\right.$$

$$\left. - \left(4p^2\overline{V}_S^2 + 4\overline{V}_S^2\frac{\cos\theta}{\overline{V}_P}\frac{\cos\theta_S}{\overline{V}_S}\right)\frac{\Delta V_S}{\overline{V}_S}\right] \tag{3.5.29}$$

$$R_{SP}(\theta) \approx \frac{\cos\theta_S}{\overline{V}_P} \frac{\overline{V}_S}{\cos\theta} R_{PS}(\theta) \tag{3.5.30}$$

$$R_{SS}(\theta) \approx -\frac{1}{2}(1 - 4p^2\overline{V}_S^2)\frac{\Delta\rho}{\overline{\rho}} - \left(\frac{1}{2\cos^2\theta_S} - 4p^2\overline{V}_S^2\right)\frac{\Delta V_S}{\overline{V}_S} \tag{3.5.31}$$

$$T_{SP}(\theta) \approx -\frac{\cos\theta_S}{\overline{V}_P} \frac{\overline{V}_S}{\cos\theta} T_{PS} \tag{3.5.32}$$

$$T_{SS}(\theta) \approx 1 - \frac{1}{2}\frac{\Delta\rho}{\overline{\rho}} + \left(\frac{1}{2\cos^2\theta_S} - 1\right)\frac{\Delta V_S}{\overline{V}_S} \tag{3.5.33}$$

where

$$p = \frac{\sin\theta_1}{V_{P1}} = \frac{\sin\theta_{S1}}{V_{S1}} \qquad \theta = (\theta_2 + \theta_1)/2$$

$$\theta_S = (\theta_{S2} + \theta_{S1})/2$$
$$\Delta\rho = \rho_2 - \rho_1 \qquad \overline{\rho} = (\rho_2 + \rho_1)/2$$
$$\Delta V_P = V_{P2} - V_{P1} \qquad \overline{V}_P = (V_{P2} + V_{P1})/2$$
$$\Delta V_S = V_{S2} - V_{S1} \qquad \overline{V}_S = (V_{S2} + V_{S1})/2$$

Often, the mean P-wave angle θ is approximated as θ_1, the P-wave angle of incidence.

The result for P-wave reflectivity can be rewritten in the familiar form:

$$R_{PP}(\theta) \approx R_{P0} + B\sin^2\theta + C(\tan^2\theta - \sin^2\theta) \tag{3.5.34}$$

or

$$R_{PP}(\theta) \approx \frac{1}{2}\left(\frac{\Delta V_P}{\overline{V}_P} + \frac{\Delta\rho}{\overline{\rho}}\right) + \left[\frac{1}{2}\frac{\Delta V_P}{\overline{V}_P} - 2\frac{\overline{V}_S^2}{\overline{V}_P^2}\left(2\frac{\Delta V_S}{\overline{V}_S} + \frac{\Delta\rho}{\overline{\rho}}\right)\right]\sin^2\theta$$
$$+ \frac{1}{2}\frac{\Delta V_P}{\overline{V}_P}(\tan^2\theta - \sin^2\theta) \tag{3.5.35}$$

This form can be interpreted in terms of different angular ranges (Castagna, 1993). In the previous equations R_{P0} is the normal incidence reflection coefficient as expressed by

$$R_{P0} = \frac{I_{P2} - I_{P1}}{I_{P2} + I_{P1}} \approx \frac{\Delta I_P}{2I_P} \approx \frac{1}{2}\left(\frac{\Delta V_P}{\overline{V}_P} + \frac{\Delta\rho}{\overline{\rho}}\right) \tag{3.5.36}$$

The parameter B describes the variation at intermediate offsets and is often called the **AVO gradient**, and C dominates at far offsets near the critical angle.

Shuey (1985) presented a similar approximation where the AVO gradient is expressed in terms of the Poisson ratio v as follows:

$$R_{PP}(\theta_1) \approx R_{P0} + \left[ER_{P0} + \frac{\Delta v}{(1 - \overline{v})^2}\right]\sin^2\theta_1 + \frac{1}{2}\frac{\Delta V_P}{\overline{V}_P}(\tan^2\theta_1 - \sin^2\theta_1) \tag{3.5.37}$$

where

$$R_{P0} \approx \frac{1}{2}\left(\frac{\Delta V_P}{\overline{V}_P} + \frac{\Delta\rho}{\overline{\rho}}\right)$$

(3.5.38)

$$E = F - 2(1+F)\left(\frac{1-2\overline{v}}{1-\overline{v}}\right)$$

(3.5.39)

$$F = \frac{\Delta V_P/\overline{V}_P}{\Delta V_P/\overline{V}_P + \Delta\rho/\overline{\rho}}$$

(3.5.40)

and

$$\Delta v = v_2 - v_1$$

(3.5.41)

$$\overline{v} = (v_2 + v_1)/2$$

(3.5.42)

The coefficients E and F used here in Shuey's equation are not the same as those defined earlier in the solutions to the Zoeppritz equations.

Smith and Gidlow (1987) offered a further simplification to the Aki-Richards equation by removing the dependence on density using Gardner's equation (see Section 7.10) as follows:

$$\rho \propto V^{1/4}$$

(3.5.43)

giving

$$R_{PP}(\theta) \approx c\frac{\Delta V_P}{\overline{V}_P} + d\frac{\Delta V_S}{\overline{V}_S}$$

(3.5.44)

where

$$c = \frac{5}{8} - \frac{1}{2}\frac{\overline{V}_S^2}{\overline{V}_P^2}\sin^2\theta + \frac{1}{2}\tan^2\theta$$

(3.5.45)

$$d = -4\frac{\overline{V}_S^2}{\overline{V}_P^2}\sin^2\theta$$

(3.5.46)

Wiggins et al. (1983) showed that when $V_P \approx 2V_S$, the AVO gradient is approximately (Spratt et al., 1993)

$$B \approx R_{P0} - 2R_{S0}$$

(3.5.47)

given that the P and S normal incident reflection coefficients are

$$R_{P0} \approx \frac{1}{2}\left(\frac{\Delta V_P}{\overline{V}_P} + \frac{\Delta\rho}{\overline{\rho}}\right)$$

(3.5.48)

$$R_{S0} \approx \frac{1}{2}\left(\frac{\Delta V_S}{\overline{V}_S} + \frac{\Delta\rho}{\overline{\rho}}\right)$$

(3.5.49)

Hilterman (1989) suggested the following slightly modified form:

$$R_{PP}(\theta) \approx R_{P0} \cos^2\theta + PR \sin^2\theta \tag{3.5.50}$$

where R_{P0} is the normal incidence reflection coefficient and

$$PR = \frac{v_2 - v_1}{(1 - \bar{v})^2} \tag{3.5.51}$$

This modified form has the interpretation that the near-offset traces reveal the P-wave impedance, and the intermediate-offset traces image contrasts in Poisson ratio (Castagna, 1993).

Gray *et al.* (1999) derived linearized expressions for P–P reflectivity in terms of the angle of incidence, θ, and the contrast in bulk modulus, K, shear modulus, μ, and bulk density, ρ:

$$R(\theta) = \left(\frac{1}{4} - \frac{1}{3}\frac{\bar{V}_S^2}{\bar{V}_P^2} \right) (\sec^2\theta) \frac{\Delta K}{K} + \left(\frac{\bar{V}_S^2}{\bar{V}_P^2} \right) \left(\frac{1}{3}\sec^2\theta - 2\sin^2\theta \right) \frac{\Delta\mu}{\mu} + \left(\frac{1}{2} - \frac{1}{4}\sec^2\theta \right) \frac{\Delta\rho}{\rho} \tag{3.5.52}$$

Similarly, their expression in terms of Lame's coefficient, λ, shear modulus, and bulk density is

$$R(\theta) = \left(\frac{1}{4} - \frac{1}{2}\frac{\bar{V}_S^2}{\bar{V}_P^2} \right) (\sec^2\theta) \frac{\Delta\lambda}{\lambda} + \left(\frac{\bar{V}_S^2}{\bar{V}_P^2} \right) \left(\frac{1}{2}\sec^2\theta - 2\sin^2\theta \right) \frac{\Delta\mu}{\mu} + \left(\frac{1}{2} - \frac{1}{4}\sec^2\theta \right) \frac{\Delta\rho}{\rho} \tag{3.5.53}$$

In the same assumptions of small layer contrast and limited angle of incidence, we can write the linearized SV-to-SV reflection (Rüger, 2001):

$$R_{SV-iso}(\theta_S) \approx -\frac{1}{2}\frac{\Delta I_S}{\bar{I}_S} + \left(\frac{7}{2}\frac{\Delta V_S}{\bar{V}_S} + 2\frac{\Delta\rho}{\rho} \right) \sin^2\theta_S - \frac{1}{2}\frac{\Delta V_S}{\bar{V}_S} \sin^2\theta_S\tan^2\theta_S \tag{3.5.54}$$

where θ_S is the SV-wave phase angle of incidence and $I_S = \rho V_S$ is the shear impedance. Similarly, for SH-to-SH reflection (Rüger, 2001):

$$R_{SH}(\theta_S) = -\frac{1}{2}\frac{\Delta I_S}{\bar{I}_S} + \frac{1}{2}\left(\frac{\Delta V_S}{\bar{V}_S} \right) \tan^2\theta_S \tag{3.5.55}$$

where θ_S is the SH-wave phase angle of incidence.

For P-to-SV converted shear-wave reflection (Aki and Richards, 1980):

$$R_{PS}(\theta_S) \approx \frac{-\tan\theta_S}{2\bar{V}_S/\bar{V}_P} \left(1 - 2\sin^2\theta_S + 2\cos\theta_S\sqrt{\left(\frac{\bar{V}_S}{\bar{V}_P} \right)^2 - \sin^2\theta_S} \right) \frac{\Delta\rho}{\bar{\rho}}$$
$$+ \frac{\tan\theta_S}{2\bar{V}_S/\bar{V}_P} \left(4\sin^2\theta_S - 4\cos\theta_S\sqrt{\left(\frac{\bar{V}_S}{\bar{V}_P} \right)^2 - \sin^2\theta_S} \right) \frac{\Delta V_S}{\bar{V}_S} \tag{3.5.56}$$

where θ_S is the S-wave phase angle of reflection. This can be rewritten as (González, 2006)

$$R_{PS}(\theta) \approx \frac{-\sin\theta}{(\overline{V}_P/\overline{V}_S)\sqrt{(\overline{V}_P/\overline{V}_S)^2 - \sin^2\theta}}$$

$$\cdot \left\{ \left[\frac{1}{2}\left(\frac{\overline{V}_P}{\overline{V}_S}\right)^2 - \sin^2\theta + \cos\theta\sqrt{\left(\frac{\overline{V}_P}{\overline{V}_S}\right)^2 - \sin^2\theta} \right] \frac{\Delta\rho}{\overline{\rho}} \right.$$

$$\left. - \left[2\sin^2\theta - 2\cos\theta\sqrt{\left(\frac{\overline{V}_P}{\overline{V}_S}\right)^2 - \sin^2\theta} \right] \frac{\Delta V_S}{\overline{V}_S} \right\} \tag{3.5.57}$$

where θ is the P-wave phase angle of incidence.

For small angles, Duffaut *et al.* (2000) give the following expression for $R_{PS}(\theta)$:

$$R_{PS}(\theta) \approx \left[-\frac{1}{2}\left(1 + 2\frac{\overline{V}_S}{\overline{V}_P}\right)\frac{\Delta\rho}{\overline{\rho}} - 2\frac{\overline{V}_S}{\overline{V}_P}\frac{\Delta V_S}{\overline{V}_S} \right]\sin\theta$$

$$+ \frac{\overline{V}_S}{\overline{V}_P}\left[\left(\frac{\overline{V}_S}{\overline{V}_P} + \frac{1}{2}\right)\left(\frac{\Delta\rho}{\overline{\rho}} + \frac{2\Delta V_S}{\overline{V}_S}\right) - \frac{1}{4}\frac{\overline{V}_S}{\overline{V}_P}\frac{\Delta\rho}{\overline{\rho}} \right]\sin^3\theta \tag{3.5.58}$$

which can be simplified further (Jílek, 2002b) to

$$R_{PS}(\theta) \approx \left[-\frac{1}{2}\left(1 + 2\frac{\overline{V}_S}{\overline{V}_P}\right)\frac{\Delta\rho}{\rho} - 2\frac{\overline{V}_S}{\overline{V}_P}\frac{\Delta V_S}{\overline{V}_S} \right]\sin\theta \tag{3.5.59}$$

Assumptions and Limitations

The equations presented in this section apply in the following cases:
- the rock is linear, isotropic, and elastic;
- plane-wave propagation is assumed; and
- most of the simplified forms assume small contrasts in material properties across the boundary and angles of incidence of less than about 30°. The simplified form for P-to-S reflection given by González (2006) is valid for large angles of incidence.

3.6   Plane-Wave Reflectivity in Anisotropic Media

Synopsis

An incident wave at a boundary between two anisotropic media (Figure 3.6.1) can generate reflected quasi-P-waves and quasi-S-waves as well as transmitted quasi-P-waves and quasi-S-waves (Auld, 1990). In general, the reflection and transmission

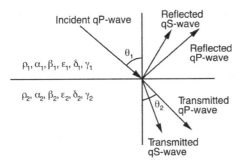

Figure 3.6.1 Reflected and transmitted rays caused by a P-wave incident at a boundary between two anisotropic media

coefficients vary with offset and azimuth. The **AVOA** (amplitude variation with offset and azimuth) can be detected by three-dimensional seismic surveys and is a useful seismic attribute for reservoir characterization.

Brute-force modeling of AVOA by solving the Zoeppritz (1919) equations can be complicated and unintuitive for several reasons: for anisotropic media in general, the two shear waves are separate (shear-wave birefringence); the slowness surfaces are nonspherical and are not necessarily convex; and the polarization vectors are neither parallel nor perpendicular to the propagation vectors.

Schoenberg and Protázio (1992) give explicit solutions for the plane-wave reflection and transmission problem in terms of submatrices of the coefficient matrix of the Zoeppritz equations. The most general case of the explicit solutions is applicable to monoclinic media with a mirror plane of symmetry parallel to the reflecting plane. Let **R** and **T** represent the **reflection** and **transmission matrices**, respectively,

$$\mathbf{R} = \begin{bmatrix} R_{\mathrm{PP}} & R_{\mathrm{SP}} & R_{\mathrm{TP}} \\ R_{\mathrm{PS}} & R_{\mathrm{SS}} & R_{\mathrm{TS}} \\ R_{\mathrm{PT}} & R_{\mathrm{ST}} & R_{\mathrm{TT}} \end{bmatrix} \tag{3.6.1}$$

$$\mathbf{T} = \begin{bmatrix} T_{\mathrm{PP}} & T_{\mathrm{SP}} & T_{\mathrm{TP}} \\ T_{\mathrm{PS}} & T_{\mathrm{SS}} & T_{\mathrm{TS}} \\ T_{\mathrm{PT}} & T_{\mathrm{ST}} & T_{\mathrm{TT}} \end{bmatrix} \tag{3.6.2}$$

where the first subscript denotes the type of incident wave and the second subscript denotes the type of reflected or transmitted wave. For "weakly" anisotropic media, the subscript P denotes the P-wave, S denotes one quasi-S-wave, and T denotes the other quasi-S-wave (i.e., the tertiary or third wave). As a convention for real $s_{3\mathrm{P}}^2$, $s_{3\mathrm{S}}^2$, and $s_{3\mathrm{T}}^2$,

$$s_{3\mathrm{P}}^2 < s_{3\mathrm{S}}^2 < s_{3\mathrm{T}}^2 \tag{3.6.3}$$

where s_{3i} is the vertical component of the phase slowness of the ith wave type when the reflecting plane is horizontal. An imaginary value for any of the vertical slownesses implies that the corresponding wave is inhomogeneous or evanescent. The impedance matrices are defined as

$$\mathbf{X} = \begin{bmatrix} e_{P1} & e_{S1} & e_{T1} \\ e_{P2} & e_{S2} & e_{T2} \\ \{-(C_{13}e_{P1}+C_{36}e_{P2})s_1 & \{-(C_{13}e_{S1}+C_{36}e_{S2})s_1 & \{-(C_{13}e_{T1}+C_{36}e_{T2})s_1 \\ -(C_{23}e_{P2}+C_{36}e_{P1})s_2 & -(C_{23}e_{S2}+C_{36}e_{S1})s_2 & -(C_{23}e_{T2}+C_{36}e_{T1})s_2 \\ -C_{33}e_{P3}s_{3P}\} & -C_{33}e_{S3}s_{3S}\} & -C_{33}e_{T3}s_{3T}\} \end{bmatrix}$$

(3.6.4)

$$\mathbf{Y} = \begin{bmatrix} \{-(C_{55}s_1+C_{45}s_2)e_{P3} & \{-(C_{55}s_1+C_{45}s_2)e_{S3} & \{-(C_{55}s_1+C_{45}s_2)e_{T3} \\ -(C_{55}e_{P1}+C_{45}e_{P2})s_{3P}\} & -(C_{55}e_{S1}+C_{45}e_{S2})s_{3S}\} & -(C_{55}e_{T1}+C_{45}e_{T2})s_{3T}\} \\ \{-(C_{45}s_1+C_{44}s_2)e_{P3} & \{-(C_{45}s_1+C_{44}s_2)e_{S3} & \{-(C_{45}s_1+C_{44}s_2)e_{T3} \\ -(C_{45}e_{P1}+C_{44}e_{P2})s_{3P}\} & -(C_{45}e_{S1}+C_{44}e_{S2})s_{3S}\} & -(C_{45}e_{T1}+C_{44}e_{T2})s_{3T}\} \\ e_{P3} & e_{S3} & e_{T3} \end{bmatrix}$$

(3.6.5)

where s_1 and s_2 are the horizontal components of the phase slowness vector; $\mathbf{e}_P$, $\mathbf{e}_S$, and $\mathbf{e}_T$ are the associated eigenvectors evaluated from the Christoffel equations (see Section 3.2), and C_{IJ} denotes elements of the stiffness matrix of the incident medium. $\mathbf{X}'$ and $\mathbf{Y}'$ are the same as previously given except that primed parameters (transmission medium) replace unprimed parameters (incidence medium). When neither $\mathbf{X}$ nor $\mathbf{Y}$ is singular and $(\mathbf{X}^{-1}\mathbf{X}' + \mathbf{Y}^{-1}\mathbf{Y}')$ is invertible, the reflection and transmission coefficients can be written as

$$\mathbf{T} = 2(\mathbf{X}^{-1}\mathbf{X}' + \mathbf{Y}^{-1}\mathbf{Y}')^{-1} \tag{3.6.6}$$

$$\mathbf{R} = (\mathbf{X}^{-1}\mathbf{X}' - \mathbf{Y}^{-1}\mathbf{Y}')(\mathbf{X}^{-1}\mathbf{X}' + \mathbf{Y}^{-1}\mathbf{Y}')^{-1} \tag{3.6.7}$$

Schoenberg and Protázio (1992) point out that a singularity occurs at a horizontal slowness for which an interface wave (e.g., a Stoneley wave) exists. When $\mathbf{Y}$ is singular, straightforward matrix manipulations yield

$$\mathbf{T} = 2\mathbf{Y}'^{-1}\mathbf{Y}(\mathbf{X}^{-1}\mathbf{X}'\mathbf{Y}'^{-1}\mathbf{Y} + \mathbf{I})^{-1} \tag{3.6.8}$$

$$\mathbf{R} = (\mathbf{X}^{-1}\mathbf{X}'\mathbf{Y}'^{-1}\mathbf{Y} + \mathbf{I})^{-1}(\mathbf{X}^{-1}\mathbf{X}'\mathbf{Y}'^{-1}\mathbf{Y} - \mathbf{I}) \tag{3.6.9}$$

Similarly, $\mathbf{T}$ and $\mathbf{R}$ can also be written without $\mathbf{X}^{-1}$ when $\mathbf{X}$ is singular as

$$\mathbf{T} = 2\mathbf{X}'^{-1}\mathbf{X}(\mathbf{I} + \mathbf{Y}^{-1}\mathbf{Y}'\mathbf{X}'^{-1}\mathbf{X})^{-1} \tag{3.6.10}$$

$$\mathbf{R} = (\mathbf{I} + \mathbf{Y}^{-1}\mathbf{Y}'\mathbf{X}'^{-1}\mathbf{X})^{-1}(\mathbf{I} - \mathbf{Y}^{-1}\mathbf{Y}'\mathbf{X}'^{-1}\mathbf{X}) \tag{3.6.11}$$

Alternative solutions can be found by assuming that $\mathbf{X}'$ and $\mathbf{Y}'$ are invertible

$$\mathbf{R} = (\mathbf{Y'}^{-1}\mathbf{Y} + \mathbf{X'}^{-1}\mathbf{X})^{-1}(\mathbf{Y'}^{-1}\mathbf{Y} - \mathbf{X'}^{-1}\mathbf{X}) \tag{3.6.12}$$

$$\begin{aligned}\mathbf{T} &= 2\mathbf{X'}^{-1}\mathbf{X}(\mathbf{Y'}^{-1}\mathbf{Y} + \mathbf{X'}^{-1}\mathbf{X})^{-1}\mathbf{Y'}^{-1}\mathbf{Y} \\ &= 2\mathbf{Y'}^{-1}\mathbf{Y}(\mathbf{Y'}^{-1}\mathbf{Y} + \mathbf{X'}^{-1}\mathbf{X})^{-1}\mathbf{X'}^{-1}\mathbf{X}\end{aligned} \tag{3.6.13}$$

These formulas allow more straightforward calculations when the media have at least monoclinic symmetry with a horizontal symmetry plane.

For a wave traveling in anisotropic media, there will generally be out-of-plane motion unless the wave path is in a symmetry plane. These symmetry planes include all vertical planes in VTI (transversely isotropic with vertical symmetry axis) media and the symmetry planes in HTI (transversely isotropic with horizontal symmetry axis) and orthorhombic media. In this case, the quasi-P- and the quasi-S-waves in the symmetry plane uncouple from the quasi-S-wave polarized transversely to the symmetry plane. For weakly anisotropic media, we can use simple analytical formulas (Banik, 1987; Thomsen, 1993; Chen, 1995; Rüger, 1995, 1996) to compute AVOA (amplitude variation with offset and azimuth) responses at the interface of anisotropic media that can be either VTI, HTI, or orthorhombic. The analytical formulas give more insight into the dependence of AVOA on anisotropy. Vavryčuk and Pšenčík (1998) and Pšenčík and Martins (2001) provide formulas for arbitrary weak anisotropy.

Transversely Isotropic Media: VTI

Thomsen (1986) introduced the following notation for weakly anisotropic VTI media with density ρ:

$$\alpha = \sqrt{\frac{C_{33}}{\rho}}; \quad \varepsilon = \frac{C_{11} - C_{33}}{2C_{33}} \tag{3.6.14}$$

$$\beta = \sqrt{\frac{C_{44}}{\rho}}; \quad \gamma = \frac{C_{66} - C_{44}}{2C_{44}} \tag{3.6.15}$$

$$\delta = \frac{(C_{13} + C_{44})^2 - (C_{33} - C_{44})^2}{2C_{33}(C_{33} - C_{44})} \tag{3.6.16}$$

In this chapter we use $\alpha = \sqrt{C_{33}/\rho}$ to refer to the P-wave velocity along the x_3-(vertical) axis in anisotropic media. We use β to refer to the S-wave velocity along the x_3-axis. The P-wave reflection coefficient for weakly anisotropic VTI media in the limit of small impedance contrast was derived by Thomsen (1993) and corrected by Rüger (1997):

$$R_{PP}(\theta) = R_{PP-iso}(\theta) + R_{PP-aniso}(\theta) \tag{3.6.17}$$

$$R_{\text{PP-iso}}(\theta) \approx \frac{1}{2}\left(\frac{\Delta Z}{\overline{Z}}\right) + \frac{1}{2}\left[\frac{\Delta\alpha}{\overline{\alpha}} - \left(\frac{2\overline{\beta}}{\overline{\alpha}}\right)^2 \frac{\Delta\mu}{\overline{\mu}}\right]\sin^2\theta + \frac{1}{2}\left(\frac{\Delta\alpha}{\overline{\alpha}}\right)\sin^2\theta\tan^2\theta \quad (3.6.18)$$

$$R_{\text{PP-aniso}}(\theta) \approx \frac{\Delta\delta}{2}\sin^2\theta + \frac{\Delta\varepsilon}{2}\sin^2\theta\tan^2\theta \quad (3.6.19)$$

where

$$\theta = (\theta_1 + \theta_2)/2 \qquad \Delta\varepsilon = \varepsilon_2 - \varepsilon_1$$

$$\overline{\rho} = (\rho_1 + \rho_2)/2 \qquad \Delta\rho = \rho_2 - \rho_1 \qquad \Delta\gamma = \gamma_2 - \gamma_1$$

$$\overline{\alpha} = (\alpha_1 + \alpha_2)/2 \qquad \Delta\alpha = \alpha_2 - \alpha_1 \qquad \Delta\delta = \delta_2 - \delta_1$$

$$\overline{\beta} = (\beta_1 + \beta_2)/2 \qquad \Delta\beta = \beta_2 - \beta_1 \qquad \mu = \rho\beta^2$$

$$\overline{Z} = (Z_1 + Z_2)/2 \qquad \Delta Z = Z_2 - Z_1 \qquad Z = \rho\alpha$$

$$\overline{Z}_S = (Z_{S1} + Z_{S2})/2 \qquad \Delta Z_S = Z_{S2} - Z_{S1} \qquad Z_S = \rho\beta$$

$$\overline{\mu} = (\mu_1 + \mu_2)/2 \qquad \Delta\mu = \mu_2 - \mu_1$$

In the preceding and following equations, Δ indicates a difference in properties between layer 1 (above) and layer 2 (below) and an overbar indicates an average of the corresponding quantities in the two layers. θ is the P-wave phase angle of incidence, $Z = \rho\alpha$ is the vertical P-wave impedance, $Z_S = \rho\beta$ is the vertical S-wave impedance, and $\mu = \rho\beta^2$ is the vertical shear modulus. Note that the VTI anisotropy causes perturbations of both the AVO gradient and far-offset terms.

Rüger (2001) derived the corresponding SV-to-SV reflectivity in VTI media, again under the assumption of weak anisotropy and small-layer contrast:

$$R_{\text{SV}}(\theta_S) = R_{\text{SV-iso}}(\theta_S) + R_{\text{SV-aniso}}(\theta_S) \quad (3.6.20)$$

$$R_{\text{SV-iso}}(\theta_S) \approx -\frac{1}{2}\frac{\Delta Z_S}{\overline{Z}_S} + \left(\frac{7}{2}\frac{\Delta\beta}{\overline{\beta}} + 2\frac{\Delta\rho}{\overline{\rho}}\right)\sin^2\theta_S - \frac{1}{2}\frac{\Delta\beta}{\overline{\beta}}\tan^2\theta_S\sin^2\theta_S \quad (3.6.21)$$

$$R_{\text{SV-aniso}}(\theta_S) = \frac{1}{2}\frac{\overline{\alpha}^2}{\overline{\beta}^2}(\Delta\varepsilon - \Delta\delta)\sin^2\theta_S \quad (3.6.22)$$

In the S-wave reflectivity equations, note that θ_S is the S-wave angle of incidence and $Z_S = \rho\beta$ is the shear impedance. The weak anisotropy, weak contrast SH-to-SH reflectivity is (Rüger, 2001)

$$R_{\text{SH}}(\theta_S) = -\frac{1}{2}\frac{\Delta Z_S}{\overline{Z}_S} + \frac{1}{2}\left(\frac{\Delta\beta}{\overline{\beta}} + \Delta\gamma\right)\tan^2\theta_S \quad (3.6.23)$$

The P–S converted wave reflectivity in VTI media is given by (Rüger, 2001)

$$R_{\text{PS}}(\theta_S) = R_{\text{PS-iso}}(\theta_S) + R_{\text{PS-aniso}}(\theta_S) \quad (3.6.24)$$

$$R_{PS-iso}(\theta_S) \approx -\frac{1}{2}\frac{\Delta\rho}{\overline{\rho}}\frac{\sin\theta_P}{\sin\theta_S} - \frac{\overline{\beta}}{\overline{\alpha}}\left(2\frac{\Delta\beta}{\overline{\beta}} + \frac{\Delta\rho}{\overline{\rho}}\right)\sin\theta_P \cos\theta_P$$

$$+ \left(\frac{\overline{\beta}}{\overline{\alpha}}\right)^2\left(2\frac{\Delta\beta}{\overline{\beta}} + \frac{\Delta\rho}{\overline{\rho}}\right)\frac{\sin^3\theta_P}{\cos\theta_S}$$

(3.6.25)

$$R_{PS-aniso}(\theta_S) = \left[\left(\frac{\overline{\alpha}^2}{2(\overline{\alpha}^2-\overline{\beta}^2)\cos\theta_S} - \frac{\overline{\alpha}\overline{\beta}\cos\theta_P}{2(\overline{\alpha}^2-\overline{\beta}^2)}\right)(\delta_2-\delta_1)\right]\sin\theta_P$$

$$+ \left[\frac{\overline{\alpha}\overline{\beta}\cos\theta_P}{(\overline{\alpha}^2-\overline{\beta}^2)}(\delta_2-\delta_1+\varepsilon_1-\varepsilon_2)\right]\sin^3\theta_P$$

$$- \left[\frac{\overline{\alpha}^2}{(\overline{\alpha}^2-\overline{\beta}^2)\cos\theta_S}(\delta_2-\delta_1+\varepsilon_1-\varepsilon_2)\right]\sin^3\theta_P$$ (3.6.26)

$$+ \left[\frac{\overline{\beta}^2}{2(\overline{\alpha}^2-\overline{\beta}^2)\cos\theta_S}(\delta_1-\delta_2)\right]\sin^3\theta_P$$

$$+ \left[\frac{\overline{\beta}^2}{(\overline{\alpha}^2-\overline{\beta}^2)\cos\theta_S}(\delta_2-\delta_1+\varepsilon_1-\varepsilon_2)\right]\sin^5\theta_P$$

Transversely Isotropic Media: HTI

In HTI (transversely isotropic with horizontal symmetry axis) media, reflectivity will vary with azimuth, ζ, as well as offset or incident angle θ. Rüger (1995, 1996) and Chen (1995) derived the P-wave reflection coefficient in the symmetry planes for reflections at the boundary of two HTI media sharing the same symmetry axis. At a horizontal interface between two HTI media with horizontal symmetry axis x_1 and vertical axis x_3, the P-wave reflectivity for propagation in the vertical symmetry axis plane (x_1–x_3 plane) parallel to the x_1 symmetry axis can be written as

$$R_{PP}(\zeta = 0, \theta) \approx \frac{1}{2}\frac{\Delta Z}{\overline{Z}} + \frac{1}{2}\left[\frac{\Delta\alpha}{\overline{\alpha}} - \left(\frac{2\overline{\beta}^\perp}{\overline{\alpha}}\right)^2\frac{\Delta\mu^\perp}{\overline{\mu}^\perp} + \Delta\delta^{(V)}\right]$$

$$\times \sin^2\theta + \frac{1}{2}\left[\frac{\Delta\alpha}{\overline{\alpha}} + \Delta\varepsilon^{(V)}\right]\sin^2\theta\tan^2\theta$$

(3.6.27)

$$\approx \frac{1}{2}\frac{\Delta Z}{\overline{Z}} + \frac{1}{2}\left[\frac{\Delta\alpha}{\overline{\alpha}} - \left(\frac{2\overline{\beta}}{\overline{\alpha}}\right)^2\left(\frac{\Delta\mu}{\overline{\mu}} - 2\Delta\gamma\right) + \Delta\delta^{(V)}\right]$$

$$\times \sin^2\theta + \frac{1}{2}\left[\frac{\Delta\alpha}{\overline{\alpha}} + \Delta\varepsilon^{(V)}\right]\sin^2\theta\tan^2\theta$$

where the azimuth angle ζ is measured from the x_1-axis and the incident angle θ is defined with respect to x_3. The isotropic part $R_{PP-iso}(\theta)$ is the same as before. In the preceding expression

$$\alpha = \sqrt{\frac{C_{33}}{\rho}};$$

$$\varepsilon^{(V)} = \frac{C_{11} - C_{33}}{2C_{33}}$$

$$\beta = \sqrt{\frac{C_{44}}{\rho}};$$

$$\delta^{(V)} = \frac{(C_{13} + C_{55})^2 - (C_{33} - C_{55})^2}{2C_{33}(C_{33} - C_{55})}$$

$$\beta^{\perp} = \sqrt{\frac{C_{55}}{\rho}};$$

$$\gamma^{(V)} = \frac{C_{66} - C_{44}}{2C_{44}} = -\frac{\gamma}{1 + 2\gamma}$$

$$\gamma = \frac{C_{44} - C_{66}}{2C_{66}};$$

where β is the velocity of a vertically propagating shear wave polarized in the x_2 direction and $\beta^{\perp}$ is the velocity of a vertically propagating shear wave polarized in the x_1 direction. $\mu = c_{44}$ and $\mu^{\perp} = c_{55}$ are the shear moduli corresponding to the shear velocities β and $\beta^{\perp}$. In the vertical symmetry plane (x_2–x_3 plane), perpendicular to the symmetry axis, the P-wave reflectivity resembles the isotropic solution:

$$R_{PP}\left(\zeta = \frac{\pi}{2}, \theta\right) \approx \frac{1}{2}\left(\frac{\Delta Z}{\overline{Z}}\right) + \frac{1}{2}\left[\frac{\Delta \alpha}{\overline{\alpha}} - \left(\frac{2\overline{\beta}}{\overline{\alpha}}\right)^2 \frac{\Delta \mu}{\overline{\mu}}\right]\sin^2\theta + \frac{1}{2}\left(\frac{\Delta \alpha}{\overline{\alpha}}\right)\sin^2\theta\tan^2\theta \qquad (3.6.28)$$

In nonsymmetry planes, Rüger (1996) derived the P-wave reflectivity $R_{PP}(\zeta, \theta)$ using a perturbation technique as follows:

$$R_{PP}(\zeta, \theta) \approx \frac{1}{2}\frac{\Delta Z}{\overline{Z}} + \frac{1}{2}\left[\frac{\Delta \alpha}{\overline{\alpha}} - \left(\frac{2\overline{\beta}}{\overline{\alpha}}\right)^2 \frac{\Delta \mu}{\overline{\mu}} + \left(\Delta \delta^{(V)} + 2\left(\frac{2\overline{\beta}}{\overline{\alpha}}\right)^2 \Delta \gamma\right)\cos^2\zeta\right]\sin^2\theta$$
$$+ \frac{1}{2}\left[\frac{\Delta \alpha}{\overline{\alpha}} + \Delta \varepsilon^{(V)} \cos^4\zeta + \Delta \delta^{(V)} \sin^2\zeta \cos^2\zeta\right]\sin^2\theta\tan^2\theta \qquad (3.6.29)$$

where ζ is the azimuth relative to the x_1-axis.

The reflectivity for SV- and SH-waves in the symmetry axis plane (x_1–x_3 plane) of the HTI medium is given by (Rüger, 2001)

$$R_{SV}(\theta_S) \approx R_{SV-iso}(\theta_S) + \frac{1}{2}\left(\frac{\overline{\alpha}}{\overline{\beta}^{\perp}}\right)^2 (\Delta\varepsilon^{(V)} - \Delta\delta^{(V)})\sin^2\theta_S \qquad (3.6.30)$$

$$R_{SH}(\theta_S) \approx R_{SH-iso}(\theta_S) + \frac{1}{2}\Delta\gamma^{(V)}\tan^2\theta_S \qquad (3.6.31)$$

where

$$R_{SV-iso}(\theta_S) \approx -\frac{1}{2}\frac{\Delta Z_S^{\perp}}{\overline{Z}_S^{\perp}} + \left(\frac{7}{2}\frac{\Delta\beta^{\perp}}{\overline{\beta}^{\perp}} + 2\frac{\Delta\rho}{\overline{\rho}}\right)\sin^2\theta_S - \frac{1}{2}\frac{\Delta\beta^{\perp}}{\overline{\beta}^{\perp}}\tan^2\theta_S \sin^2\theta_S \qquad (3.6.32)$$

$$R_{SH-iso}(\theta_S) = -\frac{1}{2}\frac{\Delta Z_S}{\overline{Z}_S} + \frac{1}{2}\left(\frac{\Delta\beta}{\overline{\beta}}\right)\tan^2\theta_S \qquad (3.6.33)$$

For waves propagating in the (x_2–x_3) plane

$$R_{SV}(\theta_S) \approx -\frac{1}{2}\frac{\Delta Z_S}{\overline{Z}_S} + \left(\frac{7}{2}\frac{\Delta\beta}{\overline{\beta}} + 2\frac{\Delta\rho}{\overline{\rho}}\right)\sin^2\theta_S - \frac{1}{2}\frac{\Delta\beta}{\overline{\beta}}\tan^2\theta_S \sin^2\theta_S \tag{3.6.34}$$

$$R_{SH}(\theta_S) = -\frac{1}{2}\frac{\Delta Z_S^\perp}{\overline{Z}_S^\perp} + \frac{1}{2}\left(\frac{\Delta\beta^\perp}{\overline{\beta}^\perp}\right)\tan^2\theta_S \tag{3.6.35}$$

Orthorhombic Media

The anisotropic parameters in orthorhombic media are given by Chen (1995) and Tsvankin (1997) as follows:

$$\alpha = \sqrt{\frac{C_{33}}{\rho}}; \quad \beta = \sqrt{\frac{C_{44}}{\rho}}; \quad \beta^\perp = \sqrt{\frac{C_{55}}{\rho}} \tag{3.6.36}$$

$$\varepsilon^{(1)} = \frac{C_{22} - C_{33}}{2C_{33}}; \quad \varepsilon^{(2)} = \frac{C_{11} - C_{33}}{2C_{33}} \tag{3.6.37}$$

$$\delta^{(1)} = \frac{(C_{23} + C_{44})^2 - (C_{33} - C_{44})^2}{2C_{33}(C_{33} - C_{44})} \tag{3.6.38}$$

$$\delta^{(2)} = \frac{(C_{13} + C_{55})^2 - (C_{33} - C_{55})^2}{2C_{33}(C_{33} - C_{55})} \tag{3.6.39}$$

$$\gamma = \frac{C_{44} - C_{55}}{2C_{55}}; \quad \gamma^{(1)} = \frac{C_{66} - C_{55}}{2C_{55}}; \quad \gamma^{(2)} = \frac{C_{66} - C_{44}}{2C_{44}} \tag{3.6.40}$$

The parameters $\varepsilon^{(1)}$ and $\delta^{(1)}$ are Thomsen's parameters for the equivalent TIV media in the x_2–x_3 plane. Similarly, $\varepsilon^{(2)}$ and $\delta^{(2)}$ are Thomsen's parameters for the equivalent TIV media in the x_1–x_3 plane; γ_2 represents the velocity anisotropy between two shear-wave modes traveling along the z-axis. The difference in the approximate P-wave reflection coefficient in the two vertical symmetry planes (with x_3 as the vertical axis) of orthorhombic media is given by Rüger (1995, 1996) in the following form:

$$R_{PP}^{(x_1,x_3)}(\theta) - R_{PP}^{(x_2,x_3)}(\theta) \approx \left[\frac{\Delta\delta^{(2)} - \Delta\delta^{(1)}}{2} + \left(\frac{2\overline{\beta}}{\overline{\alpha}}\right)^2 \Delta\gamma\right]\sin^2\theta + \left[\frac{\Delta\varepsilon^{(2)} - \Delta\varepsilon^{(1)}}{2}\right]\sin^2\theta\tan^2\theta \tag{3.6.41}$$

The equations are good approximations for angles of incidence up to 30°–40°.

Rüger (2001) gives the following expressions for P-wave reflectivity in the orthorhombic symmetry planes:

$$R_{PP}^{(x_1,x_3)}(\theta) \approx \frac{1}{2}\frac{\Delta Z}{\overline{Z}} + \frac{1}{2}\left[\frac{\Delta\alpha}{\overline{\alpha}} - \left(\frac{2\overline{\beta}^\perp}{\overline{\alpha}}\right)^2\frac{\Delta\mu^\perp}{\overline{\mu}^\perp} + \Delta\delta^{(2)}\right]\sin^2\theta + \frac{1}{2}\left(\frac{\Delta\alpha}{\overline{\alpha}} + \Delta\varepsilon^{(2)}\right)\sin^2\theta\tan^2\theta \tag{3.6.42}$$

$$R_{PP}^{(x_2,x_3)}(\theta) \approx \frac{1}{2}\frac{\Delta Z}{\overline{Z}} + \frac{1}{2}\left[\frac{\Delta\alpha}{\overline{\alpha}} - \left(\frac{2\overline{\beta}}{\overline{\alpha}}\right)^2\frac{\Delta\mu}{\overline{\mu}} + \Delta\delta^{(1)}\right]\sin^2\theta + \frac{1}{2}\left(\frac{\Delta\alpha}{\overline{\alpha}} + \Delta\varepsilon^{(1)}\right)\sin^2\theta\tan^2\theta \tag{3.6.43}$$

Jílek (2002a, 2002b) derived expressions for P-to-SV reflectivity for arbitrary weak anisotropy and weak layer contrasts. In the case of orthorhombic and higher symmetries with aligned vertical symmetry planes, the expressions for P–SV in the vertical symmetry planes simplify to

$$R_{\text{P-SV}} = B_1 \frac{\sin\phi_{\text{P}}}{\cos\phi_{\text{S}}} + B_2 \cos\phi_{\text{P}} \sin\phi_{\text{P}} + B_3 \frac{\sin^3\phi_{\text{P}}}{\cos\phi_{\text{S}}} + B_4 \cos\phi_{\text{P}} \sin^3\phi_{\text{P}} + B_5 \frac{\sin^5\phi_{\text{P}}}{\cos\phi_{\text{S}}} \qquad (3.6.44)$$

where ϕ_{P} is the P-wave incident phase angle and ϕ_{S} is the converted S-wave reflection phase angle.

In the x–z symmetry plane (the z-axis is vertical), the coefficients B_i are

$$B_1 = -\frac{1}{2}\frac{\Delta\rho}{\bar{\rho}} + \frac{\bar{\alpha}^2}{2(\bar{\alpha}^2 - \bar{\beta}^2)}\left(\tilde{\delta}_2^{(2)} - \tilde{\delta}_1^{(2)}\right) \qquad (3.6.45)$$

$$B_2 = -\frac{\bar{\beta}}{\bar{\alpha}}\frac{\Delta\rho}{\bar{\rho}} - 2\frac{\bar{\beta}}{\bar{\alpha}}\frac{\Delta\beta}{\bar{\beta}} - \frac{\bar{\alpha}\bar{\beta}}{2(\bar{\alpha}^2 - \bar{\beta}^2)}\left(\tilde{\delta}_2^{(2)} - \tilde{\delta}_1^{(2)}\right) \qquad (3.6.46)$$

$$B_3 = \frac{\bar{\beta}^2}{\bar{\alpha}^2}\frac{\Delta\rho}{\bar{\rho}} + 2\frac{\bar{\beta}^2}{\bar{\alpha}^2}\frac{\Delta\beta}{\bar{\beta}} + \frac{\bar{\beta}^2}{2(\bar{\alpha}^2 - \bar{\beta}^2)}\left(\tilde{\delta}_1^{(2)} - \tilde{\delta}_2^{(2)}\right)$$

$$+ \frac{\bar{\alpha}^2}{(\bar{\alpha}^2 - \bar{\beta}^2)}\left[\left(\varepsilon_2^{(2)} - \varepsilon_1^{(2)}\right) + \left(\tilde{\delta}_1^{(2)} - \tilde{\delta}_2^{(2)}\right)\right] \qquad (3.6.47)$$

$$B_4 = -\frac{\bar{\alpha}\bar{\beta}}{(\bar{\alpha}^2 - \bar{\beta}^2)}\left[\left(\varepsilon_2^{(2)} - \varepsilon_1^{(2)}\right) + \left(\tilde{\delta}_1^{(2)} - \tilde{\delta}_2^{(2)}\right)\right] \qquad (3.6.48)$$

$$B_5 = -\frac{\bar{\beta}^2}{(\bar{\alpha}^2 - \bar{\beta}^2)}\left[\left(\varepsilon_2^{(2)} - \varepsilon_1^{(2)}\right) + \left(\tilde{\delta}_1^{(2)} - \tilde{\delta}_2^{(2)}\right)\right] \qquad (3.6.49)$$

and in the y–z symmetry plane, the coefficients are

$$B_1 = -\frac{1}{2}\frac{\Delta\rho}{\bar{\rho}} + \frac{\bar{\alpha}^2}{2(\bar{\alpha}^2 - \bar{\beta}^2)}\left(\tilde{\delta}_2^{(1)} - \tilde{\delta}_1^{(1)}\right) \qquad (3.6.50)$$

$$B_2 = -\frac{\bar{\beta}}{\bar{\alpha}}\frac{\Delta\rho}{\bar{\rho}} - 2\frac{\bar{\beta}}{\bar{\alpha}}\frac{\Delta\beta}{\bar{\beta}} - \frac{\bar{\alpha}\bar{\beta}}{2(\bar{\alpha}^2 - \bar{\beta}^2)}\left(\tilde{\delta}_2^{(1)} - \tilde{\delta}_1^{(1)}\right) - 2\frac{\bar{\beta}}{\bar{\alpha}}\left(\gamma_2^{(S)} - \gamma_1^{(S)}\right) \qquad (3.6.51)$$

$$B_3 = \frac{\bar{\beta}^2}{\bar{\alpha}^2}\frac{\Delta\rho}{\bar{\rho}} + 2\frac{\bar{\beta}^2}{\bar{\alpha}^2}\frac{\Delta\beta}{\bar{\beta}} + \frac{\bar{\beta}^2}{2(\bar{\alpha}^2 - \bar{\beta}^2)}\left(\tilde{\delta}_1^{(1)} - \tilde{\delta}_2^{(1)}\right)$$

$$+ \frac{\bar{\alpha}^2}{(\bar{\alpha}^2 - \bar{\beta}^2)}\left[\left(\varepsilon_2^{(1)} - \varepsilon_1^{(1)}\right) + \left(\tilde{\delta}_1^{(1)} - \tilde{\delta}_2^{(1)}\right) + 2\frac{\bar{\beta}^2}{\bar{\alpha}^2}\left(\gamma_2^{(S)} - \gamma_1^{(S)}\right)\right] \qquad (3.6.52)$$

$$B_4 = -\frac{\overline{\alpha}\overline{\beta}}{(\overline{\alpha}^2 - \overline{\beta}^2)}\left[\left(\varepsilon_2^{(1)} - \varepsilon_1^{(1)}\right) + \left(\widetilde{\delta}_1^{(1)} - \widetilde{\delta}_2^{(1)}\right)\right] \tag{3.6.53}$$

$$B_5 = -\frac{\overline{\beta}^2}{(\overline{\alpha}^2 - \overline{\beta}^2)}\left[\left(\varepsilon_2^{(1)} - \varepsilon_1^{(1)}\right) + \left(\widetilde{\delta}_1^{(1)} - \widetilde{\delta}_2^{(1)}\right)\right] \tag{3.6.54}$$

In these expressions, a subscript 1 on $\varepsilon_1^{(m)}, \widetilde{\delta}_1^{(m)}$ and $\gamma_1^{(S)}$ refers to properties of the incident layer and a subscript 2 on $\varepsilon_2^{(m)}, \widetilde{\delta}_2^{(m)}$ and $\gamma_2^{(S)}$ refers to properties of the reflecting layer. "Background" isotropic P- and S-wave velocities are defined within each layer ($I = 1, 2$ refer to the incident and reflecting layers, respectively) and can be taken to be equal to the actual vertical velocities within those layers:

$$\alpha_I = \sqrt{A_{33}^{(I)}}; \quad \beta_I = \sqrt{A_{55}^{(I)}} \tag{3.6.55}$$

In addition, the overall isotropic "background" P-wave velocity, S-wave velocity, and density are defined as averages over the two layers:

$$\overline{\alpha} = \frac{1}{2}(\alpha_1 + \alpha_2); \quad \overline{\beta} = \frac{1}{2}(\beta_1 + \beta_2); \quad \overline{\rho} = \frac{1}{2}(\rho_1 + \rho_2) \tag{3.6.56}$$

and contrasts across the boundary are defined as

$$\Delta\alpha = \alpha_2 - \alpha_1 \quad \Delta\beta = \beta_2 - \beta_1 \quad \Delta\rho = \rho_2 - \rho_1 \tag{3.6.57}$$

The anisotropic parameters of each layer are defined as

$$\varepsilon_I^{(1)} = \frac{A_{22}^{(I)} - A_{33}^{(I)}}{2A_{33}^{(I)}}; \quad \widetilde{\delta}_I^{(1)} = \frac{A_{23}^{(I)} + 2A_{44}^{(I)} - A_{33}^{(I)}}{A_{33}^{(I)}} \tag{3.6.58}$$

$$\varepsilon_I^{(2)} = \frac{A_{11}^{(I)} - A_{33}^{(I)}}{2A_{33}^{(I)}}; \quad \widetilde{\delta}_I^{(2)} = \frac{A_{13}^{(I)} + 2A_{55}^{(I)} - A_{33}^{(I)}}{A_{33}^{(I)}} \tag{3.6.59}$$

$$\gamma_I^{(S)} = \frac{A_{44}^{(I)} - A_{55}^{(I)}}{2A_{55}^{(I)}}; \quad \widetilde{\delta}_I^{(3)} = \frac{A_{12}^{(I)} + 2A_{66}^{(I)} - A_{11}^{(I)}}{A_{11}^{(I)}} \tag{3.6.60}$$

where $A_{jk}^{(I)}$ are the density-normalized Voigt-notation elastic constants of each layer.

In the approximation of weak anisotropy and weak layer contrasts, ϕ_S can be related approximately to ϕ_P using the linearized isotropic Snell's law:

$$\frac{1}{\cos\phi_S} \approx \frac{1}{\sqrt{1 - (\overline{\beta}^2/\overline{\alpha}^2)\sin^2\phi_P}} \approx 1 + \frac{1}{2}\frac{\overline{\beta}^2}{\overline{\alpha}^2}\sin^2\phi_P \tag{3.6.61}$$

Arbitrary Anisotropy

Vavryčuk and Pšenčik (1998) derived P-wave reflectivity for arbitrary weak anisotropy and weak layer contrast:

$$R_{PP}(\zeta, \theta_P) = \frac{1}{2}\frac{\Delta Z}{\overline{Z}} + \frac{1}{2}\left[\frac{\Delta\alpha}{\overline{\alpha}} - 4\left(\frac{\overline{\beta}}{\overline{\alpha}}\right)^2\frac{\Delta\mu}{\overline{\mu}}\right]\sin^2\theta_P + \frac{1}{2}\frac{\Delta\alpha}{\overline{\alpha}}\tan^2\theta_P\sin^2\theta_P$$

$$+ \frac{1}{2}\left\{\Delta\left(\frac{A_{13} + 2A_{55} - A_{33}}{A_{33}}\right)\cos^2\zeta\right.$$

$$+ \left[\Delta\left(\frac{A_{23} + 2A_{44} - A_{33}}{A_{33}}\right) - 8\Delta\left(\frac{A_{44} - A_{55}}{2A_{33}}\right)\right]\sin^2\zeta$$

$$+ 2\left[\Delta\left(\frac{A_{36} + 2A_{45}}{A_{33}}\right) - 4\Delta\left(\frac{A_{45}}{A_{33}}\right)\right]\cos\zeta\,\sin\zeta\right\}\sin^2\theta_P$$

$$+ \frac{1}{2}\left[\Delta\left(\frac{A_{11} - A_{33}}{2A_{33}}\right)\cos^4\zeta + \Delta\left(\frac{A_{22} - A_{33}}{2A_{33}}\right)\sin^4\zeta\right.$$

$$+ \Delta\left(\frac{A_{12} + 2A_{66} - A_{33}}{A_{33}}\right)\cos^2\zeta\,\sin^2\zeta + 2\Delta\left(\frac{A_{16}}{A_{33}}\right)\cos^3\zeta\,\sin\zeta$$

$$\left.+ 2\Delta\left(\frac{A_{26}}{A_{33}}\right)\cos\zeta\,\sin^3\zeta\right]\sin^2\theta_P\tan^2\theta_P \tag{3.6.62}$$

where $A_{ij} = c_{ij}/\rho$ is the Voigt-notation elastic constant divided by density. θ_P is the P-wave angle of incidence and ζ is the azimuth relative to the x_1-axis. Δ signifies the contrast of any parameter across the reflector, e.g., $\Delta\psi = \psi_2 - \psi_1$. $\overline{\alpha}, \overline{\beta}$, and ρ are the background average isotropic P-wave velocity, S-wave velocity, and density, respectively. The accuracy of the result varies slightly with the choice of the background values, but a reasonable choice is the average of the vertical P-wave velocities across the two layers for $\overline{\alpha}$, the average of the vertical S-wave velocities across the reflector for $\overline{\beta}$, and the average of the densities across the two layers for ρ. The corresponding expression for the P-wave transmission coefficient for arbitrary anisotropy is given by Pšenčik and Vavryčuk (1998).

Jílek (2002a, 2002b) derived expressions for P-to-SV reflectivities for arbitrary weak anisotropy and weak layer contrasts.

Assumptions and Limitations

The equations presented in this section apply under the following conditions:
- the rock is linear elastic;
- approximate forms apply to the P-P reflection at near offset for slightly contrasting, weakly anisotropic media.

3.7   Elastic Impedance

The elastic impedance is a **pseudo-impedance attribute** (Connolly, 1998; Mukerji et al., 1998) and is a far-offset equivalent of the more conventional zero-offset acoustic impedance. This far-offset impedance has been called the "elastic impedance" (EI), as it contains information about the V_P/V_S ratio. One can obtain the elastic impedance cube from a far-offset stack using the same trace-based one-dimensional inversion algorithm used to invert the near-offset stack. Though only approximate, the inversion for elastic impedance is economical and simple compared to full prestack inversion. As with any impedance inversion, the key to effectively using this extracted attribute for quantitative reservoir characterization is calibration with log data.

The acoustic impedance, $I_a = \rho V_P$, can be expressed as

$$I_a = \exp\left(2\int R_{PP}(0)dt\right) \tag{3.7.1}$$

where $R_{PP}(0)$ is the time sequence of normal-incidence P-to-P reflection coefficients. Similarly, the **elastic impedance** may be defined in terms of the sequence of non-normal P-P reflection coefficients, $R_{PP}(\theta)$, at incidence angle θ as

$$I_e(\theta) = \exp\left(2\int R_{PP}(\theta)\,dt\right) \tag{3.7.2}$$

Substituting into this equation one of the well-known approximations for $R_{PP}(\theta)$ (e.g., Aki and Richards, 1980),

$$R_{PP}(\theta) \approx R_{PP}(0) + M\,\sin^2\theta + N\tan^2\theta \tag{3.7.3}$$

$$R_{PP}(0) = \frac{1}{2}\left(\frac{\Delta V_P}{\overline{V}_P} + \frac{\Delta\rho}{\overline{\rho}}\right) \tag{3.7.4}$$

$$M = -2\left(\frac{\overline{V}_S}{\overline{V}_P}\right)^2\left(\frac{2\Delta V_S}{\overline{V}_S} + \frac{\Delta\rho}{\overline{\rho}}\right) \tag{3.7.5}$$

$$N = \frac{1}{2}\frac{\Delta V_P}{\overline{V}_P} \tag{3.7.6}$$

leads to the following expression for I_e:

$$I_e(\theta) = \rho V_P \exp\left[\tan^2\theta\int d(\ln V_P)\right]\exp\left[-4\sin^2\theta(\overline{V}_S/\overline{V}_P)^2\int 2d(\ln V_S)\right]$$
$$\times\exp\left[-4\sin^2\theta(\overline{V}_S/\overline{V}_P)^2\int d(\ln\rho)\right] \tag{3.7.7}$$

or

$$I_e(\theta) = V_P^{(1+\tan^2\theta)}\rho^{(1-4K\sin^2\theta)}V_S^{(-8K\sin^2\theta)} \tag{3.7.8}$$

where $K = (\overline{V}_S/\overline{V}_P)^2$ is taken to be a constant. The elastic impedance reduces to the usual acoustic impedance, $I_a = \rho V_P$, when $\theta = 0$. Unlike the acoustic impedance, the elastic impedance is not a function of the rock properties alone but also depends on the angle of incidence. Using only the first two terms in the approximation for $R(\theta)$ gives a similar expression for I_e with the $\tan^2\theta$ terms being replaced by $\sin^2\theta$,

$$I_e(\theta) = V_P^{(1+\sin^2\theta)}\rho^{(1-4K\sin^2\theta)}V_S^{(-8K\sin^2\theta)} \tag{3.7.9}$$

Expressions for elastic impedance for modes P-P, P-SV, SV-SV, SV-P, SH-SH are given in Tables 3.7.1, 3.7.2, 3.7.3, and 3.7.4, for isotropic, VTI, and orthorhombic symmetries. The expression in equation 3.7.9 has been referred to as the **first-order elastic impedance** (Connolly, 1999), and it goes to $(V_P/V_S)^2$ at $\theta = 90°$, assuming $K = \dfrac{1}{4}$. Expressions for P-to-S converted-wave elastic impedance have been given by Duffaut *et al.* (2000) and González (2006). The elastic impedance has strange units and dimensions that change with angle. Whitcombe (2002) defines a useful normalization for the elastic impedance:

$$I_e(\theta) = [V_{P0}\rho_0]\left(\frac{V_P}{V_{P0}}\right)^{(1+\tan^2\theta)}\left(\frac{\rho}{\rho_0}\right)^{(1-4K\sin^2\theta)}\left(\frac{V_S}{V_{S0}}\right)^{(-8K\sin^2\theta)} \tag{3.7.10}$$

where the normalizing constants V_{P0}, V_{S0}, and ρ_0 may be taken to be either the average values of velocities and densities over the zone of interest, or the values at the top of the target zone. With this normalization, the elastic impedance has the same dimensionality as the acoustic impedance. The **extended elastic impedance** (EEI) (Whitcombe *et al.*, 2002) is defined over an angle χ, ranging from $-90°$ to $+90°$. The angle χ should not be interpreted as the actual reflection angle, but rather as the independent input variable in the definition of EEI. The EEI is expressed as

$$I_e(\chi) = [V_{P0}\rho_0]\left(\frac{V_P}{V_{P0}}\right)^{(\cos\chi+\sin\chi)}\left(\frac{\rho}{\rho_0}\right)^{(\cos\chi-4K\sin\chi)}\left(\frac{V_S}{V_{S0}}\right)^{(-8K\sin\chi)} \tag{3.7.11}$$

Under certain approximations, the EEI for specific values of χ becomes proportional to rock elastic parameters such as the bulk modulus and the shear modulus.

In applications to reservoir characterization, care must be taken to filter the logs to match the seismic frequencies, as well as to account for the differences in frequency content in near- and far-offset data. Separate wavelets should be extracted for the near- and far-offset angle stacks. The inverted acoustic and elastic impedances co-located at the wells must be calibrated with the known facies and fluid types in the well, before classifying the seismic cube in the interwell region.

Major limitations in using partial-stack elastic-impedance inversion arise from the assumptions of the one-dimensional convolutional model for far offsets, the assumption of a constant value for K, and errors in estimates of the incidence angle. The convolutional model does not properly handle all the reflections at far offsets, because the primary reflections get mixed with other events. The approximations used to derive the expressions for elastic impedance become less accurate at larger angles. The first-order, two-term elastic impedance has been found to give stabler results than the three-term elastic impedance (Mallick, 2001). Mallick compares prestack inversion versus partial-stack elastic impedance inversions and recommends a hybrid approach. In this approach, full prestack inversion is done at a few control points to get reliable estimates of P and S impedance. These prestack inversions are used as anchors for cheaper, one-dimensional, trace-based inversions over large data volumes. Based on these results, small zones may be selected for detailed analysis by prestack inversions.

Elastic Impedance Expressions: Isotropic

Table 3.7.1 *Exponents in expressions for elastic impedance of an isotropic, flat-layered Earth, $\text{EI} \approx V_P^A V_S^B \rho^C$. In all expressions, θ is the P-wave angle of incidence and θ_S is the shear-wave angle of incidence. Expressions are given for P–P reflection, converted wave reflections P–S and S–P, and shear-wave reflections SV–SV and SH–SH. $\langle \overline{V}_S^2 / \overline{V}_P^2 \rangle$ and $\langle \overline{V}_S / \overline{V}_P \rangle$ are assumed to be constant, often taken as the average of the corresponding quantities.*

P–P	A	$1 + \tan^2\theta$
	B	$-8 \left\langle \dfrac{\overline{V}_S^2}{\overline{V}_P^2} \right\rangle \sin^2\theta$
	C	$1 - 4 \left\langle \dfrac{\overline{V}_S^2}{\overline{V}_P^2} \right\rangle \sin^2\theta$
P–SV	A	0
	B	$\dfrac{\sin\theta}{\sqrt{1 - \langle \overline{V}_S^2/\overline{V}_P^2 \rangle \sin^2\theta}} \left(4 \sin^2\theta \left\langle \dfrac{\overline{V}_S^2}{\overline{V}_P^2} \right\rangle - 4 \left\langle \dfrac{\overline{V}_S}{\overline{V}_P} \right\rangle \cos\theta \sqrt{1 - \left\langle \dfrac{\overline{V}_S^2}{\overline{V}_P^2} \right\rangle \sin^2\theta} \right)$
	C	$-\dfrac{\sin\theta}{\sqrt{1 - \langle \overline{V}_S^2/\overline{V}_P^2 \rangle \sin^2\theta}} \left(1 - 2 \left\langle \dfrac{\overline{V}_S^2}{\overline{V}_P^2} \right\rangle \sin^2\theta + 2 \left\langle \dfrac{\overline{V}_S}{\overline{V}_P} \right\rangle \cos\theta \sqrt{1 - \left\langle \dfrac{\overline{V}_S^2}{\overline{V}_P^2} \right\rangle \sin^2\theta} \right)$
SV–P	A	0
	B	$\dfrac{\sin\theta_S}{\sqrt{1 - \langle \overline{V}_S^2/\overline{V}_P^2 \rangle^{-1} \sin^2\theta_S}} \left(4 \sin^2\theta_S - 4 \left\langle \dfrac{\overline{V}_S}{\overline{V}_P} \right\rangle \cos\theta_S \sqrt{1 - \left\langle \dfrac{\overline{V}_S^2}{\overline{V}_P^2} \right\rangle^{-1} \sin^2\theta_S} \right)$
	C	$-\dfrac{\sin\theta_S}{\sqrt{1 - \langle \overline{V}_S^2/\overline{V}_P^2 \rangle^{-1} \sin^2\theta_S}} \left(1 - 2 \sin^2\theta_S + 2 \left\langle \dfrac{\overline{V}_S}{\overline{V}_P} \right\rangle \cos\theta_S \sqrt{1 - \left\langle \dfrac{\overline{V}_S^2}{\overline{V}_P^2} \right\rangle^{-1} \sin^2\theta_S} \right)$
SV–SV	A	0
	B	$-\dfrac{1}{\cos^2\theta_S} + 8 \sin^2\theta_S$
	C	$-1 + 4 \sin^2\theta_S$
SH–SH	A	0
	B	$-1 + \tan^2\theta_S$
	C	-1

Elastic Impedance Expressions: VTI Anisotropic

Table 3.7.2 *Exponents in expressions for elastic impedance of a VTI, flat-layered Earth,* $EI \approx V_P^A V_S^B \rho^C exp(D\varepsilon + E\delta + F\gamma^{(V)})$. *$\theta$ is the P-wave angle of incidence, and θ_S is the shear-wave angle of incidence. Expressions are given for P–P reflection, and shear-wave reflections SV–SV and SH–SH. $\langle \overline{V}_S^2 / \overline{V}_P^2 \rangle$ is assumed to be constant, often taken as the average of the squared velocity ratio.*

P–P	A	$1 + \tan^2\theta$
	B	$-8 \left\langle \dfrac{\overline{V}_S^2}{\overline{V}_P^2} \right\rangle \sin^2\theta$
	C	$1 - 4 \left\langle \dfrac{\overline{V}_S^2}{\overline{V}_P^2} \right\rangle \sin^2\theta$
	D	$\sin^2\theta \tan^2\theta$
	E	$\sin^2\theta$
	F	0
SV–SV	A	0
	B	$-\dfrac{1}{\cos^2\theta_S} + 8\sin^2\theta_S$
	C	$-1 + 4\sin^2\theta_S$
	D	$\sin^2\theta_S / \left\langle \dfrac{\overline{V}_S^2}{\overline{V}_P^2} \right\rangle$
	E	$-\sin^2\theta_S / \left\langle \dfrac{\overline{V}_S^2}{\overline{V}_P^2} \right\rangle$
	F	0
SH–SH	A	0
	B	$-1 + \tan^2\theta_S$
	C	-1
	D	0
	E	0
	F	$\tan^2\theta_S$

$$V_P = \alpha = \sqrt{c_{33}/\rho}; \quad V_S = \beta = \sqrt{c_{44}/\rho}$$

$$\varepsilon = \frac{c_{11} - c_{33}}{2c_{33}}; \quad \gamma^{(V)} = \frac{c_{66} - c_{44}}{2c_{44}}; \quad \delta = \frac{(c_{13} + c_{44})^2 - (c_{33} - c_{44})^2}{2c_{33}(c_{33} - c_{44})}$$

Elastic Impedance Expressions: Orthorhombic x_1–x_3 Symmetry Plane

Table 3.7.3 *Exponents in expressions for P–P elastic impedance of an orthorhombic, flat-layered Earth, $EI \approx V_P^A V_S^{\perp B} \rho^C exp(D\varepsilon^{(2)} + E\delta^{(2)})$. θ is the P-wave angle of incidence. $\langle \overline{V}_S^2 / \overline{V}_P^2 \rangle$ is assumed to be constant, often taken as the average of the squared velocity ratio. $V_P = \sqrt{c_{33}/\rho}$; $V_S^{\perp} = \beta^{\perp} = \sqrt{c_{55}/\rho}$*

P–P	A	$1 + \tan^2 \theta$
	B	$-8 \left\langle \dfrac{\overline{V}_S^{\perp 2}}{\overline{V}_P^2} \right\rangle \sin^2\theta$
	C	$1 - 4 \left\langle \dfrac{\overline{V}_S^{\perp 2}}{\overline{V}_P^2} \right\rangle \sin^2\theta$
	D	$\sin^2 \theta \tan^2 \theta$
	E	$\sin^2 \theta$

Elastic Impedance Expressions: Orthorhombic x_2–x_3 Symmetry Plane

Table 3.7.4 *Exponents in expressions for P–P elastic impedance of an orthorhombic, flat-layered Earth, $EI \approx V_P^A V_S^B \rho^C exp(D\varepsilon^{(1)} + E\delta^{(1)})$. θ is the P-wave angle of incidence. $\langle \overline{V}_S^2 / \overline{V}_P^2 \rangle$ is assumed to be constant, often taken as the average of the squared velocity ratio. $V_P = \sqrt{c_{33}/\rho}$; $V_S = \beta = \sqrt{c_{44}/\rho}$*

P–P	A	$1 + \tan^2 \theta$
	B	$-8 \left\langle \dfrac{\overline{V}_S^2}{\overline{V}_P^2} \right\rangle \sin^2\theta$
	C	$1 - 4 \left\langle \dfrac{\overline{V}_S^2}{\overline{V}_P^2} \right\rangle \sin^2\theta$
	D	$\sin^2 \theta \tan^2 \theta$
	E	$\sin^2 \theta$

$$\varepsilon^{(2)} = \frac{c_{11} - c_{33}}{2c_{33}}; \qquad\qquad \varepsilon^{(1)} = \frac{c_{22} - c_{33}}{2c_{33}}$$

$$\delta^{(2)} = \frac{(c_{13} + c_{55})^2 - (c_{33} - c_{55})^2}{2c_{33}(c_{33} - c_{55})}; \quad \delta^{(1)} = \frac{(c_{23} + c_{44})^2 - (c_{33} - c_{44})^2}{2c_{33}(c_{33} - c_{44})}$$

$$\gamma^{(2)} = \frac{c_{66} - c_{44}}{2c_{44}}; \qquad\qquad \gamma^{(1)} = \frac{c_{66} - c_{55}}{2c_{55}}$$

$$\delta^{(3)} = \frac{(c_{12} + c_{66})^2 - (c_{11} - c_{66})^2}{2c_{11}(c_{11} - c_{66})}$$

Uses

The elastic impedance provides an alternative for interpreting far-offset data. As with any inversion, the elastic impedance removes some wavelet effects and attempts to determine interval properties from the band-limited reflectivity (seismic traces).

Assumptions and Limitations

Limitations of using partial-stack elastic impedance inversion arise from the assumptions of the one-dimensional convolutional model for far offsets, the assumption of a constant value for $K = (V_S/V_P)^2$, and errors in estimates of the incidence angle. The convolutional model does not properly handle all the reflections at far offsets because the primary reflections get mixed with other events. Furthermore, approximations used to derive the expressions for elastic impedance get worse at larger angles. The first-order two-term elastic impedance has been found to give more stable results than the three-term elastic impedance (Mallick, 2001).

3.8   Viscoelasticity and Q

Synopsis

Materials are **linear elastic** when the stress is proportional to the strain:

$\dfrac{\sigma_{11} + \sigma_{22} + \sigma_{33}}{3}$	$=$	$K(\varepsilon_{11} + \varepsilon_{22} + \varepsilon_{33})$	volumetric	(3.8.1)
σ_{ij}	$=$	$2\mu\varepsilon_{ij}, \quad i \neq j$	shear	(3.8.2)
σ_{ij}	$=$	$\lambda\delta_{ij}\varepsilon_{kk} + 2\mu\varepsilon_{ij}$	general isotropic	(3.8.3)

where σ_{ij} and ε_{ij}, are the stress and strain, K is the bulk modulus, μ is the shear modulus, and λ is Lamé's coefficient. Or, in terms of the time-independent elastic stiffness tensor, $C_{ijkl}^{(el)}$,

$$\sigma_{ij} = C_{ijkl}^{(el)} \varepsilon_{kl}. \tag{3.8.4}$$

Real materials, especially rocks, almost always deviate from perfect elastic behavior – their stress–strain relations have hysteresis and are rate-dependent. In geophysics, this rate dependence is often encountered as seismic wave velocity dispersion or creep. For small-strain applications, rocks are often described as **linear viscoelastic materials**, with rate or history dependence in the stress–strain behavior. Constitutive equations can be expressed by introducing time derivatives. For example, shear stress and shear strain might be related by using one of the following simple models:

$$\dot{\varepsilon}_{ij} = \frac{\dot{\sigma}_{ij}}{2\mu} + \frac{\sigma_{ij}}{2\eta} \qquad\qquad \text{Maxwell solid} \qquad\qquad (3.8.5)$$

$$\sigma_{ij} = 2\eta\dot{\varepsilon}_{ij} + 2\mu\varepsilon_{ij} \qquad\qquad \text{Voigt solid} \qquad\qquad (3.8.6)$$

$$\eta\dot{\sigma}_{ij} + (E_1 + E_2)\sigma_{ij} = E_2(\eta\dot{\varepsilon}_{ij} + E_1\varepsilon_{ij}) \quad \text{Standard linear solid} \qquad (3.8.7)$$

where E_1 and E_2 are additional elastic moduli and η is a material constant resembling viscosity.

More complicated linear viscoelastic behavior can be represented by incorporating higher-order time derivatives, for example

$$\sum_{k=0}^{N} a_k \frac{\partial^k \varepsilon}{\partial t^k} = \sum_{j=0}^{M} b_j \frac{\partial^j \sigma}{\partial t^j} \qquad\qquad (3.8.8)$$

where a_i and b_j are material constants. In the frequency domain, equation 3.8.8 becomes

$$\sum_{k=0}^{N} a_k (i\omega)^k \tilde{\varepsilon}(\omega) = \sum_{j=0}^{M} b_j (i\omega)^j \tilde{\sigma}(\omega), \qquad\qquad (3.8.9)$$

which can be rearranged as

$$\tilde{\sigma}(\omega) = \frac{\displaystyle\sum_{k=0}^{N} a_k (i\omega)^k}{\displaystyle\sum_{j=0}^{M} b_j (i\omega)^j} \tilde{\varepsilon}(\omega) = M^*(\omega)\tilde{\varepsilon}(\omega). \qquad\qquad (3.8.10)$$

In equation 3.8.10, $M^*(\omega)$ is a complex modulus, and tilde indicates the Fourier transform. Similar equations would be necessary to describe the generalizations of other elastic constants such as bulk modulus. A discussion of commonly used viscoelastic material models can be found later in this section.

Consider a plane wave propagating in a viscoelastic solid so that the displacement is given by

$$u(x,t) = u_0 \exp[-a(\omega)x]\exp[i(\omega t - kx)] \qquad\qquad (3.8.11)$$

At any point in the solid, the stress and strain are out of phase.

$$\sigma = \sigma_0 \exp[i(\omega t - kx)] \tag{3.8.12}$$

$$\varepsilon = \varepsilon_0 \exp[i(\omega t - kx - \varphi)] \tag{3.8.13}$$

The phase difference is an indication that $M^*(\omega)$ is complex.

The **quality factor**, Q, is a measure of how dissipative the material is. The lower Q, the larger the dissipation. There are several ways to express Q. One precise way is as the ratio of the imaginary and real parts of the complex modulus:

$$\frac{1}{Q} = \frac{M_I}{M_R} \tag{3.8.14}$$

In terms of energies, Q can be expressed as

$$\frac{1}{Q} = \frac{\Delta W}{2\pi W} \tag{3.8.15}$$

where ΔW is the energy dissipated per cycle of oscillation and W is the peak strain energy during the cycle. In terms of the spatial attenuation factor, α, in equation 3.8.11,

$$\frac{1}{Q} \approx \frac{\alpha V}{\pi f} \tag{3.8.16}$$

where V is the velocity and f is the frequency. In terms of the wave amplitudes of an oscillatory signal with period τ,

$$\frac{1}{Q} \approx \frac{1}{\pi} \ln\left[\frac{u(t)}{u(t+\tau)}\right] \tag{3.8.17}$$

which measures the amplitude loss per cycle. This is sometimes called the **logarithmic decrement**. Finally, in terms of the phase delay φ between the stress and strain,

$$\frac{1}{Q} \approx \tan(\varphi) \tag{3.8.18}$$

Winkler and Nur (1979) showed that if we define $Q_E = E_R/E_I$, $Q_K = K_R/K_I$, and $Q_\mu = \mu_R/\mu_I$, where the subscripts R and I denote real and imaginary parts of the Young's, bulk, and shear moduli, respectively, and if the attenuation is small, then the various Q factors can be related through the following equations:

$$\frac{(1-v)(1-2v)}{Q_P} = \frac{(1+v)}{Q_E} - \frac{2v(2-v)}{Q_S}; \quad \frac{Q_P}{Q_S} = \frac{\mu_I}{K_I + \frac{4}{3}\mu_I} \frac{V_P^2}{V_S^2}$$

$$\frac{(1+v)}{Q_K} = \frac{3(1-v)}{Q_P} - \frac{2(1-2v)}{Q_S}; \quad \frac{3}{Q_E} = \frac{(1-2v)}{Q_K} + \frac{2(1+v)}{Q_S} \tag{3.8.19}$$

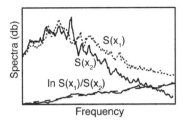

Figure 3.8.1 The slope of the log of the spectral ratio (difference of the spectra in dB) can be interpreted in terms of Q.

One of the following relations always occurs (Bourbié et al., 1987):

$$Q_K > Q_P > Q_E > Q_S \text{ for high } V_P/V_S \text{ ratios} \tag{3.8.20}$$

$$Q_K < Q_P < Q_E < Q_S \text{ for low } V_P/V_S \text{ ratios} \tag{3.8.21}$$

$$Q_K = Q_P = Q_E = Q_S \tag{3.8.22}$$

The **spectral ratio** method is a popular way to estimate Q in both the laboratory and the field. Because $1/Q$ is a measure of the fractional loss of energy per cycle of oscillation, after a fixed distance of propagation there is a tendency for shorter wavelengths to be attenuated more than longer wavelengths. If the amplitude of the propagating wave is

$$u(x,t) = u_0 \exp[-\alpha(\omega)x] = u_0 \exp\left(-\frac{\pi f}{VQ}x\right), \tag{3.8.23}$$

we can compare the spectral amplitudes at two different distances and determine Q from the slope of the logarithmic decrement (Figure 3.8.1):

$$\ln\left[\frac{S(f, x_2)}{S(f, x_1)}\right] = -\frac{\pi f}{QV}(x_2 - x_1) \tag{3.8.24}$$

Time Domain Description of Viscoelastic Creep and Relaxation

In the time domain, and especially when looking at quasi-static problems, the stress–strain relation of **linear viscoelastic** materials is often presented in the form of a convolution (Lakes, 2010)

$$\sigma_{ij}(t) = \int_0^t C_{ijkl}(t - \tau) \frac{\partial \varepsilon_{kl}}{\partial \tau} d\tau, \tag{3.8.25}$$

where each modulus tensor element $C_{ijkl}(t)$ is the causal **relaxation function** for that component. The tensor $\mathbf{C}(t)$ is a material property. Each independent component of the modulus tensor can have a different time (or frequency) dependence.

For the special case of an *isotropic* linear viscoelastic material, only two independent relaxation functions are required, and the stress–strain relation, equation 3.8.25, reduces to

$$\sigma_{ij}(t) = \int_0^t \lambda(t-\tau)\delta_{ij}\frac{\partial \varepsilon_{kk}}{\partial \tau}d\tau + \int_0^t 2\mu(t-\tau)\frac{\partial \varepsilon_{ij}}{\partial \tau}d\tau \,, \tag{3.8.26}$$

where δ_{ij} is the Kronecker delta. The isotropic relaxation behavior can also be written as

$$\sigma_{kk}(t) = 3\int_0^t K(t-\tau)\frac{\partial \varepsilon_{kk}}{\partial \tau}d\tau \,, \quad \sigma_{ij}^{dev}(t) = 2\int_0^t \mu(t-\tau)\frac{\partial \varepsilon_{ij}^{dev}}{\partial \tau}d\tau \,, \tag{3.8.27}$$

where ε_{kk} is the volume strain, σ_{kk} is the trace of the stress tensor, ε_{ij}^{dev} is the deviatoric strain, and σ_{ij}^{dev} is the deviatoric stress.

In equations 3.8.26 and 3.8.27, $K(t)$, $\mu(t)$, and $\lambda(t)$ are the relaxation function analogs of the familiar elastic bulk modulus, shear modulus, and Lamé coefficient. In homogeneous solids, the bulk modulus is often nearly constant with time, while the shear modulus is viscoelastic. For composites (most rocks), the bulk modulus can more easily be viscoelastic as well.

An equivalent way to describe the stress–strain response of a linear viscoelastic material is

$$\varepsilon_{ij}(t) = \int_0^t J_{ijkl}(t-\tau)\frac{\partial \sigma_{kl}}{\partial \tau}d\tau \,, \tag{3.8.28}$$

where $J_{ijkl}(t)$ is the tensor of **creep functions**, the viscoelastic analog of the compliance tensor. (Note that equations 3.8.25 and 3.8.28 are inverses of each other.) For an isotropic material, the creep behavior can be expressed as

$$\varepsilon_{kk}(t) = \frac{1}{3}\int_0^t B(t-\tau)\frac{\partial \sigma_{kk}}{\partial \tau}d\tau \,, \quad \varepsilon_{ij}^{dev}(t) = \frac{1}{2}\int_0^t J_S(t-\tau)\frac{\partial \sigma_{ij}^{dev}}{\partial \tau}d\tau \,, \tag{3.8.29}$$

In equation 3.8.29, $B(t)/3$ is the volume creep compliance function, and $J_S(t)/2$ is the shear creep compliance function.

The relaxation function (equation 3.8.25) and creep function (equation 3.8.28) are generally different from each other but are related. In the Laplace transform domain we can write $\widetilde{\mathbf{C}}(s) = [s^2\widetilde{\mathbf{J}}(s)]^{-1}$, where s is the Laplace transform variable. For the isotropic case in the Laplace domain, $s^2\widetilde{K}(s) = 1/\widetilde{B}(s)$ and $s^2\widetilde{\mu}(s) = 1/\widetilde{J}_S(s)$.

Under oscillatory loading (e.g., as associated with seismic waves), when stress has the form $\sigma = \sigma_{ij}^0 e^{i\omega t}$, and strain, $\varepsilon = \varepsilon_{kl}^0 e^{i\omega t}$, the convolution in equations 3.8.25 and 3.8.28 can be written in the frequency domain as

$$\widetilde{\sigma}_{ij}(\omega) = C_{ijkl}^*(\omega)\,\widetilde{\varepsilon}_{ij}(\omega) \tag{3.8.30}$$

where $C_{ijkl}^*(\omega)$ is a frequency-dependent, complex modulus, and the tilde on stress and strain indicates the Fourier transform of these quantities. For the isotropic case the volumetric and deviatoric problems can be decoupled:

$$\widetilde{\sigma}_{aa}(\omega) = 3K^*(\omega)\,\widetilde{\varepsilon}_{\beta\beta}(\omega)\,,\ \ \widetilde{\sigma}_{kl}^{dev}(\omega) = 2\mu^*(\omega)\,\widetilde{\varepsilon}_{kl}^{dev}(\omega) \tag{3.8.31}$$

where $K^*(\omega)$ and $\mu^*(\omega)$ are the complex frequency-dependent bulk and shear moduli. The complex shear modulus in equation 3.8.31 can be written as $\mu^*(\omega) = \mu_R + i\mu_I$. The real part, μ_R, is often called the **storage modulus**, and it represents the linear elastic (in-phase) part of the relation between stress and strain. The imaginary part, μ_I, is called the **loss modulus** and represents the part of the relation where stress and strain are out of phase by 90°.

An alternative description is to define the complex, frequency-dependent **viscosity**, $\eta^* = \eta_R - i\eta_I$, which relates the stress to the shear *strain rate*:

$$\widetilde{\sigma}_{ij}(\omega) = 2\eta^*(\omega)\,i\omega\widetilde{\varepsilon}_{ij}(\omega) \tag{3.8.32}$$

Comparing equations 3.8.31 and 3.8.32, we see the relation between the complex modulus and complex viscosity:

$$\eta^* = \frac{\mu^*(\omega)}{i\omega} \tag{3.8.33}$$

Comprehensive descriptions of viscoelasticity theory can be found in Fung (1965), Zener (1948), Christensen (1982), and Lakes (2010).

Common Viscoelastic Models

One of the simplest viscoelastic models, the **standard linear solid (SLS)**, has a stress–strain relation that can be defined by a differential equation (Zener, 1948) of the form:

$$\frac{\partial\sigma}{\partial t} + \omega_s\,\sigma = M_\infty\,\frac{\partial\varepsilon}{\partial t} + M_0\omega_s\,\varepsilon \tag{3.8.34}$$

where M_0, M_∞, and ω_s are material constants. M_0 is the (real) modulus (ratio of stress to strain) at extremely low stress and strain rates, M_∞ is the (real) modulus at extremely high stress and strain rates, and ω_s is a characteristic frequency. It is customary to visualize the mechanical behavior of linear viscoelastic equations with analog systems of springs and dashpots. Figure 3.8.2 illustrates, in fact, two different systems, both of

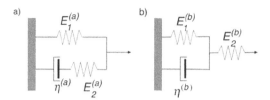

Figure 3.8.2 Two distinct mechanical analogs of the Standard Linear Solid, equation 3.8.33. Model on the left has also been referred to as the standard three-parameter Maxwell model, and the model on the right, the standard three-parameter Voigt model (Tschoegl, 1989) or Kelvin form.

which are described by equation 3.8.34 (Tschoegl, 1989). We refer to the model on the left as SLS(a) (also referred to as the Maxwell form of the Standard Linear solid), and the model on the right, SLS(b) (also referred to as the Voigt or Kelvin form of the Standard Linear Solid). Both systems have the same viscoelastic response, provided that the spring constants $E_i^{(j)}$ and viscosities $\eta^{(j)}$ are given the appropriate values shown in Table 3.8.1. *Caution: the "viscosity" in the spring-dashpot analogs is a parameter that introduces rate dependence into the model; this viscosity is not the same as the viscosity of a fluid within a poroelastic material.*

The creep and relaxation functions (transient step function responses) for the SLS are exponentials with characteristic times τ_c and τ_r, respectively (see Table 3.8.1). The relaxation function is monotonically decreasing, and the creep function is monotonically increasing. Both times are inversely proportional to characteristic frequency ω_s. Note also that $\tau_c/\tau_r = M_\infty/M_0 > 1$, so that the ratio of creep to relaxation times is proportional to the ratio of high and low frequency limiting moduli, and the characteristic creep time is always longer than the characteristic relaxation time.

Note that the SLS viscoelastic functions described by equation 3.8.34 and in Table 3.8.1 can refer to any one of the $C_{ijkl}(t)$ components of the tensor of relaxation functions or to the eigenfunctions of $\mathbf{C}(t)$.

Figure 3.8.3 Illustrates the real and imaginary parts of the Fourier-domain complex modulus $M_{SLS}^*(\omega)$ for a material where $M_\infty = 1.2M_0$. The frequency at which the real modulus is changing most rapidly is ω_s.

$$\omega_s = \frac{\sqrt{E_1(E_1 + E_2)}}{\eta} \tag{3.8.35}$$

The attenuation as a function of frequency, $1/Q_{SLS} = \mathrm{Im}(M(\omega))/\mathrm{Re}(M(\omega))$, is also given in Table 3.8.1 and plotted in Figure 3.8.3. The maximum attenuation occurs at characteristic frequency ω_s, with a value

$$\begin{aligned}
\left(\frac{1}{Q}\right)_{max} &= \frac{1}{2}\frac{E_2}{\sqrt{E_1(E_1+E_2)}} \\
&= \frac{1}{2}\frac{M_\infty - M_0}{\sqrt{M_\infty M_0}}
\end{aligned} \tag{3.8.36}$$

Table 3.8.1 *Transient and Fourier responses of the Standard Linear Solid model in equation 3.8.32 and Figure 3.8.2. Models SLS (a) and SLS (b) have identical response if the spring stiffnesses $E_i^{(j)}$ and viscosities $\eta^{(j)}$ are related to M_0, M_∞, and ω_s as shown. Expressions labeled with * are the same for both SLS(a) and SLS(b)*

	SLS (a)	SLS (b)
Differential equation*	$\dfrac{\partial\sigma}{\partial t} + \omega_s\,\sigma = M_\infty\,\dfrac{\partial\varepsilon}{\partial t} + M_0\omega_s\,\varepsilon$	
First spring stiffness	$E_1^{(a)} = M_\infty - M_0$	$E_1^{(b)} = \dfrac{M_\infty M_0}{M_\infty - M_0}$
Second spring stiffness	$E_2^{(a)} = M_0$	$E_2^{(b)} = M_\infty$
Viscosity	$\eta^{(a)} = \dfrac{M_\infty - M_0}{\omega_s}$	$\eta^{(b)} = \dfrac{M_\infty^2}{\omega_s(M_\infty - M_0)}$
Relaxation time*	$\tau_r = \dfrac{1}{\omega_s}$	
Creep time*	$\tau_c = \dfrac{M_\infty}{\omega_s M_0}$	
Ratio of times*	$\dfrac{\tau_r}{\tau_c} = \dfrac{M_0}{M_\infty}$	
Relaxation function*	$C(t) = M_0 + (M_\infty - M_0)e^{-t/\tau_r} = \begin{cases} M_\infty,\ t=0 \\ M_0,\ t\to\infty \end{cases}$	
Creep function*	$J(t) = \dfrac{1}{M_0} + \left(\dfrac{1}{M_\infty} - \dfrac{1}{M_0}\right)e^{-t/\tau_c} = \begin{cases} 1/M_\infty,\ t=0 \\ 1/M_0,\ t\to\infty \end{cases}$	
Complex modulus*	$M^*(\omega) = \dfrac{M_0 + i\,M_\infty(\omega/\omega_s)}{1 + i\,(\omega/\omega_s)} = \begin{cases} M_0,\ \omega=0 \\ M_\infty,\ \omega\to\infty \end{cases}$	
Real part of modulus*	$M_R(\omega) = \dfrac{M_0 + M_\infty(\omega/\omega_s)^2}{1 + (\omega/\omega_s)^2}$	
Imaginary part of modulus*	$M_I(\omega) = \dfrac{(M_\infty - M_0)(\omega/\omega_s)}{1 + (\omega/\omega_s)^2}$	
$\dfrac{1}{Q} = \dfrac{M_I}{M_R}$ *	$\dfrac{1}{Q} = \dfrac{(M_\infty - M_0)(\omega/\omega_s)}{M_0 + M_\infty(\omega/\omega_s)^2}$	

This peak attenuation is sometimes written as

$$\left(\frac{1}{Q}\right)_{\max} = \frac{1}{2}\frac{\Delta M}{\overline{M}} \tag{3.8.37}$$

where $\Delta M/\overline{M} = (M_\infty - M_0)/\overline{M}$ is the **modulus defect** and $\overline{M} = \sqrt{M_\infty M_0}$.

Liu *et al.* (1976) considered the **nearly constant Q model** in which simple attenuation mechanisms are combined so that the attenuation is nearly constant over a finite range of frequencies (Figure 3.8.4). One can then write

$$\frac{V(\omega)}{V(\omega_0)} = 1 + \frac{1}{\pi Q}\ln\left(\frac{\omega}{\omega_0}\right) \tag{3.8.38}$$

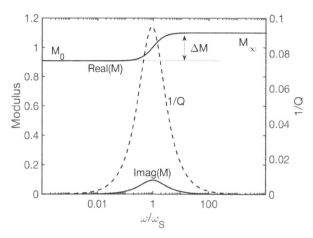

Figure 3.8.3 Example plot of real and imaginary parts of $M_{SLS}^*(\omega)$ versus frequency and $1/Q$ versus frequency for a Standard Linear Solid.

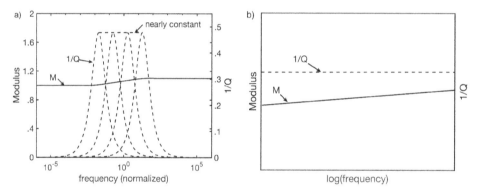

Figure 3.8.4 Schematic of the nearly constant Q model (left) and constant Q model (right).

which relates the velocity dispersion within the band of constant Q to the value of Q and the frequency. For large Q, this can be approximated as

$$\left(\frac{1}{Q}\right)_{max} \approx \frac{\pi}{\log(\omega/\omega_0)}\left(\frac{1}{2}\frac{M-M_0}{M_0}\right) \tag{3.8.39}$$

where M and M_0 are the moduli at two different frequencies ω and ω_0 within the band where Q is nearly constant. Note the resemblance of this expression to equation 3.8.36 for the standard linear solid.

Kjartansson (1979) considered the **constant Q model** in which Q is strictly constant (Figure 3.8.5). In this case, the complex modulus and Q are related by

$$M(\omega) = M_0\left(\frac{i\omega}{\omega_0}\right)^{2\gamma}, \tag{3.8.40}$$

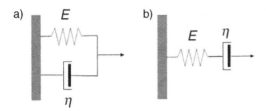

Figure 3.8.5 Spring-dashpot analogs of the Voigt (left) and Maxwell (right) viscoelastic models

where

$$\gamma = \frac{1}{\pi} \tan^{-1}\left(\frac{1}{Q}\right) \qquad (3.8.41)$$

For large Q, this can be approximated as

$$\left(\frac{1}{Q}\right)_{\text{max}} \approx \frac{\pi}{\log(\omega/\omega_0)} \left(\frac{1}{2} \frac{M - M_0}{M_0}\right) \qquad (3.8.42)$$

where M and M_0 are the moduli at two different frequencies ω and ω_0. Note the resemblance of this expression to equation 3.8.36 for the standard linear solid and equation (3.8.39) for the nearly constant Q model.

If $M_0 \to 0$ in equation 3.8.34, then the viscoelastic description becomes a **Maxwell model** defined by the following equation:

$$\frac{\partial \sigma}{\partial t} + \omega_s \, \sigma = M_\infty \, \frac{\partial \varepsilon}{\partial t}. \qquad (3.8.43)$$

The spring and dashpot representation of equation 3.8.43 is shown in Figure 3.8.5b. At high strain rates, the dashpot "locks" and the Maxwell material has a real elastic modulus M_∞. At low strain rates, the material approaches "viscous" behavior. Again, the Maxwell viscoelastic function might describe any of the stiffness tensor elements or eigenfunction. If the Maxwell model is used to describe the shear stiffness, then the low strain-rate behavior can be thought of as that of a Newtonian viscous fluid.

The relaxation function of a Maxwell material is a decaying exponential, with characteristic time $\tau_r = 1/\omega_s = \eta/M_\infty$ (see Table 3.8.2) and η is viscosity. The creep function grows without bound, linearly with time. The complex modulus relating stress and strain under oscillatory loading is

$$M^*_{MAX}(\omega) = \frac{i \, M_\infty(\omega/\omega_s)}{1 + i \, (\omega/\omega_s)} = \begin{cases} 0, & \omega = 0 \\ M_\infty, & \omega \to \infty \end{cases} \qquad (3.8.44)$$

where the subscript 'MAX' denotes Maxwell, and not maximum. The real and imaginary parts of the Maxwell modulus are plotted in Figure 3.8.6. The attenuation, $1/Q_{MAX} = \text{Im}(M(\omega))/\text{Re}(M(\omega))$, is also plotted.

The Maxwell material modulus can also be rewritten as

Table 3.8.2 *Transient and Fourier responses of Maxwell and Voigt (Kelvin) models. Subscripts* R *and* I *indicate real and imaginary parts.*

	Maxwell Model	Voigt Model
Differential equation	$\dfrac{\partial \sigma}{\partial t} + \omega_s\, \sigma = M_\infty\, \dfrac{\partial \varepsilon}{\partial t}$	$\omega_s\, \sigma = M_0\, \dfrac{\partial \varepsilon}{\partial t} + M_0 \omega_s\, \varepsilon$
Spring stiffness	$E = M_\infty$	$E = M_0$
Viscosity	$\eta = \dfrac{M_\infty}{\omega_s}$	$\eta = \dfrac{M_0}{\omega_s}$
Relaxation time	$\tau_r = \dfrac{1}{\omega_s} = \dfrac{\eta}{M_\infty}$	$-$
Creep time	$-$	$\tau_c = \dfrac{1}{\omega_s} = \dfrac{\eta}{M_0}$
Relaxation function	$C(t) = M_\infty e^{-t/\tau_r} = \begin{cases} M_\infty, & t=0 \\ 0, & t\to\infty \end{cases}$	$C(t) = M_0\, H(t) + \eta\, \delta(t)$
Creep function	$J(t) = \left(\dfrac{1}{M_\infty} + \dfrac{t}{\eta} \right) H(t) = \begin{cases} 1/M_\infty, & t=0 \\ t/\eta, & t\to\infty \end{cases}$	$J(t) = \dfrac{1}{M_0}\left(1 - e^{-t/\tau_c}\right)$
Complex modulus	$M^* = \dfrac{i\, M_\infty(\omega/\omega_s)}{1 + i\,(\omega/\omega_s)} = \begin{cases} 0, & \omega=0 \\ M_\infty, & \omega\to\infty \end{cases}$	$M^* = M_0 + iM_0(\omega/\omega_s)$ $\quad = \begin{cases} M_0, & \omega=0 \\ i\omega\eta, & \omega\to\infty \end{cases}$ $\quad = M_0 + i\omega\eta$
Real part of modulus	$M_R(\omega) = \dfrac{M_\infty(\omega/\omega_s)^2}{1 + (\omega/\omega_s)^2}$	$M_R = M_0$
Imaginary part of modulus	$M_I(\omega) = \dfrac{M_\infty(\omega/\omega_s)}{1 + (\omega/\omega_s)^2}$	$M_I = M_0(\omega/\omega_s)$
$\dfrac{1}{Q} = \dfrac{M_I}{M_R}$	$\dfrac{1}{Q} = \dfrac{1}{\omega/\omega_s}$	$\dfrac{1}{Q} = \dfrac{\omega}{\omega_s}$

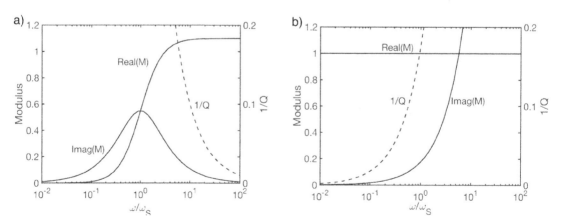

Figure 3.8.6 Example plot of real and imaginary parts of $M^*_{MAX}(\omega)$ vs. frequency for a Maxwell material (left) and Voigt material (right). The curve $1/Q = M_I/M_R$.

$$\frac{M^*_{MAX}(\omega)}{M_\infty} = 1 - \left(1 + \frac{i\omega\eta}{M_\infty}\right)^{-1}$$

$$= 1 - (1 + i\omega\tau_r)^{-1} \tag{3.8.45}$$

The **Voigt solid** (also sometimes called the Kelvin solid) is described by the following equation:

$$\sigma = \eta\frac{\partial\varepsilon}{\partial t} + M_0\varepsilon$$

$$= \frac{M_0}{\omega_s}\frac{\partial\varepsilon}{\partial t} + M_0\varepsilon \tag{3.8.46}$$

The creep function is exponential with characteristic time $\tau_c = 1/\omega_s = \eta/M_0$. The relaxation function is a delta function followed by constant stress. Under oscillatory loading, the complex Voigt modulus is

$$M^*_V(\omega) = M_0 + i\omega\eta . \tag{3.8.47}$$

When $\omega\eta \ll M_0$, the Voigt material approaches an elastic solid, and when $\omega\eta \gg M_0$ the Voigt material behaves as a Newtonian viscous fluid. The real and imaginary parts of the Voigt modulus are shown in Figure 3.8.6. The spring-dashpot analog of equation 3.8.47 is shown in Figure 3.8.5a. Creep and relaxation properties of the Voigt model are given in Table 3.8.2.

A **Newtonian linear viscous** fluid is a material in which the shear stress is proportional to the shear strain rate:

$$\sigma_{ij} = 2\eta\frac{\partial\varepsilon_{ij}}{\partial t}; \ i \neq j \tag{3.8.48}$$

The complex shear modulus is

$$M^*_N(\omega) = i\omega\eta . \tag{3.8.49}$$

The frequency responses of Newtonian and Maxwell materials are compared in Figure 3.8.7.

Many real viscoelastic materials display behavior that can be approximated with a spectrum of relaxation times, $S(\tau)$,

$$M^*(\omega) = \int_0^\infty \frac{S(\tau)}{\tau}\frac{i\omega\tau}{1 + i\omega\tau}d\tau . \tag{3.8.50}$$

When the distribution of relaxation times is a delta function, $S(\tau)/\tau = M_\infty\delta(\tau - \tau_1)$, then equation 3.8.50 reverts to equation 3.8.43, the Maxwell model. When the distribution is, for example, a boxcar function with a log-uniform distribution of

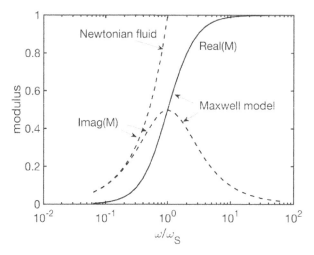

Figure 3.8.7 Example plot of real and imaginary parts of $M^*(\omega)$ vs. frequency for Newtonian and Maxwell materials

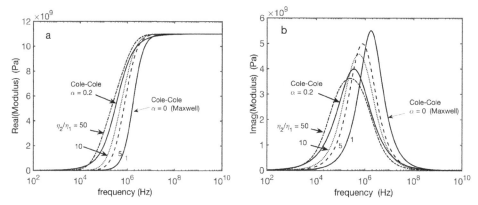

Figure 3.8.8 Example plot of real and imaginary parts of $M^*_{CC}(\omega)$ vs. ω comparing the boxcar distribution of relaxation times for values $\tau_2/\tau_1 = \eta_2/\eta_1 = 1$, 5, 10, 50, a Maxwell material, a Cole–Cole material with $\alpha = 0$, 0.2.

relaxation times between τ_1 and τ_2, then equation 3.8.50 becomes (O'Connell and Budiansky, 1977; Lakes, 2009)

$$M^* = \frac{M_\infty}{\ln(\tau_2/\tau_1)} \left\{ \frac{1}{2} \ln\left[\frac{1 + \omega^2\tau_2^2}{1 + \omega^2\tau_1^2}\right] + i\left[\tan^{-1}(\omega\tau_2) - \tan^{-1}(\omega\tau_1)\right] \right\}. \qquad (3.8.51)$$

Other distributions, such as log-normal have also been previously discussed (Nowick and Berry, 1972; O'Connell and Budiansky, 1977; Lakes, 2009).

Figure 3.8.8 Compares the real and imaginary parts of the viscoelastic moduli for four different boxcar relaxation spectra, $\tau_2/\tau_1 = 1$, 5, 10, 50 (equivalent to

$\eta_2/\eta_1 = 1, 5, 10, 50$, if M_∞ is kept constant). Most prominent in the figure is *the frequency spreading* of both the real and imaginary parts of the moduli, compared with those of the Maxwell model, which has a single relaxation time.

The ***Cole–Cole*** model is a generalization of the Maxwell model in equation 3.8.43:

$$\frac{M^*_{CC}(\omega)}{M_\infty} = 1 - \left[1 + \left(\frac{i\omega\eta}{M_\infty}\right)^{1-\alpha}\right]^{-1}$$
$$= 1 - [1 + (i\omega\tau)^{1-\alpha}]^{-1} \tag{3.8.52}$$

where $0 \le \alpha \le 1$ is an empirical constant. Figure 3.8.8 illustrates the real and imaginary parts of $M^*_{CC}(\omega)$. The slope of modulus versus frequency decreases as α increases; hence, α is sometimes referred to as the ***spreading factor***. It can also be interpreted as describing a distribution or spectrum of relaxation times. In the case $\alpha = 0$, the Cole–Cole model is exactly the same as the Maxwell model.

The ***Havriliak–Negami*** model is a further generalization of the Maxwell and Cole–Cole models:

$$\frac{M^*_{HN}(\omega)}{M_\infty} = 1 - \left[1 + \left(\frac{i\omega\eta}{M_\infty}\right)^{1-\alpha}\right]^{-\gamma},$$
$$= 1 - [1 + (i\omega\tau)^{1-\alpha}]^{-\gamma} \tag{3.8.53}$$

where γ is an additional empirical parameter. When $\gamma = 1$, the Havriliak–Negami model becomes the same as the Cole–Cole model.

The **fractional Maxwell model** (FMM) (Jaishankar and McKinley, 2012) is a particularly useful description, with creep function

$$J(t) = \frac{1}{\mathbf{V}}\frac{t^\alpha}{\Gamma(1+\alpha)} + \frac{1}{\mathbf{G}}\frac{t^\beta}{\Gamma(1+\beta)}. \tag{3.8.54}$$

The parameters $\mathbf{V}$ and $\mathbf{G}$ in equation 3.8.54 have been called "quasi-properties" of the material. $\mathbf{V}$ has dimensions $Pa \cdot s^\alpha$. Note that if $\alpha = 0$, the stress is proportional to the zeroth derivative of strain as in a linear elastic spring, and $\mathbf{V}$ is an elastic modulus. When $\alpha = 1$, the stress is proportional to the first derivative of strain, as in a dashpot, and $\mathbf{V}$ is a viscosity. Similarly $\mathbf{G}$ has dimensions $Pa \cdot s^\beta$, and also lies between the limiting cases of a spring and a dashpot. Γ is the **gamma function**.

When $\beta = 0$, the FMM creep function becomes **the power law creep** function, which is often observed in ductile rocks:

$$J(t) = \frac{1}{\mathbf{V}}\frac{t^\alpha}{\Gamma(1+\alpha)} + \frac{1}{\mathbf{G}}. \tag{3.8.55}$$

In the limit of $\alpha = 1$ and $\beta = 0$, the FMM reduces to the Maxwell response, with $\mathbf{V}$ becoming the viscosity and $\mathbf{G}$ the high frequency limiting modulus; if in addition

Table 3.8.3 *Expressions for three viscoelastic models as a creep function in time, the complex shear modulus in the Fourier domain, and the transformed modulus in the Laplace domain.*

	Shear Creep Function	Fourier Domain Shear Modulus	Laplace Domain Shear Modulus
Newtonian Viscous	$J(t) = \dfrac{t}{\eta}$	$G^*(\omega) = i\omega\eta$	$\hat{G}(s) = s\eta$
Maxwell	$J(t) = \left(\dfrac{1}{G_\infty} + \dfrac{t}{\eta}\right)$	$G^*(\omega) = \dfrac{i\,\omega\eta}{1 + i\,\omega\eta/G_\infty}$	$\hat{G}(s) = \dfrac{s\eta}{1 + s\eta/G_\infty}$
Fractional Maxwell (Power Law)	$J(t) = \dfrac{1}{V}\dfrac{t^\alpha}{\Gamma(1+\alpha)} + \dfrac{1}{G}\dfrac{t^\beta}{\Gamma(1+\beta)}$	$G^*(\omega) = \dfrac{\mathbf{V}(i\omega)^\alpha\mathbf{G}(i\omega)^\beta}{\mathbf{V}(i\omega)^\alpha + \mathbf{G}(i\omega)^\beta}$	$\hat{G}(s) = \dfrac{\mathbf{V}s^\alpha\mathbf{G}s^\beta}{\mathbf{V}s^\alpha + \mathbf{G}s^\beta}$

$G \to \infty$, the FMM reduces to the Newtonian viscous response. The Newtonian Viscous, Maxwell, and Fractional Maxwell models are compared in Table 3.8.3.

Transient and Fourier responses for the Fractional Maxwell model are shown in Table 3.8.4.

Wave Velocity Dispersion and Attenuation in a Linear Viscoelastic Medium

In a homogeneous isotropic linear viscoelastic medium, the phase velocity $V(\omega)$ and attenuation coefficient $\alpha(\omega)$ of a harmonic plane wave can be written as (Christensen, 1989; Lakes, 2009)

$$V(\omega) = \sqrt{\frac{|M^*|}{\rho}}\sec\left(\frac{\delta}{2}\right), \tag{3.8.56}$$

$$\alpha(\omega) = \frac{\omega}{V(\omega)}\tan\left(\frac{\delta}{2}\right), \tag{3.8.57}$$

where $M^* = M_R + iM_I$ is the complex wave modulus of the material (e.g., shear modulus for shear waves, P-wave modulus for P-waves), ρ is the bulk density, ω is the angular frequency, and δ is the viscoelastic phase angle, such that $\tan(\delta) = M_I/M_R$. We can also write the dispersion relation as

$$\frac{k^*(\omega)}{\omega} = \sqrt{\frac{\rho}{M_R(\omega) + iM_I(\omega)}}, \tag{3.8.58}$$

where $k^* = k_R + ik_I$ is the **complex wave number**. With some manipulation, the wave number in equation 3.8.58 can be separated into real and imaginary parts:

$$\frac{k^*(\omega)}{\omega} = \frac{1}{V_0}\left[\left(\frac{\sqrt{1+z^2}+1}{2(1+z^2)}\right)^{1/2} - i\left(\frac{\sqrt{1+z^2}-1}{2(1+z^2)}\right)^{1/2}\right], \tag{3.8.59}$$

Table 3.8.4 *Transient and Fourier responses of the Fractional Maxwell model. Subscripts R and* I *indicate real and imaginary parts.*

	Fractional Maxwell Model
Differential equation	$\sigma + \dfrac{\mathbf{V}}{\mathbf{G}}\dfrac{d^{\alpha-\beta}\sigma}{dt^{\alpha-\beta}} = \mathbf{V}\dfrac{\partial^{\alpha}\varepsilon}{\partial t^{\alpha}}$
Relaxation function	$G(t) = \mathbf{G}t^{-\beta}E_{\alpha-\beta,\,1-\beta}\left(-\dfrac{\mathbf{G}}{\mathbf{V}}t^{\alpha-\beta}\right)$
	where $E_{a,b}(z) = \displaystyle\sum_{k=0}^{\infty}\dfrac{z^{k}}{\Gamma(ak+b)},\ a>0,\ b>0$
Creep function	$J(t) = \dfrac{1}{\mathbf{V}}\dfrac{t^{\alpha}}{\Gamma(1+\alpha)} + \dfrac{1}{\mathbf{G}}\dfrac{t^{\beta}}{\Gamma(1+\beta)}$
Modulus Laplace domain	$\hat{G}(s) = \dfrac{\mathbf{V}s^{\alpha}\mathbf{G}s^{\beta}}{\mathbf{V}s^{\alpha} + \mathbf{G}s^{\beta}}$
Complex modulus Fourier domain	$G^{*}(\omega) = \dfrac{\mathbf{V}(i\omega)^{\alpha}\mathbf{G}(i\omega)^{\beta}}{\mathbf{V}(i\omega)^{\alpha} + \mathbf{G}(i\omega)^{\beta}}$
Real part of modulus	$G_{R}(\omega) = \dfrac{(\mathbf{G}\omega^{\beta})^{2}(\mathbf{V}\omega^{\alpha})\cos(\pi\alpha/2) + (\mathbf{G}\omega^{\beta})(\mathbf{V}\omega^{\alpha})^{2}\,\cos(\pi\beta/2)}{(\mathbf{V}\omega^{\alpha})^{2} + (\mathbf{G}\omega^{\beta})^{2} + 2(\mathbf{V}\omega^{\alpha})(\mathbf{G}\omega^{\beta})\cos(\pi(\alpha-\beta)/2)}$
Imaginary part of modulus	$G_{I}(\omega) = \dfrac{(\mathbf{G}\omega^{\beta})^{2}(\mathbf{V}\omega^{\alpha})\sin(\pi\alpha/2) + (\mathbf{G}\omega^{\beta})(\mathbf{V}\omega^{\alpha})^{2}\,\sin(\pi\beta/2)}{(\mathbf{V}\omega^{\alpha})^{2} + (\mathbf{G}\omega^{\beta})^{2} + 2(\mathbf{V}\omega^{\alpha})(\mathbf{G}\omega^{\beta})\cos(\pi(\alpha-\beta)/2)}$
$\dfrac{1}{Q} = \dfrac{M_{I}}{M_{R}}$	$\dfrac{(\mathbf{G}\omega^{\beta})^{2}(\mathbf{V}\omega^{\alpha})\sin(\pi\alpha/2) + (\mathbf{G}\omega^{\beta})(\mathbf{V}\omega^{\alpha})^{2}\,\sin(\pi\beta/2)}{(\mathbf{G}\omega^{\beta})^{2}(\mathbf{V}\omega^{\alpha})\cos(\pi\alpha/2) + (\mathbf{G}\omega^{\beta})(\mathbf{V}\omega^{\alpha})^{2}\,\cos(\pi\beta/2)}$

where $z = M_{I}/M_{R}$ and $V_{0} = \sqrt{M_{R}/\rho}$. The **phase velocity** is given by

$$V(\omega) = \frac{\omega}{k_{R}} = V_{0}\left(\frac{2(1+z^{2})}{\sqrt{1+z^{2}}+1}\right)^{1/2}. \qquad (3.8.60)$$

The attenuation coefficient, $\alpha = k_{I}$, of the wave is

$$\alpha = \frac{\omega}{V_{0}}\left(\frac{\sqrt{1+z^{2}}-1}{2(1+z^{2})}\right)^{1/2} \qquad (3.8.61)$$

For **P-waves**, we use the following parameters in equations 3.8.56–3.8.61.

$$M_{P}^{*} = K^{*} + (4/3)\mu^{*};\ V_{0} = \sqrt{\frac{\mathrm{Re}(M_{P}^{*})}{\rho}};\ z = \frac{\mathrm{Im}(M_{P}^{*})}{\mathrm{Re}(M_{P}^{*})} \qquad (3.8.62)$$

where K^{*} is the complex bulk modulus and μ^{*} is the complex shear modulus.

For **S-waves** in a viscoelastic medium, we use the following parameters in equations 3.8.56–3.8.61, as long as $\mathrm{Re}(M_{S}) \neq 0$:

Table 3.8.5 *Wave properties in a linear viscoelastic material*

General linear viscoelastic	
$M^*(\omega)$	$M_R + iM_I$
z	M_I/M_R
k^*/ω	$\left[\dfrac{\rho}{M_R}\dfrac{\sqrt{1+z^2}+1}{2(1+z^2)}\right]^{1/2} - i\left[\dfrac{\rho}{M_R}\dfrac{\sqrt{1+z^2}-1}{2(1+z^2)}\right]^{1/2}$
$V = \dfrac{\omega}{\mathrm{Re}(k^*)}$	$V_0\left[\dfrac{2(1+z^2)}{\sqrt{1+z^2}+1}\right]^{1/2}; \; V_0 = \sqrt{\dfrac{M_R}{\rho}}$
$Z^* = \sqrt{\rho M^*}$	$\sqrt{\dfrac{\rho M_R}{2}}\left\{\left[\sqrt{1+z^2}+1\right]^{1/2} - i\left[\sqrt{1+z^2}-1\right]^{1/2}\right\}$
$\lvert Z^* \rvert$	$\sqrt{\rho}\,[M_R^2 + M_I^2]^{1/4}$

$$M_S^* = \mu^*; \; V_0 = \sqrt{\frac{\mathrm{Re}(M_S^*)}{\rho}}; \; z = \frac{\mathrm{Im}(M_S^*)}{\mathrm{Re}(M_S^*)}; \; \mathrm{Re}(M_S^*)\neq 0 . \tag{3.8.63}$$

Wave properties in a general linear viscoelastic material are summarized in Table 3.8.5.

In the special case of a **P-wave in a pure Newtonian viscous fluid**, $K = K_f$ and $\mu = i\omega\eta$, where K_f is the real fluid bulk modulus and η is the fluid viscosity. The parameters in equations 3.8.56–3.8.61 become

$$M_{PN}^* = K_f + i\frac{4}{3}\omega\eta; \; V_0 = \sqrt{\frac{K_f}{\rho}}; \; z = \frac{4\eta\omega}{3K_f} . \tag{3.8.64}$$

In this case, all of the attenuation results from the shear dissipation associated with the P-wave propagation.

In the special case of *S-waves in a pure Newtonian viscous fluid*, $\mu^* = i\omega\eta$. Then equation 3.8.59 is replaced by

$$\begin{aligned}
\frac{k_{SN}^*}{\omega} &= \sqrt{\frac{\rho}{i\omega\eta}}. \\
&= \sqrt{\frac{\rho}{2\omega\eta}}(1-i)
\end{aligned} \tag{3.8.65}$$

We get a degenerate case of an over-damped wave with "phase velocity"

$$V_{SN} = \frac{\omega}{\mathrm{Re}(k^*)} = \sqrt{\frac{2\omega\eta}{\rho}} \tag{3.8.66}$$

and viscous skin depth

$$\xi_{SN} = \frac{1}{\mathrm{Im}(k_{SN}^*)} = \sqrt{\frac{2\eta}{\omega\rho}} = \frac{\lambda}{2\pi} . \tag{3.8.67}$$

Table 3.8.6 *Wave properties in a standard linear solid*

Standard linear solid	
$M^*(\omega)$	$\dfrac{M_0 + M_\infty(\omega/\omega_S)^2}{1 + (\omega/\omega_S)^2} + i\dfrac{(M_\infty - M_0)(\omega/\omega_S)}{1 + (\omega/\omega_S)^2};\quad \omega_S = \dfrac{M_\infty - M_0}{\eta}$
z	$\dfrac{(M_\infty - M_0)(\omega/\omega_S)}{M_0 + M_\infty(\omega/\omega_S)^2};\quad \omega_S = \dfrac{M_\infty - M_0}{\eta}$
k^*/ω	$\left[\dfrac{\rho}{M_R}\dfrac{\sqrt{1+z^2}+1}{2(1+z^2)}\right]^{1/2} - i\left[\dfrac{\rho}{M_R}\dfrac{\sqrt{1+z^2}-1}{2(1+z^2)}\right]^{1/2}$
V	$V_0\left[\dfrac{2(1+z^2)}{\sqrt{1+z^2}+1}\right]^{1/2};\ V_0 = \sqrt{\dfrac{M_R}{\rho}}$
Z^*	$\sqrt{\dfrac{\rho M_R}{2}}\left\{\left[\sqrt{1+z^2}+1\right]^{1/2} - i\left[\sqrt{1+z^2}-1\right]^{1/2}\right\}$
$|Z^*|$	$\sqrt{\rho}\,[M_R^2 + M_I^2]^{1/4}$

Table 3.8.7 *Wave properties in a Maxwell solid*

Maxwell solid	
$M^*(\omega)$	$\dfrac{i\omega\eta}{1 + i\omega\eta/M_\infty} = \dfrac{\omega^2\eta^2/M_\infty}{1 + \omega^2\eta^2/M_\infty^2} + i\dfrac{\omega\eta}{1 + \omega^2\eta^2/M_\infty^2}$
z	$\dfrac{M_\infty}{\omega\eta}$
k^*/ω	$\sqrt{\dfrac{\rho}{2M_\infty}}\left\{\left[\sqrt{1 + \dfrac{M_\infty^2}{\omega^2\eta^2}}+1\right]^{1/2} - i\left[\sqrt{1 + \dfrac{M_\infty^2}{\omega^2\eta^2}}-1\right]^{1/2}\right\}$
V	$\sqrt{\dfrac{M_\infty}{\rho}}\sqrt{\dfrac{2\omega\eta}{\omega\eta + \sqrt{M_\infty^2 + \omega^2\eta^2}}}$
Z^*	$\sqrt{\dfrac{\rho M_\infty}{2}}\left\{\left[\dfrac{\omega\eta\sqrt{M_\infty^2 + \omega^2\eta^2} + \omega^2\eta^2}{M_\infty^2 + \omega^2\eta^2}\right]^{1/2} - i\left[\dfrac{\omega\eta\sqrt{M_\infty^2 + \omega^2\eta^2} - \omega^2\eta^2}{M_\infty^2 + \omega^2\eta^2}\right]^{1/2}\right\}$
$|Z^*|$	$\left[\dfrac{\omega^2\eta^2\rho^2}{1 + \omega^2\eta^2/M_\infty^2}\right]^{1/4} = \sqrt{\rho M_\infty}\left[\dfrac{\omega^2\eta^2}{M_\infty^2 + \omega^2\eta^2}\right]^{1/4}$

Phase velocities and complex moduli and wave numbers for other viscoelastic models are shown in Tables 3.8.6–3.8.9.

The complex impedance of an isotropic linear viscoelastic medium, $Z^* = Z_R + iZ_I$, is given by (Sheen *et al.*, 1996)

$$Z^* = \sqrt{\rho M^*}\,, \tag{3.8.68}$$

Table 3.8.8 *P-wave properties in a Newtonian viscous fluid*

P-wave Newtonian viscous fluid	
$M^*(\omega)$	$K + i4\omega\eta/3$
z	$\dfrac{4\omega\eta}{3K}$
k^*/ω	$\left[\dfrac{\rho}{K}\dfrac{\sqrt{1+(4\omega\eta/3/K)^2}+1}{2\left(1+(4\omega\eta/3K)^2\right)}\right]^{1/2} - i\left[\dfrac{\rho}{K}\dfrac{\sqrt{1+(4\omega\eta/3K)^2}-1}{2\left(1+(4\omega\eta/3K)^2\right)}\right]^{1/2}$
V	$V_0\left[\dfrac{2\left(1+(4\omega\eta/3K)^2\right)}{\sqrt{1+(4\omega\eta/3K)^2}+1}\right]^{1/2}$; $V_0 = \sqrt{\dfrac{K}{\rho}}$
Z^*	$\sqrt{\rho K/2}\left\{\left[\sqrt{1+(4\omega\eta/3K)^2}+1\right]^{1/2} - i\left[\sqrt{1+(4\omega\eta/3K)^2}-1\right]^{1/2}\right\}$
$\lvert Z^* \rvert$	$\sqrt{\rho}\left[K^2+(4\omega\eta/3)^2\right]^{1/4} = \sqrt{\rho K}\left[1+(4\omega\eta/3K)^2\right]^{1/4}$

Table 3.8.9 *S-wave properties in a Newtonian viscous fluid*

S-wave Newtonian viscous fluid	
$M^*(\omega)$	$i\omega\eta$
k^*/ω	$\sqrt{\dfrac{\rho}{2\omega\eta}}(1-i)$
V	$\sqrt{\dfrac{2\omega\eta}{\rho}}$
Z^*	$\sqrt{i\omega\eta\rho} = \sqrt{\dfrac{\omega\eta\rho}{2}}(1+i)$
$\lvert Z^* \rvert$	$\sqrt{\omega\eta\rho}$

where $M^* = M_R + iM_I$ is the complex modulus of the medium.

For the special case of a shear wave in a Maxwell medium, the complex impedance is

$$
\begin{aligned}
Z^*_{SM} &= \left[\dfrac{i\omega\eta\rho}{1+i\omega\eta/M_\infty}\right]^{1/2}. \\
&= \sqrt{\rho M_\infty}\left[\dfrac{i\omega\eta}{M_\infty+i\omega\eta}\right]^{1/2}
\end{aligned}
\tag{3.8.69}
$$

For the case of a shear wave in a Newtonian viscous fluid,

$$
Z_R = Z_I = \sqrt{\dfrac{\rho\omega\eta}{2}}.
\tag{3.8.70}
$$

Using equation 3.8.66 for the velocity of a shear wave in a Newtonian viscous fluid, we see that

$$\rho V_{SN} = \sqrt{2\rho\omega\eta} \quad \text{(Newtonian fluid)}$$
$$= \sqrt{2} \cdot |Z_{SN}|$$

$$(3.8.71)$$

> **PITFALL:**
>
> For viscoelastic materials, the impedance is given by $Z^* = \sqrt{\rho M^*}$, as stated in equation 3.8.68. However, the phase velocity is NOT simply the ratio of impedance and density, i.e., $V \neq |Z^*|/\rho$.

Uses

The equations presented in this section are used for phenomenological modeling of attenuation and velocity dispersion of seismic waves.

Assumptions and Limitations

The equations presented in this section assume that the material is linear, dissipative, and causal.

3.9   Kramers–Kronig Relations between Velocity Dispersion and Q

Synopsis

For linear viscoelastic systems, causality requires that there be a very specific relation between velocity or modulus dispersion and Q; that is, if the dispersion is completely characterized for all frequencies then Q is known for all frequencies and vice versa.

We can write a viscoelastic constitutive law between the stress and strain components as

$$\sigma(t) = \frac{dr}{dt} * \varepsilon(t)$$

$$(3.9.1)$$

where $r(t)$ is the relaxation function and $*$ denotes convolution. Then in the Fourier domain we can write

$$\widetilde{\sigma}(\omega) = M(\omega)\,\widetilde{\varepsilon}(\omega)$$

$$(3.9.2)$$

where $M(\omega)$ is the complex modulus. For $r(t)$ to be causal, in the frequency domain the real and imaginary parts of $M(\omega)/(I\omega)$ must be Hilbert transform pairs (Bourbié *et al.*, 1987), as discussed in Section 1.1:

$$M_{\mathrm{I}}(\omega) = \frac{\omega}{\pi} \int_{-\infty}^{+\infty} \frac{M_{\mathrm{R}}(\alpha) - M_{\mathrm{R}}(0)}{\alpha} \frac{d\alpha}{\alpha - \omega} \tag{3.9.3}$$

$$M_{\mathrm{R}}(\omega) - M_{\mathrm{R}}(0) = -\frac{\omega}{\pi} \int_{-\infty}^{+\infty} \frac{M_{\mathrm{I}}(\alpha)}{\alpha} \frac{d\alpha}{\alpha - \omega} \tag{3.9.4}$$

where $M_{\mathrm{R}}(0)$ is the real part of the modulus at zero frequency, which results because there is an instantaneous elastic response from a viscoelastic material. If we express this in terms of

$$Q^{-1} = \frac{M_{\mathrm{I}}(\omega)\mathrm{sgn}(\omega)}{M_{\mathrm{R}}(\omega)} \tag{3.9.5}$$

then we obtain

$$Q^{-1}(\omega) = \frac{|\omega|}{\pi M_{\mathrm{R}}(\omega)} \int_{-\infty}^{+\infty} \frac{M_{\mathrm{R}}(\alpha) - M_{\mathrm{R}}(0)}{\alpha} \frac{d\alpha}{\alpha - \omega} \tag{3.9.6}$$

and its inverse

$$M_{\mathrm{R}}(\omega) - M_{\mathrm{R}}(0) = \frac{-\omega}{\pi} \int_{-\infty}^{+\infty} \frac{Q^{-1}(\alpha)M_{\mathrm{R}}(\alpha)}{|\alpha|} \frac{d\alpha}{(\alpha - \omega)} \tag{3.9.7}$$

From these we see the expected result that a larger attenuation is generally associated with larger dispersion. Zero attenuation requires zero velocity dispersion.

One never has more than partial information about the frequency dependence of velocity and Q, but the Kramers–Kronig relation allows us to put some constraints on the material behavior. For example, Lucet (1989) measured velocity and attenuation at two frequencies ($\approx$ 1 kHz and 1 MHz) and used the Kramers–Kronig relations to compare the differences with various viscoelastic models, as shown schematically in Figure 3.9.1. Using equation 3.9.1, we can determine the expected ratio of low-frequency modulus or velocity, V_{lf}, and high-frequency modulus or velocity, V_{hf}, for various functional forms of Q (for example, constant Q or nearly constant Q). In all cases, linear viscoelastic behavior should lead to an intercept of $V_{\mathrm{lf}}/V_{\mathrm{hf}} = 1$ at $1/Q = 0$.

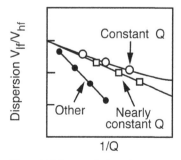

Figure 3.9.1 Lucet's (1989) use of the Kramers–Kronig relations to compare high-and low-frequency measured velocities and Q with various viscoelastic models

Mechanisms with peaked attenuation curves between the measurement points will generally cause a larger dispersion, which appears as a steeper negative slope.

Uses

The Kramers–Kronig equations can be used to relate velocity dispersion and Q in linear viscoelastic materials.

Assumptions and Limitations

The Kramers–Kronig equations apply when the material is linear and causal.

3.10   Waves in Layered Media: Full-Waveform Synthetic Seismograms

Synopsis

One of the approaches for computing wave propagation in layered media is the use of propagator matrices (Aki and Richards, 1980; Claerbout, 1985). The wave variables of interest (usually stresses and particle velocity or displacements) at the top and bottom of the stack of layers are related by a product of propagator matrices, one for each layer. The calculations are done in the frequency domain and include the effects of all multiples. For waves traveling perpendicularly to n layers with layer velocities, densities, and thicknesses V_k, ρ_k, and d_k, respectively

$$\begin{bmatrix} S \\ W \end{bmatrix}_n = \prod_{k=1}^{n} A_k \begin{bmatrix} S \\ W \end{bmatrix}_1 \tag{3.10.1}$$

where S and W are the Fourier transforms of the wave variables σ and w, respectively. For normal-incidence P-waves, σ is interpreted as the normal stress across each interface, and w is the normal component of the particle velocity. For normal-incidence S-waves, σ is the shear traction across each interface, and w is the tangential component of the particle velocity. Each layer matrix A_k has the form

$$A_k = \begin{bmatrix} \cos\left(\dfrac{\omega d_k}{V_k}\right) & i\rho_k V_k \sin\left(\dfrac{\omega d_k}{V_k}\right) \\ \dfrac{i}{\rho_k V_k} \sin\left(\dfrac{\omega d_k}{V_k}\right) & \cos\left(\dfrac{\omega d_k}{V_k}\right) \end{bmatrix} \tag{3.10.2}$$

where ω is the angular frequency.

Kennett (1974, 1983) used the invariant imbedding method to generate the response of a layered medium recursively by adding one layer at a time (Figure 3.10.1). The overall reflection and transmission matrices, $\hat{R}_D$ and $\hat{T}_D$, respectively, for downgoing waves through a stack of layers are given by the following recursion relations:

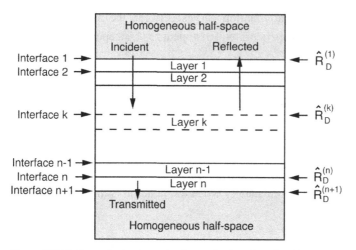

Figure 3.10.1 Recursively determined transfer functions in a layered medium

$$\hat{\mathbf{R}}_{\mathrm{D}}^{(k)} = \mathbf{R}_{\mathrm{D}}^{(k)} + \mathbf{T}_{\mathrm{U}}^{(k)}\mathbf{E}_{\mathrm{D}}^{(k)}\hat{\mathbf{R}}_{\mathrm{D}}^{(k+1)}\mathbf{E}_{\mathrm{D}}^{(k)}[\mathbf{I} - \mathbf{R}_{\mathrm{U}}^{(k)}\mathbf{E}_{\mathrm{D}}^{(k)}\hat{\mathbf{R}}_{\mathrm{D}}^{(k+1)}\mathbf{E}_{\mathrm{D}}^{(k)}]^{-1}\mathbf{T}_{\mathrm{D}}^{(k)} \tag{3.10.3}$$

$$\hat{\mathbf{T}}_{\mathrm{D}}^{(k)} = \hat{\mathbf{T}}_{\mathrm{D}}^{(k+1)}\mathbf{E}_{\mathrm{D}}^{(k)}[\mathbf{I} - \mathbf{R}_{\mathrm{U}}^{(k)}\mathbf{E}_{\mathrm{D}}^{(k)}\hat{\mathbf{R}}_{\mathrm{D}}^{(k+1)}\mathbf{E}_{\mathrm{D}}^{(k)}]^{-1}\mathbf{T}_{\mathrm{D}}^{(k)} \tag{3.10.4}$$

where $\mathbf{R}_{\mathrm{D}}^{(k)}, \mathbf{T}_{\mathrm{D}}^{(k)}, \mathbf{R}_{\mathrm{U}}^{(k)}$, and $\mathbf{T}_{\mathrm{U}}^{(k)}$ are just the single-interface downward and upward reflection and transmission matrices for the kth interface:

$$\mathbf{R}_{\mathrm{D}}^{(k)} = \begin{bmatrix} \overset{\downarrow\uparrow}{\mathrm{PP}} & \overset{\downarrow\uparrow}{\mathrm{SP}}\left(\dfrac{V_{\mathrm{P}(k-1)}\cos\theta_{k-1}}{V_{\mathrm{S}(k-1)}\cos\phi_{k-1}}\right)^{1/2} \\ \overset{\downarrow\uparrow}{\mathrm{PS}}\left(\dfrac{V_{\mathrm{S}(k-1)}\cos\phi_{k-1}}{V_{\mathrm{P}(k-1)}\cos\theta_{k-1}}\right)^{1/2} & \overset{\downarrow\uparrow}{\mathrm{SS}} \end{bmatrix} \tag{3.10.5}$$

$$\mathbf{T}_{\mathrm{D}}^{(k)} = \begin{bmatrix} \overset{\downarrow\downarrow}{\mathrm{PP}}\left(\dfrac{\rho_k V_{\mathrm{P}(k)}\cos\theta_k}{\rho_{k-1}V_{\mathrm{P}(k-1)}\cos\theta_{k-1}}\right)^{1/2} & \overset{\downarrow\downarrow}{\mathrm{SP}}\left(\dfrac{\rho_k V_{\mathrm{P}(k)}\cos\theta_k}{\rho_{k-1}V_{\mathrm{S}(k-1)}\cos\phi_{k-1}}\right)^{1/2} \\ \overset{\downarrow\downarrow}{\mathrm{PS}}\left(\dfrac{\rho_k V_{\mathrm{S}(k)}\cos\phi_k}{\rho_{k-1}V_{\mathrm{P}(k-1)}\cos\theta_{k-1}}\right)^{1/2} & \overset{\downarrow\downarrow}{\mathrm{SS}}\left(\dfrac{\rho_k V_{\mathrm{S}(k)}\cos\phi_k}{\rho_{k-1}V_{\mathrm{S}(k-1)}\cos\phi_{k-1}}\right)^{1/2} \end{bmatrix} \tag{3.10.6}$$

$$\mathbf{R}_{\mathrm{U}}^{(k)} = \begin{bmatrix} \overset{\uparrow\downarrow}{\mathrm{PP}} & \overset{\uparrow\downarrow}{\mathrm{SP}}\left(\dfrac{V_{\mathrm{P}(k)}\cos\theta_k}{V_{\mathrm{S}(k)}\cos\phi_k}\right)^{1/2} \\ \overset{\uparrow\downarrow}{\mathrm{PS}}\left(\dfrac{V_{\mathrm{S}(k)}\cos\phi_k}{V_{\mathrm{P}(k)}\cos\theta_k}\right)^{1/2} & \overset{\uparrow\downarrow}{\mathrm{SS}} \end{bmatrix} \tag{3.10.7}$$

$$\mathbf{T}_{\mathrm{U}}^{(k)} = \begin{bmatrix} \overset{\uparrow\uparrow}{\mathrm{PP}}\left(\dfrac{\rho_{k-1}V_{\mathrm{P}(k-1)}\cos\theta_{k-1}}{\rho_k V_{\mathrm{P}(k)}\cos\theta_k}\right)^{1/2} & \overset{\uparrow\uparrow}{\mathrm{SP}}\left(\dfrac{\rho_{k-1}V_{\mathrm{P}(k-1)}\cos\theta_{k-1}}{\rho_k V_{\mathrm{S}(k)}\cos\phi_k}\right)^{1/2} \\ \overset{\uparrow\uparrow}{\mathrm{PS}}\left(\dfrac{\rho_{k-1}V_{\mathrm{S}(k-1)}\cos\phi_{k-1}}{\rho_k V_{\mathrm{P}(k)}\cos\theta_k}\right)^{1/2} & \overset{\uparrow\uparrow}{\mathrm{SS}}\left(\dfrac{\rho_{k-1}V_{\mathrm{S}(k-1)}\cos\phi_{k-1}}{\rho_k V_{\mathrm{S}(k)}\cos\phi_k}\right)^{1/2} \end{bmatrix} \tag{3.10.8}$$

with

$$
\begin{pmatrix}
\overset{\downarrow\uparrow}{PP} & \overset{\downarrow\uparrow}{SP} & \overset{\uparrow\uparrow}{PP} & \overset{\uparrow\uparrow}{SP} \\
\overset{\downarrow\uparrow}{PS} & \overset{\downarrow\uparrow}{SS} & \overset{\uparrow\uparrow}{PS} & \overset{\uparrow\uparrow}{SS} \\
\overset{\downarrow\downarrow}{PP} & \overset{\downarrow\downarrow}{SP} & \overset{\uparrow\downarrow}{PP} & \overset{\uparrow\downarrow}{SP} \\
\overset{\downarrow\downarrow}{PS} & \overset{\downarrow\downarrow}{SS} & \overset{\uparrow\downarrow}{PS} & \overset{\uparrow\downarrow}{SS}
\end{pmatrix}
= \mathbf{M}^{-1}\mathbf{N}
\tag{3.10.9}
$$

$$
\mathbf{M} =
\begin{bmatrix}
-\sin\theta_{k-1} & -\cos\phi_{k-1} & \sin\theta_k & \cos\phi_k \\
\cos\theta_{k-1} & -\sin\phi_{k-1} & \cos\theta_k & -\sin\phi_k \\
2I_{S(k-1)}\sin\phi_{k-1}\cos\theta_{k-1} & I_{S(k-1)}(1-2\sin^2\phi_{k-1}) & 2I_{S(k)}\sin\phi_k\cos\theta_k & I_{S(k)}(1-2\sin^2\phi_k) \\
-I_{P(k-1)}(1-2\sin^2\phi_{k-1}) & I_{S(k-1)}\sin2\phi_{k-1} & I_{P(k)}(1-2\sin^2\phi_k) & -I_{S(k)}\sin2\phi_k
\end{bmatrix}
\tag{3.10.10}
$$

$$
\mathbf{N} =
\begin{bmatrix}
\sin\theta_{k-1} & \cos\phi_{k-1} & -\sin\theta_k & -\cos\phi_k \\
\cos\theta_{k-1} & -\sin\phi_{k-1} & \cos\theta_k & -\sin\phi_k \\
2I_{S(k-1)}\sin\phi_{k-1}\cos\theta_{k-1} & I_{S(k-1)}(1-2\sin^2\phi_{k-1}) & 2I_{S(k)}\sin\phi_k\cos\theta_k & I_{S(k)}(1-2\sin^2\phi_k) \\
I_{P(k-1)}(1-2\sin^2\phi_{k-1}) & -I_{S(k-1)}\sin2\phi_{k-1} & -I_{P(k)}(1-2\sin^2\phi_k) & I_{S(k)}\sin2\phi_k
\end{bmatrix}
\tag{3.10.11}
$$

where $I_{P(k)} = \rho_k V_{P(k)}$ and $I_{S(k)} = \rho_k V_{S(k)}$ are the P and S impedances, respectively, of the kth layer, and θ_k and ϕ_k are the angles made by the P- and S-wave vectors with the normal to the kth interface. The elements of the reflection and transmission matrices $\mathbf{R}_D^{(k)}$, $\mathbf{T}_D^{(k)}$, $\mathbf{R}_U^{(k)}$, and $\mathbf{T}_U^{(k)}$ are the reflection and transmission coefficients for scaled displacements, which are proportional to the square root of the energy flux. The scaled displacement u' is related to the displacement u by $u' = u\sqrt{\rho V\cos\theta}$.

For normal-incidence wave propagation with no mode conversions, the reflection and transmission matrices reduce to the scalar coefficients:

$$
R_D^{(k)} = \frac{\rho_{k-1}V_{k-1} - \rho_k V_k}{\rho_{k-1}V_{k-1} + \rho_k V_k}, \quad R_U^{(k)} = -R_D^{(k)}
\tag{3.10.12}
$$

$$
T_U^{(k)} = \frac{2\sqrt{\rho_{k-1}V_{k-1}\rho_k V_k}}{\rho_{k-1}V_{k-1} + \rho_k V_k}, \quad T_U^{(k)} = T_D^{(k)}
\tag{3.10.13}
$$

The phase shift operator for propagation across each new layer is given by $\mathbf{E}_D^{(k)}$

$$
\mathbf{E}_D^{(k)} =
\begin{bmatrix}
\exp\left(i\omega d_k\cos\theta_k/V_{P(k)}\right) & 0 \\
0 & \exp\left(i\omega d_k\cos\phi_k/V_{S(k)}\right)
\end{bmatrix}
\tag{3.10.14}
$$

where θ_k and ϕ_k are the angles between the normal to the layers and the directions of propagation of P- and S-waves, respectively. The terms $\hat{\mathbf{R}}_D$ and $\hat{\mathbf{T}}_D$ are functions of ω and represent the overall transfer functions of the layered medium in the frequency domain. Time-domain seismograms are obtained by multiplying the overall transfer function by the Fourier transform of the source wavelet and then performing an inverse transform. The recursion starts at the base of the layering at interface $n + 1$ (Figure 3.10.1). Setting $\mathbf{R}_D^{(n+1)} = \hat{\mathbf{R}}_D^{(n+1)} = 0$ and $\mathbf{T}_D^{(n+1)} = \hat{\mathbf{T}}_D^{(n+1)} = \mathbf{I}$ simulates a stack of layers overlying a semi-infinite homogeneous half-space with properties equal to those of the last layer, layer n. The recursion relations are stepped up through the stack of layers one at a time to finally give $\hat{\mathbf{R}}_D^{(1)}$ and $\hat{\mathbf{T}}_D^{(1)}$ the overall reflection and transmission response for the whole stack.

Example: Calculate the P-wave normal-incidence overall reflection and transmission functions $\hat{R}_D^{(1)}$ and $\hat{T}_D^{(1)}$ recursively for a three-layered medium with layer properties as follows:

$$V_{P(1)} = 4000\text{m/s}, \quad \rho_1 = 2300\text{kg/m}^3, \quad d_1 = 100\text{m}$$

$$V_{P(2)} = 3000\text{m/s}, \quad \rho_2 = 2100\text{kg/m}^3, \quad d_2 = 50\text{m}$$

$$V_{P(3)} = 5000\text{m/s}, \quad \rho_3 = 2500\text{kg/m}^3, \quad d_3 = 200\text{m}$$

The recursion starts with $\hat{R}_D^{(4)} = 0$ and $\hat{T}_D^{(4)} = 1$. The normal-incidence reflection and transmission coefficients at interface 3 are

$$R_D^{(3)} = \frac{\rho_2 V_{P(2)} - \rho_3 V_{P(3)}}{\rho_2 V_{P(2)} + \rho_3 V_{P(3)}} = -0.33, \quad R_U^{(3)} = -R_D^{(3)}$$

$$T_D^{(3)} = \frac{2\sqrt{\rho_2 V_{P(2)} \rho_3 V_{P(3)}}}{\rho_2 V_{P(2)} + \rho_3 V_{P(3)}} = 0.94, \quad T_U^{(3)} = T_D^{(3)}$$

and the phase factor for propagation across layer 3 is

$$E_D^{(3)} = \exp(i2\pi f d_3 / V_{P(3)}) = \exp(i2\pi f 200/5000)$$

where f is the frequency. The recursion relations give

$$\hat{R}_D^{(3)} = R_D^{(3)} + \frac{T_U^{(3)} E_D^{(3)} \hat{R}_D^{(4)} E_D^{(3)}}{1 - R_U^{(3)} E_D^{(3)} \hat{R}_D^{(4)} E_D^{(3)}} T_D^{(3)}$$

$$\hat{T}_D^{(3)} = \frac{\hat{T}_D^{(4)} E_D^{(3)}}{1 - R_U^{(3)} E_D^{(3)} \hat{R}_D^{(4)} E_D^{(3)}} T_D^{(3)}$$

The recursion is continued in a similar manner until finally we obtain $\hat{R}_D^{(1)}$ and $\hat{T}_D^{(1)}$.

The matrix inverse

$$[\mathbf{I} - \mathbf{R}_U^{(k)} \mathbf{E}_D^{(k)} \hat{\mathbf{R}}_D^{(k+1)} \mathbf{E}_D^{(k)}]^{-1} \tag{3.10.15}$$

is referred to as the reverberation operator and includes the response caused by all internal reverberations. In the series expansion of the matrix inverse

$$\left[\mathbf{I} - \mathbf{R}_U^{(k)} \mathbf{E}_D^{(k)} \hat{\mathbf{R}}_D^{(k+1)} \mathbf{E}_D^{(k)}\right]^{-1} = \mathbf{I} + \mathbf{R}_U^{(k)} \mathbf{E}_D^{(k)} \hat{\mathbf{R}}_D^{(k+1)} \mathbf{E}_D^{(k)}$$

$$+ \mathbf{R}_U^{(k)} \mathbf{E}_D^{(k)} \hat{\mathbf{R}}_D^{(k+1)} \mathbf{E}_D^{(k)} \mathbf{R}_U^{(k)} \mathbf{E}_D^{(k)} \hat{\mathbf{R}}_D^{(k+1)} \mathbf{E}_D^{(k)} + \ldots \qquad (3.10.16)$$

the first term represents the primaries and each successive term corresponds to higher-order multiples. Truncating the expansion to $m + 1$ terms includes m internal multiples in the approximation. The full multiple sequence is included with the exact matrix inverse.

Uses

The methods described in this section can be used to compute full-wave seismograms, which include the effects of multiples for wave propagation in layered media.

Assumptions and Limitations

The algorithms described in this section assume the following:
- layered medium with no lateral heterogeneities;
- layers are isotropic, linear, elastic; and
- plane-wave, time-harmonic propagation.

3.11   Waves in Layered Media: Stratigraphic Filtering and Velocity Dispersion

Synopsis

Waves in layered media undergo attenuation and velocity dispersion caused by multiple scattering at the layer interfaces. The effective phase slowness of normally incident waves through layered media depends on the relative scales of the wavelength and layer thicknesses and may be written as $S_{\text{eff}} = S_{\text{rt}} + S_{\text{st}}$. The term S_{rt} is the ray theory slowness of the direct ray that does not undergo any reflections and is just the thickness-weighted average of the individual layer slownesses. The individual slownesses may be complex to account for intrinsic attenuation. The excess slowness S_{st} (sometimes called the stratigraphic slowness) arises because of multiple scattering within the layers. A flexible approach to calculating the effective slowness and travel time follows from Kennett's (1974) invariant imbedding formulation for the transfer function of a layered medium. The layered medium, of total thickness L, consists of layers with velocities (inverse slownesses), densities, and thicknesses, V_j, ρ_j, and l_j, respectively.

The complex stratigraphic slowness is frequency dependent and can be calculated recursively (Frazer, 1994) by

$$S_{st} = \frac{1}{i\omega L} \sum_{j=1}^{n} \ln\left(\frac{t_j}{1 - R_j\theta_j^2 r_j}\right) \tag{3.11.1}$$

As each new layer $j + 1$ is added to the stack of j layers, R is updated according to

$$R_{j+1} = -r_{j+1} + \frac{R_j\theta_{j+1}^2 t_{j+1}^2}{1 - R_j\theta_{j+1}^2 r_{j+1}} \tag{3.11.2}$$

(with $R_0 = 0$) and the term

$$\ln[t_{j+1}(1 - R_{j+1}\theta_{j+1}^2 r_{j+1})^{-1}] \tag{3.11.3}$$

is accumulated in the sum. In the previous expressions, t_j and r_j, are the transmission and reflection coefficients defined as

$$t_j = \frac{2\sqrt{\rho_j V_j \rho_{j+1} V_{j+1}}}{\rho_j V_j + \rho_{j+1} V_{j+1}} \tag{3.11.4}$$

$$r_j = \frac{\rho_{j+1} V_{j+1} - \rho_j V_j}{\rho_j V_j + \rho_{j+1} V_{j+1}} \tag{3.11.5}$$

whereas $\theta_j = \exp(i\omega l_j/V_j)$ is the phase shift for propagation across layer j and ω is the angular frequency. The total travel time is $T = T_{rt} + T_{st}$, where T_{rt} is the ray theory travel time given by

$$T_{rt} = \sum_{j=1}^{n} \frac{l_j}{V_j} \tag{3.11.6}$$

and T_{st} is given by

$$T_{st} = \text{Re}\left[\frac{1}{i\omega} \sum_{j=1}^{n} \ln\left(\frac{t_j}{1 - R_j\theta_j^2 r_j}\right)\right] \tag{3.11.7}$$

The deterministic results previously given are not restricted to small perturbations in the material properties or statistically stationary geology.

Example: Calculate the excess stratigraphic travel time caused by multiple scattering for a normally incident P-wave traveling through a three-layered medium with layer properties as follows:

$V_{P(1)} = 4000\text{m/s}, \quad \rho_1 = 2300\text{kg/m}^3, \quad l_1 = 100\text{m}$

$V_{P(2)} = 3000\text{m/s}, \quad \rho_2 = 2100\text{kg/m}^3, \quad l_2 = 50\text{m}$

$V_{P(3)} = 5000\text{m/s}, \quad \rho_3 = 2500\text{kg/m}^3, \quad l_3 = 200\text{m}$

The excess travel time is given by

$$T_{st} = \text{Re}\left[\frac{1}{i\omega}\sum_{j=1}^{n}\ln\left(\frac{t_j}{1 - R_j\theta_j^2 r_j}\right)\right]$$

The recursion begins with $R_0 = 0$,

$$R_1 = -r_1 + \frac{R_0\theta_1^2 t_1^2}{1 - R_0\theta_1^2 r_1}$$

where

$$t_1 = \frac{2\sqrt{\rho_1 V_1 \rho_2 V_2}}{\rho_1 V_1 + \rho_2 V_2} = 0.98$$

$$r_1 = \frac{\rho_2 V_2 - \rho_1 V_1}{\rho_1 V_1 + \rho_2 V_2} = -0.19$$

$\theta_1 = \exp(i2\pi f l_1/V_1) = \exp(i2\pi f 100/4000)$ with f as the frequency.

The recursion is continued to obtain R_2 and R_3. Setting $t_3 = 1$ and $r_3 = 0$ simulates an impedance-matching homogeneous infinite half-space beneath layer 3.

Finally, the excess travel time, which is a function of the frequency, is obtained by taking the real part of the sum as follows:

$$T_{st} = \text{Re}\left\{\frac{1}{i2\pi f}\left[\ln\left(\frac{t_1}{1 - R_1\theta_1^2 r_1}\right) + \ln\left(\frac{t_2}{1 - R_2\theta_2^2 r_2}\right) + \ln\left(\frac{t_3}{1 - R_3\theta_3^2 r_3}\right)\right]\right\}$$

The effect of the layering can be thought of as a filter that attenuates the input wavelet and introduces a delay. The function

$$A(\omega) = \exp(i\omega x S_{st}) = \exp(i\omega T_{rt}S_{st}/S_{rt}) \tag{3.11.8}$$

(where S_{rt} is assumed to be real in the absence of any intrinsic attenuation) is sometimes called the stratigraphic filter.

The **O'Doherty-Anstey formula** (O'Doherty and Anstey, 1971; Banik *et al.*, 1985)

$$|A(\omega)| \approx \exp(-\hat{R}(\omega)T_{rt}) \tag{3.11.9}$$

approximately relates the amplitude of the stratigraphic filter to the power spectrum $\hat{R}(\omega)$ of the reflection coefficient time series $r(\tau)$ where

$$\tau(x) = \int_0^x dx'/V(x') \tag{3.11.10}$$

is the one-way travel time. Initially the O'Doherty-Anstey formula was obtained by a heuristic approach (O'Doherty and Anstey, 1971). Later, various authors substantiated the result using statistical ensemble averages of wavefields (Banik *et al.*, 1985), deterministic formulations (Resnick *et al.*, 1986), and the concepts of self-averaged values and wave localization (Shapiro and Zien, 1993). Resnick *et al.* (1986) showed that the O'Doherty-Anstey formula is obtained as an approximation from the exact frequency-domain theory of Resnick *et al.* by neglecting quadratic terms in the Riccatti equation of Resnick *et al.* Another equivalent way of expressing the O'Doherty-Anstey relation is

$$\frac{\text{Im}(S_{\text{st}})}{S_{\text{rt}}} \approx \frac{1}{2Q} \approx \frac{\hat{R}(\omega)}{\omega} = \frac{1}{2}\omega \hat{M}(2\omega) \tag{3.11.11}$$

Here $1/Q$ is the scattering attenuation caused by the multiples, and $\hat{M}(\omega)$ is the power spectrum of the logarithmic impedance fluctuations of the medium, $\ln[\rho(\tau)V(\tau)] - \langle \ln[\rho(\tau)V(\tau)] \rangle$, where $\langle \cdot \rangle$ denotes a stochastic ensemble average. Because the filter is minimum phase, $\omega \, \text{Re}(S_{\text{st}})$ and $\omega \, \text{Im}(S_{\text{st}})$ are a Hilbert transform pair,

$$\frac{\text{Re}(S_{\text{st}})}{S_{\text{rt}}} \approx \frac{\delta t}{T_{\text{rt}}} \approx \frac{H\{\hat{R}(\omega)\}}{\omega} \tag{3.11.12}$$

where $H\{\cdot\}$ denotes the Hilbert transform and δt is the excess travel caused by multiple reverberations.

Shapiro and Zien (1993) generalized the O'Doherty–Anstey formula for non-normal incidence. The derivation is based on a small perturbation analysis and requires the fluctuations of material parameters to be small (<30%). The generalized formula for plane pressure (scalar) waves in an acoustic medium incident at an angle θ with respect to the layer normal is

$$|A(\omega)| \approx \exp\left[-\frac{\hat{R}(\omega \cos\theta)}{\cos^4\theta} T_{\text{rt}}\right] \tag{3.11.13}$$

whereas

$$|A(\omega)| \approx \exp\left[-\frac{(2\cos^2\theta - 1)^2 \hat{R}(\omega \cos\theta)}{\cos^4\theta} T_{\text{rt}}\right] \tag{3.11.14}$$

for SH-waves in an elastic medium (Shapiro *et al.*, 1994).

For a perfectly periodic stratified medium made up of two constituents with phase velocities V_1, V_2; densities ρ_1, ρ_2; and thicknesses l_1, l_2, the velocity dispersion relation may be obtained from the **Floquet solution** (Christensen, 1991) for periodic media:

$$\cos\left[\frac{\omega(l_1 + l_2)}{V}\right] = \cos\left(\frac{\omega l_1}{V_1}\right)\cos\left(\frac{\omega l_2}{V_2}\right) - \chi \sin\left(\frac{\omega l_1}{V_1}\right)\sin\left(\frac{\omega l_2}{V_2}\right)$$

$$\chi = \frac{(\rho_1 V_1)^2 + (\rho_2 V_2)^2}{2\rho_1\rho_2 V_1 V_2} \tag{3.11.15}$$

The Floquet solution is valid for arbitrary contrasts in the layer properties. If the spatial period ($l_1 + l_2$) is an integer multiple of one-half wavelength, multiple reflections are in phase and add constructively, resulting in a large total accumulated reflection. The frequency at which this **Bragg scattering** condition is satisfied is called the Bragg frequency. Waves cannot propagate within a stop-band around the Bragg frequency.

Uses

The results described in this section can be used to estimate velocity dispersion and attenuation caused by scattering for normal-incidence wave propagation in layered media.

Assumptions and Limitations

The methods described in this section apply under the following conditions:
- layers are isotropic, linear elastic with no lateral variation;
- propagation is normal to the layers except for the generalized O'Doherty-Anstey formula; and
- plane-wave, time-harmonic propagation is assumed.

3.12 Waves in Layered Media: Frequency-Dependent Anisotropy, Dispersion, and Attenuation

Synopsis

Waves in layered media undergo attenuation and velocity dispersion caused by multiple scattering at the layer interfaces. Thinly layered media also give rise to velocity anisotropy. At low frequencies this phenomenon is usually described by the Backus average. Velocity anisotropy and dispersion in a multilayered medium are two aspects of the same phenomenon and are related to the frequency- and angle-dependent transmissivity resulting from multiple scattering in the medium. Shapiro *et al.* (1994) and Shapiro and Hubral (1995, 1996, 1999) have presented a whole-frequency-range statistical theory for the angle-dependent transmissivity of layered media for scalar waves (pressure waves in fluids) and elastic waves. The theory encompasses the Backus average in the low-frequency limit and ray theory in the high-frequency limit. The formulation avoids the problem of ensemble averaging versus measurements for a single realization by working with parameters that are averaged by the wave-propagation process itself for sufficiently long propagation paths. The results are obtained in the limit when the path length tends to infinity. Practically, this means the results are applicable when path lengths are very much longer than the characteristic correlation lengths of the medium.

The slowness (s) and density (ρ) distributions of the stack of layers (or a continuous inhomogeneous one-dimensional medium) are assumed to be realizations of random stationary processes. The fluctuations of the physical parameters are small (<30%) compared with their constant mean values (denoted by subscripts 0):

$$s^2(z) = \frac{1}{c_0^2}[1 + \varepsilon_s(z)] \tag{3.12.1}$$

$$\rho(z) = \rho_0[1 + \varepsilon_\rho(z)] \tag{3.12.2}$$

where the fluctuating parts $\varepsilon_s(z)$ (the squared slowness fluctuation) and ε_ρ (z) (the density fluctuation) have zero means by definition. The depth coordinate is denoted by z, and the x- and y-axes lie in the plane of the layers. The velocity

$$c_0 = \langle s^2 \rangle^{-1/2} \tag{3.12.3}$$

corresponds to the average squared slowness of the medium. Instead of the squared slowness fluctuations, the random medium may also be characterized by the P- and S-velocity fluctuations, α and β, respectively, as follows:

$$\alpha(z) = \langle \alpha \rangle[1 + \varepsilon_\alpha(z)] = \alpha_0[1 + \varepsilon_\alpha(z)] \tag{3.12.4}$$

$$\beta(z) = \langle \beta \rangle[1 + \varepsilon_\beta(z)] = \beta_0[1 + \varepsilon_\beta(z)] \tag{3.12.5}$$

In the case of small fluctuations, $\varepsilon_\alpha \approx -\varepsilon_s/2$ and $\langle \alpha \rangle \approx c_0(1 + 3\sigma_{\alpha\alpha}^2/2)$, where $\sigma_{\alpha\alpha}^2$ is the normalized variance (the variance divided by the square of the mean) of the velocity fluctuations. The horizontal wavenumber k_x is related to the incidence angle θ by

$$k_x = k_0 \sin\theta = \omega p \tag{3.12.6}$$

where $k_0 = \omega/c_0$, $p = \sin\theta/c_0$, the horizontal component of the slowness (also called the ray parameter), and ω is the angular frequency. For elastic media, depending on the type of the incident wave, $p = \sin\theta/\alpha_0$ or $p = \sin\theta/\beta_0$. The various autocorrelation and cross-correlation functions of the density and velocity fluctuations are denoted by

$$B_{\rho\rho}(\xi) = \langle \varepsilon_\rho(z)\varepsilon_\rho(z + \xi) \rangle \tag{3.12.7}$$

$$B_{\alpha\alpha}(\xi) = \langle \varepsilon_\alpha(z)\varepsilon_\alpha(z + \xi) \rangle \tag{3.12.8}$$

$$B_{\beta\beta}(\xi) = \langle \varepsilon_\beta(z)\varepsilon_\beta(z + \xi) \rangle \tag{3.12.9}$$

$$B_{\alpha\beta}(\xi) = \langle \varepsilon_\alpha(z)\varepsilon_\beta(z + \xi) \rangle \tag{3.12.10}$$

$$B_{\alpha\rho}(\xi) = \langle \varepsilon_\alpha(z)\varepsilon_\rho(z + \xi) \rangle \tag{3.12.11}$$

$$B_{\beta\rho}(\xi) = \langle \varepsilon_\beta(z)\varepsilon_\rho(z + \xi) \rangle \tag{3.12.12}$$

These correlation functions can often be obtained from sonic and density logs. The corresponding normalized variances and cross-variances are given by

$$\sigma_{pp}^2 = B_{pp}(0), \quad \sigma_{\alpha\alpha}^2 = B_{\alpha\alpha}(0), \quad \sigma_{\beta\beta}^2 = B_{\beta\beta}(0) \tag{3.12.13}$$

$$\sigma_{\alpha\beta}^2 = B_{\alpha\beta}(0), \quad \sigma_{\alpha p}^2 = B_{\alpha p}(0), \quad \sigma_{\beta p}^2 = B_{\beta p}(0) \tag{3.12.14}$$

The real part of the effective vertical wavenumber for pressure waves in acoustic media is (neglecting higher than second-order powers of the fluctuations)

$$k_z = k_z^{\text{stat}} - k_0^2 \cos^2\theta \int_0^\infty d\xi B(\xi)\sin(2k_0\xi \cos\theta) \tag{3.12.15}$$

$$k_z^{\text{stat}} = k_0 \cos\theta(1 + \sigma_{pp}^2/2 + \sigma_{p\alpha}^2/\cos^2\theta) \tag{3.12.16}$$

$$B(\xi) = B_{pp}(\xi) + 2B_{p\alpha}(\xi)/\cos^2\theta + B_{\alpha\alpha}(\xi)/\cos^4\theta \tag{3.12.17}$$

For waves in an elastic layered medium, the real part of the vertical wavenumber for P-, SV-, and SH-waves is given by

$$k_z^P = \lambda_a + \omega\hat{A}_P - \omega^2 \int_0^\infty d\xi[B_P(\xi)\sin(2\xi\lambda_a) + B_{BB}(\xi)\sin(\xi\lambda_-) + B_{DD}(\xi)\sin(\xi\lambda_+)] \tag{3.12.18}$$

$$k_z^{SV} = \lambda_b + \omega\hat{A}_{SV} - \omega^2 \int_0^\infty d\xi[B_{SV}(\xi)\sin(2\xi\lambda_b) - B_{BB}(\xi)\sin(\xi\lambda_-) + B_{DD}(\xi)\sin(\xi\lambda_+)] \tag{3.12.19}$$

$$k_z^{SH} = \lambda_b + \omega\hat{A}_{SH} - \omega^2 \int_0^\infty d\xi[B_{SH}(\xi)\sin(2\xi\lambda_b)] \tag{3.12.20}$$

and the imaginary part of the vertical wavenumber (which is related to the attenuation coefficient due to scattering) is

$$\gamma_P = \omega^2 \int_0^\infty d\xi[B_P(\xi)\cos(2\xi\lambda_a) + B_{BB}(\xi)\cos(\xi\lambda_-) + B_{DD}(\xi)\cos(\xi\lambda_+)] \tag{3.12.21}$$

$$\gamma_{SV} = \omega^2 \int_0^\infty d\xi[B_{SV}(\xi)\cos(2\xi\lambda_b) + B_{BB}(\xi)\cos(\xi\lambda_-) + B_{DD}(\xi)\cos(\xi\lambda_+)] \tag{3.12.22}$$

$$\gamma_{SH} = \omega^2 \int_0^\infty d\xi[B_{SH}(\xi)\cos(2\xi\lambda_b)] \tag{3.12.23}$$

where $\lambda_a = \omega\sqrt{1/\alpha_0^2 - p^2}$, $\lambda_b = \omega\sqrt{1/\beta_0^2 - p^2}$, $\lambda_+ = \lambda_a + \lambda_b$, and $\lambda_- = \lambda_b - \lambda_a$ (only real-valued $\lambda_{a,b}$ are considered). The other quantities in the preceding expressions are

$$B_P(\xi) = (X^2\alpha_0^2)^{-1}[B_{pp}(\xi)C_1^2 + 2B_{\alpha p}(\xi)C_1 + 2B_{\beta p}(\xi)C_1C_2 + 2B_{\alpha\beta}(\xi)C_2 + B_{\alpha\alpha}(\xi) + B_{\beta\beta}(\xi)C_2^2] \tag{3.12.24}$$

$$B_{SV}(\xi) = (Y^2\beta_0^2)^{-1}[B_{pp}(\xi)C_7^2 + 2B_{\beta p}(\xi)C_7C_8 + B_{\beta\beta}(\xi)C_8^2] \tag{3.12.25}$$

$$B_{SH}(\xi) = (Y^2\beta_0^2)^{-1}[B_{pp}(\xi)Y^4 + 2B_{\beta p}(\xi)C_{10}Y^2 + 2B_{\beta\beta}(\xi)C_{10}^2] \tag{3.12.26}$$

$$B_{BB}(\xi) = p^2\beta_0(4XY\alpha_0)^{-1}[B_{\rho\rho}(\xi)C_3^2 + 2B_{\beta\rho}(\xi)C_3C_4 + B_{\beta\beta}(\xi)C_4^2] \tag{3.12.27}$$

$$B_{DD}(\xi) = p^2\beta_0(4XY\alpha_0)^{-1}[B_{\rho\rho}(\xi)C_5^2 + 2B_{\beta\rho}(\xi)C_5C_6 + B_{\beta\beta}(\xi)C_6^2] \tag{3.12.28}$$

$$\hat{A}_P = (2X\alpha_0)^{-1}A_P, \quad \hat{A}_{SV} = (2Y\beta_0)^{-1}A_{SV}, \quad \hat{A}_{SH} = (2Y\beta_0)^{-1}A_{SH} \tag{3.12.29}$$

$$A_P = \sigma_{\rho\rho}^2(1 - 4p^2\beta_0^2 Z) + 2\sigma_{\alpha\rho}^2(1 - 4p^2\beta_0^2) + 8\sigma_{\beta\rho}^2 p^2\beta_0^2(1 - 2Z)$$
$$+3\sigma_{\alpha\alpha}^2 - 16p^2\beta_0^2\sigma_{\alpha\beta}^2 + 16p^2\beta_0^2\sigma_{\beta\beta}^2(1 - Z) \tag{3.12.30}$$

$$A_{SV} = \sigma_{\rho\rho}^2(1 - C_9) + \sigma_{\beta\rho}^2(2 - 4C_9) + \sigma_{\beta\beta}^2(3 - 4C_9) \tag{3.12.31}$$

$$A_{SH} = \sigma_{\rho\rho}^2 Y^2 + 2\sigma_{\beta\rho}^2(2Y^2 - 1) + \sigma_{\beta\beta}^2(4Y^2 - 1) \tag{3.12.32}$$

where

$$X = \sqrt{(1 - p^2\alpha_0^2)}, \quad Y = \sqrt{(1 - p^2\beta_0^2)}, \quad Z = p^2(\alpha_0^2 - \beta_0^2) \tag{3.12.33}$$

$$C_1 = X^2(1 - 4p^2\beta_0^2), \quad C_2 = -8p^2\beta_0^2 X^2 \tag{3.12.34}$$

$$C_3 = X(3 - 4\beta_0^2 p^2) - Y[(\alpha_0/\beta_0) + (2\beta_0/\alpha_0) - 4\alpha_0\beta_0 p^2] \tag{3.12.35}$$

$$C_4 = 4X(1 - 2\beta_0^2 p^2) - 4Y[(\beta_0/\alpha_0) - 2\alpha_0\beta_0 p^2] \tag{3.12.36}$$

$$C_5 = X(4\beta_0^2 p^2 - 3) - Y[(\alpha_0/\beta_0) + (2\beta_0/\alpha_0) - 4\alpha_0\beta_0 p^2] \tag{3.12.37}$$

$$C_6 = 4X(2\beta_0^2 p^2 - 1) - 4Y[(\beta_0/\alpha_0) - 2\alpha_0\beta_0 p^2], \quad C_7 = Y^2(1 - 4p^2\beta_0^2) \tag{3.12.38}$$

$$C_8 = 1 - 8p^2\beta_0^2 Y^2, \quad C_9 = 4Y^2 Z\beta_0^2/\alpha_0^2, \quad C_{10} = 1 - 2p^2\beta_0^2 \tag{3.12.39}$$

For multimode propagation (P–SV), neglecting higher-order terms restricts the range of applicable pathlengths L. This range is approximately given as

$$\max\{\lambda, a\} < L < Y^2 \max\{\lambda, a\}/\sigma^2 \tag{3.12.40}$$

where λ is the wavelength, a is the correlation length of the medium, and σ^2 is the variance of the fluctuations. The equations are valid for the whole frequency range, and there is no restriction on the wavelength to correlation length ratio.

The angle- and frequency-dependent phase and group velocities are given by

$$c^{\text{phase}} = \frac{\omega}{\sqrt{k_x^2 + k_z^2}} = \frac{1}{\sqrt{p^2 + k_z^2/\omega^2}} \tag{3.12.41}$$

$$c^{\text{group}} = \sqrt{\left(\frac{\partial\omega}{\partial k_x}\right)^2 + \left(\frac{\partial\omega}{\partial k_z}\right)^2} \tag{3.12.42}$$

In the low- and high-frequency limits the phase and group velocities are the same. The low-frequency limit for pressure waves in acoustic media is

$$c_{\text{fluid}}^{\text{low freq}} \approx c_0[1 - (\sigma_{\rho\rho}^2/2)\cos^2\theta - \sigma_{\rho\alpha}^2]$$

(3.12.43)

and for elastic waves

$$c_{\text{P}}^{\text{low}} = \alpha_0(1 - A_{\text{P}}/2)$$

(3.12.44)

$$c_{\text{SV}}^{\text{low}} = \beta_0(1 - A_{\text{SV}}/2)$$

(3.12.45)

$$c_{\text{SH}}^{\text{low}} = \beta_0(1 - A_{\text{SH}}/2)$$

(3.12.46)

These limits are the same as the result obtained from Backus averaging with higher-order terms in the medium fluctuations neglected. The high-frequency limit of phase and group velocities is

$$c_{\text{fluid}}^{\text{high freq}} = c_0[1 + (\sigma_{\alpha\alpha}^2/2\cos^2\theta)]$$

(3.12.47)

for fluids, and

$$c_{\text{P}}^{\text{high freq}} = \alpha_0[1 - \sigma_{\alpha\alpha}^2(1 - 3p^2\alpha_0^2/2)/(1 - p^2\alpha_0^2)]$$

(3.12.48)

$$c_{\text{SV,SH}}^{\text{high freq}} = \beta_0[1 - \sigma_{\beta\beta}^2(1 - 3p^2\beta_0^2/2)/(1 - p^2\beta_0^2)]$$

(3.12.49)

for elastic media, which are in agreement with ray theory predictions, again neglecting higher-order terms.

Shear-wave splitting or birefringence and its frequency dependence can be characterized by

$$S(\omega,p) = \frac{c_{\text{SV}}(\omega,p) - c_{\text{SH}}(\omega,p)}{c_{\text{SV}}(\omega,p)} \approx \frac{c_{\text{SV}}(\omega,p) - c_{\text{SH}}(\omega,p)}{\beta_0}$$

(3.12.50)

In the low-frequency limit $S^{\text{low freq}}(\omega, p) \approx (A_{\text{SH}} - A_{\text{SV}})/2$, which is the shear-wave splitting in the transversely isotropic medium obtained from a Backus average of the elastic moduli. In the high-frequency limit $S^{\text{high freq}}(\omega, p) = 0$.

For a medium with exponential correlation functions $\sigma^2\exp(-\xi/a)$ (where a is the correlation length) for the velocity and density fluctuations with different variances σ^2 but the same correlation length a, the complete frequency dependence of the phase and group velocities is expressed as

$$c_{\text{fluid}}^{\text{phase}} = c_0\left(1 + \frac{2k_0^2a^2\sigma_{\alpha\alpha}^2 - \sigma_{\rho\alpha}^2 - \frac{1}{2}\cos^2\theta\sigma_{\rho\rho}^2}{1 + 4k_0^2a^2\cos^2\theta}\right)$$

(3.12.51)

$$c_{\text{fluid}}^{\text{group}} = c_0\left[1 + \frac{N}{(1 + 4k_0^2a^2\cos^2\theta)^2}\right]$$

(3.12.52)

$$N = 2k_0^2a^2\sigma_{\alpha\alpha}^2(3 + 4k_0^2a^2\cos^2\theta) + \left(\frac{\sigma_{\rho\rho}^2\cos^2\theta}{2} + \sigma_{\rho\alpha}^2\right)(4k_0^2a^2\cos^2\theta - 1)$$

(3.12.53)

for pressure waves in acoustic media. For elastic P-, SV-, and SH-waves in a randomly layered medium with exponential spatial autocorrelation with correlation length a, the real part of the vertical wavenumber (obtained by Fourier sine transforms) is given as (Shapiro and Hubral, 1995, 1996, 1999)

$$k_z^P = \lambda_a + \omega A_P - \omega^2 a^2 \left[B_P(0) \frac{2\lambda_a}{1 + 4a^2\lambda_a^2} + B_{BB}(0) \frac{\lambda_-}{1 + a^2\lambda_-^2} + B_{DD}(0) \frac{\lambda_+}{1 + a^2\lambda_+^2} \right]$$

$$(3.12.54)$$

$$k_z^{SV} = \lambda_b + \omega A_{SV} - \omega^2 a^2 \left[B_{SV}(0) \frac{2\lambda_b}{1 + 4a^2\lambda_b^2} - B_{BB}(0) \frac{\lambda_-}{1 + a^2\lambda_-^2} + B_{DD}(0) \frac{\lambda_+}{1 + a^2\lambda_+^2} \right]$$

$$(3.12.55)$$

$$k_z^{SH} = \lambda_b + \omega A_{SH} - \omega^2 a^2 B_{SH}(0) \frac{2\lambda_b}{1 + 4a^2\lambda_b^2}$$

$$(3.12.56)$$

and the imaginary part of the vertical wavenumber (which is related to the attenuation coefficient due to scattering) is

$$\gamma_P = \omega^2 a \left[B_P(0) \frac{1}{1 + 4a^2\lambda_a^2} + B_{BB}(0) \frac{1}{1 + a^2\lambda_-^2} + B_{DD}(0) \frac{1}{1 + a^2\lambda_+^2} \right]$$

$$(3.12.57)$$

$$\gamma_{SV} = \omega^2 a \left[B_{SV}(0) \frac{1}{1 + 4a^2\lambda_b^2} + B_{BB}(0) \frac{1}{1 + a^2\lambda_-^2} + B_{DD}(0) \frac{1}{1 + a^2\lambda_+^2} \right]$$

$$(3.12.58)$$

$$\gamma_{SH} = \omega^2 a B_{SH}(0) \frac{1}{1 + 4a^2\lambda_b^2}$$

$$(3.12.59)$$

The shear-wave splitting for exponentially correlated randomly layered media is

$$S(\omega, p) \approx S^{\text{low freq}} + \omega a^2 \beta_0 Y \left\{ \frac{2\lambda_b}{1 + 4a^2\lambda_b^2} [B_{SV}(0) - B_{SH}(0)] \right.$$

$$\left. - \frac{\lambda_+}{1 + a^2\lambda_+^2} B_{DD}(0) + \frac{\lambda_-}{1 + a^2\lambda_-^2} B_{BB}(0) \right\}$$

$$(3.12.60)$$

These equations reveal the general feature that the *anisotropy* (change in velocity with angle) *depends on the frequency*, and the *dispersion* (change in velocity with frequency) *depends on the angle*. Stratigraphic filtering causes the transmitted amplitudes to decay as exp($-\gamma L$), where L is the path length and γ is the attenuation coefficient for the different wave modes as described in the previous equations.

One-Dimensional Layered Poroelastic Medium

The small-perturbation statistical theory has been extended to one-dimensional layered poroelastic media (Gurevich and Lopatnikov, 1995; Gelinsky and Shapiro, 1997b; Gelinsky et al., 1998). In addition to the attenuation due to multiple scattering in random elastic media, waves in a random porous saturated media cause inter-layer flow of pore fluids, leading to additional attenuation and velocity dispersion. The constituent poroelastic layers are governed by the Biot equations (see Section 6.1) and can support two P-waves: the fast and the slow P-wave. The poroelastic parameters of the random one-dimensional medium consist of a homogeneous background (denoted by subscript 0) upon which is superposed a zero-mean fluctuation. The fluctuations are characterized by their variance and a normalized spatial correlation function $B(r/a)$, such that $B(0) = 1$, where a is the correlation length. All parameters of the medium are assumed to have the same normalized correlation function and the same correlation length, but can have different variances. The poroelastic material parameters include: ϕ, porosity; ρ, saturated bulk rock density; κ, permeability; ρ_f, fluid density; η, fluid viscosity, and K_f, fluid bulk modulus. $P_d = K_d + \frac{4}{3}\mu_d$ is the dry (drained) P-wave modulus, with K_d and μ_d being the dry bulk and shear moduli, respectively. $\alpha = 1 - K_d/K_0$ is the Biot coefficient (note that in this section, α denotes the Biot coefficient, not the P-wave velocity). K_0 is the mineral bulk modulus; $M = [\phi/K_f + (\alpha - \phi)/K_0]^{-1}$; $H = P_d + \alpha^2 M$ is the saturated P-wave modulus (equivalent to Gassmann's equation). $N = MP_d/H$; $\omega_0 = \kappa N/\eta a^2$ is the characteristic frequency separating inter-layer-flow and no-flow regimes. $\omega_c = \eta\phi/\kappa\rho_f$ is the Biot critical frequency.

Plane P-waves are assumed to be propagating vertically (along the z-direction) normal to the stack of horizontal layers. The fast P-wavenumber and attenuation coefficient γ are given by (Gelinsky et al., 1998):

$$\psi = k_1^R + A - \int_0^\infty B(z/a)[\sqrt{2}D\exp(-zk_-^I)\cos(zk_-^R - \pi/4)$$

$$+ \sqrt{2}D\exp(-zk_+^I)\cos(zk_+^R - \pi/4) + C\exp(-2zk_1^I)\sin(2zk_1^R)]dz \quad (3.12.61)$$

$$\gamma = k_1^I + \int_0^\infty B(z/a)[\sqrt{2}D\exp(-zk_-^I)\cos(zk_-^R + \pi/4)$$

$$- \sqrt{2}D\exp(-zk_+^I)\cos(zk_+^R + \pi/4) + C\exp(-2zk_1^I)\cos(2zk_1^R)]dz \quad (3.12.62)$$

where superscripts R and I denote real and imaginary parts; $\tilde{k}_1 = k_1^R + ik_1^I$ and $\tilde{k}_2 = k_2^R + ik_2^I$ are the complex wavenumbers for the Biot fast and slow P-waves in the homogeneous background medium; $\tilde{k}_+ = k_+^R + ik_+^I = \tilde{k}_2 + \tilde{k}_1$; $\tilde{k}_- = k_-^R + ik_-^I = \tilde{k}_2 - \tilde{k}_1$. The quantities A, C, and D) involve complicated functions of frequency and linear combinations of the variances and covariances of the medium fluctuations. Approximations for these quantities are discussed in this section. The preceding expressions assume small fluctuations in the poroelastic parameters but are not limited by any restriction on the relation between wavelength and the correlation length of the medium fluctuations. The phase velocity for the fast P-wave is given as $V_P = \omega/\psi$. For a medium with an exponential correlation function with correlation length a, the expressions after carrying out the integrations are (Gelinsky et al., 1998):

$$
\frac{\omega}{V_P} = k_1^R + A - \frac{Da[1 + a(k_-^R + k_-^I)]}{1 + 2ak_-^I + a^2[(k_-^R)^2 + (k_-^I)^2]} - \frac{Da[1 + a(k_+^R + k_+^I)]}{1 + 2ak_+^I + a^2(k_+^{R^2} + k_+^{I^2})}
$$

$$
- \frac{2Ca^2 k_1^R}{1 + 4ak_1^I + 4a^2(k_1^{R^2} + k_1^{I^2})}
\tag{3.12.63}
$$

$$
\gamma = k_1^I + \frac{Da[1 - a(k_-^R - k_-^I)]}{1 + 2ak_-^I + a^2(k_-^{R^2} + k_-^{I^2})} + \frac{Da[1 - a(k_+^R - k_+^I)]}{1 + 2ak_+^I + a^2(k_+^{R^2} + k_+^{I^2})}
$$

$$
+ \frac{Ca(1 + 2ak_1^I)}{1 + 4ak_1^I + 4a^2(k_1^{R^2} + k_1^{I^2})}
\tag{3.12.64}
$$

Gelinsky et al. (1998) introduce approximate expressions for A, C, and D, valid in the frequency range below Biot's critical frequency. Other approximations used are:

$$
\tilde{k}_+ \approx \tilde{k}_- \approx \tilde{k}_2; \quad \tilde{k}_2 \approx (1+i)k_2; \quad \tilde{k}_1 \approx k_1^R = k_1; \quad k_1 = \omega\sqrt{\rho/H}; \text{ and } k_2 = \sqrt{\omega\omega_c\rho_f/(2\phi N)}
\tag{3.12.65}
$$

$$
A = \frac{\omega}{2}\sqrt{\frac{\rho}{H_0}}\frac{P_{d0}}{H_0}\left[\sigma_{PP}^2 + \frac{\alpha_0^2 M_0}{P_{d0}}(\sigma_{MM}^2 + 2\sigma_{P\alpha}^2) + \frac{\alpha_0^4 M_0^2}{P_{d0}^2}\sigma_{\alpha\alpha}^2\right]
\tag{3.12.66}
$$

$$
\frac{2D}{k_2} = \frac{\omega}{2}\sqrt{\frac{\rho}{H_0}}\frac{P_{d0}\alpha_0^2 M_0}{H_0^2}\left[\sigma_{PP}^2 - 2\sigma_{PM}^2 + \sigma_{MM}^2 - 2\frac{P_{d0} - \alpha_0^2 M_0}{P_{d0}}(\sigma_{P\alpha}^2 - \sigma_{M\alpha}^2)\right.
$$

$$
\left. + \frac{(P_{d0} - \alpha_0^2 M_0)^2}{P_{d0}^2}\sigma_{\alpha\alpha}^2\right]
\tag{3.12.67}
$$

$$
\frac{C}{2k_1} = \frac{\omega}{8}\sqrt{\frac{\rho}{H_0}}\frac{P_{d0}^2}{H_0^2}\left[\sigma_{PP}^2 + 2\frac{\alpha_0^2 M_0}{P_{d0}}(\sigma_{PM}^2 + 2\sigma_{P\alpha}^2) + \frac{\alpha_0^4 M_0^2}{P_{d0}^2}(\sigma_{MM}^2 + 4\sigma_{\alpha\alpha}^2 + 4\sigma_{M\alpha}^2)\right]
\tag{3.12.68}
$$

The phase velocity has three limiting values: a quasi-static value for $\omega \ll \omega_0$ and $\omega \to 0$, $V_{qs} = \omega[k_1 + A]^{-1}$; an intermediate no-flow velocity for, $\omega \gg \omega_0$, $V_{nf} = \omega[k_1 + A - 2D/k_2]^{-1}$, and a ray-theoretical limit $V_{ray} = \omega[k_1 + A - 2D/k_2 - C/(2k_1)]^{-1}$. The quasi-static limit is equivalent to the poroelastic Backus average (see Section 4.18). The inter-layer flow effect contributes significantly to the total attenuation in the seismic frequency range for highly permeable thin layers with correlation lengths of a few centimeters.

Uses

The equations described in this section can be used to estimate velocity dispersion and frequency-dependent anisotropy for plane-wave propagation at any angle in randomly layered, one-dimensional media. They can also be used to apply angle-dependent amplitude corrections to correct for the effect of stratigraphic filtering in amplitude-versus-offset (AVO) modeling of a target horizon below a thinly layered overburden. The corrected amplitudes are obtained by multiplying the transmissivity by $\exp(\gamma L)$ for the down-going and up-going ray paths (Widmaier *et al.*, 1996). The equations for poroelastic media can be used to compute velocity dispersion and attenuation in heterogeneous, fluid-saturated one-dimensional porous media.

Assumptions and Limitations

The results described in this section are based on the following assumptions:
- layers are isotropic, linear elastic or poroelastic with no lateral variation;
- the layered medium is statistically stationary with small fluctuations (<30%) in the material properties;
- the propagation path is very much longer than any characteristic correlation length of the medium; and
- incident plane-wave propagation is assumed.

The expressions for poroelastic media assume plane waves along the normal to the layers (0° angle of incidence). Numerical calculations (Gelinsky *et al.*, 1998) indicate that the expressions give reasonable results for angles of incidence up to 20°.

3.13 Scale-Dependent Seismic Velocities in Heterogeneous Media

Synopsis

Measurable travel times of seismic events propagating in heterogeneous media depend on the scale of the seismic wavelength relative to the scale of the geological heterogeneities. In general, the velocity inferred from arrival times is slower when the wavelength, λ, is longer than the scale of the heterogeneity, a, and faster when the wavelength is shorter (Mukerji *et al.*, 1995b).

Layered (One-Dimensional) Media

For normal-incidence propagation in stratified media, in the long-wavelength limit $(\lambda/a \gg 1)$, where a is the scale of the layering, the stratified medium behaves as a homogeneous effective medium with a velocity given by effective medium theory as

$$V_{\text{EMT}} = \left(\frac{M_{\text{EMT}}}{\rho_{\text{av}}}\right)^{1/2} \tag{3.13.1}$$

The effective modulus M_{EMT} is obtained from the Backus average. For normal-incidence plane-wave propagation, the effective modulus is given by the harmonic average

$$M_{EMT} = \left(\sum_k \frac{f_k}{M_k} \right)^{-1} \tag{3.13.2}$$

$$\frac{1}{\rho_{av} V_{EMT}^2} = \sum_k \frac{f_k}{\rho_k V_k^2} \tag{3.13.3}$$

$$\rho_{av} = \sum_k f_k \rho_k \tag{3.13.4}$$

where f_k, ρ_k, M_k, and V_k are the volume fractions, densities, moduli, and velocities of each constituent layer, respectively. The modulus M can be interpreted as C_{3333} or $K + 4\mu/3$ for P-waves and as C_{2323} or μ for S-waves (where K and μ are the bulk and shear moduli, respectively).

In the short-wavelength limit $\lambda/a \ll 1$, the travel time for plane waves traveling perpendicularly to the layers is given by ray theory as the sum of the travel times through each layer. The ray theory or short-wavelength velocity through the medium is, therefore,

$$\frac{1}{V_{RT}} = \sum_k \frac{f_k}{V_k} \tag{3.13.5}$$

The ray theory velocity involves averaging slownesses, whereas the effective medium velocity involves averaging compliances (slownesses squared). The result is that V_{RT} is always faster than V_{EMT}.

In **two- and three-dimensional heterogeneous media**, there is also the path effect as a result of Fermat's principle. Shorter wavelengths tend to find fast paths and diffract around slower inhomogeneities, thus biasing the travel times to lower values (Nolet, 1987; Müller et al., 1992). This is sometimes referred to as the "Wielandt effect," "fast path effect," or "velocity shift." The velocity shift was quantified by Boyse (1986) and by Roth et al. (1993) using an asymptotic ray-theoretical approach. The heterogeneous random medium is characterized by a spatially varying, statistically stationary slowness field

$$n(\mathbf{r}) = n_0 + \varepsilon n_1(\mathbf{r}) \tag{3.13.6}$$

where $\mathbf{r}$ is the position vector, $n_1(\mathbf{r})$ is a zero-mean small fluctuation superposed on the constant background slowness n_0, and $\varepsilon \ll 1$ is a small perturbation parameter. The spatial structure of the heterogeneities is described by the isotropic spatial autocorrelation function:

$$\langle \varepsilon n_1(\mathbf{r}_1) \varepsilon n_1(\mathbf{r}_2) \rangle = \varepsilon^2 \langle n_1^2 \rangle N(|\mathbf{r}_1 - \mathbf{r}_2|) \tag{3.13.7}$$

and the coefficient of variation (normalized standard deviation) is given by

$$\sigma_n = \frac{\varepsilon \sqrt{\langle n_1^2 \rangle}}{n_0} \tag{3.13.8}$$

where $\langle \cdot \rangle$ denotes the expectation operator. For short-wavelength, $\lambda/a \ll 1$, initially plane waves traveling along the x-direction, the expected travel time (spatial average of the travel time over a plane normal to x) at distance X is given as (Boyse, 1986)

$$\langle T \rangle = n_0 [X + \alpha \sigma_n^2 \int_0^x (X - \xi)^2 \frac{N'(\xi)}{\xi} d\xi] + O(\varepsilon^3) \tag{3.13.9}$$

where $\alpha = 1, \dfrac{1}{2}$, and 0 for three, two, and one dimension(s), respectively, and the prime denotes differentiation. When the wave has traveled a large distance compared with the correlation length a of the medium $X \gg a$, and when the autocorrelation function is such that $N(\xi) \ll N(0)$ for $\xi > a$ then

$$\langle T \rangle \approx n_0 (X - \alpha X^2 \sigma_n^2 D) \tag{3.13.10}$$

$$D = -\int_0^\infty \frac{N'(\xi)}{\xi} d\xi > 0 \tag{3.13.11}$$

The ray-theory slowness calculated from the average ray-theory travel time is given by $n_{RT} = \langle T \rangle / X$. Scaling the distance by the correlation length, a, results in the following equation:

$$\frac{n_{RT}}{n_0} = 1 - \alpha \sigma_n^2 \left(\frac{X}{a} \right) \hat{D} \tag{3.13.12}$$

where $\hat{D}$ is defined similarly to D but with the autocorrelation function being of unit correlation length ($a = 1$). For a Gaussian autocorrelation function $\hat{D} = \sqrt{\pi}$. In one dimension $\alpha = 0$, and the ray-theory slowness is just the average slowness. In two and three dimensions the path effect is described by the term $\alpha \sigma_n^2 (X/a) \hat{D}$. In this case, the wave arrivals are on average faster in the random medium than in a uniform medium having the same mean slowness. The expected travel time in three dimensions is less than that in two dimensions, which in turn is less than that in one dimension. This is because in higher dimensions more admissible paths are available to minimize the travel time.

Müller $et\ al.$ (1992) have established ray-theoretical results relating $N(|r|)$ to the spatial autocorrelation function $\phi(|r|)$ of the travel time fluctuations around the mean travel time. Using first-order straight-ray theory, they show

$$\phi(\zeta) = 2X \int_\zeta^\infty N(\xi) \frac{\xi}{\sqrt{\xi^2 - \zeta^2}} d\xi \tag{3.13.13}$$

For a medium with a Gaussian slowness autocorrelation function, $N(r) = \exp(-r^2/a^2)$, the variance of travel time fluctuations $\phi(0)$ is related to the variance of the slowness fluctuations by

$$\phi(0) = \sqrt{\pi} X \, a \, n_0^2 \sigma_n^2 \tag{3.13.14}$$

The expected travel time $\langle T \rangle$ of the wavefield $U(n)$ in the heterogeneous medium with random slowness n is distinct from the travel time T of the expected wavefield $\langle U(n) \rangle$ in the random medium. The expected wave, which is an ensemble average of the wavefield over all possible realizations of the random medium, travels slower than the wavefield in the average medium $\langle n \rangle$ (Keller, 1964).

Thus,

$$\langle T(U(n)) \rangle \leq T(U(\langle n \rangle)) \leq T\left(\langle U(n) \rangle \right) \tag{3.13.15}$$

Gold *et al.* (2000) analyzed wave propagation in random heterogeneous elastic media in the limit of small fluctuations and low frequencies (low-frequency limit of the elastic Bourret approximation) to obtain the effective P- and S-wave slownesses for the coherent wavefield. The random medium is characterized by zero-mean fluctuations of Lamé parameters and density superposed on a homogeneous background with constant mean values of Lamé parameters (λ_o, μ_o) and density (ρ_o):

$$\lambda(\mathbf{r}) = \lambda_0(1 + \varepsilon_\lambda(\mathbf{r})); \quad \langle \varepsilon_\lambda \rangle = 0 \tag{3.13.16}$$

$$\mu(\mathbf{r}) = \mu_0(1 + \varepsilon_\mu(\mathbf{r})); \quad \langle \varepsilon_\mu \rangle = 0 \tag{3.13.17}$$

$$\rho(\mathbf{r}) = \rho_0(1 + \varepsilon_\rho(\mathbf{r})); \quad \langle \varepsilon_\rho \rangle = 0 \tag{3.13.18}$$

The effective slownesses for P- and S-waves for three-dimensional random media in the low-frequency, small-fluctuation limit are expressed as (Gold *et al.*, 2000):

$$S_e^P = S_0^P \left[1 + \frac{1}{2} \frac{\lambda_0^2}{(\lambda_0 + 2\mu_0)^2} \sigma_{\lambda\lambda}^2 + \frac{2}{3} \frac{\lambda_0\mu_0}{(\lambda_0 + 2\mu_0)^2} \sigma_{\lambda\mu}^2 + \frac{2}{5} \frac{\mu_0^2}{(\lambda_0 + 2\mu_0)^2} \sigma_{\mu\mu}^2 + \frac{4}{15} \frac{\mu_0}{(\lambda_0 + 2\mu_0)} \sigma_{\mu\mu}^2 \right] \tag{3.13.19}$$

$$S_e^S = S_0^S \left[1 + \frac{1}{5} \sigma_{\mu\mu}^2 + \frac{2}{15} \frac{\mu_0}{(\lambda_0 + 2\mu_0)} \sigma_{\mu\mu}^2 \right] \tag{3.13.20}$$

For two-dimensional random media Gold *et al.* give the following expressions:

$$S_e^P = S_0^P \left[1 + \frac{1}{2} \frac{\lambda_0^2}{(\lambda_0 + 2\mu_0)^2} \sigma_{\lambda\lambda}^2 + \frac{\lambda_0\mu_0}{(\lambda_0 + 2\mu_0)^2} \sigma_{\lambda\mu}^2 + \frac{3}{4} \frac{\mu_0^2}{(\lambda_0 + 2\mu_0)^2} \sigma_{\mu\mu}^2 + \frac{1}{4} \frac{\mu_0}{(\lambda_0 + 2\mu_0)} \sigma_{\mu\mu}^2 \right] \tag{3.13.21}$$

$$S_e^S = S_0^S \left[1 + \frac{1}{4} \sigma_{\mu\mu}^2 + \frac{1}{4} \frac{\mu_0}{(\lambda_0 + 2\mu_0)} \sigma_{\mu\mu}^2 \right] \tag{3.13.22}$$

In the preceding equations, $S_0^{\mathrm{P}} = \sqrt{\rho_0/(\lambda_0 + 2\mu_0)}$ and $S_0^{\mathrm{S}} = \sqrt{\rho_0/\mu_0}$ are the P- and S-wave slownesses in the homogeneous background medium, respectively. The normalized variances and covariances of the fluctuations of parameters x and y are denoted by σ_{xy}. In this small-contrast, low-frequency limit, density heterogeneities do not affect the properties of the effective medium. The derivation assumes isotropic spatial correlation of the heterogeneities.

Uses

The results described in this section can be used for the following purposes:
- to estimate the velocity shift caused by fast path effects in heterogeneous media;
- to relate statistics of observed travel times to the statistics of the heterogeneities; and
- to smooth and upscale heterogeneous media in the low-frequency limit.

Assumptions and Limitations

The equations described in this section apply under the following conditions:
- small fluctuations in the material properties of the heterogeneous medium;
- isotropic spatial autocorrelation function of the fluctuations;
- ray-theory results are valid only for wavelengths much smaller than the spatial correlation length of the media; and
- low-frequency, effective-smoothing results are valid only for wavelengths much larger than the spatial correlation length of the medium, and for a smoothing window that is much smaller than the wavelength, but much larger than the spatial correlation length.

3.14 Scattering Attenuation

Synopsis

The attenuation coefficient, $\gamma_s = \pi f / QV$ (where Q is the quality factor, V is the seismic velocity, and f is frequency) that results from elastic scattering depends on the ratio of seismic wavelength, λ, to the diameter, d_s, of the scattering heterogeneity. Roughly speaking, there are three domains:
- Rayleigh scattering, where $\lambda > d_s$ and $\gamma_s \propto d_s^3 f^4$;
- stochastic/Mie scattering, where $\lambda \approx d_s$ and $\gamma_s \propto d_s f^2$; and
- diffusion scattering, where $\lambda < d_s$ and $\gamma_s \propto 1/d_s$.

When $\lambda \gg d_s$, the heterogeneous medium behaves like an effective homogeneous medium, and scattering effects may be negligible. At the other limit, when $\lambda \ll d_s$, the heterogeneous medium may be treated as a piecewise homogeneous medium.

Figure 3.14.1 shows schematically the general scale dependence (or, equivalently, frequency dependence) of wave velocity that is expected owing to scattering in

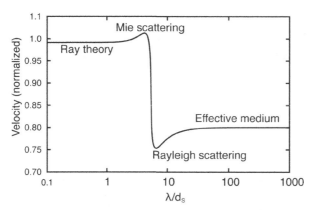

Figure 3.14.1 Scale dependence (or, equivalently, frequency dependence) of the wave velocity due to scattering in heterogeneous media

heterogeneous media. At very long wavelengths $(\lambda \gg d_s)$, the phase velocity is nondispersive and is close to the static effective medium result. As the wavelength decreases (frequency increases), scattering causes velocity dispersion. In the Rayleigh scattering domain $(\lambda/d_s \approx 2\pi)$, the velocity shows a slight decrease with increasing frequency. This is usually followed by a rapid and much larger increase in phase velocity owing to resonant (or Mie) scattering $(\lambda \approx d_s)$. When $\lambda \ll d_s$ (specular scattering or ray theory), the velocity is again nondispersive or weakly dispersive and is usually significantly higher than its long-wavelength limit.

It is usually assumed that the long-wavelength Rayleigh limit is most appropriate for analyzing laboratory rock physics results because the seismic wavelength is often much larger than the grain size. However, Lucet and Zinszner (1992) and Blair (1990), among others, have shown that the scattering heterogeneities can be *clusters* of grains that are comparable to, or larger than, the wavelength. Certainly any of the domains are possible *in situ*.

Blair (1990) suggests a simple (ad hoc) expression that is consistent with both the Rayleigh and diffusion scattering limits:

$$\gamma_s(f) = \frac{C_s}{d_s} \frac{(f/f_d)^4}{(1+f/f_d)^4}, \quad f_d = \frac{k_s V}{d_s} = \frac{k_s f \lambda}{d_s} \tag{3.14.1}$$

where C_s and k_s are constants.

Many theoretical estimates of scattering effects on velocity and attenuation have appeared (see Mehta, 1983, and Berryman, 1992b, for reviews). Most are in the long-wavelength limit, and most assume that the concentration of scatterers is small, and thus only single scattering is considered.

The attenuation of P-waves caused by a low concentration of small spherical inclusions is given by (Yamakawa, 1962; Kuster and Toksöz, 1974)

$$\gamma_{sph} = c\,\frac{3\omega}{4V_P}\left(\frac{\omega}{V_P}a\right)^3\left[2B_0^2 + \frac{2}{3}(1+2\zeta^3)B_1^2 + \frac{(2+3\zeta^5)}{5}B_2^2\right] \tag{3.14.2}$$

where

$$B_0 = \frac{K - K'}{3K' + 4\mu} \tag{3.14.3}$$

$$B_1 = \frac{\rho - \rho'}{3\rho} \tag{3.14.4}$$

$$B_2 = \frac{20}{3}\frac{\mu(\mu' - \mu)}{6\mu'(K + 2\mu) + \mu(9K + 8\mu)} \tag{3.14.5}$$

$$\zeta = V_P/V_S \tag{3.14.6}$$

The terms V_P and V_S are the P- and S-velocities of the host medium, respectively, c is the volume concentration of the spheres, a is their radius, $2\pi f = \omega$ is the frequency, ρ is the density, K is the bulk modulus, and μ is the shear modulus. The unprimed moduli refer to the background host medium and the primed moduli refer to the inclusions.

In the case of elastic spheres in a linear viscous fluid, with viscosity η, the attenuation is given by (Epstein, 1941; Epstein and Carhart, 1953; Kuster and Toksöz, 1974)

$$\gamma_{sph} = c\,\frac{\omega}{2V_P}(\rho - \rho')\,\text{Real}\left[\frac{i + b_0 - ib_0^2/3}{\rho(1 - ib_0) - (\rho + 2\rho')b_0^2/9}\right] \tag{3.14.7}$$

where

$$b_0 = (1 + i)a\sqrt{\frac{\pi f \rho}{\eta}} \tag{3.14.8}$$

Hudson (1981) gives the attenuation coefficient for elastic waves in cracked media (see Section 4.13 on Hudson). For aligned penny-shaped ellipsoidal cracks with normals along the 3-axis, the attenuation coefficients for P-, SV-, and SH-waves are

$$\gamma_P = \frac{\omega}{V_S}\varepsilon\left(\frac{\omega a}{V_P}\right)^3\frac{1}{30\pi}\left[AU_1^2\sin^2 2\theta + BU_3^2\left(\frac{V_P^2}{V_S^2} - 2\sin^2\theta\right)^2\right] \tag{3.14.9}$$

$$\gamma_{SV} = \frac{\omega}{V_S}\varepsilon\left(\frac{\omega a}{V_S}\right)^3\frac{1}{30\pi}(AU_1^2\cos^2 2\theta + BU_3^2\sin^2 2\theta) \tag{3.14.10}$$

$$\gamma_{SH} = \frac{\omega}{V_S}\varepsilon\left(\frac{\omega a}{V_S}\right)^3\frac{1}{30\pi}(AU_1^2\cos^2\theta) \tag{3.14.11}$$

$$A = \frac{3}{2} + \frac{V_S^5}{V_P^5} \tag{3.14.12}$$

$$B = 2 + \frac{15}{4}\frac{V_S}{V_P} - 10\frac{V_S^3}{V_P^3} + 8\frac{V_S^5}{V_P^5} \tag{3.14.13}$$

In these expressions, θ is the angle between the direction of propagation and the 3-axis (axis of symmetry), and ε is the crack density parameter:

$$\varepsilon = \frac{N}{V}a^3 = \frac{3\phi}{4\pi\alpha} \tag{3.14.14}$$

where N/V is the number of penny-shaped cracks of radius a per unit volume, ϕ is the crack porosity, and α is the crack aspect ratio.

U_1 and U_3 depend on the crack conditions. For dry cracks

$$U_1 = \frac{16(\lambda + 2\mu)}{3(3\lambda + 4\mu)}; \quad U_3 = \frac{4(\lambda + 2\mu)}{3(\lambda + \mu)} \tag{3.14.15}$$

For "weak" inclusions (i.e., when $\mu a/\left(K' + \frac{4}{3}\mu'\right)$ is of the order 1 and is not small enough to be neglected)

$$U_1 = \frac{16(\lambda + 2\mu)}{3(3\lambda + 4\mu)}\frac{1}{(1 + M)}; \quad U_3 = \frac{4(\lambda + 2\mu)}{3(\lambda + \mu)}\frac{1}{(1 + \kappa)} \tag{3.14.16}$$

where

$$M = \frac{4\mu'}{\pi\alpha\mu}\frac{(\lambda + 2\mu)}{(3\lambda + 4\mu)}; \quad \kappa = \frac{\left(K' + \frac{4}{3}\mu'\right)(\lambda + 2\mu)}{\pi\alpha\mu(\lambda + \mu)} \tag{3.14.17}$$

with K' and μ' equal to the bulk and shear moduli of the inclusion material. The criterion for an inclusion to be "weak" depends on its shape, or aspect ratio α, as well as on the relative moduli of the inclusion and matrix material. Dry cavities can be modeled by setting the inclusion moduli to zero. Fluid-saturated cavities are simulated by setting the inclusion shear modulus to zero. Remember that these give only the scattering losses and do not incorporate other viscous losses caused by the pore fluid.

Hudson also gives expressions for infinitely thin fluid-filled cracks:

$$U_1 = \frac{16(\lambda + 2\mu)}{3(3\lambda + 4\mu)}; \quad U_3 = 0 \tag{3.14.18}$$

These assume no discontinuity in the normal component of crack displacements and therefore predict no change in the compressional modulus with saturation. There is, however, a shear displacement discontinuity and a resulting effect on shear stiffness. This case should be used with care.

For randomly oriented cracks (isotropic distribution) the P- and S-attenuation coefficients are given as

$$\gamma_{\mathrm{P}} = \frac{\omega}{V_{\mathrm{S}}} \varepsilon \left(\frac{\omega a}{V_{\mathrm{P}}}\right)^3 \frac{4}{15^2 \pi} \left(A U_1^2 + \frac{1}{2}\frac{V_{\mathrm{P}}^5}{V_{\mathrm{S}}^5} B(B-2)U_3^2\right) \tag{3.14.19}$$

$$\gamma_{\mathrm{S}} = \frac{\omega}{V_{\mathrm{S}}} \varepsilon \left(\frac{\omega a}{V_{\mathrm{S}}}\right)^3 \frac{1}{75 \pi} \left(A U_1^2 + \frac{1}{3}B U_3^2\right) \tag{3.14.20}$$

The fourth-power dependence on ω is characteristic of Rayleigh scattering.

Random heterogeneous media with spatially varying velocity $c = c_0 + c'$ may be characterized by the autocorrelation function

$$N(r) = \frac{\langle \xi(r')\xi(r'+r)\rangle}{\langle \xi^2 \rangle} \tag{3.14.21}$$

where $\xi = -c'/c_0$ and c_0 denotes the mean background velocity. For small fluctuations, the fractional energy loss caused by scattering is given by (Aki and Richards, 1980)

$$\frac{\Delta E}{E} = \frac{8\langle \xi^2 \rangle k^4 a^3 L}{1 + 4k^2 a^2}, \quad \text{for } N(r) = e^{-r/a} \tag{3.14.22}$$

$$\frac{\Delta E}{E} = \sqrt{\pi}\langle \xi^2 \rangle k^2 aL(1 - e^{-k^2 a^2}), \quad \text{for } N(r) = e^{-r^2/a^2} \tag{3.14.23}$$

These expressions are valid for small $\Delta E/E$ values as they are derived under the Born approximation, which assumes that the primary incident waves are unchanged as they propagate through the heterogeneous medium.

Aki and Richards (1980) classify scattering phenomena in terms of two dimensionless numbers ka and kL, where $k = 2\pi/\lambda$ is the wavenumber, a is the characteristic scale of the heterogeneity, and L is the path length of the primary incident wave in the heterogeneous medium. Scattering effects are not very important for very small or very large ka, and they become increasingly important with increasing kL. Scattering problems may be classified on the basis of the fractional energy loss caused by scattering, $\Delta E/E$, and the wave parameter D defined by $D = 4L/ka^2$. The wave parameter is the ratio of the first Fresnel zone to the scale length of the heterogeneity. Ray theory is applicable when $D < 1$. In this case the inhomogeneities are smooth enough to be treated as piecewise homogeneous. Effective medium theories are appropriate when ka and $\Delta E/E$ are small. These domains are summarized in Figure 3.14.2.

Scattering becomes complex when heterogeneity scales are comparable with the wavelength and when the path lengths are long. Energy diffusion models are used for long path lengths and strong scattering.

Uses

The results described in this section can be used to estimate the seismic attenuation caused by scattering.

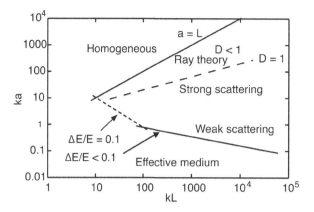

Figure 3.14.2 Domains of applicability for various scattering theories

Assumptions and Limitations

The results described in this section have the following limitations:
- formulas for spherical and ellipsoidal inclusions are limited to low pore concentrations and wavelengths much larger than the scatterer diameter; and
- formulas for fractional energy loss in random heterogeneous media are limited to weak scattering.

3.15   Waves in Cylindrical Rods: the Resonant Bar

Synopsis

Time-harmonic waves propagating in the axial direction along a circular cylindrical rod involve radial, circumferential, and axial components of displacement, u_r, u_θ, and u_z, respectively. Motions that depend on z but are independent of θ may be separated into **torsional** waves involving u_θ only and **longitudinal** waves involving u_r and u_z. **Flexural** waves consist of motions that depend on both z and θ.

Torsional Waves

Torsional waves involve purely circumferential displacements that are independent of θ. The dispersion relation (for free-surface boundary conditions) is of the form (Achenbach, 1984)

$$saJ_0(sa) - 2J_1(sa) = 0 \tag{3.15.1}$$

$$s^2 = \frac{\omega^2}{V_S^2} - k^2 \tag{3.15.2}$$

where $J_n(\cdot)$ are Bessel functions of the first kind of order n, a is the radius of the cylindrical rod, V_S is the S-wave velocity, and k is the wavenumber for torsional waves.

For practical purposes, the lowest mode of each kind of motion is important. The lowest torsional mode consists of displacement proportional to the radius, and the motion is a rotation of each cross-section of the cylinder about its center. The phase velocity of the lowest torsional mode is nondispersive and is given by

$$V_{\text{torsion}} = V_S = \sqrt{\frac{\mu}{\rho}} \qquad (3.15.3)$$

where μ and ρ are the shear modulus and density of the rod, respectively.

Longitudinal Waves

Longitudinal waves are axially symmetric and have displacement components in the axial and radial directions. The dispersion relation (for free-surface boundary conditions), known as the **Pochhammer** equation, is (Achenbach, 1984)

$$2pa[(sa)^2 + (ka)^2]J_1(pa)J_1(sa) - [(sa)^2 - (ka)^2]^2 J_0(pa)J_1(sa)$$
$$- 4(ka)^2(pa)(sa)J_1(pa)J_0(sa) = 0 \qquad (3.15.4)$$

$$p^2 = \frac{\omega^2}{V_P^2} - k^2 \qquad (3.15.5)$$

where V_P is the P-wave velocity.

The phase velocity of the lowest longitudinal mode for small ka ($ka \ll 1$) can be expressed as

$$V_{\text{long}} = \sqrt{\frac{E}{\rho}\left[1 - \frac{1}{4}v^2(ka)^2\right]} + O[(Ka)^4] \qquad (3.15.6)$$

where E is the Young modulus of the cylindrical rod and v is the Poisson ratio of the cylindrical rod.

In the limit as $(ka) \rightarrow 0$, the phase velocity tends to the bar velocity or extensional velocity $V_E = \sqrt{E/\rho}$. For very large ka ($ka \gg 1$), V_{long} approaches the Rayleigh wave velocity.

Flexural Waves

Flexural modes have all three displacement components – axial, radial, and circumferential – and involve motion that depends on both z and θ. The phase velocity of the lowest flexural mode for small values of ka ($ka \ll 1$) may be written as

$$V_{\text{flex}} = \frac{1}{2}\sqrt{\frac{E}{\rho}}(ka) + O[(ka)^3] \qquad (3.15.7)$$

The phase velocity of the lowest flexural mode goes to zero as $(ka) \rightarrow 0$ and approaches the Rayleigh wave velocity for large ka values.

Bar Resonance

Resonant modes (or standing waves) occur when the bar length is an integer number of half-wavelengths:

$$V = \lambda f = \frac{2Lf}{n} \tag{3.15.8}$$

where V is velocity, λ is wavelength, f is the resonant frequency, L is the bar length, and n is a positive integer.

In practice, the shear or extensional velocity is calculated from the observed resonant frequency, most often at the fundamental mode, where $n = 1$.

Porous, Fluid-Saturated Rods

Biot's theory has been used to extend Pochhammer's method of analysis for fluid-saturated porous rods (Gardner, 1962; Berryman, 1983). The dependence of the velocity and attenuation of longitudinal waves on the skeleton and fluid properties is rather complicated. The motions of the solid and the fluid are partly parallel to the axis of the cylinder and partly along the radius. The dispersion relations are obtained from plane-wave solutions of Biot's equations in cylindrical (r, θ, z) coordinates. For an open (unjacketed) surface boundary condition the ω–k_z dispersion relation is given by (in the notation of Berryman, 1983)

$$D_{\text{open}} = \begin{vmatrix} a_{11} & a_{12} & a_{13} \\ a_{21} & a_{22} & 0 \\ a_{31} & a_{32} & a_{33} \end{vmatrix} = 0 \tag{3.15.9}$$

$$a_{11} = \frac{[(C\Gamma_- - H)k_+^2 + 2\mu_{\text{fr}}k_z^2]J_0(k_+a) + 2\mu_{\text{fr}}k_{r+}J_1(k_+a)/a}{(\Gamma_+ - \Gamma_-)} \tag{3.15.10}$$

$$a_{12} = \frac{[(H - C\Gamma_+)k_-^2 - 2\mu_{\text{fr}}k_z^2]J_0(k_-a) - 2\mu_{\text{fr}}k_{r-}J_1(k_-a)/a}{(\Gamma_+ - \Gamma_-)} \tag{3.15.11}$$

$$a_{13} = -2\mu_{\text{fr}}k_{\text{sr}}[k_{\text{sr}}J_0(k_{\text{sr}}a) - J_1(k_{\text{sr}}a)/a] \tag{3.15.12}$$

$$a_{21} = \frac{(M\Gamma_- - C)k_+^2 J_0(k_+a)}{(\Gamma_+ - \Gamma_-)} \tag{3.15.13}$$

$$a_{22} = \frac{(C - M\Gamma_+)k_-^2 J_0(k_-a)}{(\Gamma_+ - \Gamma_-)} \tag{3.15.14}$$

$$a_{23} = 0 \tag{3.15.15}$$

$$a_{31} = \frac{-2i\mu_{\text{fr}}k_z k_{r+}J_1(k_+a)}{(\Gamma_+ - \Gamma_-)} \tag{3.15.16}$$

$$a_{32} = \frac{2i\mu_{\text{fr}}k_z k_{r-}J_1(k_-a)}{(\Gamma_+ - \Gamma_-)} \tag{3.15.17}$$

$$a_{33} = -\mu_{\text{fr}}(k_s^2 - 2k_z^2)k_{\text{sr}}J_1(k_{\text{sr}}a)/(ik_z) \tag{3.15.18}$$

$$k_{r\pm}^2 = k_{\pm}^2 - k_z^2, \quad k_{sr}^2 = k_s^2 - k_z^2 \tag{3.15.19}$$

$$k_s^2 = \omega^2(\rho - \rho_{fl}^2/q)\mu_{fr} \tag{3.15.20}$$

$$k_+^2 = \frac{1}{2}\left[b + f - \sqrt{(b-f)^2 + 4cd}\right] \tag{3.15.21}$$

$$k_-^2 = \frac{1}{2}\left[b + f + \sqrt{(b-f)^2 + 4cd}\right] \tag{3.15.22}$$

$$b = \omega^2(\rho M - \rho_{fl}C)/\Delta \tag{3.15.23}$$

$$c = \omega^2(\rho_{fl}M - qC)/\Delta \tag{3.15.24}$$

$$d = \omega^2(\rho_{fl}H - \rho C)/\Delta \tag{3.15.25}$$

$$f = \omega^2(qH - \rho_{fl}C)/\Delta \tag{3.15.26}$$

$$\Delta = MH - C^2 \tag{3.15.27}$$

$$\Gamma_\pm = d/(k_\pm^2 - b) = (k_\pm^2 - f)/c \tag{3.15.28}$$

$$H = K_{fr} + \frac{4}{3}\mu_{fr} + \frac{(K_0 - K_{fr})^2}{(D - K_{fr})} \tag{3.15.29}$$

$$C = \frac{(K_0 - K_{fr})K_0}{(D - K_{fr})} \tag{3.15.30}$$

$$M = \frac{K_0^2}{(D - K_{fr})} \tag{3.15.31}$$

$$D = K_0[1 + \phi(K_0/K_{fl} - 1)] \tag{3.15.32}$$

$$\rho = (1 - \phi)\rho_0 + \phi\rho_{fl} \tag{3.15.33}$$

$$q = \frac{\alpha\rho_{fl}}{\phi} - \frac{i\eta F(\zeta)}{\omega\kappa} \tag{3.15.34}$$

where K_{fr}, μ_{fr} = effective bulk and shear moduli of rock frame: either the dry frame or the high-frequency unrelaxed "wet frame" moduli predicted by the Mavko–Jizba squirt theory

K_0 = bulk modulus of mineral material making up rock
K_{fl} = effective bulk modulus of pore fluid
ϕ = porosity
ρ_0 = mineral density
ρ_{fl} = fluid density
α = tortuosity parameter (always greater than 1)
η = viscosity of the pore fluid
k = absolute permeability of the rock
ω = angular frequency of the plane wave.

The viscodynamic operator $F(\zeta)$ incorporates the frequency dependence of viscous drag and is defined by

$$F(\zeta) = \frac{1}{4} \frac{\zeta T(\zeta)}{1 + 2iT(\zeta)/\zeta} \qquad (3.15.35)$$

$$T(\zeta) = \frac{\text{ber}'(\zeta) + i\text{bei}'(\zeta)}{\text{ber}(\zeta) + i\text{bei}(\zeta)} = \frac{e^{i3\pi/4}J_1(\zeta e^{-i\pi/4})}{J_0(\zeta e^{-i\pi/4})} \qquad (3.15.36)$$

$$\zeta = (\omega/\omega_r)^{1/2} = \left(\frac{\omega h^2 \rho_{\text{fl}}}{\eta}\right)^{1/2} \qquad (3.15.37)$$

where ber() and bei() are the real and imaginary parts of the Kelvin function, respectively, $J_n()$ is a Bessel function of order n, and h is the pore-size parameter.

The pore-size parameter h depends on both the dimensions and shape of the pore space. Stoll (1974) found that values of between $\frac{1}{6}$ and $\frac{1}{7}$ for the mean grain diameter gave good agreement with experimental data from several investigators. For spherical grains, Hovem and Ingram (1979) obtained $h = \phi d/[3(1 - \phi)]$, where d is the grain diameter.

This dispersion relation gives the same results as Gardner (1962). When the surface pores are closed (jacketed) the resulting dispersion relation is

$$D_{\text{closed}} = \begin{vmatrix} a_{11} & a_{12} & a_{13} \\ a_{31} & a_{32} & a_{33} \\ a_{41} & a_{42} & a_{43} \end{vmatrix} = 0 \qquad (3.15.38)$$

$$a_{41} = \frac{k_{r+}\Gamma_-J_1(k_+a)}{\Gamma_+ - \Gamma_-}; \quad a_{42} = \frac{-k_{r-}\Gamma_+J_1(k_-a)}{\Gamma_+ - \Gamma_-}; \quad a_{43} = k_{\text{sr}}J_1(k_{\text{sr}}a)\rho_{\text{fl}}/q \qquad (3.15.39)$$

For open-pore surface conditions the vanishing of the fluid pressure at the surface of the cylinder causes strong radial motion of the fluid relative to the solid. This relative motion absorbs energy, causing greater attenuation than would be present in a plane longitudinal wave in an extended porous saturated medium (White, 1986). Narrow stop bands and sharp peaks in the attenuation can occur if the slow P-wave has wavelength $\lambda < 2.6a$. Such stop bands do not exist in the case of the jacketed, closed-pore surface. A slow extensional wave propagates under jacketed boundary conditions but not under the open-surface condition.

Uses

The results described in this section can be used to model wave propagation and geometric dispersion in resonant bar experiments.

Assumptions and Limitations

The results described in this section assume the following:

- an isotropic, linear, homogeneous, and elastic-poroelastic rod of solid circular cross-section;

- for elastic rods the cylindrical surface is taken to be free of tractions; and
- for porous rods, unjacketed and jacketed surface boundary conditions are assumed.

3.16 Waves in Boreholes

Synopsis

Elastic wave propagation in the presence of a cylindrical fluid-filled borehole involves different modes caused by internal refraction, constructive interference and trapping of wave energy in the borehole. The theory of borehole wave propagation has been described in the books by White (1983), Paillet and Cheng (1991), and Tang and Cheng (2004), where references to the original literature may be found. The dispersion characteristics of borehole wave modes depend strongly on the shear wave velocity of the elastic medium surrounding the borehole. Two scenarios are usually considered: "fast" formation when the S-wave velocity in the formation is greater than the borehole fluid velocity and "slow" formation when the S-wave velocity in the formation is slower than the borehole fluid velocity. Wave modes are guided by the borehole only when the formation S-wave velocity is greater than the phase velocity of the modes; otherwise the modes become leaky modes, radiating energy into the formation.

In fast formations, pseudo-Rayleigh modes or shear normal modes exist above characteristic cut-off frequencies. The pseudo-Rayleigh mode is strongly dispersive and is a combined effect of reflected waves in the fluid and critical refraction along the borehole walls. The phase velocity of the pseudo-Rayleigh wave at the cut-off frequency drops from the shear wave velocity of the formation and approaches the fluid velocity at high frequencies. In slow formations pseudo-Rayleigh modes do not exist. Stoneley waves in boreholes refer to waves along the borehole interface. At low frequencies the Stoneley waves are referred to as tube waves. Stoneley waves exist at all frequencies and in both fast and slow formations. "Leaky P" modes exist in slow formations and are dominated by critical refraction of P-waves at the borehole wall. They lose energy by conversion to shear waves. Higher-order modes include the dipole or flexural mode and the quadrupole or screw mode (Tang and Cheng, 2004).

Isotropic Elastic Formation

A low-frequency (static) analysis for a thick-walled elastic tube of inner radius b and outer radius a, with Young's modulus E, and Poisson ratio v, containing fluid with bulk modulus B, and density ρ, gives the speed of tube waves as (White, 1983)

$$c_t = \left[\rho \left(\frac{1}{B} + \frac{1}{M} \right) \right]^{-1/2} \qquad (3.16.1)$$

$$M = \frac{E(a^2 - b^2)}{2[(1 + v)(a^2 + b^2) - 2vb^2]} \qquad (3.16.2)$$

For a thin-walled tube with thickness $h = (a - b)$, $a \approx b$, and $M \approx Eh/2b$. For a borehole in an infinite solid $a \gg b$ and the speed of the tube wave in the low-frequency limit is

$$c_t = \left[\rho \left(\frac{1}{B} + \frac{1}{\mu} \right) \right]^{-1/2} \tag{3.16.3}$$

where μ is the shear modulus of the formation. For a borehole with a casing of thickness h, inner radius b, and Young's modulus E, the tube wave velocity (in the low-frequency limit) is

$$c_t = \left[\rho \left(\frac{1}{B} + \frac{1}{\mu + Eh/2b} \right) \right]^{-1/2} \tag{3.16.4}$$

The ω–k dispersion relation gives a more complete description of the modes. The dispersion relation (or period equation) yields characteristic cut-off frequencies and phase velocities as a function of frequency for the different modes (Paillet and Cheng, 1991; Tang and Cheng, 2004). For a cylindrical borehole of radius R, in an infinite, isotropic, elastic formation with Lamé constants λ and μ, and density ρ, and open-hole boundary conditions at the interface, the dispersion relation is given by (in the notation of Tang and Cheng, 2004)

$$D(\omega, k) = \begin{vmatrix} M_{11} & M_{12} & M_{13} & M_{14} \\ M_{21} & M_{22} & M_{23} & M_{24} \\ M_{31} & M_{32} & M_{33} & M_{34} \\ M_{41} & M_{42} & M_{43} & M_{44} \end{vmatrix} = 0 \tag{3.16.5}$$

$$M_{11} = -\frac{n}{R} I_n(fR) - f I_{n+1}(fR) \tag{3.16.6}$$

$$M_{12} = -p Y_1(pR) \tag{3.16.7}$$

$$M_{13} = \frac{n}{R} K_n(sR) \tag{3.16.8}$$

$$M_{14} = -iks Y_1(sR) \tag{3.16.9}$$

$$M_{21} = \rho_f \omega^2 I_n(fR) \tag{3.16.10}$$

$$M_{22} = \rho(2k^2\beta^2 - \omega^2) K_n(pR) + \frac{2p\rho\beta^2}{R} Y_2(pR) \tag{3.16.11}$$

$$M_{23} = -\frac{2n\rho s\beta^2}{R} Y_3(sR) \tag{3.16.12}$$

$$M_{24} = 2ik\rho s^2\beta^2 K_n(sR) + \frac{2ik\,\rho s\beta^2}{R} Y_2(sR) \tag{3.16.13}$$

$$M_{31} = 0 \tag{3.16.14}$$

$$M_{32} = \frac{2n\rho p\beta^2}{R} Y_3(pR) \tag{3.16.15}$$

$$M_{33} = -\rho s^2 \beta^2 Y_4(sR) \tag{3.16.16}$$

$$M_{34} = \frac{2ikns\,\rho\beta^2}{R} Y_3(sR) \tag{3.16.17}$$

$$M_{41} = 0 \tag{3.16.18}$$

$$M_{42} = -2ikp\,\rho\beta^2 Y_1(pR) \tag{3.16.19}$$

$$M_{43} = \frac{ikn\rho\beta^2}{R} K_n(sR) \tag{3.16.20}$$

$$M_{44} = (k^2 + s^2)s\,\rho\beta^2 Y_1(sR) \tag{3.16.21}$$

where I_n, and K_n are the modified Bessel functions of the first and second kind, respectively, of order n. The azimuthal order number, n, controls the azimuthal variations, with $n = 0$, 1, and 2 corresponding to monopole, dipole, and quadrupole modes, respectively. In the previous expressions Y_i denotes the following combinations of modified Bessel functions:

$$Y_1(x) = -\frac{n}{x} K_n(x) + K_{n+1}(x) \tag{3.16.22}$$

$$Y_2(x) = \frac{n(n-1)}{x} K_n(x) + K_{n+1}(x) \tag{3.16.23}$$

$$Y_3(x) = \frac{1-n}{x} K_n(x) + K_{n+1}(x) \tag{3.16.24}$$

$$Y_4(x) = \left[1 + \frac{2n(n-1)}{x^2}\right] K_n(x) + \frac{2}{x} K_{n+1}(x) \tag{3.16.25}$$

Other terms in the expressions are:

ω = angular frequency
k = axial wavenumber
$p = (k^2 - k_\alpha^2)^{1/2}$ = compressional radial wavenumber
$s = (k^2 - k_\beta^2)^{1/2}$ = shear radial wavenumber
$f = (k^2 - k_f^2)^{1/2}$ = radial wavenumber in the borehole fluid
$k_\alpha = \omega/\alpha$ = compressional wavenumber
$k_\beta = \omega/\beta$ = shear wavenumber
$k_f = \omega/\alpha_f$ = acoustic wavenumber in borehole fluid

where α and β are the P- and S-wave velocities in the formation, and α_f is the compressional wave velocity in the borehole fluid. From the roots of the dispersion relation $D(\omega, k) = 0$, the phase and group velocities of the wave modes are obtained as

$$V_{\text{phase}} = \frac{\omega}{k(\omega)} \tag{3.16.26}$$

$$V_{group} = \frac{d\omega}{dk} = -\left(\frac{\partial D}{\partial k}\right) \Big/ \left(\frac{\partial D}{\partial \omega}\right)$$

(3.16.27)

The roots and the partial derivatives have to be calculated numerically.

Transversely Isotropic (TI) Elastic Formation

The elastic properties of the transversely isotropic formation are given by the five independent components of the elastic stiffness tensor (c_{11}, c_{13}, c_{33}, c_{44}, c_{66}), and ρ denotes the density. The TI symmetry axis is along the z-axis and the borehole axis coincides with the symmetry axis. The borehole radius is R, with borehole fluid density ρ_f, and compressional wave velocity α_f. For open-hole boundary conditions the dispersion relation is given by (Tang and Cheng, 2004)

$$D(\omega, k) = \begin{vmatrix} Q_{11} & Q_{12} & Q_{13} & Q_{14} \\ Q_{21} & Q_{22} & Q_{23} & Q_{24} \\ Q_{31} & Q_{32} & Q_{33} & Q_{34} \\ Q_{41} & Q_{42} & Q_{43} & Q_{44} \end{vmatrix} = 0$$

(3.16.28)

$$Q_{11} = -\frac{n}{R} I_n(fR) - f I_{n+1}(fR)$$

(3.16.29)

$$Q_{12} = -(1 + ika') q_p Y_1(q_p R)$$

(3.16.30)

$$Q_{13} = \frac{n}{R} K_n(q_{sh} R)$$

(3.16.31)

$$Q_{14} = -(ik + b') q_{sv} Y_1(q_{sv} R)$$

(3.16.32)

$$Q_{21} = \rho_f \omega^2 I_n(fR)$$

(3.16.33)

$$Q_{22} = [c_{11} q_p^2 - c_{13} k^2 + (c_{11} - c_{13}) ika' q_p^2] K_n(q_p R) + \frac{2 c_{66} q_p}{R} (1 + ika') Y_2(q_p R)$$

(3.16.34)

$$Q_{23} = -\frac{2 c_{66} n q_{sh}}{R} Y_3(q_{sh} R)$$

(3.16.35)

$$Q_{24} = [(c_{11} q_{sv}^2 - c_{13} k^2) b' + (c_{11} - c_{13}) ik q_{sv}^2] K_n(q_{sv} R) + \frac{2 c_{66} q_{sv}}{R} (ik + b') Y_2(q_{sv} R)$$

(3.16.36)

$$Q_{31} = 0$$

(3.16.37)

$$Q_{32} = \frac{2 c_{66} n q_p}{R} (1 + ika') Y_3(q_p R)$$

(3.16.38)

$$Q_{33} = -c_{66} q_{sh}^2 Y_4(q_{sh} R)$$

(3.16.39)

$$Q_{34} = \frac{2 c_{66} n q_{sv}}{R} (ik + b') Y_3(q_{sv} R)$$

(3.16.40)

$$Q_{42} = -c_{44} q_p [2ik - a'(k^2 + q_p^2)] Y_1(q_p R)$$

(3.16.41)

$$Q_{43} = \frac{iknc_{44}}{R} K_n(q_{sh}R) \tag{3.16.42}$$

$$Q_{44} = c_{44}q_{sv}[(k^2 + q_{sv}^2) - 2ikb']Y_1(q_{sv}R) \tag{3.16.43}$$

$$a' = -\frac{1}{ik} \frac{(c_{13} + 2c_{44})k^2 - c_{11}q_p^2 - \rho\omega^2}{c_{44}k^2 - (c_{11} - c_{13} - c_{44})q_p^2 - \rho\omega^2} \tag{3.16.44}$$

$$b' = -ik \frac{c_{44}k^2 - (c_{11} - c_{13} - c_{44})q_{sv}^2 - \rho\omega^2}{(c_{13} + 2c_{44})k^2 - c_{11}q_{sv}^2 - \rho\omega^2} \tag{3.16.45}$$

$$q_{sh} = \sqrt{\frac{c_{44}k^2 - \rho\omega^2}{c_{66}}} \tag{3.16.46}$$

$$q_p = \omega\sqrt{\frac{-V + \sqrt{V^2 - 4UW}}{2U}} \tag{3.16.47}$$

$$q_{sv} = \omega\sqrt{\frac{-V - \sqrt{V^2 - 4UW}}{2U}} \tag{3.16.48}$$

$f = (k^2 - k_f^2)^{1/2} = $ radial wavenumber in the borehole fluid
$k_f = \omega/\alpha_f = $ acoustic wavenumber in the borehole fluid

$U = c_{11}c_{44}$

$$V = \rho(c_{11} + c_{44}) - (c_{11}c_{33} - c_{13}^2 - 2c_{13}c_{44})(k^2/\omega^2) \tag{3.16.49}$$

$$W = c_{33}c_{44}(\rho/c_{44} - k^2/\omega^2)(\rho/c_{33} - k^2/\omega^2) \tag{3.16.50}$$

where I_n, and K_n are the modified Bessel functions of the first and second kind, respectively, of order n. The azimuthal order number, n, controls the azimuthal variations, with $n = 0$, 1, and 2 corresponding to monopole, dipole, and quadrupole modes, respectively. In the previous expressions, Y_i denotes the same combinations of modified Bessel functions as described for the isotropic case.

Isotropic, Poroelastic, Permeable Formation

Borehole wave in permeable formations interacts with the formation permeability through the propagation of Biot slow waves in the pore fluid. The Biot–Rosenbaum theory (Rosenbaum, 1974) was the first to model monopole excitation in an open borehole within an isotropic poroelastic formation. The theory was later extended to transversely isotropic poroelastic formations with multipole excitations (Schmitt, 1989).

The Biot–Rosenbaum model solves the Biot poroelastic equations for a cylindrical borehole geometry and boundary conditions. For the monopole case in an open borehole of radius R, within a permeable formation, the dispersion relation is given by (Tang and Cheng, 2004)

$$D(\omega, k) = \begin{vmatrix} N_{11} & N_{12} & N_{13} & N_{14} \\ N_{21} & N_{22} & N_{23} & N_{24} \\ N_{31} & N_{32} & N_{33} & N_{34} \\ N_{41} & N_{42} & N_{43} & N_{44} \end{vmatrix} = 0 \tag{3.16.51}$$

$$N_{11} = -fI_1(fR) \tag{3.16.52}$$

$$N_{12} = -q_{\text{fast}}(1 + b_{\text{fast}})K_1(q_{\text{fast}}R) \tag{3.16.53}$$

$$N_{13} = -q_{\text{slow}}(1 + b_{\text{slow}})K_1(q_{\text{slow}}R) \tag{3.16.54}$$

$$N_{14} = -ik(1 + b_{\text{s}})K_1(q_{\text{s}}R) \tag{3.16.55}$$

$$N_{21} = \rho_{\text{f}}\omega^2 I_0(fR) \tag{3.16.56}$$

$$N_{22} = -2\mu[q_{\text{fast}}^2 K_0(q_{\text{fast}}R) + q_{\text{fast}}K_1(q_{\text{fast}}R)/R] + k_{\text{fast}}^2\left[\lambda + \frac{\alpha}{\beta}(\alpha + b_{\text{fast}})\right]K_0(q_{\text{fast}}R) \tag{3.16.57}$$

$$N_{23} = -2\mu[q_{\text{slow}}^2 K_0(q_{\text{slow}}R) + q_{\text{slow}}K_1(q_{\text{slow}}R)/R] + k_{\text{slow}}^2\left[\lambda + \frac{\alpha}{\beta}(\alpha + b_{\text{slow}})\right]K_0(q_{\text{slow}}R) \tag{3.16.58}$$

$$N_{24} = 2ik\mu[q_{\text{s}}^2 K_0(q_{\text{s}}R) + q_{\text{s}}K_1(q_{\text{s}}R)/R] \tag{3.16.59}$$

$$N_{31} = 0 \tag{3.16.60}$$

$$N_{32} = -2ik\mu q_{\text{fast}}K_1(q_{\text{fast}}R) \tag{3.16.61}$$

$$N_{33} = -2ik\mu q_{\text{slow}}K_1(q_{\text{slow}}R) \tag{3.16.62}$$

$$N_{34} = \mu(2k^2 - k_{\text{shear}}^2)q_{\text{s}}K_1(q_{\text{s}}R) \tag{3.16.63}$$

$$N_{41} = -\rho_{\text{f}}\omega^2 I_0(fR) \tag{3.16.64}$$

$$N_{42} = k_{\text{fast}}^2[(\alpha + b_{\text{fast}})/\beta]K_0(q_{\text{fast}}R) \tag{3.16.65}$$

$$N_{43} = k_{\text{slow}}^2[(\alpha + b_{\text{slow}})/\beta]K_0(q_{\text{slow}}R) \tag{3.16.66}$$

$$N_{44} = 0 \tag{3.16.67}$$

where I_n and K_n are the modified Bessel functions of the first and second kind, respectively, of order n, k is the axial wavenumber along the borehole axis, and ω is the angular frequency. The radial wavenumbers are given by

$$q_{\text{fast}} = \sqrt{k^2 - k_{\text{fast}}^2} \tag{3.16.68}$$

$$q_{\text{slow}} = \sqrt{k^2 - k_{\text{slow}}^2} \tag{3.16.69}$$

$$q_{\text{s}} = \sqrt{k^2 - k_{\text{shear}}^2} \tag{3.16.70}$$

$$f = (k^2 - k_{\text{f}}^2)^{1/2} = \text{the radial wavenumber in the borehole fluid}$$

$k_f = \omega/\alpha_f$ = the acoustic wavenumber in the borehole fluid

$$k_{fast} = k_{p0}\sqrt{\frac{1 + b_{fast}\rho_{pf}/\rho}{1 - b_{fast}/b_0}} \qquad (3.16.71)$$

$$k_{slow} = k_{p0}\sqrt{\frac{1 + b_{slow}\rho_{pf}/\rho}{1 - b_{slow}/b_0}} \qquad (3.16.72)$$

$$k_{p0} = \frac{\omega}{\sqrt{(\lambda + 2\mu + \alpha^2/\beta)/\rho}} \qquad (3.16.73)$$

$$b_{\substack{fast \\ slow}} = \frac{1}{2}\left[c \mp \sqrt{c^2 - 4\alpha(1 - c)/b_0}\right] \qquad (3.16.74)$$

$$b_0 = -\frac{\beta(\lambda + 2\mu + \alpha^2/\beta)}{\alpha} \qquad (3.16.75)$$

$$c = \frac{\alpha - b_s\rho/(\rho_{pf}b_0)}{\alpha + b_s} \qquad (3.16.76)$$

$$b_s = \rho_{pf}\omega^2\theta \qquad (3.16.77)$$

$$k_{shear} = \omega/\sqrt{\mu/(\rho + \rho_{pf}^2\omega^2\theta)} \qquad (3.16.78)$$

$$\alpha = 1 - K_{dry}/K_m \qquad (3.16.79)$$

$$\beta = (\alpha - \phi)/K_m + \phi/K_{pf} \qquad (3.16.80)$$

$$\rho = (1 - \phi)\rho_0 + \phi\rho_{pf} \qquad (3.16.81)$$

K_{dry} = dry (drained) bulk modulus
λ, μ = Lamé constants of the dry (drained) porous formation
K_m = mineral bulk modulus
ρ_0 = mineral density
K_{pf} = pore fluid bulk modulus
ρ_{pf} = pore fluid density
ϕ = porosity
$\theta = i\kappa(\omega)/\eta\omega$
η = fluid viscosity
$\kappa(\omega)$ = dynamic permeability

The dynamic permeability measures fluid transport in a porous medium under dynamic wave excitation. It is given by (Johnson *et al.*, 1987)

$$\kappa(\omega) = \frac{\kappa_0}{[1 - 4i\tau^2\kappa_0^2\rho_{pf}\omega/(\eta\Lambda^2\phi^2)]^{1/2} - i\tau\kappa_0\rho_{pf}\omega/(\eta\phi)} \qquad (3.16.82)$$

where κ_0 is the static Darcy permeability, τ is tortuosity, Λ is a measure of the pore size given approximately by $\Lambda \approx \sqrt{8\tau\kappa_0/\phi}$.

In the low-frequency limit, the dynamic permeability goes to the static Darcy permeability, while in the high-frequency limit,

$$\kappa(\omega) \rightarrow i\eta\phi/(\tau\rho_{pf}\omega) \tag{3.16.83}$$

After finding the roots of the dispersion equation numerically, the Stoneley wave phase velocity and attenuation are given by

$$V_{stoneley} = \omega/\text{Re}(k) \tag{3.16.84}$$

$$Q_{stoneley}^{-1} = 2 \, \text{Im}(k)/\text{Re}(k) \tag{3.16.85}$$

Tang *et al.* (1991) present a simplified Biot–Rosenbaum model by decoupling the elastic and flow effects in the full Biot–Rosenbaum model. This reduces the complexity of root finding to an approximate analytical expression for the Stoneley phase wavenumber k:

$$k = \sqrt{k_e^2 + \frac{2i\rho_{pf}\omega\kappa(\omega)R}{\eta(R^2 - a^2)} \sqrt{-i\omega/D + k_e^2} \frac{K_1(R\sqrt{-i\omega/D + k_e^2})}{K_0(R\sqrt{-i\omega/D + k_e^2})}} \tag{3.16.86}$$

$$D = \frac{\kappa(\omega)K_{pf}/(\phi\eta)}{1 + \dfrac{K_{pf}}{\phi(\lambda + 2\mu)}\left[1 + \dfrac{4\alpha\mu/3 - K_{dry} - \phi(\lambda + 2\mu)}{K_m}\right]} \tag{3.16.87}$$

a = tool radius

where D is the diffusivity modified to correct for the elasticity of the medium, and K_0 and K_1 are modified Bessel functions of the second kind. The parameter k_e in the previous expression is the elastic Stoneley wavenumber, which can be estimated from the density and effective elastic moduli (or P- and S-wave velocities) of the equivalent formation (Tang and Cheng, 2004). The tool radius a can be set to 0 to model a fluid-filled borehole without a tool. A low-frequency approximation to the Stoneley wave slowness (assuming $a = 0$) is given by (Tang and Cheng, 2004)

$$S^2 = \left(\frac{\rho_{pf}}{K_{pf}} + \frac{\rho_{pf}}{\mu}\right) + \left[\frac{2i\rho_{pf}\kappa_0}{\eta\omega R}\sqrt{-i\omega/D}\frac{K_1(R\sqrt{-i\omega/D})}{K_0(R\sqrt{-i\omega/D})}\right] \tag{3.16.88}$$

The first term is the elastic component while the second term represents the effect of formation permeability and fluid flow. The second term is usually small compared with the first term. Therefore, it is difficult to estimate formation permeability using this expression for Stoneley wave slowness.

The Stoneley wavenumber k obtained from the simplified theory matches very well the exact wavenumber for fast formations. For slow formations, a better match is obtained by applying an empirical correction to account for borehole compliance. With the empirical correction, the expression for the Stoneley wavenumber is (Tang and Cheng, 2004)

$$k = \sqrt{k_e^2 + \frac{2i\rho_{pf}\omega\kappa(\omega)R}{\eta(R^2 - a^2)} \frac{\sqrt{-i\omega/D + k_e^2}}{1 + B^\gamma} \frac{K_1(R\sqrt{-i\omega/D + k_e^2})}{K_0(R\sqrt{-i\omega/D + k_e^2})}} \qquad (3.16.89)$$

$$B = f_e R \frac{I_1(f_e R)}{I_0(f_e R)} \qquad (3.16.90)$$

$f_e = \sqrt{k_e^2 - k_f^2}$ = the radial wavenumber for an equivalent elastic formation

$k_f = \omega/\alpha_f$ = the acoustic wavenumber in the borehole fluid

$\gamma = V_S/\alpha_f$ = the ratio of the formation S-wave velocity to the acoustic velocity in the borehole fluid.

Uses

The results described in this section can be used to model wave propagation and geometric dispersion in boreholes.

Assumptions and Limitations

The results described in this section assume the following:
- isotropic or TI, linear, homogeneous, elastic or poroelastic formation;
- a borehole of circular cross-section;
- for the TI dispersion relation, the borehole axis is along the symmetry axis; and
- open-hole boundary conditions at the borehole interface.

Extensions

The Biot-Rosenbaum theory has been extended to anisotropic poroelastic formations with multipole excitations (Schmitt, 1989).

4 Effective Elastic Media: Bounds and Mixing Laws

4.1 Voigt and Reuss Bounds

Synopsis

If we wish to predict theoretically the effective elastic moduli of a mixture of grains and pores, we generally need to specify: (1) the volume fractions of the various phases, (2) the elastic moduli of the various phases, and (3) the geometric details of how the phases are arranged relative to each other. If we specify only the volume fractions and the constituent moduli, the best we can do is predict the upper and lower bounds (shown schematically in Figure 4.1.1).

At any given volume fraction of constituents, the effective modulus will fall between the bounds (somewhere along the vertical dashed line in Figure 4.1.1), but its precise value depends on the geometric details. We use, for example, terms like "stiff pore shapes" and "soft pore shapes." Stiffer shapes cause the value to be higher within the allowable range; softer shapes cause the value to be lower. The simplest, but not necessarily the best, bounds are the Voigt and Reuss bounds. (See also Section 4.2 on Hashin–Shtrikman bounds, which are narrower.)

The **Voigt upper bound** on the effective elastic *bulk or shear modulus, M_V,* of N phases is

$$M_V = \sum_{i=1}^{N} f_i M_i \tag{4.1.1}$$

where f_i is the volume fraction of the ith phase and M_i is the elastic bulk or shear modulus of the ith phase. The Voigt bound is sometimes called the **isostrain average**, because it gives the ratio of the average stress to the average strain when all constituents are assumed to have the same strain.

The **Reuss lower bound** on the effective elastic *bulk or shear modulus, M_R,* is (Reuss, 1929)

$$\frac{1}{M_R} = \sum_{i=1}^{N} \frac{f_i}{M_i} \tag{4.1.2}$$

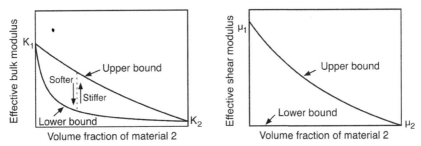

Figure 4.1.1 The effective bulk and shear moduli of a two phase composite must fall between the upper and lower bounds. At any given volume fraction, the pore and grain microgeometries determine the position of the effective modulus between the bounds.

The Reuss bound is sometimes called the **isostress average** because it gives the ratio of the average stress to the average strain when all constituents are assumed to have the same stress. When one of the constituents is a liquid or a gas with zero shear modulus, the Reuss average bulk and shear moduli for the composite are exactly the same as given by the Hashin–Shtrikman lower bound (Section 4.2).

The Reuss average exactly describes the effective moduli of a suspension of solid elastic grains in a fluid. It also describes the moduli of "shattered" materials in which solid fragments are completely surrounded by the pore fluid. When all constituents are gases or liquids, or both, with zero shear moduli, the Reuss average gives the effective moduli of the mixture exactly.

In contrast to the Reuss average, which describes a number of real physical systems, real isotropic mixtures can never be as stiff as the Voigt bound (except for the single phase end-members).

The recommended procedure for finding the Voigt and Reuss bounds on elastic moduli other than the bulk and shear moduli is to compute them from the bounds on bulk and shear moduli. For example, the Voigt and Reuss bounds on Young's modulus, $E_R \leq E \leq E_V$, can be expressed in terms of the bounds on bulk and shear moduli using the usual isotropic relations. The Reuss average is written as

$$\frac{3}{E_R} = \frac{1}{\mu_R} + \frac{1}{3K_R} \tag{4.1.3}$$

which is equivalent to

$$\frac{1}{E_R} = \frac{f_1}{E_1} + \frac{f_2}{E_2} \tag{4.1.4}$$

(Note that the harmonic average of Young's modulus yields the Reuss bound on Young's modulus). The Voigt bound on Young's modulus is

$$\frac{3}{E_V} = \frac{1}{\mu_V} + \frac{1}{3K_V} \quad \text{but} \quad E_V \geq f_1 E_1 + f_2 E_2, \tag{4.1.5}$$

That is, the Voigt bound on Young's modulus is generally larger than the arithmetic average of the phase Young's moduli, except when the Poisson's ratio is the same in both phases.

The Voigt–Reuss Bounds on the elastic stiffness tensor $\mathbf{C}$ are written as (Gibiansky, 1993)

$$\mathbf{C}_R \equiv \langle \mathbf{C}^{-1}(\mathbf{x}) \rangle^{-1} \le \mathbf{C}_0 \le \langle \mathbf{C}(\mathbf{x}) \rangle \equiv \mathbf{C}_V , \tag{4.1.6}$$

where brackets $\langle \cdot \rangle$ refers to the volume average of the enclosed property. The tensor inequality $\mathbf{A} > \mathbf{B}$ means that the difference tensor $\mathbf{D} = \mathbf{A} - \mathbf{B}$ is positive semidefinite; all the eigenvalues of $\mathbf{D}$ are greater than or equal to zero.

Uses

The methods described in this section can be used for the following purposes:
- to compute the estimated range of the average mineral modulus for a mixture of mineral grains; and
- to compute the upper and lower bounds on effective elastic moduli for a mixture of mineral and pore fluid.

Assumptions and Limitations

The methods described in this section presuppose that each constituent is isotropic, linear, and elastic.

4.2 Hashin–Shtrikman–Walpole Bounds

Synopsis

The Voigt and Reuss bounds, discussed in Section 4.1, are the easiest to remember and to implement, because they are simply arithmetic and harmonic averages. In this section and the next, we describe bounds that are often considered to be "better," because they provide narrower ranges of uncertainty between the maximum and minimum possible effective elastic moduli. We begin with the **Hashin–Shtrikman (HS) bounds**.

For an N-phase linear elastic composite with "well-ordered" bulk and shear phase moduli, such that $K_{min} = K_1 \le K_2 \cdots \le K_N = K_{max}$ and $\mu_{min} = \mu_1 \le \mu_2 \cdots \le \mu_N = \mu_{max}$, the **Hashin–Shtrikman** (1963) **bounds** on the isotropic effective bulk and shear moduli are given by

$$K^{HS+} = \Lambda(4\mu_{max}/3); \quad K^{HS-} = \Lambda(4\mu_{min}/3) \tag{4.2.1}$$

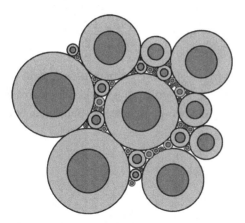

Figure 4.2.1 One of several known physical interpretations of the Hashin–Shtrikman bounds for the bulk modulus of a two-phase material

$$\mu^{HS+} = \Gamma\big(\zeta(K_{\max}, \mu_{\max})\big); \quad \mu^{HS-} = \Gamma\big(\zeta(K_{\min}, \mu_{\min})\big) \tag{4.2.2}$$

where

$$\Lambda(z) = \left\langle \frac{1}{K(r)+z} \right\rangle^{-1} - z; \quad \Gamma(z) = \left\langle \frac{1}{\mu(r)+z} \right\rangle^{-1} - z; \quad \zeta(K,\mu) = \frac{\mu}{6}\left(\frac{9K+8\mu}{K+2\mu}\right). \tag{4.2.3}$$

In equations 4.2.1 and 4.2.2, the superscript +/− indicates upper or lower bound respectively. The brackets $\langle \cdot \rangle$ in equation 4.2.3 indicate a volume average over the medium, which is the same as an average over the constituents weighted by their volume fractions.

For a two-phase "well-ordered" case, $(K_1 - K_2)(\mu_1 - \mu_2) \geq 0$, the Hashin–Shtrikman bounds can be written as

$$K^{HS\pm} = \left[\frac{f_1}{K_1 + (4/3)\mu_m} + \frac{f_2}{K_2 + (4/3)\mu_m}\right]^{-1} - (4/3)\mu_m, \tag{4.2.4}$$

$$\mu^{HS\pm} = \left[\frac{f_1}{\mu_1 + \zeta_m} + \frac{f_2}{\mu_2 + \zeta_m}\right]^{-1} - \zeta_m, \text{ with } \zeta_m = \frac{\mu_m}{6}\frac{(9K_m + 8\mu_m)}{(K_m + 2\mu_m)} \tag{4.2.5}$$

where K_1 and K_2 are the bulk moduli of individual phases; μ_1 and μ_2 are the shear moduli of individual phases; and f_1 and f_2 are the volume fractions of individual phases.

Alternative expressions, which are algebraically equivalent to equations 4.2.4 and 4.2.5, are

$$K^{HS\pm} - f_1K_1 + f_2K_2 - \frac{f_1f_2(K_2 - K_1)^2}{f_2K_1 + f_1K_2 + \frac{4}{3}\mu_m}, \tag{4.2.6}$$

$$\mu^{HS\pm} = f_1\mu_1 + f_2\mu_2 - \frac{f_1 f_2 (\mu_2 - \mu_1)^2}{f_2\mu_1 + f_1\mu_2 + \zeta_m} \ , \text{with } \zeta_m = \frac{\mu_m}{6}\frac{(9K_m + 8\mu_m)}{(K_m + 2\mu_m)} \tag{4.2.7}$$

In equations 4.2.4–4.2.7, $K_m = \max(K_1, K_2)$ and $\mu_m = \max(\mu_1, \mu_2)$ yield the upper bounds, while $K_m = \min(K_1, K_2)$ and $\mu_m = \min(\mu_1, \mu_2)$ yield the lower bounds. Another set of equivalent expressions for the two-phase Hashin–Shtrikman bounds are

$$K^{HS\pm} = K_1 + \frac{f_2}{(K_2 - K_1)^{-1} + f_1\left(K_1 + \frac{4}{3}\mu_1\right)^{-1}} \tag{4.2.8}$$

$$\mu^{HS\pm} = \mu_1 + \frac{f_2}{(\mu_2 - \mu_1)^{-1} + 2f_1(K_1 + 2\mu_1)/\left[5\mu_1\left(K_1 + \frac{4}{3}\mu_1\right)\right]} \tag{4.2.9}$$

For equations 4.2.8 and 4.2.9, upper and lower bounds are computed by interchanging which material is termed 1 and which is termed 2. The expressions yield the upper bound when the stiffest material is termed 1 and the lower bound when the softest material is termed 1.

Remember that equations 4.2.1–4.2.9 apply when the constituent with the largest bulk modulus also has the largest shear modulus, and the constituent with the smallest bulk modulus also has the smallest shear modulus. Within this range of validity, the two-phase Hashin–Shtrikman bounds are optimum, giving the narrowest possible range without specifying anything about the geometries of the constituents,

For two-phase well-ordered composites, the Hashin–Shtrikman bounds are physically attainable by a multitude of microstructures (Hashin and Shtrikman, 1963; Norris, 1985; Milton, 1984; Gibiansky and Sigmund, 2000). For example, the *HS* bound on the bulk modulus, *but not the HS shear modulus*, can be realized by a multi-scale, space-filling pack of "coated spheres," shown schematically in Figure 4.2.1. Each sphere of material 2 is surrounded by a shell of material 1. Each sphere and its shell have precisely the volume fractions f_2 and $f_1 = 1 - f_2$, respectively. This coated sphere realization requires a semi-infinite distribution of sizes down to infinitesimal, to ensure that all space is completely filled. The upper bound on bulk modulus is realized when the stiffer material forms the shell; the lower bound on bulk modulus is realized when the stiffer material is in the core. Milton (1984) found that both the bulk and shear Hashin–Shtrikman bounds can be realized simultaneously by certain multi-rank laminate geometries. For more than two phases, the Hashin–Shtrikman bounds may not be optimal (realizable) over the entire range of volume fractions. The differential effective medium model was shown by Norris (1985) to achieve the bounds on both bulk and shear moduli simultaneously when the inclusions are disk-shaped. Boucher (1976) and Norris (1985) pointed out that the nonsymmetric self-consistent scheme of Wu (1966) for disk-shaped geometries attains both bounds when $(K_1 - K_2)(G_1 - G_2) \geq 0$. The Mori–Tanaka effective medium model for bulk modulus with spherical inclusions in an elastic background predicts values that also fall on the upper Hashin–Shtrikman bound.

If a well-sorted two-phase composite is subject to an external uniform hydrostatic stress field, then its effective bulk modulus will be on one of the Hashin–Shtrikman bounds *if and only if* the stresses and strains in one of the phases (the inclusions) are uniform (Gibiansky and Sigmund, 2000). Similarly if a two-phase composite is subject to an external uniform deviatoric stress field, then its effective shear modulus falls on one of the Hashin–Shtrikman bounds if and only if the stresses and strains in one of the phases are uniform. In the analogous *multiphase* problem, the stress and strain fields in all but one phase will be uniform, though different from one phase to another. Gibiansky and Sigmund refer to this as the "field optimality condition." This is in contrast to the Voigt and Reuss bounds, which require that the stress or strain be constant throughout the composite.

Milton and Phan-Thien (1982) have shown that, in fact, the two-phase Hashin–Shtrikman bounds are valid beyond the well-ordered case under the following conditions. When $\mu_1 \geq \mu_2$ and

$$(K_1 - K_2) \geq -\frac{(3K_2 + 8\mu_2)^2}{42K_2^2} \frac{K_1 K_2}{\mu_1 \mu_2} (\mu_1 - \mu_2), \tag{4.2.10}$$

the lower Hashin–Shtrikman bound on shear modulus is still valid. Similarly, when $\mu_1 \geq \mu_2$ and

$$(K_1 - K_2) \geq -\frac{(3K_1 + 8\mu_1)^2}{42K_1^2} (\mu_1 - \mu_2), \tag{4.2.11}$$

the Hashin–Shtrikman upper bound on shear modulus is still valid. Note that when the conditions in equations 4.2.10 and 4.2.11 do not hold, then the Hashin–Shtrikman equations on shear modulus (equations 4.2.5, 4.2.7, and 4.2.9) may no longer be valid bounds.

The **Walpole bounds** are defined for the "non-well-ordered" case $(K_1 - K_2)(\mu_1 - \mu_2) < 0$. The Walpole bounds on bulk modulus are the same as the Hashin–Shtrikman bounds on bulk modulus:

$$K^{\mathrm{HSW}\pm} = K_1 + \frac{f_2}{(K_2 - K_1)^{-1} + f_1 \left(K_1 + \frac{4}{3}\mu_m \right)^{-1}} \tag{4.2.12}$$

If $\mu_1 \geq \mu_2$ and $K_2 \geq K_1$, the Walpole bounds on shear modulus can be written as

$$\mu^{\mathrm{HSW}\pm} = \mu_1 + \frac{f_2}{(\mu_2 - \mu_1)^{-1} + f_1 \left[\mu_1 + \frac{\mu_m}{6} \left(\frac{9K_m + 8\mu_m}{K_m + 2\mu_m} \right) \right]^{-1}} \tag{4.2.13}$$

(Torquato, 2002) where the subscripts 1 and 2 again refer to the properties of the two components. Equations 4.2.12 and 4.2.13 yield the upper bound when K_m and μ_m are the maximum bulk and shear moduli of the individual constituents, and the lower bound when K_m and μ_m are the minimum bulk and shear moduli of the constituents. The maximum (minimum) shear modulus might come from a different constituent than that with the maximum (minimum) bulk modulus. This would be the case, for example, for a mixture of calcite $(K = 71, \mu = 30 \text{ GPa})$ and quartz $(K = 37, \mu = 45 \text{ GPa})$.

Equation 4.2.13 reduces to the Hashin–Shtrikman bound equation 4.2.7 when the constituents are well-ordered. Therefore, equations 4.2.12 and 4.2.13 are sometimes called the **Hashin–Shtrikman–Walpole bounds**. In geophysical practice, the Hashin–Shtrikman–Walpole bounds are commonly implemented, ignoring the distinction between well-ordered and non-well-ordered constituents.

Unlike the Hashin–Shtrikman bounds, the Walpole bound on shear modulus for the non-well-ordered case is never realizable. The two-phase bounds of Milton and Phan-Thien (1982) described in Section 4.3 are always tighter than the Walpole bounds on shear modulus.

Example: Compute the Hashin–Shtrikman–Walpole upper and lower bounds on the bulk and shear moduli for a mixture of quartz, calcite, and water. The porosity (water fraction) is 27%, quartz is 80% by volume of the solid fraction, and calcite is 20% by volume of the solid fraction. The moduli of the individual constituents are:

$$K_{quartz} = 36 \text{GPa} \quad K_{calcite} = 75 \text{GPa} \quad K_{water} = 2.2 \text{GPa},$$

$$\mu_{quartz} = 45 \text{ GPa} \quad \mu_{calcite} = 31 \text{ GPa} \quad \mu_{water} = 0 \text{ GPa}$$

Hence

$$\mu_{min} = 0 \text{ GPa}, \quad \mu_{max} = 45 \text{ GPa}, \quad K_{min} = 2.2 \text{ GPa, and } K_{max} = 75 \text{ GPa}$$

$$K^{HS-} = \Lambda(\mu_{min})$$

$$= \left[\frac{\phi}{2.2} + \frac{(1-\phi)(0.8)}{36.0} + \frac{(1-\phi)(0.2)}{75.0} \right]^{-1}$$

$$= 7.10 \text{ GPa}$$

$$K^{HS+} = \Lambda(\mu_{max})$$

$$= \left[\frac{\phi}{2.2 + \left(\frac{4}{3}\right)45} + \frac{(1-\phi)(0.8)}{36.0 + \left(\frac{4}{3}\right)45} + \frac{(1-\phi)(0.2)}{75.0 + \left(\frac{4}{3}\right)45.0} \right]^{-1} - \left(\frac{4}{3}\right)45$$

$$= 26.9 \text{ GP}a$$

$$\zeta(K_{max}, \mu_{max}) = \frac{45}{6}\left(\frac{9 \cdot 75 + 8 \cdot 45}{75 + 2 \cdot 45}\right) = 47.0 \text{ GP}a$$

$$\zeta(K_{min}, \mu_{nin}) = 0 \text{ GP}a$$

$$\mu^{HS+} = \Gamma(\zeta(K_{max}, \mu_{max}))$$

$$= \left[\frac{\phi}{47.0} + \frac{(1-\phi)(0.8)}{45.0 + 47.0} + \frac{(1-\phi)(0.2)}{31.0 + 47.0}\right]^{-1} - 47.0$$

$$= 24.6 \text{ GP}a$$

$$\mu^{HS-} = \Gamma(\zeta(K_{min}, \mu_{min}))$$

$$= 0$$

Note that this example involves a non-well-ordered set of minerals, since quartz has the maximum shear modulus, but calcite has the maximum bulk modulus. Therefore, this calculation yields the bounds on shear modulus, but they are not realizable. If we modify the fractions to porosity=27%, quartz fraction = 99.99999% of the solid, and calcite fraction=0.00001%, of the solid, we find that

$$\mu^{HSW+} = 26.11.\text{GPa}$$

But, if we eliminate calcite completely, the constituents become well-ordered, and we can compute the HS bounds:
$$\mu^{HS+} = 25.30.\text{GPa}$$
When only quartz and water are present, $K_{max} = K_{quartz} = 36$ and $\mu_{max} = \mu_{quartz} = 45$. But with even a miniscule amount of calcite (e.g., 1 molecule), $K_{max} = K_{calcite} = 75$ and $\mu_{max} = \mu_{quartz} = 45$, and the predicted upper shear bound takes a nonphysical step jump.

The separation between upper and lower elastic bounds depends on how different the constituents are. As illustrated in Figure 4.2.2, the bounds are often quite similar when mixing solids, since the moduli of common minerals are usually within a factor of 2 of each other. Because many effective-medium models (e.g., Biot, Gassmann, Kuster–Toksöz, etc.) assume a homogeneous mineral modulus, it is often useful (and adequate) to represent a mixed mineralogy with an "average mineral" equal to either one of the bounds or to their average $(M^{HS+} + M^{HS-})/2$. On the other hand, when the constituents are quite different, such as minerals and pore fluids, the bounds become quite separated, and we lose some of the predictive value.

Note that when $\mu_{min} = 0$, the bounds K^{HS-}, μ^{HS-}, K^{HSW-}, and μ^{HSW-} are the same as the Reuss bound. In this case, the Reuss and Hashin–Shtrikman lower bound exactly describe the moduli of a suspension of grains in a pore fluid (see Section 4.1 on Voigt–

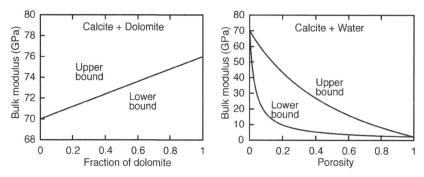

Figure 4.2.2 Hashin–Shtrikman bounds for elastically similar constituents (left) and dissimilar constituents (right)

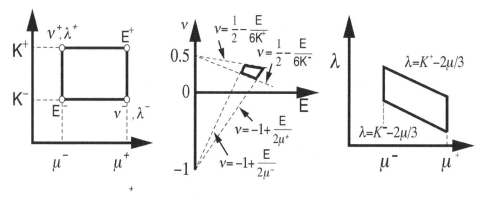

Figure 4.2.3 Schematic representation of the Hashin–Shtrikman bounds in the (K, μ), (ν, E) and (λ, μ) planes

Reuss bounds and also Section 4.4 on Wood's relation). These also describe the moduli of a mixture of fluids or gases, or both.

When all phases have the same shear modulus, $\mu = \mu_{\min} = \mu_{\max}$ the upper and lower bounds become identical, and we obtain the expression from Hill (1963) for the effective bulk modulus of a composite with uniform shear modulus (see Section 4.6 on composites with uniform shear modulus).

The Hashin–Shtrikman bounds are defined for the bulk and shear moduli. When bounds on other isotropic elastic constants are needed, the correct procedure is to compute the bounds for bulk and shear moduli and then compute the other elastic constants from them. Figure 4.2.3 shows schematically the bounds in the planes (K, μ), (ν, E), and (λ, μ). In each case, the allowed pairs of elastic constants lie within the polygons. Notice how the upper and lower Hashin–Shtrikman bounds for (K, μ) are uncoupled, suggesting that any value $K^{HS-} \leq K \leq K^{HS+}$ can exist with any value $\mu^{HS-} \leq \mu \leq \mu^{HS+}$. In contrast, the bounds on (ν, E) both depend on (K, μ) – the upper and lower bounds on ν depend on the value of E, and vice versa. Similarly, the bounds on λ are coupled to the value of μ. *K* and μ are more "orthogonal" than λ and μ *We will actually see in Section 4.3 that certain combinations of K and μ within the Hashin–Shtrikman rectangle cannot exist together.*

Uses

The bounds described in this section can be used for the following:

- to compute the estimated range of average mineral modulus for a mixture of mineral grains; and
- to compute the upper and lower bounds for a mixture of mineral and pore fluid.

Assumptions and Limitations

The bounds described in this section apply under the following conditions:

- each constituent is isotropic, linear, and elastic; and
- the rock is isotropic, linear, and elastic.

4.3   Improvements on the Hashin–Shtrikman–Walpole Bounds

Useful Canonical Form

All of the bounds discussed in this book can be conveniently expressed in terms of "canonical" functions of effective moduli (Berryman, 2005), which are defined for a 2-phase elastic composite as

$$K = \Lambda(y_K) = \left[\frac{f_1}{K_1 + y_K} + \frac{f_2}{K_2 + y_K} \right]^{-1} - y_K , \tag{4.3.1}$$

$$\mu = \Gamma(y_\mu) = \left[\frac{f_1}{\mu_1 + y_\mu} + \frac{f_2}{\mu_2 + y_\mu} \right]^{-1} - y_\mu , \tag{4.3.2}$$

where *K* and μ are the effective bulk and shear moduli, and f_i, K_i, and μ_i are the volume fractions, bulk moduli, and shear moduli of the individual phases. Functions $\Lambda(z)$ and $\Gamma(z)$ are the same as defined by equation 4.2.3. (**Caution: Some authors put a factor of (4/3) in front of y_k.**) The inverses of functions Λ and Γ are the **scalar y-transforms** of the effective moduli (Vinogradov and Milton, 2005, p. 1251):

$$y_K = -f_2 K_1 - f_1 K_2 + \frac{f_1 f_2 (K_1 - K_2)^2}{f_1 K_1 + f_2 K_2 - K} , \tag{4.3.3}$$

$$y_\mu = -f_2 \mu_1 - f_1 \mu_2 + \frac{f_1 f_2 (\mu_1 - \mu_2)^2}{f_1 \mu_1 + f_2 \mu_2 - \mu} . \tag{4.3.4}$$

Since $\Lambda(y_K)$ and $\Gamma(y_\mu)$ are monotonically increasing functions of their arguments, bounds on y_K and y_μ correspond to bounds on K and μ, respectively. Variables y_K and y_μ are always non-negative. The y-transforms in equations 4.3.3 and 4.3.4 are defined for any effective moduli K and μ that lie within their respective bounds.

The Hashin–Shtrikman bounds on bulk and shear moduli (Section 4.2) are defined for the "well-ordered" case $(K_1 - K_2)(\mu_1 - \mu_2) \geq 0$. If the phases are labeled such that $\mu_1 \geq \mu_2$, the Hashin–Shtrikman bounds can be written in the y-transform domain as (Torquato, 2002)

$$y_K^{HS-} \equiv \frac{4}{3}\mu_2 \leq y_K \leq \frac{4}{3}\mu_1 \equiv y_K^{HS+} , \tag{4.3.5}$$

$$y_\mu^{HS-} \equiv \frac{\mu_2(9K_2 + 8\mu_2)}{6(K_2 + 2\mu_2)} \leq y_\mu \leq \frac{\mu_1(9K_1 + 8\mu_1)}{6(K_1 + 2\mu_1)} \equiv y_\mu^{HS+} , \tag{4.3.6}$$

For "non-well-ordered" phases, with $K_{min} = \min(K_1, K_2)$, $K_{max} = \max(K_1, K_2)$, $\mu_{min} = \min(\mu_1, \mu_2)$, and $\mu_{max} = \max(\mu_1, \mu_2)$, the Walpole bounds can be written as

$$y_K^{HSW-} \equiv \frac{4}{3}\mu_{min} \leq y_K \leq \frac{4}{3}\mu_{max} \equiv y_K^{HSW+} , \tag{4.3.7}$$

$$y_\mu^{HSW-} \equiv \frac{\mu_{min}(9K_{min} + 8\mu_{min})}{6(K_{min} + 2\mu_{min})} \leq y_\mu \leq \frac{\mu_{max}(9K_{max} + 8\mu_{max})}{6(K_{max} + 2\mu_{max})} \equiv y_\mu^{HSW+} , \tag{4.3.8}$$

It is also noted that the Voigt and Reuss bounds (Section 4.1) can also be expressed in the y-transform domain:

$$y_\mu^R \equiv 0 \leq y_\mu \leq \infty \equiv y_\mu^V \tag{4.3.9}$$

In equation 4.3.9, the Voigt average is found as $y_\mu^V \to \infty$ in the limit.

Bounds That Are Narrower Than the Walpole Bounds

Bounds on effective media properties (electrical or thermal conductivity, elastic moduli, dielectric constant) can be found that are more restrictive than those of Hashin–Shtrikman and Walpole if information about the microgeometry is included. Microgeometric parameters can be estimated directly from three-dimensional micro images of a composite (Berryman, 1988) or by using information from measurement of one property to narrow the bounds of a second (Prager, 1969; Milton, 1981; Kantor and Bergman, 1984; Berryman and Milton, 1988).

Milton and Phan-Thien (1982) and Berryman and Milton (1988), for example, considered two-phase isotropic composites and used two geometric quantities ζ and η, which depend on the isotropic three-point correlations functions S_3 of the composite:

$$S_3(r_{12}, r_{13}, u_{12,13}) = \langle h(\mathbf{x} + \mathbf{r}_1)h(\mathbf{x} + \mathbf{r}_2)h(\mathbf{x} + \mathbf{r}_3) \rangle , \qquad (4.3.10)$$

where brackets $\langle \ \rangle$ indicate volume average. Variables $\mathbf{x}$ and $\mathbf{r}$ are the position vector and lag, respectively. Relative positions of points i and j are expressed by the quantities

$$\mathbf{r}_{ij} = \mathbf{r}_j - \mathbf{r}_i , \quad r_{ij} = |\mathbf{r}_{ij}| , \quad u_{ij,ik} = \mathbf{r}_{ij} \cdot \mathbf{r}_{ik}/|\mathbf{r}_{ij}||\mathbf{r}_{ik}| . \qquad (4.3.11)$$

Function $h(\mathbf{x})$ is a three-dimensional isotropic indicator function equal to $h(\mathbf{x}) = 1$ in material 1 and $h(\mathbf{x}) = 0$ in material 2. The microgeometry parameters are then defined as

$$\zeta_1, = \lim_{\Delta \to 0} \lim_{\Delta' \to \infty} \frac{9}{2f_1f_2} \int_\Delta^{\Delta'} dr \int_\Delta^{\Delta'} ds \int_{-1}^1 du \, \frac{S_3(r,s,u)}{rs} P_2(u), \qquad (4.3.12)$$

$$\eta_1 = \frac{5\zeta_1}{21} + \lim_{\Delta \to 0} \lim_{\Delta' \to \infty} \frac{150}{7f_1f_2} \int_\Delta^{\Delta'} dr \int_\Delta^{\Delta'} ds \int_{-1}^1 du \, \frac{S_3(r,s,u)}{rs} P_4(u), \qquad (4.3.13)$$

where f_1 and f_2 are volume fractions of the phases, such that $f_1 + f_2 = 1$, and $P_2(u)$ and $P_4(u)$ are Legendre polynomials:

$$P_2(u) = \frac{1}{2}(3u^2 - 1) \quad \text{and} \quad P_4(u) = \frac{1}{8}(35u^4 - 30u^2 + 3). \qquad (4.3.14)$$

Complementary geometric parameters can also be defined, $\zeta_2 = 1 - \zeta_1$ and $\eta_2 = 1 - \eta_1$. The microgeometry parameters have the limits $0 \le \eta_i \le 1$ and $0 \le \zeta_i \le 1$.

Milton and Phan-Thien (1982) showed that the possible range of η_1 is constrained by the value of ζ_1:

$$\eta_1^- \equiv 5\zeta_1/21 \le \eta_1 \le (16 + 5\zeta_1)/21 \equiv \eta_1^+. \qquad (4.3.15)$$

This latter relation is useful when direct measurement of η_1 is not possible.

To express various correlation-dependent bounds compactly, it is useful to define the following weighted averages:

$$\langle \Pi \rangle = f_1\Pi_1 + f_2\Pi_2 , \qquad (4.3.16)$$

$$\langle \Pi \rangle_\zeta = \zeta_1\Pi_1 + \zeta_2\Pi_2 , \qquad (4.3.17)$$

$$\langle \Pi \rangle_\eta = \eta_1\Pi_1 + \eta_2\Pi_2 , \qquad (4.3.18)$$

where Π is any bimodal scalar property associated with phases 1 and 2.

Using this notation, compact expressions for the bounds of Beran and Molyneux (1966) and McCoy (1970) can be written. For the bulk modulus, these bounds are (if ζ is known)

$$y_K^{BM-} \equiv \frac{4}{3}\langle \mu^{-1} \rangle_\zeta^{-1} \le y_K \le \frac{4}{3}\langle \mu \rangle_\zeta \equiv y_K^{BM+} \qquad (4.3.19)$$

and for the shear modulus the bounds are (if ζ and η are known)

$$y_\mu^{BM-} \equiv 1/6\Xi \leq y_\mu \leq \Theta/6 \equiv y_\mu^{BM+} \tag{4.3.20}$$

where

$$\Theta = \frac{10\langle\mu\rangle^2\langle K\rangle_\zeta + 5\langle\mu\rangle\langle 2K + 3\mu\rangle\langle\mu\rangle_\zeta + \langle 3K + \mu\rangle^2\langle\mu\rangle_\eta}{\langle K + 2\mu\rangle^2}, \tag{4.3.21}$$

$$\Xi = \frac{10\langle K\rangle^2\langle 1/K\rangle_\zeta + 5\langle\mu\rangle\langle 2K + 3\mu\rangle\langle 1/\mu\rangle_\zeta + \langle 3K + \mu\rangle^2\langle 1/\mu\rangle_\eta}{\langle 9K + 8\mu\rangle^2}. \tag{4.3.22}$$

Milton and Phan-Thien (1982) found improved bounds on shear modulus given by

$$y_\mu^{MP-} \equiv 1/6\hat{\Xi} \leq y_\mu \leq \hat{\Theta}/6 \equiv y_\mu^{MP+} \tag{4.3.23}$$

where

$$\hat{\Theta} = \frac{3\langle\mu\rangle_\eta\langle 6K + 7\mu\rangle_\zeta - 5\langle\mu\rangle_\zeta^2}{\langle 2K - \mu\rangle_\zeta + 5\langle\mu\rangle_\eta}, \tag{4.3.24}$$

$$\hat{\Xi} = \frac{5\langle 1/\mu\rangle_\zeta\langle 6/K - 1/\mu\rangle_\zeta + \langle 1/\mu\rangle_\eta\langle 2/K + 21/\mu\rangle_\zeta}{\langle 128/K + 99/\mu\rangle_\zeta + \langle 45/\mu\rangle_\eta}. \tag{4.3.25}$$

When the correlation functions are not known, correlation-independent bounds can be obtained by finding the extrema when considering all $5\zeta_1/21 \leq \eta_1 \leq (16 + 5\zeta_1)/21$ while $0 \leq \zeta_1 \leq 1$, with $\mu_1 \geq \mu_2$. In equations 4.3.26 and 4.3.27, find the minimum/maximum value of the expression over all values of ζ_1 in the specified range.

$$y_\mu^{MP2+} \geq \frac{\min}{\substack{\zeta_1 \\ 0 \leq \zeta_1 \leq 1}} \frac{8\langle 6/\mu + 7/K\rangle_\zeta + 15/\mu_2}{2(\langle 21/\mu + 2/K\rangle_\zeta/\mu_2 + 40\langle 1/\mu\rangle_\zeta\langle 1/K\rangle_\zeta)} \tag{4.3.26}$$

$$y_\mu^{MP2-} \leq \frac{\max}{\substack{\zeta_1 \\ 0 \leq \zeta_1 \leq 1}} \frac{8\mu_1\langle 6K + 7\mu\rangle_\zeta + 15\langle K\rangle_\zeta\langle\mu\rangle_\zeta}{2(\langle 21K + 2\mu\rangle_\zeta + 40\mu_1)} \tag{4.3.27}$$

Bounds $y_\mu^{MP2\pm}$ coincide with the Hashin–Shtrikman bounds over the extended range where the Hashin–Shtrikman bounds are valid (see Section 4.2). Bounds $y_\mu^{MP2\pm}$ are always more restrictive than the Walpole bounds.

Berryman and Milton (1988) discuss how to use ζ and η to create cross bounds of different composite parameters, without introducing explicit information about the microgeometry. For example, if we know the effective bulk modulus, K, of a two-phase composite, then the bounds of Beran and Molyneux (1966) can be inverted to put the following limits on ζ_1 (if $\mu_1 \geq \mu_2$):

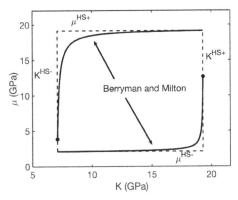

Figure 4.3.1 Example of cross-bounds between bulk and shear moduli from Berryman and Milton (1988)

$$\zeta_1^-(y_K) \equiv \frac{4/(3y_K) - 1/\mu_1}{1/\mu_2 - 1/\mu_1} \le \zeta_1 \le \frac{\mu_1 - 3y_K/4}{\mu_1 - \mu_2} \equiv \zeta_1^+(y_K) \tag{4.3.28}$$

where, in this case, y_K is the transform of the known bulk modulus.

The constrained range of ζ_1 puts limits as well on the range of η_1, as previously expressed. Finally, ζ_1 and η_1 determine the maximum value μ^+ and minimum value μ^- that are possible for the given K, and we obtain the Berryman–Milton correlations of effective bulk and shear moduli that are more restrictive than the Hashin–Shtrikman (HS) bounds.

Figure 4.3.1 shows a two-phase composite of quartz ($K_1 = 36$ GPa and $\mu_2 = 45$ GPa) with 40% volume fraction of a soft solid, such as bitumen ($K_2 = 2.9$ GPa and $\mu_2 = 0.5$ GPa). We see that the Berryman–Milton bounds are slightly narrower than the HS bounds, which form the square.

The result illustrates that while any bulk modulus on or within the HS bounds is realizable, and any shear modulus on or within the HS bounds is also realizable, not all combinations of bulk and shear modulus are possible. It is not possible for bulk modulus of a composite to be on the lower HS bound while the corresponding shear modulus is on the upper HS bound (and vice versa) (Berryman and Milton, 1988). Milton also proved that the bulk modulus of a two-phase composite can be on the upper or lower HS bound, while the shear is not on the bound; in contrast, if the shear modulus is on the HS bound, the bulk modulus must be on the HS bound as well.

If both effective moduli K and μ are known, the range of ζ_1 can be determined from K as in the previous example. Knowledge of μ constrains the range of η_1 further than if only the range of ζ_1 is known.

4.4   Wood's Formula

Synopsis

In a fluid suspension or fluid mixture, where the heterogeneities are small compared with a wavelength, the sound velocity is given exactly by Wood's (1955) relation

$$V = \sqrt{\frac{K_R}{\rho}} \qquad (4.4.1)$$

where K_R is the Reuss (isostress) average of the composite

$$\frac{1}{K_R} = \sum_{i=1}^{N} \frac{f_i}{K_i} \qquad (4.4.2)$$

and ρ is the average density defined by

$$\rho = \sum_{i=1}^{N} f_i \rho_i \qquad (4.4.3)$$

f_i, K_i, and ρ_i are the volume fraction, bulk moduli, and densities of the phases, respectively.

Example: Use Wood's relation to estimate the speed of sound in a water-saturated suspension of quartz particles at atmospheric conditions. The quartz properties are $K_{quartz} = 36$ GPa and $\rho_{quartz} = 2.65 \text{g/cm}^3$. The water properties are $K_{water} = 2.2$ GPa and $\rho_{water} = 1.0 \text{ g/cm}^3$ The porosity is $\phi = 0.40$.

The Reuss average bulk modulus of the suspension is given by

$$K_{Reuss} = \left(\frac{\phi}{K_{water}} + \frac{1-\phi}{K_{quartz}} \right)^{-1} = \left(\frac{0.4}{2.2} + \frac{0.6}{36} \right)^{-1} = 5.04 \text{GPa}$$

The density of the suspension is

$$\rho = \phi \rho_{water} + (1-\phi)\rho_{quartz} = 0.4 \times 1.0 + 0.6 \times 2.65 = 1.99 \text{ g/cm}^3$$

This gives a sound speed of

$$V = \sqrt{K/\rho} = \sqrt{5.04/1.99} = 1.59 \text{ km/s}$$

Example: Use Wood's relation to estimate the speed of sound in a suspension of quartz particles in water with 50% saturation of air at atmospheric conditions. The quartz properties are $K_{\text{quartz}} = 36$ GPa and $\rho_{\text{quartz}} = 2.65\text{g/cm}^3$. The water properties are $K_{\text{water}} = 2.2$ GPa and $\rho_{\text{water}} = 1.0\text{g/cm}^3$. The air properties are $K_{\text{alr}} = 0.000131$ GPa and $\rho_{\text{air}} = 0.00119$ g/cm^3.

The porosity is $\phi = 0.40$.

The Reuss average bulk modulus of the suspension is given by

$$K_{\text{Reuss}} = \left(\frac{0.5\phi}{K_{\text{water}}} + \frac{0.5\phi}{K_{\text{air}}} + \frac{1 - \phi}{K_{\text{quartz}}} \right)^{-1}$$

$$= \left(\frac{0.5 \times 0.4}{2.2} + \frac{0.5 \times 0.4}{0.000131} + \frac{0.6}{36} \right)^{-1} = 0.00065 \text{ GPa}$$

The density of the suspension is

$$\rho = 0.5\phi\rho_{\text{water}} + 0.5\phi\rho_{\text{air}} + (1 - \phi)\rho_{\text{quartz}}$$
$$= 0.5 \times 0.4 \times 1.0 + 0.5 \times 0.4 \times 0.00119 + 0.6 \times 2.65 = 1.79 \text{ g/cm}^3$$

This gives a sound speed of

$$V = \sqrt{K/\rho} = \sqrt{0.00065/1.79} = 0.019 \text{ km/s}$$

Uses

Wood's formula may be used to estimate the velocity in suspensions.

Assumptions and Limitations

Wood's formula presupposes that composite rock and each of its components are isotropic, linear, and elastic.

4.5   Voigt–Reuss–Hill Average Moduli Estimate

Synopsis

The Voigt–Reuss–Hill average is simply the arithmetic average of the Voigt upper bound and the Reuss lower bound. (See the discussion of the Voigt–Reuss bounds in Section 4.1.) This average is expressed as

$$M_{\text{VRH}} = \frac{M_{\text{V}} + M_{\text{R}}}{2} \tag{4.5.1}$$

where

$$M_{\text{v}} = \sum_{i=1}^{N} f_i M_j ; \quad \frac{1}{M_{\text{R}}} = \sum_{i=1}^{N} \frac{f_i}{M_i} \tag{4.5.2}$$

The terms f_i and M_j are the volume fraction and the modulus of the ith component, respectively. M can be either the shear modulus or the bulk modulus.

The Voigt–Reuss–Hill average is useful when an *estimate* of the moduli is needed, not just the allowable range of values. An obvious extension would be to average, instead, the Hashin–Shtrikman upper and lower bounds.

This resembles, but is not exactly the same as the average of the algebraic and harmonic means of velocity used by Greenberg and Castagna (1992) in their empirical V_P–V_S relation (see Section 7.9).

Uses

The Voigt–Reuss–Hill average is used to estimate the effective elastic moduli of a rock in terms of its constituents and pore space.

Assumptions and Limitations

The following limitation and assumption apply to the Voigt–Reuss–Hill average:
- The result is strictly heuristic. Hill (1952) showed that the Voigt and Reuss averages are upper and lower bounds, respectively. Several authors have shown that the average of these bounds can be a useful and sometimes accurate estimate of rock properties.
- The rock is isotropic.

4.6 Composite with Uniform Shear Modulus

Synopsis

Hill (1963) showed that when all of the phases or constituents in a composite have the same shear modulus, μ, the effective P-wave modulus, $M_{\text{eff}} = \left(K_{\text{eff}} + \frac{4}{3}\mu_{\text{eff}} \right)$ is given exactly by

$$\frac{1}{\left(K_{\text{eff}} + \frac{4}{3}\mu_{\text{eff}} \right)} = \sum_{i=1}^{N} \frac{x_i}{\left(K_i + \frac{4}{3}\mu \right)} = \left\langle \frac{1}{K + \frac{4}{3}\mu} \right\rangle \tag{4.6.1}$$

where x_i is the volume fraction of the ith component, K_i is its bulk modulus, and $\langle \cdot \rangle$ refers to the volume average. Because $\mu_{\text{eff}} = \mu_j = \mu$, any of the effective moduli can then be easily obtained from K_{eff} and μ_{eff}.

This is obviously the same as

$$\frac{1}{\left(\rho V_P^2\right)_{\text{eff}}} = \left\langle\frac{1}{\rho V_P^2}\right\rangle \tag{4.6.2}$$

This striking result states that the effective moduli of a composite with uniform shear modulus can be found *exactly* if one knows only the volume fractions of the constituents independent of the constituent geometries. There is no dependence, for example, on ellipsoids, spheres, or other idealized shapes.

Hill's equation follows simply from the expressions for Hashin–Shtrikman bounds (see Section 4.2 on Hashin–Shtrikman bounds) on the effective bulk modulus:

$$\frac{1}{K^{\text{HS}\pm} + \frac{4}{3}\mu_{\{\substack{\max \\ \min}\}}} = \left\langle\frac{1}{K + \frac{4}{3}\mu_{\{\substack{\max \\ \min}\}}}\right\rangle \tag{4.6.3}$$

where $\mu_{\min}$ and $\mu_{\max}$ are the minimum and maximum shear moduli of the various constituents, yielding, respectively, the lower and upper bounds on the bulk modulus, $K^{\text{HS}\pm}$. Any composite must have an effective bulk modulus that falls between the bounds. Because here $\mu = \mu_{\min} = \mu_{\max}$, the two bounds on the bulk modulus are equal and reduce to the previous Hill expression.

In the case of a mixture of liquids or gases, or both, where $\mu = 0$ for all the constituents, the Hill's equation becomes the well-known isostress equation or Reuss average:

$$\frac{1}{K_{\text{eff}}} = \sum_{i=1}^{N}\frac{x_i}{K_i} = \left\langle\frac{1}{K}\right\rangle \tag{4.6.4}$$

A somewhat surprising result is that a finely layered medium, where each layer is isotropic and has the same shear modulus but a different bulk modulus, is *isotropic* with a bulk modulus given by Hill's equation. (See Section 4.16 on the Backus average.)

Uses

Hill's equation can be used to calculate the effective low-frequency moduli for rocks with spatially nonuniform or *patchy* saturation. At low frequencies, Gassmann's relations predict no change in the shear modulus between dry and saturated patches, allowing this relation to be used to estimate K.

Assumptions and Limitations

Hill's equation applies when the composite rock and each of its components are isotropic and have the same shear modulus.

4.7   Rock and Pore Compressibilities and Some Pitfalls

Synopsis

This section summarizes useful relations among the compressibilities of porous materials and addresses some commonly made mistakes.

A nonporous elastic solid has a single compressibility

$$\beta = \frac{1}{V}\frac{\partial V}{\partial \sigma} \tag{4.7.1}$$

where σ is the hydrostatic stress (defined as being positive in tension) applied on the outer surface, and V is the sample bulk volume. In contrast, compressibilities for porous media are more complicated. We have to account for at least two stresses (the external confining stress σ_c and the internal pore stress σ_p) and two volumes (bulk volume V_b and pore volume v_p). Therefore, we can define at least four compressibilities. Following Zimmerman's (1991a) notation, in which the first subscript indicates the volume change (b for bulk and p for pore) and the second subscript denotes the pressure that is varied (c for confining and p for pore), these compressibilities are

$$\beta_{bc} = \frac{1}{V_b}\left(\frac{\partial V_b}{\partial \sigma_c}\right)_{\sigma_p} \tag{4.7.2}$$

$$\beta_{bp} = -\frac{1}{V_b}\left(\frac{\partial V_b}{\partial \sigma_p}\right)_{\sigma_c} \tag{4.7.3}$$

$$\beta_{pc} = \frac{1}{v_p}\left(\frac{\partial v_p}{\partial \sigma_c}\right)_{\sigma_p} \tag{4.7.4}$$

$$\beta_{pp} = -\frac{1}{v_p}\left(\frac{\partial v_p}{\partial \sigma_p}\right)_{\sigma_c} \tag{4.7.5}$$

Note that the signs are chosen to ensure that the compressibilities are positive when *tensional* stress is taken to be positive. Thus, for instance, β_{bp} is to be interpreted as the fractional change in the bulk volume with respect to change in the pore pressure while the confining pressure is held constant. The effective **dry** or **drained** bulk modulus is, $K_{dry} = 1/\beta_{bc}$ and the dry-pore-space stiffness is $K_\phi = 1/\beta_{pc}$. In addition, there is the **saturated** or **undrained** bulk compressibility when the mass of the pore fluid is kept constant as the confining pressure changes:

$$\beta_u = \frac{1}{K_{\text{sat low f}}} = \frac{1}{V_b}\left(\frac{\partial V_b}{\partial \sigma_c}\right)_{m-\text{fluid}} \tag{4.7.6}$$

This equation assumes that the pore pressure is equilibrated throughout the pore space, and the expression is therefore appropriate for very low frequencies. At high frequencies, with unequilibrated pore pressures, the appropriate bulk modulus is $K_{\text{sat hi f}}$, calculated from some high-frequency theory such as the squirt, Biot, or inclusion models, or some other viscoelastic model. The subscript *m-fluid* indicates constant mass of the pore fluid.

The moduli K_{dry}, $K_{\text{sat low f}}$, $K_{\text{sat hi f}}$, and K_ϕ are the ones most useful in wave-propagation rock physics. The other compressibilities are used in calculations of subsidence caused by fluid withdrawal and reservoir-compressibility analyses. Some of the compressibilities can be related to each other by linear superposition and reciprocity. The well-known Gassmann's equation relates K_{dry} to K_{sat} through the mineral and fluid bulk moduli K_0 and K_{fl}. A few other relations are (for simple derivations, see Zimmerman, 1991a)

$$\beta_{\text{bp}} = \beta_{\text{bc}} - \frac{1}{K_0} \tag{4.7.7}$$

$$\beta_{\text{pc}} = \frac{\beta_{\text{bp}}}{\phi} \tag{4.7.8}$$

$$\beta_{\text{pp}} = \left[\beta_{\text{bc}} - (1 + \phi)\frac{1}{K_0} \right]/\phi \tag{4.7.9}$$

More on Dry-Rock Compressibility

The effective dry-rock compressibility of a homogeneous, linear, porous, elastic solid with any arbitrarily shaped pore space (sometimes called the "drained" or "frame" compressibility) can be written as

$$\frac{1}{K_{\text{dry}}} = \frac{1}{K_0} + \frac{1}{V_b}\frac{\partial v_p}{\partial \sigma_c}\bigg|_{\sigma_p} \tag{4.7.10}$$

or

$$\frac{1}{K_{\text{dry}}} = \frac{1}{K_0} + \frac{\phi}{K_\phi} \tag{4.7.11}$$

where

$$\frac{1}{K_\phi} = \frac{1}{v_p}\frac{\partial v_p}{\partial \sigma_c}\bigg|_{\sigma_p} \tag{4.7.12}$$

is defined as the dry pore-space compressibility (K_ϕ is the dry pore-space stiffness),

$K_{\text{dry}} = 1/\beta_{\text{bc}} = $ effective bulk modulus of dry porous solid

$K_0 = $ bulk modulus of intrinsic mineral material

$V_b = $ total bulk volume

$v_p = $ pore volume

$\phi = v_p/V_b = $ porosity

$\sigma_c, \sigma_p = $ hydrostatic confining stress and pore stress (pore pressure)

We assume that no inelastic effects such as friction or viscosity are present. These equations are strictly true, regardless of pore geometry and pore concentration.

Caution:

"Dry rock" is not the same as gas-saturated rock. The dry-frame modulus refers to the incremental bulk deformation resulting from an increment of applied confining pressure while the pore pressure is held constant. This corresponds to a "drained" experiment in which pore fluids can flow freely in or out of the sample to ensure constant pore pressure. Alternatively, it can correspond to an undrained experiment in which the pore fluid has zero bulk modulus, and thus the pore compressions do not induce changes in pore pressure, which is approximately the case for an air-filled sample at standard temperature and pressure. However, at reservoir conditions (high pore pressure), the gas takes on a non-negligible bulk modulus and should be treated as a saturating fluid.

Caution:

The harmonic average of the mineral and dry-pore moduli, which resembles equation 4.7.11, is incorrect:

$$\frac{1}{K_{dry}} \overset{?}{=} \frac{1 - \phi}{K_0} + \frac{\phi}{K_\phi}. \qquad \textbf{(incorrect)} \qquad (4.7.13)$$

Equation 4.7.13 is sometimes "guessed" because it resembles the Reuss average, but it has no justification from elasticity analysis. It is also *incorrect* to write

$$\frac{1}{K_{dry}} \overset{?}{=} \frac{1}{K_0} + \frac{\partial \phi}{\partial \sigma_c}. \qquad \textbf{(incorrect)} \qquad (4.7.14)$$

The correct expression is

$$\frac{1}{K_{dry}} = \frac{1}{(1 - \phi)} \left(\frac{1}{K_0} + \frac{\partial \phi}{\partial \sigma_c} \right) \qquad (4.7.15)$$

The incorrect equation 4.7.14 appears as an intermediate result in some of the classic literature of rock physics. The notable final results are still correct, for the actual derivations are done in terms of the pore volume change, $\partial v_p / \partial \sigma_c$ and not $\partial \phi / \partial \sigma_c$.

Not distinguishing between changes in differential pressure, $\sigma_d = \sigma_c - \sigma_p$, and confining pressure, σ_c can lead to confusion. Changing σ_c while σ_p is kept constant ($\delta\sigma_p = 0$) is not the same as changing σ_c with $\delta\sigma_p = \delta\sigma_c$ (i.e., the differential stress is kept constant). In the first situation, the porous medium deforms with the effective dry modulus K_{dry}. The second situation is one of uniform hydrostatic pressure outside and inside the porous rock. For this stress state, the rock deforms with the intrinsic mineral modulus K_0 (see Table 4.7.1). Not understanding this can lead to the following erroneous results:

$$\frac{1}{K_0} \overset{?}{=} \frac{1}{K_{dry}} - \frac{1}{(1-\phi)} \frac{\partial \phi}{\partial \sigma_c} \quad \text{(incorrect)} \tag{4.7.16}$$

or

$$\frac{\partial \phi}{\partial \sigma_c} \overset{?}{=} (1 - \phi) \left(\frac{1}{K_{dry}} - \frac{1}{K_0} \right) \quad \text{(incorrect)} \tag{4.7.17}$$

Table 4.7.1 *Correct and incorrect versions of the fundamental equations*

Incorrect	Correct
$\dfrac{1}{K_{dry}} \overset{?}{=} \dfrac{1}{K_0} + \dfrac{\partial \phi}{\partial \sigma_c}$	$\dfrac{1}{K_{dry}} = \dfrac{1}{K_0} + \dfrac{1}{V_b} \dfrac{\partial v_p}{\partial \sigma_c}$
$\dfrac{1}{K_{dry}} \overset{?}{=} \dfrac{1-\phi}{K_0} + \dfrac{\phi}{K_\phi}$	$\dfrac{1}{K_{dry}} = \dfrac{1}{K_0} + \dfrac{\phi}{K_\phi}$
$\dfrac{1}{K_{dry}} \overset{?}{=} \dfrac{1}{K_0} + \dfrac{1}{(1-\phi)} \dfrac{\partial \phi}{\partial \sigma_c}$	$\dfrac{1}{K_{dry}} = \dfrac{1}{(1-\phi)} \left(\dfrac{1}{K_0} + \dfrac{\partial \phi}{\partial \sigma_c} \right)$
$\dfrac{\partial \phi}{\partial \sigma_c} \overset{?}{=} \left(\dfrac{1}{K_{dry}} - \dfrac{1}{K_0} \right)(1-\phi)$	$\dfrac{\partial \phi}{\partial \sigma_c} = \dfrac{1-\phi}{K_{dry}} - \dfrac{1}{K_0}$

Assumptions and Limitations

The following presuppositions apply to the equations presented in this section:
- They assume isotropic, linear, porous, and elastic media.
- All derivations here are in the context of linear elasticity with infinitesimal, incremental strains and stresses. Hence Eulerian and Lagrangian formulations are equivalent.
- It is assumed that the temperature is always held constant as the pressure varies.
- Inelastic effects such as friction and viscosity are neglected.

4.8   General Comments on Inclusion-Based Estimation Models

Synopsis

In contrast to theoretical *bounds*, which define the limits of physically possible effective elastic moduli, there are many models that attempt to *estimate* the effective moduli of a composite with specified composition, volume fractions, and microgeometry.

Consider an N-phase composite of linear elastic phases (including both minerals and pore-fill), each phase with elastic stiffnesss tensor $\mathbf{C}_i$, such that the stress and strain in that phase are related by $\boldsymbol{\sigma}_i = \mathbf{C}_i \boldsymbol{\varepsilon}_i$, $i = 1, \ldots N$. We define the effective stiffness (modulus) $\mathbf{C}^{eff}$ of the composite by $\bar{\boldsymbol{\sigma}} \equiv \mathbf{C}^{eff}\bar{\boldsymbol{\varepsilon}}$ where $\bar{\boldsymbol{\sigma}}$ and $\bar{\boldsymbol{\varepsilon}}$ are the stress and strain tensors, respectively, each averaged over the entire sample volume. If there are N discrete phases, then $\bar{\boldsymbol{\sigma}} = \sum_{i=1}^{N} x_i \bar{\boldsymbol{\sigma}}_i$ where $\bar{\boldsymbol{\sigma}}_i$ is the average stress tensor within the ith phase, and x_i is the fraction of the total volume occupied by the ith phase, such that $\sum_{i=1}^{N} x_i = 1$. Similarly, the average strain tensor is $\bar{\boldsymbol{\varepsilon}} = \sum_{i=1}^{N} x_i \bar{\boldsymbol{\varepsilon}}_i$. We can then write the *exact* expression

$$\sum_{1}^{N} x_i (\mathbf{C}_i - \mathbf{C}^{eff}) \mathbf{T}^{0i} = 0 \qquad (4.8.1)$$

where $\mathbf{T}^{0i}$ is the *strain concentration tensor*, which relates the average strain tensor in the *i*th phase to the overall average strain tensor of the composite $\bar{\varepsilon}_i = \mathbf{T}^{0i}\bar{\varepsilon}_0$.

When there is a random distribution of identical inclusions over all orientations, the strain concentration tensor must be summed over all rotations and becomes isotropic, and can be expressed as

$$\mathbf{T}_{iso}^{0i} = T_h^{0i} \Lambda_h + T_s^{0i} \Lambda_s \qquad (4.8.2)$$

where Λ_h and Λ_s are the hydrostatic and deviatoric isotropic projection tensors that were defined in Section 1.6. Using the 6x6 Kelvin notation defined in Section 2.2, the isotropic projections of $\hat{\mathbf{T}}^{0i}$ can be found:

$$P^{0i} \equiv T_h^{0i} = \hat{\mathbf{T}}^{0i} : \hat{\Lambda}_h$$

$$= \frac{1}{3} \begin{pmatrix} \hat{T}_{11} & \hat{T}_{12} & \hat{T}_{13} & 0 & 0 & 0 \\ \hat{T}_{21} & \hat{T}_{22} & \hat{T}_{23} & 0 & 0 & 0 \\ \hat{T}_{31} & \hat{T}_{32} & \hat{T}_{33} & 0 & 0 & 0 \\ 0 & 0 & 0 & \hat{T}_{44} & 0 & 0 \\ 0 & 0 & 0 & 0 & \hat{T}_{55} & 0 \\ 0 & 0 & 0 & 0 & 0 & \hat{T}_{66} \end{pmatrix} : \begin{pmatrix} 1 & 1 & 1 & 0 & 0 & 0 \\ 1 & 1 & 1 & 0 & 0 & 0 \\ 1 & 1 & 1 & 0 & 0 & 0 \\ 0 & 0 & 0 & 0 & 0 & 0 \\ 0 & 0 & 0 & 0 & 0 & 0 \\ 0 & 0 & 0 & 0 & 0 & 0 \end{pmatrix} = \frac{1}{3} T_{iijj}^{0i}$$

$$= \frac{1}{3} \begin{pmatrix} T_{1111} & T_{1122} & T_{1133} & 0 & 0 & 0 \\ T_{2211} & T_{2222} & T_{2233} & 0 & 0 & 0 \\ T_{3311} & T_{3322} & T_{3333} & 0 & 0 & 0 \\ 0 & 0 & 0 & 2T_{2323} & 0 & 0 \\ 0 & 0 & 0 & 0 & 2T_{1313} & 0 \\ 0 & 0 & 0 & 0 & 0 & 2T_{1212} \end{pmatrix} : \begin{pmatrix} 1 & 1 & 1 & 0 & 0 & 0 \\ 1 & 1 & 1 & 0 & 0 & 0 \\ 1 & 1 & 1 & 0 & 0 & 0 \\ 0 & 0 & 0 & 0 & 0 & 0 \\ 0 & 0 & 0 & 0 & 0 & 0 \\ 0 & 0 & 0 & 0 & 0 & 0 \end{pmatrix} = \frac{1}{3} T_{iijj}^{0i}$$

$$(4.8.3)$$

$$Q^{0i} \equiv T_s^{0i} = \hat{\mathbf{T}}^{0i} \hat{\Lambda}_s / \|\hat{\Lambda}_s\|^2$$

$$= \frac{1}{15} \begin{pmatrix} \hat{T}_{11} & \hat{T}_{12} & \hat{T}_{13} & 0 & 0 & 0 \\ \hat{T}_{21} & \hat{T}_{22} & \hat{T}_{23} & 0 & 0 & 0 \\ \hat{T}_{31} & \hat{T}_{32} & \hat{T}_{33} & 0 & 0 & 0 \\ 0 & 0 & 0 & \hat{T}_{44} & 0 & 0 \\ 0 & 0 & 0 & 0 & \hat{T}_{55} & 0 \\ 0 & 0 & 0 & 0 & 0 & \hat{T}_{66} \end{pmatrix} : \begin{pmatrix} 2 & -1 & -1 & 0 & 0 & 0 \\ -1 & 2 & -1 & 0 & 0 & 0 \\ -1 & -1 & 2 & 0 & 0 & 0 \\ 0 & 0 & 0 & 3 & 0 & 0 \\ 0 & 0 & 0 & 0 & 3 & 0 \\ 0 & 0 & 0 & 0 & 0 & 3 \end{pmatrix}$$

$$= \frac{1}{5}\left(T^{0i}_{ijij} - \frac{1}{3}T^{0i}_{iijj}\right)$$

$$= \frac{1}{15}\begin{pmatrix} T_{1111} & T_{1122} & T_{1133} & 0 & 0 & 0 \\ T_{2211} & T_{2222} & T_{2233} & 0 & 0 & 0 \\ T_{3311} & T_{3322} & T_{3333} & 0 & 0 & 0 \\ 0 & 0 & 0 & 2T_{2323} & 0 & 0 \\ 0 & 0 & 0 & 0 & 2T_{1313} & 0 \\ 0 & 0 & 0 & 0 & 0 & 2T_{1212} \end{pmatrix} : \begin{pmatrix} 2 & -1 & -1 & 0 & 0 & 0 \\ -1 & 2 & -1 & 0 & 0 & 0 \\ -1 & -1 & 2 & 0 & 0 & 0 \\ 0 & 0 & 0 & 3 & 0 & 0 \\ 0 & 0 & 0 & 0 & 3 & 0 \\ 0 & 0 & 0 & 0 & 0 & 3 \end{pmatrix}$$

$$= \frac{1}{5}\left(T^{0i}_{ijij} - \frac{1}{3}T^{0i}_{iijj}\right) \tag{4.8.4}$$

In equations 4.8.3 and 4.8.4, the symbol $\hat{}$ indicates Kelvin notation. The products are inner products of the two matrices.

For an isotropic composite, equation 4.8.1 for effective stiffness reduces to two scalar equations:

$$\sum_1^N x_i(K_i - K^{eff})P^{0i} = 0 \tag{4.8.5}$$

$$\sum_1^N x_i(\mu_i - \mu^{eff})Q^{0i} = 0 \tag{4.8.6}$$

where K^{eff} and μ^{eff} are the effective bulk and shear moduli, and P^{0i} and Q^{0i} are the scalar volumetric and deviatoric strain concentration coefficients, respectively, from equations 4.8.3 and 4.8.4. Note that $\sum_{i=1}^{N} x_i\mathbf{T}^{0i} = \mathbf{I}$, $\sum_{i=1}^{N} x_iP^{0i} = 1$, and $\sum_{i=1}^{N} x_iQ^{0i} = 1$. P^{0i} and Q^{0i} are the same shape factors that appear in the Mori–Tanaka, Self-Consistent, DEM, and Kuster–Toksöz models. For a two-phase isotropic material,

$$K^{eff} = K_1 + x_2(K_2 - K_1)P^{02} \tag{4.8.7}$$

$$\mu^{eff} = \mu_1 + x_2(\mu_2 - \mu_1)Q^{02} \tag{4.8.8}$$

If we know how strain is heterogeneously distributed within a composite when stress is applied, i.e., P^{0i}, Q^{0i}, and $\mathbf{T}^{0i}$, then we know the effective moduli *exactly*. However, exact values of P^{0i}, Q^{0i}, and $\mathbf{T}^{0i}$ are rarely known, since they can be complicated functions of composition and microgeometry. The models discussed in the following sections all aim to make plausible estimates of strain concentrations (and therefore effective moduli) using Eshelby's (1957) analytical results for ellipsoidal heterogeneities (see Section 2.12).

An Abundance of Inclusion Models

A widely adopted strategy for modeling composites is to represent heterogeneities in a rock as ellipsoidal inclusions, and then use the exact expressions of Eshelby (1957) for the strain concentration tensor $\mathbf{T}^{mi}$ in a single ellipsoid i sitting in an infinite uniform background of material m. Aside from the shortcoming that ellipsoidal microgeometries are almost never observed in rocks, a major challenge for modeling a statistically homogeneous composite is somehow *extrapolating* Eshelby's exact single-ellipsoid result to situations where there are many heterogeneities. No single modeling approach is ideal and universally applicable, and so different estimation strategies have proliferated. Some models take the "noninteracting assumption," in which the change in effective stiffness or compliance resulting from many inclusions is approximated as simply the sum over the change in effective stiffness of many isolated Eshelby problems. These models are most easily justified for very dilute concentrations of inclusions.

Other models attempt to approximate the elastic interaction of the inclusions. Differential Effective Medium (DEM) models exploit the accuracy of very dilute concentration models, by recursively adding small fractions of inclusions, each time updating the background effective medium (Section 4.12). Self-Consistent (SC) models follow the heuristic argument that each inclusion deforms as though it sits in the as-yet-unknown effective medium, rather than in the original uniform background phase (Section 4.11). Mori–Tanaka (MT) models (Section 4.9) follow from the argument that each inclusion deforms as though it sits in an effective strain field.

One important distinction among models is whether or not the predicted elastic moduli are physically realizable. For example, predictions of the symmetric *n-phase* Coherent Potential Approximation model (Berryman, 1980a, 1980b), one of the Self-Consistent formulations, always falls within the Hashin–Shtrikman bounds, while other nonsymmetric formulations might not. The *two-phase* DEM model of Norris (1985) always lies within the HS bounds, while the formulation of Zimmerman (1984a, 1984b) will violate the bounds for some inclusion types (Norris, 1985). For *two isotropic phases*, the Mori–Tanaka (1973) model (MT model) always predicts moduli that fall within the Hashin–Shtrikman bounds. However, the Mori–Tanaka can violate the bounds when there are three or more phases. The Kuster–Toksöz (1974) model (KT model) can violate the HS bounds when the inclusions are either disks at any finite concentration of needles at volume fractions great than ~60% (Berryman, 1980a, 1980b; Berryman and Berge, 1996).

Kachanov *et al.* (2004) and Berge *et al.* (1993) point out that among the several realizable models, there is no answer to which model is *best*, without more information about the microgeometry of the heterogeneity (including spatial distributions of sizes, orientations, and spacing of inclusions) that we wish to describe. For example, is there a clearly defined background or host material? The DEM model can be realized by well-separated inclusions, surrounded by a sea of much smaller well-separated inclusions, ad infinitum (Milton, 2002). The original background phase always remains connected. On the other hand, with the symmetric SC formulation, there is no preferred background and it is possible for multiple phases to be interconnected. In the case where one of the phases is a fluid, the SC model predicts a threshold fraction of the fluid phase at which the effective shear modulus goes to zero. In contrast, the DEM and MT formulations only predict loss of rigidity at 100% fluid fraction. Mukerji *et al.* (1995) suggest a modified DEM in which the fluid phase is replaced by the critical phase, thus creating a shear percolation less than one. The Hashin–Shtrikman bounds on bulk modulus can be realized by Hashin's coated sphere assemblage, in which spherical inclusions of one phase are completely surrounded by a shell of the stiffer phase, with the special features that inclusion sizes range from finite to infinitesimal, and each inclusion with its background coating have precisely the overall volume fractions. The Mori–Tanaka, Maxwell, and Kuster–Toksöz models also fall on the HS bound for spherical inclusions. In contrast, the SC and DEM models both deviate from the bounds and from each other. Granular materials, such as sandstones, might be represented well with the SC model, while the DEM model will overpredict the moduli (Berge *et al.*, 1993). On the other hand, low porosity materials with somewhat isolated cracks and pores could be described with SC, DEM, MT, or KT. Very high porosity cellular materials can sometimes be represented with the MT and KT models with spherical pores, which agree exactly with the upper HS bounds.

An important practical consideration is that many models are implicit, requiring a recursive or iterative computation of the effective properties. Only a few popular models, including the MT, Maxwell, and KT formulations are explicit, allowing for rapid computation and inversion, and analytic extension to viscoelastic and elastoplastic materials (Hu and Weng, 2000). In a commercial geophysical environment, consistent use of a model choice is sometimes the preferred strategy. Recognizing that no model is ideal, comparison of results might be more useful if collaborators all use the same one.

Finally, it is important to remember that physically realizable models can describe more than one microstructure. For example, the Hashin–Shtrikman bounds can be realized by laminate composites, coated spheres (for bulk modulus), and disks of one phase surrounded by the other. Materials lying within the HS bounds can be realized by an infinite number of microstructures (Mavko and Saxena, 2013). Hence, agreement of a model with an observation does not guarantee that the microstructure is the one intended by the modeler.

Eshelby Strain Concentration Factors Used in Inclusion Models

Many inclusion-based models for the isotropic effective moduli incorporate the isotropic volumetric and deviatoric strain concentration factors P^{mn} and Q^{mn} (equations 4.8.5 and 4.8.6) based on Eshelby's results. The P^{mn} and Q^{mn} for a distribution of randomly oriented identical ellipsoids are estimated by averaging the single ellipsoid strain concentration tensor $\mathbf{T}^{mn}$ over all orientations:

$$P^{mn} = \frac{1}{3} T_{iijj}^{mn} \qquad (4.8.9)$$

$$Q^{mn} = \frac{1}{5}\left(T^{mn}_{ijij} - \frac{1}{3}T^{mn}_{iijj}\right)$$

(4.8.10)

where the tensor T^{mn}_{ijkl} (equation 4.8.1) relates the uniform far-field strain field (m) to the strain within the nth ellipsoidal inclusion (Wu, 1966). Berryman (1980b) gives the pertinent scalars required for computing P^{mn} and Q^{mn} for spheroids of arbitrary aspect ratio:

$$T^{mn}_{iijj} = \frac{3F_1}{F_2}$$

(4.8.11)

$$T^{mn}_{ijij} - \frac{1}{3}T^{mn}_{iijj} = \frac{2}{F_3} + \frac{1}{F_4} + \frac{F_4F_5 + F_6F_7 - F_8F_9}{F_2F_4}$$

(4.8.12)

where

$$F_1 = 1 + A\left[\frac{3}{2}(f + \theta) - R\left(\frac{3}{2}f + \frac{5}{2}\theta - \frac{4}{3}\right)\right]$$

(4.8.13)

$$F_2 = 1 + A\left[1 + \frac{3}{2}(f + \theta) - \frac{1}{2}R(3f + 5\theta)\right] + B(3 - 4R)$$
$$+ \frac{1}{2}A(A + 3B)(3 - 4R)[f + \theta - R(f - \theta + 2\theta^2)]$$

(4.8.14)

$$F_3 = 1 + A\left[1 - \left(f + \frac{3}{2}\theta\right) + R(f + \theta)\right]$$

(4.8.15)

$$F_4 = 1 + \frac{1}{4}A[f + 3\theta - R(f - \theta)]$$

(4.8.16)

$$F_5 = A\left[-f + R\left(f + \theta - \frac{4}{3}\right)\right] + B\theta(3 - 4R)$$

(4.8.17)

$$F_6 = 1 + A[1 + f - R(f + \theta)] + B(1 - \theta)(3 - 4R)$$

(4.8.18)

$$F_7 = 2 + \frac{1}{4}A[3f + 9\theta - R(3f + 5\theta)] + B\theta(3 - 4R)$$

(4.8.19)

$$F_8 = A\left[1 - 2R + \frac{1}{2}f(R - 1) + \frac{1}{2}\theta(5R - 3)\right] + B(1 - \theta)(3 - 4R)$$

(4.8.20)

$$F_9 = A[(R - 1)f - R\theta] + B\theta(3 - 4R)$$

(4.8.21)

with A, B, and R given by

$$A = \mu_i/\mu_m - 1$$

(4.8.22)

$$B = \frac{1}{3}\left(\frac{K_i}{K_m} - \frac{\mu_i}{\mu_m}\right)$$

(4.8.23)

Table 4.8.1 *Scalar volumetric and deviatoric strain concentration factors* P^{mi} *and* Q^{mi} *for some specific spheroidal shapes, randomly oriented in the composite. With the Mori–Tanaka and Kuster–Toksöz models, the subscripts* m *and* i *refer to the background and inclusion phases, respectively. With the symmetric self-consistent formulation,* m *refers to the effective medium and* i *refers to each phase (both solid and pore). With the DEM model,* m *refers to the effective medium from the previous integration step and* i *refers to the inclusion phase (from Berryman 1995).*

Inclusion shape	P^{mi}	Q^{mi}
Spheres	$\dfrac{K_m + \frac{4}{3}\mu_m}{K_i + \frac{4}{3}\mu_m}$	$\dfrac{\mu_m + \zeta_m}{\mu_i + \zeta_m}$
Needles	$\dfrac{K_m + \mu_m + \frac{1}{3}\mu_i}{K_i + \mu_m + \frac{1}{3}\mu_i}$	$\dfrac{1}{5}\left(\dfrac{4\mu_m}{\mu_m + \mu_i} + 2\dfrac{\mu_m + \gamma_m}{\mu_i + \gamma_m} + \dfrac{K_i + \frac{4}{3}\mu_m}{K_i + \mu_m + \frac{1}{3}\mu_i}\right)$
Disks	$\dfrac{K_m + \frac{4}{3}\mu_i}{K_i + \frac{4}{3}\mu_i}$	$\dfrac{\mu_m + \zeta_i}{\mu_i + \zeta_i}$
Penny cracks	$\dfrac{K_m + \frac{4}{3}\mu_i}{K_i + \frac{4}{3}\mu_i + \pi\alpha\beta_m}$	$\dfrac{1}{5}\left[1 + \dfrac{8\mu_m}{4\mu_i + \pi\alpha(\mu_m + 2\beta_m)} + 2\dfrac{K_i + \frac{2}{3}(\mu_i + \mu_m)}{K_i + \frac{4}{3}\mu_i + \pi\alpha\beta_m}\right]$

$\beta = \mu\dfrac{(3K+\mu)}{(3K+4\mu)}$, $\gamma = \mu\dfrac{(3K+\mu)}{(3K+7\mu)}$, $\zeta = \dfrac{\mu}{6}\dfrac{(9K+8\mu)}{(K+2\mu)}$, α = crack aspect ratio. A

disk is a crack of zero thickness

and

$$R = \frac{(1 - 2v_m)}{2(1 - v_m)} \tag{4.8.24}$$

The functions θ and f are given by

$$\theta = \begin{cases} \dfrac{\alpha}{(\alpha^2 - 1)^{3/2}}[\alpha(\alpha^2 - 1)^{1/2} - \cosh^{-1}\alpha] \\[2ex] \dfrac{\alpha}{(1 - \alpha^2)^{3/2}}[\cos^{-1}\alpha - \alpha(1 - \alpha^2)^{1/2}] \end{cases} \tag{4.8.25}$$

for prolate and oblate spheroids, respectively, and

$$f = \frac{\alpha^2}{1 - \alpha^2}(3\theta - 2) \tag{4.8.26}$$

Note that $\alpha < 1$ for oblate spheroids and $\alpha > 1$ for prolate spheroids.

Table 4.8.1 gives expressions for P^{mi} and Q^{mi} for some simple spheroidal inclusion shapes. The subscripts m and i in the table refer to the background and inclusion materials that are used when applying the Eshelby results. The Mori–Tanaka (Section 4.9), Kuster–Toksöz (Section 4.10), Self-Consistent (Section 4.11), and DEM (Section 4.12) models all employ the same expressions with different strategies for estimating the role of the background.

4.9 Mori–Tanaka Formulation for Effective Moduli

The Mori–Tanaka (1973) approximation is one of the few inclusion-based effective medium methods that yield explicit, closed-form expressions for effective medium properties.

The general definition of the fourth-order effective linear elastic stiffness tensor, $\mathbf{C}^{eff}$, of an elastic composite of phases 0, 1, ... N is given by

$$\bar{\sigma} = \mathbf{C}^{eff}\bar{\varepsilon} \tag{4.9.1}$$

where $\bar{\sigma} = \sum_{j=0}^{N} x_j \bar{\sigma}_j$ is the mean stress in the (N+1)-phase composite, and $\bar{\varepsilon} = \sum_{j=0}^{N} x_j \bar{\varepsilon}_j$ is the mean strain. In these expressions x_j is the volume fraction of each constituent, such that $\sum_{j=0}^{N} x_j = 1$, and $\bar{\sigma}_j$ and $\bar{\varepsilon}_j$ are the mean stress and strain within each jth constituent. The exact expression for the fourth-order effective modulus tensor, $\mathbf{C}^{eff}$ can be written as

$$\mathbf{C}^{eff} = \mathbf{C}_0 + \left(\sum_{j=1}^{N} x_j (\mathbf{C}_j - \mathbf{C}_0) \mathbf{T}^{0j} \right) \left(x_0 \mathbf{I} + \sum_{j=1}^{N} x_j \mathbf{T}^{0j} \right)^{-1} \tag{4.9.2}$$

where $\mathbf{I}$ is the fourth-rank isotropic identity tensor, and $\mathbf{C}_j$ are the fourth-rank elastic stiffness tensors for each of the constituent materials. Tensor $\mathbf{T}^{0j}$ is the strain concentration tensor that relates the average strain in the jth inclusion to the average strain $\bar{\varepsilon}_0$ in phase 0:

$$\bar{\varepsilon}_j = T^{0j}\bar{\varepsilon}_0. \tag{4.9.3}$$

As with many effective medium problems, the challenge is in estimating $\mathbf{T}^{0j}$. In the Mori–Tanaka method, $\mathbf{T}^{0j}$ is approximated as the analogous expression $\mathbf{T}^{0j}$ for a single inclusion of material j in an infinite background of material 0. This is equivalent to the dilute limit of the exact $\mathbf{T}^{0j}$.

When dilute sets of inclusions are identical, aligned ellipsoids $\mathbf{T}^{0j}$ can be written in terms of the Eshelby (1957) tensor $\mathbf{S}^{0j}$ of a single ellipsoid of material j in a background of material 0.

$$\mathbf{T}^{0j} = [\mathbf{I} + \mathbf{S}^{0j}\mathbf{C}_0^{-1}(\mathbf{C}_j - \mathbf{C}_0)]^{-1} . \tag{4.9.4}$$

For a statistically isotropic composite of isotropic constituents, $\mathbf{T}^{0j}$ is replaced by its isotropic form, and the Mori–Tanaka estimates of effective bulk and shear moduli are

$$\sum_{j=0}^{N} x_j (K_j - K_{MT}^{eff}) P^{0j} = 0, \qquad (4.9.5)$$

$$\sum_{j=0}^{N} x_j (\mu_j - \mu_{MT}^{eff}) Q^{0j} = 0, \qquad (4.9.6)$$

Expressions for the volumetric and deviatoric strain concentration factors P^{0j} and Q^{0j} are given in Table 4.8.1 and equations 4.8.9–4.8.12, for certain specific types of spheroids. Dry cavities can be modeled by setting the inclusion moduli to zero. Fluid saturated cavities are simulated by setting the inclusion shear modulus to zero.

Equations 4.9.5 and 4.9.6 can be written more explicitly as

$$K_{MT}^{eff} = K_0 + \left(\sum_{j=1}^{N} x_j (K_j - K_0) P^{0j} \right) \left(\sum_{j=0}^{N} x_j P^{0j} \right)^{-1}, \qquad (4.9.7)$$

$$\mu_{MT}^{eff} = \mu_0 + \left(\sum_{j=1}^{N} x_j (\mu_j - \mu_0) Q^{0j} \right) \left(\sum_{j=0}^{N} x_j Q^{0j} \right)^{-1}, \qquad (4.9.8)$$

The Mori–Tanaka approach works best for spheres and aligned ellipsoids when the number of phases is greater than or equal to two, and for randomly oriented ellipsoids when there are two phases. In other instances, the approximations can be problematic (Torquato, 2002).

The Mori–Tanaka bulk and shear moduli for soft spherical inclusions in a stiffer background lie on the upper Hashin–Shtrikman bound (Weng, 1984); similarly, moduli for stiff spherical inclusions in a softer background lie on the lower HS bound. Norris (1985) showed that the predictions for disk-shaped inclusions also fall on the HS bounds (soft inclusions yield the lower bound, and stiff inclusions yield the upper bound). The Mori–Tanaka bulk and shear moduli for randomly oriented ellipsoidal inclusions always fall within the HS bounds (Benveniste, 1987), though for more than two phases, the predicted moduli can fall outside of the HS bounds for some inclusion shapes.

4.10   Kuster and Toksöz Formulation for Effective Moduli

Synopsis

Kuster and Toksöz (1974) derived expressions for P- and S-wave velocities by using a long-wavelength first-order scattering theory. A generalization of their expressions for the effective moduli K_{KT}^{eff} and μ_{KT}^{eff} for a variety of inclusion shapes can be written as (Kuster and Toksöz, 1974; Berryman, 1980b)

$$(K_{KT}^{eff} - K_m) \frac{\left(K_m + \frac{4}{3}\mu_m\right)}{\left(K_{KT}^{eff} + \frac{4}{3}\mu_m\right)} = \sum_{i=1}^{N} x_i(K_i - K_m)P^{mi} \tag{4.10.1}$$

$$(\mu_{KT}^{eff} - \mu_m) \frac{(\mu_m + \zeta_m)}{(\mu_{KT}^{eff} + \zeta_m)} = \sum_{i=1}^{N} x_i(\mu_i - \mu_m)Q^{mi} \tag{4.10.2}$$

where the summation is over the different inclusion types with volume concentration x_i, and

$$\zeta = \frac{\mu (9K + 8\mu)}{6 (K + 2\mu)}, \tag{4.10.3}$$

The coefficients P^{mi} and Q^{mi} describe the effect of an inclusion of material i in a background medium m. (Most often, the mineral is assumed to be the background medium m.) For example, a two-phase material with a single type of inclusion embedded within a background medium has a single term on the right-hand side of equations 4.10.1 and 4.10.2. Inclusions with different material properties or different shapes require separate terms in the summation. Each set of inclusions must be randomly oriented, and thus its effect is isotropic. Expressions for P^{mi} and Q^{mi} are given in Table 4.8.1 and equations 4.8.9–4.8.12, for certain specific types of spheroids.

Dry cavities can be modeled by setting the inclusion moduli to zero. Fluid saturated cavities are simulated by setting the inclusion shear modulus to zero.

> **Caution:**
>
> Because the cavities are isolated with respect to flow, this approach simulates very high-frequency saturated rock behavior appropriate to ultrasonic laboratory conditions. At low frequencies, when there is time for wave-induced pore-pressure increments to flow and equilibrate, it is better to find the effective moduli for dry cavities and then saturate them with the Gassmann low-frequency relations (see Section 6.3). This should not be confused with the tendency to term this approach a low-frequency theory, for crack dimensions are assumed to be much smaller than a wavelength.

Example: Calculate the effective bulk and shear moduli, K_{KT}^{eff} and μ_{KT}^{eff} for a quartz matrix with spherical, water-filled inclusions of porosity 0.1:

$K_m = 37$ GPa, $\mu_m = 44$ GPa, $K_i = 2.25$ GPa, $\mu_i = 0$ GPa.

The volume fraction of spherical inclusions is $x_1 = 0.1$, and $N = 1$. The P and Q values for spheres are obtained from Table 4.8.1 as follows:

$$P^{m1} = \frac{\left(37 + \frac{4}{3} \cdot 44\right)}{\left(2.25 + \frac{4}{3} \cdot 44\right)} = 1.57$$

$$\zeta_m = \frac{44}{6} \frac{(9 \cdot 37 + 8 \cdot 44)}{(37 + 2 \cdot 44)} = 40.2 \text{ GPa}$$

$$Q^{m1} = \frac{(44 + 40.2)}{(0 + 40.2)} = 2.095$$

Substituting these in the Kuster–Toksöz equations 4.10.1–4.10.2 gives

$$K_{KT}^{eff} = 31.84 \text{ GPa}, \quad \mu_{KT}^{eff} = 35.7 \text{ GPa}$$

Note that for spherical inclusions, the Kuster–Toksöz expressions for bulk modulus are identical to the Hashin–Shtrikman upper bound, even though the Kuster–Toksöz expressions are formally limited to low porosity.

Assumptions and Limitations

The following presuppositions and limitations apply to the Kuster–Toksöz formulations:

- they assume isotropic, linear, and elastic media;
- they are limited to dilute concentrations of the inclusions; and
- they assume idealized ellipsoidal inclusion shapes.

4.11 Self-Consistent Approximations of Effective Moduli

Synopsis

The **self-consistent approximation** (Budiansky, 1965; Hill, 1965; Wu, 1966) again uses the mathematical solution for the deformation of isolated inclusions, but the interaction of inclusions is approximated by replacing the background medium with the as-yet-unknown effective medium. These methods were made popular following a series of papers by O'Connell and Budiansky (see, for example, O'Connell and Budiansky, 1974). Their equations for effective bulk and shear moduli, K_{SC}^{eff} and μ_{SC}^{eff}, respectively, of a cracked medium with randomly oriented dry penny-shaped cracks (in the limiting case when the aspect ratio α goes to 0) are

$$\frac{K_{SC}^{eff}}{K} = 1 - \frac{16}{9}\left(\frac{1 - (v_{SC}^{eff})^2}{1 - 2v_{SC}^{eff}}\right)\varepsilon \tag{4.11.1}$$

$$\frac{\mu_{SC}^{eff}}{\mu} = 1 - \frac{32}{45}\frac{(1 - v_{SC}^{eff})(5 - v_{SC}^{eff})}{(2 - v_{SC}^{eff})}\varepsilon \tag{4.11.2}$$

where K and μ are the bulk and shear moduli, respectively, of the uncracked medium, and ε is the crack density parameter, which is defined as the number of cracks per unit volume times the crack radius cubed. The effective Poisson ratio v_{SC}^{eff} is related to ε and the Poisson ratio v of the uncracked solid by

$$\varepsilon = \frac{45}{16}\frac{(v - v_{SC}^{eff})(2 - v_{SC}^{eff})}{\left(1 - (v_{SC}^{eff})^2\right)(10v - 3vv_{SC}^{eff} - v_{SC}^{eff})} \tag{4.11.3}$$

Equation 4.11.3 must first be solved for v_{SC}^{eff} for a given ε, after which K_{SC}^{eff} and μ_{SC}^{eff} can be evaluated. The nearly linear dependence of v_{SC}^{eff} on ε is well approximated by

$$v_{SC}^{eff} \approx v\left(1 - \frac{16}{9}\varepsilon\right) \tag{4.11.4}$$

and this simplifies the calculation of the effective moduli. For fluid-saturated, infinitely thin penny-shaped cracks

$$\frac{K_{SC}^{eff}}{K} = 1 \tag{4.11.5}$$

$$\frac{\mu_{SC}^{eff}}{\mu} = 1 - \frac{32}{15}\left(\frac{1 - v_{SC}^{eff}}{2 - v_{SC}^{eff}}\right)\varepsilon \tag{4.11.6}$$

$$\varepsilon = \frac{45}{32}\frac{(v_{SC}^{eff} - v)(2 - v_{SC}^{eff})}{\left(1 - (v_{SC}^{eff})^2\right)(1 - 2v)} \tag{4.11.7}$$

However, equations 4.11.5–4.11.7 are inadequate for cracks with small aspect ratio α with soft-fluid saturation, such as when the parameter $\omega = K_{fluid}/\alpha K$ is of the order of 1. In this latter case, the appropriate equations given by O'Connell and Budiansky are

$$\frac{K_{SC}^{eff}}{K} = 1 - \frac{16}{9}\frac{\left(1 - (v_{SC}^{eff})^2\right)}{(1 - 2v_{SC}^{eff})}D\varepsilon \tag{4.11.8}$$

$$\frac{\mu_{SC}^{eff}}{\mu} = 1 - \frac{32}{45}(1 - v_{SC}^{eff})\left[D + \frac{3}{(2 - v_{SC}^{eff})}\right]\varepsilon \tag{4.11.9}$$

$$\varepsilon = \frac{45}{16} \frac{(v - v_{SC}^{eff})}{\left(1 - (v_{SC}^{eff})^2\right)} \frac{(2 - v_{SC}^{eff})}{[D(1 + 3v)(2 - v_{SC}^{eff}) - 2(1 - 2v)]} \tag{4.11.10}$$

$$D = \left[1 + \frac{4}{3\pi} \frac{\left(1 - (v_{SC}^{eff})^2\right)}{(1 - 2v_{SC}^{eff})} \frac{K}{K_{SC}^{eff}} \omega\right]^{-1} \tag{4.11.11}$$

Wu's self-consistent modulus estimates for two-phase composites may be expressed as (m = matrix, i = inclusion)

$$K_{SC}^{eff} = K_m + x_i(K_i - K_m)P^{eff\ i} \tag{4.11.12}$$

$$\mu_{SC}^{eff} = \mu_m + x_i(\mu_i - \mu_m)Q^{eff\ i} \tag{4.11.13}$$

Berryman (1980b, 1995) gives a more general form of the self-consistent approximations for N-phase composites:

$$\sum_{i=1}^{N} x_i(K_i - K_{SC}^{eff})P^{eff\ i} = 0 \tag{4.11.14}$$

$$\sum_{i=1}^{N} x_i(\mu_i - \mu_{SC}^{eff})Q^{eff\ i} = 0 \tag{4.11.15}$$

where i refers to the ith material, x_i is its volume fraction, $P^{eff\ i}$ and $Q^{eff\ i}$ are geometric strain concentration factors given in Table 4.8.1 and equations 4.8.9 – 4.8.12. The superscript $^{eff\ i}$ on $P^{eff\ i}$ and $Q^{eff\ i}$ indicates that the factors are for an inclusion of material i in a background medium with self-consistent effective moduli K_{SC}^{eff} and μ_{SC}^{eff}. The summation is over all phases, including minerals and pores. Equations 4.11.14 and 4.11.15 are referred to as the "symmetric" self –consistent formulation, since all phases (solid and pore) are treated as inclusions in the effective medium background. In contrast, the formulations in equations 4.11.1–4.11.13 all single out one of the phases as the host background.

Equations 4.11.14 and 4.11.15 are coupled and must be solved by simultaneous iteration. Although Berryman's self-consistent method does not converge for fluid disks ($\mu_2 = 0$), the formulas for penny-shaped fluid-filled cracks are generally not singular and converge rapidly. However, his estimates for needles, disks, and penny cracks should be used cautiously for fluid-saturated composite materials.

Dry cavities can be modeled by setting the inclusion moduli to zero. Fluid saturated cavities are simulated by setting the inclusion shear modulus to zero.

The generalization of equations 4.11.14 and 4.11.15 to full tensor notation is

$$\sum_{i=1}^{N} x_i(\mathbf{C}_i - \mathbf{C}_{SC}^{eff}) : \mathbf{T}^{eff\ i} = 0 \tag{4.11.16}$$

Caution:

Because the cavities are isolated with respect to flow, this approach simulates very high-frequency saturated rock behavior appropriate to ultrasonic laboratory conditions. At low frequencies, when there is time for wave-induced pore-pressure increments to flow and equilibrate, it is better to find the effective moduli for dry cavities and then saturate them with the Gassmann low-frequency relations. This should not be confused with the tendency to term this approach a low-frequency theory, for crack dimensions are assumed to be much smaller than a wavelength.

Example: Calculate the self-consistent effective bulk and shear moduli, K_{SC}^{eff} and μ_{SC}^{eff}, for a water-saturated rock consisting of spherical quartz grains (aspect ratio $\alpha = 1$) and total porosity 0.3. The pore space consists of spherical pores $\alpha = 1$ and thin penny-shaped cracks ($\alpha = 10^{-2}$). The thin cracks have a porosity of 0.01, whereas the remaining porosity (0.29) is made up of the spherical pores.

The total number of phases, $N = 3$.

K_1(quartz)$= 37$ GPa, μ_1 (quartz) $= 44$ GPa,

$\alpha_1 = 1$, x_1 (volume fraction) $= 0.7$

K_2 (water, spherical pores) $= 2.25$ GPa,

μ_2 (water, spherical pores) $= 0$ GPa,

α_2 (spherical pores) $= 1$, x_2 (volume fraction) $= 0.29$

K_3 (water, thin cracks) $= 2.25$ GPa,

μ_3 (water, thin cracks) $= 0$ GPa,

α_3(thin cracks) $= 10^{-2}$, x_3 (volume fraction) $= 0.01$

The coupled equations for K_{SC}^{eff} and μ_{SC}^{eff} are

$$x_1(K_1 - K_{SC}^{eff})P^{eff1} + x_2(K_2 - K_{SC}^{eff})P^{eff2} + x_3(K_3 - K_{SC}^{eff})P^{eff3} = 0$$

$$x_1(\mu_1 - \mu_{SC}^{eff})Q^{eff1} + x_2(\mu_2 - \mu_{SC}^{eff})Q^{eff2} + x_3(\mu_3 - \mu_{SC}^{eff})Q^{eff3} = 0$$

The P's and Q's are obtained from Table 4.8.1 or from the more general equation for ellipsoids of arbitrary aspect ratio. In the equations for P and Q, K_m and μ_m are replaced everywhere by K_{SC}^{eff} and μ_{SC}^{eff}. The Voigt average may be taken as the starting point. The converged solutions (known as the fixed points of the coupled equations) are $K_{SC}^{eff} = 16.8$ GPa and $\mu_{SC}^{eff} = 11.6$ GPa.

Assumptions and Limitations

The approach described in this section has the following presuppositions:
- idealized ellipsoidal inclusion shapes;
- isotropic, linear, elastic media;
- cracks are isolated with respect to fluid flow; and

- pore pressures are unequilibrated and adiabatic, which is appropriate for high-frequency laboratory conditions. For low-frequency field situations, use dry inclusions and then saturate by using Gassmann relations. This should not be confused with the tendency to term this approach a low-frequency theory, for crack dimensions are assumed to be much smaller than a wavelength.

4.12 Differential Effective Medium Model

Synopsis

The DEM theory models two-phase composites by incrementally adding inclusions of one phase (phase 2) to the matrix phase (Cleary et al., 1980; Norris, 1985; Zimmerman, 1991a). The matrix begins as phase 1 (when the concentration of phase 2 is zero) and is changed at each step as a new increment of phase 2 material is added. The process is continued until the desired proportion of the constituents is reached. The DEM formulation does not treat each constituent symmetrically. There is a preferred matrix or host material, and the effective moduli depend on the construction path taken to reach the final composite. Starting with material 1 as the host and incrementally adding inclusions of material 2 will not, in general, lead to the same effective properties as starting with phase 2 as the host and adding inclusions of material 1. For multiple inclusion shapes or multiple constituents, the effective moduli depend not only on the final volume fractions of the constituents but also on the order in which the incremental additions are made. The process of incrementally adding inclusions to the matrix is really a thought experiment and should not be taken to provide an accurate description of the true evolution of rock porosity in nature.

The coupled system of ordinary differential equations for the effective bulk and shear moduli, K_{DEM}^{eff} and μ_{DEM}^{eff}, respectively, are (Berryman, 1992b)

$$(1 - y)\frac{d}{dy}[K_{DEM}^{eff}(y)] = (K_2 - K_{DEM}^{eff})P^{eff2}(y) \tag{4.12.1}$$

$$(1 - y)\frac{d}{dy}[\mu_{DEM}^{eff}(y)] = (\mu_2 - \mu_{DEM}^{eff})Q^{eff2}(y) \tag{4.12.2}$$

with initial conditions $K_{DEM}^{eff}(0) = K_1$ and $\mu_{DEM}^{eff}(0) = \mu_1$, where K_1 and μ_1 are the bulk and shear moduli of the initial host material (phase 1), K_2 and μ_2 are the bulk and shear moduli of the incrementally added inclusions (phase 2), and y is the concentration of phase 2.

For fluid inclusions and voids, y equals the porosity, ϕ. Expressions for the volumetric and deviatoric strain concentration factors, P^{eff2} and Q^{eff2}, are given in Table 4.8.1 and equations 4.8.9 – 4.8.12, for certain specific types of ellipsoids. The superscript ^{eff2} on P and Q indicates that the factors are for an inclusion of material 2 in a background medium with effective moduli $K_{DEM}^{eff}(y)$ and $\mu_{DEM}^{eff}(y)$. Dry cavities can be modeled by setting the inclusion moduli to zero. Fluid-saturated cavities are simulated by setting the inclusion shear modulus to zero.

The generalization of equations 4.12.1 and 4.12.2 to full tensor notation is

$$(1-y)\frac{d\,\mathbf{C}^{eff}_{DEM}(y)}{dy} = (\mathbf{C}_2 - \mathbf{C}^{eff}_{DEM}) : \mathbf{T}^{eff\,2} \qquad (4.12.3)$$

Caution:

Because the cavities are isolated with respect to flow, this approach simulates very high-frequency saturated rock behavior appropriate to ultrasonic laboratory conditions. At low frequencies, when there is time for wave-induced pore-pressure increments to flow and equilibrate, it is better to find the effective moduli for dry cavities and then saturate them with the Gassmann low-frequency relations. This should not be confused with the tendency to term this approach a low-frequency theory, for inclusion dimensions are assumed to be much smaller than a wavelength.

Norris *et al.* (1985) have shown that the DEM is realizable and therefore is always consistent with the Hashin–Shtrikman upper and lower bounds.

The DEM equations as previously given (Norris, 1985; Zimmerman, 1991b; Berryman *et al.*, 1992) assume that, as each new inclusion (or pore) is introduced, it displaces on average either the host matrix material or the inclusion material, with probabilities $(1-y)$ and y, respectively. A slightly different derivation by Zimmerman (1984a) (superseded by Zimmerman (1991a)) assumed that when a new inclusion is introduced, it always displaces the host material alone. This leads to similar differential equations with $dy\,(1-y)$ replaced by dy. The effective moduli predicted by this version of DEM are always slightly stiffer (for the same inclusion geometry and concentration) than the DEM equations previously given. They both predict the same first-order terms in y but begin to diverge at concentrations above 10%. The dependence of effective moduli on concentration goes as $e^{-2y} = (1 - 2y + 2y^2 - \cdots)$ for the version without $(1-y)$. whereas it behaves as $(1-y)^2 = (1 - 2y + y^2)$ for the version with $(1-y)$. Including the $(1-y)$ term makes the results of Zimmerman (1991a) consistent with the Hashin–Shtrikman bounds. In general, for a fixed inclusion geometry and porosity, the Kuster–Toksöz effective moduli are stiffer than the DEM predictions, which in turn are stiffer than the Berryman self-consistent effective moduli.

An important conceptual difference between the DEM and self-consistent schemes for calculating effective moduli of composites is that the DEM scheme identifies one of the constituents as a host or matrix material in which inclusions of the other constituent(s) are embedded, whereas the self-consistent scheme does not identify any specific host material but treats the composite as an aggregate of all the constituents.

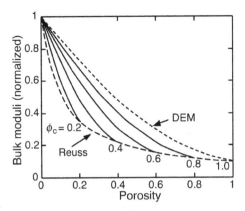

Figure 4.12.1 Normalized bulk-modulus curves for the conventional DEM theory (percolation at $\phi = 1$) and for the modified DEM (percolation at $\phi_c < 1$) for a range of ϕ_c values

Modified DEM with Critical Porosity Constraints

In the usual DEM model, starting from a solid initial host, a porous material stays intact at all porosities and falls apart only at the very end when $y = 1$ (100% porosity). This is because the solid host remains connected, and therefore load bearing.

Although DEM is a good model for materials such as glass foam (Berge *et al.*, 1993) and oceanic basalts (Berge *et al.*, 1992), most reservoir rocks fall apart at a critical porosity, ϕ_c, significantly less than 1.0 and are not represented very well by the conventional DEM theory. The modified DEM model (Mukerji *et al.*, 1995a) incorporates percolation behavior at any desired ϕ_c by redefining the phase 2 end member. The inclusions are now no longer made up of pure fluid (the original phase 2 material) but are composite inclusions of the critical phase at ϕ_c with elastic moduli (K_c, μ_c). With this definition, y denotes the concentration of the critical phase in the matrix. The total porosity is given by $\phi = y\phi_c$.

The computations are implemented by replacing (K_2, μ_2) with (K_c, μ_c) everywhere in the equations. Integrating along the reverse path, from $\phi = \phi_c$ to $\phi = 0$, gives lower moduli, for now the softer critical phase is the matrix. The moduli of the critical phase may be taken as the Reuss average value at ϕ_c of the pure end-member moduli. Because the critical phase consists of grains just barely touching each other, better estimates of K_c and μ_c may be obtained from measurements on loose sands, or from models of granular material. For porosities greater than ϕ_c, the material is a suspension and is best characterized by the Reuss average (or Wood's equation).

Figure 4.12.1 shows normalized bulk-modulus curves for the conventional DEM theory (percolation at $\phi = 1$ and for the modified DEM (percolation at $\phi_c < 1$) for a range of ϕ_c values. When $\phi_c = 1$, the modified DEM coincides with the conventional DEM curve. The shapes of the inclusions were taken to be spheres. The path was from 0 to ϕ_c, and (K_c, μ_c) were taken as the Reuss-average values at ϕ_c. For this choice of (K_c, μ_c) estimates along the reversed path coincide with the Reuss curve.

Uses

The purpose of the differential effective-medium model is to estimate the effective elastic moduli of a rock in terms of its constituents and pore space.

Assumptions and Limitations

The following assumptions and limitations apply to the differential effective-medium model:

- the rock is isotropic, linear, and elastic;
- the process of incrementally adding inclusions to the matrix is a thought experiment and should not be taken to provide an accurate description of the true evolution of rock porosity in nature;
- idealized ellipsoidal inclusion shapes are assumed; and
- cracks are isolated with respect to fluid flow. Pore pressures are unequilibrated and adiabatic. The model is appropriate for high-frequency laboratory conditions. For low-frequency field situations, use dry inclusions and then saturate using Gassmann relations. This should not be confused with the tendency to term this approach a low-frequency theory, for crack dimensions are assumed to be much smaller than a wavelength.

4.13   Hudson's Model for Cracked Media

Synopsis

Hudson's model is based on a scattering-theory analysis of the mean wave field in an elastic solid with thin, penny-shaped ellipsoidal cracks or inclusions (Hudson, 1980, 1981). The effective moduli c_{ij}^{eff} are given as

$$c_{ij}^{\text{eff}} = c_{ij}^{(0)} + c_{ij}^{(1)} + c_{ij}^{(2)} \tag{4.13.1}$$

where $c_{ij}^{(0)}$ are the isotropic background moduli, and $c_{ij}^{(1)}$, $c_{ij}^{(2)}$ are the first- and second-order corrections, respectively. (See Section 2.2 on anisotropy for the two-index notation of elastic moduli. Note also that Hudson uses a slightly different definition, and that there is an extra factor of 2 in his c_{44}, c_{55}, and c_{66}. This makes the equations given in his paper for $c_{44}^{(1)}$ and $c_{44}^{(2)}$ slightly different from those given below, which are consistent with the more standard Voigt notation described in Section 2.2 on anisotropy.)

For a single crack set with crack normals aligned along the 3-axis, the cracked media show transverse isotropic symmetry, and the corrections are

$$c_{11}^{(1)} = -\frac{\lambda^2}{\mu}\varepsilon U_3 \tag{4.13.2}$$

$$c_{13}^{(1)} = -\frac{\lambda(\lambda + 2\mu)}{\mu}\varepsilon U_3 \tag{4.13.3}$$

$$c_{33}^{(1)} = -\frac{(\lambda + 2\mu)^2}{\mu}\varepsilon U_3 \tag{4.13.4}$$

$$c_{44}^{(1)} = -\mu\varepsilon U_1 \tag{4.13.5}$$

$$c_{66}^{(1)} = 0 \tag{4.13.6}$$

and

$$c_{11}^{(2)} = \frac{q}{15}\frac{\lambda^2}{(\lambda + 2\mu)}(\varepsilon U_3)^2 \tag{4.13.7}$$

$$c_{13}^{(2)} = \frac{q}{15}\lambda(\varepsilon U_3)^2 \tag{4.13.8}$$

$$c_{33}^{(2)} = \frac{q}{15}(\lambda + 2\mu)(\varepsilon U_3)^2 \tag{4.13.9}$$

$$c_{44}^{(2)} = \frac{2}{15}\frac{\mu(3\lambda + 8\mu)}{\lambda + 2\mu}(\varepsilon U_1)^2 \tag{4.13.10}$$

$$c_{66}^{(2)} = 0 \tag{4.13.11}$$

where

$$q = 15\frac{\lambda^2}{\mu^2} + 28\frac{\lambda}{\mu} + 28 \tag{4.13.12}$$

$$\varepsilon = \frac{N}{V}a^3 = \frac{3\phi}{4\pi\alpha} = \text{crack density} \tag{4.13.13}$$

The isotropic background elastic moduli are λ and μ, and a and α are the crack radius and aspect ratio, respectively. The corrections $c_{ij}^{(1)}$ and $c_{ij}^{(2)}$ obey the usual symmetry properties for transverse isotropy or hexagonal symmetry (see Section 2.2 on anisotropy).

Caution:

The second-order expansion is not a uniformly converging series and predicts increasing moduli with crack density beyond the formal limit (Cheng, 1993). Better results will be obtained by using just the first-order correction rather than inappropriately using the second-order correction. Cheng gives a new expansion based on the Padé approximation, which avoids this problem. See Section 4.14.

The terms U_1 and U_3 depend on the crack conditions. For dry cracks,

$$U_1 = \frac{16(\lambda + 2\mu)}{3(3\lambda + 4\mu)} \qquad U_3 = \frac{4(\lambda + 2\mu)}{3(\lambda + \mu)}$$
(4.13.14)

For "weak" inclusions (i.e., when $\mu a / \left(K' + \frac{4}{3}\mu' \right)$ is of the order of 1 and is not small enough to be neglected)

$$U_1 = \frac{16(\lambda + 2\mu)}{3(3\lambda + 4\mu)} \frac{1}{(1 + M)} \qquad U_3 = \frac{4(\lambda + 2\mu)}{3(\lambda + \mu)} \frac{1}{(1 + \kappa)}$$
(4.13.15)

where

$$M = \frac{4\mu'}{\pi a \mu} \frac{(\lambda + 2\mu)}{(3\lambda + 4\mu)} \qquad \kappa = \frac{\left(K' + \frac{4}{3}\mu' \right)(\lambda + 2\mu)}{\pi a \mu(\lambda + \mu)}$$
(4.13.16)

where K' and μ' are the bulk and shear modulus of the inclusion material. The criteria for an inclusion to be "weak" depend on its shape or aspect ratio α as well as on the relative moduli of the inclusion and matrix material. Dry cavities can be modeled by setting the inclusion moduli to zero. Fluid-saturated cavities are simulated by setting the inclusion shear modulus to zero.

Caution:

Because the cavities are isolated with respect to flow, this approach simulates very high-frequency behavior appropriate to ultrasonic laboratory conditions. At low frequencies, when there is time for wave-induced pore-pressure increments to flow and equilibrate, it is better to find the effective moduli for dry cavities and then saturate them with the Brown and Korringa low-frequency relations (Section 6.5). This should not be confused with the tendency to term this approach a low-frequency theory, for crack dimensions are assumed to be much smaller than a wavelength.

Hudson also gives expressions for infinitely thin, fluid-filled cracks:

$$U_1 = \frac{16(\lambda + 2\mu)}{3(3\lambda + 4\mu)} \qquad U_3 = 0$$
(4.13.17)

These assume no discontinuity in the normal component of crack displacements and therefore predict no change in the compressional modulus with saturation. There is, however, a shear displacement discontinuity and a resulting effect on shear stiffness. This case should be used with care.

The first-order changes λ_1 and μ_1 in the isotropic elastic moduli λ and μ of a material containing randomly oriented inclusions are given by

$$\mu_1 = -\frac{2\mu}{15}\varepsilon(3U_1 + 2U_3)$$
(4.13.18)

$$3\lambda_1 + 2\mu_1 = -\frac{(3\lambda + 2\mu)^2}{3\mu}\varepsilon U_3 \qquad (4.13.19)$$

These results agree with the self-consistent results of Budiansky and O'Connell (1976).

For two or more crack sets aligned in different directions, corrections for each crack set are calculated separately in a crack-local coordinate system with the 3-axis normal to the crack plane and then rotated or transformed back (see Section 1.5 on coordinate transformations) into the coordinates of c_{ij}^{eff}; finally, the results are added to obtain the overall correction. Thus, for three crack sets with crack densities ε_1, ε_2, and ε_3 with crack normals aligned along the 1-, 2-, and 3-axes, respectively, the overall first-order corrections to c_{ij}^0, $c_{ij}^{(1)(3sets)}$, may be given in terms of linear combinations of the corrections for a single set, with the appropriate crack densities as follows (where we have taken into account the symmetry properties of $c_{ij}^{(1)}$):

$$c_{11}^{(1)(3sets)} = c_{33}^{(1)}(\varepsilon_1) + c_{11}^{(1)}(\varepsilon_2) + c_{11}^{(1)}(\varepsilon_3) \qquad (4.13.20)$$

$$c_{12}^{(1)(3sets)} = c_{13}^{(1)}(\varepsilon_1) + c_{13}^{(1)}(\varepsilon_2) + c_{12}^{(1)}(\varepsilon_3) \qquad (4.13.21)$$

$$c_{13}^{(1)(3sets)} = c_{13}^{(1)}(\varepsilon_1) + c_{12}^{(1)}(\varepsilon_2) + c_{13}^{(1)}(\varepsilon_3) \qquad (4.13.22)$$

$$c_{22}^{(1)(3sets)} = c_{11}^{(1)}(\varepsilon_1) + c_{33}^{(1)}(\varepsilon_2) + c_{11}^{(1)}(\varepsilon_3) \qquad (4.13.23)$$

$$c_{23}^{(1)(3sets)} = c_{12}^{(1)}(\varepsilon_1) + c_{13}^{(1)}(\varepsilon_2) + c_{13}^{(1)}(\varepsilon_3) \qquad (4.13.24)$$

$$c_{33}^{(1)(3sets)} = c_{11}^{(1)}(\varepsilon_1) + c_{11}^{(1)}(\varepsilon_2) + c_{33}^{(1)}(\varepsilon_3) \qquad (4.13.25)$$

$$c_{44}^{(1)(3sets)} = c_{44}^{(1)}(\varepsilon_2) + c_{44}^{(1)}(\varepsilon_3) \qquad (4.13.26)$$

$$c_{55}^{(1)(3sets)} = c_{44}^{(1)}(\varepsilon_1) + c_{44}^{(1)}(\varepsilon_3) \qquad (4.13.27)$$

$$c_{66}^{(1)(3sets)} = c_{44}^{(1)}(\varepsilon_1) + c_{44}^{(1)}(\varepsilon_2) \qquad (4.13.28)$$

Note that $c_{66}^{(1)} = 0$ and $c_{12}^{(1)} = c_{11}^{(1)} - 2c_{66}^{(1)} = c_{11}^{(1)}$

Hudson (1981) also gives the attenuation coefficient $(\gamma = \omega Q^{-1}/2V)$ for elastic waves in cracked media. For aligned cracks with normals along the 3-axis, the attenuation coefficients for P-, SV-, and SH-waves are

$$\gamma_P = \frac{\omega}{V_S}\varepsilon\left(\frac{\omega a}{V_P}\right)^3\frac{1}{30\pi}\left[AU_1^2\sin^2 2\theta + BU_3^2\left(\frac{V_P^2}{V_S^2} - 2\sin^2\theta\right)^2\right] \qquad (4.13.29)$$

$$\gamma_{SV} = \frac{\omega}{V_S}\varepsilon\left(\frac{\omega a}{V_S}\right)^3\frac{1}{30\pi}(AU_1^2\cos^2 2\theta + BU_3^2\sin^2 2\theta) \qquad (4.13.30)$$

$$\gamma_{SH} = \frac{\omega}{V_S}\varepsilon\left(\frac{\omega a}{V_S}\right)^3\frac{1}{30\pi}(AU_1^2\cos^2\theta) \qquad (4.13.31)$$

$$A = \frac{3}{2} + \frac{V_S^5}{V_P^5} \tag{4.13.32}$$

$$B = 2 + \frac{15}{4} \frac{V_S}{V_P} - 10 \frac{V_S^3}{V_P^3} + 8 \frac{V_S^5}{V_P^5} \tag{4.13.33}$$

In the preceding expressions, V_P and V_S are the P and S velocities in the uncracked isotropic background matrix, ω is the angular frequency, and θ is the angle between the direction of propagation and the 3-axis (axis of symmetry).

For randomly oriented cracks (isotropic distribution), the P and S attenuation coefficients are given as

$$\gamma_P = \frac{\omega}{V_S} \varepsilon \left(\frac{\omega a}{V_P} \right)^3 \frac{4}{15^2 \pi} \left[A U_1^2 + \frac{1}{2} \frac{V_P^5}{V_S^5} B(B-2) U_3^2 \right] \tag{4.13.34}$$

$$\gamma_S = \frac{\omega}{V_S} \varepsilon \left(\frac{\omega a}{V_S} \right)^3 \frac{1}{75\pi} \left(A U_1^2 + \frac{1}{3} B U_3^2 \right) \tag{4.13.35}$$

The fourth-power dependence on ω is characteristic of Rayleigh scattering.

Hudson (1990) gives results for overall elastic moduli of material with various distributions of penny-shaped cracks. If conditions at the cracks are taken to be uniform, so that U_1 and U_3 do not depend on the polar and azimuthal angles θ and ϕ, the first-order correction is given as

$$c_{ijpq}^{(1)} = -\frac{A}{\mu} U_3 [\lambda^2 \delta_{ij} \delta_{pq} + 2\lambda\mu(\delta_{ij}\tilde{\varepsilon}_{pq} + \delta_{pq}\tilde{\varepsilon}_{ij}) + 4\mu^2 \tilde{\varepsilon}_{ijpq}]$$

$$- A\mu U_1 (\delta_{jq}\tilde{\varepsilon}_{ip} + \delta_{jp}\tilde{\varepsilon}_{iq} + \delta_{iq}\tilde{\varepsilon}_{jp} + \delta_{ip}\tilde{\varepsilon}_{jq} - 4\tilde{\varepsilon}_{ijpq}) \tag{4.13.36}$$

where

$$A = \int_0^{2\pi} \int_0^{\pi/2} \varepsilon(\theta, \phi) \sin\theta \, d\theta \, d\phi \tag{4.13.37}$$

$$\tilde{\varepsilon}_{ij} = \frac{1}{A} \int_0^{2\pi} \int_0^{\pi/2} \varepsilon(\theta, \phi) \, n_i n_j \sin\theta \, d\theta \, d\phi \tag{4.13.38}$$

$$\tilde{\varepsilon}_{ijpq} = \frac{1}{A} \int_0^{2\pi} \int_0^{\pi/2} \varepsilon(\theta, \phi) \, n_i n_j n_p n_q \sin\theta \, d\theta \, d\phi \tag{4.13.39}$$

and n_i are the components of the unit vector along the crack normal, $\mathbf{n} = (\sin\theta\cos\phi, \sin\theta\sin\phi, \cos\theta)$, whereas $\varepsilon(\theta, \phi)$ is the crack density distribution function, so that $\varepsilon(\theta, \phi)\sin\theta d\theta d\phi$ is the density of cracks with normals lying in the solid angle between $(\theta, \theta + d\theta)$ and $(\phi, \phi + d\phi)$.

Special Cases of Crack Distributions

(a) Cracks with total crack density ε_t, which have all their normals aligned along $\theta = \theta_0$, $\phi = \phi_0$:

$$\varepsilon(\theta, \phi) = \varepsilon_t \frac{\delta(\theta - \theta_0)}{\sin\theta} \delta(\phi - \phi_0)$$
$$A = \varepsilon_t$$
$$\tilde{\varepsilon}_{ij} = n_i^0 n_j^0 \tag{4.13.40}$$
$$\tilde{\varepsilon}_{ijpq} = n_i^0 n_j^0 n_p^0 n_q^0$$

where $n_1^0 = \sin\theta_0 \cos\phi_0$, $n_2^0 = \sin\theta_0 \sin\phi_0$, $n_3^0 = \cos\theta_0$, and $\delta(\ )$ is the delta function.
(b) Rotationally symmetric crack distributions with normals symmetrically distributed about $\theta = 0$, that is, where ε is a function of θ only:

$$A = 2\pi \int_0^{\pi/2} \varepsilon(\theta)\sin\theta \, d\theta \tag{4.13.41}$$

$$\tilde{\varepsilon}_{12} = \tilde{\varepsilon}_{23} = \tilde{\varepsilon}_{31} = 0 \tag{4.13.42}$$

$$\tilde{\varepsilon}_{11} = \tilde{\varepsilon}_{22} = \frac{\pi}{A} \int_0^{\pi/2} \varepsilon(\theta) \sin^3\theta \, d\theta = \frac{1}{2}(1 - \tilde{\varepsilon}_{33}) \tag{4.13.43}$$

$$\tilde{\varepsilon}_{1111} = \tilde{\varepsilon}_{2222} = \frac{3\pi}{4A} \int_0^{\pi/2} \varepsilon(\theta) \sin^5\theta \, d\theta = 3\tilde{\varepsilon}_{1122} = 3\tilde{\varepsilon}_{1212} \text{ etc.} \tag{4.13.44}$$

$$\tilde{\varepsilon}_{3333} = \frac{8}{3}\tilde{\varepsilon}_{1111} - 4\tilde{\varepsilon}_{11} + 1 \tag{4.13.45}$$

$$\tilde{\varepsilon}_{1133} = \tilde{\varepsilon}_{11} - \frac{4}{3}\tilde{\varepsilon}_{1111} = \tilde{\varepsilon}_{2233} = \tilde{\varepsilon}_{1313} = \tilde{\varepsilon}_{2323} \text{ etc.} \tag{4.13.46}$$

Elements other than those related to the preceding elements by symmetry are zero. A particular rotationally symmetric distribution is the Fisher distribution, for which $\varepsilon(\theta)$ is

$$\varepsilon(\theta) = \frac{\varepsilon_t}{2\pi \sigma^2 (e^{1/\sigma^2} - 1)} e^{(\cos\theta)/\sigma^2} \tag{4.13.47}$$

For small σ^2, this is approximately a model for a Gaussian distribution on the sphere

$$\varepsilon(\theta) \approx \varepsilon_t \frac{e^{-\theta^2/2\sigma^2}}{2\pi\sigma^2} \tag{4.13.48}$$

The proportion of crack normals outside the range $0 \leq \theta \leq 2\sigma$ is approximately $1/e^2$. For this distribution,

$$A = \varepsilon_t \tag{4.13.49}$$

$$\tilde{\varepsilon}_{11} = \frac{-1 + 2\sigma^2 e^{1/\sigma^2} - 2\sigma^4(e^{1/\sigma^2} - 1)}{2(e^{1/\sigma^2} - 1)} \approx \sigma^2 \tag{4.13.50}$$

$$\tilde{\varepsilon}_{1111} = \frac{3}{8}\left[\frac{-1 + 4\sigma^4(2e^{1/\sigma^2} + 1) - 24\sigma^6 e^{1/\sigma^2} + 24\sigma^8(e^{1/\sigma^2} - 1)}{(e^{1/\sigma^2} - 1)}\right] \approx 3\sigma^4 \tag{4.13.51}$$

This distribution is suitable when crack normals are oriented randomly with a small variance about a mean direction along the 3-axis.

(c) Cracks with normals randomly distributed at a fixed angle from the 3-axis forming a cone. In this case ε is independent of ϕ and is zero unless $\theta = \theta_0$, $0 \le \phi \le 2\pi$.

$$\varepsilon(\theta) = \varepsilon_t \frac{\delta(\theta - \theta_0)}{2\pi \sin\theta} \tag{4.13.52}$$

which gives

$$A = \varepsilon_t \tag{4.13.53}$$

$$\widetilde{\varepsilon}_{11} = \frac{1}{2} \sin^2\theta_0 \tag{4.13.54}$$

$$\widetilde{\varepsilon}_{1111} = \frac{3}{8} \sin^4\theta_0 \tag{4.13.55}$$

and the first-order corrections are

$$c_{1111}^{(1)} = -\frac{\varepsilon_t}{2\mu} \left[U_3 \left(2\lambda^2 + 4\lambda\mu \sin^2\theta_0 + 3\mu^2 \sin^4\theta_0 \right) + U_1\mu^2 \sin^2\theta_0 \left(4 - 3 \sin^2\theta_0 \right) \right]$$
$$= C_{2222}^{(1)}$$
$$\tag{4.13.56}$$

$$c_{3333}^{(1)} = -\frac{\varepsilon_t}{\mu} \left[U_3 (\lambda + 2\mu \cos^2\theta_0)^2 + U_1\mu^2 4 \cos^2\theta_0 \sin^2\theta_0 \right] \tag{4.13.57}$$

$$c_{1122}^{(1)} = -\frac{\varepsilon_t}{2\mu} \left[U_3 \left(2\lambda^2 + 4\lambda\mu \sin^2\theta_0 + \mu^2 \sin^4\theta_0 \right) - U_1\mu^2 \sin^4\theta_0 \right] \tag{4.13.58}$$

$$c_{1133}^{(1)} = -\frac{\varepsilon_t}{\mu} \left[U_3 (\lambda + \mu \sin^2\theta_0)(\lambda + 2\mu \cos^2\theta_0) - U_1\mu^2 2 \sin^2\theta_0 \cos^2\theta_0 \right]$$
$$= c_{2233}^{(1)}$$
$$\tag{4.13.59}$$

$$c_{2323}^{(1)} = -\frac{\varepsilon_t}{2}\mu \left[U_3 4 \sin^2\theta_0 \cos^2\theta_0 + U_1 \left(\sin^2\theta_0 + 2 \cos^2\theta_0 - 4 \sin^2\theta_0 \cos^2\theta_0 \right) \right]$$
$$= C_{1313}^{(1)}$$
$$\tag{4.13.60}$$

$$c_{1212}^{(1)} = -\frac{\varepsilon_t}{2}\mu \left[U_3 \sin^4\theta_0 + U_1 \sin^2\theta_0 \left(2 - \sin^2\theta_0 \right) \right] \tag{4.13.61}$$

Heavily Faulted Structures

Hudson and Liu (1999) provide a model for the effective elastic properties of rocks with an array of parallel faults based on the averaging process of Schoenberg and Douma (1988). The individual faults themselves are modeled using two different approaches: model 1 as a planar distribution of small circular cracks; model 2 as a planar distribution of small circular welded contacts. Hudson and Liu also give expressions where both models 1 and 2 are replaced by an equivalent layer of constant thickness and appropriate infill material, which can then be averaged using the Backus average. Model 2 (circular contacts) provides expressions for effective elastic properties of heavily cracked media with cracks aligned and confined within the fault planes.

Assuming the fault normals to be aligned along the 3-axis, the elements of the effective stiffness tensor are (Hudson and Liu, 1999)

$$c_{1111} = c_{2222} = \lambda + 2\mu \tag{4.13.62}$$

$$c_{3333} = \frac{\lambda + 2\mu}{1 + E_N} \tag{4.13.63}$$

$$c_{1122} = \lambda \tag{4.13.64}$$

$$c_{1133} = c_{2233} = \frac{\lambda}{1 + E_N} \tag{4.13.65}$$

$$c_{2323} = c_{1313} = \frac{\mu}{1 + E_T} \tag{4.13.66}$$

$$c_{1212} = \mu \tag{4.13.67}$$

The quantities E_N and E_T depend on the fault model. For **model 1**, where the fault plane consists of an array of circular cracks,

$$E_N = \left(\frac{v^s a^3}{H}\right)\left(\frac{\lambda + 2\mu}{\mu}\right) U_3 \left[1 + \pi U_3 (v^s a^2)^{3/2}\left(1 - \frac{\mu}{\lambda + 2\mu}\right)\right] \tag{4.13.68}$$

$$E_T = \left(\frac{v^s a^3}{H}\right) U_1 \left[1 + \frac{\pi}{4} U_1 (v^s a^2)^{3/2}\left(3 - 2\frac{\mu}{\lambda + 2\mu}\right)\right] \tag{4.13.69}$$

where a is the crack radius, v^s is the number density of cracks on the fault surface, H is the spacing between parallel faults, and λ and μ are the Lamé parameters of the uncracked background material. The overall number density of cracks, $v = v^s/H$ and the overall crack density $\varepsilon = va^3 = v^s a^3/H$. The relative area of cracking on the fault is given by $r = v^s \pi a^2$.

Alternatively, the faults may be replaced by an elastically equivalent layer with bulk and shear moduli given by

$$K^* = K' - \frac{\pi a \mu}{4}\left(\frac{\mu}{\lambda + 2\mu}\right)\left[1 - \frac{4}{3}\left(\frac{r^3}{\pi}\right)^{1/2}\right] \tag{4.13.70}$$

$$\mu^* = \mu' + \frac{3\pi a \mu}{16}\left(\frac{3\lambda + 4\mu}{\lambda + 2\mu}\right)\left[1 - \frac{4}{3}\left(\frac{r^3}{\pi}\right)^{1/2}\right] \tag{4.13.71}$$

In the previous expressions, K' and μ' are, respectively, the bulk and shear moduli of the material filling the cracks, of aspect ratio a, within the faults. Viscous fluid may be modeled by setting μ' to be imaginary. Setting both moduli to zero represents dry cracks. The equivalent layers can be used in Backus averaging to estimate the overall stiffness. These give results very close (but not completely identical, because of the approximations involved) to the results obtained from the equations for c_{ijkl} previously given.

For **model 2**, which assumes a planar distribution of small, circular, welded contacts:

$$E_N = \frac{(\lambda + 2\mu)^2}{4\mu(\lambda + \mu)} \frac{1}{(v^w Hb)} (1 + 2\sqrt{v^w b^2})^{-1} \tag{4.13.72}$$

$$E_T = \frac{(3\lambda + 4\mu)}{8(\lambda + \mu)} \frac{1}{(v^w Hb)} (1 + 2\sqrt{v^w b^2})^{-1} \tag{4.13.73}$$

where b is the radius of the welded contact region and v^w is the number density of welded contacts on the fault surface.

In this case, the relative area of cracking (noncontact) on the fault is given by $r = 1 - v^w \pi b^2$.

Again, as an alternate representation, the faults in model 2 may be replaced by an elastically equivalent layer with bulk and shear moduli given by

$$K^* = K' - \frac{4\mu(d/b)r(1-r)}{3\pi} \frac{(\lambda + \mu)(4\mu - \lambda)}{(3\lambda + 4\mu)(\lambda + 2\mu)} \left[1 + 2\left(\frac{1-r}{\pi}\right)^{1/2} \right] \tag{4.13.74}$$

$$\mu^* = \mu' + \frac{8\mu(d/b)r(1-r)}{\pi} \left(\frac{\lambda + \mu}{3\lambda + 4\mu} \right) \left[1 + 2\left(\frac{1-r}{\pi}\right)^{1/2} \right] \tag{4.13.75}$$

where d is the aperture of the cracked (noncontact) area. Note that d/b is not the aspect ratio of cracks since b is the radius of the welded regions. The previous equations are valid for small d/b. The equivalent layers can be used in Backus averaging to estimate the overall stiffness.

The equations for model 1 are valid for $r \ll 1$ (dilute cracks, predominantly welded) while the equations for model 2 are valid for $(1 - r) \ll 1$ (dilute contacts, heavily cracked). Hudson and Liu (1999) show that, to first order, the elastic response of heavily cracked faults (model 2) is the same as that of a cubic packing of identical spheres with the same number density and size of contact areas. To the first order, the actual distribution of contact regions and the shape of cavities between the contact regions do not affect the overall elastic moduli.

Uses

Hudson's model is used to estimate the effective elastic moduli and attenuation of a rock in terms of its constituents and pore space.

Assumptions and Limitations

The use of Hudson's model requires the following considerations:

- an idealized crack shape (penny-shaped) with small aspect ratios and either small crack density or small contact density are assumed. The crack radius and the distance between cracks are much smaller than a wavelength. The formal limit quoted by Hudson for both first- and second-order terms is ε less than 0.1;
- the second-order expansion is not a uniformly converging series and predicts increasing moduli with crack density beyond the formal limit (Cheng, 1993). Better results will be obtained by using just the first-order correction rather than inappropriately using the second-order correction. Cheng gives a new expansion based on the Padé approximation, which avoids this problem;
- cracks are isolated with respect to fluid flow. Pore pressures are unequilibrated and adiabatic. The model is appropriate for high-frequency laboratory conditions. For low-frequency field situations, use Hudson's dry equations and then saturate by using the Brown and Korringa relations (Section 6.5). This should not be confused with the tendency to think of this approach as a low-frequency theory, because crack dimensions are assumed to be much smaller than a wavelength; and
- sometimes a single crack set may not be an adequate representation of crack-induced anisotropy. In this case we need to superpose several crack sets with angular distributions.

4.14   Eshelby–Cheng Model for Cracked Anisotropic Media

Synopsis

Cheng (1978, 1993) has given a model for the effective moduli of cracked, transversely isotropic rocks based on Eshelby's (1957) static solution for the strain inside an ellipsoidal inclusion in an isotropic matrix. The effective moduli C_{ij}^{eff} for a rock containing fluid-filled ellipsoidal cracks with their normals aligned along the 3-axis are given as

$$c_{ij}^{eff} = c_{ij}^{(0)} - \phi c_{ij}^{(1)} \tag{4.14.1}$$

where ϕ is the porosity and $c_{ij}^{(0)}$ are the moduli of the uncracked isotropic rock. The corrections $c_{ij}^{(1)}$ are

$$c_{11}^{(1)} = \lambda(S_{31} - S_{33} + 1) + \frac{2\mu E}{D(S_{12} - S_{11} + 1)} \tag{4.14.2}$$

$$c_{33}^{(1)} = \frac{(\lambda + 2\mu)(-S_{12} - S_{11} + 1) + 2\lambda S_{13} + 4\mu C}{D} \tag{4.14.3}$$

$$c_{13}^{(1)} = \frac{(\lambda + 2\mu)(S_{13} + S_{31}) - 4\mu C + \lambda(S_{13} - S_{12} - S_{11} - S_{33} + 2)}{2D} \tag{4.14.4}$$

$$c_{44}^{(1)} = \frac{\mu}{1 - 2S_{1313}} \qquad c_{66}^{(1)} = \frac{\mu}{1 - 2S_{1212}} \tag{4.14.5}$$

with

$$C = \frac{K_{fl}}{3(K - K_{fl})} \tag{4.14.6}$$

$$\begin{aligned} D = &S_{33}S_{11} + S_{33}S_{12} - 2S_{31}S_{13} - (S_{11} + S_{12} + S_{33} - 1 - 3C) \\ &- C[S_{11} + S_{12} + 2(S_{33} - S_{13} - S_{31})] \end{aligned} \tag{4.14.7}$$

$$E = S_{33}S_{11} - S_{31}S_{13} - (S_{33} + S_{11} - 2C - 1) + C(S_{31} + S_{13} - S_{11} - S_{33}) \tag{4.14.8}$$

$$S_{11} = QI_{aa} + RI_a \qquad S_{33} = Q\left(\frac{4\pi}{3} - 2I_{ac}\alpha^2\right) + I_c R \tag{4.14.9}$$

$$S_{12} = QI_{ab} - RI_a \qquad S_{13} = QI_{ac}\alpha^2 - RI_a \tag{4.14.10}$$

$$S_{31} = QI_{ac} - RI_c \qquad S_{1212} = QI_{ab} + RI_a \tag{4.14.11}$$

$$S_{1313} = \frac{Q(1 + \alpha^2)I_{ac}}{2} + \frac{R(I_a + I_c)}{2} \tag{4.14.12}$$

$$I_a = \frac{2\pi\alpha(\cos^{-1}\alpha - \alpha S_a)}{S_a^3} \qquad I_c = 4\pi - 2I_a \tag{4.14.13}$$

$$I_{ac} = \frac{I_c - I_a}{3S_a^2} \qquad I_{aa} = \pi - \frac{3I_{ac}}{4} \qquad I_{ab} = \frac{I_{aa}}{3} \tag{4.14.14}$$

$$\sigma = \frac{3K - 2\mu}{6K + 2\mu} \qquad S_a = \sqrt{1 - \alpha^2} \tag{4.14.15}$$

$$R = \frac{1 - 2\sigma}{8\pi(1 - \sigma)} \qquad Q = \frac{3R}{1 - 2\sigma} \tag{4.14.16}$$

In the preceding equations, K and μ are the bulk and shear moduli of the isotropic matrix, respectively; K_{fl} is the bulk modulus of the fluid; and α is the crack aspect ratio. Dry cavities can be modeled by setting the inclusion moduli to zero. Do not confuse S with the anisotropic compliance tensor. This model is valid for arbitrary aspect ratios, unlike the Hudson model, which assumes very small aspect ratio cracks (see Section 4.13 on Hudson's model). The results of the two models are essentially the same for small aspect ratios and low crack densities (< 0.1), as long as the "weak inclusion" form of Hudson's theory is used (Cheng, 1993).

Uses

The Eshelby–Cheng model is used to obtain the effective anisotropic stiffness tensor for transversely isotropic, cracked rocks.

Assumptions and Limitations

The following presuppositions and limitations apply to the Eshelby-Cheng model:
- the model assumes an isotropic, homogeneous, elastic background matrix and an idealized ellipsoidal crack shape;
- the model assumes low crack concentrations but can handle all aspect ratios; and
- because the cavities are isolated with respect to flow, this approach simulates very-high-frequency behavior appropriate to ultrasonic laboratory conditions.

At low frequencies, when there is time for wave-induced pore-pressure increments to flow and equilibrate, it is better to find the effective moduli for dry cavities and then saturate them with the Brown and Korringa low-frequency relations. This should not be confused with the tendency to term this approach a low-frequency theory, for crack dimensions are assumed to be much smaller than a wavelength.

Extensions

The model has been extended to a transversely isotropic background by Nishizawa (1982).

4.15   *T*-Matrix Inclusion Models for Effective Moduli

Synopsis

One approach for estimating effective elastic constants for composites is based on the integral equation or *T*-matrix approach of quantum scattering theory. This approach takes into account interactions between inclusions based on multiple-point correlation functions. The integral equation for effective elastic constants of macroscopically homogeneous materials with statistical fluctuation of properties at the microscopic level is very similar to the Lippmann-Schwinger-Dyson equation of multiple scattering in quantum mechanics. The theory in the context of elastic composites was developed by Eimer (1967, 1968), Kröner (1967, 1977, 1986), Zeller and Dederichs (1973), Korringa (1973), and Gubernatis and Krumhansl (1975). Willis (1977) used a *T*-matrix approach to obtain bounds and estimates for anisotropic composites. Middya and Basu (1986) applied the *T*-matrix formalism for polycrystalline aggregates. Jakobsen *et al.* (2003a) synthesized many of the existing effective medium approximations and placed them on a common footing using the *T*-matrix language. They also applied the formalism to model elastic properties of anisotropic shales.

A homogeneous anisotropic matrix with elastic stiffness tensor $\mathbf{C}^{(0)}$ has embedded inclusions, divided into families $r = 1, 2, \ldots, N$ having concentrations $v^{(r)}$ and shapes $\alpha^{(r)}$ For ellipsoidal inclusions $\alpha^{(r)}$ denotes the aspect ratio of the rth family of inclusions. While the general theory is not limited to ellipsoidal inclusions, most practical calculations are carried out assuming idealized ellipsoidal shapes. The effective elastic stiffness is given by Jakobsen *et al.* (2003a) as

$$\mathbf{C}_T^* = \mathbf{C}^{(0)} + \langle \mathbf{T}_1 \rangle (\mathbf{I} - \langle \mathbf{T}_1 \rangle^{-1} \mathbf{X})^{-1} \tag{4.15.1}$$

where

$$\langle \mathbf{T}_1 \rangle = \sum_r v^{(r)} \mathbf{t}^{(r)} \tag{4.15.2}$$

$$\mathbf{X} = -\sum_r \sum_s v^{(r)} \mathbf{t}^{(r)} \mathbf{G}_d^{(rs)} \mathbf{t}^{(s)} v^{(s)} \tag{4.15.3}$$

and $\mathbf{I} = I_{jklm} = \dfrac{1}{2}\left(\delta_{jl}\delta_{km} + \delta_{jm}\delta_{kl}\right)$ is the fourth-rank identity tensor. The *T*-matrix $\mathbf{t}^{(r)}$ for a single inclusion of elastic stiffness $\mathbf{C}^{(r)}$ is given by

$$\mathbf{t}^{(r)} = \delta\mathbf{C}^{(r)}(\mathbf{I} - \mathbf{G}^{(r)}\delta\mathbf{C}^{(r)})^{-1} \tag{4.15.4}$$

where

$$\delta\mathbf{C}^{(r)} = \mathbf{C}^{(r)} - \mathbf{C}^{(0)} \tag{4.15.5}$$

The fourth-rank tensor $\mathbf{G}^{(r)}$ (not the same as the tensor $\mathbf{G}_d^{(rs)}$) is a function of only $\mathbf{C}^{(0)}$ and the inclusion shape. It is computed using the following equations:

$$G_{pqrs}^{(r)} = -\frac{1}{4}[E_{pqrs}^{(r)} + E_{pqsr}^{(r)} + E_{qprs}^{(r)} + E_{qpsr}^{(r)}] \tag{4.15.6}$$

$$E_{pqrs}^{(r)} = \int_0^\pi d\theta \, \sin\theta \int_0^{2\pi} d\phi \, D_{qs}^{-1}(k) \, k_p k_r A^{(r)}(\theta, \phi) \tag{4.15.7}$$

$$D_{qs} = C_{qmsn}^{(0)} k_m k_n \tag{4.15.8}$$

where k, θ, and ϕ are spherical coordinates in *k*-space (Fourier space), and the components of the unit *k*-vector are given as usual by $[\sin\theta \cos\phi, \sin\theta \sin\phi, \cos\theta]$. The term $A^{(r)}(\theta, \phi)$ is a shape factor for the *r*th inclusion and is independent of elastic constants. For an oblate spheroidal inclusion with short axis b_3 aligned along the x_3-axis, long axes $b_1 = b_2$, and aspect ratio $\alpha^{(r)} = b_3/b_1$, $A^{(r)}$ is independent of ϕ and is given by

$$A^{(r)}(\theta) = \frac{1}{4\pi}\frac{\alpha^{(r)}}{[\sin^2\theta + (\alpha^{(r)})^2 \cos^2\theta]^{3/2}} \tag{4.15.9}$$

For a spherical inclusion, $A^{(r)} = 1/(4\pi)$. For the case when $\mathbf{C}^{(0)}$ has hexagonal symmetry and the principal axes of the oblate spheroidal inclusions are aligned along the symmetry axes of $\mathbf{C}^{(0)}$, the nonzero elements of E_{pqrs} are given by definite integrals of polynomial functions (Mura, 1982):

$$E_{1111} = E_{2222} = \frac{\pi}{2} \int_0^1 \Delta(1-x^2)\{ [f(1-x^2) + h\gamma^2 x^2] \cdot [(3e+d)(1-x^2)$$

$$+ 4f\gamma^2 x^2] - g^2\gamma^2 x^2 (1-x^2)\} dx \tag{4.5.10}$$

$$E_{3333} = 4\pi \int_0^1 \Delta\gamma^2 x^2 [d(1-x^2) + f\gamma^2 x^2][e(1-x^2) + f\gamma^2 x^2] dx \tag{4.15.11}$$

$$E_{1122} = E_{2211} = \frac{\pi}{2} \int_0^1 \Delta(1-x^2)\{ [f(1-x^2) + h\gamma^2 x^2] \cdot [(e+3d)(1-x^2)$$

$$+ 4f\gamma^2 x^2] - 3g^2\gamma^2 x^2 (1-x^2)\} dx \tag{4.15.12}$$

$$E_{1133} = E_{2233} = 2\pi \int_0^1 \Delta\gamma^2 x^2 \{ [(d+e)(1-x^2) + 2f\gamma^2 x^2] \cdot [f(1-x^2) + h\gamma^2 x^2]$$

$$- g^2\gamma^2 x^2 (1-x^2)\} dx \tag{4.15.13}$$

$$E_{3311} = E_{3322} = 2\pi \int_0^1 \Delta(1-x^2)[d(1-x^2) + f\gamma^2 x^2] \cdot [e(1-x^2) + f\gamma^2 x^2] dx \tag{4.15.14}$$

$$E_{1212} = \frac{\pi}{2} \int_0^1 \Delta(1-x^2)^2 \{ g^2\gamma^2 x^2 - (d-e)[f(1-x^2) + h\gamma^2 x^2]\} dx \tag{4.15.15}$$

$$E_{1313} = E_{2323} = -2\pi \int_0^1 \Delta g\gamma^2 x^2 (1-x^2)[e(1-x^2) + f\gamma^2 x^2] dx \tag{4.15.16}$$

where

$$\Delta^{-1} = [e(1-x^2) + f\gamma^2 x^2]\{ [d(1-x^2) + f\gamma^2 x^2][f(1-x^2) + h\gamma^2 x^2] - g^2\gamma^2 x^2 (1-x^2)\} \tag{4.15.17}$$

$$d = C_{11}^{(0)} \quad e = [C_{11}^{(0)} - C_{12}^{(0)}]/2 \tag{4.15.18}$$

$$f = C_{44}^{(0)} \quad g = C_{13}^{(0)} + C_{44}^{(0)} \quad h = C_{33}^{(0)} \tag{4.15.19}$$

and x is the integration variable while $\gamma = \alpha$ is the inclusion aspect ratio.

Using these equations, the tensor E_{pqrs} (and hence G_{pqrs}) can be obtained by numerical or symbolic integration. For numerical computations, instead of the more common Voigt notation, the Kelvin notation of fourth-rank tensors in terms of 6 x 6 matrices is more efficient, since matrix operations can be performed on the Kelvin 6x6 matrices according to the usual matrix rules (see also Section 2.2).

In the equation for the effective elastic stiffness, the term $\mathbf{X}$ involving the tensor $\mathbf{G}_d^{(rs)}$ represents the two-point interaction between the *rth* set and the *sth* set of inclusions. The tensor $\mathbf{G}_d^{(rs)}$ is computed in the same way as the tensor $\mathbf{G}_{ijkl}^{(r)}$ except that the aspect ratio of the associated inclusion is set to be the aspect ratio, α_d of the ellipsoidal two-point spatial correlation function, representing the symmetry of $p^{(s|r)}(\mathbf{x}-\mathbf{x}')$ the conditional probability of finding an inclusion of type s at $\mathbf{x}'$, given that there is an inclusion of type r at $\mathbf{x}$. The aspect ratio of the spatial correlation function can be different from the aspect ratio of the inclusions. The effect of spatial distribution is of higher order in the volume concentrations than the effect of inclusion shapes. The effect of spatial distribution becomes important when the inclusion concentration increases beyond the dilute limit. However, even the T-matrix approximation given by equation 4.12.1 may be invalid at very high concentration, as it neglects the higher-order interactions beyond two-point interactions (Jakobsen $et\ al.$, 2003a). According to Jakobsen $et\ al.$ (2003a) it appears to be difficult to determine the range of validity for the various T-matrix approximations. For the case of a matrix phase with just one type of inclusion, Ponte-Castaneda and Willis (1995) found that the T-matrix approximation is consistent only for certain combinations of α_d (aspect ratio of the correlation function), α (inclusion aspect ratio), and v (inclusion concentration). When $\alpha_d > \alpha$, the minimum possible value for inclusion aspect ratio is $\alpha_{\min} = \alpha_d v$.

Using a series expansion of the reciprocal, the equation for the effective stiffness may be written as

$$\mathbf{C}_T^* = \mathbf{C}^{(0)} + \sum_r v^{(r)} \mathbf{t}^{(r)} - \sum_r \sum_s v^{(r)} \mathbf{t}^{(r)} \mathbf{G}_d^{(rs)} \mathbf{t}^{(s)} v^{(s)} + O[(v^{(r)})^3] \qquad (4.15.20)$$

which agrees with Eshelby's (1957) dilute estimate to first order in concentration. When the spatial distribution is assumed to be the same for all pairs of interacting inclusions, $\mathbf{G}_d^{(rs)} = \mathbf{G}_d$ for all r and s. Under this assumption, the equation for $\mathbf{C}_T^*$ becomes

$$\mathbf{C}_T^* = \mathbf{C}^0 + \left[\sum_r v^{(r)} \mathbf{t}^{(r)} \right] \left\{ I + G_d \left[\sum_s v^{(s)} \mathbf{t}^{(s)} \right] \right\}^{-1} \qquad (4.15.21)$$

or

$$\mathbf{C}_T^* = \mathbf{C}^{(0)} + \left(\sum_r v^{(r)} \mathbf{t}^{(r)} \right) - \left(\sum_r v^{(r)} \mathbf{t}^{(r)} \right) G_d \left(\sum_s v^{(s)} \mathbf{t}^{(s)} \right) + O\left[(v^{(r)})^3 \right] \qquad (4.15.22)$$

These non-self-consistent, second-order T-matrix approximations with a well-defined $\mathbf{C}^{(0)}$ are sometimes termed the generalized **optical potential approximation** (OPA) (Jakobsen $et\ al.$, 2003a). The generalized OPA equations reduce to Nishizawa's (1982) differential effective medium formulation when the second-order correction terms are dropped, and the calculations are done incrementally, increasing the inclusion concentration in small steps, and taking $\mathbf{C}^{(0)}$ to be the effective stiffness from the previous incremental step. This formulation can be written as a system of ordinary differential equations for the effective stiffness C_{DEM}^* (Jakobsen $et\ al.$, 2000):

$$[1 - v^{(r)}]\frac{d\mathbf{C}^*_{\text{DEM}}}{dv^{(r)}} = [\mathbf{C}^{(r)} - \mathbf{C}^*_{\text{DEM}}][\mathbf{I} - \mathbf{G}^{(\text{DEM})}(\mathbf{C}^{(r)} - \mathbf{C}^*_{\text{DEM}})]^{-1} \qquad (4.15.23)$$

with initial conditions

$$\mathbf{C}^*_{\text{DEM}}(v^{(r)} = 0) = \mathbf{C}^{(0)} \qquad (4.15.24)$$

The superscript (DEM) on the tensor $\mathbf{G}$ indicates that it is computed for a medium having elastic stiffness $\mathbf{C}^*_{\text{DEM}}$.

When the multiphase aggregate does not have a clearly defined matrix material and each phase is in the form of inclusions embedded in a sea of inclusions, one can set $\mathbf{C}^{(0)} = \mathbf{C}^*_T$ to give the self-consistent *T*-matrix approximation, also called the generalized **coherent potential approximation** (CPA):

$$\left[\sum_r v^{(r)} \mathbf{t}^{(r)}_*\right]\left\{\mathbf{I} + \left[\sum_s v^{(s)} \mathbf{t}^{(s)}_*\right]^{-1}\sum_p\sum_q v^{(p)} \mathbf{t}_*^{(p)} \mathbf{G}^{(pq)}_{d^*} v^{(q)} \mathbf{t}^{(q)}_*\right\}^{-1} = 0 \qquad (4.15.25)$$

or

$$\left[\sum_r v^{(r)} \mathbf{t}^{(r)}_*\right] - \sum_p\sum_q v^{(p)} \mathbf{t}_*^{(p)} \mathbf{G}^{(pq)}_{d^*} v^{(q)} \mathbf{t}^{(q)}_* \approx 0 \qquad (4.15.26)$$

where the subscript $*$ on the tensor quantities $\mathbf{t}$ and $\mathbf{G}$ indicates that they are computed for a reference medium having the (as yet unknown) elastic stiffness $\mathbf{C}^*_T$. The generalized CPA equations are an implicit set of equations for the effective stiffness and must be solved iteratively. The generalized CPA reduces to the first-order CPA, ignoring higher-order terms:

$$\sum_r v^{(r)} \mathbf{t}^{(r)}_* \approx 0 \qquad (4.15.27)$$

which is equivalent to Berryman's symmetric self-consistent approximation (see Section 4.11).

When the *r*th set of inclusions have a spatial distribution described by an orientation distribution function $O^{(r)}(\theta, \psi, \phi)$, the *T*-matrix for the individual inclusion, $\mathbf{t}^{(r)}$ in the preceding equations for the first- and second-order corrections have to be replaced by the orientation averaged tensor $\bar{\mathbf{t}}^{(r)}$ given by

$$\bar{\mathbf{t}}^{(r)} = \int_0^\pi d\theta\,\sin\theta\int_0^{2\pi} d\psi\int_0^{2\pi} d\phi\, O^{(r)}(\theta, \psi, \phi)\, \mathbf{t}^{(r)}(\theta, \psi, \phi) \qquad (4.15.28)$$

The three Euler angle (θ, ψ, ϕ) define the orientation of the ellipsoid with principal axes $X_1 X_2 X_3$ with respect to the fixed global coordinates $x_1 x_2 x_3$, where θ is the angle between the short axis of the ellipsoid and the x_3-axis. Computing the orientation average *T*-matrix involves coordinate transformation of the inclusion stiffness tensor

$\mathbf{C}^{(r)}$ from the local coordinate system of the inclusion to the global coordinates using the usual transformation laws for the stiffness tensor.

Jakobsen *et al.* (2003b) have extended the *T*-matrix formulation to take into account fluid effects in porous rocks. For a single fully saturated ellipsoidal inclusion of the *r*th set that is allowed to communicate and exchange fluid mass with other cavities, the *T*-matrix, $\mathbf{t}_{\text{sat}}^{(r)}$ is given as the dry *T*-matrix $\mathbf{t}_{\text{dry}}^{(r)}$ plus an extra term accounting for fluid effects. The *T*-matrix for dry cavities is obtained by setting $\mathbf{C}^{(r)}$ to zero in the preceding equations. The saturated *T*-matrix is expressed as

$$\mathbf{t}_{\text{sat}}^{(r)} = \mathbf{t}_{\text{dry}}^{(r)} + \frac{\Theta \mathbf{Z}^{(r)} + i\omega\tau\kappa_{\text{f}}\mathbf{X}^{(r)}}{1 + i\omega\tau\gamma^{(r)}} \tag{4.15.29}$$

$$\mathbf{X}^{(r)} = \mathbf{t}_{\text{dry}}^{(r)}\mathbf{S}^{(0)}(\mathbf{I}_2 \otimes \mathbf{I}_2)\mathbf{S}^{(0)}\mathbf{t}_{\text{dry}}^{(r)} \tag{4.15.30}$$

$$\mathbf{Z}^{(r)} = \mathbf{t}_{\text{dry}}^{(r)}\mathbf{S}^{(0)}(\mathbf{I}_2 \otimes \mathbf{I}_2)\mathbf{S}^{(0)}\left(\sum_r \frac{v^{(r)}\mathbf{t}_{\text{dry}}^{(r)}}{1 + i\omega\tau\gamma^{(r)}}\right) \tag{4.15.31}$$

$$\Theta = \kappa_{\text{f}}\left\{\left(1 - \kappa_{\text{f}}S_{uuvv}^{(0)}\right)\left(\sum_r \frac{v^{(r)}}{1 + i\omega\tau\gamma^{(r)}}\right) + \kappa_{\text{f}}\left(\sum_r \frac{v^{(r)}\left(K_d^{(r)}\right)_{uuvv}}{1 + i\omega\tau\gamma^{(r)}}\right) - \frac{ik_uk_v\Gamma_{uv}\kappa_{\text{f}}}{\omega\eta_{\text{f}}}\right\}^{-1} \tag{4.15.32}$$

$$\mathbf{K}_d^{(r)} = (\mathbf{I}_4 + \mathbf{G}^{(r)}\mathbf{C}^{(0)})^{-1}\mathbf{S}^{(0)} \tag{4.15.33}$$

$$\gamma^{(r)} = 1 + \kappa_f\left(\mathbf{K}_d^{(r)} - \mathbf{S}^{(0)}\right)_{uuvv} \tag{4.15.34}$$

In the preceding equations, $\mathbf{S}^0 = [\mathbf{C}^{(0)}]^{-1}$ is the elastic compliance of the background medium, κ_{f} and η_{f} are the bulk modulus and viscosity of the fluid, ω is the frequency, τ is a flow-related relaxation-time constant, k_u, k_v are components of the wave vector, and Γ_{uv} is the (possibly anisotropic) permeability tensor of the rock. Repeated subscripts *u* and *v* imply summation over *u, v* = 1, 2, 3. The tensors $\mathbf{I}_2$ and $\mathbf{I}_4$ are second- and fourth-rank identity tensors, respectively. The dyadic product is denoted by $\otimes$ and $(\mathbf{I}_2 \otimes \mathbf{I}_2)_{ijkl} = \delta_{ij}\delta_{kl}$.

Usually the relaxation-time constant τ is poorly defined and has to be calibrated from experimental data. Jakobsen *et al.* (2003b) show that the first-order approximations of these expressions reduce to the Brown and Korringa results (see Section 6.5) in the low-frequency limit.

Uses

The results described in this section can be used to model effective-medium elastic properties of composites made up of anisotropic ellipsoidal inclusions in an anisotropic background. The formulation for saturated, communicating cavities can be used to model fluid-related velocity dispersion and attenuation in porous rocks.

Assumptions and Limitations

The results described in this section assume the following:
- idealized ellipsoidal inclusion shape; and
- linear elasticity and a Newtonian fluid.

4.16 Elastic Constants in Finely Layered Media: Backus Average

Synopsis

A transversely isotropic medium with the symmetry axis in the x_3-direction has an elastic stiffness tensor that can be written in the condensed Voigt matrix form (see Section 2.2 on anisotropy):

$$\begin{bmatrix} a & b & f & 0 & 0 & 0 \\ b & a & f & 0 & 0 & 0 \\ f & f & c & 0 & 0 & 0 \\ 0 & 0 & 0 & d & 0 & 0 \\ 0 & 0 & 0 & 0 & d & 0 \\ 0 & 0 & 0 & 0 & 0 & m \end{bmatrix}, \quad m = \frac{1}{2}(a - b) \tag{4.16.1}$$

where a, b, c, d, and f are five independent elastic constants. Backus (1962) showed that in the long-wavelength limit a stratified medium composed of **layers of transversely isotropic materials** (each with its symmetry axis normal to the strata) is also effectively anisotropic, with effective stiffness as follows:

$$\begin{bmatrix} A & B & F & 0 & 0 & 0 \\ B & A & F & 0 & 0 & 0 \\ F & F & C & 0 & 0 & 0 \\ 0 & 0 & 0 & D & 0 & 0 \\ 0 & 0 & 0 & 0 & D & 0 \\ 0 & 0 & 0 & 0 & 0 & M \end{bmatrix}, \quad M = \frac{1}{2}(A - B) \tag{4.16.2}$$

where

$$A = \langle a - f^2 c^{-1} \rangle + \langle c^{-1} \rangle^{-1} \langle f c^{-1} \rangle^2 \tag{4.16.3}$$

$$B = \langle b - f^2 c^{-1} \rangle + \langle c^{-1} \rangle^{-1} \langle f c^{-1} \rangle^2 \tag{4.16.4}$$

$$C = \langle c^{-1} \rangle^{-1} \tag{4.16.5}$$

$$F = \langle c^{-1} \rangle^{-1} \langle f c^{-1} \rangle \tag{4.16.6}$$

$$D = \langle d^{-1} \rangle^{-1} \tag{4.16.7}$$

$$M = \langle m \rangle \tag{4.16.8}$$

The brackets $\langle \cdot \rangle$ indicate averages of the enclosed properties weighted by their volumetric proportions. This is often called the *Backus average*.

If the **individual layers are isotropic**, the effective medium is still transversely isotropic, but the number of independent constants needed to describe each individual layer is reduced to 2:

$$a = c = \lambda + 2\mu, \ b = f = \lambda, \ d = m = \mu \tag{4.16.9}$$

giving for the effective medium

$$A = \left\langle \frac{4\mu(\lambda + \mu)}{\lambda + 2\mu} \right\rangle + \left\langle \frac{1}{\lambda + 2\mu} \right\rangle^{-1} \left\langle \frac{\lambda}{\lambda + 2\mu} \right\rangle^{2} \tag{4.16.10}$$

$$B = \left\langle \frac{2\mu\lambda}{\lambda + 2\mu} \right\rangle + \left\langle \frac{1}{\lambda + 2\mu} \right\rangle^{-1} \left\langle \frac{\lambda}{\lambda + 2\mu} \right\rangle^{2} \tag{4.16.11}$$

$$C = \left\langle \frac{1}{\lambda + 2\mu} \right\rangle^{-1} \tag{4.16.12}$$

$$F = \left\langle \frac{1}{\lambda + 2\mu} \right\rangle^{-1} \left\langle \frac{\lambda}{\lambda + 2\mu} \right\rangle \tag{4.16.13}$$

$$D = \left\langle \frac{1}{\mu} \right\rangle^{-1} \tag{4.16.14}$$

$$M = \langle \mu \rangle \tag{4.16.15}$$

In terms of the P- and S-wave velocities and densities in the isotropic layers (Levin, 1979),

$$a = \rho V_{\mathrm{P}}^{2} \tag{4.16.16}$$
$$d = \rho V_{\mathrm{S}}^{2} \tag{4.16.17}$$
$$f = \rho(V_{\mathrm{P}}^{2} - 2V_{\mathrm{S}}^{2}) \tag{4.16.18}$$

the effective parameters can be rewritten as

$$A = \left\langle 4\rho V_{\mathrm{S}}^{2} \left(1 - \frac{V_{\mathrm{S}}^{2}}{V_{\mathrm{P}}^{2}}\right) \right\rangle + \left\langle 1 - 2\frac{V_{\mathrm{S}}^{2}}{V_{\mathrm{P}}^{2}} \right\rangle^{2} \langle (\rho V_{\mathrm{P}}^{2})^{-1} \rangle^{-1} \tag{4.16.19}$$

$$B = \left\langle 2\rho V_{\mathrm{S}}^{2} \left(1 - \frac{2V_{\mathrm{S}}^{2}}{V_{\mathrm{P}}^{2}}\right) \right\rangle + \left\langle 1 - 2\frac{V_{\mathrm{S}}^{2}}{V_{\mathrm{P}}^{2}} \right\rangle^{2} \langle (\rho V_{\mathrm{P}}^{2})^{-1} \rangle^{-1} \tag{4.16.20}$$

$$C = \langle (\rho V_{\mathrm{P}}^{2})^{-1} \rangle^{-1} \tag{4.16.21}$$

$$F = \left\langle 1 - 2\frac{V_{\mathrm{S}}^{2}}{V_{\mathrm{P}}^{2}} \right\rangle \langle (\rho V_{\mathrm{P}}^{2})^{-1} \rangle^{-1} \tag{4.16.22}$$

$$D = \langle (\rho V_{\mathrm{S}}^{2})^{-1} \rangle^{-1} \tag{4.16.23}$$

$$M = \langle \rho V_{\mathrm{S}}^{2} \rangle \tag{4.16.24}$$

The P- and S-wave velocities in the effective anisotropic medium can be written as

$$V_{\mathrm{SH,h}} = \sqrt{M/\rho} \tag{4.16.25}$$

$$V_{SH,v} = V_{SV,h} = V_{SV,v} = \sqrt{D/\rho}$$ (4.16.26)

$$V_{P,h} = \sqrt{A/\rho}$$ (4.16.27)

$$V_{P,v} = \sqrt{C/\rho}$$ (4.16.28)

where ρ is the average density; $V_{P,v}$ is for the vertically propagating P-wave; $V_{P,h}$ is for the horizontally propagating P-wave; $V_{SH,h}$ is for the horizontally propagating, horizontally polarized S-wave; $V_{SV,h}$ is for the horizontally propagating, vertically polarized S-wave; and $V_{SV,v}$ and $V_{SH,v}$ are for the vertically propagating S-waves of any polarization (vertical is defined as being normal to the layering).

Example: Calculate the effective anisotropic elastic constants and the velocity anisotropy for a thinly layered sequence of dolomite and shale with the following layer properties:

$$V_{P(1)} = 5200 \text{ m/s}, \quad V_{S(1)} = 2700 \text{ m/s}, \quad \rho_1 = 2450 \text{ kg/m}^3, \quad h_1 = 0.75 \text{ m}$$
$$V_{P(2)} = 2900 \text{ m/s}, \quad V_{S(2)} = 1400 \text{ m/s}, \quad \rho_2 = 2340 \text{ kg/m}^3, \quad h_2 = 0.5 \text{ m}$$

The volumetric fractions are
$$f_1 = h_1/(h_1 + h_2) = 0.6, \quad f_2 = h_2/(h_1 + h_2) = 0.4$$
If one takes the volumetric weighted averages of the appropriate properties,

$$A = f_1 4\rho_1 V_{S(1)}^2 \left(1 - \frac{V_{S(1)}^2}{V_{P(1)}^2}\right) + f_2 4\rho_2 V_{S(2)}^2 \left(1 - \frac{V_{S(2)}^2}{V_{P(2)}^2}\right)$$
$$+ \left[f_1 \left(1 - 2\frac{V_{S(1)}^2}{V_{P(1)}^2}\right) + f_2 \left(1 - 2\frac{V_{S(2)}^2}{V_{P(2)}^2}\right)\right]^2 \frac{1}{f_1/\left(\rho_1 V_{P(1)}^2\right) + f_2/\left(\rho_2 V_{P(2)}^2\right)}$$
$$A = 45.1 \text{ GPa}$$

Similarly, computing the other averages, we obtain $C = 34.03$ GPa, $D = 8.28$ GPa, $M = 12.55$ GPa, $F = 16.7$ GPa, and $B = A - 2M = 20$ GPa.

The average density $\rho = f_1\rho_1 + f_2\rho_2 = 2406 \text{ kg/m}^3$. The anisotropic velocities are:

$$V_{SH,h} = \sqrt{M/\rho} = 2284.0 \text{ m/s}$$
$$V_{SH,v} = V_{SV,h} = V_{SV,v} = \sqrt{D/\rho} = 1854.8 \text{ m/s}$$
$$V_{P,v} = \sqrt{A/\rho} = 4329.5 \text{ m/s}$$
$$V_{P,v} = \sqrt{C/\rho} = 3761.0 \text{ m/s}$$

P-wave anisotropy = $(4329.5 - 3761.0)/3761.0 \approx 15\%$
S-wave anisotropy = $(2284.0 - 1854.8)/1854.8 \approx 23\%$

Finally, consider the case in which **each layer is isotropic with the same shear modulus** but with a different bulk modulus. This might be the situation, for example, for a massive, homogeneous rock with fine layers of different fluids or saturations. Then, the elastic constants of the medium become

$$A = C = \left\langle \frac{1}{\rho V_P^2} \right\rangle^{-1} = \left\langle \frac{1}{K + \frac{4}{3}\mu} \right\rangle^{-1} \tag{4.16.29}$$

$$B = F = \left\langle \frac{1}{\rho V_P^2} \right\rangle^{-1} - 2\mu = \left\langle \frac{1}{K + \frac{4}{3}\mu} \right\rangle^{-1} - 2\mu = A - 2\mu \tag{4.16.30}$$

$$D = M = \mu \tag{4.16.31}$$

A finely layered medium of isotropic layers, all having the same shear modulus, is isotropic. (see Section 4.6)

Bakulin and Grechka (2003) show that any effective property, $\mathbf{m}^{\text{eff}}$, of a finely-layered medium (in fact, any heterogeneous medium) can be written as

$$\mathbf{F}[\mathbf{m}^{\text{eff}}] = \frac{1}{V}\int_V \mathbf{F}[\mathbf{m}(\mathbf{x})]d\mathbf{x} \tag{4.16.32}$$

where $\mathbf{F}$ is the appropriate averaging operator, $\mathbf{m}$ is the parameter (tensor, vector, or scalar) of interest (e.g., the elastic stiffnesses, c_{ij}, or Thomsen parameters, ε, δ, γ), V is the representative volume, and $\mathbf{x}$ denotes the Cartesian coordinates over which $\mathbf{m}$ varies. For example, from the Backus average one can write

$$c_{66}^{\text{eff}} = \frac{1}{V}\int_V c_{66}(\mathbf{x})d\mathbf{x} \tag{4.16.33}$$

where $\mathbf{F}$ is the scalar identity function,

$$c_{44}^{\text{eff}} = \left[\frac{1}{V}\int_V [c_{44}(\mathbf{x})]^{-1}d\mathbf{x} \right]^{-1} \tag{4.16.34}$$

where $\mathbf{F}[\mathbf{m}] = \mathbf{m}^{-1}$, and so on. Some other examples of averaging common in rock physics are shown in Table 4.16.1.

An important consequence is that

$$\mathbf{m}^{\text{eff}} = \overline{\mathbf{m}} + O(\widetilde{\mathbf{m}}^2) \tag{4.16.35}$$

where

$$\overline{\mathbf{m}} = \frac{1}{V}\int_V \mathbf{m}(\mathbf{x})d\mathbf{x} \qquad \widetilde{\mathbf{m}} = \mathbf{m} - \overline{\mathbf{m}} \tag{4.16.36}$$

In other words, the effective property of a finely layered medium is, to first order, just the arithmetic average of the individual layer properties. The effect of the layer contrasts is second order in the deviations from the mean.

A composite of fine isotropic layers with weak layer contrasts is, to first order, isotropic (Bakulin, 2003). The well-known anisotropy resulting from the Backus average is second order. When the layers are intrinsically anisotropic, as with shales,

Table 4.16.1 *Examples of averaging equations common in rock physics*

	operator	Effective quantity
Arithmetic (Voigt)	$\mathbf{F}_V(\mathbf{m}) = \mathbf{m}$	$\mathbf{C}_V = \dfrac{1}{V}\displaystyle\int_V \mathbf{C}(\mathbf{x})d\mathbf{x}$
Harmonic (Reuss)	$\mathbf{F}_R(\mathbf{m}) = \mathbf{m}^{-1}$	$\mathbf{C}_R = \left[\dfrac{1}{V}\displaystyle\int_V \mathbf{C}^{-1}(\mathbf{x})d\mathbf{x}\right]^{-1}$
Dix equation	$\mathbf{F}_D(V) = V^2$	$V_D = \left[\dfrac{1}{T}\displaystyle\int_0^T V^2(t)dt\right]^{1/2}$
Hashin-Shtrikman (bulk modulus)	$F_{HS} = \dfrac{1}{K+z}$	$K_{HS} = \left[\dfrac{1}{V}\displaystyle\int_V [K+z]^{-1}d\mathbf{x}\right]^{-1} - z$

the first-order anisotropy is the average of the layer intrinsic anisotropies, and the anisotropy due to the layer contrasts is often less important, especially when the layer contrasts are small.

Uses

The Backus average is used to model a finely stratified medium as a single homogeneous medium.

Assumptions and Limitations

The following presuppositions and conditions apply to the Backus average:

- all materials are linearly elastic;
- there are no sources of intrinsic energy dissipation, such as friction or viscosity; and
- the layer thickness must be much smaller than the seismic wavelength. How small is still a subject of disagreement and research, but a rule of thumb is that the wavelength must be at least ten times the layer thickness.

4.17   Elastic Constants in Finely Layered Media: General Layer Anisotropy

Techniques for calculation of the effective elastic moduli of a finely layered one-dimensional medium were first developed by Riznichenko (1949) and Postma (1955). Backus (1962) further developed the method and introduced the well-known Backus average of isotropic or VTI layers (see Section 4.16). Helbig and Schoenberg (1987) and Schoenberg and Muir (1989) presented the results for layers with arbitrary anisotropy in convenient matrix form, shown here. Assuming that the $x3$-axis is normal to the layering, one can define the matrices

$$
\mathbf{C}_{NN}^{(i)} = \begin{bmatrix} c_{33}^{(i)} & c_{34}^{(i)} & c_{35}^{(i)} \\ c_{34}^{(i)} & c_{44}^{(i)} & c_{45}^{(i)} \\ c_{35}^{(i)} & c_{45}^{(i)} & c_{55}^{(i)} \end{bmatrix} \tag{4.17.1}
$$

$$
\mathbf{C}_{TN}^{(i)} = \begin{bmatrix} c_{13}^{(i)} & c_{14}^{(i)} & c_{15}^{(i)} \\ c_{23}^{(i)} & c_{24}^{(i)} & c_{25}^{(i)} \\ c_{36}^{(i)} & c_{46}^{(i)} & c_{56}^{(i)} \end{bmatrix} \tag{4.17.2}
$$

$$
\mathbf{C}_{TT}^{(i)} = \begin{bmatrix} c_{11}^{(i)} & c_{12}^{(i)} & c_{16}^{(i)} \\ c_{12}^{(i)} & c_{22}^{(i)} & c_{26}^{(i)} \\ c_{16}^{(i)} & c_{26}^{(i)} & c_{66}^{(i)} \end{bmatrix} \tag{4.17.3}
$$

where the $c_{kl}^{(i)}$ are the Voigt-notation elastic constants of the ith layer. Then the effective medium having the same stress–strain behavior of the composite has the corresponding matrices

$$
\overline{\mathbf{C}}_{NN} = \begin{bmatrix} \overline{c}_{33} & \overline{c}_{34} & \overline{c}_{35} \\ \overline{c}_{34} & \overline{c}_{44} & \overline{c}_{45} \\ \overline{c}_{35} & \overline{c}_{45} & \overline{c}_{55} \end{bmatrix} = \langle \mathbf{C}_{NN}^{-1} \rangle^{-1} \tag{4.17.4}
$$

$$
\overline{\mathbf{C}}_{TN} = \begin{bmatrix} \overline{c}_{13} & \overline{c}_{14} & \overline{c}_{15} \\ \overline{c}_{23} & \overline{c}_{24} & \overline{c}_{25} \\ \overline{c}_{36} & \overline{c}_{46} & \overline{c}_{56} \end{bmatrix} = \langle \mathbf{C}_{TN} \mathbf{C}_{NN}^{-1} \rangle \overline{\mathbf{C}}_{NN} \tag{4.17.5}
$$

$$
\overline{\mathbf{C}}_{TT} = \begin{bmatrix} \overline{c}_{11} & \overline{c}_{12} & \overline{c}_{16} \\ \overline{c}_{12} & \overline{c}_{22} & \overline{c}_{26} \\ \overline{c}_{16} & \overline{c}_{26} & \overline{c}_{66} \end{bmatrix} = \langle \mathbf{C}_{TT} \rangle - \langle \mathbf{C}_{TN} \mathbf{C}_{NN}^{-1} \mathbf{C}_{NT} \rangle + \langle \mathbf{C}_{TN} \mathbf{C}_{NN}^{-1} \rangle \overline{\mathbf{C}}_{NN} \langle \mathbf{C}_{NN}^{-1} \mathbf{C}_{NT} \rangle
$$

$$
\tag{4.17.6}
$$

where the $\overline{c}_{kl}$ terms are the effective elastic constants. In the preceding equations, the operator $\langle \cdot \rangle$ denotes a thickness-weighted average over the layers.

4.18 Poroelastic and Viscoelastic Backus Average

Synopsis

For a medium consisting of thin poroelastic layers with different saturating fluids, Gelinsky and Shapiro (1997a) give the expressions for the poroelastic constants of the effective homogeneous anisotropic medium. This is equivalent to the Backus average for thin elastic layers (see Section 4.16). At very low frequencies (quasi-static limit) seismic waves cause interlayer flow across the layer boundaries. At higher frequencies, layers behave as if they are sealed with no flow across the boundaries, and this scenario is equivalent to elastic Backus averaging with saturated rock moduli obtained from Gassmann's relations for individual layers.

When the individual poroelastic layers are themselves transversely anisotropic with a vertical axis of symmetry, x_3, normal to the layers, each layer is described by five independent stiffness constants, b_d, c_d, f_d, d_d, and m_d, and three additional poroelastic constants, p, q, and r, describing the coupling between fluid flow and elastic deformation. The five stiffness constants correspond to the components of the dry-frame stiffness tensor. In the usual two-index Voigt notation, they are: $b_d = C_{12}^d$, $c_d = C_{33}^d$, $f_d = C_{13}^d$, $d_d = C_{44}^d = C_{55}^d$, where the sub- and superscript "d" indicates dry frame. The constants p, q, and r for the individual layers can be expressed in terms of the dry anisotropic frame moduli and fluid properties as follows:

$$p = \alpha_1 r \qquad q = \alpha_2 r \tag{4.18.1}$$

$$r = K_0 \left[\left(1 - \frac{K_d}{K_0} \right) - \phi \left(1 - \frac{K_0}{K_f} \right) \right]^{-1} \tag{4.18.2}$$

$$\alpha_1 = 1 - (C_{11}^d + C_{12}^d + C_{13}^d)/(3K_0) \tag{4.18.3}$$

$$\alpha_2 = 1 - (2C_{13}^d + C_{33}^d)/(3K_0) \tag{4.18.4}$$

$$K_d = \frac{1}{9} C_{iijj} = \text{dry bulk modulus}$$

$K_0 = $ mineral bulk modulus

$K_f = $ pore fluid bulk modulus

$\phi = $ porosity

In the quasi-static limit, the effective homogeneous poroelastic medium is itself transversely anisotropic with effective constants given by (Gelinsky and Shapiro, 1997a):

$$B = \langle b_d - f_d^2 c_d^{-1} \rangle + \langle f_d c_d^{-1} \rangle^2 \langle c_d^{-1} \rangle^{-1} + \frac{P^2}{R} \tag{4.18.5}$$

$$C = \langle c_d^{-1} \rangle^{-1} + \frac{Q^2}{R} \tag{4.18.6}$$

$$F = \langle c_d^{-1} \rangle^{-1} \langle f_d c_d^{-1} \rangle + \frac{PQ}{R} \tag{4.18.7}$$

$$D = \langle d_d^{-1} \rangle^{-1} \qquad M = \langle m_d \rangle \tag{4.18.8}$$

$$P = R \left(\left\langle \frac{p}{r} \right\rangle - \left\langle \frac{q}{r} f_d c_d^{-1} \right\rangle + \left\langle \frac{q}{r} c_d^{-1} \right\rangle \left\langle c_d^{-1} \right\rangle^{-1} \left\langle f_d c_d^{-1} \right\rangle \right) \tag{4.18.9}$$

$$Q = R \left\langle \frac{q}{r} c_d^{-1} \right\rangle \langle c_d^{-1} \rangle^{-1} \tag{4.18.10}$$

$$R = \left[\langle r^{-1} \rangle + \left\langle \frac{q^2}{r^2} c_{\mathrm{d}}^{-1} \right\rangle - \left\langle \frac{q}{r} c_{\mathrm{d}}^{-1} \right\rangle^2 \langle c_{\mathrm{d}}^{-1} \rangle^{-1} \right]^{-1} \tag{4.18.11}$$

The symbol $\langle \cdot \rangle$ indicates volumetric averages of the enclosed properties. When the individual thin layers are transversely isotropic with a horizontal symmetry axis (x_1), the effective medium has orthorhombic symmetry and is described by the stiffness tensor A_{ij} with nine independent constants and four additional Biot parameters, P_1, P_2, P_3, and R. The layers themselves are characterized by the transversely isotropic coefficients, $b_{\mathrm{d}}, c_{\mathrm{d}}, f_{\mathrm{d}}, d_{\mathrm{d}}$ and m_{d} (and $a_{\mathrm{d}} = 2m_{\mathrm{d}} + b_{\mathrm{d}}$), and three additional poroelastic constants, $p, q,$ and r, with x_1 as the symmetry axis instead of x_3. The effective constants are given by (Gelinsky and Shapiro, 1997a):

$$A_{11} = \langle c_{\mathrm{d}} \rangle + \left\langle \frac{f_{\mathrm{d}}}{a_{\mathrm{d}}} \right\rangle^2 \langle a_{\mathrm{d}}^{-1} \rangle^{-1} - \left\langle \frac{f_{\mathrm{d}}^2}{a_{\mathrm{d}}} \right\rangle + \frac{P_1^2}{R} \tag{4.18.12}$$

$$A_{22} = \langle a_{\mathrm{d}} \rangle + \left\langle \frac{b_{\mathrm{d}}}{a_{\mathrm{d}}} \right\rangle^2 \langle a_{\mathrm{d}}^{-1} \rangle^{-1} - \left\langle \frac{b_{\mathrm{d}}^2}{a_{\mathrm{d}}} \right\rangle + \frac{P_2^2}{R} \tag{4.18.13}$$

$$A_{33} = \langle a_{\mathrm{d}}^{-1} \rangle^{-1} + \frac{P_3^2}{R} \tag{4.18.14}$$

$$A_{12} = \langle f_{\mathrm{d}} \rangle + \left\langle \frac{f_{\mathrm{d}}}{a_{\mathrm{d}}} \right\rangle \left\langle \frac{b_{\mathrm{d}}}{a_{\mathrm{d}}} \right\rangle \langle a_{\mathrm{d}}^{-1} \rangle^{-1} - \left\langle \frac{f_{\mathrm{d}} b_{\mathrm{d}}}{a_{\mathrm{d}}} \right\rangle + \frac{P_1 P_2}{R} \tag{4.18.15}$$

$$A_{13} = \left\langle \frac{f_{\mathrm{d}}}{a_{\mathrm{d}}} \right\rangle \langle a_{\mathrm{d}}^{-1} \rangle^{-1} + \frac{P_1 P_3}{R} \tag{4.18.16}$$

$$A_{23} = \left\langle \frac{b_{\mathrm{d}}}{a_{\mathrm{d}}} \right\rangle \langle a_{\mathrm{d}}^{-1} \rangle^{-1} + \frac{P_2 P_3}{R} \tag{4.18.17}$$

$$A_{44} = \left\langle \frac{1}{m_{\mathrm{d}}} \right\rangle^{-1} \tag{4.18.18}$$

$$A_{55} = \left\langle \frac{1}{d_{\mathrm{d}}} \right\rangle^{-1} \tag{4.18.19}$$

$$A_{66} = \langle d_{\mathrm{d}} \rangle \tag{4.18.20}$$

$$P_1 = R \left(\left\langle \frac{q}{r} \right\rangle - \left\langle \frac{p f_{\mathrm{d}}}{r a_{\mathrm{d}}} \right\rangle + \left\langle \frac{p}{r a_{\mathrm{d}}} \right\rangle \left\langle \frac{f_{\mathrm{d}}}{a_{\mathrm{d}}} \right\rangle \langle a_{\mathrm{d}}^{-1} \rangle^{-1} \right) \tag{4.18.21}$$

$$P_2 = R \left(\left\langle \frac{p}{r} \right\rangle - \left\langle \frac{p b_{\mathrm{d}}}{r a_{\mathrm{d}}} \right\rangle + \left\langle \frac{p}{r a_{\mathrm{d}}} \right\rangle \left\langle \frac{b_{\mathrm{d}}}{a_{\mathrm{d}}} \right\rangle \langle a_{\mathrm{d}}^{-1} \rangle^{-1} \right) \tag{4.18.22}$$

$$P_3 = R \left(\left\langle \frac{p}{r a_{\mathrm{d}}} \right\rangle \langle a_{\mathrm{d}}^{-1} \rangle^{-1} \right) \tag{4.18.23}$$

$$R = \left(\left\langle \frac{1}{r} \right\rangle + \left\langle \frac{p^2}{r^2 a_d} \right\rangle - \left\langle \frac{p}{r a_d} \right\rangle^2 \langle a_d^{-1} \rangle^{-1} \right)^{-1}$$ (4.18.24)

When the individual thin layers are isotropic, the effective constants for the equivalent transversely isotropic homogeneous poroelastic medium can be expressed in terms of averages of the Lamé constants λ_d and μ_d of the dry frame of the layers. For isotropic layers, $\alpha_1 = \alpha_2 = \alpha = 1 - K_d/K_0$. In the quasi-static limit, the effective poroelastic constants are:

$$B = 2 \left\langle \frac{\lambda_d \mu_d}{\lambda_d + 2\mu_d} \right\rangle + \left\langle \frac{\lambda_d}{\lambda_d + 2\mu_d} \right\rangle^2 \left\langle \frac{1}{\lambda_d + 2\mu_d} \right\rangle^{-1} + \frac{P^2}{R}$$ (4.18.25)

$$C = \left\langle \frac{1}{\lambda_d + 2\mu_d} \right\rangle^{-1} + \frac{Q^2}{R}$$ (4.18.26)

$$F = \left\langle \frac{1}{\lambda_d + 2\mu_d} \right\rangle^{-1} \left\langle \frac{\lambda_d}{\lambda_d + 2\mu_d} \right\rangle + \frac{PQ}{R}$$ (4.18.27)

$$D = \langle \mu_d^{-1} \rangle^{-1} \qquad M = \langle \mu_d \rangle$$ (4.18.28)

$$P = -R \left(2 \left\langle \frac{\alpha \mu_d}{\lambda_d + 2\mu_d} \right\rangle + \left\langle \frac{\alpha}{\lambda_d + 2\mu_d} \right\rangle \left\langle \frac{\lambda_d}{\lambda_d + 2\mu_d} \right\rangle \left\langle \frac{1}{\lambda_d + 2\mu_d} \right\rangle^{-1} \right)$$ (4.18.29)

$$Q = -R \left\langle \frac{\alpha}{\lambda_d + 2\mu_d} \right\rangle \left\langle \frac{1}{\lambda_d + 2\mu_d} \right\rangle^{-1}$$ (4.18.30)

$$R = \left[\langle r^{-1} \rangle + \left\langle \frac{\alpha^2}{\lambda_d + 2\mu_d} \right\rangle - \left\langle \frac{\alpha}{\lambda_d + 2\mu_d} \right\rangle^2 \left\langle \frac{1}{\lambda_d + 2\mu_d} \right\rangle^{-1} \right]^{-1}$$ (4.18.31)

Long Wavelength Viscoelastic Backus Average

In the long-wavelength limit, the effective attenuation (Q^{-1}), P-wave modulus, and velocity for waves travelling normal to a stratified medium composed of linear viscoelastic layers can be obtained by applying the correspondence principle of viscoelasticity theory to the Backus average for linear elastic stratified media. The resulting complex frequency-dependent effective P-wave modulus in the long-wavelength limit (wavelength much longer than layer thickness) is given as (Das *et al.*, 2019):

$$\frac{1}{M_{eff}^*(\omega)} = \sum_{j=1}^{N} \frac{f_j}{M_j^*(\omega)} \quad j = 1, 2 \ldots, N \tag{4.18.32}$$

where

$M_{eff}^*(\omega)$ = complex P-wave effective modulus at frequency ω

f_j = volume fraction of the jth layer

$M_j^*(\omega)$ = complex P-wave modulus for j^{th} layer at frequency ω

N = total number of layers

The effective viscoelastic attenuation, expressed by the inverse quality factor $Q_{eff}^{-1}(\omega)$ in the long wavelength limit, is given as:

$$Q_{eff}^{-1}(\omega) = \frac{\displaystyle\sum_{j=1}^{N} \frac{f_j Q_j^{-1}(\omega)}{Re(M_j(\omega))(1 + (Q_j^{-1}(\omega))^2)}}{\displaystyle\sum_{j=1}^{N} \frac{f_j}{Re(M_j(\omega))(1 + (Q_j^{-1}(\omega))^2)}} \tag{4.18.33}$$

where

$$Q_j^{-1}(\omega) = \frac{Im(M_j(\omega))}{Re(M_j(\omega))} \tag{4.18.34}$$

is the intrinsic attenuation of the jth constituent layer.

The effective velocity for propagation normal to the layers is given as:

$$V_{eff}(\omega) = V_o \sqrt{\frac{2\left(1 + \left(Q_{eff}^{-1}(\omega)\right)^2\right)}{1 + \sqrt{1 + \left(Q_{eff}^{-1}(\omega)\right)^2}}} \tag{4.19.35}$$

where $V_o = \sqrt{\dfrac{Re(M_{eff}(\omega))}{\sum f_k \rho_k}}$ and ρ_k is the density of each layer.

Reformulating the results in terms of the velocity ($V_j(\omega)$), density (ρ_j) and inverse quality factor ($Q_j^{-1}(\omega)$) of the constituent viscoelastic layers, the effective attenuation and velocity for normal-incidence propagation are given as:

$$Q_{\mathit{eff}}^{-1}(\omega) = \frac{\sum\limits_{j=1}^{N} \dfrac{f_j Q_j^{-1}(\omega)}{\rho_j (Re(V_j(\omega)))^2 (1 + (Q_j^{-1}(\omega))^2)}}{\sum\limits_{j=1}^{N} \dfrac{f_j}{\rho_j (Re(V_j(\omega)))^2 (1 + (Q_j^{-1}(\omega))^2)}}$$

(4.19.36)

$$V_{\mathit{eff}}(\omega) = V_o \sqrt{\frac{2(1 + (Q_{\mathit{eff}}^{-1}(\omega))^2)}{1 + \sqrt{1 + (Q_{\mathit{eff}}^{-1}(\omega))^2}}}$$

(4.19.37)

$$V_o = \sqrt{\frac{\left[\left(\sum\limits_{j=1}^{N} \left(\dfrac{f_j}{\rho_j (Re(V_j(\omega)))^2 (1 + (Q_j^{-1}(\omega))^2)} \right) \right) \left(1 + \left(Q_{\mathit{eff}}^{-1}(\omega) \right)^2 \right) \right]^{-1}}{\sum\limits_{j=1}^{N} f_j \rho_j}}$$

(4.19.38)

In the limit when the layers become perfectly elastic, $Q^{-1} = 0$. These equations reduce to Backus' results for elastic layered media.

These equations apply in the effective medium- or long-wavelength limit, and are hence valid only when layer thicknesses are much smaller than the wavelength. Typical rule of thumb would require the wavelength to be greater than 10 times the layer thickness. These equations apply for normal-incidence propagation. All constituent layers are assumed to be isotropic, linear viscoelastic media.

Uses

These equations can be used to model a finely stratified heterogeneous poroelastic or viscoelastic medium as an effective homogeneous anisotropic medium.

Assumptions and Limitations

- All materials are linear and poroelastic or viscoelastic. The viscoelastic equations are for waves propagating normal to the layering.
- Layer thicknesses are much smaller than the seismic wavelength. A rule of thumb is that the wavelength must be at least 10 times the layer thickness.

4.19   Seismic Response to Fractures

Synopsis

One approach to modeling the effects of fractures on the elastic properties of rocks is to represent them as thin layers for which the elastic moduli are smaller (softer) than those of the unfractured background rock. The soft fracture layer modulus represents the

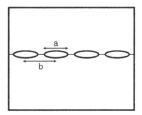

Figure 4.19.1 Representation of a partially open fracture with coplanar two-dimensional elliptical cracks

extra compliance of the fracture relative to the background rock. When the seismic wavelength is long compared with the layer thickness and fracture spacing, then the overall elastic response of a parallel array of such fracture layers can be estimated using the Backus (1962) average (Section 4.16) (Hudson *et al.*, 1996; Schoenberg, 1980; Schoenberg and Douma, 1988; Hudson and Liu, 1999).

If the thickness of the weak fracture layer is h, and the fracture layer P-wave modulus and shear modulus are M_f and μ_f, respectively, then the total normal and shear displacements across the thin layer under normal and shear stresses, σ_n and σ_τ, are

$$\delta_n = \sigma_n h / M_f \qquad \delta_\tau = \sigma_\tau h / (2\mu_f) \tag{4.19.1}$$

The fracture normal and shear compliances, Z_n and Z_τ, can be defined as

$$\frac{1}{Z_n} = \frac{\partial \sigma_n}{\partial \delta_n} = \frac{M_f}{h}, \quad \frac{1}{Z_\tau} = \frac{\partial \sigma_\tau}{\partial \delta_\tau} = \frac{2\mu_f}{h} \tag{4.19.2}$$

where Z_n and Z_τ have units of (stress/length)$^{-1}$. By defining these fracture compliances, then at long wavelengths the fracture layer can be thought of as a plane of weakness, across which displacement discontinuities, δ_n and δ_τ, occur when stress is applied. Hence, these are often referred to as *displacement discontinuity models*.

Quasi-Static Fracture Compliance

Both empirical and model-based methods have been applied to estimating values of fracture compliance (and its reciprocal the *fracture stiffness*). Jaeger *et al.* (2007) give an excellent summary. Myer (2000) represented joint deformation using the solution for the deformation of coplanar two-dimensional elliptical cavities (Figure 4.19.1) from Sneddon and Lowengrub (1969):

$$Z_n = \frac{d\delta_n}{d\sigma_n} = \frac{4(1-\nu)a}{\pi\mu(1-r)} \ln \sec\left[\frac{\pi}{2}(1-r)\right] \tag{4.19.3}$$

where μ is the rock shear modulus, ν is the Poisson ratio, r is the fractional area of contact of the crack faces, a is the crack length, and $b = a/(1-r)$ is the center-to-center crack spacing. With this model, crack closure under stress can be modeled using a

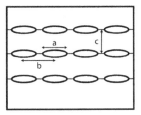

Figure 4.19.2 Representation of parallel partially open fractures with a doubly periodic array of two-dimensional elliptical cracks

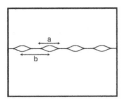

Figure 4.19.3 Representation of a partially open fracture with nonelliptical two-dimensional cracks with tapered ends

distribution of crack aspect ratios, α, with a corresponding distribution of closing stresses (Section 2.10)

$$\sigma_{\text{close}} \approx \frac{\pi}{2(1-v)} \alpha \mu_0 \tag{4.19.4}$$

Hence, as normal compressive stress increases, a fraction of the cracks close, the contact area grows, and the fracture stiffens. Delameter *et al.* (1974) solved the related problem of the deformation of a doubly periodic array of cracks (Figure 4.19.2), which represents an extension of the Myer model to a set of parallel fractures. Mavko and Nur (1978) modeled the deformation of two-dimensional nonelliptical cracks (Figure 4.19.3). They showed that the deformation under infinitesimal increments of stress is identical to the deformation of an elliptical crack of the same length; hence, the crack stiffness of the models in Figures 4.19.1 and 4.19.3 will be the same. An important difference of the nonelliptical cracks is that under finite compressive stress, the cracks shorten, leading to increased contact area and fracture stiffening. Mavko and Nur also gave the solution for a nonelliptical crack that makes contact at a midpoint (Figure 4.19.4) before closing.

Another set of theoretical models represents the fracture as a distribution of Hertzian contacts (Section 5.5) of random heights (Greenwood and Williamson, 1966; White, 1983; Brown and Scholz, 1986; Jaeger *et al.*, 2007).

Based on laboratory measurements of joint closure, Goodman (1976) proposed an empirical stress–displacement function

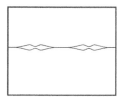

Figure 4.19.4 Representation of a partially open fracture with nonelliptical two-dimensional cracks that make multiple contacts under normal compression

$$\sigma_n = \sigma_0 \left[1 + \left(\frac{\delta_n}{\delta_m - \delta_n} \right)^t \right]; \quad 0 \le \delta_n < \delta_m \tag{4.19.5}$$

where σ_0 is an initial reference stress at $\delta_n = 0$, δ_m is the maximum possible closure, and t is an empirical exponent. Bandis *et al.* (1983) proposed a simpler form

$$\sigma_n = \frac{1}{Z_0} \frac{\delta_m \delta_n}{\delta_m - \delta_n} \tag{4.19.6}$$

leading to the fracture compliance

$$\frac{1}{Z_n} = \frac{d\sigma_n}{d\delta_n} = \frac{1}{Z_0} \left(1 + \frac{Z_0 \sigma_n}{\delta_m} \right)^2 \tag{4.19.7}$$

where Z_0 is the compliance at zero stress.

As discussed in Section 4.13, Hudson and Liu (1999) estimated the compliance of rough joint faces in contact using two models. In one model they represent the fracture as a planar distribution of small circular cracks with first-order normal and shear compliances equal to

$$Z_n = \frac{\gamma^S a^3}{\mu} U_3 = \frac{ra}{\pi\mu} U_3 \tag{4.19.8}$$

$$Z_\tau = \frac{\gamma^S a^3}{\mu} U_1 = \frac{ra}{\pi\mu} U_1 \tag{4.19.9}$$

where γ^S is the number of cracks per unit area of the fracture plane, a is the crack radius, r is the fraction of the fracture plane that is not in contact, and μ is the unfractured rock shear modulus. For dry cracks

$$U_1 = \frac{16(\lambda + 2\mu)}{3(3\lambda + 4\mu)} \qquad U_3 = \frac{4(\lambda + 2\mu)}{3(\lambda + \mu)} \tag{4.19.10}$$

Hence for dry fractures,

$$\left. \frac{Z_n}{Z_\tau} \right|_{dry} = 1 - \frac{\nu}{2} \tag{4.19.11}$$

where v is the Poisson ratio of the unfractured rock, or $Z_{n-dry} \approx Z_{\tau-dry}$. For "weak"

inclusions (i.e., when $\mu a / \left[K' + \dfrac{4}{3}\mu' \right]$ is of the order of 1 and is not small enough to be

neglected)

$$U_1 = \frac{16(\lambda + 2\mu)}{3(3\lambda + 4\mu)} \frac{1}{(1 + M)} \tag{4.19.12}$$

$$U_3 = \frac{4(\lambda + 2\mu)}{3(\lambda + \mu)} \frac{1}{(1 + \kappa)} \tag{4.19.13}$$

where

$$M = \frac{4\mu'}{\pi a \mu} \frac{(\lambda + 2\mu)}{(3\lambda + 4\mu)} \tag{4.19.14}$$

$$\kappa = \frac{\left[K' + \frac{4}{3}\mu' \right](\lambda + 2\mu)}{\pi a \mu (\lambda + \mu)} \tag{4.19.15}$$

and K' and μ' are the bulk and shear modulus of the inclusion material. Hence, for fluid-filled fractures ($\mu' = 0$)

$$\left. \frac{Z_\tau}{Z_n} \right|_{wet} = \left. \frac{Z_\tau}{Z_n} \right|_{dry} \frac{1 + \kappa}{1 + M} \tag{4.19.16}$$

In general, shear fracture compliance models depend very much on the amount of shear displacement. Large shear displacement is thought to involve interference of asperities on opposite fracture faces (Jaeger *et al.*, 2007). At small normal stresses, asperities might slide over one another, causing dilatancy in the fracture zone. At larger normal stresses, asperities need to fail in either brittle or plastic fashion to accommodate slip.

Infinitesimal shear deformation in the fault zone can be modeled analogously to the normal deformation discussed above. Open cracks undergo shear deformation, while welded or frictional contacts have zero slip. The two-dimensional models for normal fracture deformation illustrated in Figures 4.19.1–4.19.4 can be scaled for shear deformation. For example, using dislocation analysis (Section 2.10), both normal and in-plane shear of two-dimensional cracks can be represented using edge dislocations, and hence the normal and shear fracture stiffnesses are equal for dry fractures. In general, a reasonable estimate of the infinitesimal shear compliance can be obtained from the infinitesimal normal compliance using the Hudson and Liu expressions for Z_τ / Z_n.

Dynamic Response

The displacement discontinuity model has been applied to derive the dynamic response of waves impinging on the fracture plane (Kendall and Tabor, 1971; Schoenberg, 1980; Angel and Achenbach, 1985; Pyrak-Nolte *et al.*, 1990). Schoenberg (1980) gives expressions for the frequency-dependent P-wave reflection $R(\omega)$ and transmission $T(\omega)$ coefficients for normal incidence:

$$R(\omega) = \frac{-i\varpi}{1 - i\varpi} \qquad (4.19.17)$$

$$T(\omega) = \frac{1}{1 - i\varpi} \qquad (4.19.18)$$

$$\varpi = \left(\frac{\omega \rho Z_n V_P}{2} \right) \qquad (4.19.19)$$

where ϖ is a dimensionless frequency, ω is the angular frequency, ρ is the density, V_P is the P-wave velocity, and Z_n is the infinitesimal fracture normal compliance. $R(\omega)$ and $T(\omega)$ are frequency-dependent and complex. An excellent review is given by Jaeger *et al.* (2007).

Uses

The models presented here can be used to predict the effects of joints and fractures on the static and dynamic elastic properties of rocks.

Assumptions and Limitations

The models presented here assume that the unfractured rock mass is isotropic, linear, and elastic. The joints and fractures are assumed to be planar.

4.20 Bound-Filling Models

Synopsis

The **Voigt–Reuss** (Section 4.1) and **Hashin–Shtrikman–Walpole** (see Section 4.2) bounds yield precise limits on the maximum and minimum possible values for the effective bulk and shear moduli of an isotropic, linear, elastic composite. Specifically, for a mixture of two components, the Hashin–Shtrikman–Walpole bounds can be written as

$$K^{HS\pm} = K_1 + \frac{f_2}{(K_2 - K_1)^{-1} + f_1 \left(K_1 + \frac{4}{3}\mu_m \right)^{-1}} \qquad (4.20.1)$$

$$\mu^{HS\pm} = \mu_1 + \frac{f_2}{(\mu_2 - \mu_1)^{-1} + f_1 \left[\mu_1 + \frac{1}{6}\mu_m \left(\frac{9K_m + 8\mu_m}{K_m + 2\mu_m} \right) \right]^{-1}} \qquad (4.20.2)$$

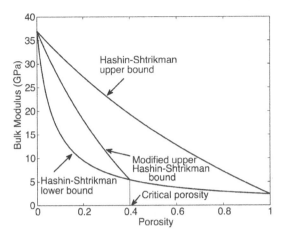

Figure 4.20.1 Hashin–Shtrikman and modified Hashin–Shtrikman bounds for bulk modulus in a quartz–water system

where subscripts 1 and 2 refer to properties of the two components, having bulk moduli, K_1, K_2, shear moduli, μ_1, μ_2, and volume fractions f_1 and f_2. Most commonly, these bounds are applied to describe mixtures of mineral and pore fluid, as illustrated in Figure 4.20.1. Equations 4.20.1 and 4.20.2 yield the upper bounds when K_m and μ_m are the maximum bulk and shear moduli of the individual constituents, and the lower bounds when K_m and μ_m are the minimum bulk and shear moduli of the constituents, respectively. The maximum (minimum) shear modulus might come from a different constituent from the maximum (minimum) bulk modulus. For example, this would be the case for a mixture of calcite ($K = 71$ GPa; $\mu = 30$ GPa) and quartz ($K = 37$ GPa; $\mu = 45$ GPa).

Equations 4.20.1 and 4.20.2 simplify to the Hashin–Shtrikman bounds (Section 4.2) when one constituent has both the maximum bulk and shear moduli, while the other constituent has the minimum bulk and shear moduli.

The **modified Hashin–Shtrikman bounds** use exactly the same equations previously shown. However, with modified bounds, the constituent end members are selected differently, such as a mineral mixed with a fluid-solid suspension (Figure 4.20.1) or a stiffly packed sediment mixed with a fluid-solid suspension. The critical-porosity model (Nur *et al.*, 1991, 1995), described in Section 7.1, identifies a critical porosity, ϕ_c, that separates load-bearing sediments at porosities $\phi < \phi_c$ and suspensions at porosities $\phi > \phi_c$. Modified Hashin–Shtrikman equations (or modified Voigt–Reuss equations) can be constructed to describe mixtures of mineral with the unconsolidated fluid-solid suspension at critical porosity. The modified upper Hashin–Shtrikman curve has been observed empirically to be a useful trend line describing, for example, how the elastic moduli of clean sandstones evolve from deposition through compaction and cementation. The

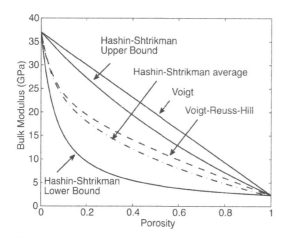

Figure 4.20.2 The Voigt–Reuss–Hill average is halfway between the Voigt upper bound and Reuss lower bound. A similar average can be found between the Hashin–Shtrikman bounds.

modified upper Hashin–Shtrikman curve, constructed as such, is not a rigorous bound on the elastic properties of clean sand, although sandstone moduli are almost always observed to lie on or below it. (See Sections 5.3 and 7.1 on stiff and soft interpolators.)

The **Voigt–Reuss–Hill average** (Section 4.5) is an estimate of elastic modulus defined to lie exactly halfway between the Voigt upper and Reuss lower bounds. A similar estimate can be constructed to lie halfway between the upper and lower Hashin–Shtrikman bounds (Figure 4.20.2). These two estimates are most practical in cases where the constituent end-members are elastically similar, as with a mixture of minerals without pore space. In this latter case, an average of upper and lower bounds yields a useful estimate of the average mineral moduli.

Marion's (1990) bounding average method (BAM) of fluid substitution (Section 6.10) uses the position of porosity–modulus data between the bounds as an indication of rock stiffness. In Figure 4.20.3, the Hashin–Shtrikman upper and lower bounds are displayed for mixtures of mineral and water. The data point A lies a distance d above the lower bound, while D is the spacing between the bounds at the same porosity. In the bounding average method, it is assumed that the ratio d/D remains constant if the pore fluid in the rock is changed, without changing the pore geometry or the stiffness of the dry frame. Though not theoretically justified, the BAM method sometimes gives a reasonable estimate of high-frequency fluid-substitution behavior.

Fabricius (2003) introduced the **isoframe model** to describe behavior between the modified upper and lower Hashin–Shtrikman bounds. In this model, the modified upper Hashin–Shtrikman curve is assumed to describe the trend of sediments that become progressively more compacted and cemented as they trend away from the lower Reuss average (Figure 4.20.4) – an empirical result. It is assumed that these

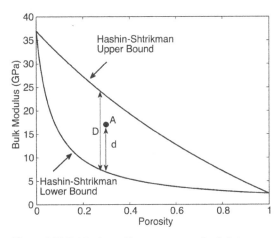

Figure 4.20.3 The bounding average method. It is assumed that the position of a data point A, described as d/D relative to bounds, is a measure of the pore stiffness.

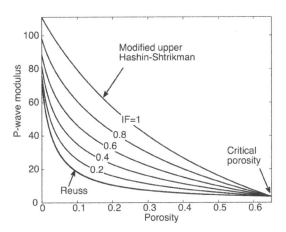

Figure 4.20.4 The isoframe model. The modified upper Hashin–Shtrikman curve is assumed to describe a strong frame of grains in good contact. The Reuss average curve describes a suspension of grains in a fluid. Each isoframe curve is a Hashin–Shtrikman mix of a fraction *IF* of frame with (1 −*IF*) of suspension.

rocks contain only grains that are load-bearing with good grain-to-grain contacts. Sediments that fall below the modified upper bound are assumed to contain inclusions of grain–fluid suspensions, in which the grains are not load-bearing. A family of curves can be generated (Figure 4.20.4), computed as an upper Hashin–Shtrikman mix of "frame," taken from the modified upper bound, and suspension, taken from the lower Reuss bound. *IF* is the volume fraction of load-bearing frame and 1 − *IF* is the fraction of suspension. The isoframe modulus at each porosity is computed from the frame and suspension moduli at the same porosity. The calculation is carried out separately for bulk and shear moduli, from which the P-wave modulus can be calculated.

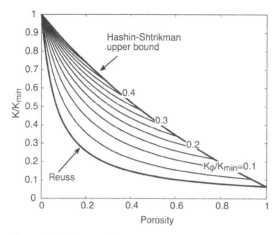

Figure 4.20.5 Normalized bulk modulus versus pressure showing contours of constant K_ϕ

The bulk modulus K of an elastic porous medium can be expressed as

$$\frac{1}{K} = \frac{1}{K_{min}} + \frac{\phi}{\widetilde{K}_\phi}$$

(4.20.3)

where $\widetilde{K}_\phi$ is the saturated pore-space stiffness (Section 2.10) given by

$$\widetilde{K}_\phi = K_\phi + \frac{K_{min}K_{fluid}}{K_{min} + K_{fluid}}$$

(4.20.4)

and K_ϕ is the dry-rock pore-space stiffness defined in terms of the pore volume v and confining stress σ_c:

$$\frac{1}{K_\phi} = \frac{1}{v}\frac{\partial v}{\partial \sigma_c}$$

(4.20.5)

Figure 4.20.5 shows a plot of bulk modulus versus porosity with contours of constant K_ϕ A large value of K_ϕ indicates a stiff pore space, while small K_ϕ indicates a soft pore space. $K_\phi = 0$ corresponds to a suspension.

The y-transform discussed in Section 4.3 yields another strategy for interpolating between bounds. The "Canonical" form of the effective elastic moduli (Berryman, 2004) can be written as

$$K = \Lambda(y_K) = \left[\frac{f_1}{K_1 + y_K} + \frac{f_2}{K_2 + y_K}\right]^{-1} - y_K$$

(4.20.6)

$$\mu = \Gamma(y_\mu) = \left[\frac{f_1}{\mu_1 + y_\mu} + \frac{f_2}{\mu_2 + y_\mu}\right]^{-1} - y_\mu$$

(4.20.7)

where y_K and y_μ are the **scalar y-transforms** of the bulk and shear moduli. For the Hashin–Shtrikman–Walpole bounds

$$y_K^{HSW-} \equiv \frac{4}{3}\mu_{\min} \leq y_K \leq \frac{4}{3}\mu_{\max} \equiv y_K^{HSW+} \tag{4.20.8}$$

$$y_\mu^{HSW-} \equiv \frac{\mu_{\min}(9K_{\min} + 8\mu_{\min})}{6(K_{\min} + 2\mu_{\min})} \leq y_\mu \leq \frac{\mu_{\max}(9K_{\max} + 8\mu_{\max})}{6(K_{\max} + 2\mu_{\max})} \equiv y_\mu^{HSW+} \tag{4.20.9}$$

In equations 4.20.6–4.20.9, K is a monotonic function of y_K and μ is a monotonic function of y_μ. Bounds on y_K and y_μ are bounds on K and μ. Therefore, values of y_K and y_μ within their allowable ranges define corresponding curves of K and μ lying in the space between the bounds. Berryman (2004) suggests, for example, that *estimates* of the moduli could be made by averaging the bounding values of the y-transforms $\overline{y}_k = (y_K^{HSW-} + y_K^{HSW+})/2$ and $\overline{y}_\mu = (y_\mu^{HSW-} + y_\mu^{HSW+})/2$, as an alternative to averaging the HSW bounds on K and μ. Other estimates can be had by taking the geometric means of the y-transforms.

Uses

The bound-filling models described here provide a means to describe the stiffness of rocks that fall in the range between upper and lower bounds. The modified Hashin–Shtrikman and modified Voigt averages have been found to be useful depth-trend lines for sand and chalk sediments. The BAM model provides a heuristic fluid-substitution strategy that seems to work best at high frequency. The isoframe model allows one to estimate the moduli of rocks composed of a consolidated grain framework with inclusions of nonload-bearing grain suspensions.

Assumptions and Limitations

These models are based on isotropic linear elasticity. All of the bound-filling models provided here contain at least some heuristic elements, such as the interpretation of the modified upper bounds.

4.21   Effective Moduli of Polycrystalline Aggregates

An important set of composites is one made of many small anisotropic grains, fully welded to one another, with no gaps. This is often referred to as a **polycrystal.** The individual grains might also be referred to as "crystals" or "domains." If the grains are linear elastic, the effective moduli of the composite, at scales much larger than that of the constituent grains, will also be linear elastic; the precise value of the effective elastic stiffness tensor and its anisotropy will depend on the shapes and volume fractions of the grains, the elastic stiffness tensors of each grain, and the spatial orientation of each grain. In this section we discuss bounds for the effective elastic moduli without considering the shapes of each grain. (Analogous treatments can be applied to electrical conductivity of a polycrystal.)

For any spatially variable tensor, $\mathbf{D}(\mathbf{x})$, defined over a region Ω, the Voigt arithmetic average of the tensor over Ω is given by

$$\mathbf{D}_V = \langle \mathbf{D}(\mathbf{x}) \rangle_\Omega = \sum_{i=1}^{m} x_i \mathbf{D}_i \tag{4.21.1}$$

The latter summation is appropriate if $\mathbf{D}(\mathbf{x})$ is piecewise homogeneous over m sub-regions spanning Ω, each with volume fraction x_i and tensor property $\mathbf{D}_i$. Similarly, the Reuss harmonic average of $\mathbf{D}(\mathbf{x})$ is given by

$$\mathbf{D}_R = \langle \mathbf{D}^{-1}(\mathbf{x}) \rangle^{-1} = \left[\sum_{i=1}^{m} x_i \mathbf{D}_i^{-1} \right]^{-1} . \tag{4.21.2}$$

The Voigt and Reuss averages have been shown to be bounds on the effective tensor $\mathbf{D}_{eff}$ (Hill, 1952); in the context of elasticity, where $\mathbf{D}$ is the stiffness tensor, these are sometimes referred to as Hill's bounds (Gibiansky, 1993):

$$\mathbf{D}_R \equiv \langle \mathbf{D}^{-1}(\mathbf{x}) \rangle^{-1} \leq \mathbf{D} \leq \langle \mathbf{D}(\mathbf{x}) \rangle \equiv \mathbf{D}_V . \tag{4.21.3}$$

The meaning of the tensor inequalities in the previous equation is that $\mathbf{A} \geq \mathbf{B}$ if $\mathbf{C} = \mathbf{A} - \mathbf{B}$ is a positive semi-definite tensor (Gibiansky, 1993). The inequality *does not* mean that each element of $\mathbf{A}$ will be greater than the corresponding element in $\mathbf{B}$.

Single Crystal Phase at Different Orientations

We consider the special case where each grain in the polycrystal is composed of the same crystalline material with elastic stiffness tensor $\mathbf{C}_0$, expressed in its crystal-fixed coordinate system. Each grain can have a different orientation described by the orientation distribution function (ODF), $W(\xi, \psi, \phi,)$, where $\xi = \cos\theta$, and θ, ψ, ϕ are the Euler angles of the crystal-fixed coordinate axes relative to the global coordinate axes (Roe, 1965; Sayers, 1987). The volume fraction of grains whose axes lie in the angular ranges ϕ to $\phi + d\phi$, ψ to $\psi + d\psi$, and ξ to $\xi + d\xi$ is given by $W(\xi, \psi, \phi) \, d\xi \, d\psi \, d\phi$, subject to the normalization

$$\int_{\phi=0}^{2\pi} \int_{\psi=0}^{2\pi} \int_{\xi=-1}^{1} W(\xi, \psi, \phi) \, d\xi \, d\psi \, d\phi = 1 . \tag{4.21.4}$$

For an isotropic distribution, the normalization leads to $W(\xi, \psi, \phi) = 1/8\pi^2$.

Even though we consider a composite of identical minerals, the rotations of the crystals cause the elastic stiffness tensor in the global coordinates $\mathbf{C}(x)$ to vary spatially. Specifically, $\mathbf{C}(x) = \mathcal{R}(x) \cdot \mathbf{C}_0 \cdot \mathcal{R}^T(x)$, where $\mathcal{R}_{ijkl} = R_{ik}(x)R_{jl}(x)$, and $R_{ij}(x)$ is the local coordinate rotation matrix through angles $\theta(x)$, $\psi(x)$, and $\phi(x)$ (Milton, 2002). The Voigt average stiffness tensor of the polycrystal can be written as

$$\mathbf{C}_V = \int_{\phi=0}^{2\pi} \int_{\psi=0}^{2\pi} \int_{\xi=-1}^{1} W(\xi,\psi,\phi)[\mathcal{R} \cdot \mathbf{C}_0 \cdot \mathcal{R}^T] \, d\xi \, d\psi \, d\phi \, . \tag{4.21.5}$$

Similarly, the Reuss stiffness tensor can be written as

$$\mathbf{C}_R^{-1} = \int_{\phi=0}^{2\pi} \int_{\psi=0}^{2\pi} \int_{\xi=-1}^{1} W(\xi,\psi,\phi)[\mathcal{R} \cdot \mathbf{C}_0^{-1} \cdot \mathcal{R}^T] \, d\xi \, d\psi \, d\phi \, . \tag{4.21.6}$$

It is sometimes convenient to express the ODF, $W(\xi,\psi,\phi)$, in terms of generalized spherical harmonics (Roe, 1969; Sayers, 1994; Johansen et al., 2002):

$$W(\xi,\psi,\phi) = \sum_{l=0}^{\infty} \sum_{m=-l}^{l} \sum_{n=-l}^{l} W_{lmn} Z_{lmn}(\xi) e^{-im\psi} e^{-in\phi}, \tag{4.21.7}$$

where $Z_{lmn}(\xi)$ are normalized Legendre functions (Roe, 1965).

For a given ODF, the coefficients W_{lmn} can be determined using the orthogonality property of the spherical harmonics:

$$W_{lmn} = \frac{1}{4\pi^2} \int_{\phi=0}^{2\pi} \int_{\psi=0}^{2\pi} \int_{\xi=-1}^{1} W(\xi,\psi,\phi) Z_{lmn}(\xi) e^{-im\psi} e^{-in\phi} \, d\xi \, d\psi \, d\phi \tag{4.21.8}$$

When spatially averaging over a fourth-order tensor (e.g., elastic stiffness or compliance), only coefficients W_{lmn} with $l \leq 4$ impact the effective tensor. In this case, the normalization requires that $W_{000} = \sqrt{2}/8\pi^2$. For a second-order tensor (e.g., electrical conductivity) only the W_{lmn} with $l \leq 2$, affect the effective tensor. When the individual crystals have hexagonal symmetry and the overall material has orthorhombic symmetry, the W_{lmn} are real and restricted to $n = 0$ and even values of l and m. For this latter case, the non-zero coefficients are W_{000}, W_{200}, W_{220}, W_{400}, W_{420}, and W_{440} for elastic stiffness. When individual crystals have hexagonal symmetry and the overall material has hexagonal symmetry, $m = n = 0$. In this latter case, only W_{200} and W_{400} are needed to determine the elastic stiffness tensor, and only W_{200} is needed for the conductivity tensor.

The subset Z_{l00} are related to the Legendre functions, $P_l(\xi) = \sqrt{2/(2l+1)} \, Z_{l00}(\xi)$, such that

$$\int_{-1}^{1} P_m P_n d\xi = \frac{2}{(2n+1)} \delta_{mn} \tag{4.21.9}$$

Some examples are $P_0(\xi) = 1$, $P_2(\xi) = (3\xi^2 - 1)/2$, and $P_4(\xi) = (35\xi^4 - 30\xi^2 + 3)/8$, or $Z_{000}(\xi) = \sqrt{1/2}$, $Z_{200}(\xi) = \sqrt{5/2}(3\xi^2 - 1)/2$, $Z_{400}(\xi) = \sqrt{9/2}(35\xi^4 - 30\xi^2 + 3)/8$.

Voigt and Reuss Bounds of an Orthorhombic Distribution of Hexagonal Crystals

Sayers (1987) found expressions for the lower bound (Reuss average) of the effective elastic moduli of a distribution of hexagonal crystals, oriented to give overall orthorhombic sample symmetry (As pointed out previously, only W_{200} and W_{400} are needed). These are conveniently expressed in terms of the effective compliance tensor $\mathbf{S} = \mathbf{C}^{-1}$ as

$$S_{ijkl} = S_{ijkl}^{iso} + \Delta S_{ijkl}, \tag{4.21.10}$$

where S_{ijkl}^{iso} is the effective compliance of an isotropic distribution of crystals, and ΔS_{ijkl} is the anisotropic difference. The non-zero components of the isotropic compliance are (Sayers, 1987)

$$S_{11}^{iso} = S_{22}^{iso} = S_{33}^{iso} = \frac{1}{15}(8s_{11} + 3s_{33} + 4s_{13} + 2s_{44}), \tag{4.21.11}$$

$$S_{12}^{iso} = S_{23}^{iso} = S_{31}^{iso} = \frac{1}{15}(s_{11} + s_{33} + 5s_{12} + 8s_{13} - s_{44}), \tag{4.21.12}$$

$$S_{44}^{iso} = S_{55}^{iso} = S_{66}^{iso} = \frac{2}{15}(7s_{11} + 2s_{33} - 5s_{12} - 4s_{13} + 3s_{44}), \tag{4.21.13}$$

where s_{ij} are the single hexagonal crystal compliances in the Voigt notation, with the symmetry axis along the crystal-centered x-3 direction. The nonzero components of the anisotropic portion of the tensor in the Voigt notation are (Sayers, 1987)

$$\Delta S_{11} = \frac{8\sqrt{10}}{105}\pi^2 s_3(W_{200} - \sqrt{6}\,W_{220}) + \frac{4\sqrt{2}}{35}\pi^2 s_1\left(W_{400} - \frac{2\sqrt{10}}{3}W_{420} + \frac{\sqrt{70}}{3}W_{440}\right),$$
$$\tag{4.21.14a}$$

$$\Delta S_{22} = \frac{8\sqrt{10}}{105}\pi^2 s_3(W_{200} + \sqrt{6}\,W_{220}) + \frac{4\sqrt{2}}{35}\pi^2 s_1\left(W_{400} + \frac{2\sqrt{10}}{3}W_{420} + \frac{\sqrt{70}}{3}W_{440}\right),$$
$$\tag{4.21.14b}$$

$$\Delta S_{33} = \frac{16\sqrt{2}}{105}\pi^2(2s_1 W_{400} - \sqrt{5}\,s_3 W_{200}), \tag{4.21.14c}$$

$$\Delta S_{12} = -\frac{8\sqrt{10}}{315}\pi^2(7s_2 - s_3)W_{200} + \frac{4\sqrt{2}}{105}\pi^2 s_1(W_{400} - \sqrt{70}\,W_{440}), \tag{4.21.14d}$$

$$\Delta S_{13} = \frac{4\sqrt{10}}{315}\pi^2(7s_2 - s_3)(W_{200} + \sqrt{6}\,W_{220}) - \frac{16\sqrt{2}}{105}\pi^2 s_1\left(W_{400} - \sqrt{\frac{5}{2}}W_{420}\right),$$
$$\tag{4.21.14e}$$

$$\Delta S_{23} = \frac{4\sqrt{10}}{315}\pi^2(7s_2 - s_3)(W_{200} - \sqrt{6}\,W_{220}) - \frac{16\sqrt{2}}{105}\pi^2 s_1\left(W_{400} + \sqrt{\frac{5}{2}}W_{420}\right),$$
$$\tag{4.21.14f}$$

$$\Delta S_{44} = -\frac{8\sqrt{10}}{315}\pi^2(7s_2 + 2s_3)(W_{200} - \sqrt{6}\,W_{220}) - \frac{64\sqrt{2}}{105}\pi^2 s_1\left(W_{400} + \sqrt{\frac{5}{2}}\,W_{420}\right),$$

$$(4.21.14g)$$

$$\Delta S_{55} = -\frac{8\sqrt{10}}{315}\pi^2(7s_2 + 2s_3)(W_{200} + \sqrt{6}\,W_{220}) - \frac{64\sqrt{2}}{105}\pi^2 s_1\left(W_{400} - \sqrt{\frac{5}{2}}W_{420}\right),$$

$$(4.21.14h)$$

$$\Delta S_{66} = \frac{16\sqrt{10}}{315}\pi^2(7s_2 + 2s_3)W_{200} + \frac{16\sqrt{2}}{105}\pi^2 s_1(W_{400} - \sqrt{70}\,W_{440})$$

The scalar constants in these expressions are

$$s_1 = s_{11} + s_{33} - 2s_{13} - s_{44}, \qquad (4.21.15)$$

$$s_2 = s_{11} - 3s_{12} + 2s_{13} - s_{44}/2, \qquad (4.21.16)$$

$$s_3 = 4s_{11} - 3s_{33} - s_{13} - s_{44}/2. \qquad (4.21.17)$$

Similarly, the upper bound (Voigt average) of the effective stiffness tensor C_{ijkl} can be written as

$$C_{ijkl} = C_{ijkl}^{iso} + \Delta C_{ijkl}, \qquad (4.21.18)$$

where C_{ijkl}^{iso} is the effective elastic stiffness of an isotropic distribution of crystals, and ΔC_{ijkl} is the anisotropic deviation from it. The non-zero components of the isotropic stiffness are (Johansen *et al.*, 2004)

$$C_{11}^{iso} = C_{22}^{iso} = C_{33}^{iso} = \frac{1}{15}(8c_{11} + 3c_{33} + 4c_{13} + 8c_{44}), \qquad (4.21.19)$$

$$C_{12}^{iso} = C_{23}^{iso} = C_{31}^{iso} = \frac{1}{15}(c_{11} + c_{33} + 5c_{12} + 8c_{13} - 4c_{44}), \qquad (4.21.20)$$

$$C_{44}^{iso} = C_{55}^{iso} = C_{66}^{iso} = \frac{1}{30}(7c_{11} + 2c_{33} - 5c_{12} - 4c_{13} + 12c_{44}), \qquad (4.21.21)$$

where c_{ij} are the single hexagonal crystal stiffnesses in the Voigt notation, with the symmetry axis along the crystal-centered x-3 direction. The nonzero components of the anisotropic portion of the tensor are

$$\Delta C_{11} = \frac{8\sqrt{10}}{105}\pi^2 c_3(W_{200} - \sqrt{6}\,W_{220}) + \frac{4\sqrt{2}}{35}\pi^2 c_1\left(W_{400} - \frac{2\sqrt{10}}{3}W_{420} + \frac{\sqrt{70}}{3}W_{440}\right),$$

$$(4.21.22a)$$

$$\Delta C_{22} = \frac{8\sqrt{10}}{105}\pi^2 c_3(W_{200} + \sqrt{6}\,W_{220}) + \frac{4\sqrt{2}}{35}\pi^2 c_1\left(W_{400} + \frac{2\sqrt{10}}{3}W_{420} + \frac{\sqrt{70}}{3}W_{440}\right),$$

$$(4.21.22b)$$

$$\Delta C_{33} = \frac{16\sqrt{2}}{105} \pi^2 (2c_1 W_{400} - \sqrt{5} \, c_3 W_{200}) , \tag{4.21.22c}$$

$$\Delta C_{12} = -\frac{8\sqrt{10}}{315} \pi^2 (7c_2 - c_3) W_{200} + \frac{4\sqrt{2}}{105} \pi^2 c_1 (W_{400} - \sqrt{70} \, W_{440}), \tag{4.21.22d}$$

$$\Delta C_{13} = \frac{4\sqrt{10}}{315} \pi^2 (7c_2 - c_3)(W_{200} + \sqrt{6} \, W_{220}) - \frac{16\sqrt{2}}{105} \pi^2 c_1 \left(W_{400} - \sqrt{\frac{5}{2}} \, W_{420} \right), \tag{4.21.22e}$$

$$\Delta C_{23} = \frac{4\sqrt{10}}{315} \pi^2 (7c_2 - c_3)(W_{200} - \sqrt{6} \, W_{220}) - \frac{16\sqrt{2}}{105} \pi^2 c_1 \left(W_{400} + \sqrt{\frac{5}{2}} \, W_{420} \right), \tag{4.21.22f}$$

$$\Delta C_{44} = -\frac{2\sqrt{10}}{315} \pi^2 (7c_2 + 2c_3)(W_{200} - \sqrt{6} \, W_{220}) - \frac{16\sqrt{2}}{105} \pi^2 c_1 \left(W_{400} + \sqrt{\frac{5}{2}} \, W_{420} \right), \tag{4.21.22g}$$

$$\Delta C_{55} = -\frac{2\sqrt{10}}{315} \pi^2 (7c_2 + 2c_3)(W_{200} + \sqrt{6} \, W_{220}) - \frac{16\sqrt{2}}{105} \pi^2 c_1 \left(W_{400} - \sqrt{\frac{5}{2}} \, W_{420} \right), \tag{4.21.22h}$$

$$\Delta C_{66} = \frac{4\sqrt{10}}{315} \pi^2 (7c_2 + 2c_3) W_{200} + \frac{4\sqrt{2}}{105} \pi^2 c_1 (W_{400} - \sqrt{70} \, W_{440}), \tag{4.21.22i}$$

where

$$c_1 = c_{11} + c_{33} - 2c_{13} - 4c_{44}, \tag{4.21.23}$$

$$c_2 = c_{11} - 3c_{12} + 2c_{13} - 2c_{44}, \tag{4.21.24}$$

$$c_3 = 4c_{11} - 3c_{33} - c_{13} - 2c_{44}. \tag{4.21.25}$$

Voigt and Reuss Bounds of a Hexagonal Distribution of Hexagonal Crystals

For a hexagonal distribution of hexagonal crystals, the Reuss lower bound on the effective stiffness can be expressed in terms of the effective compliance tensor $\mathbf{S} = \mathbf{C}^{-1}$ as

$$S_{11} = S_{11}^{iso} + \frac{4\sqrt{2}}{105} \pi^2 [2\sqrt{5} s_3 W_{200} + 3s_1 W_{400}], \tag{4.21.26a}$$

$$S_{33} = S_{33}^{iso} - \frac{16\sqrt{2}}{105} \pi^2 [\sqrt{5} s_3 W_{200} - 2s_1 W_{400}], \tag{4.21.26b}$$

$$S_{12} = S_{12}^{iso} - \frac{4\sqrt{2}}{315} \pi^2 [2\sqrt{5}(7s_2 - s_3) W_{200} - 3s_1 W_{400}], \tag{4.21.26c}$$

$$S_{13} = S_{13}^{iso} + \frac{4\sqrt{2}}{315} \pi^2 [\sqrt{5}(7s_2 - s_3) W_{200} - 12s_1 W_{400}] , \tag{4.21.26d}$$

$$S_{44} = S_{44}^{iso} - \frac{8\sqrt{2}}{315}\pi^2[\sqrt{5}(7s_2 + 2s_3)W_{200} + 24s_1 W_{400}] , \tag{4.21.26e}$$

$$S_{66} = 2(S_{11} - S_{12}), \tag{4.21.26f}$$

where

$$s_1 = s_{11} + s_{33} - 2s_{13} - s_{44}, \tag{4.21.27}$$

$$s_2 = s_{11} - 3s_{12} + 2s_{13} - s_{44}/2, \tag{4.21.28}$$

$$s_3 = 4s_{11} - 3s_{33} - s_{13} - s_{44}/2. \tag{4.21.29}$$

The constants are defined such that $s_1 = s_2 = s_3 = 0$ if the particles are isotropic ($W_{200} = W_{400} = 0$). When the particles are completely aligned, such that $W(\xi, \psi, \phi) = \delta(\xi - 1)/(4\pi^2)$ then

$$W_{200} = \frac{\sqrt{5/2}}{4\pi^2}, \text{ and } W_{400} = \frac{\sqrt{9/2}}{4\pi^2}, \tag{4.21.30}$$

and $S_{ij} = s_{ij}$.

The resulting stiffness coefficients estimated from the Voigt average are (Morris, 1969; Sayers, 1995a)

$$C_{11} = C_{11}^{iso} + \frac{4\sqrt{2}}{105}\pi^2[2\sqrt{5}c_3 W_{200} + 3c_1 W_{400}], \tag{4.21.31a}$$

$$C_{33} = C_{33}^{iso} - \frac{16\sqrt{2}}{105}\pi^2[\sqrt{5}c_3 W_{200} - 2c_1 W_{400}], \tag{4.21.31b}$$

$$C_{12} = C_{12}^{iso} - \frac{4\sqrt{2}}{315}\pi^2[2\sqrt{5}(7c_2 - c_3)W_{200} - 3c_1 W_{400}], \tag{4.21.31c}$$

$$C_{13} = C_{13}^{iso} + \frac{4\sqrt{2}}{315}\pi^2[\sqrt{5}(7c_2 - c_3)W_{200} - 12c_1 W_{400}], \tag{4.21.31d}$$

$$C_{44} = C_{44}^{iso} - \frac{2\sqrt{2}}{315}\pi^2[\sqrt{5}(7c_2 + 2c_3)W_{200} + 24c_1 W_{400}], \tag{4.21.31e}$$

$$C_{66} = (C_{11} - C_{12})/2, \tag{4.21.31f}$$

where

$$c_1 = c_{11} + c_{33} - 2c_{13} - 4c_{44}, \tag{4.21.32}$$

$$c_2 = c_{11} - 3c_{12} + 2c_{13} - 2c_{44}, \tag{4.21.33}$$

$$c_3 = 4c_{11} - 3c_{33} - c_{13} - 2c_{44}. \tag{4.21.34}$$

The constants are defined such that $c_1 = c_2 = c_3 = 0$ if the particles are isotropic ($W_{200} = W_{400} = 0$). Note that the expressions for a hexagonal distribution of hexagonal crystals can be obtained from the expressions for an orthorhombic distribution of hexagonal crystals by dropping the dependence on W_{220}, W_{420}, and W_{440}.

When the particles are completely aligned,

$$W_{200} = \frac{\sqrt{5/2}}{4\pi^2}, \text{ and } W_{400} = \frac{\sqrt{9/2}}{4\pi^2}, \tag{4.21.35}$$

and $C_{ij} = c_{ij}$.

Thomsen Parameters for Polycrystals

The Thomsen (1986) parameters of transversely isotropic (hexagonal) symmetry are defined in terms of the components of the elastic stiffness tensor. We can use the previous results to write the effective parameters of the polycrystal, $\bar{\varepsilon}$, $\bar{\gamma}$, $\bar{\delta}$, and $\bar{\eta} = (\bar{\varepsilon} - \bar{\delta})/(1 - 2\bar{\delta})$, in terms of the corresponding parameters of the individual crystals, ε, γ, δ, and $\eta = (\varepsilon - \delta)/(1 - 2\delta)$. For a hexagonal distribution of hexagonal crystals, the Thomsen parameters take a relatively simple form (Bandyopadhyay, 2009):

$$\bar{\varepsilon} = \frac{15[(8\varepsilon - \delta)W_{200} - (\varepsilon - \delta)W_{400}]}{105 + 28(4\varepsilon + \delta) - 20(8\varepsilon - \delta)W_{200} + 48(\varepsilon - \delta)W_{400}} \tag{4.21.36}$$

$$\bar{\delta} = \frac{15[(8\varepsilon - \delta)W_{200} - 8(\varepsilon - \delta)W_{400}]}{105 + 28(4\varepsilon + \delta) - 20(8\varepsilon - \delta)W_{200} + 48(\varepsilon - \delta)W_{400}} \tag{4.21.37}$$

$$\bar{\gamma} = \frac{15\left[\left(-\varepsilon + 7\frac{\mu}{M}\gamma + \delta\right)W_{200} - (\varepsilon - \delta)W_{400}\right]}{105\frac{\mu}{M} + 14\left(\varepsilon + 5\frac{\mu}{M}\gamma - \delta\right) + 10(\varepsilon - \delta)W_{200} - 24(\varepsilon - \delta)W_{400}} \tag{4.21.38}$$

$$\bar{\eta} = \frac{\bar{\varepsilon} - \bar{\delta}}{1 + 2\bar{\delta}} = \frac{105(\varepsilon - \delta)W_{400}}{105 + 28(4\varepsilon + \delta) + 10(8\varepsilon - \delta)W_{200} - 192(\varepsilon - \delta)W_{400}^1} \tag{4.21.39}$$

where $M = c_{33}$ and $\mu = c_{44}$ are the P- and S-wave moduli for waves propagating along the symmetry axis of the individual crystal. Note that the effective anellipticity, $\bar{\eta}$, is proportional to W_{400}. Therefore, $\bar{\eta}$ goes to zero if W_{400} goes to zero. The effective anellipticity, $\bar{\eta}$, is also proportional to $(\varepsilon - \delta)$ of the crystal; therefore, the effective anellipticity goes to zero if the crystal anellipticity goes to zero.

Bounds on the Effective Bulk and Shear Moduli of an Isotropic Distribution of Crystals

The Voigt and Reuss bounds on effective bulk modulus, K, and shear modulus, μ, of an isotropic distribution of crystals with **triclinic** symmetry are (Hill, 1952; Cowin et al., 1999)

$$K_V^T = \frac{1}{9}[c_{11} + c_{22} + c_{33} + 2(c_{12} + c_{23} + c_{13})], \tag{4.21.40}$$

$$\frac{1}{K_R^T} = (s_{11} + s_{22} + s_{33}) + 2(s_{12} + s_{23} + s_{13}), \tag{4.21.41}$$

$$\mu_V^T = \frac{1}{15}[c_{11} + c_{22} + c_{33} - c_{12} - c_{23} - c_{13} + 3(c_{44} + c_{55} + c_{66})], \qquad (4.21.42)$$

$$\frac{1}{\mu_R^T} = \frac{1}{15}[4(s_{11} + s_{22} + s_{33}) - 4(s_{12} + s_{23} + s_{13}) + 3(s_{44} + s_{55} + s_{66})], \qquad (4.21.43)$$

where c_{ij} are the single crystal stiffnesses, s_{ij} are the single crystal compliances in the Voigt notation, and superscript T indicates triclinic crystal symmetry. Note that these results only depend on nine of the 21 independent elastic moduli of a triclinic crystal.

Crystals with **trigonal** symmetry (crystal classes 3 and $\bar{3}$) have seven independent elastic stiffnesses $c_{11}, c_{12}, c_{13}, c_{14}, c_{25}, c_{33}, c_{44}$, with $c_{66} = (c_{11} - c_{12})/2$. (Note that for trigonal symmetry $c_{22} = c_{11}, c_{23} = c_{13}, c_{55} = c_{44}, c_{24} = -c_{14}, c_{25} = -c_{15}, c_{46} = c_{25}, c_{56} = c_{14}$. All remaining constants are zero.) Trigonal classes $\bar{3}m$, 32 and $3m$ have only six independent constants ($c_{25} = 0$). For classes 3 and $\bar{3}$ the Voigt and Reuss bounds on effective bulk and shear moduli of an isotropic distribution of crystals are (Watt, 1986)

$$K_V^{trig} = \frac{1}{9}[2(c_{11} + c_{12}) + c_{33} + 4c_{13}], \qquad (4.21.44)$$

$$K_R^{trig} = c^2/B, \qquad (4.21.45)$$

$$\mu_V^{trig} = \frac{1}{30}(B + 12c_{44} + 12c_{66}), \qquad (4.21.46)$$

$$\mu_R^{trig} = \frac{5}{2}\frac{c^2[c_{44}c_{66} - (c_{14}^2 + c_{25}^2)]}{3K_V^{trig}[c_{44}c_{66} - (c_{14}^2 + c_{25}^2)] + c^2(c_{44} + c_{66})}, \qquad (4.21.47)$$

where superscript *trig* indicates trigonal crystal symmetry, and

$$B = c_{11} + c_{12} + 2c_{33} - 4c_{13}, \qquad (4.21.48)$$

$$c^2 = (c_{11} + c_{12})c_{33} - 2c_{13}^2. \qquad (4.21.49)$$

The previous expressions for K_V^{trig}, K_R^{trig}, and μ_V^{trig} apply also to trigonal cases with only six independent constants, while μ_R reduces to these simpler classes when $c_{25} = 0$.

Crystals with **tetragonal** symmetry (classes 4, $\bar{4}$, $4m$) have seven independent elastic stiffness $c_{11}, c_{12}, c_{13}, c_{16}, c_{33}, c_{44}, c_{66}$. (For tetragonal symmetry $c_{22} = c_{11}$, $c_{23} = c_{13}, c_{55} = c_{44}$, and $c_{26} = -c_{16}$. All remaining constants are zero.) Classes $4/mmm$, $\bar{4}2m$, $4mm$, and 422 have only six independent constants ($c_{16} = 0$). For classes 4, $\bar{4}$, $4m$ the Voigt and Reuss bounds on bulk modulus are the same as above for trigonal, $K_V^{tet} = K_V^{trig}$ and $K_R^{tet} = K_R^{trig}$, while the Voigt and Reuss bounds on shear modulus are (Watt, 1986)

$$\mu_V^{tet} = \frac{1}{30}\left(B + 3(c_{11} - c_{12}) + 12c_{44} + 6c_{66}\right), \qquad (4.21.50)$$

$$\mu_R^{tet} = 15\left(\frac{18K_V^{tet}}{c^2} + \frac{6}{c_{11} - c_{12}} + \frac{6}{c_{44}} + \frac{3(c_{11} - c_{12})^2 + 12c_{16}^2}{(c_{11} - c_{12})d^2}\right)^{-1}, \qquad (4.21.51)$$

$$d^2 = (c_{11} - c_{12})c_{66} - 2c_{16}^2.$$

For tetragonal cases with only six independent constants, the expressions for K_V^{tet}, K_R^{tet}, and μ_V^{tet} remain the same, while μ_R^{tet} simplifies by noting that $c_{16} = 0$.

Crystals with **cubic** symmetry have only three independent elastic constants. (For cubic symmetry $c_{33} = c_{22} = c_{11}$, $c_{12} = c_{23} = c_{13}$, $c_{55} = c_{44} = c_{66}$. All remaining constants are zero.) The Voigt and Reuss bounds on bulk and shear moduli for an isotropic distribution of **cubic** crystals are (Watt, 1986)

$$K_V^{cub} = \frac{1}{3}(c_{11} + 2c_{12}),$$

(4.21.52)

$$K_R^{cub} = K_V^{cub},$$

(4.21.53)

$$\mu_V^{cub} = \frac{1}{5}(c_{11} - c_{12} + 3c_{44}),$$

(4.21.54)

$$\frac{1}{\mu_R^{cub}} = \frac{4}{5}(s_{11} - s_{12}) + \frac{3}{5}s_{44}.$$

(4.21.55)

The Voigt and Reuss bounds on bulk modulus are only equal when the grains have cubic or isotropic symmetry.

Until recently, the Hashin–Shtrikman bounds on the effective bulk and shear moduli could be calculated for polycrystals of all symmetries except triclinic (Berryman, 2011; Hashin, 1983; Hashin and Shtrikman, 1962a, 1962b; Hill, 1952; Meister and Peselnick, 1966; Middya *et al.*, 1984; Peselnick and Meister, 1965; Watt and Peselnick, 1980; Watt, 1986). Brown (2015) published a method and a MATLAB code for computing the bounds for *any* crystal symmetry.

4.22   Comments on the Representative Volume Element

The models described in this book attempt to quantify and predict "effective properties" of rocks (elastic, electrical, dielectric, and transport) in terms of the heterogeneous mix of minerals and pore fluids. Many of the models are based upon or tested against laboratory measurements on rock samples or computational simulations on high-resolution images of rock samples. An implicit goal of effective medium modeling is to extrapolate what is learned to larger scales in the field. A "representative" laboratory or digital sample is one that behaves as a homogeneous medium when the scale of inclusions (grains, pores, fractures) is much smaller than the sample and the wavelength of the externally applied fields. Such a sample is called a "Representative Elementary Volume" (REV) or "Representative Volume Element" (RVE).

Caution:

Practicalities of laboratory measurement, imaging, and computation put limits on the sample size that can be handled. Some reported results are taken from samples that are below representative volume, and therefore, the results may be difficult to relate to larger scales.

To elaborate, consider a mechanical measurement on a linear elastic sample in which we apply surfaces displacements **u** that are consistent with a uniform strain field $\bar{\varepsilon}$:

$$\mathbf{u} = \bar{\varepsilon}\mathbf{x} \tag{4.22.1}$$

where **x** is the position vector. If the sample is heterogeneous, the displacement fields throughout the sample will not be uniform, but the average strain will still be $\bar{\varepsilon}$. The mean stress $\bar{\sigma}$ in the body can be found from the resulting surface tractions. Because the system is linear, an *apparent stiffness tensor*, $\mathbf{C}^{app,d}$ (Huet, 1990), relating the mean stress to mean strain can be defined as

$$\bar{\sigma} \equiv \mathbf{C}^{app,\,d}\,\bar{\varepsilon} \tag{4.22.2}$$

with the corresponding *apparent compliance tensor* $\mathbf{S}^{app,\,d} = (\mathbf{C}^{app,\,d})^{-1}$. The super-script d in equation 4.22.2 indicates results from the experiment with displacement boundary conditions. The term "apparent" indicates that the sample might not be sufficiently large or statistically homogeneous enough for $\mathbf{C}^{app,\,d}$ and $\mathbf{S}^{app,\,d}$ to be true effective medium properties.

In a second experiment we apply surface tractions **T** that are consistent with a uniform stress field $\bar{\sigma}$:

$$\mathbf{T} = \bar{\sigma} \cdot \mathbf{n} \tag{4.22.3}$$

where **n** is the unit normal to the surface. The resulting mean strain in the body can be found from the resulting surface deformation. An *apparent compliance tensor* $\mathbf{S}^{(app,\,t)}$ relating the mean stress to mean strain can be defined as

$$\bar{\varepsilon} \equiv \mathbf{S}^{app,\,t}\bar{\sigma} \tag{4.22.4}$$

with the corresponding apparent modulus $\mathbf{C}^{(app,\,t)} = \left(\mathbf{S}^{(app,\,t)}\right)^{-1}$. In equation 4.22.4, superscript t indicates results from the experiment with traction boundary conditions.

For a heterogeneous sample, $\mathbf{C}^{(app,\,d)}$ may not be equal to $\mathbf{C}^{(app,\,t)}$, and $\mathbf{S}^{(app,\,d)}$ may not be equal to $\mathbf{S}^{(app,\,t)}$; furthermore $\mathbf{C}^{(app,\,t)} \leq \mathbf{C}^{(app,\,d)}$. In this case the apparent stiffnesses are not purely material properties, since they also depend on the boundary conditions. However, if the sample is sufficiently large relative to the microstructure, i.e., the representative volume, then the apparent moduli under stress and displacement boundary conditions will coincide, at least to some level of accuracy. In this case the superscripts d and t become irrelevant, and the apparent moduli are the effective moduli (Huet, 1990). The Hill condition for the existence of effective moduli states that when uniform displacement (equation 4.22.1) or traction boundary conditions (equation 4.22.3) are applied, the total strain energy found by integrating over the entire volume of the sample is equal to the strain energy found from the average stress and average strain:

$$\frac{1}{2}\overline{\sigma_{ij}\varepsilon_{ij}} = \frac{1}{2}\overline{\sigma_{ij}}\,\overline{\varepsilon_{ij}} \tag{4.22.5}$$

Huet (1990) proved the following **partition theorem** for samples either larger than or smaller than the representative volume. Imagine a sample D_0 with apparent stiffness $\mathbf{C}_0^{(app,\ d)}$ under displacement boundary conditions as defined in equation 10.22.2. Next, divide the sample into n coarse subsamples of equal size and shape. Each subsample has apparent moduli determined under displacement boundary conditions $\mathbf{C}_{c,\ i}^{(app,\ d)}$, $i = 1, 2, \ldots n$, and apparent compliance determined under traction boundary conditions $\mathbf{S}_{c,\ i}^{(app,\ t)}$, $i = 1, 2, \ldots n$, where "c" designates these as "coarse" subsamples. The arithmetic average stiffness $\overline{\mathbf{C}_c^{(app,\ d)}} = \dfrac{1}{n}\sum_{i=1}^{n} \mathbf{C}_{c,\ i}^{(app,\ d)}$ and average compliance $\left(\overline{\mathbf{C}_c^{(app,\ t)}}\right)^{-1} = \overline{\mathbf{S}_c^{(app,\ t)}} = \dfrac{1}{n}\sum_{i=1}^{n} \mathbf{S}_{c,\ i}^{(app,\ t)}$ over all of the coarse subsamples satisfy the relations $\overline{C_c^{(app,\ t)}} \le \mathbf{C}_0^{(app,\ t)} \le \mathbf{C}_0^{(app,\ d)} \le \overline{\mathbf{C}_c^{(app,\ d)}}$. Each of the coarse subsamples can be further divided into m "fine" subsamples of equal size and shape, which have apparent displacement moduli $\mathbf{C}_{f,\ j}^{(app,\ d)}$ and traction compliances $S_{f,\ j}^{(app,\ t)}$, $j = 1, 2, \ldots nm$. Averaging over all of the fine subsample moduli and compliances yields $\left(\overline{\mathbf{C}_f^{(app,\ t)}}\right)^{-1} = \overline{\mathbf{S}_f^{(app,\ t)}} = \dfrac{1}{nm}\sum_{j=1}^{nm} \mathbf{S}_{f,\ j}^{(app,\ t)}$ and $\left(\overline{\mathbf{S}_f^{(app,\ d)}}\right)^{-1} = \overline{\mathbf{C}_f^{(app,\ d)}}$ $= \dfrac{1}{nm}\sum_{j=1}^{nm} \mathbf{C}_{f,\ j}^{(app,\ d)}$, which satisfy the inequalities:

$$\overline{\mathbf{C}_f^{(app,\ t)}} \le \overline{\mathbf{C}_c^{(app,\ t)}} \le \mathbf{C}_0^{(app,\ t)} \le \mathbf{C}_0^{(app,\ d)} \le \overline{\mathbf{C}_c^{(app,\ d)}} \le \overline{\mathbf{C}_f^{(app,\ d)}} \tag{4.22.6}$$

Subsampling can continue until each piece contains only a single phase, so that averaging over the subsamples yields the Voigt and Reuss bounds on modulus. We then find

$$\overline{\mathbf{C}_{0,\ Reuss}} \le \overline{\mathbf{C}_f^{(app,\ t)}} \le \overline{\mathbf{C}_c^{(app,\ t)}} \le \mathbf{C}_0^{(app,\ t)} \le \mathbf{C}_0^{(app,\ d)} \le \overline{\mathbf{C}_c^{(app,\ d)}} \le \overline{\mathbf{C}_f^{(app,\ d)}} \le \overline{\mathbf{C}_{0,\ Voigt}} \tag{4.22.7}$$

Of course, if the sample D_0 is large enough, then $\mathbf{C}_0^{(app,\ t)} = \mathbf{C}_0^{(d)} = \mathbf{C}_0^{(eff)}$.

The partition theorem helps us better understand how measurements or simulations on multiple subsamples lead to bounds on the desired effective medium property.

4.23   A Few Theorems on Strain in an Effective Medium

For a two-phase isotropic elastic composite one can write

$$P_1 = \frac{K_{eff} - K_2}{f_1(K_1 - K_2)} \tag{4.23.1}$$

$$P_2 = \frac{1 - f_1 P_1}{f_2} = \frac{K_{eff} - K_1}{f_2(K_2 - K_1)} \tag{4.23.2}$$

A similar statement can be made for the effective shear modulus:

$$Q_1 = \frac{\mu_{eff} - \mu_2}{f_1(\mu_1 - \mu_2)} \tag{4.23.3}$$

$$Q_2 = \frac{1 - f_1 Q_2}{f_2} = \frac{\mu_{eff} - \mu_1}{f_1(\mu_2 - \mu_1)} \tag{4.23.4}$$

where μ_{eff} is the effective shear modulus, and μ_i is the shear modulus of the ith phase. Parameter $Q_i = \|\langle\gamma\rangle_i\| / \|\langle\gamma\rangle\|$ is the scalar *deviatoric strain concentration factor*, the mean deviatoric strain norm in the ith phase divided by the mean deviatoric strain norm in the entire sample, when remote uniform deviatoric strain is applied. Thus, from equations 4.23.2–4.23.4, we see that the *mean* hydrostatic and deviatoric strain concentrations in each component will generally be different, and can be found from the effective moduli. (We are not aware of a comparable way to find the deviatoric field induced by remote volumetric strain and the volumetric field induced by remote deviatoric strain.) Values of the effective moduli can come from measurements, rock physics models, or numerical simulations.

If the applied strain field has *no deviatoric part*, it can be shown that the induced squared hydrostatic and deviatoric strains in each phase can be related to the derivatives of the effective moduli as follows (Bergman, 1978; Beran, 1981; and Bobeth and Diener, 1986):

$$\xi_{KK(i)} = \frac{\langle Tr(\boldsymbol{\varepsilon})^2\rangle_i}{\langle Tr(\boldsymbol{\varepsilon})\rangle^2} = \frac{1}{f_i} \frac{\partial K_{eff}}{\partial K_i} \tag{4.23.5}$$

$$\xi_{K\mu(i)} = \frac{\langle\|\boldsymbol{\gamma}\|^2\rangle_i}{\langle Tr(\boldsymbol{\varepsilon})\rangle^2} = \frac{\langle\gamma_{jk}\gamma_{jk}\rangle_i}{\langle Tr(\boldsymbol{\varepsilon})\rangle^2} = \frac{1}{2f_i} \frac{\partial K_{eff}}{\partial \mu_i} \tag{4.23.6}$$

In equations 4.23.5 and 4.23.6, $Tr(\boldsymbol{\varepsilon})$ and γ are the hydrostatic and deviatoric strains induced by a *purely hydrostatic remote field*.

If the applied strain is purely *deviatoric*,

$$\eta_{\mu K(i)} = \frac{\langle Tr(\boldsymbol{\varepsilon})^2\rangle_i}{\|\langle\boldsymbol{\gamma}\rangle\|^2} = \frac{2}{f_i} \frac{\partial \mu_{eff}}{\partial K_i} \tag{4.23.7}$$

$$\eta_{\mu\mu(i)} = \frac{\langle\|\boldsymbol{\gamma}\|^2\rangle_i}{\|\langle\boldsymbol{\gamma}\rangle\|^2} = \frac{\langle\gamma_{jk}\gamma_{jk}\rangle_i}{\|\langle\boldsymbol{\gamma}\rangle\|^2} = \frac{1}{f_i} \frac{\partial \mu_{eff}}{\partial \mu_i} \tag{4.23.8}$$

In equations 4.237 and 4.238, $Tr(\varepsilon)$ and γ are the hydrostatic and deviatoric strains induced by a *purely deviatoric remote field*.

The normalized variances of the volumetric and deviatoric strain fields in the *ith* phase under hydrostatic and deviatoric remote loading, respectively, are

$$\left(\sigma_{\varepsilon(i)}\right)^2 / \langle Tr(\boldsymbol{\varepsilon}) \rangle^2 = \langle [Tr(\boldsymbol{\varepsilon}) - \langle Tr(\boldsymbol{\varepsilon})_i \rangle]^2 \rangle_i = \langle Tr(\boldsymbol{\varepsilon})^2 \rangle_i - \langle Tr(\boldsymbol{\varepsilon}) \rangle_i^2 = \zeta_{KK}^{(i)} - P_i^2$$

$$(4.23.9)$$

$$\left(\sigma_{\gamma(i)}\right)^2 / \|\langle \boldsymbol{\gamma} \rangle\|^2 = \langle [\gamma_{jk} - \langle \gamma_{jk} \rangle_i]^2 \rangle = \langle \|\boldsymbol{\gamma}\|^2 \rangle_i - \|\boldsymbol{\gamma}\|_i^2 = \eta_{\mu\mu}^{(i)} / f_i - Q_i^2 \qquad (4.23.10)$$

So we see that from the effective modulus and the derivative of the effective moduli, we can get the mean and variance of the load-induced strains in each phase.

5 Granular Media

5.1 Packing and Sorting of Spheres and Irregular Particles

Synopsis

Spheres are often used as idealized representations of grains in unconsolidated and poorly consolidated sands. They provide a means of quantifying geometric relations, such as the porosity and the coordination number, as functions of packing and sorting. Using spheres also allows an analytical treatment of mechanical grain interactions under stress. Recent research also started to reveal how the properties of the granular medium change as particle shapes differ from perfect spheres.

Packings of Identical Spheres

Porosity

Packing of identical spheres has been studied most, often as a means to represent very well-sorted sand. Table 5.1.1 lists geometric properties of various packings of identical spheres (summarized from Cumberland and Crawford, 1987, and Bourbié *et al.*, 1987). The first four rows describe perfectly ordered sphere packs. These are directly analogous to the atomic arrangements in crystals having the same symmetries, although natural sands will never achieve such order. The last row in Table 5.1.1 is for a random, close packing of identical spheres. (Note that complementary interpretations are possible, depending on whether the grains or the pores are considered to be spheres.)

Random packing of spheres has been studied both experimentally and theoretically (Bernal and Mason, 1960; Makse *et al.*, 1999, 2004; Garcia and Medina, 2006; Sain, 2010; Baule and Makse, 2014). Denton (1957) performed 116 repeated experiments of moderately dense, random packings of spheres. His results highlight the statistical nature of describing grain packs (hence the use of the term **random packings**). He found a mean porosity of 0.391 with a standard deviation of 0.0016. Cumberland and Crawford (1987) found that Denton's porosity distribution over the many realizations could be described well using a beta distribution, and reasonably well with a normal distribution.

Smith *et al.* (1929) determined the porosity of random sphere packs experimentally using several preparation techniques, summarized in Table 5.1.2. They observed that loose random packings resulting from pouring spheres into a container had a porosity

Table 5.1.1 *Geometric properties of packs of identical spheres. Radii of maximum inscribable sphere and maximum sphere fitting in narrowest channel can be analyzed using Descarte's theorem, discussed later in this section.*

Packing type	Porosity (nonspheres)	Solid fraction (spheres)	Void ratio[d]	Specific surface area[a]	Number of contacts per sphere	Radius of maximum inscribable sphere[b,c]	Radius of maximum sphere fitting in narrowest channels[b,c]
Simple cubic	$1 - \pi/6 = 0.476$	$\pi/6 = 0.524$	0.910	$\pi/2R$	6	0.732	0.414
Simple hexagonal	$1 - 4\pi \cos(\pi/6)/18 = 0.395$	$4\pi \cos(\pi/6)/18 = 0.605$	0.654	$\pi/\sqrt{3}R$	8	0.528	0.414 and 0.155
Tetragonal	$1 - \pi/4.5 = 0.3019$	0.6981	0.432	$2\pi/3R$	10		
Hexagonal close pack	0.259	0.741	0.350	$\pi/\sqrt{2}R$	12	0.225 and 0.414	0.155
Face centered cubic	0.259	0.741	0.350	$\pi/\sqrt{2}R$			
Dense random pack	~0.36	~0.64	~0.56	~1.92/R	~9		

Notes:

[a] Specific surface area, S, is defined as the pore surface area in a sample divided by the total volume of the sample. If the grains are spherical, $S = 3(1 - \varphi)/R$.

[b] Expressed in units of the radius of the packed spheres.

[c] Note that if the pore space is modeled as a packing of spherical pores, the inscribable spheres always have radius equal to 1.

[d] Void ratio is the volume of voids divided by volume of solids.

Table 5.1.2 *Experimental data on coordination numbers in random packings of spheres, from Smith et al. (1929)*

Method of random packing	Total number of spheres counted	Distribution of contacts per sphere										Mean coordination number*	Porosity
		4	5	6	7	8	9	10	11	12		
Poured into beaker	905	6	78	243	328	200	48	2	0	0	6.87±1.05	0.447
Poured and shaken	906	3	54	173	309	233	118	14	2	0	7.25±1.16	0.440
Poured and shaken	887	0	14	69	182	316	212	87	7	0	8.05±1.17	0.426
Shaken to maximum density	1494	0	14	86	192	233	193	161	226	389	9.57±2.02	0.372
Added in small quantities and intermittently tamped	1562	1	13	77	245	322	310	208	194	192	9.05±1.78	0.359

Note:

* Column shows mean ±1 standard deviation.

of ~0.447. Shaking systematically decreased porosity, to a minimum of 0.372. Slight tamping reduced the porosity further to a random close-packed value of 0.359. More accurate (Cumberland and Crawford, 1987) experimental values for the porosity of random close packs of 0.3634 ± 0.008 and 0.3634 ± 0.004 were found by Scott and Kilgour (1969) and Finney (1970), respectively.

Coordination Number

The **coordination number** of a grain pack is the average number of contacts that each grain has with surrounding grains. Table 5.1.1 shows that the coordination numbers for regular packings of identical spheres range from a low of 6 for a simple cubic packing, to a high of 12 for hexagonal close packing. Coordination numbers in random packings have been determined by tediously counting the contacts in experimentally prepared samples (e.g., Smith *et al.*, 1929; Wadsworth, 1960; Bernal and Mason, 1960). Table 5.1.2 shows results from Smith *et al.* (1929), after counting more than 5000 spheres in five different packings. There are several important conclusions.

- The coordination number increases with decreasing porosity, both being the result of tighter packing.
- Random packs of identical spheres have coordination numbers ranging from ~6.9 (loose packings) to ~9.1 (tight packings).

The coordination number can also refer to the number of contacts that *each* grain has with neighboring grains – we might call it the *local* coordination number. In a random packing of spheres, the local coordination number varies widely throughout each sample. The data of Smith *et al.* (1929), shown in Table 5.1.2 reveal a distribution of contacts per sphere ranging from 4 to 12. Many models for the effective elastic moduli of grain packs are formulated only in terms of the mean. Numerical results indicate that effective elastic moduli of random packs, especially the shear modulus, depend on both the mean and the variance of the contacts numbers. (see Section 5.5)

Tables 5.1.3–5.1.5 show data illustrating a correlation between average coordination number and porosity. Manegold and von Engelhardt (1933) described theoretically the arrangements of identical spheres. They chose a criterion of the most regular packing possible for porosities ranging from 0.2595 to 0.7766. Their results are summarized in Table 5.1.4. Murphy (1982) also compiled coordination number data from the literature, summarized in Table 5.1.5.

Figure 5.1.1 compares coordination number versus porosity from Smith *et al.* (1929), Manegold and von Engelhardt (1933), and Murphy (1982). The data from various sources are consistent with each other and with the exact values for hexagonal close packing, simple hexagonal packing, and simple cubic packing. The thin lines show the approximate range of ± one standard deviation of coordination number observed by Smith *et al.* (1929). The standard deviation describes the spatial variation of coordination number throughout the packing, and not the standard deviation of the mean.

Table 5.1.3 *Porosity and coordination number data for randomly packed identical spheres, from* Smith *et al.* *(1929)*

Porosity	Coordination number
0.447	6.87 ± 1.05
0.440	7.25 ± 1.16
0.426	8.05 ± 1.17
0.372	9.57 ± 2.02
0.359	9.05 ± 1.78

Table 5.1.4 *Porosity and coordination number data for randomly packed identical spheres, from Manegold and von Engelhardt (1933)*

Porosity	Coordination number
0.7766	3
0.6599	4
0.5969	5
0.4764	6
0.4388	7
0.3955	8
0.3866	9
0.3019	10
0.2817	11
0.2595	12

One source of uncertainty in the experimental estimation of coordination number is the difficulty of distinguishing between actual grain contacts and near-grain contacts. For the purposes of understanding porosity and transport properties, this distinction might not be important. However, mechanical and elastic properties in granular media are determined entirely by load-bearing grain contacts. For this reason, one might speculate that the equivalent coordination numbers for mechanical applications might be smaller than those shown in Tables 5.1.1–5.1.5. Wadsworth (1960) argued that coordination numbers in random packings may be smaller than those discussed so far, reduced by about 1.

Garcia and Medina (2006) derived a power-law relation between the mean coordination number and the porosity based on fitting data from numerical simulations of granular media:

$$C = C_0 + 9.7(\phi_0 - \phi)^{0.48} \tag{5.1.1}$$

Table 5.1.5 *Porosity and coordination number data for randomly packed spheres, compiled by Murphy (1982). Not all sphere sizes are identical at the two smallest porosities.*

Porosity	Coordination number
0.20	14.007
0.25	12.336
0.3	10.843
0.35	9.508
0.40	8.315
0.45	7.252
0.50	6.311
0.55	5.488
0.60	4.783

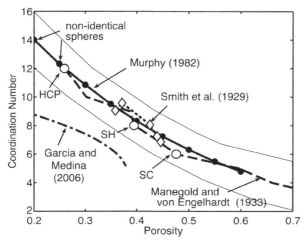

Figure 5.1.1 Coordination number versus porosity in random sphere packs, from Smith *et al.* (1929), Manegold and von Engelhardt (1933), Murphy (1982), and Garcia and Medina (2006). HCP = hexagonal close packed; SH = simple hexagonal; and SC = simple cubic. The thin lines approximately indicate ±1 standard deviation from the data of Smith *et al.* (1929).

where $C_0 = 4.46$ and $\phi_0 = 0.384$. The relation holds for $\phi \leq \phi_0$. These relations between coordination number and porosity can be used in effective medium models such as those described later in Section 5.4.

Makse *et al.* (2004) derived a similar relation using numerical simulation of frictionless spheres:

$$C = C_0 + 9.1(\phi_0 - \phi)^{0.48} \tag{5.1.2}$$

where $C_0 = 6$ and $\phi_0 = 0.37$. The Makse *et al.* relation parallels that of Garcia and Medina, shifted upward by about 1.5.

Pores and Throats: Descarte's Theorem and Soddy Circles

Descarte's theorem allows us to analyze the sizes of the maximum inscribed sphere and the maximum sphere fitting in the narrowest throats, as listed in Table 5.1.1. Descarte's theorem states that in two-dimensional Euclidean space, if four circles are mutually tangent at six distinct points, then the radii r_i are related by (Soddy, 1936; Assouline and Rouault, 1997):

$$\left(\frac{1}{r_1} + \frac{1}{r_2} + \frac{1}{r_3} + \frac{1}{r_4}\right)^2 = 2\left(\frac{1}{r_1^2} + \frac{1}{r_2^2} + \frac{1}{r_3^2} + \frac{1}{r_4^2}\right). \tag{5.1.3}$$

In other words, given three circles that are tangent at three distinct points, there are exactly two circles that are mutually tangent to the three, with radii given by

$$\frac{1}{r_4} = \left|\frac{1}{r_1} + \frac{1}{r_2} + \frac{1}{r_3} \pm 2\sqrt{\frac{1}{r_1 r_2} + \frac{1}{r_2 r_3} + \frac{1}{r_3 r_1}}\right|. \tag{5.1.4}$$

One solution is for the inscribed circle, and one is for the circumscribed circle (see Figure 5.1.2).

If three circles have equal radii, $r_1 = r_2 = r_3$, the two solutions for the mutually tangent circles are

$$\frac{1}{r_4} = \left|\frac{3}{r_1} \pm \frac{2\sqrt{3}}{r_1}\right| \tag{5.1.5}$$

In three-dimensional Euclidean space, the maximum number of mutually tangent *spheres* is five. The radii are related by

$$\left(\frac{1}{r_1} + \frac{1}{r_2} + \frac{1}{r_3} + \frac{1}{r_4} + \frac{1}{r_5}\right)^2 = 3\left(\frac{1}{r_1^2} + \frac{1}{r_2^2} + \frac{1}{r_3^2} + \frac{1}{r_4^2} + \frac{1}{r_5^2}\right). \tag{5.1.6}$$

Given four spheres that are tangent at four distinct points, there are exactly two spheres that are mutually tangent to the four, with radii given by

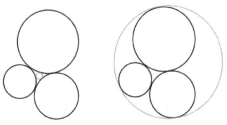

Figure 5.1.2 Inscribed and circumscribed circles given by equation 5.1.4

$$\frac{2}{r_5} = \left| \left(\frac{1}{r_1} + \frac{1}{r_2} + \frac{1}{r_3} + \frac{1}{r_4} \right) \pm \sqrt{3} \sqrt{ \left(\frac{1}{r_1} + \frac{1}{r_2} + \frac{1}{r_3} + \frac{1}{r_4} \right)^2 - 2 \left(\frac{1}{r_1^2} + \frac{1}{r_2^2} + \frac{1}{r_3^2} + \frac{1}{r_4^2} \right) } \right| \qquad (5.1.7)$$

If four of the spheres have equal radii, $r_1 = r_2 = r_3 = r_4$, the solutions for the inscribed and circumscribed mutually tangent spheres are

$$\frac{1}{r_5} = \left| \frac{2}{r_1} \pm \frac{\sqrt{6}}{r_1} \right|. \qquad (5.1.8)$$

The solution for finding the maximum inscribed and minimum circumscribed spheres to four spheres 1, 2, 3, 4 that are not necessarily in contact, has been found by Mackay (1973). The spheres have radii r_1, r_2, r_3, and r_4, and the fifth sphere has unknown radius r_5. The relative positions of the centers of the five spheres are given by vectors $\mathbf{v}_{12}, \mathbf{v}_{13}$, $\mathbf{v}_{14}$, and $\mathbf{v}_{15}$. The distances between the centers of the ith and jth spheres are denoted as d_{ij}. The radius of the fifth sphere can be found as the roots of the determinant relation

$$D(r_5) = \begin{vmatrix} 0 & d_{12}^2 & d_{13}^2 & d_{14}^2 & (r_1 + r_5)^2 & 1 \\ d_{12}^2 & 0 & d_{23}^2 & d_{24}^2 & (r_2 + r_5)^2 & 1 \\ d_{13}^2 & d_{23}^2 & 0 & d_{34}^2 & (r_3 + r_5)^2 & 1 \\ d_{14}^2 & d_{24}^2 & d_{34}^2 & 0 & (r_4 + r_5)^2 & 1 \\ (r_1 + r_5)^2 & (r_2 + r_5)^2 & (r_3 + r_5)^2 & (r_4 + r_5)^2 & 0 & 1 \\ 1 & 1 & 1 & 1 & 1 & 0 \end{vmatrix} = 0 \qquad (5.1.9)$$

The positive root corresponds to the inscribed circle and the negative root corresponds to the circumscribed circle.

Binary Mixtures of Spheres – Ideal Mixture Model

Binary mixtures of spheres (i.e., two different sphere sizes) add additional complexity to random packings. Cumberland and Crawford (1987) review the specific problem of predicting porosity. Figure 5.1.3 compares theoretical models for porosity (Cumberland and Crawford, 1987) in an **ideal binary mixture**. The ideal mixture is composed of two different sphere sizes, where R is the ratio of the large sphere diameter to the small sphere diameter. The horizontal axis is the volume fraction of large grains relative to the total solid volume. Point A represents a random close packing of only small spheres, while point B represents a random packing of only large spheres. Since A and B each correspond to random packings of identical spheres, the porosity in each case should be close to 0.36. The lower curves, ACB, correspond to mixtures where R approaches infinity (practically, $R > 30$). Moving from B toward C corresponds to gradually adding small spheres to the pore space of the large sphere pack, gradually decreasing porosity. In the ideal model, the packing of each set of spheres is assumed to be undisturbed by the presence of the other set (*the definition of an ideal binary mixture*). The minimum

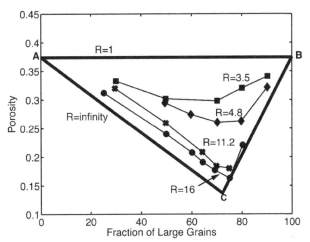

Figure 5.1.3 Porosity in an ideal binary mixture. R is the ratio of the large grain diameter to the small grain diameter. Curves ABC are bounds; highest porosity occurs when $R = 1$, and the lowest occurs when R becomes very large. Data for carefully prepared laboratory binary mixtures are from McGeary (1967).

porosity occurs at point C, where the pore space of the large sphere pack is completely filled by the smaller sphere pack. The total porosity at point C is $\phi = (0.36)^2 = 0.13$ and occurs at a volume fraction of large grains ~0.73. Moving from A toward C corresponds to gradually adding more large grains to the packing of small grains. Each solid large grain replaces a spherical region of the porous small grain packing, hence decreasing porosity. Curve AB corresponds to $R = 1$. In this case, all spheres are identical, so the porosity is always ~0.36. All other ideal binary mixtures should fall within the triangle ABC. Data from McGeary (1967) for ratios $R = 16, 11.2, 4.8$, and 3.5, are plotted and are observed to lie within the expected bounds. It should be pointed out that McGeary's experiments were designed to approach an ideal mixture. First, a close random packing of large spheres was created, and then small particles were added, while keeping a weight on the packing to minimize disturbing the large grain pack. The ideal binary mixture model has been used successfully to describe porosities in mixtures of sand and clay (Marion and Nur, 1991; Marion et al., 1992; Yin, 1992; Avseth, 2000). The binary mixture model is a key component of the Thomas–Stieber (1975, 1977) model discussed in Section 5.3.

Binary Mixtures of Spheres – Non-Ideal Mixing

We now look at deviations from ideal binary mixing. The porosity of a *bimodal* random sphere pack depends on the ratio of small to large particle diameters, $\delta = d/D = 1/R$, the volume fraction, x_D, of large particles, the volume fraction, $x_d = 1 - x_D$, of small particles, and the packing. (Note that x_D and x_d are the fractions of the solid phase.)

The absolute minimum porosity of a random bimodal sphere pack occurs when $\delta \rightarrow 0$, ($R \rightarrow \infty$) and when the small particle pack is placed within the pore spaces of an undisturbed large particle pack (Figure 5.1.4, and Table 5.1.6 row b). The total

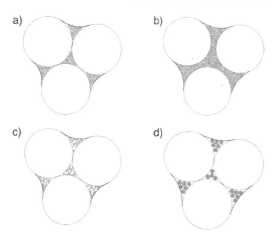

Figure 5.1.4 Examples of bimodal sphere packs. a: large particles touching as though no small particles are present; small particles fill the spaces; in this case $\delta \approx 0.03$; absolute minimum porosity occurs as $\delta \to 0$. b, c: porosity larger than the absolute minimum; in this case $\delta \approx 0.03$. d: porosity larger than absolute minimum as δ grows and small particle packing is disturbed.

porosity at this point is $\phi^0_{min} = \phi^0_D \phi^0_d$, where ϕ^0_d is the porosity of the small grain pack in the absence of large grains, and ϕ^0_D is the porosity of the large grain pack in the absence of small grains. The volume fraction of large grains at this state is $x^0_{Dmin} = (1 - \phi^0_D)/(1 - \phi^0_D \phi^0_d)$. The porosity grows larger than ϕ^0_{min} if any of the following occurs:

(1) The solid fraction of large particles x_D becomes less than x^0_{Dmin} as the large particles separate and remain filled by small particles (Figure 5.1.4b, and Table 5.1.6 row a). If $\delta \to 0$, the porosity is

$$\phi = \frac{\phi^0_d(1 - x_D)}{(1 - x_D \phi^0_d)}, \text{ and } \phi^0_D \phi^0_d \leq \phi \leq \phi^0_d. \tag{5.1.10}$$

(2) The solid fraction of large particles becomes larger than x^0_{Dmin} as the large particles stay in contact but the pore space is only partially filled by small particles (Figure 5.1.4c, and Table 5.1.6, row c). If $\delta \to 0$, the porosity is

$$\phi = \frac{x_D + \phi^0_D - 1}{x_D}, \text{ and } \phi^0_D \phi^0_d \leq \phi \leq \phi^0_D. \tag{5.1.11}$$

(3) The particle size ratio δ grows and the packing efficiency of small particles between the large particles decreases (Figure 5.1.4d, and Table 5.1.6, rows d, e, and f). If the large particles remain in contact with the maximum possible number of small particles then (Table 5.1.6, row e), the porosity is estimated to be (Dias et al., 2006)

$$\phi'_{min} = \phi^0_D \phi^0_d + \phi^0_d(1 - \phi^0_D)\exp\left(0.25(1 - 1/\delta)\right). \tag{5.1.12}$$

Table 5.1.6 *Estimated parameters of bimodal random sphere packs. The ratio of small to large particle radii is δ. Total porosity is φ. Large particle fraction is x_D.*

	δ	ϕ	x_D
a) Very fine grains filling separated large grains.	$\delta \to 0$	$\phi_D^0\phi_d^0 \leq \dfrac{\phi_d^0(1-x_D)}{(1-x_D\phi_d^0)} \leq \phi_d^0$	$\dfrac{1-\phi_D^0}{1-\phi_D^0\phi_d^0} \geq \dfrac{\phi_d^0 - \phi}{\phi_d^0(1-\phi)} \geq 0$
b) Absolute minimum ϕ: very fine grains filling tight large grain pack.	$\delta \to 0$	$\phi_{\min} = \phi_D^0\phi_d^0$	$x_{D\min}^0 = \dfrac{1-\phi_D^0}{1-\phi_D^0\phi_d^0}$
c) Very fine grains under-fill tight large grain pack.	$\delta \to 0$	$\phi_D^0\phi_d^0 \leq \dfrac{x_D + \phi_D^0 - 1}{x_D} \leq \phi_D^0$	$\dfrac{1-\phi_D^0}{1-\phi_D^0\phi_d^0} \leq \dfrac{1-\phi_D^0}{1-\phi} \leq 1$
d) Tight pack of medium grains within separated large grains.	$0 < \delta < .3$	$\phi'_{\min} \leq \dfrac{\phi_d^0(1-x_D)}{(1-x_D\phi_d^0)} F \leq \phi_d^0;\; F = \exp\left(1.2264\,x_D^{1/\sqrt{\delta}}\right)$	$x'_{D\min} \geq \dfrac{(\phi_d^0 F - \phi)}{\phi_d^0(F - \phi)} \geq 0$
e) Best pack of medium grains in tight large grain pack.	$0 < \delta < .3$	$\phi'_{\min} = \phi_D^0\phi_d^0 + \phi_d^0(1-\phi_D^0)E;$ $E = \exp\big((1-1/\delta)/4\big)$	$x'_{D\min} = \dfrac{1-\phi_D}{1-\phi_d^0\phi_D^0 - \phi_d^0(1-\phi_D^0)E};$ $E = \exp\big((1-1/\delta)/4\big)$
f) Sparse medium grains in tight large grain pack.	$0 < \delta < .3$	$\phi'_{\min} < \dfrac{x_D + \phi_D^0 - 1}{x_D} \leq \phi_D^0$	$x'_{D\min} \geq \dfrac{1-\phi_D^0}{1-\phi} \geq 0$
g) Large monosphere pack.	$\delta = 1$	ϕ_D^0	1

ϕ_D^0 = porosity of large grain pack without small particles

ϕ_d^0 = porosity of small grain pack without large particles

x_D = fraction of solid phase composed of large particles.

x_d = fraction of solid phase composed of small particles $= 1 - x_D$

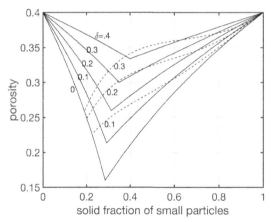

Figure 5.1.5 Porosity models for bimodal packs of spheres as a function of volume fraction of small particles, x_d, and ratio of small to large particle radii, δ. Dashed curves are from equations 5.1.13–5.1.14. Solid curves are from equations 5.1.19–5.1.23.

If the large particles separate with maximum number of small particles between them (Table 5.1.6, row d), the porosity is estimated to be (Dias *et al.*, 2006)

$$\phi = \frac{\phi_d^0 (1 - x_D)}{(1 - x_D \phi_d^0)} \exp\left(1.2264 x_D^{1/\sqrt{\delta}}\right), \text{ and } \phi'_{min} \leq \phi \leq \phi_d^0. \tag{5.1.13}$$

If the large particles remain in contact, but the number of small particles decreases (Table 5.1.6, row f), the porosity is

$$\phi = \frac{x_D + \phi_D^0 - 1}{x_D}, \text{ and } \phi'_{min} < \phi \leq \phi_D^0. \tag{5.1.14}$$

The dashed curves in Figure 5.1.5 show estimated porosity trends from equations 5.1.13 and 5.1.14 as functions of δ and x_d. The lowest set of curves correspond to $\delta \to 0$ and are sometimes referred to as the "ideal binary mixture model," (Cumberland and Crawford, 1987). Curves corresponding to $0 < \delta$ deviate from the ideal mixture due to distortions of the fine particle pack caused by the large particles and distortions of the large particle pack caused by the fine particles. These distortions increase the porosity.

An alternative description of porosity increase due to non-ideal mixtures is the "fractional packing model" (Kolterman and Gorelick, 1995), which attempts to approximate non-optimal packing in real bimodal mixtures. Their empirical expressions for porosity are equivalent to

$$\phi = \phi_{ideal} + \Delta\phi_{KG}\left(\frac{c}{\phi_D^0}\right)^2 \; ; \; c \leq \phi_D^0, \tag{5.1.15}$$

$$\phi = \phi_{ideal} + \Delta\phi_{KG}\left(\frac{1 - c}{1 - \phi_D^0}\right) \; ; \; c \geq \phi_D^0, \tag{5.1.16}$$

where ϕ_{ideal} is the ideal mixture porosity (Table 5.1.6, rows a, b, c), c is the volume fraction of porous fine particles (solid plus pore space), and $\Delta\phi_{KG} > 0$ is an empirical factor, corresponding to the porosity correction when $c = \phi_D^0$. These can be rewritten replacing c in terms of x_D:

$$\phi = \phi_{ideal} + \Delta\phi_{KG}\left(\frac{1 - x_D}{\phi_D^0(1 - x_D\phi_D^0)}\right)^2 \; ; \; c \leq \phi_D^0 , \tag{5.1.17}$$

$$\phi = \phi_{ideal} + \Delta\phi_{KG}\left(\frac{x_D(1 - \phi_d^0)}{(1 - \phi_D^0)(1 - x_D\phi_d^0)}\right) \; ; \; c \geq \phi_D^0 . \tag{5.1.18}$$

The *linear packing model* for binary mixtures (Yu and Standish, 1987) of spheres begins with an empirical relation (Ridgway and Tarbuck, 1968) for the minimum porosity ϕ_{min} at the maximum packing of large and small particles:

$$\phi_{min} = \begin{cases} \phi_0 - \phi_0(1 - \phi_0)(1 - 2.35\delta + 1.35\delta^2), & \delta \leq 0.741 \\ \phi_0, & \delta > 0.741 \end{cases} \tag{5.1.19}$$

where it is assumed that $\phi_d^0 = \phi_D^0 = \phi_0$. The corresponding maximum packing is

$$p_{max} = 1 - \phi_{min} = (1 - \phi_0) + \phi_0(1 - \phi_0)(1 - 2.35\delta + 1.35\delta^2) , \tag{5.1.20}$$

The relative solid fractions at maximum packing are

$$X_{L\,max} = \frac{1 - \delta^2}{1 + \phi_0} \tag{5.1.21}$$

$$X_{S\,max} = 1 - \frac{1 - \delta^2}{1 + \phi_0} \tag{5.1.22}$$

At any other fractions, the inverse packing is a linear interpolation between the inverse end-member packings, $1/p_0 = 1/(1 - \phi_0)$, and the minimum inverse packing, $1/p_{max}$:

$$\frac{1}{p} = \begin{cases} \frac{1}{p_0} - \left(\frac{1}{p_0} - \frac{1}{p\,\max(\delta)}\right)\frac{X_L}{X_{L\,max}}, & 0 \leq X_L \leq X_{L\,max} \\ \frac{1}{p_0} - \left(\frac{1}{p_0} - \frac{1}{p\,\max(\delta)}\right)\frac{X_S}{X_{S\,max}}, & 0 \leq X_S \leq X_{S\,max} \end{cases} \tag{5.1.23}$$

The porosity of the mixture $\phi = 1 - p$ is plotted in Figure 5.1.5 for different values of δ.

Non-Spherical Particles

Particle shape is known to affect the porosity, permeability, and mechanical properties of random packings. Table 5.1.7 compares measurements from Richardson *et al.* (2002) for **regular geometric grain shapes**. Table 5.1.8 shows additional values of

Table 5.1.7 *Specific surface area, porosity, permeability of beds of some regular-shaped materials (Richardson* et al., *2002)*

Solid constituents	Porous mass		
Description	Specific surface area $S(m^2/m^3)$	Fractional voidage, $e(-)$	Permeability coefficient B (m^2)
Spheres			
0.794 mm diam.	7600	0.393	6.2×10^{-10}
1.588 mm diam.	3759	0.405	2.8×10^{-9}
3.175 mm diam.	1895	0.393	9.4×10^{-9}
6.35 mm diam.	948	0.405	4.9×10^{-8}
7.94 mm diam.	756	0.416	9.4×10^{-8}
Cubes			
3.175 mm	1860	0.190	4.6×10^{-10}
3.175 mm	1860	0.425	1.5×10^{-8}
6.35 mm	1078	0.318	1.4×10^{-8}
6.35 mm	1078	0.455	6.9×10^{-8}
Hexagonal prisms			
4.76 mm x 4.76 mm thick	1262	0.355	1.3×10^{-8}
4.76 mm x 4.76 mm thick	1262	0.472	5.9×10^{-8}
Triangular pyramids			
6.35 mm length x 2.87 mm ht.	2410	0.361	6.0×10^{-9}
6.35 mm length x 2.87 mm ht.	2410	0.518	1.9×10^{-8}
Cylinders			
3.175 mm x 3.175 mm diam.	1840	0.401	1.1×10^{-8}
3.175 mm x 6.35 mm diam.	1585	0.397	1.2×10^{-8}
6.35 mm x 6.35 mm diam.	945	0.410	4.6×10^{-8}
Plates			
6.35 mm x 6.35 mm x 0.794 mm	3033	0.410	5.0×10^{-9}
6.35 mm x 6.35 mm x 1.59 mm	1984	0.409	1.1×10^{-8}
Discs			
3.175 mm diam. x 1.59 mm	2540	0.398	6.3×10^{-9}

measured and modeled minimum porosities of random packings of regular-shaped particles (Baule and Makse, 2014).

One measure of particles shape is **sphericity** ψ (Waddell, 1932), defined as:

$$\psi = \frac{S_{sphere}}{S_{grain}} \tag{5.1.24}$$

Table 5.1.8 *Measured and modeled close-packed porosities of random packings of particles with regular shapes (modified from a compilation of Baule and Makse, 2014)*

Shape	ϕ_{min} simulation	ϕ_{min} experiment	ϕ_{min} theory
Sphere	0.355[a]	0.360[g]	0.366[j]
M&M candy		0.335[c]	
Dimer	0.297[b]		0.293[k]
Oblate ellipsoid	0.293[c]		
Prolate ellipsoid	0.284[c]		
Spherocylinder	0.278[d]		0.269[k]
Lens-shaped particle			0.264[k]
Octahedron	0.303[e]		
Icosahedron	0.293[e]		
Dodecahedron	0.284[e]		
General ellipsoid	0.265[c]	0.26[h]	
Tetrahedron	0.2142[f]	0.24[i]	

[a]Skoge *et al.* (2006); [b]Faure *et al.* (2009); [c]Donov *et al.* (2004); [d]Zhao *et al.* (2012); [e]Jiao and Torquato (2011); [f]Haji-Akbar *et al.* (2009); [g]Bernal and Mason (1960); [h]Man *et al.* (2005); [i]Jaoshvili *et al.* (2010); [j]Song *et al.* (2008); [k]Baule *et al.* (2013)

where S_{sphere} is the surface area of a sphere having the same volume as the particle, and S_{grain} is the actual surface area of the particle. Brown *et al.* (1966) observed experimentally that small deviations from a spherical shape (decreasing sphericity) tend to result in increased porosity, as shown in Figure 5.1.6.

Kerimov *et al.* (2018) found, using numerical simulations, that porosities in random packings of spheroidal grains decrease to a minimum with slight deviation from spherical shape, followed by an increase in porosity with larger deviations from spherical (Figure 5.1.7). The magnitude of the variation varies when the aspect ratio of the spheroids, α, deviate from 1. When *irregularly* shaped grains are generated by stochastically deviating from regular spheroids, the porosity and permeability trends mimic those observed with regular shapes, but the magnitude of the deviations from spherical are larger with the irregular grains than with the corresponding regular spheroids (Figure 5.1.8).

Other Practical Parameters Affecting Packing

Cumberland and Crawford (1987) list additional factors affecting the efficiency of packing, and consequently the porosity and coordination number:

Energy of Deposition. Experiments with spheres indicate that when spheres are poured (sometimes called *pluviated*), the packing density increases with the height of the drop; for many materials, an optimum drop height exists, above and below which the packing is looser (Figure 5.1.6).

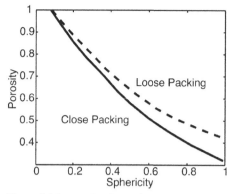

Figure 5.1.6 Trends of porosity versus grain sphericity observed by Waddell (1932)

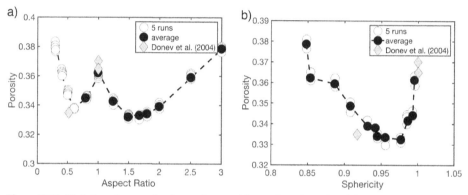

Figure 5.1.7 Variations in porosity for random packings of regular spheroids versus aspect ratio and sphericity. For spheres, $\alpha = 1$ and $\psi = 1$. Circles are values from numerical simulations by Kerimov *et al.* (2018). Diamonds are data from measurements by Donev *et al.* (2004); after Kerimov *et al.* (2018).

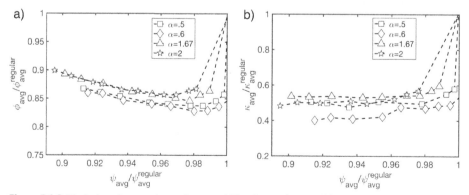

Figure 5.1.8 Variations in porosity and permeability for random packings of irregular grain shapes versus aspect ratio and sphericity (numerical simulations). The simulated values are from Kerimov *et al.* (2018). All values are normalized by the values for the underlying regular shapes, from which the irregular particles are perturbed.

Absolute Particle Size. Experimental evidence suggests that porosity sometimes increases with decreasing absolute particle size. This could result from the increased importance of surface forces (e.g., friction and adhesion) as the surface-to-volume ratio grows.

Uses

These results can be used to estimate the geometric relations of the packing of granular materials, which is useful for understanding porosity, transport properties, and effective elastic constants.

5.2   Percolation of Random Ellipsoidal Packs

Percolation refers to the formation of long-range connectivity in a random medium. In a two-phase system, the percolation threshold is the volume fraction of the inclusion phase at which connectivity occurs. Below the threshold, no large-scale connected clusters exist; above the threshold, connected clusters exist on the scale of the overall system. This description is strictly geometrical. Percolation sometimes also refers to the asymptotic variation of a physical property of the medium (e.g., electrical conductivity, fluid permeability, mechanical strength) near a critical concentration, which might or might not be the same concentration at which geometrical connection is made (Garboczi *et al.*, 2005).

In a finite sized random system the geometrical percolation point is probabilistic – connectivity will occur in some realizations, but not in others. In an infinite system, however, the percolation point can be precisely defined. Figure 5.2.1 (after Yi and Sastry, 2004) illustrates the scale effect based on numerical simulations of randomly

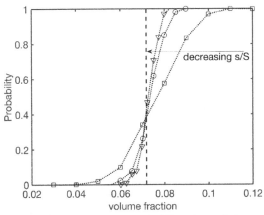

Figure 5.2.1 Simulations of probability of percolation as a function of inclusion volume fraction. From data by Yi and Sastry (2004) showing the influence of scale on geometric percolation. Inclusions are randomly oriented, randomly positioned, overlapping, prolate ellipsoids of aspect ratio 10. As ratio of inclusion size to system size (s/S) decreases, the observed probability function steepens.

positioned and oriented prolate spheroids of aspect ratio 10. The horizontal axis is the volume fraction of inclusions in a simulation. The vertical axis is the probability of geometric percolation, estimated as the fraction of realizations that achieve connection at a given inclusion concentration. When the inclusion size, s, is comparable to the simulated system size, S, there is a broad range of thresholds over which connectivity might occur. As s/S decreases, the probability function steepens, ultimately becoming a step function as $s/S \to 0$.

Geometric percolation has been studied extensively both theoretically and numerically by many authors. Excellent discussions can be found in Balberg et al. (1984), Balberg (1985), Yi and Sastry (2004), Yi et al. (2004), Garboczi et al. (1995), Torquato (2002), Consiglio et al. (2003). The number density, ρ, of particles or inclusions can be defined as

$$\rho = \frac{N}{V} \tag{5.2.1}$$

where N is the number of particles in system volume V. The dimensionless density η is

$$\eta = \frac{Nv}{V} = \rho v \tag{5.2.2}$$

where v is the particle volume. For non-overlapping objects the volume fraction of inclusions at a given density is

$$\varphi = \eta . \tag{5.2.3}$$

For overlapping objects, the average volume fraction φ occupied by objects of general shape (Garboczi et al., 1991) is

$$\varphi = 1 - e^{-\eta} \tag{5.2.4}$$

For mixtures of different objects A and B (Consiglio et al., 2003) the dimensionless density is

$$\eta \equiv \frac{N_A v_A}{V} + \frac{N_A v_B}{V} = \eta_A + \eta_B \tag{5.2.5}$$

and

$$\varphi = 1 - e^{-(\eta_A + \eta_B)} = 1 - e^{-\eta} \tag{5.2.6}$$

For a binary mixture of overlapping spheres the volume fraction of spheres is

$$\varphi = 1 - e^{-\langle n \rangle (\pi/6)[x(1-\lambda^3)+\lambda^3]} \tag{5.2.7}$$

where λ is the ratio of the small to large radii, $x = N_A/(N_A + N_B)$ is the number fraction of larger spheres, and $\langle n \rangle$ is the mean number of spheres per unit volume.

Table 5.2.1 (Consiglio *et al.*, 2003) shows simulated percolation thresholds, φ_c, for randomly positioned spheres with $\lambda = 0.5$.

Table 5.2.2 shows simulation results for randomly placed and randomly oriented three-dimensional distributions of *overlapping* spheroids (Garboczi *et al.*, 1995). The percolation volume fraction is φ_c. The ellipsoid aspect ratio is $\alpha = a/b$, where b is dimension in the direction of the symmetry axis, and a is the dimension perpendicular

Table 5.2.1 *Simulations from Consiglio* et al. *(2003). Overlapping spheres with ratio of radii 0.5. Parameter* $x = N_{large}/(N_{large} + N_{small})$ *is the number fraction of large spheres.* φ_c *is the volume fraction of spheres at percolation.*

x	0	0.1	0.11	0.15	0.25	0.5	0.75	1
φ_c	0.28955	0.29730	0.29731	0.29700	0.29537	0.29233	0.29065	0.28955

Table 5.2.2 *Simulations of percolation volume fraction* φ_c *for overlapping ellipsoids with a range of aspect ratios from thin disks (*$\alpha \ll 1$*) to needles (*$\alpha \gg 1$*) by Garboczi* et al. *(1995). The average number of ellipsoids at percolation in each simulation is* n_c.

α	a	b	n_c	φ_c
1/2000	0.000012	0.024	22005	0.000637
1/1000	0.000024	0.024	22028	0.001275
1/100	0.00024	0.024	21691	0.01248
1/10	0.0025	0.025	17089	0.1058
1/8	0.0030	0.024	18637	0.1262
1/5	0.0044	0.022	21659	0.1757
1/4	0.0055	0.022	20046	0.2003
1/3	0.0070	0.021	20103	0.2289
1/2	0.010	0.020	18209	0.2629
3/4	0.015	0.020	13243	0.2831
1	0.025	0.025	5134	0.2854
3/2	0.030	0.0200	6521	0.2795
2	0.020	0.0100	36235	0.2618
3	0.030	0.0100	20219	0.2244
4	0.040	0.0100	12581	0.1901
5	0.040	0.0080	16557	0.1627
10	0.050	0.0050	17389	0.08703
20	0.060	0.0030	18740	0.04150
30	0.060	0.0020	26679	0.02646
50	0.060	0.0012	41827	0.01502
100	0.060	0.0006	77069	0.006949
200	0.060	0.0003	141458	0.003195
300	0.060	0.0002	204373	0.002052
500	0.060	0.00012	333258	0.001205

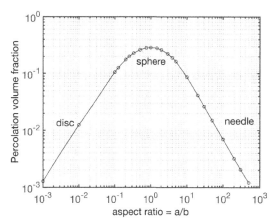

Figure 5.2.2 Simulations of percolation volume fraction φ_c for ellipsoids with a range of aspect ratios from thin disks (aspect ratio $\ll 1$) to needles (aspect ratio $\gg 1$), by Garboczi et al. (1995)

to the symmetry axis. The results from Table 5.2.2 are also plotted in Figures 5.2.2 and 5.2.3.

Garbozi et al. observed that in the limits of very small aspect ratio and very large aspect ratio, the simulations are consistent with the asymptotic approximations

$$\varphi_p \approx \begin{cases} 0.6/\alpha & \text{(prolate)} \\ 1.27\alpha & \text{(oblate)} \end{cases}. \tag{5.2.8}$$

The excluded volume for a given object is defined as that volume surrounding and including the object that cannot be occupied by a second object. For identical ellipsoids, Ogston and Winzor (1975) obtained the following expression for the excluded volume normalized by the volume of a single particle:

$$c = 2 + \frac{3}{2}\alpha\left[1 + \frac{1 - \zeta^2}{2\zeta}\ln\left(\frac{1+\zeta}{1-\zeta}\right)\right]\left[\sqrt{1 - \zeta^2} + \frac{\sin^{-1}\zeta}{\zeta}\right] \tag{5.2.9}$$

where

$$\zeta = 1 - \alpha^{-2}. \tag{5.2.10}$$

Yi and Sastry (2004) approximated the relation between the dimensionless particle density η_p at percolation and the excluded volume using functions of the form: $\eta_p = 1/c$, $\eta_p = 2/c$, $\eta_p = 3/c$, and $\eta_p = 4/c$. These, in turn, allow estimates of the inclusion volume fraction at percolation, $\varphi_p = 1 - e^{-\eta_p}$. Figure 5.2.3 compares percolation thresholds simulated by Garboczi et al. (1995) with the model curves based on excluded volume. At low aspect ratios, the percolation thresholds are close to $\eta = 3/c$, while at larger aspect ratios, the values are between $\eta = 1/c$ and $\eta = 2/c$.

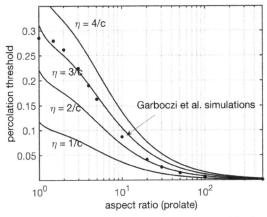

Figure 5.2.3 Simulations of percolation threshold volume fraction for prolate ellipsoids (Garboczi *et al.*, 1995), compared with simple functions of the excluded volume (Yi and Sastry, 2004)

5.3   Thomas–Stieber–Yin–Marion Model for Sand–Shale Systems

Synopsis

Thomas–Stieber Volumetric Model

Thomas and Stieber (1975, 1977) developed a model for the porosity of thinly bedded sands and shales, under the assumption that all rocks in the interval, including dirty sands, can be constructed by mixing clean, high-porosity sand and low-porosity shale (see also Juhasz, 1986; Pedersen and Nordal, 1999). Figure 5.3.1 illustrates some of the mixtures considered, and Figure 5.3.2 shows the associated porosities. The approach to analyzing porosity can be illustrated as a form of ternary diagram with the following limiting points and boundaries (Figure 5.3.2):

1. Shale (layer 1 in Figure 5.3.1; point C in Figure 5.3.2), with total porosity $\phi_{\text{total}} = \phi_{\text{shale}}$.

2. Clean sand (layer 2 in Figure 5.3.1; point A in Figure 5.3.2), with total porosity $\phi_{\text{total}} = \phi_{\text{clean sand}}$.

3. Sand with some dispersed shale in the sand pore space (layer 3 in Figure 5.3.1; line A-B in Figure 5.3.2), with total porosity

$$\phi_{\text{total}} = \phi_{\text{clean sand}} - (1 - \phi_{\text{shale}})V_{\text{disp shale}}$$

where $V_{\text{disp shale}}$ is the volume fraction of microporous dispersed shale (clay) in the sand. It is assumed that the sand packing is undisturbed by the shale in the pore space, as long as $V_{\text{disp shale}} \leq \phi_{\text{clean sand}}$. The total porosity is reduced by the solid fraction of the pore-filling shale, $(1 - \phi_{\text{shale}})V_{\text{disp shale}}$. This is sometimes called

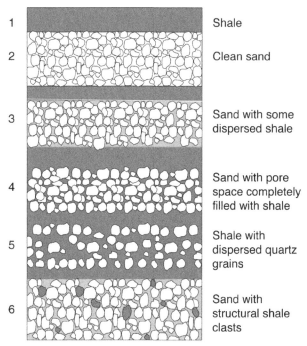

Figure 5.3.1 Laminations of shale, clean sand, sand with dispersed shale, and sand with structural shale clasts

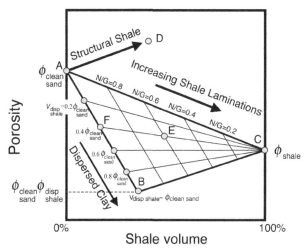

Figure 5.3.2 Porosity versus shale volume in the Thomas–Stieber model. $V_{disp\ shale}$ is the volume of dispersed shale in the sand pore space. The minimum porosity occurs when the sand porosity is completely filled with shale, $V_{disp\ shale} = \phi_{clean\ sand}$. Net-to-gross, N/G, is the thickness fraction of sand in the laminated composite.

an *ideal binary mixture model* (Cumberland and Crawford, 1987; Marion *et al.*, 1992) and corresponds to the limiting case of very large particles (sand) mixed with very small particles (clay). See Section 5.1.

4. Sand with pore space completely filled with shale (layer 4 in Figure 5.3.1; point B in Figure 5.3.2), with total porosity

$$\phi_{total} = \phi_{clean\ sand}\phi_{shale}.$$

We sometimes also refer to this as the "V-point."
In this case, $V_{disp\ shale} = \phi_{clean\ sand}.$

5. Shale with dispersed quartz grains (layer 5 in Figure 5.3.1; line B-C in Figure 5.3.2), with total porosity

$$\phi_{total} = V_{shale}\phi_{shale}.$$

In this case, regions of microporous shale are replaced by solid quartz grains, reducing the total porosity.

6. An additional case falls outside the triangle A-B-C: sand with structural shale clasts (layer 6 in Figure 5.3.1; line A-D in Figure 5.3.2), with porosity

$$\phi_{total} = \phi_{clean\ sand} + \phi_{shale}V_{shale\ clast}.$$

In this case, it is assumed that some of the solid quartz grains are replaced by porous shale clasts, resulting in a net increase in total porosity (see also Florez, 2005).

A thinly interbedded or *laminated* system of rocks has total porosity

$$\phi_{total} = \sum_i X_i\phi_i$$

where ϕ_i is the total porosity of the *i*th layer, and X_i is the thickness fraction of the *i*th layer. For a laminated sequence of shale and dirty sand (sand with dispersed shale), the total porosity is

$$\phi = N/G[\phi_{clean\ sand} - (1 - \phi_{shale})V_{disp\ shale}] + (1 - N/G)\phi_{shale}$$

where net-to-gross, *N/G*, is the thickness fraction of sand layers. *N/G* is *not* identical to the shale fraction, since some dispersed shale can be within the sand. The laminated systems fall within the triangle A-B-C in Figure 5.3.2. As previously mentioned, the line A-C corresponds to laminations of *clean sand* and shale. Points inside the triangle correspond to laminations of shale with *dirty sand* (sand with dispersed shale).

Figure 5.3.3 is similar to Figure 5.3.2 except that the horizontal axis is the approximate gamma ray response of the thinly laminated system. A practical workflow is to construct the model curves shown in Figure 5.3.3 and then superimpose well-log-derived total

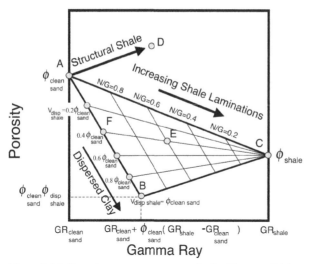

Figure 5.3.3 Porosity versus gamma ray in the Thomas–Stieber model, similar to Figure 5.3.2

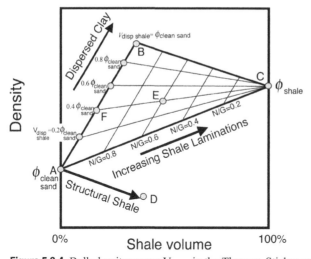

Figure 5.3.4 Bulk density versus V_{shale} in the Thomas–Stieber model, similar to Figure 5.3.2

porosity and gamma ray data for interpretation to identify laminar, dispersed, and structural shale. For example, point E in Figures 5.3.2 and 5.3.3 can be interpreted as a laminar sequence where the net-to-gross is 60%. Dispersed clay fills 40% of the pore space in the sand layers.

Figure 5.3.4 is similar to Figure 5.3.2, except that the vertical axis now represents bulk density. Table 5.3.1 summarizes total porosity and density at several key locations in the model.

Table 5.3.1 *Total porosity and bulk density at key locations in Figures 5.3.2 and 5.3.4, corresponding to the Thomas–Stieber model*

	Total porosity	Bulk density
Point A	$\phi_{\text{clean sand}}$	$\rho_{\text{clean sand}}$
Point C	ϕ_{shale}	ρ_{shale}
Line A–C	$\phi = (N/G)\phi_{\text{clean sand}} + (1 - N/G)\phi_{\text{shale}}$	$(N/G)\rho_{\text{clean sand}} + (1 - N/G)\rho_{\text{shale}}$
Point B	$\phi_{\text{clean sand}}\phi_{\text{shale}}$	$\rho_{\text{clean sand}} + \phi_{\text{clean sand}}(\rho_{\text{shale}} - \rho_{\text{water}})$
		$= \phi_{\text{clean sand}}\rho_{\text{shale}} + (1 - \phi_{\text{clean sand}})\rho_{\text{quartz}}$
Line A–B	$\phi_{\text{total}} = \phi_{\text{clean sand}} - (1 - \phi_{\text{shale}})V_{\text{disp shale}}$	$\rho_{\text{clean sand}} + V_{\text{shale}}(\rho_{\text{shale}} - \rho_{\text{water}})$
Line B–C	$\phi_{\text{total}} = V_{\text{shale}}\phi_{\text{shale}}$	$V_{\text{shale}}\rho_{\text{shale}} + (1 - V_{\text{shale}})\rho_{\text{quartz}}$

Notes:

$\phi_{\text{clean sand}}$, ϕ_{shale} = porosities of clean sand and shale; N/G = fractional thickness of sand layers; $V_{\text{disp shale}}$ = volume fraction of dispersed shale in sand; V_{shale} = total volume fraction of shale in the laminated stack; and, $\rho_{\text{clean sand}}$, ρ_{shale}, ρ_{water}, ρ_{quartz} = densities of clean sand, water-saturated shale, water, and quartz.

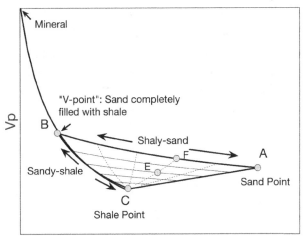

Figure 5.3.5 Relation between P-wave velocity and total porosity curves for the dispersed and interbedded sand-shale systems. Each dashed line within the triangle represents constant sand layer properties with varying fractions of interbedded shale. Each solid line within the triangle represents constant fraction of shale layer, with varying amounts of dispersed shale in sand layers.

Yin–Marion–Dvorkin–Gutierrez–Avseth Elastic Model

Several authors (e.g., Marion, 1990; Yin, 1992; Dvorkin and Gutierrez, 2002; Avseth *et al.*, 2010) have extended the Thomas–Stieber approach to model *elastic* properties of clastics composed of sand and shale end members (Figure 5.3.5). The velocity-porosity trend of "sandy shale" (line B-C in Figure 5.3.5) has been modeled using a Hashin–Shtrikman lower bound (HSLB) extending from the shale point to the sand mineral point, passing through point B (sometimes call the "V-point"), which might be physically realized as a mixture of elastically stiff grains (e.g., quartz) enveloped by softer microporous shale. One can similarly model the "shaly-sand" trend (line A-B in Figure 5.3.5) as a HSLB extending from the clean sand point to the V-point. (Heuristically, the role of the HSLB is that of a "soft interpolator" between specified end points, as discussed in Section 7.1. For the sandy-shale trend, we imagine quartz mineral suspended in the much softer load-bearing shale matrix. For the shaly-sand trend, we imagine that the quartz sand is load bearing, and that a substantial fraction of the clay mineral sits passively in the pore space. Neither realization is strictly correct, but the model predictions are very often consistent with observed trends. When the elastic contrast between the porous clean sand point and V-point is relatively small, the HSLB, the Reuss lower bound, and the harmonic average of P-wave moduli give similar results (Dvorkin and Gutierrez, 2002). The curve A-C corresponds to thinly inter-bedded ("laminated") sand and shale, which can be modeled using the Backus average (Section 4.16).

When plotting elastic properties against total porosity, the shaly-sand and sandy-shale legs of the dispersed sand-shale system form an "inverted-V" trend (Figure 5.3.5) analogous to the Thomas–Stieber "V" (Figure 5.3.2). The inverted-V trend has been shown by several authors to be a good approximation for velocity-porosity from this type of sand-shale system (e.g., Dvorkin and Gutierrez, 2002; Avseth et al., 2003, 2005; Flórez-Niño, 2005; Simm and Bacon, 2014). Flòrez-Niño (2005) compared the inverted-V trend with data from four different depositional environments. He observed, for example, a pronounced dispersed "V" pattern in fluvial deposits, a linear trend associated with interbedded mixtures in a fining-upward sequence of mud-rich deep-water sediments, and both dispersed and laminated mixtures of lithologies in coarsening-upward shallow water sediments. The laminated mixtures were at the decimeter scale. Avseth et al. (2003) observed sand-clay laminations below the cm scale.

Note that the curves in Figure 5.3.5 correspond to sediments at the same depth or same effective pressure. With further compaction, the sand point (A) and shale point (C) are expected to move toward smaller porosity and larger Vp, taking the connecting curves with them. We often observe that the sand point falls approximately along the Modified Upper Hashin–Shtrikman curve connecting mineral with clean sand on the Reuss average at the critical porosity (Section 7.1).

The elastic modulus of the V-point material cannot be determined precisely, without knowing the exact microstructure of the V-point material. One approach is to estimate the V-point elastic modulus as a harmonic average of the P-wave moduli of shale and quartz, assuming that the V-point is a suspension of quartz in soft shale. Dvorkin et al. (2007) estimated the effective mineral modulus of shaly sand (for their fluid substitution work) as a Voigt–Reuss–Hill estimate of quartz and shale, which yields a slightly stiffer value. The fluid substitution equation for these sand-shale systems that we present in Section 6.9 is independent of the V-point modulus, though for physical consistency the V-point modulus and Dvorkin's effective mineral modulus should be the same.

Assumptions and Limitations

There are several important assumptions made in the Thomas–Stieber model:

- clean sand is homogeneous with constant porosity;

- shale is the only destroyer of porosity (i.e., reduction of porosity by cementation and/ or sorting is ignored); and
- the shale content in sands is primarily detrital, having essentially the same properties as the shale laminae.

Extensions

The Backus average allows an exact prediction of elastic properties of a thinly laminated composite in terms of the individual layer properties.

5.4   Particle Size and Sorting

Synopsis

Particle Size

Particle size is one of the most fundamental parameters of clastic rocks. The **Udden–Wentworth scale** uses a geometric progression of grain diameters in millimeters, $d = 1$, 2, 4, 8, 16, ..., and divides sediments into seven grain-size grades: clay, silt, sand, granules, pebbles, cobbles, and boulders. The Udden–Wentworth scale is summarized in Table 5.4.1. The third column in the table is a subclassification: vc = very coarse,

Table 5.4.1 *Grain sizes and classes, according to the Udden–Wentworth scale*

Diameter (mm)	ϕ		Class
4096	−12		Block
2048	−11	vc	Boulder
1024	−10	c	
512	−9	m	
256	−8	f	
128	−7	c	Cobble
64	−6	f	
32	−5	vc	Pebble
16	−4	c	
8	−3	m	
4	−2	f	
2	−1		Granule
1	0	vc	Sand
0.50	1	c	
0.25	2	m	
0.125	3	f	
0.063	4	vf	
0.031	5	c	Silt
0.015	6	m	
0.008	7	f	
0.004	8	vf	
			Clay

c = coarse, m = medium, f = fine, and vf = very fine. Krumbein (1934, 1938) introduced a logarithmic transformation of the Udden–Wentworth scale, known as the **phi-scale**, which is the most-used measure of grain size:

$$\phi = -\log_2 d. \tag{5.4.1}$$

A lesser-known description is the **psi-scale**, $\psi = -\phi$.

Sorting

Sorting is the term that geologists use to describe the size distributions, which are a fundamental property of all naturally occurring packings of particles or grains. Sorting is no more than a measure of the spread of the grain sizes, and there are standard statistical methods to describe the spread of any population. Two of the main parameters to measure the spread of a population are the standard deviation and the coefficient of variation (the standard deviation normalized by the mean).

In the **phi-scale**, sorting is usually expressed as being equal to the **standard deviation** (Boggs, 2001). Sorting is often estimated graphically by plotting a frequency histogram of grain sizes (measured by weight) expressed in the phi-scale, and normalized by the total weight of grains (see Figure 5.4.1). The cumulative distribution function (CDF) of grain sizes, ranging from 0% to 100%, is derived by successively adding and smoothing the bars in the histogram (from small to large sizes). The CDF makes it easy to identify the percentiles of the distribution. For example, the 25th percentile, ϕ_{25}, is the value of ϕ such that 25% of the grains by weight are smaller.

Folk and Ward (1957) suggested various measures of the grain size distribution that can be computed from the graphically determined percentiles (after Tucker, 2001):

$$\text{Median} \quad MD = \phi_{50} \tag{5.4.2}$$

$$\text{Mean} \quad M = \frac{\phi_{16} + \phi_{50} + \phi_{84}}{3} \tag{5.4.3}$$

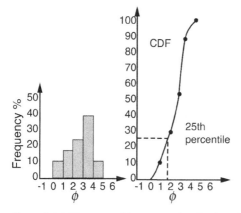

Figure 5.4.1 Frequency histogram of grain sizes (left) and the corresponding cumulative distribution function (after Tucker, 2001)

$$\text{Standard deviation} \quad \sigma\phi = \frac{\phi_{84} - \phi_{16}}{4} + \frac{\phi_{95} - \phi_{5}}{6.6} \tag{5.4.4}$$

$$\text{Skewness} \quad SK = \frac{\phi_{16} + \phi_{84} - 2\phi_{50}}{2(\phi_{84} - \phi_{16})} + \frac{\phi_{5} + \phi_{95} - 2\phi_{50}}{2(\phi_{95} - \phi_{5})} \tag{5.4.5}$$

Statistics such as these, measured in phi units, should have ϕ attached to them. The **Folk and Ward formula** for **phi standard deviation**, $\sigma\phi$, follows from assuming a log-normal distribution of grain diameters in millimeters, which corresponds to a normal distribution of sizes in ϕ. If grain sizes have a normal distribution, then 68% of the grains fall between ± 1 standard deviation, or between ϕ_{16} and ϕ_{84}. Then we can write $\sigma\phi_{68} = (\phi_{84} - \phi_{16})/2$. Similarly, in a normal distribution, 90% of the grains fall between ± 1.65 standard deviations, or between ϕ_{05} and ϕ_{95}. Hence $\sigma\phi_{90} = (\phi_{95} - \phi_{05})/3.30$. The Folk and Ward phi standard deviation is the average of these two estimates: $\sigma\phi = [(\phi_{84} - \phi_{16})/2 + (\phi_{95} - \phi_{05})/3.30]/2$. *The sorting constant S_0 is equal to the phi standard deviation:*

$$S_0 = \frac{\phi_{84} - \phi_{16}}{4} + \frac{\phi_{95} - \phi_{5}}{6.6} \approx \frac{\phi_{84} - \phi_{16}}{2} \tag{5.4.6}$$

It is interesting to convert this expression back to the millimeter scale:

$$S_0 \approx \frac{\phi_{84} - \phi_{16}}{2} \quad \Rightarrow \quad 2^{S_0} = \sqrt{\frac{d_{16}}{d_{84}}} \tag{5.4.7}$$

Another popular way to express the variation of the population in millimeters is using the **coefficient of variation**, which is the standard deviation divided by the mean.

It is worth pointing out that many serious sedimentologists have studied distributions of grain sizes using standard statistical methods, without attempting to define or redefine the term sorting.

5.5　Random Spherical Grain Packings: Contact Models and Effective Moduli

Synopsis

The Hertz Contact Sphere Problem

When two *frictionless*, linear elastic spheres are pressed into contact by a **normal force** F (Figure 5.5.1), a circular area of contact is formed, with radius, a, given by (Hertz, 1881)

$$a = \left[\frac{3F \left(\dfrac{1 - v_1^2}{E_1} + \dfrac{1 - v_2^2}{E_2} \right)}{4 \left(\dfrac{1}{R_1} + \dfrac{1}{R_2} \right)} \right]^{1/3}, \tag{5.5.1}$$

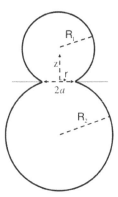

Figure 5.5.1 Two elastic spheres in compressional contact

where v_1, E_1, and R_1 are the Poisson's ratio, Young's modulus, and radius of sphere 1, and v_2, E_2, and R_2 are the Poisson's ratio, Young's modulus, and radius of sphere 2. It is convenient to define the *relative curvature*, $1/R^*$,

$$\frac{1}{R^*} = \frac{1}{R_1} + \frac{1}{R_2}, \tag{5.5.2}$$

and an effective modulus (not to be confused with the effective modulus of a composite, as discussed in Chapter 4), E^*,

$$\frac{1}{E^*} = \frac{1 - v_1^2}{E_1} + \frac{1 - v_2^2}{E_2}. \tag{5.5.3}$$

Note that when the two spheres are identical, then $R^* = R/2$, and $E^* = E/2(1 - v^2)$.

Caution:

Some authors use alternate definitions for R^* and E^* which might differ from those in equations 5.5.2 and 5.5.3 by a factor of 2.

The pressure distribution $p(r)$ at the contact is rotationally symmetric, and takes the form of an ellipse:

$$p(r) = \frac{3F}{2\pi a^2} \sqrt{1 - (r/a)^2}, \ 0 \leq r < a, \tag{5.5.4}$$

where the maximum pressure, $p_{max} = 3F/2\pi a^2$, occurs at the center of the contact, and r is the radial distance from the contact center, $0 \leq r < a$. The average pressure at the contact is $p_{avg} = 2p_{max}/3$. The normal stresses at the contact are the same in both spheres ($\sigma_{zz} = -p$). In terms of the effective modulus and curvature, we can also write

$$p_{avg} = \frac{F^{1/3}}{\pi} \left[\frac{4E^*}{3R^*} \right]^{2/3}. \tag{5.5.5}$$

Notice that the contact pressure increases with increasing contact force and decreasing effective radius.

The horizontal stresses (radial stress σ_{rr} and hoop stress $\sigma_{\theta\theta}$) at the contact are also radially symmetric, but depend on the Poisson's ratio and can therefore vary from one sphere to the other:

$$\frac{\sigma_{rr}}{p_{\max}} = \left(\frac{1-2v_i}{3}\right)\left(\frac{a^2}{r^2}\right)\left[1 - \left(1 - \frac{r^2}{a^2}\right)^{3/2}\right] - \left(1 - \frac{r^2}{a^2}\right)^{1/2}, \quad i = 1, 2, \quad (5.5.6)$$

$$\frac{\sigma_{\theta\theta}}{p_{\max}} = -\left(\frac{1-2v_i}{3}\right)\left(\frac{a^2}{r^2}\right)\left[1 - \left(1 - \frac{r^2}{a^2}\right)^{3/2}\right] - 2v_i\left(1 - \frac{r^2}{a^2}\right)^{1/2}, \quad i = 1, 2.$$

$$(5.5.7)$$

(Note that if $v = 1/2$ the stress becomes hydrostatic, $\sigma_{rr} = \sigma_{\theta\theta} = \sigma_{zz}$). *The radial stress is tensional near and beyond the edge of the contact*, reaching its maximum value of $(\sigma_{rr})_{\max} = (1 - 2v_i)p_{\max}/3$ at $r = a$. *This is the largest tensional stress anywhere in the medium under pure normal loading.* Tensile ring fractures are often observed around the contact in brittle materials. At the center of the contact ($r = 0$ and $z = 0$)   equations   5.5.6   and   5.5.7   reduce   to $\sigma_{rr} = \sigma_{\theta\theta} = -p_{\max}(1 + 2v_i)/2$. The von Mises stress (see Section 2.1) at the center of the contact is $\sigma_{vM} = p_{\max}(1 - 2v_i)/2$.

The radial and vertical displacements within the contact circle are

$$u_r = -\frac{(1 - 2v_i)(1 + v_i)a^2}{3E_i}\frac{1}{r}p_{\max}\left[1 - \left(1 - \frac{r^2}{a^2}\right)^{3/2}\right], \quad r \le a \quad (5.5.8)$$

$$u_z = \frac{1 - v_i^2}{E_i}\frac{\pi}{4a}p_{\max}(2a^2 - r^2). \quad (5.5.9)$$

The mutual approach of the centers of the two spheres is 2δ:

$$2\delta = \frac{\pi a p_{\max}}{2E^*} = \frac{a^2}{R^*}. \quad (5.5.10)$$

Note that δ is the average displacement of each sphere center toward the plane of contact.

Caution:

Some authors define δ as the displacement of one grain center relative to the other grain center, thus differing by a factor of 2 from the definition that we use here.

Within each sphere, the principal stresses along the z-axis of symmetry are

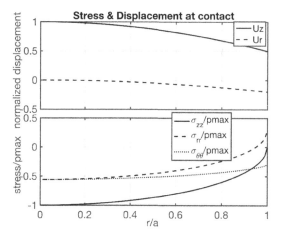

Figure 5.5.2 Stresses and displacements versus radius at the contact under pure compression loading, for Poisson's ratio 0.06. The contact radius is a.

$$\sigma_{rr} = \sigma_{\theta\theta} = \sigma_{xx} = \sigma_{yy} = -p_{\max}\left[(1+v_i)\left(1 - \left|\frac{z}{a}\right|\tan^{-1}\left|\frac{a}{z}\right|\right) - \frac{1}{2}\left(\frac{z^2}{a^2}+1\right)^{-1}\right],$$

$$(5.5.11)$$

$$\sigma_{zz} = -p_{\max}\left(\frac{z^2}{a^2}+1\right)^{-1},$$

$$(5.5.12)$$

where z is the distance from the contact (Figure 5.5.2). Remember that stresses are defined as positive in tension, so that σ_{zz} is compressive everywhere along the z-axis, and $\sigma_{\theta\theta}$ varies from compressive at the contact ($z = 0$) to tensional at larger z. The maximum shear stresses occur on planes oriented at 45°From the axis of symmetry and achieve a maximum value

$$\tau_{\max} = \left|\frac{\sigma_{rr} - \sigma_{\theta\theta}}{2}\right|.$$

$$(5.5.13)$$

The von Mises stress is

$$\sigma_{vM} = \sqrt{\frac{(\sigma_1 - \sigma_2)^2 + (\sigma_2 - \sigma_3)^2 + (\sigma_3 - \sigma_1)^2}{2}},$$

$$(5.5.14)$$

where $\sigma_1 = \sigma_{xx}$, $\sigma_2 = \sigma_{yy}$, and $\sigma_3 = \sigma_{zz}$ are the principal stresses. The largest shear stress and the maximum von Mises stress, $\sigma_{vM}^{(\max)} \approx p_{\max}(0.05v^2 - 0.10v + 0.77)$, occurs below the contact at a depth $z_{vM}^{(\max)} \approx a(0.2v + 0.38)$

The vertical displacements along the symmetry axis are:

$$u_z(z) = p_{\max}\frac{(1+v_i)}{aE_i}\left\{[(1-v_i)a^2 - v_i z^2]\sin^{-1}\left(\frac{a}{\sqrt{a^2 + z^2}}\right) + v_i az\right\}.$$

$$(5.5.15)$$

The stresses within the sphere near the contact can be written as follows (Hertz, 1881; Sneddon, 1948; Hanson, 1992; Kachanov et al., 2003)

$$\sigma_{zz} = -\frac{3F}{2\pi a^3}\frac{(a^2 - l_1^2)^{3/2}}{l_2^2 - l_1^2} \,,$$ (5.5.16)

$$\sigma_{rr} = (\sigma_1 + \sigma_2)/2 \,,$$ (5.5.17)

$$\sigma_{\theta\theta} = (\sigma_1 - \sigma_2)/2 \,,$$ (5.5.18)

where

$$\sigma_1 = \frac{3F}{2\pi a^3}\left\{2(1+v)z\sin^{-1}\left(\frac{l_1}{r}\right) - (1+2v)\sqrt{a^2 - l_1^2} - \frac{za\sqrt{l_2^2 - a^2}}{l_2^2 - l_1^2}\right\} \,,$$ (5.5.19)

$$\sigma_2 = \frac{F}{2\pi a^3}\frac{1}{r^2}\left\{(1-2v)[2a^3 - (l_1^2 + 2a^2)\sqrt{a^2 - l_1^2}] - \frac{3zl_1^3\sqrt{r^2 - l_1^2}}{l_2^2 - l_1^2}\right\} \,,$$ (5.5.20)

and

$$2l_1 = \sqrt{(r+a)^2 + z^2} - \sqrt{(r-a)^2 + z^2} \,,$$ (5.5.21)

$$2l_2 = \sqrt{(r+a)^2 + z^2} + \sqrt{(r-a)^2 + z^2} \,.$$ (5.5.22)

The displacement within the sphere can be written as

$$u_z = \frac{3F(1+v)}{8\pi a^3 E}\left\{[2(1-v)(2a^2 - r^2) - 4vz^2]\sin^{-1}\left(\frac{l_1}{r}\right)\right.$$

$$\left. + 4z\sqrt{a^2 - l_1^2} + 2(1-v)\frac{(3l_1^2 - 2a^2)\sqrt{l_2^2 - a^2}}{a}\right\}$$

$$u_r = \frac{3F(1+v)}{4\pi a^3 E}r\left\{2(1-v)z\sin^{-1}\left(\frac{l_1}{r}\right) - \frac{za\sqrt{l_2^2 - a^2}}{l_2^2}\right.$$

$$\left. - (1-2v)\frac{2a^3 + (3r^2 - 2a^2 - l_1^2)\sqrt{a^2 - l_1^2}}{3r^2}\right\}$$ (5.5.23) and (5.5.24)

When evaluating these expressions for stress and displacement, it is useful to note that for small r:

$$l_1 \approx ar/\sqrt{a^2 + z^2}$$ (5.5.25)

$$l_2 \approx \sqrt{a^2 + z^2}$$ (5.5.26)

$$l_1/r \approx a/\sqrt{a^2 + z^2}$$ (5.5.27)

If the contacts have coefficient of friction, f, then a **tangential force**, $Q = fF$, acting in the x-direction can cause sliding. When the entire surface is sliding, the surface shear stresses in the x-direction are proportional to the normal stress

$$\sigma_{zx} = -\frac{3Q}{2\pi a^2}\left(1 - \frac{r^2}{a^2}\right)^{1/2} \quad r \le a; \ y = 0 \tag{5.5.28}$$
$$= fp(r)$$

The radial stress at the contact ($z = 0$) along the x-axis (from tangential loading) is

$$\sigma_{rr} = \sigma_{xx} = -\frac{3Q}{2\pi a^3}\frac{\pi x(4 + v)}{8} \ , \ r \le a; \ y = 0 \ , \tag{5.5.29}$$

which is compressive in front of the sliding contact and tensional behind. When superimposed on the radial stress induced by the normal compression the total radial stress along the x-axis (tangential loading) is

$$\sigma_{rr} = \frac{3F}{2\pi a^2}\left[\frac{1 - 2v}{3}\left(\frac{x}{a}\right)^2 - \pi f\left(\frac{x}{a}\right)\frac{(4 + v)}{8}\right], r \le a; \ y = 0 \tag{5.5.30}$$

which leads to maximum tension at $x = -a$. (Even though the contacts are now frictional, we approximate the effect of normal pressure using the frictionless Hertz results previously described.) The von Mises stress at the surface point, $z = 0$; $x = -a$; $y = 0$, is

$$\frac{\sigma_{vM}}{p_{max}} = \left\{\frac{(1 - 2v)^2}{3} + \frac{(1 - 2v)(2 - v)f\pi}{4} + \frac{(16 - 4v + 7v^2)f^2\pi^2}{64}\right\}^{1/2}. \tag{5.5.31}$$

With increasing friction, the point of maximum von Mises stress moves from below the center of contact (when frictionless) to $x = -a$ at the surface.

The Hertz–Mindlin Model for Effective Moduli of Random Identical Sphere Packs

When there are two identical spheres in normal contact, the **Hertz model**, equation 5.5.1, simplifies, giving the radius of the contact area, a, and the normal displacement, δ of each sphere center toward the contact plane as:

$$a = \left[\frac{3FR}{8\mu}(1 - v)\right]^{1/3}, \quad \delta = \frac{a^2}{R} \tag{5.5.32}$$

where μ and v are the shear modulus and the Poisson's ratio of the grain material, respectively (Figure 5.5.4).

If hydrostatic confining pressure P is applied to a random, identical-sphere pack, the mean confining force acting between two particles is often approximated as

$$F = \frac{4\pi R^2 P}{C(1 - \phi)} \tag{5.5.33}$$

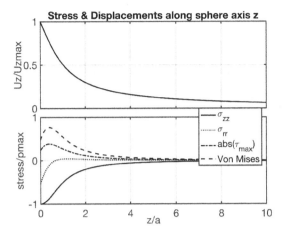

Figure 5.5.3 Stresses and displacements along the symmetry z-axis for Poisson's ratio 0.06 (pure compressional loading). z is the distance from the contact plane.

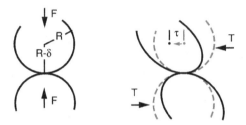

Figure 5.5.4 Normal and tangential displacement in a two-particle system

where ϕ is the porosity, C is the average coordination number (average number of contacts per sphere), and R is the grain radius. Equation 5.5.33 assumes that the magnitude of normal force is the same on all contacts.

Combining equations 5.5.32 and 5.5.33, the average contact radius is expressed as

$$a = R\left[\frac{3\pi(1-v)}{2C(1-\phi)\mu}P\right]^{1/3} \tag{5.5.34}$$

The normal and tangential stiffnesses of two identical spheres in contact are defined by

$$S_n = \frac{\partial F}{\partial \delta}, \quad S_\tau = \frac{\partial T}{\partial \tau} \tag{5.5.35}$$

where S_n and S_τ are the normal and tangential stiffnesses, respectively, F is the normal force, T is the tangential force, δ is the normal displacement of one sphere center relative to the plane of contact, and τ is the tangential displacement of one sphere center relative to the contact center.

Caution:

Some authors define δ and τ as the displacements of one sphere center relative to the other sphere center, rather than the definitions used here, which are the displacements of one sphere center relative to the contact center on the plane of contact. This introduces factors of 2 to the displacements and to the stiffnesses in equation 5.5.35.

The normal stiffness can be found from equations 5.5.32 and 5.5.34:

$$S_n = \frac{4\mu a}{1 - v} \tag{5.5.36}$$

Hertz estimated the *effective bulk modulus* of a dry, random, identical-sphere pack as

$$K_{\text{eff}} = \frac{C(1 - \phi)}{12\pi R} S_n = \left[\frac{C^2 (1 - \phi)^2 \mu^2}{18\pi^2 (1 - v)^2} P \right]^{1/3} \tag{5.5.37}$$

Implicit in this result is the assumption that for a remote uniform volumetric strain field ε_0 the displacement of the center of each sphere at position $\mathbf{x}$ is the same as for a homogeneous material, $\mathbf{u} = \varepsilon_0 \mathbf{x}$. This is equivalent to assuming that each contact feels the same normal compressive force.

Mindlin (1949) showed that if the spheres are first pressed together, and a tangential force is applied *afterward*, slip may occur at the edges of the contact (Figure 5.5.4). The extent of the slip depends on the coefficient of friction η between the contacting bodies. The normal stiffness of the two-grain system remains the same as that given by the Hertz solution. The shear stiffness is

$$S_\tau = \frac{8a\mu(1 - F_t/\eta F_n)^{1/3}}{2 - v}, \quad F_t \le \eta F_n \tag{5.5.38}$$

where v and μ are the Poisson's ratio and shear modulus of the solid grains, respectively, and F_n and F_t are the normal and tangential forces applied to the spheres, respectively. Typically, if we are only concerned with acoustic wave propagation, the normal force, which is due to the confining (overburden) stress imposed on the granular system, is much larger than the tangential force, which is imposed by the very small oscillatory stress due to the propagating wave: $F_t \ll F_n$. As a result, the shear and normal stiffnesses are as follows (the latter being the same as in the Hertz solution):

$$S_\tau = \frac{8a\mu}{2 - v}, \quad S_n = \frac{4a\mu}{1 - v} \tag{5.5.39}$$

The effective shear modulus of a dry, random, identical sphere pack is then

$$\mu_{\text{eff}} = \frac{C(1-\phi)}{20\pi R}\left(S_n + \frac{3}{2}S_\tau\right)$$
$$= \frac{5-4v}{5(2-v)}\left[\frac{3C^2(1-\phi)^2\mu^2}{2\pi^2(1-v)^2}P\right]^{1/3} \tag{5.5.40}$$

where again, the displacement of the sphere centers is assumed to be consistent with a uniform deviatoric strain field. Both equations 5.5.37 and 5.5.40 assume that every grain has the same coordination number. The ratio of the effective bulk modulus to effective shear modulus and the effective Poisson's ratio of the dry frame of this system are

$$\frac{K_{\text{eff}}}{\mu_{\text{eff}}} = \frac{5(2-v)}{3(5-4v)}, \quad v_{\text{eff}} = \frac{v}{2(5-3v)} \tag{5.5.41}$$

Note that the ratio $K_{\text{eff}}/\mu_{\text{eff}}$ and the effective Poisson's ratio depend only on the grain Poisson's ratio. The effective P- and S-wave velocities for the dry sphere packs are:

$$V_P^2 = \frac{3C}{20\pi R\rho}\left(S_n + \frac{2}{3}S_\tau\right) \tag{5.5.42}$$

$$V_S^2 = \frac{C}{20\pi R\rho}\left(S_n + \frac{3}{2}S_\tau\right) \tag{5.5.43}$$

where ρ is the grain material density. Observed values for coordination number C are discussed in Section 5.1.

One may also consider the case of **absolutely frictionless spheres**, where $\eta = 0$ ($\eta \to 0$, as $(F_\tau/\eta F_n) \to 1$) and therefore $S_\tau = 0$. In this case, the effective bulk and shear moduli of the pack become

$$K_{\text{eff}} = \frac{C(1-\phi)}{12\pi R}S_n, \quad \mu_{\text{eff}} = \frac{C(1-\phi)}{20\pi R}S_n, \quad \frac{K_{\text{eff}}}{\mu_{\text{eff}}} = \frac{5}{3} \tag{5.5.44}$$

and the effective Poisson's ratio becomes exactly 0.25.

Figure 5.5.5 demonstrates that there is a large difference between the effective Poisson's ratio of a dry frictionless pack and that of a dry pack with perfect adhesion between the particles. In the latter case, the effective Poisson's ratio does not exceed 0.1 regardless of the grain material.

The absence of friction between particles may conceivably occur in an unconsolidated system because of the presence of a lubricant at some contacts. It is virtually impossible to predict in advance which fraction of the individual contacts is frictionless ($S_\tau = 0$), and which has perfect adhesion ($S_\tau = 8a\mu/(2-v)$). To account for all possibilities, an ad hoc coefficient $f\,(0 \le f \le 1)$ can be introduced such that

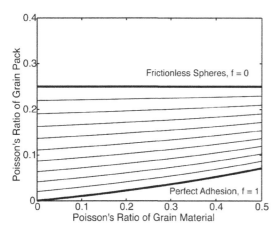

Figure 5.5.5 Poisson's ratio of a grain pack versus that of the grain material. The upper bold curve is for frictionless spheres while the lower bold curve is for spheres with perfect adhesion. The thin curves in between are for values of f varying between 0 and 1, with a step size of 0.1.

$$S_n = \frac{4a\mu}{1 - \nu}, \quad S_\tau = f\frac{8a\mu}{2 - \nu} \tag{5.5.45}$$

and the grains have perfect adhesion for $f = 1$ and no friction for $f = 0$. As a result, the effective bulk modulus of the pack remains the same, but the expression for the effective shear modulus becomes

$$\mu_{\text{eff}} = \frac{2 + 3f - \nu(1 + 3f)}{5(2 - \nu)}\left[\frac{3C^2(1 - \phi)^2\mu^2}{2\pi^2(1 - \nu)^2}P\right]^{1/3} \tag{5.5.46}$$

and

$$\frac{K_{\text{eff}}}{\mu_{\text{eff}}} = \frac{5(2 - \nu)}{3[2 + 3f - \nu(1 + 3f)]}, \quad \nu_{\text{eff}} = \frac{2 - 2f + \nu(2f - 1)}{2[4 + f - \nu(2 + f)]} \tag{5.5.47}$$

Equations 5.5.46 and 5.5.47 are the same as those introduced by Bachrach and Avseth (2008). Figure 5.5.5 demonstrates how the Poisson's ratio of the dry pack gradually moves from the frictionless line down to the perfect-adhesion curve.

The Walton Model

It is assumed in the Walton model (Walton, 1987) *that normal and shear deformation of a two-grain combination occur simultaneously.* This assumption leads to results somewhat different from those given by the Hertz–Mindlin model. Specifically, there is no partial slip in the contact area. The slip occurs across the whole area once applied tractions exceed the friction resistance. The results discussed in the following paragraphs are given for two special cases: infinitely rough spheres (where the friction coefficient is very large) and ideally smooth spheres (where the friction coefficient is zero).

Under *hydrostatic pressure P*, an identical-sphere random packing is isotropic. Its effective bulk and shear moduli for the rough-spheres case (dry pack) are described by

$$K_{eff} = \frac{1}{6}\left[\frac{3(1-\phi)^2 C^2 P}{\pi^4 B^2}\right]^{1/3}, \quad \mu_{eff} = \frac{3}{5}K_{eff}\frac{5B+A}{2B+A} \tag{5.5.48}$$

$$A = \frac{1}{4\pi}\left(\frac{1}{\mu} - \frac{1}{\mu+\lambda}\right), \quad B = \frac{1}{4\pi}\left(\frac{1}{\mu} + \frac{1}{\mu+\lambda}\right) \tag{5.5.49}$$

where λ is Lamé's coefficient of the grain material. For the smooth-spheres case (dry pack),

$$\mu_{eff} = \frac{1}{10}\left[\frac{3(1-\phi)^2 C^2 P}{\pi^4 B^2}\right]^{1/3}, \quad K_{eff} = \frac{5}{3}\mu_{eff} \tag{5.5.50}$$

It is clear that the effective density of the aggregate is

$$\rho_{eff} = (1-\phi)\rho \tag{5.5.51}$$

Under *uniaxial pressure σ_1*, a dry, identical-sphere packing is transversely isotropic, and if the spheres are infinitely rough, it can be described by the following five constants:

$$\begin{matrix} C_{11} = 3(\alpha+2\beta), & C_{12} = \alpha - 2\beta, & C_{13} = 2C_{12} \\ C_{33} = 8(\alpha+\beta), & C_{44} = \alpha + 7\beta \end{matrix} \tag{5.5.52}$$

where

$$\alpha = \frac{(1-\phi)Ce}{32\pi^2 B}, \quad \beta = \frac{(1-\phi)Ce}{32\pi^2(2B+A)} \tag{5.5.53}$$

$$e = \left[\frac{24\pi^2 B(2B+A)\sigma_1}{(1-\phi)AC}\right]^{1/3} \tag{5.5.54}$$

The Digby Model

The Digby model gives effective moduli for a dry, random packing of identical elastic spherical particles. Neighboring particles are initially firmly bonded across small, flat, circular regions of radius a. Outside these adhesion surfaces, the shape of each particle is assumed to be ideally smooth (with a continuous first derivative). Notice that this condition differs from that of Hertz, where the shape of a particle is not smooth at the intersection of the spherical surface and the plane of contact. Digby's normal and shear stiffnesses under hydrostatic pressure P are (Digby, 1981)

$$S_n = \frac{4\mu b}{1-v}, \quad S_\tau = \frac{8\mu a}{2-v} \tag{5.5.55}$$

where v and μ are the Poisson's ratio and shear modulus of the grain material, respectively. Parameter b can be found from the relation

$$\frac{b}{R} = \left[d^2 + \left(\frac{a}{R} \right)^2 \right]^{1/2} \tag{5.5.56}$$

where d satisfies the cubic equation

$$d^3 + \frac{3}{2} \left(\frac{a}{R} \right)^2 d - \frac{3\pi(1-v)P}{2C(1-\phi)\mu} = 0 \tag{5.5.57}$$

Example: Use the Digby model to estimate the effective bulk and shear moduli for a dry random pack of spherical grains under a confining pressure of 10 MPa. The ratio of the radius of the initially bonded area to the grain radius a/R is 0.01. The bulk and shear moduli of the grain material are $K = 37$ GPa and $\mu = 44$ GPa, respectively. The porosity of the grain pack is 0.36.

The Poisson's ratio v for the grain material is calculated from K and μ:

$$v = \frac{3K - 2\mu}{2(3K + \mu)} = 0.07$$

The coordination number is $C = 9$. Solving the cubic equation for d

$$d^3 + \frac{3}{2} \left(\frac{a}{R} \right)^2 d - \frac{3\pi(1-v)P}{2C(1-\phi)\mu} = 0$$

and taking the real root, neglecting the pair of complex conjugate roots, we obtain $d = 0.0547$. Next, we calculate b/R as

$$\frac{b}{R} = \sqrt{d^2 + \left(\frac{a}{R} \right)^2} = 0.0556$$

The values of a/R and b/R are used to compute S_n/R and S_τ/R:

$$\frac{S_n}{R} = \frac{4\mu(b/R)}{1-v} = 10.5 \qquad \frac{S_\tau}{R} = \frac{8\mu(a/R)}{2-v} = 1.8$$

which then finally give us

$$K_{\text{eff}} = \frac{C(1-\phi)}{12\pi}(S_n/R) = 1.6 \text{ GPa}$$

and

$$\mu_{\text{eff}} = \frac{C(1-\phi)}{20\pi}[(S_n/R) + 1.5(S_\tau/R)] = 1.2 \text{ GPa}$$

The Jenkins *et al.* Model

Jenkins *et al.* (2005) derive expressions for the effective moduli of a random packing of identical frictionless spheres. Their approach differs from that of Digby (1981), Walton (1987), and Hertz–Mindlin in that under applied deviatoric strain, the particle motion of each sphere relative to its neighbors is allowed to deviate from the mean homogeneous strain field of a corresponding homogeneous effective medium. Under hydrostatic compression, the grain motion is consistent with homogeneous strain. The additional degrees of freedom of particle motion result in calculated shear moduli that are smaller than predicted by the previously mentioned models.

The expressions for dry effective shear modulus, μ_{eff}, and effective Lamé constant, λ_{eff}, in terms of the coordination number, C, grain diameter, d, and porosity, ϕ, are

$$\mu_{\text{eff}} = \frac{C(1-\phi)}{5\pi d}\widetilde{S}_n\{1 - 2[\psi^{-1}(\omega_1 + 2\omega_2) - \psi^{-2}(\kappa_1 + 2\kappa_2) + \psi^{-3}(\xi_1 + 2\xi_2)]\}$$

(5.5.58)

$$\lambda_{\text{eff}} = \frac{C(1-\phi)}{5\pi d}\widetilde{S}_n\{1 - 2\psi^{-1}(\omega_1 + 7\omega_2) + 2\psi^{-2}(\kappa_1 + 2\kappa_2 + 5\kappa_3) - 2\psi^{-3}(\xi_1 + 2\xi_2 + 5\xi_3)\}$$

(5.5.59)

where

$$\psi = C/3 \tag{5.5.60}$$

$$\omega_1 = (166 - 11C)/128 \tag{5.5.61}$$

$$\widetilde{\omega}_1 = (38 - 11C)/128 \tag{5.5.62}$$

$$\omega_2 = -(C + 14)/128 \tag{5.5.63}$$

$$\alpha_1 = (19C - 22)/48 \tag{5.5.64}$$

$$\alpha_2 = (22 - 3C)/16 \tag{5.5.65}$$

$$\widetilde{\alpha}_2 = (18 - 9C)/48 \tag{5.5.66}$$

$$\kappa_1 = -(\alpha_1\widetilde{\omega}_1 + \widetilde{\alpha}_2\widetilde{\omega}_1 + 2\widetilde{\alpha}_2\omega_2) + \widetilde{\omega}_1 - [0.52(C-2)(C-4)$$
$$+0.10C(C-2) - 0.13C(C-4) - 0.01C^2]/16\pi \tag{5.5.67}$$

$$\kappa_2 = -\alpha_1\omega_2 + \omega_2 + [0.44(C-2)(C-4) - 0.24C(C-2)$$
$$-0.11C(C-4) - 0.14C^2]/16\pi \tag{5.5.68}$$

$$\kappa_3 = -(\alpha_1\omega_2 + \widetilde{\alpha}_2\omega_2) + \omega_2 - [0.44(C-2)(C-4) - 0.42C(C-2)$$
$$-0.11C(C-4) + 0.04C^2]/16\pi \tag{5.5.69}$$

$$\eta_1 = -\alpha_1^2 + \alpha_1 + [1.96(C-2)(C-4) + 3.30C(C-2)$$
$$+0.49C(C-4) + 0.32C^2]/16\pi \tag{5.5.70}$$

$$\eta_2 = -(2\alpha_1\widetilde{\alpha}_2 + \widetilde{\alpha}_2^2) + \widetilde{\alpha}_2 - [2.16(C-2)(C-4) + 2.30C(C-2)$$
$$+0.54C(C-4) - 0.06C^2]/16\pi \tag{5.5.71}$$

$$\xi_1 = (\eta_1\omega_1 + \eta_2\omega_1 + 2\eta_2\omega_2) \tag{5.5.72}$$

$$\xi_2 = \eta_1\omega_2 \tag{5.5.73}$$

$$\xi_3 = (\eta_1\omega_2 + \eta_2\omega_2) \tag{5.5.74}$$

In the previous expressions, $\widetilde{S}_n$ is the contact normal stiffness determined from Hertz–Mindlin theory:

$$\widetilde{S}_n = \frac{2\mu R}{(1-v)}\left[\frac{3\pi}{2C}\frac{(1-v)\,P}{(1-\phi)\,\mu}\right]^{1/3} \tag{5.5.75}$$

where μ and v are the shear modulus and Poisson's ratio of the solid material making up the grains, P is the pressure, and R is the grain radius. Note that the normal stiffness $\widetilde{S}_n$ defined in equation 5.5.61 is related to the S_n, defined in the Hertz–Mindlin model, as $\widetilde{S}_n = S_n/2$. The difference occurs because S_n is defined in terms of the incremental change in grain *radius*, R, under compression, while $\widetilde{S}_n$ is defined in terms of the incremental change in grain *diameter, d*, under compression.

The Brandt Model

The Brandt model allows one to calculate the bulk modulus of randomly packed elastic spheres of identical mechanical properties but different sizes. This packing is subject to external and internal hydrostatic pressures. The effective pressure P is the difference between these two pressures. The effective bulk modulus is (Brandt, 1955)

$$K_{\text{eff}} = \frac{2P^{1/3}}{9\phi}\left[\frac{E}{1.75(1-v^2)}\right]^{2/3} Z - 1.5PZ \tag{5.5.76}$$

$$Z = \frac{(1+30.75z)^{5/3}}{1+46.13z}, \quad z = \frac{K^{3/2}(1-v^2)}{E\sqrt{P}} \tag{5.5.77}$$

In this case E and v are the mineral Young's modulus and Poisson's ratio, and K is the fluid bulk modulus.

The Johnson *et al.* Model

Norris and Johnson (1997) and Johnson *et al.* (1998) have developed an effective medium theory for the nonlinear elasticity of granular sphere packs, generalizing the earlier results of Walton, based on the underlying Hertz–Mindlin theory of grain-to-grain contacts. Johnson *et al.* (1998) show that for this model, the second-order elastic constants, C_{ijkl}, are unique path-independent functions of an arbitrary strain environment ε_{ij}. However, the stress tensor σ_{ij}, depends on the strain path, and consequently C_{ijkl} (considered as a function of applied stresses) are path dependent. For a pack of identical spheres of radius R with no-slip Hertz–Mindlin contacts, the elastic constants in the effective medium approximation are given by Johnson *et al.* (1997) as

$$C_{ijkl} = \frac{3n(1 - \phi)}{4\pi^2 R^{1/2} B_W (2B_W + C_W)}$$
$$\times \langle \zeta^{1/2} [2C_W N_i N_j N_k N_l + B_W (\delta_{ik} N_j N_l + \delta_{il} N_j N_k + \delta_{jl} N_i N_k + \delta_{jk} N_i N_l)] \rangle$$

$$(5.5.78)$$

where $\langle \dots \rangle$ denotes an average over all solid angles. ϕ is porosity, n is the coordination number or average number of contacts per grain, N_i is the unit vector along sphere centers, and $\zeta \equiv - N_i \varepsilon_{ij} N_j R$ is the normal component of displacement at the contacts:

$$B_W = \frac{2}{\pi C_n} \tag{5.5.79}$$

$$C_W = \frac{4}{\pi} \left[\frac{1}{C_t} - \frac{1}{C_n} \right] \tag{5.5.80}$$

$$C_n = \frac{4\mu_S}{1 - \nu_S} \tag{5.5.81}$$

$$C_t = \frac{8\mu_S}{2 - \nu_S} \tag{5.5.82}$$

where μ_S and ν_S are the shear modulus and the Poisson's ratio of individual grains, respectively.

When the strain is a combination of hydrostatic compression and uniaxial compression (along the 3-axis),

$$\varepsilon_{ij} = \varepsilon \delta_{ij} + \varepsilon_3 \delta_{i3} \delta_{j3}, \tag{5.5.83}$$

the sphere pack exhibits transversely isotropic symmetry. Assuming the angular distribution of contacts to be isotropic, explicit expressions for the five independent elements of the stiffness tensor are given as (Johnson *et al.*, 1998):

$$C_{11} \equiv C_{1111} = \frac{\gamma}{\alpha} \left\{ 2B_W [I_0(\alpha) - I_2(\alpha)] + \frac{3}{4} C_W [I_0(\alpha) - 2I_2(\alpha) + I_4(\alpha)] \right\} \tag{5.5.84}$$

$$C_{13} \equiv C_{1133} = \frac{\gamma}{\alpha} \{ C_W [I_2(\alpha) - I_4(\alpha)] \} \tag{5.5.85}$$

$$C_{33} \equiv C_{3333} = \frac{\gamma}{\alpha} [4B_W I_2(\alpha) + 2C_W I_4(\alpha)] \tag{5.5.86}$$

$$C_{44} \equiv C_{2323} = \frac{\gamma}{\alpha} \left\{ \frac{1}{2} B_W [I_0(\alpha) + I_2(\alpha)] + C_W [I_2(\alpha) - I_4(\alpha)] \right\} \tag{5.5.87}$$

$$C_{66} \equiv C_{1212} = \frac{\gamma}{\alpha} \left\{ B_W [I_0(\alpha) - I_2(\alpha)] + \frac{1}{4} C_W [I_0(\alpha) - 2I_2(\alpha) + I_4(\alpha)] \right\} \tag{5.5.88}$$

where

$$\alpha = \sqrt{\varepsilon/\varepsilon_3} \tag{5.5.89}$$

$$\gamma = \frac{3n(1-\phi)(-\varepsilon)^{1/2}}{4\pi^2 B_W (2B_W + C_W)} = \frac{3}{32} n C_n C_t (1-\phi)(-\varepsilon)^{1/2} \tag{5.5.90}$$

The integral $I_n(\alpha) \equiv \int_0^1 x^n \sqrt{\alpha^2 + x^2} \, dx$ can be evaluated analytically as

$$I_0(\alpha) = \frac{1}{2} \left[\sqrt{1+\alpha^2} + \alpha^2 \ln \left(\frac{1 + \sqrt{1+\alpha^2}}{\alpha} \right) \right] \tag{5.5.91}$$

$$I_2(\alpha) = \frac{1}{4} [(1+\alpha^2)^{3/2} - \alpha^2 I_0(\alpha)] \tag{5.5.92}$$

$$I_4(\alpha) = \frac{1}{6} \left[(1+\alpha^2)^{3/2} - 3\alpha^2 I_2(\alpha) \right] \tag{5.5.93}$$

The results of the Johnson *et al.*, model are consistent with the Walton model in the limiting cases of pure hydrostatic strain $(\varepsilon_3 \to 0)$ and pure uniaxial strain $(\varepsilon \to 0)$. When a small uniaxial strain is superimposed on a large isotropic strain, the equations can be expanded and simplified to the first order in $\varepsilon_3/\varepsilon$ (Johnson *et al.*, 1998), giving

$$C_{11} = \gamma \left[\frac{4B_W}{3} + \frac{2C_W}{5} + \frac{\varepsilon_3}{\varepsilon} \left(\frac{2B_W}{15} + \frac{C_W}{35} \right) \right] \tag{5.5.94}$$

$$C_{13} = \gamma \left[\frac{2C_W}{15} + \frac{\varepsilon_3}{\varepsilon} \left(\frac{C_W}{35} \right) \right] \tag{5.5.95}$$

$$C_{33} = \gamma \left[\frac{4B_W}{3} + \frac{2C_W}{5} + \frac{\varepsilon_3}{\varepsilon} \left(\frac{2B_W}{5} + \frac{C_W}{7} \right) \right] \tag{5.5.96}$$

$$C_{44} = \gamma \left[\frac{2B_W}{3} + \frac{2C_W}{15} + \frac{\varepsilon_3}{\varepsilon} \left(\frac{2B_W}{15} + \frac{C_W}{35} \right) \right] \tag{5.5.97}$$

$$C_{66} = \gamma \left[\frac{2B_W}{3} + \frac{2C_W}{15} + \frac{\varepsilon_3}{\varepsilon} \left(\frac{B_W}{15} + \frac{C_W}{105} \right) \right] \tag{5.5.98}$$

The components of the stress tensor σ_{ij} depend on the strain path. When the system first undergoes hydrostatic strain followed by a uniaxial strain, the nonzero components of the stress tensor, σ_{33} and $\sigma_{11} = \sigma_{22}$ are given by

$$\sigma_{33} = -\frac{2(-\varepsilon)^{3/2}(1-\phi)n}{4\pi a^3} \tag{5.5.99}$$

$$\times \left\{ C_t \left[\alpha^2 I_0(\alpha) + (1-\alpha^2) I_2(\alpha) - I_4(\alpha) - \frac{2}{3}\alpha^3 \right] + C_n[\alpha^2 I_2(\alpha) + I_4(\alpha)] \right\} \tag{5.5.100}$$

$$\sigma_{11} = -\frac{(-\varepsilon)^{3/2}(1-\phi)n}{4\pi a^3} \left\{ -C_t \left[\alpha^2 I_0(\alpha) + (1-\alpha^2) I_2(\alpha) - I_4(\alpha) - \frac{2}{3}\alpha^3 \right] \right.$$

$$\left. + C_n[\alpha^2 I_0(\alpha) + (1-\alpha^2) I_2(\alpha) - I_4(\alpha)] \right\} \tag{5.5.101}$$

When a uniaxial compression is followed by an isotropic compression, the stresses are

$$\sigma_{33} = -\frac{2(-\varepsilon)^{3/2}(1-\phi)n}{4\pi a^3} \left\{ \frac{1}{12} C_t + C_n[\alpha^2 I_2(\alpha) + I_4(\alpha)] \right\} \tag{5.5.102}$$

$$\sigma_{11} = -\frac{(-\varepsilon)^{3/2}(1-\phi)n}{4\pi a^3} \left\{ -\frac{1}{12} C_t + C_n[\alpha^2 I_0(\alpha) + (1-\alpha^2) I_2(\alpha) - I_4(\alpha)] \right\} \tag{5.5.103}$$

and finally when the two strain components are applied simultaneously the stress components are

$$\sigma_{33} = -\frac{2(-\varepsilon)^{3/2}(1-\phi)n}{4\pi a^3} \left\{ C_t[I_2(\alpha) - I_4(\alpha)] + C_n[\alpha^2 I_2(\alpha) + I_4(\alpha)] \right\} \tag{5.5.104}$$

$$\sigma_{11} = -\frac{(-\varepsilon)^{3/2}(1-\phi)n}{4\pi a^3} \left\{ -C_t[I_2(\alpha) - I_4(\alpha)] + C_n[\alpha^2 I_0(\alpha) + (1-\alpha^2) I_2(\alpha) - I_4(\alpha)] \right\} \tag{5.5.105}$$

The differences in the stresses for the three different strain paths are quite small (Johnson et al., 1998).

The concept of third-order elastic constants described by nonlinear hyperelasticity with three third-order constants breaks down for unconsolidated granular media with uniaxial strain superposed on a large hydrostatic strain. The changes in second-order elastic constants due to an imposed strain are governed (to the first order) by four "third-order" moduli and not just the usual three third-order moduli (Norris and Johnson, 1997). In granular media, the contact forces are path dependent and are not derivable from a potential energy function. As a result, the system cannot be described by an equivalent nonlinear hyperelastic medium.

Dvorkin's Cemented-Sand Model

The cemented-sand model allows one to calculate the bulk and shear moduli of dry sand in which cement is deposited *at grain contacts*. The cement is elastic and its properties may differ from those of the spheres.

It is assumed that the starting framework of cemented sand is a dense, random pack of identical spherical grains with porosity $\phi_0 \approx 0.36$ and an average number of contacts per grain $C = 9$. Adding cement to the grains reduces porosity and increases the effective elastic moduli of the aggregate. Then, these effective dry-rock bulk and shear moduli are (Dvorkin and Nur, 1996)

$$K_{\text{eff}} = \frac{1}{6}C(1 - \phi_0)M_c\hat{S}_n \tag{5.5.106}$$

$$\mu_{\text{eff}} = \frac{3}{5}K_{\text{eff}} + \frac{3}{20}C(1 - \phi_0)\mu_c\hat{S}_\tau \tag{5.5.107}$$

$$M_c = \rho_c V_{\text{Pc}}^2 \tag{5.5.108}$$

$$\mu_c = \rho_c V_{\text{Sc}}^2 \tag{5.5.109}$$

where ρ_c is the density of the cement and V_{Pc} and V_{Sc} are its P- and S-wave velocities, respectively. Parameters $\hat{S}_n$ and $\hat{S}_\tau$ are proportional to the normal and shear stiffnesses, respectively, of a cemented two-grain combination. They depend on the amount of contact cement and on the properties of the cement and the grains as defined in the following relations:

$$\hat{S}_n = A_n\alpha^2 + B_n\alpha + C_n \tag{5.5.110}$$

$$A_n = -0.024153\Lambda_n^{-1.3646} \tag{5.5.111}$$

$$B_n = 0.20405\Lambda_n^{-0.89008} \tag{5.5.112}$$

$$C_n = 0.00024649\Lambda_n^{-1.9864} \tag{5.5.113}$$

$$\hat{S}_\tau = A_\tau\alpha^2 + B_\tau\alpha + C_\tau \tag{5.5.114}$$

$$A_\tau = -10^{-2}(2.26v^2 + 2.07v + 2.3)\Lambda_\tau^{0.079v^2+0.1754v-1.342} \tag{5.5.115}$$

$$B_\tau = (0.0573v^2 + 0.0937v + 0.202)\Lambda_\tau^{0.0274v^2+0.0529v-0.8765} \tag{5.5.116}$$

$$C_\tau = 10^{-4}(9.654v^2 + 4.945v + 3.1)\Lambda_\tau^{0.01867v^2+0.4011v-1.8186} \tag{5.5.117}$$

$$\Lambda_n = \frac{2\mu_c}{\pi\mu}\frac{(1 - v)(1 - v_c)}{(1 - 2v_c)} \tag{5.5.118}$$

$$\Lambda_\tau = \frac{\mu_c}{\pi\mu} \tag{5.5.119}$$

$$\alpha = \frac{a}{R} \tag{5.5.120}$$

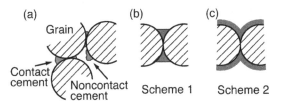

Figure 5.5.6 (a) Schematic representation of types of cement deposition. (b) All cement deposited at grain contacts. (c) Cement deposited in uniform layer around grains.

where μ and v are the shear modulus and the Poisson's ratio of the grains, respectively; μ_c and v_c are the shear modulus and the Poisson's ratio of the cement, respectively; a is the radius of the contact cement layer; and R is the grain radius. The theory is limited to values of $a \ll R$.

The amount of contact cement can be expressed through the ratio α of the radius of the cement layer a to the grain radius R:

$$\alpha = \frac{a}{R} \tag{5.5.121}$$

The radius a of the contact cement layer is not necessarily directly related to the total amount of cement; part of the cement may be deposited away from the intergranular contacts. However, by assuming that porosity reduction in sands is due to cementation only and by adopting certain schemes of cement deposition, we can relate the parameter α to the current porosity of cemented sand ϕ. For example, we can use Scheme 1 (see Figure 5.5.6), in which all cement is deposited at grain contacts, to obtain the formula

$$\alpha = 2\left[\frac{\phi_0 - \phi}{3C(1 - \phi_0)}\right]^{1/4} = 2\left[\frac{S\phi_0}{3C(1 - \phi_0)}\right]^{1/4} \tag{5.5.122}$$

or we can use Scheme 2, in which cement is evenly deposited on the grain surface:

$$\alpha = \left[\frac{2(\phi_0 - \phi)}{3(1 - \phi_0)}\right]^{1/2} = \left[\frac{2S\phi_0}{3(1 - \phi_0)}\right]^{1/2} \tag{5.5.123}$$

In these formulas, S is the cement saturation of the pore space. It is the fraction of the pore space (of the uncemented sand) occupied by cement (in the cemented sand).

If the properties of the cement are identical to those of the grains, the cementation theory gives results that are very close to those of the Digby model. The cementation theory allows one to **diagnose** a rock by determining what type of cement prevails. For example, the theory helps to distinguish between quartz and clay cement (Figure 5.5.7). Generally, the model predicts V_P much more reliably than V_S.

Xia and Mavko (2015) generalized the contact cement model to larger contact areas by using finite element (FE) calculations to simulate normal and shear contact stiffnesses. The contact geometry is shown in Figure 5.5.8. Grain radius is R; radius of the

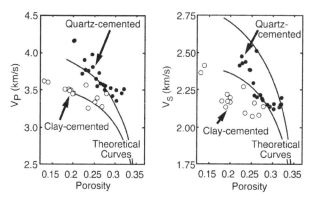

Figure 5.5.7 Predictions of V_P and V_S using the Scheme 2 model for quartz and clay cement, compared with data from quartz- and clay-cemented rocks from the North Sea

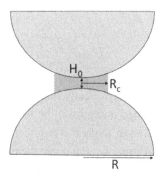

Figure 5.5.8 Geometry of cemented sphere contact. Sphere radius is R; radius of cemented patch is R_c; separation of spheres before cement is H_0.

cemented patch is R_c; separation of spheres before cement is H_0. If pressure is applied before cementation, a Hertz indentation is formed, and is described by $H_0 < 0$. Normalized parameters are $h = H_0/R$ and $r_c = R_c/R$. Pre-cement contact compression was not considered in the original Dvorkin model.

Empirical polynomial fits to the computed contact stiffnesses with varying amounts of pre-compaction and cementation are given as follows. Note that grain radius is in meters, computed contact stiffness is in N/m.

If both grain and cement are quartz (shear modulus $\mu = 45 \cdot 10^9$ Pa and Poisson's ratio $v = 0.08$), the stiffnesses are:

$$S_n = 10^{10} R(10.85r_c^2 + 17.73r_c - 11.72h_0 + 0.09) \text{ (N/m)} \qquad (5.5.124)$$

$$S_\tau = 10^{10} R(-7.127r_c^2 + 12.34r_c - 18.06h_0 + 0.2061) \text{ (N/m)} \qquad (5.5.125)$$

If the grain is quartz ($\mu = 45$ GPa and $v = 0.08$) and cement is calcite ($\mu = 32$ GPa and $v = 0.32$), the stiffnesses are:

$$S_n = 10^{10} R(9.632r_c^2 + 18.76r_c - 8.828h_0 + 0.0097) \text{ (N/m)} \qquad (5.5.126)$$

$$S_\tau = 10^{10}R(-7.674r_c^2 + 12.32r_c - 20.4h_0 + 0.195) \ (\text{N/m}) \qquad (5.5.127)$$

If the grain is quartz ($\mu = 45$ GPa and $v = 0.08$) and cement is clay ($\mu = 9$ GPa and $v = 0.34$), the stiffnesses are:

$$S_n = 10^{10}R(0.6775r_c^2 + 18.21r_c - 28.22h_0 + 0.0509) \ (\text{N/m}) \qquad (5.5.128)$$

$$S_\tau = 10^{10}R(-5.68r_c^2 + 9.55r_c + 388.3h_0^2 - 39.07h_0 + 0.339) \ (\text{N/m}) \qquad (5.5.129)$$

Figure 5.5.9 Compares the simulated quartz–quartz contact stiffness predicted by equations 5.5.124 and 5.5.125 $R=1$ m, with Dvorkin's analytical contact stiffness,

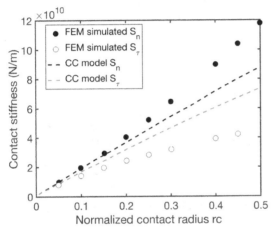

Figure 5.5.9 Comparison of simulated normal and shear cemented contact stiffness (Xia and Mavko, 2015) with analytical values from the contact cement model of Dvorkin and Nur (1996). Parameters are quartz cement and quartz grains; R = 1 m.

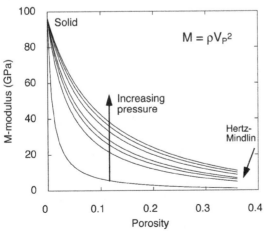

Figure 5.5.10 Illustration of the modified lower Hashin–Shtrikman bound for various effective pressures. The pressure dependence follows from the Hertz–Mindlin theory incorporated into the right end-member. All grain contacts have perfect adhesion. The grains are pure quartz.

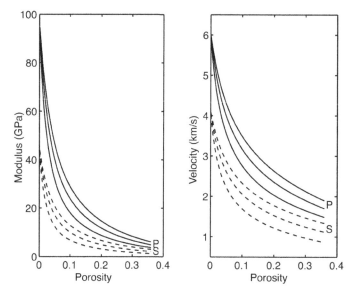

Figure 5.5.11 Illustration of the modified lower Hashin–Shtrikman bound for a varying fractions of contact with perfect adhesion. The grains are pure quartz and the pack is dry. The effective pressure is 20 MPa and the critical porosity is 0.36. Left: shear and compressional modulus versus porosity. Right: V_P and V_S versus porosity. Solid curves are for the P-wave velocity and modulus, while dashed lines are for the S-wave velocity and modulus. For each computed parameter (modulus and velocity) the upper curves are for perfect adhesion while the lower curves are for perfect slip. The curve in between is for $f = 0.5$.

equations 5.5.106–5.5.120. At small r_c the models converge, but as r_c grows, the contact cement model underpredicts the simulated S_n and overpredicts the simulated S_τ.

Substituting the new simulated contact stiffnesses (equations 5.5.124–5.5.129) into Dvorkin's contact cement model yields the following approximations for effective moduli.

If both grain and cement are quartz ($\mu = 45$ GPa and $v = 0.08$), the stiffnesses are:

$$K_{eff} = \frac{5}{6\pi} C(1 - \phi_0)(10.85r_c^2 + 17.73r_c - 11.72h_0 + 0.09) \text{ (GPa)} \tag{5.5.130}$$

$$\mu_{eff} = \frac{3}{5} K_{eff} + \frac{3}{4\pi} C(1 - \phi_0)(-7.127r_c^2 + 12.34r_c - 18.06h_0 + 0.2061) \text{ (GPa)} \tag{5.5.131}$$

If the grain is quartz ($\mu = 45$ GPa and $v = 0.08$) and cement is calcite ($\mu = 32$ GPa and $v = 0.32$), the stiffnesses are:

$$K_{eff} = \frac{5}{6\pi} C(1 - \phi_0)(9.632r_c^2 + 18.76r_c - 8.828h_0 + 0.0097) \text{ (GPa)} \tag{5.5.132}$$

$$\mu_{eff} = \frac{3}{5} K_{eff} + \frac{3}{4\pi} C(1 - \phi_0)(-7.674r_c^2 + 12.32r_c - 20.4h_0 + 0.195) \text{ (GPa)} \tag{5.5.133}$$

If the grain is quartz ($\mu = 45$ GPa and $v = 0.08$) and cement is clay ($\mu = 9$ GPa and $v = 0.34$), the stiffnesses are:

$$K_{eff} = \frac{5}{6\pi} C(1 - \phi_0)(0.6775r_c^2 + 18.21r_c - 28.22h_0 + 0.0509) \text{ (GPa)} \qquad (5.5.134)$$

$$\mu_{eff} = \frac{3}{5} K_{eff} + \frac{3}{4\pi} C(1 - \phi_0)(-5.68r_c^2 + 9.546r_c + 388.3h_0^2 - 39.07h_0 + 0.339) \text{ (GPa)}$$

$$(5.5.135)$$

Figure 5.5.9 compares the contact stiffness predicted by equations 5.5.124 and 5.5.125 with Dvorkin's contact stiffness, equations for quartz grains ($\mu = 45 \cdot 10^9$ Pa and $v = .08$) and R = 1 m.

The Uncemented (Soft) Sand Model

The uncemented-sand (or "soft-sand") model allows one to calculate the bulk and shear moduli of dry sand in which cement is deposited *away from grain contacts*. It is assumed that the starting framework of uncemented sand is a dense random pack of identical spherical grains with porosity ϕ_0 about 0.36 and average number of contacts per grain $C = 5$ to 9. At this porosity, the contact Hertz–Mindlin theory gives the following expressions for the effective bulk (K_{HM}) and shear (μ_{HM}) moduli of a dry, dense, random pack of identical spherical grains subject to a hydrostatic pressure P:

$$K_{HM} = \left[\frac{C^2(1 - \phi_0)^2 \mu^2}{18\pi^2(1 - v)^2} P \right]^{1/3} \qquad (5.5.136)$$

$$\mu_{HM} = \frac{5 - 4v}{5(2 - v)} \left[\frac{3C^2(1 - \phi_0)^2 \mu^2}{2\pi^2(1 - v)^2} P \right]^{1/3} \qquad (5.5.137)$$

where v is the grain Poisson's ratio and μ is the grain shear modulus. A version of the same model allows a fraction f of grain contacts to have perfect adhesion and the rest to be frictionless; it uses

$$\mu_{HM} = \frac{2 + 3f - v(1 + 3f)}{5(2 - v)} \left[\frac{3C^2(1 - \phi)^2 \mu^2}{2\pi^2(1 - v)^2} P \right]^{1/3} \qquad (5.5.138)$$

To find the effective moduli (K_{eff} and μ_{eff}) at a different porosity ϕ, a heuristic modified Hashin–Shtrikman lower bound is used as an interpolator between the Hertz–Mindlin moduli at porosity ϕ_0 and the solid grain at zero porosity:

$$K_{eff} = \left[\frac{\phi/\phi_0}{K_{HM} + \frac{4}{3}\mu_{HM}} + \frac{1 - \phi/\phi_0}{K + \frac{4}{3}\mu_{HM}} \right]^{-1} - \frac{4}{3}\mu_{HM} \qquad (5.5.139)$$

$$\mu_{\text{eff}} = \left[\frac{\phi/\phi_0}{\mu_{\text{HM}} + \frac{\mu_{\text{HM}}}{6}\left(\frac{9K_{\text{HM}} + 8\mu_{\text{HM}}}{K_{\text{HM}} + 2\mu_{\text{HM}}}\right)} + \frac{1 - \phi/\phi_0}{\mu + \frac{\mu_{\text{HM}}}{6}\left(\frac{9K_{\text{HM}} + 8\mu_{\text{HM}}}{K_{\text{HM}} + 2\mu_{\text{HM}}}\right)} \right]^{-1}$$

$$- \frac{\mu_{\text{HM}}}{6}\left(\frac{9K_{\text{HM}} + 8\mu_{\text{HM}}}{K_{\text{HM}} + 2\mu_{\text{HM}}}\right) \tag{5.5.140}$$

where K is the grain bulk modulus. Figure 5.5.10 shows the curves from this model in terms of the P-wave modulus, $M = K + \frac{4}{3}\mu$. Figure 5.5.11 illustrates the effect of the parameter f (the fraction of the perfect-adhesion contacts) on the elastic moduli and velocity in a dry pack of identical quartz grains.

Example: Calculate V_P and V_S in uncemented dry quartz sand of porosity 0.3, at 40 MPa overburden and 20 MPa pore pressure. Use the uncemented sand model. For pure quartz, $\mu = 45$ GPa, $K = 36.6$ GPa, and $v = 0.06$. Then, for effective pressure 20 MPa $= 0.02$ GPa,

$$K_{\text{HM}} = \left[\frac{9^2(1 - 0.36)^2 45^2}{18 \cdot 3.14^2(1 - 0.06)^2} 0.02 \right]^{1/3} = 2 \text{ GPa}$$

$$\mu_{\text{HM}} = \frac{5 - 4 \cdot 0.06}{5(2 - 0.06)}\left[\frac{3 \cdot 9^2(1 - 0.36)^2 45^2}{2 \cdot 3.14^2(1 - 0.06)^2} 0.02 \right]^{1/3} = 3 \text{ GPa}$$

Next,

$$K_{\text{eff}} = \left(\frac{0.3/0.36}{2 + \frac{4}{3} \cdot 3} + \frac{1 - 0.3/0.36}{36.6 + \frac{4}{3} \cdot 3} \right)^{-1} - \frac{4}{3} \cdot 3 = 3 \text{ GPa}$$

$$\mu_{\text{eff}} = \left(\frac{0.3/0.36}{3 + 2.625} + \frac{1 - 0.3/0.36}{45 + 2.265} \right)^{-1} - 2.625 = 3.97 \text{ GPa}$$

Pure quartz density is 2.65 g/cm^3; then, the density of the sandstone is

$$2.65 \cdot (1 - 0.3) = 1.855 \text{ g/cm}^3$$

The P-wave velocity is

$$V_P = \sqrt{\frac{3 + \frac{4}{3} \cdot 3.97}{1.855}} = 2.11 \text{ km/s}$$

and the S-wave velocity is

$$V_S = \sqrt{\frac{3.97}{1.855}} = 1.46 \text{ km/s}$$

The Stiff-Sand and Intermediate Stiff-Sand Models

A counterpart to the soft-sand model is the "stiff-sand" model, which uses precisely the same end-members in the porosity–elastic–modulus plane but connects them with a heuristic modified Hashin–Shtrikman *upper* bound as a stiff interpolator (see Section 7.1):

$$K_{\text{eff}} = \left[\frac{\phi/\phi_0}{K_{\text{HM}} + \frac{4}{3}\mu} + \frac{1 - \phi/\phi_0}{K + \frac{4}{3}\mu} \right]^{-1} - \frac{4}{3}\mu \qquad (5.5.141)$$

$$\mu_{\text{eff}} = \left[\frac{\phi/\phi_0}{\mu_{\text{HM}} + \frac{\mu}{6}\left(\frac{9K + 8\mu}{K + 2\mu}\right)} + \frac{1 - \phi/\phi_0}{\mu + \frac{\mu}{6}\left(\frac{9K + 8\mu}{K + 2\mu}\right)} \right]^{-1} - \frac{\mu}{6}\left(\frac{9K + 8\mu}{K + 2\mu}\right) \qquad (5.5.142)$$

where K_{HM} and μ_{HM} are precisely the same as in the soft-sand model. Figure 5.5.12 compares the elastic moduli and velocity in a dry pack of identical quartz grains as calculated by the soft- and stiff-sand models. Notice that the stiff-sand model produces velocity–porosity curves that are essentially identical to those from the Raymer–Hunt–Gardner empirical model.

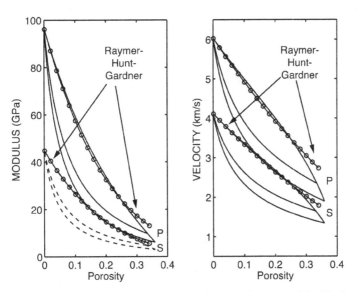

Figure 5.5.12 Illustration of the modified lower and upper Hashin–Shtrikman bounds (soft- and stiff-sand models, respectively). The grains are pure quartz and the pack is dry. The effective pressure is 20 MPa, the coordination number is 9, and the critical porosity is 0.36. Left: shear and compressional moduli versus porosity. Right: V_P and V_S versus porosity. Solid curves are for the P-wave velocity and modulus while dashed lines are for the S-wave velocity and modulus. The curves between the two bounds are for the intermediate stiff-sand model that uses the soft-sand equation with an artificial coordination number of 15.

The intermediate stiff-sand model uses the functional form of the soft-sand model (the modified Hashin–Shtrikman lower bound) but with the high-porosity end-point situated on the stiff-sand model curve. The easiest way to generate such curves is by simply increasing the coordination number in the soft-sand model (Figure 5.5.12). This artificially increased coordination number may not be representative of the actual coordination number of the grain pack at the high-porosity end-point.

These models connect two end-members; one has zero porosity and the modulus of the solid phase, and the other has high porosity and a pressure-dependent modulus, as given by the Hertz–Mindlin theory. This contact theory allows one to describe the noticeable pressure dependence normally observed in sands. Notice that in many experiments on natural sands and artificial granular packs, the observed dependence of the elastic moduli on pressure is different from that given by the Hertz–Mindlin theory. This is because the grains are not perfect spheres, and the contacts have configurations different from those between perfectly spherical particles. Hertz–Mindlin theory also fails to incorporate the spatial heterogeneity of stress and strain within the random grain pack.

The high-porosity end-member does not necessarily have to be calculated from the Hertz–Mindlin theory. The end-member can be measured experimentally on high porosity sands from a given reservoir. Then, to estimate the moduli of sands of different porosities, the modified Hashin–Shtrikman lower bound formulas can be used, where K_{HM} and μ_{HM} are set at the measured values.

This method provides estimates for velocities in uncemented sands. In Figures 5.5.13 and 5.5.14, the curves are from the theory.

This method can also be used for estimating velocities in sands of porosities exceeding 0.36.

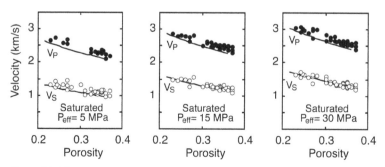

Figure 5.5.13 Prediction of V_P and V_S using the modified lower Hashin–Shtrikman bound, compared with measured velocities from unconsolidated North Sea samples.

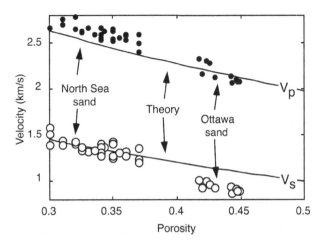

Figure 5.5.14 Prediction of V_P and V_S using the modified lower Hashin–Shtrikman bound, compared with measured velocities from North Sea and Ottawa sand samples.

Granular Dynamics Simulations and Caveat on the Use of Effective Medium Models for Granular Media

Granular media have properties lying somewhat between solids and liquids and are sometimes considered to be a distinct form of matter. Complex behavior arises from the ability of grains to move relative to each other, modify their packing and coordination numbers, and rotate. Observed behavior, depending on the stress and strain conditions, is sometimes approximately nonlinearly elastic, sometimes viscoelastic, and sometimes somewhat fluid-like. A number of authors (Goddard, 1990; Makse *et al.*, 1999, 2000, 2004) have shown that this complex behavior causes effective medium theory to fail in cohesionless granular assemblies. Closed-form effective medium theories tend to predict the incorrect (relative to laboratory observations) dependence of effective moduli on pressure, and poor estimates of the bulk to shear moduli ratio. Numerical methods, referred to as "molecular dynamics" or "discrete element modeling," which simulate the motions and interactions of thousands of grains, appear to come closer to predicting observed behavior (e.g., Makse *et al.*, 2004; Garcia and Medina, 2006; Sain, 2010; Sain *et al.*, 2016). Closed-form effective medium models can be useful, because it is not always practical to run a numerical simulation; however, model predictions of the types presented in this section must be used with care.

Uses

The methods can be used to model granular high-porosity rocks, as well as rocks in the entire porosity range, depending on what type of diagenetic transformation reduced the original high porosity.

Assumptions and Limitations

The grain contact models presuppose the following:
- the strains are small;
- most models assume that grains are homogeneous, isotropic, elastic spheres;
- except for the models for regular packings, packings are assumed to be random and statistically isotropic; and
- the effective elastic constants are relevant for long-wavelength propagation, where wavelengths are much longer (more than 10 times) compared with the grain radius.

Extensions

To calculate the effective elastic moduli of saturated rocks (and their low-frequency acoustic velocities), Gassmann's formula should be applied.

5.6   Ordered Spherical Grain Packings: Effective Moduli

Synopsis

Ordered packings of identical spherical particles are generally anisotropic; thus their effective elastic properties can be described through stiffness matrices.

Simple Cubic Packing

The coordination number is 6, and the porosity is 47%. The stiffness matrix is

$$
\begin{pmatrix}
C_{11} & C_{12} & C_{12} & 0 & 0 & 0 \\
C_{12} & C_{11} & C_{12} & 0 & 0 & 0 \\
C_{12} & C_{12} & C_{11} & 0 & 0 & 0 \\
0 & 0 & 0 & \frac{1}{2}(C_{11} - C_{12}) & 0 & 0 \\
0 & 0 & 0 & 0 & \frac{1}{2}(C_{11} - C_{12}) & 0 \\
0 & 0 & 0 & 0 & 0 & \frac{1}{2}(C_{11} - C_{12})
\end{pmatrix}
\tag{5.6.1}
$$

where

$$
C_{11} = C_0 \quad C_{12} = \frac{v}{2(2 - v)} C_0 \quad C_0 = \left[\frac{3\mu^2 P}{2(1 - v)^2} \right]^{1/3}
\tag{5.6.2}
$$

P is the hydrostatic pressure, and μ and v are the shear modulus and Poisson's ratio of the grain material, respectively.

Hexagonal Close Packing

The coordination number is 12, and the porosity is about 26%. The stiffness matrix is where

$$\begin{pmatrix} C_{11} & C_{12} & C_{13} & 0 & 0 & 0 \\ C_{12} & C_{11} & C_{13} & 0 & 0 & 0 \\ C_{13} & C_{13} & C_{33} & 0 & 0 & 0 \\ 0 & 0 & 0 & C_{44} & 0 & 0 \\ 0 & 0 & 0 & 0 & C_{44} & 0 \\ 0 & 0 & 0 & 0 & 0 & C_{66} \end{pmatrix} \tag{5.6.3}$$

where

$$C_{11} = \frac{1152 - 1848v + 725v^2}{24(2-v)(12-11v)} C_0 \tag{5.6.4}$$

$$C_{12} = \frac{v(120 - 109v)}{24(2-v)(12-11v)} C_0 \tag{5.6.5}$$

$$C_{13} = \frac{v}{3(2-v)} C_0 \tag{5.6.6}$$

$$C_{33} = \frac{4(3-2v)}{3(2-v)} C_0 \tag{5.6.7}$$

$$C_{44} = C_{55} = \frac{6-5v}{3(2-v)} C_0 \tag{5.6.8}$$

$$C_{66} = \frac{576 - 948v + 417v^2}{24(2-v)(12-11v)} C_0 \tag{5.6.9}$$

Face-Centered Cubic Packing

The coordination number is 12, and the porosity is about 26%. The stiffness matrix is

$$\begin{pmatrix} C_{11} & C_{12} & C_{12} & 0 & 0 & 0 \\ C_{12} & C_{11} & C_{12} & 0 & 0 & 0 \\ C_{12} & C_{12} & C_{11} & 0 & 0 & 0 \\ 0 & 0 & 0 & C_{44} & 0 & 0 \\ 0 & 0 & 0 & 0 & C_{44} & 0 \\ 0 & 0 & 0 & 0 & 0 & C_{44} \end{pmatrix} \tag{5.6.10}$$

where

$$C_{11} = 2C_{44} = \frac{4-3v}{2-v} C_0 \tag{5.6.11}$$

$$C_{12} = \frac{v}{2(2-v)} C_0 \tag{5.6.12}$$

Uses

The results of this section are sometimes used to estimate the elastic properties of granular materials.

Assumptions and Limitations

These models assume identical, elastic, spherical grains under small-strain conditions.

6 Fluid Effects on Wave Propagation

6.1 Biot's Velocity Relations

Synopsis

Biot (1956) derived theoretical formulas for predicting the frequency-dependent seismic velocities of saturated rocks in terms of the dry-rock properties. His formulation incorporates some, but not all, of the mechanisms of viscous and inertial interaction between the pore fluid and the mineral matrix of the rock. The low-frequency limiting velocities, V_{P0} and V_{S0} are the same as those predicted by Gassmann's relations (see the discussion of Gassmann in Section 6.3). The high-frequency limiting velocities, $V_{P\infty}$ and $V_{S\infty}$ (cast in the notation of Johnson and Plona, 1982), are given by

$$V_{P\infty}(\text{fast, slow}) = \left\{ \frac{\Delta \pm [\Delta^2 - 4(\rho_{11}\rho_{22} - \rho_{12}^2)(PR - Q^2)]^{1/2}}{2(\rho_{11}\rho_{22} - \rho_{12}^2)} \right\}^{1/2} \tag{6.1.1}$$

$$V_{S\infty} = \left(\frac{\mu_{\text{fr}}}{\rho - \phi\rho_{\text{fl}}\alpha^{-1}} \right)^{1/2} \tag{6.1.2}$$

$$\Delta = P\rho_{22} + R\rho_{11} - 2Q\rho_{12} \tag{6.1.3}$$

$$P = \frac{(1-\phi)(1-\phi-K_{\text{fr}}/K_0)K_0 + \phi K_0 K_{\text{fr}}/K_{\text{fl}}}{1-\phi-K_{\text{fr}}/K_0 + \phi K_0/K_{\text{fl}}} + \frac{4}{3}\mu_{\text{fr}} \tag{6.1.4}$$

$$Q = \frac{(1-\phi-K_{\text{fr}}/K_0)\phi K_0}{1-\phi-K_{\text{fr}}/K_0 + \phi K_0/K_{\text{fl}}} \tag{6.1.5}$$

$$R = \frac{\phi^2 K_0}{1-\phi-K_{\text{fr}}/K_0 + \phi K_0/K_{\text{fl}}} \tag{6.1.6}$$

$$\rho_{11} = (1-\phi)\rho_0 - (1-\alpha)\phi\rho_{\text{fl}} \tag{6.1.7}$$

$$\rho_{22} = \alpha\phi\rho_{\text{fl}} \tag{6.1.8}$$

$$\rho_{12} = (1-\alpha)\phi\rho_{\text{fl}} \tag{6.1.9}$$

$$\rho = \rho_0(1-\phi) + \rho_{\text{fl}}\phi \tag{6.1.10}$$

where K_{fr} and μ_{fr} are the effective bulk and shear moduli of the rock frame, respectively – either the dry-frame moduli or the high-frequency, unrelaxed, "wet-frame" moduli predicted by the Mavko–Jizba squirt theory (Section 6.11); K_0 is the bulk modulus of the mineral material making up the rock; K_{fl} is the effective bulk modulus of the pore fluid; ϕ is the porosity; ρ_0 is the mineral density; ρ_{fl} is the fluid density; and α is the tortuosity parameter, which is always greater than or equal to 1.

The term ρ_{12} describes the induced mass resulting from inertial drag caused by the relative acceleration of the solid frame and the pore fluid. The tortuosity, α (sometimes called the structure factor), is a purely geometrical factor independent of the solid or fluid densities. Berryman (1981) obtained the relation

$$\alpha = 1 - r\left(1 - \frac{1}{\phi}\right) \tag{6.1.11}$$

where $r = \dfrac{1}{2}$ for spheres, and lies between 0 and 1 for other ellipsoids. For uniform cylindrical pores with axes parallel to the pore pressure gradient, α equals 1 (the minimum possible value), whereas for a random system of pores with all possible orientations, $\alpha = 3$ (Stoll, 1977). The high-frequency limiting velocities depend quite strongly on α, with higher fast P-wave velocities for lower α values.

The two solutions previously given for the high-frequency limiting P-wave velocity, designated by $\pm$, correspond to the "fast" and "slow" waves. The fast wave is the compressional body-wave most easily observed in the laboratory and the field, and it corresponds to overall fluid and solid motions that are in phase. The slow wave is a highly dissipative wave in which the overall solid and fluid motions are out of phase.

Another approximate expression for the high-frequency limit of the fast P-wave velocity is (Geertsma and Smit, 1961)

$$V_{P\infty} = \left\{ \frac{1}{\rho_0(1 - \phi) + \phi\rho_{fl}(1 - \alpha^{-1})} \right.$$

$$\left. \times \left[\left(K_{fr} + \frac{4}{3}\mu_{fr}\right) + \frac{\phi\dfrac{\rho}{\rho_{fl}}\alpha^{-1} + \left(1 - \dfrac{K_{fr}}{K_0}\right)\left(1 - \dfrac{K_{fr}}{K_0} - 2\phi\alpha^{-1}\right)}{\left(1 - \dfrac{K_{fr}}{K_0} - \phi\right)\dfrac{1}{K_0} + \dfrac{\phi}{K_{fl}}} \right] \right\}^{1/2} \tag{6.1.12}$$

Caution:

This form predicts velocities that are too high (by about 3%–6%) compared with the actual high-frequency limit.

The complete frequency dependence can be obtained from the roots of the dispersion relations (Biot, 1956; Stoll, 1977; Berryman, 1980a):

$$\begin{vmatrix} H/V_P^2 - \rho & \rho_{fl} - C/V_P^2 \\ C/V_P^2 - \rho_{fi} & q - M/V_P^2 \end{vmatrix} = 0 \tag{6.1.13}$$

$$\begin{vmatrix} \rho - \mu_{fr}/V_S^2 & \rho_{fl} \\ \rho_{fl} & q \end{vmatrix} = 0 \tag{6.1.14}$$

The complex roots are

$$\frac{1}{V_P^2} = \frac{-(Hq + M\rho - 2C\rho_{fl}) \pm \sqrt{(Hq + M\rho - 2C\rho_{fl})^2 - 4(C^2 - MH)(\rho_{fl} - \rho q)}}{2(C^2 - MH)} \tag{6.1.15}$$

$$\frac{1}{V_S^2} = \frac{q\rho - \rho_{fl}}{q\mu_{fr}} \tag{6.1.16}$$

The real and imaginary parts of the roots give the velocity and the attenuation, respectively. Again, the two solutions correspond to the fast and slow P-waves. The various terms are

$$H = K_{fr} + \frac{4}{3}\mu_{fr} + \frac{(K_0 - K_{fr})^2}{(D - K_{fr})} \tag{6.1.17}$$

$$C = \frac{(K_0 - K_{fr})K_0}{(D - K_{fr})} \tag{6.1.18}$$

$$M = \frac{K_0^2}{(D - K_{fr})} \tag{6.1.19}$$

$$D = K_0 \left[1 + \phi \left(\frac{K_0}{K_{fl}} - 1 \right) \right] \tag{6.1.20}$$

$$\rho = (1 - \phi)\rho_0 + \phi\rho_{fl} \tag{6.1.21}$$

$$q = \frac{\alpha\rho_{fl}}{\phi} - \frac{i\eta F(\zeta)}{\omega\kappa} \tag{6.1.22}$$

where η is the viscosity of the pore fluid, κ is the absolute permeability of the rock, and ω is the angular frequency of the plane wave.

The viscodynamic operator $F(\zeta)$ incorporates the frequency dependence of viscous drag and is defined by

$$F(\zeta) = \frac{1}{4} \left(\frac{\zeta T(\zeta)}{1 + 2iT(\zeta)/\zeta} \right) \tag{6.1.23}$$

$$T(\zeta) = \frac{\text{ber}'(\zeta) + i\text{bei}'(\zeta)}{\text{ber}(\zeta) + i\text{bei}(\zeta)} = \frac{e^{i3\pi/4}J_1(\zeta e^{-i\pi/4})}{J_0(\zeta e^{-i\pi/4})} \tag{6.1.24}$$

$$\zeta = (\omega/\omega_r)^{1/2} = \left(\frac{\omega a^2 \rho_{fl}}{\eta} \right)^{1/2} \tag{6.1.25}$$

where ber( ) and bei( ) are real and imaginary parts of the Kelvin function, respectively, $J_n(\ )$ is a Bessel function of order n, and a is the pore-size parameter.

The pore-size parameter a depends on both the dimensions and the shape of the pore space. Stoll (1974) found that values between $\frac{1}{6}$ and $\frac{1}{7}$ of the mean grain diameter gave good agreement with experimental data from several investigators. For spherical grains, Hovem and Ingram (1979) obtained $a = \phi d/[3(1 - \phi)]$, where d is the grain diameter. The velocity dispersion curve for fast P-waves can be closely approximated by a standard linear solid viscoelastic model when $\kappa/a^2 \geq 1$ (see Sections 3.8 and 6.13). However, for most consolidated crustal rocks, κ/a^2 is usually less than 1.

At very low frequencies, $F(\zeta) \rightarrow 1$ and at very high frequencies (large ζ), the asymptotic values are $T(\zeta) \rightarrow (1 + i)/\sqrt{2}$ and $F(\zeta) \rightarrow (\kappa/4)(1 + i)/\sqrt{2}$.

The reference frequency, f_c, which determines the low-frequency range, $f \ll f_c$, and the high-frequency range, $f \gg f_c$, is given by

$$f_c = \frac{\phi\eta}{2\pi\rho_{fl}\kappa} \tag{6.1.26}$$

One interpretation of this relation is that it is the frequency where viscous forces acting on the pore fluid approximately equal the inertial forces acting on it. In the high-frequency limit, the fluid motion is dominated by inertial effects, and in the low-frequency limit, the fluid motion is dominated by viscous effects.

As previously mentioned, Biot's theory predicts the existence of a slow, highly attenuated P-wave in addition to the usual fast P- and S-waves. The slow P-wave has been observed in the laboratory, and it is sometimes invoked to explain diffusional loss mechanisms.

Slow S-wave

In the Biot theory, the only loss mechanism is the average motion of the fluid with respect to the solid frame, ignoring viscous losses within the pore fluid. The fluid strain-rate term is not incorporated into the constitutive equations. As a result, the Biot relaxation term includes only a part of the drag force involving permeability but does not account for the dissipation due to shear drag within the fluid. Incorporation of the viscous term is achieved by volume averaging of the pore-scale constitutive relations of the solid and fluid constituents (de al Cruz and Spanos, 1985; Sahay *et al.*, 2001; Spanos, 2002). This gives rise to two propagating shear processes, corresponding to in-phase and out-of-phase shear motion of the phases, with fast and slow S-wave velocities V_{SI} and V_{SII}, respectively. The slow S-wave has the characteristics of a rapidly decaying viscous wave in a Newtonian fluid. The slow S-wave may play a role in attenuation of fast P- and S-waves by drawing energy from fast waves due to mode conversion at interfaces and discontinuities. Incorporation of the fluid strain-rate term introduces an additional relaxation frequency, the saturated shear frame relaxation frequency (Sahay, 2008), given by $\omega_\beta = \mu_{fr}/\eta$, which is typically above Biot's peak relaxation frequency. Expressions for the complex fast and slow S-wave velocities derived from the extended theory are given by Sahay (2008) as

$$V_{\text{SI,SII}} = \left(\frac{T \pm \sqrt{T^2 - 4\Delta}}{2} \right)^{1/2} \tag{6.1.27}$$

$$T = \left(\frac{\mu_{\text{fr}}}{\rho} \right) \left\{ 1 + \frac{1}{1 + i(\omega_i/\omega)} d_{\text{f}} m_{\text{f}} - i \frac{\omega}{\omega_\beta} \left[\gamma + \frac{1}{1 + i(\omega_i/\omega)} d_{\text{f}} m_{\text{s}} \left(\frac{\phi}{m_{\text{f}}} - \gamma \right) \right] \right\} \tag{6.1.28}$$

$$\Delta = -i\omega \frac{1}{1 + i(\omega_i/\omega)} \left(\frac{\mu_{\text{fr}}}{\rho} \right) d_{\text{f}} v \tag{6.1.29}$$

$$v = \eta/\rho_{\text{fl}} = \text{kinematic shear viscosity of pore fluid} \tag{6.1.30}$$

$$\gamma = 1 - \mu_{\text{fr}}/\mu_0 \tag{6.1.31}$$

$\mu_{\text{fr}}, \mu_0 = \text{shear moduli of dry rock frame and solid mineral, respectively} \tag{6.1.32}$

$$m_{\text{s}} = (1 - \phi)\rho_0/\rho = \text{solid mass fraction} \tag{6.1.33}$$

$$m_{\text{f}} = \phi\rho_{\text{fl}}/\rho = \text{fluid mass fraction} \tag{6.1.34}$$

$$d_{\text{f}} = \frac{1}{\alpha - m_{\text{f}}} \tag{6.1.35}$$

$$\omega_j = d_{\text{f}}\omega_c = d_{\text{f}}2\pi f_c = \frac{1}{\alpha - m_{\text{f}}} \frac{\phi\eta}{\rho_{\text{fl}}\kappa} \tag{6.1.36}$$

The frequency ω_i, interpreted as the Biot critical frequency scaled by tortuosity, is the peak frequency associated with Biot relaxation. Asymptotic approximations for $\omega \ll \omega_i$, and $\omega_i \ll \omega < \omega_\beta$ are derived by Sahay (2008):

$$V_{\text{SI}} \approx \begin{cases} \sqrt{\frac{\mu_{\text{fr}}}{\rho}} \left(1 - \frac{i}{2} \frac{\omega}{\omega_i} d_{\text{f}} m_{\text{f}} \right) & \omega \ll \omega_i \\ \sqrt{\frac{\mu_{\text{fr}}}{\rho(1 - m_{\text{f}}/\alpha)}} \left\{ 1 - \frac{i}{2} \left[\frac{\omega_i}{\omega} \frac{m_{\text{f}}}{\alpha} + \frac{\omega}{\omega_\beta} \left(\gamma(1 - \alpha^{-1}) + \frac{m_{\text{s}}\phi}{m_{\text{f}}\alpha} \right) \right] \right\} & \omega_i \ll \omega < \omega_\beta \end{cases} \tag{6.1.37}$$

$$V_{\text{SII}} \approx \begin{cases} \omega\sqrt{\frac{v}{\omega_i}} \left(\frac{\omega}{\omega_i} \frac{\alpha}{2} - i \right) & \omega \ll \omega_i \\ \omega^{1/2}\sqrt{\frac{v}{2\alpha}}(1 - i) & \omega_i \ll \omega < \omega_\beta \end{cases} \tag{6.1.38}$$

The slow S-wave mode in the regime above the Biot relaxation frequency is a diffusive viscous wave with a phase velocity that has a square-root frequency dependence, and a diffusion constant given by the kinematic shear viscosity scaled by the tortuosity. Below the Biot relaxation frequency the viscous drag of the fluid on the solid frame dominates over inertial effects. The slow S-wave is highly attenuating with a linear and a quadratic frequency dependence of its attenuation and phase velocity, respectively (Sahay, 2008).

Uses

Biot's theory can be used for the following purposes:
• estimating saturated-rock velocities from dry-rock velocities;
• estimating frequency dependence of velocities; and
• estimating reservoir compaction caused by pumping using the quasi-static limit of Biot's poroelasticity theory.

Assumptions and Limitations

The use of Biot's equations presented in this section requires the following considerations:
• The rock is isotropic.
• All minerals making up the rock have the same bulk and shear moduli.
• The fluid-bearing rock is completely saturated.
• The pore fluid is Newtonian.
• The wavelength, even in the high-frequency limit, is much larger than the grain or pore scale.

Caution:

For most crustal rocks the amount of squirt dispersion (which is not included in Biot's formulation) is comparable to or greater than Biot's dispersion, and thus using Biot's theory alone will lead to poor predictions of high-frequency saturated velocities. Exceptions include very-high-permeability materials such as ocean sediments and glass beads, materials at very high effective pressure, or near open boundaries, such as at a borehole or at the surfaces of a laboratory sample. The recommended procedure is to use the Mavko–Jizba squirt theory (Section 6.11) first to estimate the high-frequency wet-frame moduli and then to substitute them into Biot's equations

Extensions

Biot's theory has been extended to anisotropic media (Biot, 1962).

6.2 Geertsma–Smit Approximations of Biot's Relations

Synopsis

Biot's theoretical formulas predict the frequency-dependent velocities of saturated rocks in terms of the dry-rock properties (see also Biot's relations, Section 6.1).

Low and middle-frequency approximations (Geertsma and Smit, 1961) of his relations may be expressed as

$$V_P^2 = \frac{V_{P\infty}^4 + V_{P0}^4 (f_c/f)^2}{V_{P\infty}^2 + V_{P0}^2 (f_c/f)^2} \qquad (6.2.1)$$

where V_P is the frequency-dependent P-wave velocity of saturated rock, V_{P0} is the Biot–Gassmann low-frequency limiting P-wave velocity, $V_{P\infty}$ is the Biot high-frequency limiting P-wave velocity, f is the frequency, and f_c is Biot's reference frequency, which determines the low-frequency range, $f \ll f_c$, and the high-frequency range, $f \gg f_c$, given by

$$f_c = \frac{\phi \eta}{2\pi \rho_{fl} \kappa} \qquad (6.2.2)$$

where ϕ is porosity, ρ_{fl} is fluid density, η is the viscosity of the pore fluid, and κ is the absolute permeability of the rock.

Uses

The Geertsma–Smit approximations can be used for the following:
• estimating saturated-rock velocities from dry-rock velocities; and
• estimating the frequency dependence of velocities.

Assumptions and Limitations

The use of the Geertsma–Smit approximations presented in this section requires the following considerations:
• mathematical approximations are valid at moderate-to-low seismic frequencies, so that $f < f_c$. This generally means moderate-to-low permeabilities, but it is in this range of permeabilities that squirt dispersion may dominate the Biot effect;
• the rock is isotropic;
• all minerals making up the rock have the same bulk and shear moduli; and
• fluid-bearing rock is completely saturated.

Caution:

For most crustal rocks the amount of squirt dispersion (not included in Biot's theory) is comparable to or greater than Biot's dispersion, and thus using Biot's theory alone will lead to poor predictions of high-frequency saturated velocities. Exceptions include very high-permeability materials such as ocean sediments and glass beads, or materials at very high effective pressure. The recommended procedure is to use the Mavko–Jizba squirt theory (Section 6.11), first to estimate the high-frequency wet-frame moduli and then to substitute them into the Biot or Geertsma–Smit equations.

6.3   Gassmann's Relations: Isotropic Form

Synopsis

One of the most important problems in the rock physics analysis of logs, cores, and seismic data is using seismic velocities in rocks saturated with one fluid to predict those of rocks saturated with a second fluid, or equivalently, predicting saturated-rock velocities from dry-rock velocities, and vice versa. This is the seismic *fluid substitution problem*.

Generally, when a rock is loaded under an increment of compression, such as from a passing seismic wave, an increment of pore-pressure change is induced, which resists the compression and therefore stiffens the rock. The low-frequency Gassmann–Biot (Gassmann, 1951; Biot, 1956) theory predicts the resulting increase in effective bulk modulus, K_{sat}, of the saturated rock using the following equation:

$$\frac{K_{sat}}{K_0 - K_{sat}} = \frac{K_{dry}}{K_0 - K_{dry}} + \frac{K_{fl}}{\phi(K_0 - K_{fl})}, \quad \mu_{sat} = \mu_{dry} \tag{6.3.1}$$

where K_{dry} is the effective bulk modulus of dry rock, K_{sat} is the effective bulk modulus of the rock saturated with pore fluid, K_0 is the bulk modulus of mineral material making up rock, K_{fl} is the effective bulk modulus of pore fluid, ϕ is the porosity, μ_{dry} is the effective shear modulus of dry rock, and μ_{sat} is the effective shear modulus of rock with pore fluid.

Gassmann's equations assume a homogeneous mineral modulus and statistical isotropy of the pore space but is free of assumptions about the pore geometry, other than the pore space is well-connected. Most importantly, it is valid only at sufficiently **low frequencies** such that the induced pore pressures are equilibrated throughout the pore space (i.e., there is sufficient time for the pore fluid to flow and eliminate wave-induced pore-pressure gradients). This limitation to low frequencies explains why Gassmann's relation works best for very low-frequency *in-situ* seismic data (<100 Hz) and may perform less well as frequencies increase toward sonic logging ($\approx 10^4$ Hz) and laboratory ultrasonic measurements ($\approx 10^6$ Hz). Gassmann's equations only describe situations where the pore filling is a fluid with zero rigidity. The analogous problem of substituting solids into the pore space is discussed in Section 6.8.

Caution:

"Dry rock" is not the same as gas-saturated rock. The "dry-rock" or "dry-frame" modulus refers to the incremental bulk deformation resulting from an increment of applied confining pressure with the pore pressure held constant. This corresponds to a "drained" experiment in which pore fluids can flow freely in or out of the sample to ensure constant pore pressure. Alternatively, the "dry-frame" modulus can correspond to an undrained experiment in which the pore fluid has zero bulk modulus, and in which pore compression therefore does not induce changes in pore pressure. This is approximately the case for an air-filled sample at standard temperature and pressure. However, at reservoir conditions (high pore pressure), gas takes on a non-negligible bulk modulus and should be treated as a saturating fluid.

Caution:

Laboratory measurements on very dry rocks, such as those prepared in a vacuum oven, are sometimes *too dry*. Several investigators have found that the first few percent of fluid saturation added to an extremely dry rock will lower the frame moduli, probably as a result of disrupting surface forces acting on the pore surfaces. It is this slightly wet or moist rock modulus that should be used as the "dry-rock" modulus in Gassmann's relations. A more thorough discussion is given in Section 6.18.

Although we often describe Gassmann's relations as allowing us to predict saturated-rock moduli from dry-rock moduli, and vice versa, the most common *in-situ* problem is to predict the changes that result when one fluid is replaced with another. One procedure is simply to apply the equation twice: transform the moduli from the initial fluid saturation to the dry state, and then immediately transform from the dry moduli to the new fluid-saturated state. Equivalently, we can algebraically eliminate the dry-rock moduli from the equation and relate the saturated-rock moduli $K_{\text{sat 1}}$ and $K_{\text{sat 2}}$ in terms of the two fluid bulk moduli $K_{\text{fl 1}}$ and $K_{\text{fl 2}}$ as follows:

$$\frac{K_{\text{sat 1}}}{K_0 - K_{\text{sat 1}}} - \frac{K_{\text{fl 1}}}{\phi(K_0 - K_{\text{fl 1}})} = \frac{K_{\text{sat 2}}}{K_0 - K_{\text{sat 2}}} - \frac{K_{\text{fl 2}}}{\phi(K_0 - K_{\text{fl 2}})} \tag{6.3.2}$$

Caution:

It is *not correct* simply to replace K_{dry} in Gassmann's equation by $K_{\text{sat 2}}$.

A few more explicit, but entirely equivalent, forms of equations 6.3.1 are

$$K_{\text{sat}} = K_{\text{dry}} + \frac{(1 - K_{\text{dry}}/K_0)^2}{\phi/K_{\text{fl}} + (1 - \phi)/K_0 - K_{\text{dry}}/K_0^2} \tag{6.3.3}$$

$$K_{\text{sat}} = \frac{\phi(1/K_0 - 1/K_{\text{fl}}) + 1/K_0 - 1/K_{\text{dry}}}{(\phi/K_{\text{dry}})(1/K_0 - 1/K_{\text{fl}}) + (1/K_0)(1/K_0 - 1/K_{\text{dry}})} \tag{6.3.4}$$

$$\frac{1}{K_{\text{sat}}} = \frac{1}{K_0} + \frac{\phi}{K_\phi + K_0 K_{\text{fl}}/(K_0 - K_{\text{fl}})} \tag{6.3.5}$$

$$K_{\text{dry}} = \frac{K_{\text{sat}}(\phi K_0/K_{\text{fl}} + 1 - \phi) - K_0}{\phi K_0/K_{\text{fl}} + K_{\text{sat}}/K_0 - 1 - \phi} \tag{6.3.6}$$

Note that the dry pore-space compressibility is

$$\frac{1}{K_\phi} = \frac{1}{v_p}\frac{\partial v_p}{\partial \sigma}\bigg|_P \qquad (6.3.7)$$

where v_p is the pore volume, σ is the confining pressure, and P is the pore pressure.

Example: Use Gassmann's relation to compare the bulk modulus of a dry quartz sandstone having the properties $\phi = 0.20$, $K_{dry} = 12$ GPa, and, $K_0 = 36$ GPa with the bulk moduli when the rock is saturated with gas and with water at $T = 80°C$ and pore pressure $P_p = 300$ bar.

Calculate the bulk moduli and density for gas and for water using the Batzle–Wang formulas discussed in Section 6.22. A gas with gravity=1 will have the properties $K_{gas} = 0.133$ GPa and $\rho_{gas} = 0.336$ g/cm³, and water with salinity 50000 ppm will have properties $K_{water} = 3.013$ GPa and $\rho_{water} = 1.055$ g/cm³.

Next, substitute these into Gassmann's relations

$$\frac{K_{sat-gas}}{36 - K_{sat-gas}} = \frac{12}{36 - 12} + \frac{0.133}{0.20(36 - 0.133)}$$

to yield a value of $K_{sat-gas} = 12.29$ GPa for the gas-saturated rock. Similarly,

$$\frac{K_{sat-water}}{36 - K_{sat-water}} = \frac{12}{36 - 12} + \frac{3.013}{0.20(36 - 3.013)}$$

which yields a value of $K_{sat-water} = 17.6$ GPa for the water-saturated rock.

Compressibility Form

An equivalent form of Gassmann's isotropic relation can be written compactly in terms of compressibilities:

$$(C_{sat} - C_0)^{-1} = (C_{dry} - C_0)^{-1} + [\phi(C_{fl} - C_0)]^{-1} \qquad (6.3.8)$$

where

$$C_{sat} = \frac{1}{K_{sat}}; \quad C_{dry} = \frac{1}{K_{dry}}; \quad C_{fl} = \frac{1}{K_{fl}}; \quad C_0 = \frac{1}{K_0} \qquad (6.3.9)$$

Reuss Average Form

An equivalent form of Gassmann's equation can be written as

$$\frac{K_{sat}}{K_0 - K_{sat}} = \frac{K_{dry}}{K_0 - K_{dry}} + \frac{K_R}{K_0 - K_R} \qquad (6.3.10)$$

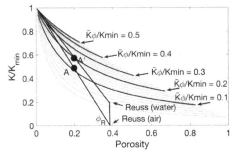

Figure 6.3.1 ϕ_R is the porosity where a straight line drawn from the mineral modulus on the left-hand axis, through the data point (A or A'), intersects the Reuss average for that corresponding pore fluid. The change in rock bulk modulus between any two pore fluids (A–A') is proportional to the change in Reuss average for the same two fluids evaluated at the intercept porosity ϕ_R.

where

$$K_R = \left(\frac{\phi}{K_{fl}} + \frac{1-\phi}{K_0} \right)^{-1} \tag{6.3.11}$$

is the Reuss average modulus of the fluid and mineral at porosity ϕ. This is consistent with the obvious result that when the dry-frame modulus of the rock goes to zero, the fluid-saturated sample will behave as a suspension and lie on the Reuss bound.

Linear Form

A particularly useful exact *linear* form of Gassmann's relation follows from the simple graphical construction shown in Figure 6.3.1 (Mavko and Mukerji, 1995). Draw a straight line from the mineral modulus on the left-hand axis (at $\phi = 0$) through the data point (A) corresponding to the rock modulus with the initial pore fluid (in this case, air). The line intersects the Reuss average for that pore fluid at some porosity, ϕ_R, which is a measure of the pore-space stiffness. Then the rock modulus (point A') for a new pore fluid (in this case, water) falls along a second straight line from the mineral modulus, intersecting the Reuss average for the new pore fluid at ϕ_R. Then we can write, *exactly,*

$$\Delta K_{Gass}(\phi) = \frac{\phi}{\phi_R} \Delta K_R(\phi_R) \tag{6.3.12}$$

where $\Delta K_{Gass} = K_{sat\,2} - K_{sat\,1}$ is the Gassmann-predicted change of saturated-rock bulk modulus between any two pore fluids (including gas), and $\Delta K_R(\phi_R)$ is the difference in the Reuss average for the two fluids *evaluated at the intercept porosity* ϕ_R.

Because pore-fluid moduli are usually much smaller than mineral moduli, we can approximate the Reuss average as

$$K_R(\phi_R) = \frac{K_{fl}K_0}{\phi_R K_0 + (1 - \phi_R)K_{fl}} \approx \frac{K_{fl}}{\phi_R} \tag{6.3.13}$$

Then, the linear form of Gassmann's relations can be approximated as

$$\Delta K_{Gass}(\phi) \approx \frac{\phi}{\phi_R^2} \Delta K_{fl} \tag{6.3.14}$$

P-Wave Modulus Form

Because Gassmann's relations predict no change in the shear modulus, we can also write the linear form of Gassmann's relation as

$$\Delta M_{Gass}(\phi) = \Delta K_{Gass}(\phi) \approx \frac{\phi}{\phi_R^2} \Delta K_{fl} \tag{6.3.15}$$

where $\Delta M_{Gass}(\phi)$ is the predicted change in the P-wave modulus $M = K + \frac{4}{3}\mu$. Similarly, we can write

$$\Delta K_{Gass} = \Delta M_{Gass} = \Delta \lambda_{Gass} \tag{6.3.16}$$

where λ is Lamé's parameter.

Velocity Form

Murphy *et al.* (1991) suggested a velocity form of Gassmann's relation

$$\rho_{sat} V_{Psat}^2 = K_p + K_{dry} + \frac{4}{3}\mu \tag{6.3.17}$$

$$\rho_{sat} V_{Ssat}^2 = \mu \tag{6.3.18}$$

which is easily written as

$$\frac{V_{Psat}^2}{V_{Ssat}^2} = \frac{K_p}{\mu} + \frac{K_{dry}}{\mu} + \frac{4}{3} \tag{6.3.19}$$

where ρ_{sat} is the density of the saturated rock, V_{Psat} is the P-wave saturated rock velocity, V_{Ssat} is the S-wave saturated rock velocity, K_{dry} is the dry-rock bulk modulus, $\mu = \mu_{dry} = \mu_{sat}$ at is the rock shear modulus, and

$$K_p = \frac{(1 - K_{dry}/K_0)^2}{\phi/K_{fl} + (1 - \phi)/K_0 - K_{dry}/K_0^2} \tag{6.3.20}$$

Pore Stiffness Interpretation

One can write the dry-rock compressibility at constant pore pressure, K_{dry}^{-1}, as (Walsh, 1965; Zimmerman, 1991a)

$$\frac{1}{K_{dry}} = \frac{1}{K_0} + \frac{\phi}{K_\phi} \tag{6.3.21}$$

where

$$\frac{1}{K_\phi} = \frac{1}{v_p}\frac{\partial v_p}{\partial \sigma}\bigg|_P \tag{6.3.22}$$

is the effective dry-rock pore-space compressibility, defined as the ratio of the fractional change in pore volume, v_p, to an increment of applied external hydrostatic stress, σ, at *constant pore pressure* (see Section 2.10). This is related to another pore compressibility, $K_{\phi P}^{-1}$, which is more familiar to reservoir engineers and hydrogeologists:

$$\frac{1}{K_{\phi P}} = -\frac{1}{v_p}\frac{\partial v_p}{\partial P}\bigg|_\sigma \tag{6.3.23}$$

which is the ratio of the fractional change in pore volume to an increment of applied pore pressure, at *constant confining pressure,* also expressed by (Zimmerman, 1991a)

$$\frac{1}{K_{\phi P}} = \frac{1}{K_\phi} - \frac{1}{K_0} \tag{6.3.24}$$

Figure 6.3.2 shows a plot of normalized dry bulk modulus, K_{dry}/K_0, versus porosity, computed for various values of normalized pore-space stiffness, K_ϕ/K_0.

The data points plotted in Figure 6.3.2 are dynamic bulk moduli (calculated from ultrasonic velocities) for: (1) ten clean sandstones, all at a pressure of 40MPa (Han, 1986); (2) a single clean sandstone at pressures ranging from 5 to 40MPa (Han, 1986); and (3) porous glass (Walsh *et al.,* 1965). Effective pore-space compressibilities for each point can be read directly from the contours.

Similarly, the low-frequency *saturated* rock compressibility, K_{sat}^{-1}, can be written as

$$\frac{1}{K_{sat}} = \frac{1}{K_0} + \frac{\phi}{\widetilde{K}_\phi} \tag{6.3.25}$$

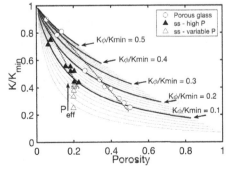

Figure 6.3.2 Normalized dry bulk modulus, K_{dry}/K_0, versus porosity, computed for various values of normalized pore-space stiffness, K_ϕ/K_0.

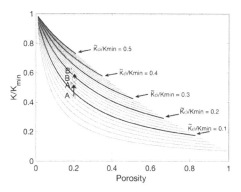

Figure 6.3.3 Normalized low-frequency saturated bulk modulus, K/K_0, versus porosity, computed for various values of the normalized modified pore-space stiffness, $\widetilde{K}_\phi/K_0$.

where

$$\widetilde{K}_\phi = K_\phi + \frac{K_0 K_{\text{fl}}}{K_0 - K_{\text{fl}}} \approx K_\phi + K_{\text{fl}} \qquad (6.3.26)$$

Here K_{fl} is the pore-fluid bulk modulus and K_ϕ is the same *dry* pore-space stiffness previously defined. The functional form for K_{sat} (equation 6.3.25) is exactly the same as for K_{dry} (equation 6.3.22). The difference is only in the term $\widetilde{K}_\phi$, which is equal to the dry-pore stiffness, K_ϕ, incremented by a fluid term $F = (K_0 K_{\text{fl}})/(K_0 - K_{\text{fl}}) \approx K_{\text{fl}}$. Hence, the only effect of a change in fluid is a change in the modified pore stiffness $\widetilde{K}_\phi$.

Therefore, fluid substitution, as predicted by Gassmann's equation, can be applied by computing this change, $\Delta \widetilde{K}_\phi$, and then jumping the appropriate number of contours, as illustrated in Figure 6.3.3.

For the example shown, the starting point A was one of Han's (1986) data points for an effective dry-rock bulk modulus, $K_{\text{dry}}/K_0 = 0.44$, and porosity, $\phi = 0.20$. Because the rock is dry, $F^{(1)}/K_0 \approx 0$. To saturate with water ($K_{\text{f}}/K_0 \approx 0.056$), we move up the amount $\left(F^{(2)} - F^{(1)} \right)/K_0 = 0.06$, or three contours. The water-saturated modulus can be read off directly as $K_{\text{sat}}/K_0 = 0.52$, point A'. Obviously, the technique works equally well for going from dry to saturated, saturated to dry, or from one fluid to another. All we do is compute $\Delta F/K_0$ and count the contours.

The second example shown (points $B - B'$) is for the same two pore fluids and the same porosity. However, the change in rock stiffness during fluid substitution is much larger.

This illustrates the important point that a softer rock will have a larger sensitivity to fluid substitution than a stiffer rock at the same porosity. In fact, anything that causes the velocity input to Gassmann's equation to be higher (such as a stiffer rock, a measurement error, or velocities contaminated by velocity dispersion) will result in a smaller predicted sensitivity to changes in pore fluids. Similarly, lower input velocities will lead to a larger predicted sensitivity.

The contours in Figure 6.3.3 have an additional interpretation. For a fixed pore fluid, jumping from one contour to another indicates a rock with a stiffer or softer dry rock bulk modulus.

Biot Coefficient

The dry-rock modulus can be written as

$$K_{dry} = K_0(1 - \beta) \tag{6.3.27}$$

where K_{dry} and K_0 are the bulk moduli of the dry rock and the mineral, and β is sometimes called the **Biot coefficient**, defined as the ratio of pore-volume change Δv_p to bulk-volume change, ΔV, under applied hydrostatic confining stress at constant pore pressure:

$$\beta = \frac{\Delta v_p}{\Delta V}\Big|_{dry} = \frac{\phi K_{dry}}{K_\phi} = 1 - \frac{K_{dry}}{K_0} \tag{6.3.28}$$

Then, Gassmann's equation can be expressed as

$$K_{sat} = K_{dry} + \beta^2 M \tag{6.3.29}$$

where

$$\frac{1}{M} = \frac{\beta - \phi}{K_0} + \frac{\phi}{K_{fl}} \tag{6.3.30}$$

V_P but no V_S

In practice, fluid substitution is performed by starting with bulk density and compressional- and shear-wave velocities measured on rocks saturated with the *initial* pore fluid (or gas) and then extracting the bulk and shear moduli. Then the bulk modulus of the rock saturated with the *new* pore fluid is calculated by using Gassmann's relations, and the velocities are reconstructed.

A practical problem arises when we wish to estimate the change of V_P during fluid substitution, but the shear velocity is unknown, which is often the case *in situ*. Then, strictly speaking, K cannot be extracted from V_P, and Gassmann's relations cannot be applied. To get around this problem, a common approach is to estimate V_S from an empirical V_S–V_P relation or to assume a dry-rock Poisson's ratio (Castagna *et al.*, 1985; Greenberg and Castagna, 1992; see Section 7.9).

Mavko *et al.* (1995) have suggested an approximate method that operates directly on the P-wave modulus, $M = \rho V_P^2$. The method is equivalent to replacing the bulk moduli of the rock and mineral in Gassmann's relation with the corresponding P-wave moduli. For example,

$$\frac{M_{\text{sat}}}{M_0 - M_{\text{sat}}} \approx \frac{M_{\text{dry}}}{M_0 - M_{\text{dry}}} + \frac{M_{\text{fl}}}{\phi(M_0 - M_{\text{fl}})} \tag{6.3.31}$$

where M_{sat}, M_{dry}, M_0, and M_{fl} are the P-wave moduli of the saturated rock, the dry rock, the mineral, and the pore fluid, respectively. The approximate method performs the same operation with the P-wave modulus, M, as is performed with the bulk modulus, K, in any of the various exact forms of Gassmann's relations previously listed.

Saxena and Mavko (2014a) derive a more exact form that depends somewhat on pore shape and Poisson's ratio. Details can be found in Section 6.8.

By analogy with equation 6.3.31, one can also write

$$\frac{\mu_{sat}}{\mu_0 - \mu_{sat}} = \frac{\mu_{dry}}{\mu_0 - \mu_{dry}} + \frac{\mu_{fl}}{\phi(\mu_0 - \mu_{fl})} \tag{6.3.32}$$

which is nothing more than Gassmann's result $\mu_{sat} = \mu_{dry}$ if $\mu_{fluid} = 0$.

Dvorkin's Fluid Substitution in Sandstone with Dispersed Clay

Although Gassmann's equations are intended for monomineralic rocks with a single, well-defined connected pore space, Dvorkin et al. (2007) proposed a strategy to extend Gassmann to sandstones containing dispersed clay. The method treats the water-saturated microporous clay as part of the "modified" mineral frame. Accordingly the mineral phase bulk modulus in Gassmann's equation is taken as the Voigt–Reuss–Hill average of the moduli of the non-clay minerals plus the modulus of the porous clay. The effective porosity – the void space other than microporosity in the clay – is substituted into Gassmann's equation in place of the total porosity.

If the volume fraction of microporous clay in the rock is C, and the intrinsic microporosity of the clay is ϕ_{clay}, the total volume occupied by the clay microporosity is $C\phi_{clay}$. The effective porosity is $\phi_e = \phi_t - C\phi_{clay}$, where ϕ_t is the total porosity. The fraction of the rock treated as the modified mineral frame is then $f_s = 1 - \phi_e = 1 - \phi_t + C\phi_{clay}$.

The fraction of modified mineral consisting of microporous clay is $f_{clay} = C/f_s$ and the fraction of remaining non-clay minerals is $1 - f_{clay} = 1 - C/f_s$. The Voigt–Reuss–Hill effective mineral bulk modulus is estimated as

$$K_s = \frac{1}{2}\left[\left(f_{clay}K_{clay} + (1 - f_{clay})\sum x_i K_i\right) + \left(f_{clay}/K_{clay} + (1 - f_{clay})\sum x_i/K_i\right)^{-1}\right] \tag{6.3.33}$$

where x_i is the volume fraction of the non-clay mineral with bulk modulus K_i. For the case of quartz sandstone with dispersed clay, the mineral modulus is

$$K_s = \frac{1}{2}\left[\left(f_{clay}K_{clay} + (1 - f_{clay})K_{qtz}\right) + \left(f_{clay}/K_{clay} + (1 - f_{clay})/K_{qtz}\right)^{-1}\right] \quad (6.3.34)$$

It is important to remember that equation 6.3.33 is just one of several mixing laws that can be used to estimate the effective mineral modulus. In specific situations, such as where the softer component of the solid frame – usually clay – is load bearing, the use of the lower Hashin–Shtrikman bound may be more appropriate.

Uses

Gassmann's relations are used to estimate the change of low-frequency elastic moduli of porous media caused by a change of pore fluids.

Assumptions and Limitations

The following considerations apply to the use of Gassmann's relations:
- Low seismic frequencies are assumed, so that pore pressures are equilibrated throughout the pore space. *In-situ* seismic conditions should generally be acceptable. Ultrasonic laboratory conditions will generally **not** be described well with Gassmann's equation. Sonic-logging frequencies may or may not be within the range of validity, depending on the rock type and the fluid viscosity.
- The rock is isotropic.
- All minerals making up the rock have the same bulk and shear moduli.
- Fluid-bearing rock is completely saturated.

Extensions

Gassmann's relations can be extended in the following ways:
- For mixed mineralogy, it is common to still use Gassmann's equation, but with the mineral bulk modulus K_0, taken as an average of the mineral moduli, most typically the Voigt–Reuss–Hill average (Section 4.5).
- For clay-rich rocks in which the clay is partially load-bearing, Dvorkin *et al.* (2007) suggest that Gassmann equation be used by taking the wet microporous clay to be part of the effective mineral mix, and the total porosity replaced by effective porosity.
- For clay-rich rocks, it sometimes works best to consider the "soft" clay to be part of the pore-filling phase rather than part of the mineral matrix. Then the pore fluid is "mud," and its modulus can be estimated with an isostress calculation, as in the next item. If this strategy is taken, the "porosity" used in Gassmann's equation should include the entire volume of the fluid-clay mixture.
- For partially saturated rocks at sufficiently low frequencies, one can usually use an effective modulus for the pore fluid that is an isostress average of the moduli of the liquid and gaseous phases (see Section 6.18):

$$\frac{1}{K_{fl}} = \frac{S}{K_L} + \frac{1 - S}{K_G}$$

where K_L is the bulk modulus of the liquid phase, K_G is the bulk modulus of the gas phase, and S is the saturation of the liquid phase.

- For rocks with mixed mineralogy, use Brown and Korringa's extension of Gassmann's relations (see Section 6.5).
- For anisotropic rocks, use the anisotropic form of Gassmann's relations (see Section 6.6).
- Berryman and Milton (1991) have extended Gassmann's relation to include multiple porous constituents (see Section 6.7).

6.4   Bounds on Fluid Substitution

When two-phase materials do not fall on the Hashin–Shtrikman bounds, fluid substitution may not be Gassmann-consistent. In the geophysics literature, deviation from Gassmann's predictions has been extensively discussed in the context of load-induced unequilibrated pore pressure as occurs in disconnected pores or during unrelaxed "squirt flow" (Biot, 1962; Stoll and Bryan, 1970; Mavko and Nur, 1975; Budiansky and O'Connell, 1977; Mavko and Jizba, 1991; Chapman *et al.*, 2002; Gurevich *et al.*, 2009). Fluid substitution with unequilibrated pore fluids yields larger changes in bulk modulus than those predicted by Gassmann.

Gibiansky and Torquato (2000) discuss the non-uniqueness of fluid substitution and present rigorous bounds on the change in bulk modulus that can occur upon fluid substitution in an isotropic two-phase composite, with mineral bulk and shear moduli $K_{\min}$ and $\mu_{\min}$. If the initial effective bulk modulus is $K_{sat}^{(1)}$ when the rock is saturated with fluid of bulk modulus $K_f^{(1)}$, then upon substituting a new fluid with bulk modulus $K_f^{(2)}$, the new saturated effective bulk modulus $K_{sat}^{(2)}$ must lie in the interval

$$F_1 \leq K_{sat}^{(2)} \leq F_2 \quad \text{if } K_f^{(2)} > K_f^{(1)}, \tag{6.4.1}$$

$$F_1 \geq K_{sat}^{(2)} \geq F_2 \quad \text{if } K_f^{(2)} < K_f^{(1)}, \tag{6.4.2}$$

where

$$F_1 = \frac{\alpha_1 K_{1*}^{(2)} A + K_h^{(2)} B}{\alpha_1 A + B}; \quad F_2 = \frac{\alpha_2 K_{1*}^{(2)} A + K_h^{(2)} B}{\alpha_2 A + B}; \tag{6.4.3}$$

$$A = \left(K_h^{(1)} - K_{sat}^{(1)}\right)\left(K_{1*}^{(1)} - K_h^{(1)}\right); \quad B = \left(K_{1*}^{(1)} - K_{sat}^{(1)}\right)\left(K_h^{(2)} - K_{1*}^{(2)}\right); \tag{6.4.4}$$

$$\alpha_1 = \frac{\left(K_{\min} - K_f^{(2)}\right)^2 \left((1 - \phi)K_f^{(1)} - \phi K_{\min}\right)^2}{\left(K_{\min} - K_f^{(1)}\right)^2 \left((1 - \phi)K_f^{(2)} - \phi K_{\min}\right)^2}; \quad \alpha_2 = \alpha_1 \frac{K_f^{(2)}\left(3K_f^{(1)} + 4G_{\min}\right)}{K_f^{(1)}\left(3K_f^{(2)} + 4G_{\min}\right)};$$

$$\tag{6.4.5}$$

$$K_h^{(1)} = \left[\frac{1-\phi}{K_{min}} + \frac{\phi}{K_f^{(1)}}\right]^{-1} \; ; \quad K_h^{(2)} = \left[\frac{1-\phi}{K_{min}} + \frac{\phi}{K_f^{(2)}}\right]^{-1} \; ; \tag{6.4.6}$$

$$K_{1*}^{(1)} = (1-\phi)K_{min} + \phi K_f^{(1)} - \frac{\phi(1-\phi)\left(K_{min} - K_f^{(1)}\right)^2}{\phi K_{min} + (1-\phi)K_f^{(1)} + 4\mu_{min}/3} \; ; \tag{6.4.7}$$

$$K_{1*}^{(2)} = (1-\phi)K_{min} + \phi K_f^{(2)} - \frac{\phi(1-\phi)\left(K_{min} - K_f^{(2)}\right)^2}{\phi K_{min} + (1-\phi)K_f^{(2)} + 4\mu_{min}/3} \; . \tag{6.4.8}$$

Some properties of the Gibiansky and Torquato bounds are as follows.

1. The modulus F_1 is equivalent to Gassmann's prediction of saturated bulk modulus. F_1 corresponds to the *smallest* possible change in bulk modulus upon fluid substitution, which occurs when the rock pore space is well connected and the pore pressure can equilibrate under bulk compression.
2. The modulus F_2 corresponds to the *largest* possible change in bulk modulus upon fluid substitution. This occurs when the pore pressure is not uniform throughout the pore space under bulk compression – for example, when the rock pore space is disconnected or tight and has very heterogeneous pore stiffness (i.e., unrelaxed squirt flow).
3. If $K_f^{(1)} > 0$ and $K_{sat}^{(1)}$ is on either the *HS+* or *HS-* bound, then upon fluid substitution to $K_f^{(2)} > 0$, $K_{sat}^{(2)} = F_1 = F_2$, is on the same corresponding bound for the new fluid; hence, fluid substitution for an initially saturated rock whose bulk modulus falls on a bound is unique.
4. If the initial case is dry ($K_f^{(1)} = 0$), and if $K_{sat}^{(1)}$ is on *HS+* or *HS-* bound, then upon fluid substitution, F_1 is still Gassmann's prediction of the saturated rock bulk modulus and will lie on the same corresponding *HS+* or *HS-* bound computed with the new fluid. If the dry $K_{sat}^{(1)}$ is on the *HS-* bound, then F_2 will be on the *HS+* bound, and if the dry $K_{sat}^{(1)}$ is on the *HS+* bound, then F_2 will stay on the *HS+* bound.
5. If the initial case is dry ($K_f^{(1)} = 0$), and if $K_{sat}^{(1)}$ lies anywhere between (but not on) the *HS-* and *HS+* bounds, then F_2 is the *HS+* bound computed with the new fluid. If the initial rock is saturated, the prediction for dry modulus F_2 lies on *HS-*.

6.5 Brown and Korringa's Generalized Gassmann Equations for Mixed Mineralogy

Synopsis

Brown and Korringa (1975) generalized Gassman's (1951) isotropic fluid substitution equations to the case where the solid phase of the rock is heterogeneous (i.e., mixed mineralogy).

$$\frac{K_{\text{sat}}}{K_{\text{S}} - K_{\text{sat}}} = \frac{K_{\text{dry}}}{K_{\text{S}} - K_{\text{dry}}} + \frac{K_{\phi\text{S}}}{K_{\text{S}}} \frac{K_{\text{fl}}}{\phi(K_{\phi\text{S}} - K_{\text{fl}})}, \quad \mu_{\text{sat}} = \mu_{\text{dry}} \tag{6.5.1}$$

where K_{dry} is the dry-rock bulk modulus, K_{sat} is the bulk modulus of the rock saturated with a pore fluid, K_{fl} is the fluid bulk modulus, μ_{dry} is the dry-rock shear modulus, μ_{sat} is the saturated-rock shear modulus, and ϕ is porosity.

Two additional moduli K_{S} and $K_{\phi\text{S}}$ are defined as

$$\frac{1}{K_{\text{S}}} = -\frac{1}{V}\left(\frac{\partial V}{\partial P_{\text{c}}}\right)\Bigg|_{P_d = constant} \tag{6.5.2}$$

$$\frac{1}{K_{\phi\text{S}}} = -\frac{1}{v_\phi}\left(\frac{\partial v_\phi}{\partial P_{\text{c}}}\right)\Bigg|_{P_d = constant} \tag{6.5.3}$$

where V is the total volume of a sample of rock, v_ϕ is the volume of pore space within the sample, P_{c} is the confining pressure, $P_{\text{d}} = P_{\text{c}} - P_{\text{p}}$ is the differential pressure, and P_{p} is the pore pressure. These special moduli (see also Zimmerman, 1991a) describe the effect on sample volume and pore volume of equally incrementing confining pressure and pore pressure – sometimes called the "unjacketed" case.

Caution:

The notation here is different from that in Brown and Korringa (1975) and Zimmerman (1991a).

The Brown and Korringa fluid substitution equation can also be written in terms of compressibilities $(C_{\text{sat}} - C_{\text{S}})^{-1} = (C_{\text{dry}} - C_{\text{S}})^{-1} + [\phi(C_{\text{fl}} - C_{\phi\text{S}})]^{-1}$ where $C_{\text{sat}} = 1/K_{\text{sat}}$, $C_{\text{dry}} = 1/K_{\text{dry}}$, $C_{\text{S}} = 1/K_{\text{S}}$, and $C_{\phi\text{S}} = 1/K_{\phi\text{S}}$. When the mineral phase is homogeneous, then $K_{\text{S}} = K_{\phi\text{S}} = K_{\text{mineral}}$, and $C_{\text{S}} = C_{\phi\text{S}} = C_{\text{mineral}}$, where $C_{\text{mineral}} = 1/K_{\text{mineral}}$, and K_{mineral} is the mineral bulk modulus. Thus, if there is only a single mineral, both the modulus and compressibility forms of fluid substitution presented here reduce exactly to the isotropic Gassmann equations.

A practical problem with using the Brown and Korringa equations for fluid substitution is that we seldom have knowledge of C_{S} and $C_{\phi\text{S}}$ (or K_{S} and $K_{\phi\text{S}}$). One useful constraint comes from the relation

$$C_{\phi\text{S}} = \frac{1}{\phi}C_{\text{S}} - \frac{1 - \phi}{\phi}\overline{C}_{\text{mineral}}$$

where $\overline{C}_{\text{mineral}} = 1/\overline{K}_{\text{mineral}}$ is the true effective compressibility of the mineral mix. $\overline{K}_{\text{mineral}}$ must satisfy the Hashin–Shtrikman–Walpole bounds, $K_{\text{mineral}}^{\text{HSW}\pm}$, and might be estimated, for example, using a Voigt–Reuss–Hill average of the individual mineral bulk moduli. Insights into the value of K_{S} come from knowledge of the grain micro-structure. K_{S} might lie close to the upper bound of the mineral bulk modulus, if the softest phases, such as clay, are pore-filling. On the other hand, if the softest minerals are load-bearing, then K_S might be closer to the lower bound. Once an estimate of K_S is made, a consistent estimate of $K_{\phi S}$ can be found from the preceding relation.

Uses

The Gassmann and Brown–Korringa relations are used to estimate the change of low-frequency elastic moduli of porous media caused by a change of pore fluids.

Assumptions and Limitations

The following considerations apply to the use of Gassmann and Brown–Korringa relations.

- **Low seismic frequencies** are assumed, so that pore pressures are equilibrated throughout the pore space. *In-situ* seismic conditions should generally be acceptable. Ultrasonic laboratory conditions will generally **not** be described well with Gassmann or Brown–Korringa equations. Sonic-logging frequencies may or may not be within the range of validity, depending on the rock type and fluid viscosity.
- The rock is isotropic.
- Fluid-bearing rock is completely saturated.

6.6   Fluid Substitution in Anisotropic Rocks

Synopsis

In addition to his famous isotropic equations, Gassmann (1951) published fluid-substitution equations for *anisotropic* porous rocks. In terms of the anisotropic linear elastic stiffness components, c_{ijkl}, his result is written as

$$c_{ijkl}^{\text{sat}} = c_{ijkl}^{\text{dry}} + \frac{(K_0\delta_{ij} - c_{ijaa}^{\text{dry}}/3)(K_0\delta_{kl} - c_{bbkl}^{\text{dry}}/3)}{(K_0/K_{\text{fl}})\phi(K_0 - K_{\text{fl}}) + (K_0 - c_{ccdd}^{\text{dry}}/9)} \tag{6.6.1}$$

where c_{ijkl}^{dry} is the effective elastic stiffness element of dry rock, c_{ijkl}^{sat} is the effective elastic stiffness element of rock saturated with pore fluid, K_0 is the mineral bulk modulus, K_{fl} is the fluid bulk modulus, and ϕ is porosity:

$$\delta_{ij} = \begin{cases} 1 & \text{for } i = j \\ 0 & \text{for } i \neq j \end{cases} \tag{6.6.2}$$

The inverse of Gassmann's anisotropic equation (not given in Gassmann's paper) can be written as

$$c_{ijkl}^{\text{dry}} = c_{ijkl}^{\text{sat}} - \frac{(K_0\delta_{ij} - c_{ijaa}^{\text{sat}}/3)(K_0\delta_{kl} - c_{bbkl}^{\text{sat}}/3)}{(K_0/K_{\text{fl}})\phi(K_0 - K_{\text{fl}}) - (K_0 - c_{ccdd}^{\text{sat}}/9)} \tag{6.6.3}$$

Successive application of equation 6.6.3 followed by equation 6.6.1 allows us to apply fluid substitution from any initial fluid to any final fluid.

Similarly, Gassmann (1951) gave anisotropic fluid-substitution expressions in terms of the elastic compliances:

$$s_{ijkl}^{\text{dry}} = s_{ijkl}^{\text{sat}} + \frac{(s_{ijaa}^{\text{sat}} - \delta_{ij}/3K_0)(s_{bbkl}^{\text{sat}} - \delta_{kl}/3K_0)}{(s_{ccdd}^{\text{sat}} - 1/K_0) + \phi(1/K_{\text{fl}} - 1/K_0)} \tag{6.6.4}$$

where s_{ijkl}^{dry} is the effective elastic compliance element of dry rock, s_{ijkl}^{sat} is the effective elastic compliance element of rock saturated with pore fluid, K_0 is the bulk modulus of the solid mineral, K_{fl} is the bulk modulus of the fluid, and ϕ is porosity.

The inverse expression (not in Gassmann's original paper) is written as

$$s_{ijkl}^{\text{dry}} = s_{ijkl}^{\text{sat}} + \frac{(s_{ijaa}^{\text{sat}} - \delta_{ij}/3K_0)(s_{bbkl}^{\text{sat}} - \delta_{kl}/3K_0)}{(s_{ccdd}^{\text{sat}} - 1/K_0) - \phi(1/K_{\text{fl}} - 1/K_0)} \tag{6.6.5}$$

Gassmann's equations assume that the mineral is homogeneous and isotropic, although the dry and saturated rocks can have arbitrary anisotropy. In all of the equations of this section, a repeated index within a term implies a sum over 1–3 (e.g., $c_{ijaa} = c_{ij11} + c_{ij22} + c_{ij33}$).

Brown and Korringa (1975) also gave expressions for fluid substitution in anisotropic rocks, expressed in the compliance domain. Their result is very similar to Gassmann's, except that they allow for an **anisotropic mineral**. (Brown and Korringa also made a generalization to allow for mixed mineralogy. The latter generalization is shown in Section 6.5 for isotropic rocks.)

$$s_{ijkl}^{\text{sat}} = s_{ijkl}^{\text{dry}} + \frac{(s_{ijaa}^{\text{dry}} - s_{ijaa}^0)(s_{bbkl}^{\text{dry}} - s_{bbkl}^0)}{(s_{ccdd}^{\text{dry}} - s_{ccdd}^0) + \phi(\beta_{\text{fl}} - \beta_0)} \tag{6.6.6}$$

where s_{ijkl}^{dry} is the effective elastic compliance tensor element of dry rock, s_{ijkl}^{sat} is the effective elastic compliance element of rock saturated with pore fluid, s_{ijkl}^0 is the effective elastic compliance element of the solid mineral β_{fl} is the fluid compressibility, $\beta_0 = s_{aa\gamma\gamma}^0$ is the mineral compressibility, and ϕ is porosity.

The inverse is given by

$$s_{ijkl}^{\text{dry}} = s_{ijkl}^{\text{sat}} + \frac{(s_{ijaa}^{\text{sat}} - s_{ijaa}^0)(s_{bbkl}^{\text{sat}} - s_{bbkl}^0)}{(s_{ccdd}^{\text{sat}} - s_{ccdd}^0) - \phi(1/K_{\text{fl}} - 1/K_0)} \tag{6.6.7}$$

The practical differences between the Gassmann and Brown–Korringa equations presented here appear to be inconsequential. Even though mineral grains are usually highly anisotropic, Brown and Korringa's equations imply that all grains in the rock are crystallographically aligned, which will not be the case in real rocks. Furthermore, complete elastic tensors for rock-forming minerals might not be well known. In most cases, our only practical choice will be to assume an average isotropic mineral modulus.

One of the major disadvantages in the application of the anisotropic Gassmann equation is that in the field we seldom measure enough parameters to completely characterize the stiffness tensor of a rock. Mavko and Bandyopadhyay (2009) find an approximate form of fluid substitution, for vertically propagating seismic waves in a transversely isotropic medium with a vertical axis of symmetry (VTI). The anisotropic correction for the fluid effect depends only on Thomsen's parameter δ:

$$c_{3333}^{\text{sat}} \approx c_{3333}^{\text{dry}} + \left[\frac{(K_{\text{fl}}/K_0)(K_0 - K_{\text{iso}}^{\text{dry}})^2}{\phi(K_0 - K_{\text{fl}}) + (K_{\text{fl}}/K_0)(K_0 - K_{\text{iso}}^{\text{dry}})}\right]\left(1 - \delta\frac{\frac{4}{3}c_{3333}^{\text{dry}}}{K_0 - K_{\text{iso}}^{\text{dry}}}\right) \qquad (6.6.8)$$

$$c_{3333}^{\text{dry}} \approx c_{3333}^{\text{sat}} - \left[\frac{(K_{\text{fl}}/K_0)(K_0 - K_{\text{iso}}^{\text{sat}})^2}{\phi(K_0 - K_{\text{fl}}) + (K_{\text{fl}}/K_0)(K_0 - K_{\text{iso}}^{\text{sat}})}\right]\left(1 - \delta\frac{\frac{4}{3}c_{3333}^{\text{sat}}}{K_0 - K_{\text{iso}}^{\text{sat}}}\right) \qquad (6.6.9)$$

where $K_{\text{iso}}^{\text{dry}}$ and $K_{\text{iso}}^{\text{sat}}$ are the *apparent* isotropic bulk moduli of the dry and saturated rock, respectively, computed from the vertical P– and S-wave velocities: $K_{\text{iso}} = \rho\left(V_{\text{P}}^2 - \frac{4}{3}V_{\text{S}}^2\right)$. In equations 6.6.7 and 6.6.8, the term within the square brackets is just the difference between the dry and saturated apparent bulk moduli as predicted by the isotropic Gassmann equation. The approximation is good for porosities larger than the magnitude of the Thomsen parameters (ε, γ, δ).

Uses

The equations of Gassmann and Brown and Korringa are applicable to the fluid substitution problem in anisotropic rocks.

Assumptions and Limitations

The following considerations apply to the use of Brown and Korringa's relations:
- **Low seismic frequencies** (equilibrated pore pressures). *In-situ* seismic frequencies generally should be acceptable. Ultrasonic laboratory conditions will generally **not** be described well with Brown and Korringa's equations. Sonic-logging frequencies may or may not be within the range of validity, depending on the rock type and fluid viscosity.
- All minerals making up the rock have the same moduli.
- Fluid-bearing rock is completely saturated.

Extensions

The following extensions of the anisotropic fluid substitution relations can be made:
- for mixed mineralogy, one can usually use an effective average set of mineral compliances for S_{ijkl}^0;

- for clay-filled rocks, it often works best to consider the "soft" clay to be part of the pore-filling phase rather than part of the mineral matrix. Then, the pore fluid is "mud," and its modulus can be estimated using an isostress calculation, as in the next item; and
- for partially saturated rocks at sufficiently low frequencies, one can usually use an effective modulus for the pore fluid that is an isostress average of the moduli of the liquid and gaseous phases:

$$\beta_{fi} = S\beta_L + (1 - S)\beta_G$$

where β_L is the compressibility of the liquid phase, β_G is the compressibility of the gas phase, and S is the liquid saturation.

6.7   Generalized Gassmann's Equations for Composite Porous Media

Synopsis

The generalized Gassmann's equation (Berryman and Milton, 1991) describes the static or low-frequency effective bulk modulus of a fluid-filled porous medium when the porous medium is a composite of two porous phases, each of which could be separately described by the more conventional Gassmann relations (Figure 6.7.1). This is a generalization of the usual Gassmann equation, which assumes that the porous medium is composed of a single, statistically homogeneous porous constituent with a single pore-space stiffness and a single solid mineral. Like Gassmann's equation, the generalized Gassmann formulation is completely independent of the pore geometry; other than that, it is well-connected. The generalized formulation assumes that the two porous constituents are bonded at points of contact and fill all the space of the composite porous medium. Furthermore, like the Gassmann formulation, it is assumed that the frequency is low enough that viscous and inertial effects are negligible, and that any stress-induced increments of pore pressure are uniform within each constituent, although they could be different from one constituent to another (Berryman and Milton extend this to include dynamic poroelasticity, which is not presented here).

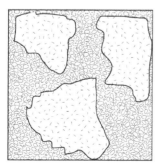

Figure 6.7.1 Composite porous medium with two porous phases

The pore microstructure within each phase is statistically homogeneous and much smaller than the size of the inclusions of each phase, which in turn are smaller than the size of the macroscopic sample. The inclusions of each porous phase are large enough to have effective dry-frame bulk moduli, $K_{\text{dry}}^{(1)}$ and $K_{\text{dry}}^{(2)}$; porosities, $\phi^{(1)}$ and $\phi^{(2)}$; and solid mineral moduli, $K_0^{(1)}$ and $K_0^{(2)}$, respectively. The volume fractions of the two porous phases are $f^{(1)}$ and $f^{(2)}$, where $f^{(1)} + f^{(2)} = 1$. The generalized Gassmann equation relates the effective saturated bulk modulus of the macroscopic sample, K_{sat}^*, to its dry-frame bulk modulus, K_{dry}^*, through two other elastic constants defined by

$$\frac{1}{K_s^*} = -\frac{1}{V}\left(\frac{\partial V}{\partial p_f}\right)_{p_d = \text{const}} \tag{6.7.1}$$

and

$$\frac{1}{K_\phi^*} = -\frac{1}{v_\phi}\left(\frac{\partial v_\phi}{\partial p_f}\right)_{p_d = \text{const}} \tag{6.7.2}$$

where V is the total sample volume, v_ϕ is the total pore volume, p_f is the pore pressure, and $p_d = p - p_f$ is the differential pressure, with p as the confining pressure. (Caution: Use here of the symbol K_ϕ is not the same as defined in Chapters 2 and 4). For a single-phase porous medium made up of a single solid-mineral constituent with modulus K_0, the two moduli are equal to the mineral modulus: $K_s = K_\phi = K_0$. The relation between K_{sat}^* and K_{dry}^* for a composite porous medium is

$$K_{\text{sat}}^* = K_{\text{dry}}^* + \alpha^* C \tag{6.7.3}$$

$$C = \frac{\alpha^*}{\alpha^*/K_s^* + \phi(1/K_f - 1/K_\phi^*)} \tag{6.7.4}$$

$$\alpha^* = 1 - \frac{K_{\text{dry}}^*}{K_s^*} \tag{6.7.5}$$

where K_f is the fluid bulk modulus. The constants K_s^* and K_ϕ^* can be expressed in terms of the moduli of the two porous constituents making up the composite medium. The key idea leading to the results is that whenever two scalar fields, such as p_d and p_f, can be varied independently in a linear composite with only two constituents, there exists a special value of the increment ratio $\delta p_d / \delta p$ that corresponds to an overall expansion or contraction of the medium without any relative shape change. This guarantees the existence of a set of consistency relations, allowing K_s^* and K_ϕ^* to be written in terms of the dry-frame modulus and the constituent moduli. By linearity, the coefficients for the special value of $\delta p_d / \delta p$ are also the coefficients for any other arbitrary ratio. The relation for K_s^* is

$$\frac{1/K_0^{(1)} - 1/K_0^{(2)}}{1/K_{\text{dry}}^{(2)} - 1/K_{\text{dry}}^{(1)}} = \frac{1/K_0^{(1)} - 1/K_s^*}{1/K_{\text{dry}}^* - 1/K_{\text{dry}}^{(1)}} = \frac{1/K_s^* - 1/K_0^{(2)}}{1/K_{\text{dry}}^{(2)} - 1/K_{\text{dry}}^*} \tag{6.7.6}$$

or, equivalently, in terms of α^*

$$\frac{\alpha^* - \alpha^{(1)}}{\alpha^{(2)} - \alpha^{(1)}} = \frac{K^*_{\text{dry}} - K^{(1)}_{\text{dry}}}{K^{(2)}_{\text{dry}} - K^{(1)}_{\text{dry}}} \tag{6.7.7}$$

where $\alpha^{(1)} = 1 - K^{(1)}_{\text{dry}}/K^{(1)}_0$ and $\alpha^{(2)} = 1 - K^{(2)}_{\text{dry}}/K^{(2)}_0$. Other equivalent expressions are

$$\frac{1}{K^*_s} = \frac{1}{K^{(1)}_0} - \frac{1/K^{(1)}_0 - 1/K^{(2)}_0}{1/K^{(2)}_{\text{dry}} - 1/K^{(1)}_{\text{dry}}}\left[1/K^*_{\text{dry}} - 1/K^{(1)}_{\text{dry}}\right] \tag{6.7.8}$$

and

$$\frac{1}{K^*_s} = \frac{1}{K^{(2)}_0} - \frac{1/K^{(1)}_0 - 1/K^{(2)}_0}{1/K^{(2)}_{\text{dry}} - 1/K^{(1)}_{\text{dry}}}\left[1/K^*_{\text{dry}} - 1/K^{(2)}_{\text{dry}}\right] \tag{6.7.9}$$

The relation for K^*_ϕ is given by

$$\frac{\langle\phi\rangle}{K^*_\phi} = \frac{\alpha^*}{K^*_s} - \left\langle\frac{\alpha(x) - \phi(x)}{K_0(x)}\right\rangle - \left(\langle\alpha(x)\rangle - \alpha^*\right)\left(\frac{\alpha^{(1)} - \alpha^{(2)}}{K^{(1)}_{\text{dry}} - K^{(2)}_{\text{dry}}}\right) \tag{6.7.10}$$

where $\langle q(x)\rangle = f^{(1)}q^{(1)} + f^{(2)}q^{(2)}$ denotes the volume average of any quantity q. Gassmann's equation for a single porous medium is recovered correctly from the generalized equations when $K^{(1)}_0 = K^{(2)}_0 = K_0 = K^*_s = K^*_\phi$ and $K^{(1)}_{\text{dry}} = K^{(2)}_{\text{dry}} = K^*_{\text{dry}}$.

Uses

The generalized Gassmann equations can be used to calculate low-frequency saturated velocities from dry velocities for composite porous media made of two porous constituents. Examples include shaly patches embedded within a sand, microporous grains within a rock with macroporosity, or large nonporous inclusions within an otherwise porous rock.

Assumptions and Limitations

The preceding equations imply the following assumptions:
- the rock is isotropic and made of up to two porous constituents;
- all minerals making up the rock are linearly elastic;
- fluid-bearing rock is completely saturated;
- the porosity in each phase is uniform, and the pore structure in each phase is smaller than the size of inclusions of each porous phase; and
- the size of the porous inclusions is big enough to have a well-defined dry-frame modulus but is much smaller than the wavelength and the macroscopic sample.

Extensions

For more than two porous constituents, the composite may be modeled by dividing it into separate regions, each containing only two phases. This approach is restrictive and not always possible.

6.8   Solid Substitution of Frame or Pore-Filling Phases

Synopsis

The solid substitution problem consists of predicting the change in effective elastic moduli of a composite when one or more of the solid constituent phases is changed, while keeping the microgeometry and volume fractions the same. Gassmann's equations for fluid substitution cannot be applied to substitution of solids in the pore space, except for a few very specific cases (see Section 6.3).

For two-phase composites whose effective bulk and/or shear moduli fall on the Hashin–Shtrikman (HS) bounds, computation of the change in effective modulus is straightforward and unique – we simply change the end-member properties and recompute the HS bounds for the new composition, while keeping the volume fractions fixed. (See Section 4.2 on Hashin–Shtrikman bounds for their range of applicability.)

For composites whose moduli fall between the bounds, neither fluid nor solid substitution is unique, unless information about the microgeometry is incorporated. Gibiansky and Torquato (1998) gave upper and lower bounds on *fluid* substitution (Section 6.4) and showed that Gassmann's (1951) equations predict the smallest possible change in effective modulus (i.e., the lower bound on the *change*). Changes in modulus larger than predicted by Gassmann can be caused by disconnected pores or unrelaxed pore pressure gradients that can be associated with large fluid viscosities and/or high measurement frequencies.

Vinogradov and Milton Bounds

Vinogradov and Milton (2005) derived bounds on *solid* substitution of the pore fill material. (They posed the problem as bounds on the total linear viscoelastic creep of a two-phase composite between the immediate and fully relaxed phase moduli of the material, both of which are elastic.) They considered a linear elastic two-phase composite with initial effective bulk modulus $K_{eff}^{(1)}$, individual phase bulk and shear moduli $(K_1^{(1)}, \mu_1^{(1)})$, $(K_2^{(1)}, \mu_2^{(1)})$, and volume fractions (f_1, f_2). In these expressions, superscripts (1) and (2) indicate the initial and final states, while subscripts 1 and 2 indicate the individual phases. Their results can be presented compactly in terms of the Y-transform, y_k, which is uniquely related to the bulk modulus K_{eff} by

$$y_k^{(i)} = -f_2 K_1^{(i)} - f_1 K_2^{(i)} - \frac{f_1 f_2 \left(K_1^{(i)} - K_2^{(i)} \right)^2}{K_{eff}^{(i)} - f_1 K_1^{(i)} - f_2 K_2^{(i)}} \tag{6.8.1}$$

and the inverse:

$$K_{eff}^{(i)} = \left[\frac{f_1}{K_1^{(i)} + y_k^{(i)}} + \frac{f_2}{K_2^{(i)} + y_k^{(i)}} \right]^{-1} - y_k^{(i)} \tag{6.8.2}$$

(More on the Y-transform can be found in Section 4.3.)

If $(\mu_2^{(1)} - \mu_1^{(1)})(\mu_2^{(2)} - \mu_1^{(2)}) \leq 0$ (Vinogradov and Milton call this the *badly ordered* case), then upon replacing phase moduli $(K_1^{(1)}, \mu_1^{(1)})$ and $(K_2^{(1)}, \mu_2^{(1)})$ with $(K_1^{(2)}, \mu_1^{(2)})$ and $(K_2^{(2)}, \mu_2^{(2)})$ one bound on the substituted effective bulk modulus $K_{eff}^{(2)}$ is defined by a straight line between points $P_1 = \frac{4}{3} \left(\mu_1^{(1)}, \mu_1^{(2)} \right)$ and $P_2 = \frac{4}{3} \left(\mu_2^{(1)}, \mu_2^{(2)} \right)$ in the Y-transform plane, $\left(y_k^{(1)}, y_k^{(2)} \right)$:

$$\frac{y_k^{(1)}}{c_1} + \frac{y_k^{(2)}}{c_2} = 1, \quad \frac{4}{3} \min(\mu_1^{(i)}, \mu_2^{(i)}) \leq y_k^{(i)} \leq \frac{4}{3} \max(\mu_1^{(i)}, \mu_2^{(i)}), \tag{6.8.3}$$

where

$$c_1 = \frac{4}{3} \frac{\mu_1^{(1)} \mu_2^{(2)} - \mu_2^{(1)} \mu_1^{(2)}}{\mu_2^{(2)} - \mu_1^{(2)}} \quad \text{and} \quad c_2 = -\frac{4}{3} \frac{\mu_1^{(1)} \mu_2^{(2)} - \mu_2^{(1)} \mu_1^{(2)}}{\mu_2^{(1)} - \mu_1^{(1)}}. \tag{6.8.4}$$

Another bound on $K_{eff}^{(2)}$ is defined by a hyperbola between points P_1 and P_2 in the Y-transform plane:

$$\left(\frac{y_k^{(1)}}{d_1} - 1 \right) \left(\frac{y_k^{(2)}}{d_2} - 1 \right) = 1, \quad \frac{4}{3} \min(\mu_1^{(i)}, \mu_2^{(i)}) \leq y_k^{(i)} \leq \frac{4}{3} \max(\mu_1^{(i)}, \mu_2^{(i)}), \tag{6.8.5}$$

where

$$d_1 = \frac{4}{3} \frac{\mu_1^{(1)} \mu_2^{(1)} \left(\mu_2^{(2)} - \mu_1^{(2)} \right)}{\mu_1^{(1)} \mu_2^{(2)} - \mu_2^{(1)} \mu_1^{(2)}} \quad \text{and} \quad d_2 = -\frac{4}{3} \frac{\mu_1^{(2)} \mu_2^{(2)} \left(\mu_2^{(1)} - \mu_1^{(1)} \right)}{\mu_1^{(1)} \mu_2^{(2)} - \mu_2^{(1)} \mu_1^{(2)}}. \tag{6.8.6}$$

If $\left(\mu_2^{(1)} - \mu_1^{(1)} \right) \left(\mu_2^{(2)} - \mu_1^{(2)} \right) \geq 0$ (Vinogradov and Milton call this the *well-ordered* case), then two additional curves must be considered (in addition to the two just given) as possible bounds. One curve is defined parametrically by

$$y_k^{(1)}(\xi_k) = \frac{4}{3} \frac{f_2 \mu_1^{(1)} \left(3K_1^{(1)} + 4\mu_2^{(1)} \right)(1 - \xi_k) + \left(3K_1^{(1)} + 4\mu_1^{(1)} \right) \mu_2^{(1)} \xi_k}{f_2 \left(3K_1^{(1)} + 4\mu_2^{(1)} \right)(1 - \xi_k) + \left(3K_1^{(1)} + 4\mu_1^{(1)} \right) \xi_k}, \tag{6.8.7}$$

$$y_k^{(2)}(\xi_k) = \frac{4f_2\mu_1^{(2)}\left(3K_1^{(2)} + 4\mu_2^{(2)}\right)(1-\xi_k) + \left(3K_1^{(2)} + 4\mu_1^{(2)}\right)\mu_2^{(2)}\xi_k}{3 \quad f_2\left(3K_1^{(2)} + 4\mu_2^{(2)}\right)(1-\xi_k) + \left(3K_1^{(2)} + 4\mu_1^{(2)}\right)\xi_k}, \tag{6.8.8}$$

$$\xi_k = \frac{\left(\frac{4}{3}\mu_1^{(1)} - y_k\right)f_2\left(k_1^{(1)} + \frac{4}{3}\mu_2^{(1)}\right)}{\left(\frac{4}{3}\mu_1^{(1)} - y_k\right)f_2\left(k_1^{(1)} + \frac{4}{3}\mu_2^{(1)}\right) - \left(\frac{4}{3}\mu_2^{(1)} - y_k\right)\left(k_1^{(1)} + \frac{4}{3}\mu_1^{(1)}\right)} \tag{6.8.9}$$

where parameter $0 \le \xi_k \le 1$ varies as $\frac{4}{3}\min(\mu_1^{(i)}, \mu_2^{(i)}) \le y_k^{(i)} \le \frac{4}{3}\max(\mu_1^{(i)}, \mu_2^{(i)})$. A fourth curve has the same form as equations 6.8.7–6.8.9 with subscripts 1 and 2 swapped. The practical recipe for using these last two curves would be to start with the initial effective modulus, $K_{eff}^{(1)}$; solve for the corresponding $y_k^{(1)}$ from equation 6.8.1 and ξ_k from equation 6.8.9; get $y_k^{(2)}$ from equations 6.8.3, 6.8.5, and 6.8.8; finally, get $K_{eff}^{(2)}$ from equation 6.8.2.

These four curves (equations 6.8.3–6.8.9) define envelopes in the plane $K_{eff}^{(1)}$ vs. $K_{eff}^{(2)}$, or equivalently, the plane $y_{eff}^{(1)}$ vs. $y_{eff}^{(2)}$. The bounds on the bulk modulus upon substitution are given by the outermost two of the four curves, which will vary with composition and microstructure. Points P_1 and P_2, through which all of the curves pass, correspond to the Hashin–Shtrikman bounds for the initial and final compositions, where substitution is unique. Three additional discrete points, P_3, P_4, and P_5, on the curve defined by equation 6.8.3, and another one, P_6, on the curved defined by equation 6.8.5, are known to be realizable by special microgeometries of polycrystalline assemblages; however, it is not known if any of the remainder of these curves is realizable. The curves defined by equations 6.8.7 and 6.8.8 are realizable by doubly coated spheres, where ξ_k is the proportion of phase 1 in the core of the doubly coated sphere (Figure 6.8.4a, d).

Example: Figure 6.8.1 shows an example of the four bounding curves for parameters $f_1 = f_2 = 0.5$, $K_1^{(1)} = K_1^{(2)} = 36$ GPa, $\mu_1^{(1)} = \mu_1^{(2)} = 45$ GPa, $K_2^{(1)} = 2$ GPa, $\mu_2^{(1)} = 1$ GPa, $K_2^{(2)} = 5$ GPa, and $\mu_2^{(2)} = 3$ GPa. The horizontal axis shows the range of possible effective bulk moduli for the initial composition, $K_{P1}^{(1)HS-} \le K^{(1)} \le K_{P2}^{(1)HS+}$, where K_{P1}^{HS-} and K_{P2}^{HS+} are the Hashin–Shtrikman bounds at points P_1 and P_2. The vertical axis shows the range of possible effective bulk moduli with the second composition, $K_{P1}^{(2)HS-} \le K^{(2)} \le K_{P2}^{(2)HS+}$. If, for example, the initial effective modulus is $K^{(1)} = 10$ GPa, as marked, then after substituting the pore fill, the final effective modulus must be in the interval $\sim 13.75 \le K^{(2)} \le \sim 15.7$.

Embedded Bounds of Mavko and Saxena

Mavko and Saxena (2013) developed a method based on nested ("embedded") Hashin–Shtrikman (HS) bounds to estimate the change in effective bulk modulus upon

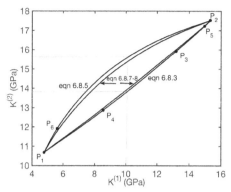

Figure 6.8.1 The four curves of Vinogradov and Milton (2005), given by equations 6.8.3–6.8.9. The outer two curves bound the relation between initial effective bulk modulus $K_{eff}^{(1)}$ and the effective modulus after substitution, $K_{eff}^{(2)}$.

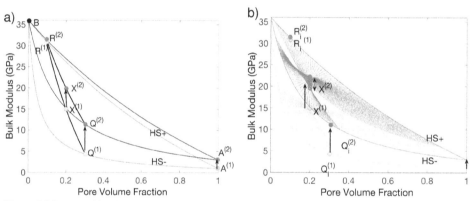

Figure 6.8.2 Embedded bound method for solid substitution

substitution of either fluid or solid in the pore space. As illustrated in Figure 6.8.2a, any point $X^{(1)}$ (for the well-ordered case) can be realized by a Hashin–Shtrikman upper or lower bound composite of materials $R^{(1)}$ and $Q^{(1)}$, which are themselves Hashin–Shtrikman composites of the original end member materials $B^{(1)}$ and $A^{(1)}$. If material $A^{(1)}$ is replaced by material $A^{(2)}$, then point $R^{(1)}$ moves to $R^{(2)}$, $Q^{(1)}$ moves to $Q^{(2)}$, and $X^{(1)}$ moves to $X^{(2)}$. While the embedded Hashin–Shtrikman equations give an exact prediction of $X^{(2)}$ from $X^{(1)}$, the prediction is not unique. As illustrated in Figure 6.8.2b, any material whose bulk modulus lies between the HS bounds can be realized by an infinite number of microstructures, each constructed from a different pair of materials at $R_i^{(1)}$ and $Q_i^{(1)}$. Each microstructure yields a different modulus upon solid substitution of the pore-filling material. Four limiting cases of the embedded bounds are shown in Figure 6.8.3 (labeled HS_{min}^{+}, HS_{min}^{-}, HS_f^{+}, HS_f^{-}) which bound the change in effective bulk modulus predicted by all of the other embedded bound

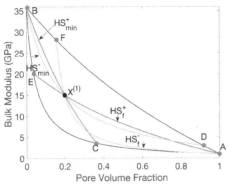

Figure 6.8.3 Four limiting cases of the embedded bounds. Curves BC and AF bound all of the other constructions and are realizable by doubly coated spheres.

Figure 6.8.4 Microstructures constructed from coated spheres that realize the embedded bound constructions of Figures 6.8.2 and 6.8.3. Gray shows mineral phase 1, and black shows pore-filling phase 2.

constructions (as in Figure 6.8.2b). Note that HS_{min}^{+} and HS_{f}^{-} are equivalent to the curves defined by equations 6.8.7–6.8.9 (Vinogradov and Milton, 2005).

Figure 6.8.4 shows examples of microstructures constructed from coated spheres. Phase 1 is the "mineral phase," shown in gray; phase 2 is the pore-filling phase, shown in black. Structure 6.8.4c is an embedded lower bound mix of a material R_i on the absolute upper bound and point Q_i on the absolute lower bound. Structure 6.8.4b is an embedded *upper* bound mix of materials R_i and Q_i. Structures 6.8.4a and 6.8.4d are the doubly coated spheres corresponding to the limiting cases when R_i is the end-member phase 1 and Q_i is the end member phase 2 (curves HS_{min}^{+} and HS_{f}^{-} in Figure 6.8.3), respectively. The doubly coated spheres are also the same as two of Vinogradov and Milton's bounding curves (defined by equations 6.8.7–6.8.9).

Since equations 6.8.7 and 6.8.8 represent embedded Hashin–Shtrikman bounds for bulk modulus, we can extend them into analogous embedded Hashin–Shtrikman constructions for shear modulus (Saxena and Mavko, 2014) by noting the Y-transform expressions for shear modulus and its inverse:

$$y_{\mu}^{(i)} = -f_2\mu_1^{(i)} - f_1\mu_2^{(i)} - \frac{f_1 f_2 \left(\mu_1^{(i)} - \mu_2^{(i)}\right)^2}{\mu_{eff}^{(i)} - f_1\mu_1^{(i)} - f_2\mu_2^{(i)}}, \tag{6.8.10}$$

$$\mu_{eff}^{(i)} = \left[\frac{f_1}{\mu_1^{(i)} + y_\mu^{(i)}} + \frac{f_2}{\mu_2^{(i)} + y_\mu^{(i)}} \right]^{-1} - y_\mu^{(i)}. \tag{6.8.11}$$

For the well-ordered case, the embedded bound curve is defined by

$$y_\mu^{(1)}(\xi_\mu) = \frac{f_2 \varepsilon_1^{(1)} \left(\mu_1^{(1)} + \varepsilon_2^{(1)} \right)(1 - \xi_\mu) + \left(\mu_1^{(1)} + \varepsilon_1^{(1)} \right) \varepsilon_2^{(1)} \xi_\mu}{f_2 \left(\mu_1^{(1)} + \varepsilon_2^{(1)} \right)(1 - \xi_\mu) + \left(\mu_1^{(1)} + \varepsilon_1^{(1)} \right) \xi_\mu} \tag{6.8.12}$$

$$y_\mu^{(2)}(\xi_\mu) = \frac{f_2 \varepsilon_1^{(2)} \left(\mu_1^{(2)} + \varepsilon_2^{(2)} \right)(1 - \xi_\mu) + \left(\mu_1^{(2)} + \varepsilon_1^{(2)} \right) \varepsilon_2^{(2)} \xi_\mu}{f_2 \left(\mu_1^{(2)} + \varepsilon_2^{(2)} \right)(1 - \xi_\mu) + \left(\mu_1^{(2)} + \varepsilon_1^{(2)} \right) \xi_\mu} \tag{6.8.13}$$

$$\xi_\mu = \frac{\left(\varepsilon_1^{(1)} - y_\mu \right) f_2 (\mu_1^{(1)} + \varepsilon_2^{(1)})}{\left(\varepsilon_1^{(1)} - y_\mu \right) f_2 (\mu_1^{(1)} + \varepsilon_2^{(1)}) - (\varepsilon_2^{(1)} - y_\mu)(\mu_1^{(1)} + \varepsilon_1^{(1)})} \tag{6.8.14}$$

where $0 \le \xi_\mu \le 1$ and

$$\varepsilon_i^{(j)} = \frac{\mu_i^{(j)} \left(9K_i^{(j)} + 8\mu_i^{(j)} \right)}{6 \left(K_i^{(j)} + 2\mu_i^{(j)} \right)}. \tag{6.8.15}$$

The additional embedded bound for shear modulus is defined using the same form as equations 6.8.12–6.8.15 with subscripts 1 and 2 swapped. Note that the Hashin–Shtrikman-based equations 6.8.12–6.8.15 are *not* realized by doubly coated spheres as was the case for the bulk modulus problem. Recall also that the Hashin–Shtrikmann–Walpole bounds (Section 4.2) are never realizable for the shear modulus.

Saxena *et al.* (2016) give the following algebraically equivalent forms for the HS_{min}^+ construction for substitution of the effective bulk modulus:

$$K_{eff}^{(2)} = K_{bc}^{(2)} + \frac{\left(1 - \dfrac{K_{bc}^{(2)}}{K_1^{(2)}} \right)^2}{\dfrac{\phi}{K_2^{(2)}} + \dfrac{1 - \phi}{K_1^{(2)}} - \dfrac{K_{bc}^{(2)}}{\left(K_1^{(2)} \right)^2}}, \tag{6.8.16}$$

$$K_{bc}^{(2)} = \frac{(1 - \phi)\left(\dfrac{1}{K_1^{(2)}} - \dfrac{1}{K_{dry}^{(2)}} \right) + \dfrac{3\phi}{4}\left(\dfrac{1}{\mu_1^{(2)}} - \dfrac{1}{\mu_2^{(2)}} \right)}{\dfrac{1}{K_1^{(2)}}\left(\dfrac{1}{K_1^{(2)}} - \dfrac{1}{K_{dry}^{(2)}} \right) + \dfrac{3\phi}{4}\left(\dfrac{1}{K_1^{(2)}\mu_1^{(2)}} - \dfrac{1}{\mu_2^{(2)}K_{dry}^{(2)}} \right)}. \tag{6.8.17}$$

Note that for fluids, $\mu_2^{(i)} = 0$ and $K_{bc}^{(i)} = K_{dry}^{(i)}$, so that equations 6.8.16 and 6.8.17 revert to Gassmann's equation. If the mineral is changed as part of the substitution, then $K_{dry}^{(1)}$ is related to $K_{dry}^{(2)}$ with:

$$K_{dry}^{(2)} = \frac{\left(M_1^{(2)} - K_1^{(2)}\right) K_1^{(2)} M_1^{(1)} K_{dry}^{(1)}}{\left(K_1^{(1)} M_1^{(2)} - K_1^{(2)} M_1^{(1)}\right) K_{dry}^{(1)} + \left(M_1^{(1)} - K_1^{(1)}\right) M_1^{(2)} K_1^{(1)}}, \tag{6.8.18}$$

where $M_i^{(j)} = K_i^{(j)} + (4/3)\mu_i^{(j)}$. If there are solids in the pore space, then the initial dry modulus $K_{dry}^{(1)}$ is obtained from the initial saturated modulus $K_{eff}^{(1)}$ using equations 6.8.19–6.8.20:

$$K_{dry}^{(1)} = \frac{\left(\dfrac{K_{bc}^{(1)}}{K_1^{(1)}} + \dfrac{3\phi}{4} \dfrac{K_{bc}^{(1)}}{\mu_2^{(2)}} - 1 + \phi\right)}{\dfrac{K_{bc}^{(1)}}{\left(K_1^{(1)}\right)^2} + \dfrac{3\phi}{4}\left(\dfrac{K_{bc}^{(1)}}{K_1^{(1)}\mu_1^{(1)}} - \dfrac{1}{\mu_1^{(1)}} + \dfrac{1}{\mu_2^{(1)}}\right) - \dfrac{1-\phi}{K_1^{(1)}}}, \tag{6.8.19}$$

$$K_{bc}^{(1)} = \frac{K_{eff}^{(1)}\left(\phi\dfrac{K_1^{(1)}}{K_2^{(1)}} + 1 - \phi\right) - K_1^{(1)}}{\phi\dfrac{K_1^{(1)}}{K_2^{(1)}} + \dfrac{K_{eff}^{(1)}}{K_1^{(1)}} - 1 - \phi}. \tag{6.8.20}$$

Saxena et al. (2016) give the following two equations for the effective shear modulus upon substitution, which represent the HS_{min}^+ construction:

$$\mu_{eff}^{(2)} = \mu_{bc}^{(2)} + \frac{\left(1 - \dfrac{\mu_{bc}^{(2)}}{\mu_1^{(2)}}\right)^2}{\dfrac{\phi}{\mu_2^{(2)}} + \dfrac{1-\phi}{\mu_1^{(2)}} - \dfrac{\mu_{bc}^{(2)}}{\left(\mu_1^{(2)}\right)^2}} \tag{6.8.21}$$

$$\mu_{bc}^{(2)} = \frac{(1-\phi)\left(\dfrac{1}{\mu_1^{(2)}} - \dfrac{1}{\mu_{dry}^{(2)}}\right) + \phi\left(\dfrac{1}{\varepsilon_1^{(2)}} - \dfrac{1}{\varepsilon_2^{(2)}}\right)}{\dfrac{1}{\mu_1^{(2)}}\left(\dfrac{1}{\mu_1^{(2)}} - \dfrac{1}{\mu_{dry}^{(2)}}\right) + \phi\left(\dfrac{1}{\mu_1^{(2)}\varepsilon_1^{(2)}} - \dfrac{1}{\varepsilon_2^{(2)}\mu_{dry}^{(2)}}\right)} \tag{6.8.22}$$

Note that for fluids, $\mu_2^{(i)} = 0$, $\mu_{bc}^{(i)} = \mu_{dry}^{(i)}$, and equations 6.8.21 and 6.8.22 revert to Gassmann's equation.

If the pore fill is not a fluid, $\mu_{dry}^{(1)}$ and $\mu_{dry}^{(2)}$ are related by

$$\mu_{dry}^{(2)} = \frac{\left(Y_1^{(2)} - \mu_1^{(2)}\right)\mu_1^{(2)}Y_1^{(1)}\mu_{dry}^{(1)}}{\left(\mu_1^{(1)}Y_1^{(2)} - \mu_1^{(2)}Y_1^{(1)}\right)\mu_{dry}^{(1)} + \left(Y_1^{(1)} - \mu_1^{(1)}\right)Y_1^{(2)}\mu_1^{(1)}}$$

(6.8.23)

where $Y_i^{(j)} = \mu_i^{(j)} + \varepsilon_i^{(j)}$. If there are solids in the pore space, then the initial dry modulus $\mu_{dry}^{(1)}$ is obtained from the initial saturated modulus $\mu_{eff}^{(1)}$ using equations 6.8.24 and 6.8.25:

$$\mu_{dry}^{(1)} = \frac{\left(\dfrac{\mu_{bc}^{(1)}}{\mu_1^{(1)}} + \phi\dfrac{\mu_{bc}^{(1)}}{\varepsilon_2^{(1)}} - 1 + \phi\right)}{\dfrac{\mu_{bc}^{(1)}}{\left(\mu_1^{(1)}\right)^2} + \phi\left(\dfrac{\mu_{bc}^{(1)}}{\mu_1^{(1)}\varepsilon_1^{(1)}} - \dfrac{1}{\varepsilon_1^{(1)}} + \dfrac{1}{\varepsilon_2^{(1)}}\right) - \dfrac{1 - \phi}{\mu_1^{(1)}}}$$

(6.8.24)

$$\mu_{bc}^{(1)} = \frac{\mu_{eff}^{(1)}\left(\phi\dfrac{\mu_1^{(1)}}{\mu_2^{(1)}} + 1 - \phi\right) - \mu_1^{(1)}}{\phi\dfrac{\mu_1^{(1)}}{\mu_2^{(1)}} + \dfrac{\mu_{eff}^{(1)}}{\mu_1^{(1)}} - 1 - \phi}$$

(6.8.25)

Caution:

The expressions in equations 6.8.16–6.8.25 describe only the lower embedded bound HS_{min}^+. This can often yield a good estimate, but not a unique one.

BAM, HSBAM, and Aliyeva Estimates

Two of the embedded Hashin–Shtrikman realizations shown in Figure 6.8.2 provide conceptually simple estimates (though not necessarily bounds) on solid substitution. In Figure 6.8.5, the material at point $X^{(1)}$, with porosity ϕ_X and effective modulus $K_X^{(1)}$, can be constructed as Hashin–Shtrikman mixtures of materials $U^{(1)}$, (with moduli $K_U^{(1)}, \mu_U^{(1)}$) and $L^{(1)}$, (with moduli $K_L^{(1)}, \mu_L^{(1)}$) which lie on the absolute upper and lower Hashin–Shtrikman bounds, respectively, at the same porosity ϕ_x. Two particular constructions are the modified upper and lower bound mixtures of $U^{(1)}$ and $L^{(1)}$, where the fractions of material $U^{(1)}$ are, respectively,

$$f_U^+ = \frac{\left[K_X^{(1)} + (4/3)\mu_U^{(1)}\right]^{-1} - \left[K_L^{(1)} + (4/3)\mu_U^{(1)}\right]^{-1}}{\left[K_U^{(1)} + (4/3)\mu_U^{(1)}\right]^{-1} - \left[K_L^{(1)} + (4/3)\mu_U^{(1)}\right]^{-1}}; \quad f_U^+ = 1 - f_L^+,$$

(6.8.26)

$$f_U^- = \frac{\left[K_X^{(1)} + (4/3)\mu_L^{(1)}\right]^{-1} - \left[K_L^{(1)} + (4/3)\mu_L^{(1)}\right]^{-1}}{\left[K_U^{(1)} + (4/3)\mu_L^{(1)}\right]^{-1} - \left[K_L^{(1)} + (4/3)\mu_L^{(1)}\right]^{-1}}; \quad f_U^- = 1 - f_L^-,$$

(6.8.27)

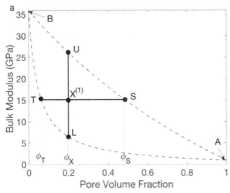

Figure 6.8.5 Two of the embedded bound constructions from Figure 6.8.2b that are algebraically simple to apply

The substitution recipe is to compute f_U^+ and f_U^- from the initial composition and effective modulus $K_X^{(1)}$ using equations 6.8.26 and 6.8.27. Then, re-compute the new moduli at points $U^{(2)}$ $(K_U^{(2)}, \mu_U^{(2)})$ and $L^{(2)}$ $(K_L^{(2)}, \mu_L^{(2)})$, using the new composition. Finally, the new effective modulus $K_X^{(2)}$ is evaluated from Hashin–Shtrikman mixes of the materials at $U^{(2)}$ and $L^{(2)}$ using the same fractions f_U^+ and f_U^-. This is the Hashin–Shtrikman extension of the linear BAM model discussed in Section 6.10.

Alternatively, using a method suggested by Aliyeva (2017), material $X^{(1)}$ can be constructed as a mixture of materials at points $S^{(1)}$ and $T^{(1)}$, which have the same bulk moduli as K_X^1. Volume fractions of materials $S^{(1)}$ and $T^{(1)}$ are:

$$f_T = \frac{\phi_S - \phi_X}{\phi_S - \phi_T} \quad \text{and} \quad f_S = \frac{\phi_X - \phi_T}{\phi_S - \phi_T}.$$

The substitution recipe is to compute f_T and f_S from the initial composition and effective modulus $K_X^{(1)}$. Then, compute the absolute and modified bounds for the new composition, and the new moduli at points $T^{(2)}$ $(K_T^{(2)}, \mu_T^{(2)})$ and $S^{(2)}$ $(K_S^{(2)}, \mu_S^{(2)})$. Finally the new effective modulus $K_X^{(2)}$ is evaluated using modified Hashin–Shtrikman upper and lower mixes of materials $T^{(2)}$ and $S^{(2)}$ using the same volume fractions f_T and f_S.

Note that after the substitution, usually, $K_T^{(2)} \neq K_S^{(2)}$.

Saxena and Mavko Equations

Saxena and Mavko (2014a) derived exact equations for solid-to-solid substitution for bulk and shear moduli for a two-phase composite. The method is based on reciprocity, but does not assume the embedded bound construction. Using the same notation as previously given, mineral phase moduli are (K_1, μ_1), initial pore-filling material moduli are $(K_2^{(1)}, \mu_2^{(1)})$, and substituted pore-filling material moduli are $(K_2^{(2)}, \mu_2^{(2)})$. The substitution relation for effective bulk modulus is

$$
\left(\frac{1}{K_2^{(1)}} - \frac{1}{K_2^{(2)}}\right)\alpha_1 + \left(\frac{1}{\mu_2^{(1)}} - \frac{1}{\mu_2^{(2)}}\right)\alpha_2 = \phi \frac{\left(\frac{1}{K_2^{(2)}} - \frac{1}{K_1}\right)\left(\frac{1}{K_2^{(1)}} - \frac{1}{K_1}\right)}{\left(\frac{1}{K_{eff}^{(1)}} - \frac{1}{K_1}\right)\left(\frac{1}{K_{eff}^{(2)}} - \frac{1}{K_1}\right)}\left(\frac{1}{K_{eff}^{(1)}} - \frac{1}{K_{eff}^{(2)}}\right).
$$

$$(6.8.28)$$

The geometric information is embedded in the two parameters

$$
\alpha_1 = \frac{\overline{P^{(1)}P^{(2)}}}{\overline{P^{(1)}}\,\overline{P^{(2)}}} = \frac{\overline{e^{(1)}e^{(2)}}}{\overline{e^{(1)}}\,\overline{e^{(2)}}},
$$

$$(6.8.29)$$

$$
\alpha_2 = \frac{1}{2}\frac{\overline{\tau_{ij}^{(1)}\tau_{ij}^{(2)}}}{\overline{P^{(1)}}\,\overline{P^{(2)}}} = 2\frac{\overline{\gamma_{ij}^{(1)}\gamma_{ij}^{(2)}}}{\overline{e^{(1)}}\,\overline{e^{(2)}}}\frac{\mu_2^{(1)}\mu_2^{(2)}}{K_2^{(1)}K_2^{(2)}} = \alpha_2'\frac{\mu_2^{(1)}\mu_2^{(2)}}{K_2^{(1)}K_2^{(2)}}.
$$

$$(6.8.30)$$

In equations 6.8.29 and 6.8.30, $P^{(1)}$ and $P^{(2)}$ are the compression-induced mean stress (trace of the stress tensor) in the initial and substituted pore-filling materials, $\gamma_{ij}^{(1)}$ and $\gamma_{ij}^{(2)}$ are the compression-induced deviatoric strains, and $e^{(1)} = P^{(1)}/K_2^{(1)}$ and $e^{(2)} = P^{(2)}/K_2^{(1)}$ are the compression-induced volumetric strains. The operator ⁻ indicates volume average over the pore-filling phase. Parameter α_1 describes the heterogeneity of compression-induced mean stress (pressure) in the initial and pore-filling materials. If $K_2^{(1)} \le K_2^{(2)}$ and $\mu_2^{(1)} \le \mu_2^{(2)}$, then α_1, α_2, and α_2' must be nonnegative. For fluid-to-fluid substitution, $\alpha_1 \ge 1$ and $\alpha_2 = 0$; for fluid-to-fluid substitution with equilibrated pore pressure, $\alpha_1 = 1$, $\alpha_2 = 0$, and equation 6.8.29 becomes Gassmann's equation for fluid substitution. Similarly, if $\alpha_1 = 1$ and $\mu_2^{(1)} = \mu_2^{(2)}$ then equation 6.8.29 also becomes Gassmann's equation. For solid-to-solid substitution, $\alpha_1 \ge 1$. Parameter α_2 is a measure of the deviatoric strain induced in the pore-filling phase caused by remote hydrostatic loading.

The substitution relation for effective shear modulus is

$$
\left(\frac{1}{\mu_2^{(1)}} - \frac{1}{\mu_2^{(2)}}\right)\beta_1 + \left(\frac{1}{K_2^{(1)}} - \frac{1}{K_2^{(2)}}\right)\beta_2 = \phi \frac{\left(\frac{1}{\mu_2^{(2)}} - \frac{1}{\mu_1}\right)\left(\frac{1}{\mu_2^{(1)}} - \frac{1}{\mu_1}\right)}{\left(\frac{1}{\mu_{eff}^{(1)}} - \frac{1}{\mu_1}\right)\left(\frac{1}{\mu_{eff}^{(2)}} - \frac{1}{\mu_1}\right)}\left(\frac{1}{\mu_{eff}^{(1)}} - \frac{1}{\mu_{eff}^{(2)}}\right).
$$

$$(6.8.31)$$

where

$$\beta_1 = \frac{1}{2}\frac{\overline{\tau_{ij}^{(1)}\tau_{ij}^{(2)}}}{\tau_{12}^{(1)}\tau_{12}^{(2)}} = \frac{1}{2}\frac{\overline{\gamma_{ij}^{(1)}\gamma_{ij}^{(2)}}}{\gamma_{12}^{(1)}\gamma_{12}^{(2)}} \tag{6.8.32}$$

$$\beta_2 = \frac{\overline{P^{(1)}P^{(2)}}}{\tau_{12}^{(1)}\tau_{12}^{(2)}} = \frac{1}{4}\frac{\overline{e^{(1)}e^{(2)}}}{\gamma_{12}^{(1)}\gamma_{12}^{(2)}}\frac{K_2^{(1)}K_2^{(2)}}{\mu_2^{(1)}\mu_2^{(2)}} = \beta_2'\frac{K_2^{(1)}K_2^{(2)}}{\mu_2^{(1)}\mu_2^{(2)}} \tag{6.8.33}$$

In equations 6.8.32 and 6.8.33, $P^{(n)}$, $\tau_{ij}^{(n)}$, and $e^{(n)}$ refer to the mean stress, deviatoric stress, and volumetric strain induced in the pore-filling materials by remote deviatoric loading. Parameter β_1 is a measure of the deviatoric strain heterogeneity in the pore-filling material induced by remote deviatoric loading. Parameter β_2 is a measure of the average mean stress in the pore-filling material induced by remote deviatoric loading.

When a material has an effective bulk modulus on a HS bound, $\alpha_1 = 1$ and $\alpha_2 = 0$, for any substitution of the pore-filling material. Similarly, when a material has an effective shear modulus on a HS bound, $\beta_1 = 1$ and $\beta_1 = 0$ for any substitution of the pore-filling material.

P-Wave Only Bounds

Saxena and Mavko (2014b) developed exact expressions for the change of effective P-wave modulus, $M_{eff} = K_{eff}^{(1)} + (4/3)\mu_{eff}$, without knowledge of the shear modulus, under either fluid or solid substitution:

$$\frac{\left(M_{eff}^{(2)} - M_1\right)\left(M_{eff}^{(1)} - M_1\right)}{\left(M_{ud}^{(2)} - M_{ud}^{(1)}\right)}$$

$$= \frac{\phi\left(K_1 - K_2^{(2)} + \left(\mu_1 - \mu_2^{(2)}\right)\frac{4}{3}\Omega_2^{(2)}\right)\left(K_1 - K_2^{(1)} + \left(\mu_1 - \mu_2^{(1)}\right)\frac{4}{3}\Omega_2^{(1)}\right)}{\left(K_2^{(2)} - K_2^{(1)}\right)\delta_1 + \left(\mu_2^{(2)} - \mu_2^{(1)}\right)\delta_2'} \tag{6.8.34}$$

where

$$\delta_1 = \frac{\overline{P^{(1)}P^{(2)}}}{\overline{P^{(1)}}\ \overline{P^{(2)}}} = \frac{\overline{e^{(1)}e^{(2)}}}{\overline{e^{(1)}}\ \overline{e^{(2)}}}, \tag{6.8.35}$$

$$\delta_2' = \frac{K_2^{(1)}K_2^{(2)}}{\mu_2^{(1)}\mu_2^{(2)}}, \tag{6.8.36}$$

$$\delta_2 - \frac{K_2^{(1)}K_2^{(2)}}{\mu_2^{(1)}\mu_2^{(2)}}\left(\frac{1}{2}\frac{\overline{\tau_{ij}^{(1)}\tau_{ij}^{(2)}}}{\overline{P^{(1)}}\ \overline{P^{(2)}}}\right) = 2\frac{\overline{\gamma_{ij}^{(1)}\gamma_{ij}^{(2)}}}{\overline{e^{(1)}}\ \overline{e^{(2)}}}, \tag{6.8.37}$$

and

$$\Omega_2^{(1)} = \frac{3\overline{\varepsilon_{11(2)}}}{2\overline{e_2}} - \frac{1}{2} \tag{6.8.38}$$

In equations 6.8.34–6.8.38, $P^{(n)}$, $\tau_{ij}^{(n)}$, $e^{(n)} = P^{(n)}/K_2^{(n)}$, and $\gamma_{ij}^{(n)} = \tau_{ij}^{(n)}/2\mu_2^{(n)}$ are the mean stress, deviatoric stress, volumetric strain, and deviatoric strain induced in the pore-filling material by a remote *uniaxial strain* $\varepsilon_{11}^{(0)}$ loading. Parameter $\Omega_2^{(n)}$ compares the uniaxial and volumetric strains in the pore-filling material under the remote uniaxial strain loading.

In the special case of fluid substitution,

$$\frac{\phi\left(M_{eff}^{(1)} - M_{eff}^{(2)}\right)}{\left(M_{eff}^{(2)} - M_1\right)\left(M_{eff}^{(1)} - M_1\right)} = \frac{\left(K_2^{(1)} - K_2^{(2)}\right)\delta_1}{\left(K_1 - K_2^{(2)} + \mu_1\frac{4}{3}\Omega_2^{(2)}\right)\left(K_1 - K_2^{(1)} + \mu_1\frac{4}{3}\Omega_2^{(1)}\right)} \tag{6.8.39}$$

Under the condition of interconnected pore space and equilibrated pore pressure (the Gassmann conditions), $\delta_1 = 1$. The P-wave only substitution approximation suggested by Mavko *et al.* (1995) is equivalent to assuming that $\Omega_2^{(n)} \approx 1$, which is almost never true. Saxena and Mavko (2014b) show that $\Omega_2^{(n)}$ depends on pore geometry and mineral Poisson's ratio. For a quartz-rich rock (low mineral Poisson's ratio), the self-consistent model estimates that $\Omega_2^{(n)} \approx 1.25$ for pore aspect ratio > 0.05 and $\Omega_2^{(n)} > 1.5$ for aspect ratio < 0.03; for a calcite-rich rock (high mineral Poisson's ratio), the self-consistent model estimates that $\Omega_2^{(n)} \approx 0.75$ for aspect ratio > 0.03 and $\Omega_2^{(n)} > 1$ for aspect ratios <0.02. The Mavko *et al.* assumption of $\Omega_2^{(n)} \approx 1$ lies within these ranges, but is not strictly constant (see also Section 6.3).

Mineral Substitution

Note that all of the solid substitution equations in this section are symmetric in the sense that either the rock frame mineral or the pore-filling material, or both, can be substituted. Saxena *et al.* (2016) refer to the problem of changing the mineral, while keeping the poregeometry and pore-filling composition fixed as **mineral substitution**. Figure 6.8.6 shows examples of mineral substitution applied to laboratory measurements of ultrasonic *Vp* and *Vs* of water-saturated sandstones, dolomites, and limestones. Figure 6.8.6a shows the initial data, with empirical curves by Castagna *et al.* (1993) superimposed. Figure 6.8.6b shows predictions (equations 6.8.16 and 6.8.17) when all of the original minerals (calcite in limestone and dolomite in dolomites) are replaced with quartz, while keeping porosity and microgeometry fixed. All transformed data fall near the empirical wet sandstone line. Figure 6.8.6c shows predictions when all of the original minerals are replaced by calcite. All data fall near the empirical wet limestone line. Deviations from a perfect prediction can be attributed to

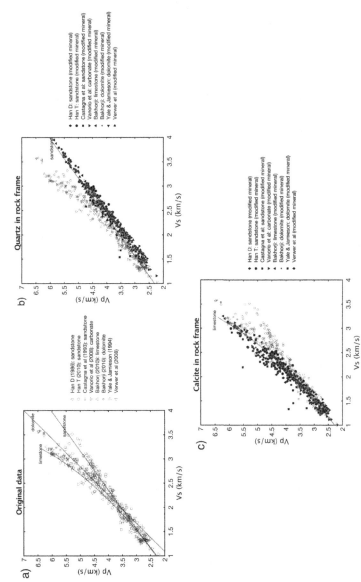

Figure 6.8.6 Examples of mineral substitution on V_p and V_s. (a) Laboratory ultrasonic velocities for dry sandstones, carbonates, limestones, and dolomites, showing the distinctive mineral-dependent trends. (b) All rocks substituted from their original mineral to quartz (shown by filled dots). (c) All rocks substituted from their original mineral to calcite.

changes in pore micro-geometry and uncertainty of the original compositions. As with all of the substitution problems discussed in this section, mineral substitution is non-unique. However Saxena *et al.* (2016) find that the uncertainty for mineral substitution is often small, so that a single prediction is often sufficient.

Mineral substitution offers a strategy to separate effects of mineralogy from effects of microstructure. For example, if a limestone becomes dolomitized, what part of the change in effective elastic moduli is due to the change of mineralogy, and what part is due to change in pore structure? Similarly, if feldspar grains alter to clay, what are the effects of changes in grain stiffness versus microstructure?

Ciz and Shapiro Approximation

Ciz and Shapiro (2007) introduced solid substitution equations for effective bulk and shear moduli:

$$\frac{1}{K_{sat}^{CS}} = \frac{1}{K_{dry}} - \frac{\left(\dfrac{1}{K_{dry}} - \dfrac{1}{K_0}\right)^2}{\phi\left(\dfrac{1}{K_{if}} - \dfrac{1}{K_0}\right) + \left(\dfrac{1}{K_{dry}} - \dfrac{1}{K_0}\right)} \tag{6.8.40}$$

$$\frac{1}{\mu_{sat}^{CS}} = \frac{1}{\mu_{dry}} - \frac{\left(\dfrac{1}{\mu_{dry}} - \dfrac{1}{\mu_0}\right)^2}{\phi\left(\dfrac{1}{\mu_{if}} - \dfrac{1}{\mu_0}\right) + \left(\dfrac{1}{\mu_{dry}} - \dfrac{1}{\mu_0}\right)} \tag{6.8.41}$$

where subscript *0* refers to the moduli of the mineral material of the porous frame, *dry* refers to the dry frame moduli, and *sat* refers to bulk and shear moduli of the solid (-or liquid) saturated porous material. The constants K_{if} and μ_{if} represent measures of pore stiffness, which are generally not known when the wave-induced stress is non-uniform in the pore-filling material. Ciz and Shapiro approximated the parameters with the moduli of the pore-filling material: $K_{if} \approx K_f$ and $\mu_{if} \approx \mu_f$. These assumptions ignore stress heterogeneity in the pore-filling material and therefore are most valid for homogeneous, equidimensional pores. Saxena *et al.* (2013) point out that equations 6.8.40 and 6.8.41 tend to underestimate the change in effective moduli upon solid substitution, and sometimes predict values that violate the Hashin–Shtrikman bounds. Equations 6.8.40 and 6.8.41 reduce to Gassmann's equations when the pore-filling material is a liquid with zero shear modulus.

Uses

The solid substitution equations can be used to estimate the change in the elastic moduli of porous rock with a change in pore-filling material when the pore-filling material has a finite shear modulus; such is the case with heavy oils, which behave like quasi-solids at low temperatures.

Assumptions and Limitations

- The derivation assumes all materials are linear and elastic.
- The pore-filling material completely saturates the pore space.

6.9   Fluid Substitution in Thinly Laminated Reservoirs

Synopsis

In laminated sand-shale sequences, fluid changes, if any, are likely to occur only in the sandy layers, while fluid changes in the shale are prevented by the capillary-held water and extremely low permeability (Katahara, 2004; Skelt, 2004). When the laminations are subresolution, measurements (seismic, sonic, or ultrasonic) yield average properties of the *composite,* rather than of the individual layers. If the subresolution heterogeneity is ignored when computing fluid substitution, the velocity changes are likely to be overpredicted (Skelt, 2004). One approach for fluid substitution is first to downscale the measured composite values to individual sand and shale properties, and then to apply Gassmann fluid substitution only to the sand.

For P- and S-waves propagating normal to the layering, the composite (upscaled) wave velocity is related to the individual layer properties via the Backus (1962) average:

$$\frac{1}{\rho V^2} = \frac{(1 - f_{\text{shale}})}{\rho_{\text{sand}} V^2{}_{\text{sand}}} + \frac{f_{\text{shale}}}{\rho_{\text{shale}} V^2{}_{\text{shale}}} \tag{6.9.1}$$

where V is the measured composite velocity, $\rho = (1 - f_{\text{shale}})\rho_{\text{sand}} + f_{\text{shale}}\rho_{\text{shale}}$ is the measured composite density, $M = \rho V^2$ is the wave modulus of the composite (P- or S-wave), ρ_{sand} is the sand bulk density, V_{sand} is the sand velocity (V_P or V_S), f_{shale}, is the shale *thickness* fraction (1 − net/gross), ρ_{shale} is the shale bulk density, and V_{shale} is the shale velocity (V_P or V_S). The Backus average assumes that layers are very thin relative to the wavelength. If expressed in terms of *wave compliance,* $C = 1/M = 1/\rho V^2$, the Backus average becomes linear for both P- and S-waves:

$$C_{\text{P}} = (1 - f_{\text{shale}})C_{\text{P-sand}} + f_{\text{shale}}C_{\text{P-shale}} \tag{6.9.2}$$

$$C_{\text{S}} = (1 - f_{\text{shale}})C_{\text{S-sand}} + f_{\text{shale}}C_{\text{S-shale}} \tag{6.9.3}$$

Solving the preceding equations for sand properties yields

$$\rho_{\text{sand}} = \frac{\rho - f_{\text{shale}}\rho_{\text{shale}}}{(1 - f_{\text{shale}})} \tag{6.9.4}$$

$$C_{\text{P-sand}} = \frac{C_{\text{P}} - f_{\text{shale}}C_{\text{P-shale}}}{(1 - f_{\text{shale}})} \tag{6.9.5}$$

$$C_{\text{S-sand}} = \frac{C_{\text{S}} - f_{\text{shale}}C_{\text{S-shale}}}{(1 - f_{\text{shale}})} \tag{6.9.6}$$

where $C_{\text{P-sand}}$ and $C_{\text{S-sand}}$ are the wave compliances measured from P- and S-wave velocities, respectively. Hence, sand properties can be estimated if the shale thickness fraction can be estimated (e.g., from a Thomas–Stieber analysis), and if the shale compliance and density can be estimated from shaly log intervals or regional trends.

The sand P-wave modulus, $M_{\text{P-sand}}$, shear modulus, μ_{sand}, and bulk modulus, K_{sand}, are given by

$$M_{\text{P-sand}} = \frac{1}{C_{\text{P-sand}}} \tag{6.9.7}$$

$$\mu_{\text{sand}} = \frac{1}{C_{\text{S-sand}}} \tag{6.9.8}$$

$$K_{\text{sand}} = M_{\text{P-sand}} - \frac{4}{3}\mu_{\text{sand}} \tag{6.9.9}$$

Finally, Gassmann fluid substitution can be applied to the sand bulk modulus, and the sand-shale composite can be upscaled again using the Backus average. If shear data are not available, the approximate form of Gassmann's equation (Section 6.3) can be applied to the sand P-wave modulus, $M_{\text{P-sand}}$.

Caution:

When the fraction of sand becomes very small, the inversion for sand properties, as outlined in equations 6.9.4–6.9.9, becomes unreliable (Skelt, 2004). Small errors in the shale properties become magnified, and predicted sand properties can be incorrect.

Katahara (2004) presented a more stable approach for downscaling, illustrated in Figure 6.9.1. Regional shale and sand trends are plotted in the plane of bulk density ρ versus effective P-wave compliance C_P. Multiple sand trends can be plotted, representing water, gas, and oil sands. The observed trend of laminated sands is also plotted,

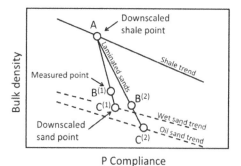

P Compliance

Figure 6.9.1 Katahara's graphical method for down scaling laminated sand data into estimates of the constituent sand and shale end-members.

extending between the sand and shale trends (i.e., the graphical interpretation of the Backus average). Because both the density and compliance of the composite are linear functions of the sand and shale densities and compliances, any measured point, $B^{(1)}$, should fall along a straight line between a shale point, A, and a sand point, $C^{(1)}$. The shale volume fraction is related to the distances $|BC|$ and $|AC|$ by

$$f_{\text{shale}} = \frac{|BC|}{|AC|} \tag{6.9.10}$$

Plausible sand and shale end-member properties corresponding to any composite point, B, are found by drawing a line through $B^{(1)}$, parallel to the laminated-sand trend (Katahara, 2004). The intersections of the line with the shale trend (point A) and sand trend (point $C^{(1)}$) are the end-members. Fluid substitution can be applied to the sand point, yielding $C^{(2)}$. Finally, a line is constructed between the original shale point A and the new fluid-substituted sand point $C^{(2)}$; the fluid-substituted point $B^{(2)}$ is along the new line at the same fractional distance f_{shale} from $C^{(1)}$ to A. The method guarantees that the downscaled end-members are always consistent with the established trends.

Caution:

Before applying Gassmann fluid substitution to the sand point, remember to estimate the downscaled sand porosity (e.g., from the sand density), and the sand fluid saturation.

Subresolution shale laminations within a reservoir can cause several effects:
- fluid changes will only occur within the permeable layers, not within the shales;
- because fluid changes only occur within the permeable fraction of the interval, the observed fluid-related changes in density and velocity are less than if the entire interval were permeable – approximately $(1 - f_{\text{shale}})$ of the effect; and
- shale laminations tend to increase the V_P/V_S ratio – both because the shale fractions have higher V_P/V_S and also because the effects of hydrocarbons on V_P/V_S are diminished.

Dejtrakulwong (2012) and Dejtrakulwong and Mavko (2016) generalized the fluid substitution methods of Katahara (2004) and Skelt (2004) by building upon the Thomas–Stieber–Yin–Marion (TSYM) model for rocks constructed from sand-clay mixtures (Section 5.3). We refer to this as the DM model. With the TSYM model, a subresolution sequence of sand (or shaly sand) interbedded with shale can be represented as a point inside the triangular mesh (see Figures 5.3.2 and 5.3.5). The mesh is controlled by the three corner points: clean sand, pure shale, and the V-point. The V-point, where the clean sand pore space is completely filled with shale is, in turn, controlled by the sand and shale points. The dispersed shale lines use modified lower bounds (Sections 5.3 and 7.1) as "soft interpolators" to connect the corners (i.e., the shaly sand line A-B connects the clean sand point with the V-point, and the sandy shale line B-C connects the shale point with the V-point). The mesh is filled by using the Backus average to inter-bed materials along the shaly sand line with shale at point C. For example, point E is a thinly bedded mix of shale (point C) and shaly sand (point F). These Backus lines are exactly straight, and the modified lower bounds are approximately straight, when plotted in the compliance $C = (\rho V_P^2)^{-1}$ versus effective porosity domain (Figure 6.9.2a).

Under fluid substitution, the sandy-shale line (line B-C), the pure shale point (C), and the V-point (B) remain unchanged due to their zero effective porosity. Dejtrakulwong and Mavko find that the shaly-sand line (A-B) remains (approximately) straight under fluid substitution and moves with the clean sand point. The Backus average lines (F-C) move with the shaly sand points (F). Therefore, fluid substitution anywhere on or within the mesh corresponds to a linear distortion of the mesh, as the clean sand point moves (Figure 6.9.2b). For example, point E goes to E' upon substitution. One can then write fluid substitution anywhere on or within the mesh as

$$C_X^{(2)} - C_X^{(1)} = \frac{\phi_{eff}}{\phi_{sand}}\left(C_{sand}^{(2)} - C_{sand}^{(1)}\right) \tag{6.9.11}$$

where $C_X^{(1)}$ and $C_X^{(2)}$ are the P-wave compliances at any point X with fluid 1 and fluid 2, respectively, $C_{sand}^{(1)}$ and $C_{sand}^{(2)}$ are the compliances of the clean sand point with fluid 1 and fluid 2, ϕ_{eff} is the effective porosity at point X, and ϕ_{sand} is the porosity of the clean sand point.

The change in P-compliance after fluid substitution for an interbedded sand-shale sequence is directly proportional to the change in P-compliance of the clean sand point, scaled by the ratio of the effective porosity of the interbedded sand-shale sequence to the porosity of clean sand. Note from equation 6.9.11 that upon fluid substitution, the

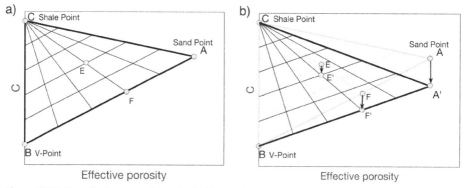

Figure 6.9.2 Detrajkulwong and Mavko fluid substitution model represented in the P-wave compliance vs. effective porosity domain. (a) Model with initial pore fluid. (b) Distortion of the model when effective porosity is substituted with a less compliant pore fluid. Gassmann fluid substitution is applied to the clean sand point A, and remainder of mesh distorts linearly.

change of compliance anywhere in the model is always less than or equal to the change of compliance of the clean sand point. This helps to make the prediction relatively robust, since fluid substitution on clean sand is well defined.

The starting fluid in the clean sand must be consistent with the fluid in the effective pore space of the laminated (interbedded) sequence. For example, let us assume that a data point represents interbedding of 50% shale and 50% shaly sand, whose effective pore space is only half saturated with oil. In this case, the effective water saturation in the shaly sand layer is 0.5. If we want to apply fluid substitution to this interbedded sequence by replacing oil with brine, we need to compute the change in P-compliance of the clean sand point by going from (effective) water saturation of 0.5 to water saturation of 1.

The mesh method is directly applicable at the measurement scale, without the need to downscale the measurements, but its fluid substitution results still agree well with those predicted when fluid substitution is applied only to the fine-scale permeable sand layers. Input parameters of the mesh such as clean-sand P-wave velocity can be constrained by rock physics trends. The method is only robust for rocks that can be represented as sand–shale mixtures, in which any porosity reduction is caused only by filling of shale in the pore space. Therefore, the method does not account for additional porosity reduction due to cementation or changes in the clean sand texture. Because the mesh method does not cover all possible variations in rocks, this method should not be universally applied to a whole dataset without checking for its applicability. To alleviate the problem of non-universality of the mesh method, Dejtrakulwong and Mavko recommend applying both the mesh method and the traditional Gassmann's equation to the data. The traditional Gassmann's equation tends to overestimate the change of elastic moduli after fluid substitution. Of the DM and Gassmann predictions, the more accurate fluid substitution result is the one that predicts the smaller change in elastic moduli.

Uses

This approach can be used for fluid substitution when subresolution impermeable layers lie within the interval.

Assumptions and Limitations

- As with any Gassman-related fluid substitution method, the seismic frequency must be low enough that wave-induced increments of pore pressure can equilibrate throughout the pore space during a seismic period.
- In the case discussed here, the equilibration only needs to take place within each sand layer.
- Downscaling to sand properties requires that reasonable estimates of shale properties and the shale fraction can be made.
- Describing the laminated composite using the Backus average implies that the layer thicknesses are very much smaller than the wavelength.

Extensions

Fluid substitution within the sand can be done using either the Gassmann or Brown–Korringa equations, depending on the sand mineralogy. If the sand itself is anisotropic, then the corresponding anisotropic fluid substitution algorithms can be applied. The anisotropy resulting from the lamination of shales with sands is not relevant to choosing the fluid substitution algorithm; the choice depends only on the properties of the permeable intervals where fluids are changing.

6.10   BAM: Marion's Bounding Average Method

Synopsis

Marion (1990) developed a heuristic method based on theoretical bounds for estimating how elastic moduli and velocities change when one pore-filling phase is substituted for another. The Hashin–Shtrikman (1963) bounds (see Sections 4.2 and 4.3) define the range of elastic moduli (velocities) possible for a given volume mix of two (well-ordered) phases, either liquid or solid (see Figure 6.10.1). At any given volume fraction of constituents, the effective modulus will fall between the bounds (somewhere along the vertical dashed line in the top figure), but its precise value depends on the geometric details of the grains and pores. We use, for example, terms such as "stiff pore shapes" and "soft pore shapes," when describing rocks. Stiffer pore shapes cause the moduli to be higher within the allowable range; softer pore shapes cause the moduli to be lower.

Marion reasoned that the fractional vertical position within the bounds, $w = d/D$, where $0 \leq w \leq 1$, is, therefore, a measure of the pore geometry and is independent of the pore-filling properties – a reasonable assumption, but one that has not been proven. Because changing the pore-filling material does not change the geometry, w might remain (nearly) constant, $w = d/D = d'/D'$, with any change in pore fluids. His method is as follows:

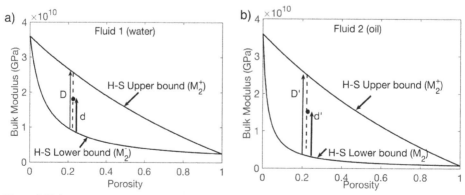

Figure 6.10.1 Hashin–Shtrikman bounds. The fractional vertical position within the bounds is $w = d/D$.

1. Begin with the mineral modulus and measurements of the modulus with the first pore-filling material (liquid, gas, or solid). Calculate the theoretical upper bound, M_1^+, and lower bound, M_1^-, corresponding to this state. (The subscript 1 refers to this first state with the first pore-filling material.) Plot the measured data value, M_1, and measure w relative to these bounds.

$$w = \frac{M_1 - M_1^-}{M_1^+ - M_1^-}$$

2. Recalculate the theoretical upper and lower bounds, M_2^+ and M_2^-, corresponding to the second pore-filling material of interest. Plot a point at the same position w relative to the new bounds. This is the new estimated modulus, M_2,

$$M_2 = M_2^- + w(M_2^+ - M_2^-)$$

Marion and others (Marion and Nur, 1991; Marion *et al.*, 1992) showed that this method works quite well for several examples: predicting water-saturated rock velocities from dry-rock velocities and predicting frozen-rock (ice-filled) velocities from water-saturated velocities.

Uses

Marion's bounding average method is applicable to the fluid substitution problem.

Assumptions and Limitations

Marion's bounding average method is primarily heuristic. Therefore, it needs to be tested empirically. A stronger theoretical basis would also be desirable. Theoretical work on bounds on fluid and solid substitution, Sections 6.4 and 6.8, shows that no estimate will be unique without more information about the rock microgeometry.

Extensions

The simplicity of the method suggests that it be tried for comparing the effects of water-filled pores with clay-filled pores, altered clay grains versus the original crystalline grain, and so forth.

6.11 Mavko–Jizba Squirt Relations

Synopsis

The *squirt* or *local flow* model suggests that the fluctuating stresses in a rock caused by a passing seismic wave induce pore-pressure gradients at virtually all scales of pore-space heterogeneity – particularly on the scale of individual grains and pores. These gradients impact the viscoelastic behavior of the rock; at high frequencies, when the gradients are unrelaxed, all elastic moduli (including the shear modulus) will be stiffer than at low frequencies, when the gradients are relaxed (the latter case being modeled by Gassmann).

Mavko and Jizba (1991) derived simple theoretical formulas for predicting the very high-frequency moduli of saturated rocks by using the pressure dependence of dry rocks as a measure of pore heterogeneity. The prediction is made in two steps: first, the squirt effect (i.e., the frame stiffening caused by liquid trapped in the thinnest cracks at very high frequency) is incorporated as high-frequency "wet-frame moduli" K_{uf} and μ_{uf}. The pore space in the "wet-frame" is dry, except in the thinnest cracks. The pressure-dependent moduli $K_{uf}(P)$ and $\mu_{uf}(P)$ are derived from the dry moduli as

$$\frac{1}{K_{uf}(P)} \approx \frac{1}{K_{dry-hiP}} + \left(\frac{1}{K_{fluid}} - \frac{1}{K_0}\right)\phi_{soft}(P) \tag{6.11.1}$$

$$\left(\frac{1}{\mu_{uf}(P)} - \frac{1}{\mu_{dry}(P)}\right) = \frac{4}{15}\left(\frac{1}{K_{uf}(P)} - \frac{1}{K_{dry}(P)}\right) \tag{6.11.2}$$

where $K_{dry}(P)$ is the effective bulk modulus of the dry rock at effective pressure P, K_0 is the bulk modulus of the mineral material making up the rock, and K_{fluid} is the effective bulk modulus of the pore fluid. Modulus $K_{dry-HiP}$ is the effective bulk modulus of the dry rock at very high effective pressure when cracks are closed. The "soft" porosity $\phi_{soft}(P)$ is the porosity of cracks that are open when $K_{dry}(P) < K_{dry-HiP}$. μ_{uf} is the effective high-frequency, unrelaxed, wet-frame shear modulus, and μ_{dry} is the effective shear modulus of dry rock.

The interpretation of equation 6.11.1 is as follows. At very high effective pressure, cracks close and the dry frame stiffens to moduli $K_{dry-hiP}$ and $\mu_{dry-hiP}$. For elastic crack models, closing a crack with high effective pressure is conceptually the same as filling the crack with mineral – in both cases the crack "disappears." If the crack aspect ratio is small, satisfying $\alpha \ll K_{fl}/K_0$, then filling the crack with liquid is approximately the same as filling it with mineral (in terms of crack-normal stiffness). Hence, the first term on the right side of equation 6.11.1 is simply the "crack-free" high pressure compliance. The second term on the right side is a first order correction for the difference in fluid and mineral compressibilities. Equation 6.11.2 expresses the wet frame shear modulus in terms of the wet frame bulk modulus, recognizing that fluids affect the normal compliance of cracks, but not their shear compliance.

The wet frame moduli are then substituted into Gassmann's or Biot's relations (in place of dry frame moduli) to incorporate the remaining fluid-saturation effects. For most crustal rocks the amount of squirt dispersion is comparable to or greater than Biot's dispersion, and thus using Biot's theory alone will lead to poor predictions of high-frequency saturated velocities. (Exceptions include very high-permeability materials such as ocean sediments and glass beads; rocks at very high effective pressure when most of the soft, crack-like porosity is closed; or rocks near free boundaries such as borehole walls.)

A more detailed analysis of the frequency dependence of the squirt mechanism is presented in Section 6.11.

Although the formulation presented here is independent of any idealized crack shape, the squirt behavior is also implicit in virtually all published formulations for effective moduli based on elliptical cracks (see Sections 4.8–4.15). In most of those models, the cavities are treated as being isolated with respect to flow, thus simulating the high-frequency limit of the squirt model.

Gurevich et al. (2009) extended the Mavko–Jizba expression for the unrelaxed frame bulk modulus to the case of a highly compressible pore fluid (e.g., gas):

$$\frac{1}{K_{uf}(P)} = \frac{1}{K_{dry-hiP}} + \cfrac{1}{\cfrac{1}{\cfrac{1}{K_{dry}(P)} - \cfrac{1}{K_{dry-hiP}}} + \cfrac{1}{\left(\cfrac{1}{K_{fluid}} - \cfrac{1}{K_0}\right)\phi_{soft}(P)}}, \tag{6.11.3}$$

where $K_{dry-hiP}$ is the dry bulk modulus at the highest effective pressure available, K_{fluid} is the fluid bulk modulus, K_0 is the mineral bulk modulus, and $\phi_{soft}(P)$ is the soft, crack porosity at effective pressure P.

The Gurevich et al. equation goes to the Mavko–Jizba equation when K_{fluid} increases, such that

$$\left(\frac{K_0 - K_{fluid}}{K_{fluid}K_0}\right)\phi_{soft} \gg \left(\frac{K_{dry-hiP} - K_{dry}(P)}{K_{dry(P)}K_{dry-hiP}}\right). \tag{6.11.4}$$

It reduces appropriately to the dry rock modulus when $K_{fluid} = 0$.

Uses

The Mavko–Jizba–Gurevich squirt relations can be used to calculate high-frequency saturated rock velocities from dry-rock velocities.

Assumptions and Limitations

The use of the Mavko–Jizba squirt relations requires the following considerations:
- High seismic frequencies that are ideally suited for ultrasonic laboratory measurements are assumed. In-situ seismic velocities generally will have neither squirt nor Biot dispersion and should be described using Gassmann's equations. Sonic-logging frequencies may or may not be within the range of validity, depending on the rock type and fluid viscosity.
- The rock is isotropic.
- All minerals making up the rock have the same bulk and shear moduli.
- Fluid-bearing rock is completely saturated with liquid.
- The Gurevich et al. modification extends the applicability to soft pore fluids and gas.

Extensions

The Mavko–Jizba–Gurevich squirt relations can be extended in the following ways:
- For mixed mineralogy, one can usually use an effective average modulus for K_0.
- For clay-filled rocks, it often works best to consider the "soft" clay to be part of the pore-filling phase rather than part of the mineral matrix. Then the pore fluid is "mud," and its modulus can be estimated with an isostress calculation.
- The anisotropic form of these squirt relations has been found by Mukerji and Mavko (1994) and is discussed in Section 6.15.

6.12   Extension of Mavko–Jizba Squirt Relations for All Frequencies

Synopsis

The Mavko and Jizba (1991) squirt relations (see Section 6.11) predict the very high-frequency moduli of saturated rocks. At a low frequency, these moduli can be calculated from Gassmann's (1951) equations. Dvorkin *et al.* (1995) introduced a model for calculating these moduli, velocities, and attenuations at any intermediate frequency. As input, the model uses such experimentally measurable parameters as the dry-rock elastic properties at a given effective pressure, the dry-rock bulk modulus at very high effective pressure, the bulk moduli of the solid and fluid phases, and the rock density and porosity. One additional parameter (Z), which determines the frequency scale of the dispersion, is proportional to the **characteristic squirt-flow length**. This parameter can be found by matching the theoretical velocity to that measured experimentally at a given frequency. Then the theory can be used to calculate velocities and attenuation at any frequency and with any pore fluid. The algorithm for calculating velocities and attenuation at a given frequency follows.

Step 1: Calculate the bulk modulus of the dry modified solid (K_{msd}) from

$$\frac{1}{K_{msd}} = \frac{1}{K_0} - \frac{1}{K_{dry-hiP}} + \frac{1}{K_{dry}} \tag{6.12.1}$$

Step 2: Calculate the ratio of the induced pore-pressure increment to the confining-stress increment ($dP/d\sigma$) from

$$\frac{dP}{d\sigma} = -\left[\alpha_0\left(1 + \frac{\phi K_{dry}}{\alpha_0^2 F_0}\right)\right]^{-1} \tag{6.12.2}$$

where

$$\frac{1}{F_0} = \frac{1}{K_{fl}} + \frac{1}{\phi Q_0} \tag{6.12.3}$$

$$\alpha_0 = 1 - \frac{K_{dry}}{K_0} \tag{6.12.4}$$

$$Q_0 = \frac{K_0}{\alpha_0 - \phi} \tag{6.12.5}$$

Step 3: Assume a certain value for the frequency-dispersion parameter Z (start with Z = 0.001) and calculate the bulk modulus of the saturated modified solid (K_{ms}) from

$$K_{ms} = \frac{K_{msd} + \alpha K_0[1 - f(\xi)]}{1 + \alpha f(\xi)dP/d\sigma} \tag{6.12.6}$$

where

$$\alpha = 1 - \frac{K_{msd}}{K_0}, \quad f(\xi) = \frac{2J_1(\xi)}{\xi J_0(\xi)}, \quad \xi = Z\sqrt{i\omega} \tag{6.12.7}$$

and where ω is angular frequency, and J_0 and J_1 are Bessel functions of zero and first order, respectively.

Step 4: Calculate the bulk modulus of the modified frame (K_m) from

$$\frac{1}{K_m} = \frac{1}{K_{ms}} + \frac{1}{K_{dry-hiP}} - \frac{1}{K_0} \tag{6.12.8}$$

Step 5: Calculate the bulk modulus of the saturated rock (K_r) from

$$K_r = \frac{K_m}{1 + \alpha_m dP/d\sigma} \tag{6.12.9}$$

where

$$\alpha_m = 1 - \frac{K_m}{K_{ms}} \tag{6.12.10}$$

Step 6: Calculate the shear modulus of the modified frame (μ_m) from

$$\frac{1}{\mu_{dry}} - \frac{1}{\mu_m} = \frac{4}{15}\left(\frac{1}{K_{dry}} - \frac{1}{K_{md}}\right) \tag{6.12.11}$$

where

$$\frac{1}{K_{md}} = \frac{1}{\tilde{K}_{ms}} + \frac{1}{K_{dry-hiP}} - \frac{1}{K_0} \tag{6.12.12}$$

$$\tilde{K}_{ms} = K_{msd} + \alpha K_0[1 - f(\xi)] \tag{6.12.13}$$

Step 7: Finally, calculate velocities V_P and V_S and inverse quality factors Q_P^{-1} and Q_S^{-1} from

$$V_P = \sqrt{\frac{\text{Re}\left(K_r + \frac{4}{3}\mu_m\right)}{\rho}}, \quad V_S = \sqrt{\frac{\text{Re}(\mu_m)}{\rho}} \tag{6.12.14}$$

$$Q_P^{-1} = \frac{\left| \text{Im}\left(K_r + \frac{4}{3}\mu_m \right) \right|}{\left| \text{Re}\left(K_r + \frac{4}{3}\mu_m \right) \right|}, \qquad Q_S^{-1} = \frac{|\text{Im}(\mu_m)|}{|\text{Re}(\mu_m)|} \tag{6.12.15}$$

Step 8: The velocities and inverse quality factors have been found for an assumed Z value (Figure 6.12.1). To find the true Z value, one has to change it until the theoretical value of one of the four parameters (V_P, V_S, Q_P^{-1}, or Q_S^{-1}) matches the experimentally measured value at a given frequency. It is preferred that V_P be used for this purpose. The Z value thus obtained should be used for calculating the velocities and quality factors at varying frequencies. The Z value can also be used for a different pore fluid. In the latter case, use the following value for Z:

$$Z_{new} = Z \sqrt{\frac{\eta_{new}}{\eta}} \tag{6.12.16}$$

where the subscript *new* indicates the new pore fluid.

The notation used is: K_{dry} is the effective bulk modulus of dry rock, $K_{dry\text{-}hiP}$ is the effective bulk modulus of dry rock at very high pressure, K_0 is the bulk modulus of the mineral material making up the rock, K_{fl} is the effective bulk modulus of the pore fluid, η is the viscosity of the pore fluid, ϕ is porosity, μ_{dry} is the effective shear modulus of dry rock, ρ is rock density, and ω is angular frequency.

Uses

The extension of the Mavko–Jizba squirt relations can be used to calculate saturated-rock velocities and attenuation at any frequency.

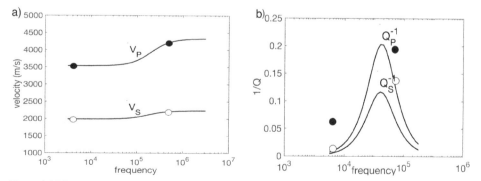

Figure 6.12.1 The V_S and inverse quality factors have been predicted by determining Z from V_P measurements (0.18 porosity limestone). Closed symbols are for P-waves; open symbols are for S-waves.

Figure 6.12.2 Schematic of a micromechanical model to describe the squirt-flow mechanism

Assumptions and Limitations

The following assumptions underlie the extension of the Mavko–Jizba squirt relations:
- the rock is isotropic;
- all minerals making up the rock have the same bulk and shear moduli; and
- fluid-bearing rock is completely saturated.

Extensions

Additional extensions of the Mavko–Jizba squirt relations include the following:
- For mixed mineralogy, one can usually use an effective average modulus for K_0.
- Murphy *et al.* (1984) introduced a micromechanical model to describe the squirt-flow mechanism. They considered a composite grain-contact stiffness, which is the parallel combination of the solid–solid contact stiffness and the stiffness of a fluid-filled gap, shown in Figure 6.12.2.

When modeling the solid–solid contact, surface energy is taken into account. The hydrodynamic contact model takes into account the squirt flow of pore fluid between a thin crack and a stiff, large pore. The results qualitatively match the observed velocity dispersion and attenuation in sandstones.

6.13 Biot–Squirt Model

Synopsis

Dvorkin and Nur (1993) and Dvorkin *et al.* (1994) introduced a unified Biot–squirt (BISQ) model. The model is applicable to rocks at high pressure with compliant cracks closed. The rock is partially saturated or apparently fully saturated (meaning that there are small, undetectable amounts of free gas left in pores). The zero-frequency velocity limit, as given by BISQ, is

$$V_{P0} = \sqrt{\frac{M_{dry}}{\rho}} \qquad (6.13.1)$$

The M_{dry} term is the dry-rock uniaxial-strain modulus ($M_{dry} = \rho_{dry}\, V_{P-dry}^2$), ρ is rock density (at saturation), and V_{P-dry} and ρ_{dry} are the dry-rock P-wave velocity and density, respectively. The BISQ high-frequency velocity limit is the same as in the Biot theory.

BISQ gives the following expressions for P-wave velocity (V_P), the attenuation coefficient (a_P), and the inverse quality factor (Q_P^{-1}) at **apparently full saturation**:

$$V_P = \frac{1}{\mathrm{Re}(\sqrt{Y})}, \quad a_P = \omega \, \mathrm{Im}(\sqrt{Y}), \quad Q_P^{-1} = \frac{2 a_P V_P}{\omega} \tag{6.13.2}$$

$$Y = -\frac{B}{2A} - \sqrt{\left(\frac{B}{2A}\right)^2 - \frac{C}{A}}, \quad A = \frac{\phi F_{sq} M_{dry}}{\rho_2^2} \tag{6.13.3}$$

$$B = \left[F_{sq}\left(2\gamma - \phi - \phi\frac{\rho_1}{\rho_2} \right) - \left(M_{dry} + F_{sq}\frac{\gamma^2}{\phi} \right)\left(1 + \frac{\rho_a}{\rho_2} + i\frac{\omega_c}{\omega} \right) \right] / \rho_2 \tag{6.13.4}$$

$$C = \frac{\rho_1}{\rho_2} + \left(1 + \frac{\rho_1}{\rho_2} \right)\left(\frac{\rho_a}{\rho_2} + i\frac{\omega_c}{\omega} \right), \quad F_{sq} = F\left[1 - \frac{2 J_1(\lambda R)}{\lambda R J_0(\lambda R)} \right] \tag{6.13.5}$$

$$\lambda^2 = \frac{\rho_{fl}\omega^2}{F}\left(\frac{\phi + \rho_a/\rho_{fl}}{\phi} + i\frac{\omega_c}{\omega} \right), \quad \rho_1 = (1 - \phi)\rho_s, \quad \rho_2 = \phi\rho_{fl} \tag{6.13.6}$$

$$\omega_c = \frac{\eta\phi}{k\rho_{fl}}, \quad \gamma = 1 - \frac{K_{dry}}{K_0}, \quad \frac{1}{F} = \frac{1}{K_{fl}} + \frac{1}{\phi K_0}\left(1 - \phi - \frac{K_{dry}}{K_0} \right) \tag{6.13.7}$$

where R is the **characteristic squirt-flow length**, ϕ is porosity, ρ_s and ρ_{fl} are the solid-phase and fluid-phase densities, respectively; $\rho_a = (1 - \alpha)\phi\rho_{fl}$ is the Biot inertial-coupling density; η is the pore-fluid viscosity; k is rock permeability; K_{dry} is the dry-rock bulk modulus; K_0 and K_{fl} are the solid-phase and the fluid-phase bulk moduli, respectively; ω is the angular frequency; and J_0 and J_1 are Bessel functions of zero and first order, respectively. The tortuosity α (sometimes called the structure factor) is a purely geometrical factor independent of the solid or fluid densities and is always greater than 1 (see Section 6.1).

All input parameters, except for the **characteristic squirt-flow length**, are experimentally measurable. The latter has to be either guessed (it should have the same order of magnitude as the average grain size or the average crack length) or adjusted by using an experimental measurement of velocity versus frequency (see Section 6.1).

For **partially saturated** rock at saturation S,

$$R_s = R\sqrt{S} \tag{6.13.8}$$

has to be used instead of R in the preceding formulas. To avoid numerical problems (caused by resonance) at high frequencies,

$$\lambda^2 = i\frac{\rho_{fl}\omega\omega_c}{F} \tag{6.13.9}$$

can be used instead of

$$\lambda^2 = \frac{\rho_{fl}\omega^2}{F}\left(\frac{\phi + \rho_a/\rho_{fl}}{\phi} + i\frac{\omega_c}{\omega} \right) \tag{6.13.10}$$

At lower frequencies, $\omega_c/\omega \gg 1$, the following simplified formulas can be used:

$$Y = \frac{(1-\phi)\rho_s + \phi\rho_{fl}}{M_{dry} + F_{sq}\gamma^2/\phi} , \qquad F_{sq} = F\left[1 - \frac{2J_1(\xi)}{\xi J_0(\xi)}\right] \qquad (6.13.11)$$

$$\xi = \sqrt{i\frac{R^2\omega}{\kappa}} \qquad \kappa = \frac{kF}{\eta\phi} \qquad (6.13.12)$$

The BISQ formulas give the Biot theory expressions for the velocity and attenuation if $F_{sq} = F$.

Uses

The BISQ formulas can be used to calculate partially saturated rock velocities and attenuation (at high pressure) at any frequency.

Assumptions and Limitations

The BISQ formulas are based on the following assumptions:
- the rock is isotropic;
- all minerals making up the rock have the same bulk and shear moduli.
    In the low-frequency limit, the BISQ formulas are not consistent with the Gassmann predictions.

Extensions

- For mixed mineralogy, one can usually use an effective average modulus for K_0.

6.14 Chapman *et al.* Squirt Model

Synopsis

Chapman *et al.* (2002) presented a *squirt* or *local flow* model that considers frequency-dependent, wave-induced exchange of fluids between pores and cracks, as well as between cracks of different orientations. The cracks are idealized as oblate spheroids with small aspect ratio r, uniform crack radius a, and crack density ε. The pores are considered to be spherical. The bulk and shear moduli and Lamé parameter of the solid mineral matrix (rock without cracks or pores) are denoted by K, μ, and λ, respectively, while the fluid bulk modulus is K_f. The total porosity is ϕ.

    The frequency-dependent effective bulk and shear moduli, K_{eff} and μ_{eff}, are expressed as follows (Chapman *et al.*, 2006):

$$K_{\text{eff}} = K - \varepsilon \left\{ \frac{4(3\lambda + 2\mu)(\lambda + 2\mu)}{\mu(\lambda + \mu)} [1 - 3A(\omega)] - 4\pi r A(\omega) \right\}$$

$$- \phi \left\{ \frac{3\lambda + 2\mu}{4\mu} \left[\frac{\lambda + 2\mu}{3\lambda + 2\mu} + B(\omega) \right] - 3B(\omega) \right\} \tag{6.14.1}$$

and

$$\mu_{\text{eff}} = \mu - \frac{16}{45} \varepsilon \frac{1}{1 + K_c} \frac{\mu(\lambda + 2\mu)}{3\lambda + 4\mu} \left(K_c + \frac{1}{1 + i\omega\tau} \right) - \frac{32}{45} \varepsilon \frac{\mu(\lambda + 2\mu)}{3\lambda + 4\mu}$$

$$- \phi \frac{15\mu(\lambda + 2\mu)}{9\lambda + 14\mu} \tag{6.14.2}$$

where ω is the angular frequency, and the other parameters are

$$A(\omega) = \frac{(1 + i\omega\gamma\tau) \frac{\lambda + 2\mu}{\lambda + \mu} \left[\frac{16\varepsilon}{27\phi(1 + K_p)} + \frac{\lambda + \mu}{3\lambda + 2\mu} \right] + i\omega\tau \left[\frac{1}{3(1 + K_c)} - \gamma' \right]}{1 + i\omega\tau + (1 + i\omega\gamma\tau) \frac{16\varepsilon(1 + K_c)}{9\phi(1 + K_p)} \frac{\lambda + 2\mu}{\lambda + \mu}} \tag{6.14.3}$$

$$B(\omega) = \frac{\frac{1}{3(1 + K_c)} + \frac{9(1 + K_p)}{16(1 + K_c)} \frac{(\lambda + \mu)}{(3\lambda + 2\mu)} + \frac{i\omega\tau}{1 + i\omega\tau} \left[\gamma' - \frac{1}{3(1 + K_c)} \right]}{\frac{9(1 + K_p)(\lambda + \mu)}{16(1 + K_c)(\lambda + 2\mu)} + \frac{1 + i\omega\gamma\tau}{1 + i\omega\tau}} \tag{6.14.4}$$

$$\gamma' = \gamma \frac{\lambda + 2\mu}{(3\lambda + 2\mu)(1 + K_p)} \tag{6.14.5}$$

$$\gamma = \frac{3\pi(\lambda + \mu)(1 + K_p)}{4(\lambda + 2\mu)(1 + K_c)} \tag{6.14.6}$$

$$K_p = \frac{4\mu}{3K_f} \tag{6.14.7}$$

and

$$K_c = \frac{\pi\mu r(\lambda + \mu)}{K_f(\lambda + 2\mu)} \tag{6.14.8}$$

The timescale parameter τ controls the frequency regime over which the dispersion occurs. It is proportional to the fluid viscosity η and inversely proportional to the permeability k, and depends on the crack radius a. For small aspect ratios ($r < 10^{-3}$), τ is approximately given by

$$\tau \approx \frac{4\eta a^3 (1 - v)}{9k\varsigma\mu} \tag{6.14.9}$$

where ς is a characteristic grain size, and v is the Poisson's ratio of the solid mineral.

One of the requirements for any model of fluid-related dispersion is that in the low-frequency limit the predictions should be consistent with the Gassmann equations. The Chapman *et al.* (2002) formulas are consistent with the Gassmann predictions in the low-frequency limit, whereas the BISQ model (Dvorkin and Nur, 1993) and the equant porosity model (Hudson *et al.*, 1996) are not.

Uses

The formulas can be used to calculate squirt-related velocity dispersion and attenuation at any frequency.

Assumptions and Limitations

The formulas assume idealized spherical pores and penny-shaped crack geometry, with all cracks having the same radius:
- the rock is isotropic;
- all minerals making up the rock have the same bulk and shear moduli.

Extensions

- The formulation has been extended to meso-scale aligned fractures with coupled fluid motion at two scales (Chapman, 2003).

6.15 Anisotropic Squirt

Synopsis

The *squirt* or *local flow* model suggests that the fluctuating stresses in a rock caused by a passing seismic wave induce pore-pressure gradients at virtually all scales of pore-space heterogeneity – particularly on the scale of individual grains and pores. These gradients impact the viscoelastic behavior of the rock; at high frequencies, when the gradients are unrelaxed, all elastic moduli will be stiffer than at low frequencies, when the gradients are relaxed. (The latter case is modeled by the anisotropic Gassmann and Brown and Korringa [1975] formalisms.) Mukerji and Mavko (1994) derived simple theoretical formulas for predicting the very high-frequency compliances of saturated anisotropic rocks in terms of the pressure dependence of dry rocks. The prediction is made in two steps: first, the squirt effect is incorporated as high-frequency "wet-frame compliances" $S_{ijkl}^{(\text{wet})}$, which are derived from the dry compliances $S_{ijkl}^{(\text{dry})}$. Then these wet-frame compliances are substituted into the Gassmann (Section 6.3), Brown and Korringa (see Section 6.5) or Biot relations (see Section 6.1) (in place of the dry compliances) to incorporate the remaining fluid saturation effects. For most crustal rocks, the amount of squirt dispersion is comparable to or greater than Biot's dispersion, and thus using Biot's theory alone will lead to poor predictions of high-frequency saturated velocities. Exceptions include very high permeability materials such as ocean sediments and glass beads, rocks at very high effective pressure, when most of the soft, crack-like porosity is closed, and rocks near permeable free boundaries such as borehole walls.

The wet-frame compliance is given by (repeated indices imply summation)

$$S_{ijkl}^{(\text{wet})} \approx S_{ijkl}^{(\text{dry})} - \frac{\Delta S_{\alpha\alpha\beta\beta}^{(\text{dry})}}{1 + \phi_{\text{soft}}(\beta_{\text{f}} - \beta_0)/\Delta S_{\gamma\gamma\delta\delta}^{(\text{dry})}} G_{ijkl} \tag{6.15.1}$$

where $\Delta S_{ijkl}^{(\text{dry})} = S_{ijkl}^{(\text{dry})} - S_{ijkl}^{(\text{dry high p})}$ is the change in dry compliance between the pressure of interest and very high confining pressure, ϕ_{soft} is the soft porosity that closes under high confining pressure, and β_{f} and β_0 are the fluid and mineral compressibilities, respectively. The soft porosity is often small enough that the second term in the denominator can be ignored. The tensor G_{ijkl} represents the fraction of the total compliance that is caused by volumetric deformation of crack-like pore space with different orientations for a given externally applied load. The tensor depends on the symmetry of the crack distribution function and is expressed as an integral over all orientations:

$$G_{ijkl} = \int f(\Omega) n_i n_j n_k n_l d\Omega \tag{6.15.2}$$

where $f(\Omega)$ is the crack-orientation distribution function normalized so that its integral over all angles equals unity, and n_i is the unit normal to the crack faces. Elements of G_{ijkl} with any permutation of a given set $ijkl$ are equal. Note that G_{ijkl} has more symmetries than the elastic compliance tensor. For **isotropic** symmetry, the elements of G_{ijkl} are given by the Table 6.14.1.

Table 6.14.1 gives exactly the same result as the isotropic equations of Mavko and Jizba (1991) presented in Section 6.11.

Table 6.14.1 *Elements of the tensor* G_{ijkl} *for isotropic symmetry*

ij	11	22	33	23	13	12
kl						
11	1/5	1/15	1/15	0	0	0
22	1/15	1/5	1/15	0	0	0
33	1/15	1/15	1/5	0	0	0
23	0	0	0	1/15	0	0
13	0	0	0	0	1/15	0
12	0	0	0	0	0	1/15

When the rock is **transversely isotropic** with the 3-axis as the axis of rotational symmetry, the five independent components of G_{ijkl} are

$$G_{1111} = \Delta\widetilde{S}_{1111}^{(dry)} - \frac{4\alpha}{1-4\alpha}\left[\Delta\widetilde{S}_{1122}^{(dry)} + \Delta\widetilde{S}_{1133}^{(dry)}\right] \tag{6.15.3}$$

$$G_{1122} = \frac{\Delta\widetilde{S}_{1122}^{(dry)}}{1-4\alpha} \tag{6.15.4}$$

$$G_{1133} = \frac{\Delta\widetilde{S}_{1133}^{(dry)}}{1-4\alpha} \tag{6.15.5}$$

$$G_{3333} = \Delta\widetilde{S}_{3333}^{(dry)} - \frac{8\alpha\Delta\widetilde{S}_{1133}^{(dry)}}{1-4\alpha} \tag{6.15.6}$$

$$G_{2323} = \frac{\Delta\widetilde{S}_{2323}^{(dry)}}{1-4\alpha} - \frac{\Delta\widetilde{S}_{1111}^{(dry)} + \Delta\widetilde{S}_{3333}^{(dry)}}{4(1-4\alpha)} + \frac{G_{1111} + G_{3333}}{4} \tag{6.15.7}$$

The nine independent components of G_{ijkl} for **orthorhombic** symmetry are

$$G_{1111} = \Delta\widetilde{S}_{1111}^{(dry)} - \frac{4\alpha}{1-4\alpha}\left[\Delta\widetilde{S}_{1122}^{(dry)} + \Delta\widetilde{S}_{1133}^{(dry)}\right] \tag{6.15.8}$$

$$G_{2222} = \Delta\widetilde{S}_{2222}^{(dry)} - \frac{4\alpha}{1-4\alpha}\left[\Delta\widetilde{S}_{1122}^{(dry)} + \Delta\widetilde{S}_{2233}^{(dry)}\right] \tag{6.15.9}$$

$$G_{3333} = \Delta\widetilde{S}_{3333}^{(dry)} - \frac{4\alpha}{1-4\alpha}\left[\Delta\widetilde{S}_{1133}^{(dry)} + \Delta\widetilde{S}_{2233}^{(dry)}\right] \tag{6.15.10}$$

$$G_{1122} = \frac{\Delta\widetilde{S}_{1122}^{(dry)}}{1-4\alpha} \tag{6.15.11}$$

$$G_{1133} = \frac{\Delta\widetilde{S}_{1133}^{(dry)}}{1-4\alpha} \tag{6.15.12}$$

$$G_{2233} = \frac{\Delta\widetilde{S}_{2233}^{(dry)}}{1-4\alpha} \tag{6.15.13}$$

$$G_{2323} = \frac{\Delta\widetilde{S}_{2323}^{(dry)}}{1-4\alpha} - \frac{\Delta\widetilde{S}_{2222}^{(dry)} + \Delta\widetilde{S}_{3333}^{(dry)}}{4(1-4\alpha)} + \frac{G_{2222} + G_{3333}}{4} \tag{6.15.14}$$

$$G_{1313} = \frac{\Delta\widetilde{S}_{1313}^{(dry)}}{1-4\alpha} - \frac{\Delta\widetilde{S}_{1111}^{(dry)} + \Delta\widetilde{S}_{3333}^{(dry)}}{4(1-4\alpha)} + \frac{G_{1111} + G_{3333}}{4} \tag{6.15.15}$$

$$G_{1212} = \frac{\Delta\widetilde{S}_{1212}^{(dry)}}{1-4\alpha} - \frac{\Delta\widetilde{S}_{1111}^{(dry)} + \Delta\widetilde{S}_{2222}^{(dry)}}{4(1-4\alpha)} + \frac{G_{1111} + G_{2222}}{4} \tag{6.15.16}$$

where

$$\Delta \widetilde{S}_{ijkl}^{(dry)} = \frac{\Delta S_{ijkl}^{(dry)}}{\Delta S_{\gamma\gamma\beta\beta}^{(dry)}} \tag{6.15.17}$$

$$\alpha = \frac{1}{4}\left(\Delta \widetilde{S}_{\gamma\beta\gamma\beta}^{(dry)} - 1\right) \tag{6.15.18}$$

Computed from the dry data, α is the ratio of the representative shear-to-normal compliance of a crack set, including all elastic interactions with other cracks. When the orthorhombic anisotropy is due to three mutually perpendicular crack sets, superposed on a general orthorhombic background, with the crack normals along the three symmetry axes, the wet-frame compliances are obtained from

$$
\begin{aligned}
S_{ijkl}^{(dry)} - S_{ijkl}^{(wet)} \approx{}& \frac{\Delta S_{1111}^{(dry)}}{1 + \phi_{soft}^{(1)}(\beta_f - \beta_0)/\Delta S_{1111}^{(dry)}}\delta_{i1}\delta_{j1}\delta_{k1}\delta_{l1} \\
&+ \frac{\Delta S_{2222}^{(dry)}}{1 + \phi_{soft}^{(2)}(\beta_f - \beta_0)/\Delta S_{2222}^{(dry)}}\delta_{i2}\delta_{j2}\delta_{k2}\delta_{l2} \\
&+ \frac{\Delta S_{3333}^{(dry)}}{1 + \phi_{soft}^{(3)}(\beta_f - \beta_0)/\Delta S_{3333}^{(dry)}}\delta_{i3}\delta_{j3}\delta_{k3}\delta_{l3}
\end{aligned}
\tag{6.15.19}
$$

where δ_{ij}, is the Kronecker delta, and $\phi_{soft}^{(i)}$ refers to the soft porosity of the ith crack set. The preceding expressions assume that the intrinsic compliance tensor of planar crack-like features is sparse, the largest components being the normal and shear compliances, whereas the other components are approximately zero. This general property of planar crack formulations reflects an approximate decoupling of normal and shear deformation of the crack and decoupling of the in-plane and out-of-plane compressive deformation. In the case of a *single crack set* (with crack normals along the 3-axis) the wet-frame compliances can be calculated from the dry compliances for a completely general, nonsparse crack compliance as

$$S_{ijkl}^{(wet)} = S_{ijkl}^{(dry)} - \frac{\Delta S_{\lambda\lambda ij}^{(dry)}\Delta S_{\eta\eta kl}^{(dry)}}{\Delta S_{\gamma\gamma\delta\delta}^{(dry)} + \phi_{soft}(\beta_f - \beta_0)} \tag{6.15.20}$$

Little or no change of the $\Delta S_{1111}^{(dry)}$ and $\Delta S_{2222}^{(dry)}$ dry compliances with stress would indicate that all the soft, crack-like porosity is aligned normal to the 3-axis. However, a rotationally symmetric distribution of cracks may often be a better model of crack-induced transversely isotropic rocks than just a single set of aligned cracks. In this case the equations in terms of G_{ijkl} should be used.

The anisotropic squirt formulation presented here does not assume any idealized crack geometries. Because the high-frequency saturated compliances are predicted entirely in terms of the measured dry compliances, the formulation automatically incorporates all elastic pore interactions, and there is no limitation to low crack density.

Although the formulation presented here is independent of any idealized crack shape, the squirt behavior is also implicit in virtually all published formulations for effective moduli based on elliptical cracks (see Sections 4.8–4.15). In most of those models, the cavities are treated as isolated with respect to flow, thus simulating the high-frequency limit of the squirt model.

Uses

The anisotropic squirt formulation can be used to calculate high-frequency, saturated-rock velocities from dry-rock velocities.

Assumptions and Limitations

The use of the anisotropic squirt formulation requires the following considerations:
- High seismic frequencies ideally suited for ultrasonic laboratory measurements are assumed. In situ seismic velocities will generally have neither squirt nor Biot dispersion and should be described by using Brown and Korringa equations. Sonic-logging frequencies may or may not be within the range of validity, depending on the rock type and fluid viscosity.
- All minerals making up rock have the same compliances.
- Fluid-bearing rock is completely saturated.

Extensions

The following extensions of the anisotropic squirt formulation can be made:
- For mixed mineralogy, one can usually use an effective average modulus for β_0.
- For clay-filled rocks, it often works best to consider the "soft" clay to be part of the pore-filling phase rather than part of the mineral matrix. The pore fluid is then "mud," and its modulus can be estimated with an isostress calculation.

6.16 Common Features of Fluid-Related Velocity Dispersion Mechanisms

Synopsis

Many physical mechanisms have been proposed and modeled to explain velocity dispersion and attenuation in rocks (Section 3.8): scattering (see Section 3.14), viscous and inertial fluid effects (see Sections 6.1, 6.2, and 6.11–6.20), hysteresis related to surface forces, thermoelastic effects, phase changes, and so forth. Scattering and surface forces appear to dominate in dry or nearly dry conditions (Tutuncu, 1992; Sharma and Tutuncu, 1994). Viscous fluid mechanisms often dominate when there is more than a trace of pore fluids, such as in the case of the poroelasticity described by Biot (1956) and the local flow or squirt mechanism (Stoll and Bryan, 1970; Mavko and

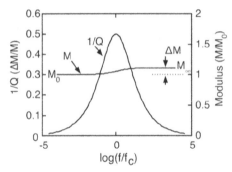

Figure 6.16.1 Key parameters in various dispersion mechanisms: M_0, the low-frequency limiting modulus, M_∞, the high-frequency limiting modulus, and f_c, the characteristic frequency separating high- and low-frequency behavior.

Nur, 1975; O'Connell and Budiansky, 1977; Stoll, 1989; Dvorkin and Nur, 1993). Extensive reviews of these were given by Knopoff (1964), Mavko, Kjartansson, and Winkler (1979, 1985, 1986), and Bourbié *et al.* (1987), among others.

This section highlights some features that attenuation–dispersion models have in common. These suggest a simple approach to analyzing dispersion, bypassing some of the complexity of the individual theories, complexity that is often not warranted by available data.

Although the various dispersion mechanisms and their mathematical descriptions are distinct, most can be described by the following three key parameters (see Figure 6.16.1):

1. a **low-frequency** limiting velocity V_0 (or modulus, M_0), often referred to as the "relaxed" state;
2. a **high-frequency** limiting velocity V_∞ (or modulus, M_∞), referred to as the "unrelaxed" state;
3. a **characteristic frequency**, f_c, that separates high-frequency behavior from low-frequency behavior and specifies the range in which velocity changes most rapidly.

High- and Low-Frequency Limits

Of the three key parameters, usually the low- and high-frequency limits can be estimated most easily. These require the fewest assumptions about the rock microgeometry and are, therefore, the most robust. In rocks, the velocity (or modulus) generally increases with frequency (though not necessarily monotonically in the case of scattering), and thus $M_\infty > M_0$ The total amount of dispersion between very low-frequency and very high-frequency $(M_\infty - M_0)/M = \Delta M/M$ is referred to as the **modulus defect** (Zener, 1948), where $M = \sqrt{M_0 M_\infty}$.

One of the first steps in analyzing any dispersion mechanism should be to estimate the modulus defect to see whether the effect is large enough to warrant any additional modeling. In most situations all but one or two mechanisms can be eliminated based on the size of the modulus defect alone.

As an example, consider the local flow or squirt mechanism, which is discussed in Sections 6.11–6.15. The squirt model recognizes that natural heterogeneities in pore stiffnessess, pore orientation, fluid compressibility, and saturation can cause spatial variations in wave-induced pore pressures at the pore scale. At sufficiently low frequencies, there is time for the fluid to flow and eliminate these variations; hence, the very low-frequency limiting bulk and shear moduli can be predicted by Gassmann's theory (Section 6.3), or by Brown and Korringa's theory (Section 6.5–6.6) if the rock is anisotropic. At high frequencies, the pore-pressure variations persist, causing the rock to be stiffer. The high-frequency limiting moduli may be estimated from dry-rock data using the Mavko and Jizba method (Section 6.11). These low- and high-frequency limits are relatively easy to estimate and require minimum assumptions about pore microgeometry. In contrast, calculating the detailed frequency variation between the low- and high-frequency regimes requires estimates of the pore aspect-ratio or throat-size distributions.

Other simple estimates of the high- and low-frequency limits associated with the squirt mechanism can be made by using ellipsoidal crack models such as the Mori–Tanaka model (Section 4.9), the Kuster and Toksöz model (Section 4.10), the self-consistent model (Section 4.11), the DEM model (Section 4.12), Hudson's model (Section 4.13), or the Eshelby–Cheng model (Section 4.14). In each case, the dry rock is modeled by setting the inclusion moduli to zero. The *high-frequency saturated* rock conditions are simulated by assigning fluid moduli to the inclusions. Because each model treats the cavities as isolated with respect to flow, this yields the unrelaxed moduli for squirt. The *low-frequency saturated* moduli are found by taking the model-predicted effective moduli for dry cavities and saturating them with the Gassmann low-frequency relations (see Section 6.3).

The characteristic frequency is also simple to estimate for most models but usually depends more on poorly determined details of grain and pore microgeometry. Hence, the estimated critical frequency is usually less robust. Table 6.16.1 summarizes approaches to estimating the low-frequency moduli, high-frequency moduli, and characteristic frequencies for five important categories of velocity dispersion and attenuation.

Each mechanism shown in Table 6.16.1 has an f_c value that depends on poorly determined parameters. The Biot (see Section 6.1) and patchy (see Section 6.18) models depend on the permeability. The squirt (see Sections 6.11–6.15) and viscous shear (Walsh, 1969) models require the crack aspect ratio. Furthermore, the formula for f_c–squirt is only a rough approximation, for the dependence on α^3 is of unknown reliability. The patchy-saturation (see Section 6.18) and scattering (see Sections 3.10–3.14) models depend on the scale of saturation and scattering heterogeneities. Unlike the other parameters, which are determined by grain and pore size, saturation and scattering can involve all scales, ranging from the pore scale to the basin scale.

Table 6.16.1 *High- and low-frequency limits and characteristic frequency of dispersion mechanisms*

Mechanism[a]	Low-frequency limit[a]	High-frequency limit[a]	Characteristic frequency (f_c)
Biot [6.1]	Gassmann's relations [6.3]	Biot's high-frequency formula [6.1]	$f_{Biot} \approx \phi\eta/2\pi\rho_{fl}\kappa$
Squirt [6.11–6.15]			
b	Gassmann's relations [6.3]	Mavko–Jizba relations [6.11]	$f_{squirt} \approx K_O\alpha^3/\eta$
c	Kuster–Toksöz dry relations [4.10] → Gassmann	Kuster–Toksöz saturated relations [4.10]	"
c	DEM dry relations [4.12] → Gassmann	DEM saturated relations [4.12]	"
c	Self-consistent dry relations [4.11] → Gassmann	Self-consistent saturated relations [4.11]	"
d	Hudson's dry relations [4.13] → Brown and Korringa [6.5]	Hudson's saturated relations [4.13]	"
Patchy saturation [6.18, 6.19]			
e	Gassmann's relations [6.3]	Hill equation [4.6]	$f_{patchy} \approx \kappa/L^2\eta(\beta_p + \beta_{fl})$
f	Gassmann's relations [6.3]	White high-frequency formula [6.19]	$f_{patchy} \approx \kappa K_s/\pi L^2\eta$
g	Generalized Gassmann's relations [6.7]	Dutta–Odé high-frequency formula [6.19]	"
Viscous shear	h	h	$f_{visc.crack} \approx \alpha\mu/2\pi\eta$
Scattering [3.10–3.14]			
i	Effective medium theory [4.1–4.15]	Ray theory [3.11–3.13]	$f_{scatter} \approx V/2\pi a$
j	Backus average [4.18]	Time average [3.11–3.13]	"

[a] Numbers in brackets [ ] refer to sections in this book.
[b] Inputs are measured dry-rock moduli; no idealized pore geometry.
[c] Pore space is modeled as idealized ellipsoidal cracks.
[d] Anisotropic rock modeled as idealized penny-shaped cracks with preferred orientations.
[e] Dry rock is homogeneous; saturation has arbitrarily shaped patches.
[f] Dry rock is homogeneous; saturation is in spherical patches. Same limits as [e].
[g] Dry rock can be heterogeneous; saturation is in spherical patches.
[h] The mechanism modeled by Walsh (1969) is related to shearing of penny-shaped cracks with viscous fluids. Interesting only for extremely high viscosity or extremely high frequency.
[i] General heterogeneous three-dimensional medium. Normal-incidence propagation through a layered medium.
ϕ = porosity, L = characteristic size (or correlation length), K_0, μ = bulk and shear moduli of mineral of saturation heterogeneity, K_s = saturated rock modulus, ρ_{fl} = density of the pore fluid, β_{fl} = compressibility of the pore fluid, η = viscosity of the pore fluid, β_P = compressibility of the pore space, α = pore aspect ratio, a = characteristic size (or correlation length) of scatterers, κ = rock permeability, V = wave velocity at f_c.

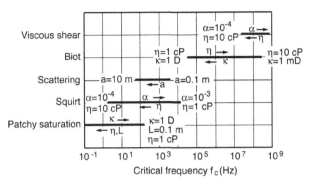

Figure 6.16.2 Comparison of characteristic frequencies for typical rock and fluid parameters. Arrows show the direction of change as the labeled parameter increases.

Figure 6.16.2 compares some values for f_c predicted by the various expressions in the table. The following parameters (or ranges of parameters) were used:

$\phi = 0.2$ = porosity
$\rho_{fl} = 1.0$ g/cm^3 = fluid density
$\eta = 1$–10 cP = fluid viscosity
$\kappa = 1$–1000 mD = permeability
$\alpha = 10^{-3}$–10^{-4} = crack aspect ratio
$V = 3500$ m/s = wave velocity
$a, L = 0.1$–10 m = characteristic scale of heterogeneity
$\mu = 17$ GPa = rock shear modulus
$K = 18$ GPa = rock bulk modulus

Caution:

Other values for rock and fluid parameters can change f_c considerably.

Complete Frequency Dependence

Figure 6.16.3 compares the complete *normalized* velocity-versus-frequency dependence predicted by the Biot, patchy-saturation, and scattering models. Although there are differences, each follows roughly the same trend from the low-frequency limits to the high-frequency limits, and the most rapid transition is in the range $f \approx f_c$. This is not always strictly true for the Biot mechanism, where certain combinations of parameters can cause the transition to occur at frequencies far from f_{Biot}. All are qualitatively similar to the dispersion predicted by the standard linear solid (see Section 3.8):

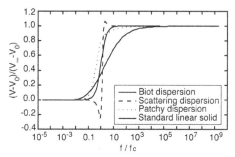

Figure 6.16.3 Comparison of the complete *normalized* velocity-versus-frequency dependence predicted by the Biot, patchy-saturation, and scattering models

$$\text{Re}(M(\omega)) = \frac{M_0 M_\infty [1 + (f/f_c)^2]}{M_\infty + (f/f_c)^2 M_0} \tag{6.16.1}$$

Because the value of f_c and the resulting curves depend on poorly determined parameters, we suggest that a simple and practical way to estimate dispersion curves is to use the standard linear solid. The uncertainty in the rock microgeometry, permeability, and heterogeneous scales – as well as the approximations in the theories themselves – often makes a more detailed analysis unwarranted. Alternatives to the standard linear solid are presented in Section 3.8.

Attenuation

The Kramers–Kronig relations (see Section 3.9) completely specify the relation between velocity dispersion and attenuation. If velocity is known for all frequencies, attenuation is determined for all frequencies. The attenuation versus frequency for the standard linear solid is given by

$$\frac{1}{Q} = \frac{M_\infty - M_0}{\sqrt{M_\infty M_0}} \frac{f/f_c}{1 + (f/f_c)^2} \tag{6.16.2}$$

Similarly, it can be argued (see Section 3.8) that, in general, the order of magnitude of attenuation can be determined from the modulus defect:

$$\frac{1}{Q} \approx \frac{M_\infty - M_0}{\sqrt{M_\infty M_0}} \tag{6.16.3}$$

Uses

The preceding simplified relations can be used to estimate velocity dispersion and attenuation in rocks.

Assumptions and Limitations

This discussion is based on the premise that difficulties in measuring attenuation and velocity dispersion, and in estimating aspect ratios, permeability, and heterogeneous scales, often make detailed analysis of dispersion unwarranted. Fortunately, much about attenuation–dispersion behavior can be estimated robustly from the high- and low-frequency limits and the characteristic frequency of each physical mechanism.

6.17 Dvorkin–Mavko Attenuation Model

Synopsis

Dvorkin and Mavko (2006) present a theory for calculating the P- and S-wave inverse quality factors ((Q_P^{-1} and Q_S^{-1}, respectively) at partial and full saturation. The basis for the quality factor estimation is the standard linear solid (SLS) model (Section 3.8) that links the inverse quality factor Q^{-1} to the corresponding elastic modulus M versus frequency / dispersion as

$$Q^{-1}(f) = \frac{(M_\infty - M_0)(f/f_c)}{\sqrt{M_0 M_\infty}[1 + (f/f_c)^2]}$$
(6.17.1)

where M_0 and M_∞ are the low- and high-frequency limits of the modulus M, respectively; and f_c is the critical frequency at which the inverse quality factor is maximum.

Partial Water Saturation

Consider rock in which K_{dry} is the bulk modulus of the dry frame of the rock; $K_{mineral}$ is the bulk modulus of the mineral phase; and ϕ is the total porosity. Its bulk modulus at partial water saturation S_W is (according to Gassmann's equation)

$$K_0 = K_{mineral} \frac{\phi K_{dry} - (1 + \phi)K_{fl}K_{dry}/K_{mineral} + K_{fl}}{(1 - \phi)K_{fl} + \phi K_{mineral} - K_{fl}K_{dry}/K_{mineral}}$$
(6.17.2)

where K_{fl} is the bulk modulus of the pore fluid which is a *uniform* mixture of water with the bulk modulus K_W and hydrocarbon (e.g., gas) with the bulk modulus K_G:

$$\frac{1}{K_{fl}} = \frac{S_W}{K_W} + \frac{1 - S_W}{K_G}$$
(6.17.3)

Then the compressional modulus M_0 is obtained from K_0 and the shear modulus of the dry frame μ_{dry} as

$$M_0 = K_0 + (4/3)\mu_{dry}$$
(6.17.4)

Alternatively, to obtain M_0, one may use the approximate V_P-only fluid-substitution equation (Section 6.3) with the dry-frame compressional modulus M_{dry} and the mineral-phase compressional modulus $M_{mineral}$:

$$M_0 \approx M_{mineral} \frac{\phi M_{dry} - (1 + \phi)K_{fl}M_{dry}/M_{mineral} + K_{fl}}{(1 - \phi)K_{fl} + \phi M_{mineral} - K_{fl}M_{dry}/M_{mineral}} \tag{6.17.5}$$

The compressional modulus M_0 thus obtained (either by Gassmann's equation or V_P-only fluid-substitution) is considered the low-frequency limit of the compressional modulus M.

The high-frequency limit of the compressional modulus M_∞ is calculated using the patchy-saturation equations (Section 6.18). It is assumed that the difference between M_0 and M_∞ is nonzero only at water saturation larger than the irreducible water saturation S_{irr}. For $S_W \leq S_{irr}$, $M_\infty = M_0$, i.e., $Q_P^{-1} = 0$.

For $S_W > S_{irr}$,

$$\frac{1}{K_\infty + \frac{4}{3}\mu_{dry}} = \frac{(S_W - S_{irr})/(1 - S_{irr})}{K_P + \frac{4}{3}\mu_{dry}} + \frac{(1 - S_W)/(1 - S_{irr})}{K_{mineral\text{-}irr} + \frac{4}{3}\mu_{dry}} \tag{6.17.6}$$

where

$$K_P = K_{mineral} \frac{\phi K_{dry} - (1 + \phi)K_W K_{dry}/K_{mineral} + K_W}{(1 - \phi)K_W + \phi K_{mineral} - K_W K_{dry}/K_{mineral}} \tag{6.17.7}$$

$$K_{mineral\text{-}irr} = K_{mineral} \frac{\phi K_{dry} - (1 + \phi)K_{fl\text{-}irr}K_{dry}/K_{mineral} + K_{fl\text{-}irr}}{(1 - \phi)K_{fl\text{-}irr} + \phi K_{mineral} - K_{fl\text{-}irr}K_{dry}/K_{mineral}} \tag{6.17.8}$$

$$\frac{1}{K_{fl\text{-}irr}} = \frac{S_{irr}}{K_W} + \frac{1 - S_{irr}}{K_G} \tag{6.17.9}$$

Then the high-frequency limit of the compressional modulus M_∞ is

$$M_\infty = K_\infty + \frac{4}{3}\mu_{dry} \tag{6.17.10}$$

and the inverse P-wave quality factor Q_P^{-1} is calculated from the SLS dispersion equation as presented at the beginning of this section.

The alternative V_P-only fluid-substitution equations are

$$\frac{1}{M_\infty} = \frac{(S_W - S_{irr})/(1 - S_{irr})}{M_P} + \frac{(1 - S_W)/(1 - S_{irr})}{M_{mineral\text{-}irr}} \tag{6.17.11}$$

where

$$M_P = M_{mineral} \frac{\phi M_{dry} - (1 + \phi)K_W M_{dry}/M_{mineral} + K_W}{(1 - \phi)K_W + \phi M_{mineral} - K_W M_{dry}/M_{mineral}} \tag{6.17.12}$$

$$M_{\text{mineral-irr}} = M_{\text{mineral}} \frac{\phi M_{\text{dry}} - (1+\phi)K_{\text{fl-irr}}M_{\text{dry}}/M_{\text{mineral}} + K_{\text{fl-irr}}}{(1-\phi)K_{\text{fl-irr}} + \phi M_{\text{mineral}} - K_{\text{fl-irr}}M_{\text{dry}}/M_{\text{mineral}}} \qquad (6.17.13)$$

Full Water Saturation (Wet Rock)

Attenuation is calculated not at a point but in an interval. The necessary condition for attenuation is elastic heterogeneity in rock. The low-frequency limit of the compressional modulus is calculated by theoretically substituting the pore fluid into the dry-frame modulus of the spatially averaged rock while the high-frequency modulus is the spatial average of the saturated-rock modulus. The difference between these two estimates may give rise to noticeable P-wave attenuation if elastic heterogeneity in rock is substantial.

Specifically, in an interval of a layered medium, the average porosity is the arithmetic average of the porosity: $\bar{\phi} = \langle \phi \rangle$; the average dry-frame compressional modulus is the harmonic average of these moduli: $\overline{M}_{dry} = \langle M_{dry}^{-1} \rangle^{-1}$; the average low-frequency compressional modulus limit is obtained from the V_P-only (or, more rigorously, Gassmann's) equation applied to these averaged values:

$$M_0 = \overline{M}_{\text{mineral}} \frac{\bar{\phi}\overline{M}_{dry} - (1+\bar{\phi})K_W\overline{M}_{dry}/\overline{M}_{\text{mineral}} + K_W}{(1-\bar{\phi})K_W + \bar{\phi}\overline{M}_{\text{mineral}} - K_W\overline{M}_{dry}/\overline{M}_{\text{mineral}}} \qquad (6.17.14)$$

where $\overline{M}_{\text{mineral}}$ can be estimated by averaging the mineral-component moduli in the interval by, for example, Hill's equation or harmonically.

The high-frequency limit is the harmonic average of the wet moduli in the interval:

$$M_\infty = \left\langle \left(M_{\text{mineral}} \frac{\phi M_{\text{dry}} - (1+\phi)K_{\text{fl}}M_{\text{dry}}/M_{\text{mineral}} + K_{\text{fl}}}{(1-\phi)K_{\text{fl}} + \phi M_{\text{mineral}} - K_{\text{fl}}M_{\text{dry}}/M_{\text{mineral}}} \right)^{-1} \right\rangle^{-1} \qquad (6.17.15)$$

Finally, the inverse P-wave quality factor Q_P^{-1} is calculated from the SLS dispersion equation as presented at the beginning of this section.

Example for wet rock. Consider a model rock that is fully water-saturated (wet) and has two parts. One part (80% of the rock volume) is shale with porosity 0.4; the clay content is 0.8 (the rest is quartz); and the P-wave velocity 1.9 km/s. The other part (the remaining 20%) is clean, high-porosity, slightly cemented sand with porosity 0.3 and P-wave velocity 3.4 km/s. The compressional modulus is 7GPa in the shale and 25 GPa in the sand. Because of the difference between the compliance of the sand and shale parts, their deformation due to a passing wave is different, which leads to macroscopic "squirt-flow." At high frequency, there is essentially no cross-flow between sand and shale, simply because the flow cannot fully develop during the short cycle of oscillation. The effective elastic modulus of the system is the harmonic (Backus) average of the moduli of the two parts: $M_\infty = 16$ GPa. At low frequency, the cross-flow can easily develop. In this case, the fluid reacts to the combined deformation of the dry frames of the sand and shale. The dry-frame compressional modulus in the shale is 2 GPa, while that in the sand is 20 GPa. The dry-frame modulus of the combined dry frames can be estimated as the harmonic average of the two: 7 GPa. The arithmetically averaged porosity of the model rock is 0.32. To estimate the effective compressional modulus of the combined dry frame with water, we theoretically substitute water into this combined frame. The result is $M_0 = 13$ GPa. The maximum inverse quality factor $Q_{P\,max}^{-1}$ is about 0.1 ($Q = 10$), which translates into a noticeable attenuation coefficient 0.05 dB/m at 50 Hz.

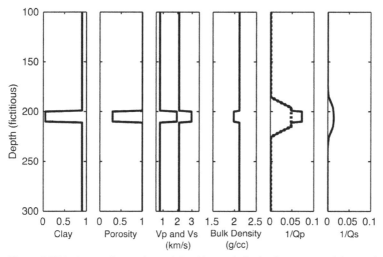

Figure 6.17.1 A pseudo-Earth model with a relatively fast gas-sand interval embedded in soft shale. From left to right: clay content; total porosity; velocity; bulk density; the inverse P-wave quality factor for wet background (dotted curve) and at the actual gas saturation (solid curve); and the inverse S-wave quality factor.

At partial saturation, the background inverse quality factor Q_{Pback}^{-1} is calculated as if the entire interval were wet. This value is added to Q_{P-Sw}^{-1} which is calculated at partial saturation (as previously shown) to (approximately) arrive at the total inverse quality factor:

$$Q_{Ptotal}^{-1} = Q_{Pback}^{-1} + Q_{P-Sw}^{-1} \tag{6.17.16}$$

An example of applying this model to a pseudo-Earth model is shown in Figure 6.17.1.

S-Wave Attenuation

To model the S-wave inverse quality factor, Dvorkin and Mavko (2006) assume that the reduction in the compressional modulus between the high-frequency and low-frequency limits in *wet rock* is due to the introduction of a hypothetical set of aligned defects or flaws (e.g., cracks). The *same set of defects* is responsible for the reduction in the shear modulus between the high- and low-frequency limits. Finally, Hudson's theory for cracked media is used to link the shear-modulus-versus-frequency dispersion to the compressional-modulus-versus-frequency dispersion and show that the proportionality coefficient between the two is a function of the P-to-S-wave velocity ratio (or Poisson's ratio). This coefficient falls between 0.5 and 3 for Poisson's ratio contained in the interval 0.25 to 0.35, typical for saturated Earth materials.

The three proposed versions for Q_S^{-1} calculation from Q_P^{-1} (as calculated using the preceding theory) are:

$$\frac{Q_P^{-1}}{Q_S^{-1}} = \frac{1}{4}\frac{(M/\mu - 2)^2(3M/\mu - 2)}{(M/\mu - 1)(M/\mu)} \tag{6.17.17}$$

$$\frac{Q_P^{-1}}{Q_S^{-1}} = \frac{5}{4}\frac{(M/\mu - 2)^2}{(M/\mu - 1)} \bigg/ \left[\frac{2M/\mu}{(3M/\mu - 2)} + \frac{M/\mu}{3(M/\mu - 1)}\right] \tag{6.17.18}$$

$$\frac{Q_P^{-1}}{Q_S^{-1}} = \frac{1}{M/\mu}\left[\frac{4}{3} + \frac{5}{4}\frac{\left(M/\mu - \frac{2}{3}\right)\left(M/\mu - \frac{4}{3}\right)^2}{M/\mu - \frac{8}{9}}\right] \tag{6.17.19}$$

where M is the compressional modulus at full water saturation. All three equations provide results of the same order of magnitude. The result according to equation 6.17.19 is displayed in Figure 6.17.1.

Assumptions and Limitations

This model estimates fluid-related dispersion and attenuation. The magnitude of attenuation is estimated as being proportional to the difference between the relaxed and unrelaxed moduli, using the dispersion–attenuation relation predicted by the standard linear solid (SLS) model. Choice of the SLS model ignores uncertainty associated with unknown distributions of fluid relaxation times. For partial saturation, the unrelaxed modulus is estimated using the patchy saturation model, while the relaxed modulus is estimated using the fine-scale saturation model. For full water saturation, the unrelaxed modulus is estimated as the spatial average of point-by-point Gassmann predictions of saturated moduli, while the relaxed modulus is estimated from the Gassmann prediction of saturated moduli applied to the spatial average of dry-rock moduli.

6.18   Partial and Multiphase Saturations

Synopsis

One of the most fundamental observations of rock physics is that seismic velocities are sensitive to pore fluids. The first-order low-frequency effects for *single fluid phases* are often described quite well with Gassmann's (1951) relations, which are discussed in Section 6.3. We summarize here variations on those fluid effects that result from *partial or mixed saturations*.

Caveat on Very Dry Rocks

Figure 6.18.1 illustrates some key features of the saturation problem. The data are for limestones measured by Cadoret (1993) using the resonant bar technique at 1 kHz. The very-dry-rock velocity is approximately 2.84 km/s. Upon initial introduction of moisture (water), the velocity drops by about 4%. This apparent softening of the rock occurs at tiny volumes of pore fluid equivalent to a few monolayers of liquid if distributed uniformly over the internal surfaces of the pore space. These amounts are hardly sufficient for a fluid dynamic description, as in the Biot–Gassmann theories. Similar behavior has been reported in sandstones by Murphy (1982), Knight and Dvorkin (1992), and Tutuncu (1992).

This velocity drop has been attributed to softening of cements (sometimes called "chemical weakening"), to clay swelling, and to surface effects. In the latter model, very dry surfaces attract each other via cohesive forces, giving a mechanical effect resembling an increase in effective stress. Water or other pore fluids disrupt these forces. A fairly thorough treatment of the subject is found in the articles of Sharma and Tutuncu (Tutuncu, 1992; Tutuncu and Sharma, 1992; Sharma and Tutuncu, 1994; Sharma et al., 1994).

After the first few percent of water saturation, additional fluid effects are primarily elastic and fluid dynamic and are amenable to analysis, for example, by the Biot–Gassmann (Sections 6.1–6.3) and squirt (Sections 6.11–6.15) models.

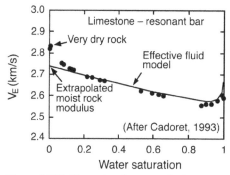

Figure 6.18.1 Extensional-wave velocity versus saturation

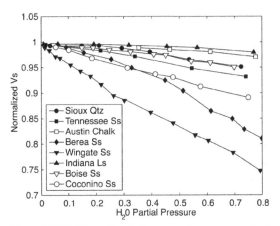

Figure 6.18.2 Normalized ultrasonic shear-wave velocity versus water partial pressure (Clark *et al.,* 1980)

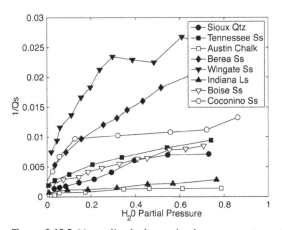

Figure 6.18.3 Normalized ultrasonic shear-wave attenuation versus water partial pressure (Clark *et al.*, 1980)

Figures 6.18.2 and 6.18.3 (Clark *et al.*, 1980; Winkler and Murphy, 1995) illustrate further the sensitivity of mechanical properties to very small amounts of water. Figure 6.18.2 shows normalized ultrasonic shear-wave velocity in sandstones and carbonates, plotted versus water vapor partial pressure. In all cases, increased exposure to water molecules systematically decreases the shear stiffness of the rocks, even though the amount of water is equivalent to only a few monolayers of molecules coating the pore surfaces. Figure 6.18.3 shows shear-wave attenuation versus water vapor partial pressure. Attenuation is very small in dry rocks but increases rapidly upon expose to moisture.

Several authors (Murphy *et al.,* 1991; Cadoret, 1993) have pointed out that classical fluid-mechanical models such as the Biot–Gassmann theories perform poorly when the measured very-dry-rock values are used for the "dry rock" or "dry frame." The models can be fairly accurate if the extrapolated "moist" rock modulus (see Figure 6.18.1) is used instead. For this reason, and to avoid the artifacts of ultra-dry rocks, it is often recommended to use samples that are at room conditions or that have been prepared in a constant-humidity environment for measuring "dry-rock" data. For the rest of this section, it is assumed that the ultra-dry artifacts have been avoided.

Caveat on Frequency

It is well known that the Gassmann theory (see Section 6.3) is valid only at sufficiently **low frequencies** that the induced pore pressures are equilibrated throughout the pore space (i.e., that there is sufficient time for the pore fluid to flow and eliminate wave-induced pore-pressure gradients). This limitation to low frequencies explains why Gassmann's relation works best for very-low-frequency, *in-situ* seismic data (< 100 Hz) and may perform less well as frequencies increase toward sonic-logging ($\approx 10^4$Hz) and laboratory ultrasonic measurements ($\approx 10^6$Hz). Knight and Nolen-Hoeksema (1990) studied the effects of pore-scale fluid distributions on ultrasonic measurements.

Effective Fluid Model

The most common approach to modeling partial saturation (air/water or gas/water) or mixed fluid saturations (gas/water/oil) is to replace the collection of phases with a single "effective fluid."

When a rock is stressed by a passing wave, pores are always elastically compressed more than the solid grains. This pore compression tends to induce increments of pore fluid pressure, which resist the compression; hence, pore phases with the largest bulk modulus K_{fl} stiffen the rock most. For single fluid phases, the effect is described quite elegantly by Gassmann's (1951) relation (see Section 6.3):

$$\frac{K_{sat}}{K_0 - K_{sat}} = \frac{K_{dry}}{K_0 - K_{dry}} + \frac{K_{fl}}{\phi(K_0 - K_{fl})}, \quad \mu_{sat} = \mu_{dry} \tag{6.18.1}$$

where K_{dry} is the effective bulk modulus of the dry rock, K_{sat} is the effective bulk modulus of the rock with pore fluid, K_0 is the bulk modulus of the mineral material making up the rock, K_{fl} is the effective bulk modulus of the pore fluid, ϕ is porosity, μ_{dry} is the effective shear modulus of the dry rock, and μ_{sat} is the effective shear modulus of the rock with the pore fluid.

Implicit in Gassmann's relations is the stress-induced pore pressure, given by Skempton's coefficient (Section 2.10)

$$\begin{aligned}
\frac{dP}{d\sigma} &= \frac{1}{1 + K_\phi(1/K_{fl} - 1/K_0)} \\
&= \frac{1}{1 + \phi(1/K_{fl} - 1/K_0)(1/K_{dry} - 1/K_0)^{-1}}
\end{aligned} \tag{6.18.2}$$

where P is the increment of pore pressure and σ is the applied hydrostatic stress. If there are multiple pore-fluid phases with different fluid bulk moduli, then there is a tendency for each to have a different induced pore pressure. However, when the phases are intimately mixed at the finest scales, these pore-pressure increments can equilibrate with each other to a single average value. This is an *isostress* situation, and therefore the effective bulk modulus of the mixture of fluids is described well by the Reuss average (see Section 4.1) as follows:

$$\frac{1}{K_{fl}} = \sum \frac{S_i}{K_i} \tag{6.18.3}$$

where K_{fl} is the effective bulk modulus of the fluid mixture, K_i denotes the bulk moduli of the individual gas and fluid phases, and S_i represents their saturations. The rock moduli can often be predicted quite accurately by inserting this effective-fluid modulus into Gassmann's relation. The effective-fluid (air + water) prediction is superimposed on the data in Figure 6.18.1 and reproduces the overall trend quite well. This approach has been discussed, for example, by Domenico (1976), Murphy (1984), Cadoret (1993), Mavko and Nolen-Hoeksema (1994), and many others.

> **Caution:**
>
> It is thought that the Reuss average effective-fluid model is valid only when all of the fluid phases are mixed at the finest scale.

Critical Relaxation Scale

A critical assumption in the effective-fluid model represented by the Reuss average is that differences in wave-induced pore pressure have time to flow and equilibrate among the various phases. As discussed in Section 8.1, the characteristic relaxation time or diffusion time for heterogeneous pore pressures of scale L is

$$\tau \approx \frac{L^2}{D} \tag{6.18.4}$$

where $D = kK_{fl}/\eta$ is the diffusivity, k is the permeability, K_{fl} is the fluid bulk modulus, and η is the viscosity. Therefore, at a seismic frequency $f = 1/\tau$, pore pressure heterogeneities caused by saturation heterogeneities will have time to relax and reach a local isostress state over scales smaller than

$$L_c \approx \sqrt{\tau D} = \sqrt{D/f} \tag{6.18.5}$$

and will be described *locally* by the effective-fluid model mentioned in the previous discussion. Spatial fluctuations on scales larger than L_c will tend to persist and will not be described well by the effective-fluid model.

Patchy Saturation

Consider the situation of a homogeneous rock type with spatially variable saturation $S_i(x, y, z)$. Each "patch" or pixel at scale $\approx L_c$ will have fluid phases equilibrated within the patch at scales smaller than L_c, but neighboring patches at scales $>L_c$ will not be equilibrated with each other. Each patch will have a different effective fluid described approximately by the Reuss average. Consequently, the rock in each patch will have a different bulk modulus describable locally with Gassmann's relations. Yet, the shear modulus will remain unchanged and spatially uniform.

The effective moduli of the rock with spatially varying bulk modulus but uniform shear modulus is described exactly by the equation of Hill (1963) discussed in Section 4.6:

$$K_{\text{eff}} = \left(\frac{1}{K + \frac{4}{3}\mu} \right)^{-1} - \frac{4}{3}\mu \qquad (6.18.6)$$

This striking result states that the effective moduli of a composite with uniform shear modulus can be found *exactly* by knowing only the volume fractions of the constituents independent of the constituent geometries. There is no dependence, for example, on ellipsoids, spheres, or other idealized shapes.

Figure 6.18.4 shows the P-wave velocity versus water saturation for another limestone (data from Cadoret, 1993). Unlike the effective-fluid behavior, which shows a small decrease in velocity with increasing saturation and then an abrupt increase as S_W approaches unity, the patchy model predicts a monotonic, almost linear increase in velocity from the dry to saturated values. The deviation of the data from the effective-fluid curve at saturations greater than 0.8 is an indication of patchy saturation (Cadoret, 1993).

The velocity-versus-saturation curves shown in Figure 6.18.4 for the effective-fluid model and patchy-saturation model represent approximate lower and upper bounds,

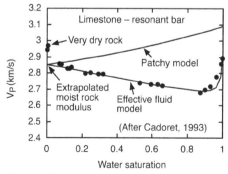

Figure 6.18.4 P-wave velocity versus saturation

respectively, at low frequencies. The lower effective-fluid curve is achieved when the fluid phases are mixed at the finest scales. The upper patchy-saturation curve is achieved when there is the greatest separation of phases: when each patch of size $>L_c$ has only a single phase. Any velocity between these "bounds" can occur when there is a range of saturation scales.

Saturation-Related Velocity Dispersion

The difference between effective-fluid behavior and patchy-saturation behavior is largely a matter of the scales at which the various phases are mixed. The critical-relaxation scale L_c, which separates the two domains, is related to the seismic frequency. Hence, spatially varying saturations can be a source of velocity dispersion. Attempts to quantify this velocity dispersion have been made by White (1975) and by Dutta and Odé (1979) (see Sections 6.19).

Voigt Average Approximation

It can be shown that an approximation to the patchy-saturation "upper bound" can be found by first computing the Voigt average (see Section 4.1) of the fluid modulus:

$$K_{fl} = \sum S_j K_j \tag{6.18.7}$$

where K_{fl} is the effective bulk modulus of the fluid mixture, K_j denotes the bulk moduli of the individual gas and fluid phases, and S_j represents their saturations. Next, this average fluid is put into Gassmann's equations to predict the overall rock moduli. This is in contrast to the effective-fluid model discussed earlier, in which the Reuss average of the fluid moduli was put into the Gassmann equations.

Brie's Fluid Mixing Equation

Brie *et al.* (1995) suggest an empirical fluid mixing law, given by

$$K_{Brie} = (K_{liquid} - K_{gas})(1 - S_{gas})^e + K_{gas} \tag{6.18.8}$$

where K_{gas} is the gas bulk modulus, $K_{liquid} = (S_{water} + S_{oil})/(S_{water}/K_{water} + S_{oil}/K_{oil})$ is the liquid bulk modulus given by the Reuss average of the water and oil moduli, K_{water} and K_{oil}, and e is an empirical constant, typically equal to about 3.

Figure 6.18.5 compares the effective fluid moduli predicted by the Reuss, Voigt, and Brie averages. While the Voigt and Reuss are bounds for the effective fluid, data rarely fall near the Voigt average. A more useful range is to assume that the effective fluid modulus will fall roughly between the Reuss and Brie averages.

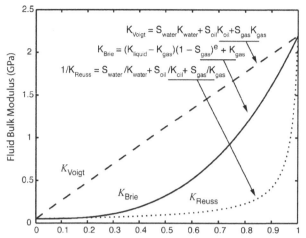

Figure 6.18.5 Comparison of effective fluid moduli, governed by the following equations: Brie's patchy mixer (solid line), $K_{Brie} = (K_{liquid} - K_{gas})(1 - K_{gas})^e + K_{gas}$; patchy mix Voigt average (dashed line), $K_{Voigt} = S_{water}K_{water} + S_{oil}K_{oil} + S_{gas}K_{gas}$; and fine-scale mix Reuss average (dotted line), $1/K_{Reuss} = S_{water}/K_{water} + S_{oil}/K_{oil} + S_{gas}/K_{gas}$.

6.19 Partial Saturation: White and Dutta–Odé Model for Velocity Dispersion and Attenuation

Synopsis

Consider the situation of a reservoir rock with spatially variable saturation $S_i(x, y, z)$. Each "patch" at scale $\approx L_c$ (where $L_c \approx \sqrt{\tau D} = \sqrt{D/f}$ is the critical fluid-diffusion relaxation scale, $D = kK_{fl}/\eta$ is the diffusivity, k is the permeability, K_{fl} is the fluid bulk modulus, and η is the viscosity) will have fluid phases equilibrated within the patch at scales smaller than L_c, but neighboring patches at scales $>L_c$ will not be equilibrated with each other. Fluid flow resulting from unequilibrated pore pressures between patches of different fluids will cause attenuation and dispersion of seismic waves traveling through the rock. White (1975) modeled the seismic effects of patchy saturation by considering porous rocks saturated with brine but containing spherical gas-filled regions. Dutta and Odé (1979) gave a more rigorous solution for White's model by using Biot's equations for poroelasticity. The patches (spheres) of heterogeneous saturation are much larger than the grain scale but are much smaller than the wavelength. The idealized geometry of a unit cell consists of a gas-filled sphere of radius a placed at the center of a brine-saturated spherical shell of outer radius b ($b > a$). Adjacent unit cells do not interact with each other. The gas saturation is $S_g = a^3/b^3$. The inner region will be denoted by subscript 1 and the outer shell by subscript 2. In the more rigorous Dutta and Odé formulation, the dry-frame properties (denoted by subscript "dry" in the following equations) in the two regions may be different. However, in White's approximate formulation, the dry-frame properties are assumed to be the same in regions 1 and 2. White's equations as given here (incorporating a correction pointed out by Dutta and Seriff, 1979) yield results that agree very well with the Dutta–Odé results. The complex bulk modulus, K^*, for the partially saturated porous rock as a function of angular frequency ω is given by

$$K^* = \frac{K_\infty}{1 - K_\infty W} = K_r^* + iK_i^* \tag{6.19.1}$$

where

$$W = \frac{3a^2(R_1 - R_2)(-Q_1 + Q_2)}{b^3 i\omega(Z_1 + Z_2)} \tag{6.19.2}$$

$$R_1 = \frac{K_1 - K_{dry_1}}{1 - K_{dry_1}/K_{0_1}} \frac{3K_2 + 4\mu_2}{K_2(3K_1 + 4\mu_2) + 4\mu_2(K_1 - K_2)S_g} \tag{6.19.3}$$

$$R_2 = \frac{K_2 - K_{dry_2}}{1 - K_{dry_2}/K_{0_2}} \frac{3K_1 + 4\mu_1}{K_2(3K_1 + 4\mu_2) + 4\mu_2(K_1 - K_2)S_g} \tag{6.19.4}$$

$$Z_1 = \frac{\eta_1 a}{\kappa_1} \left[\frac{1 - e^{-2\alpha_1 a}}{(\alpha_1 a - 1) + (\alpha_1 a + 1)e^{-2\alpha_1 a}} \right] \tag{6.19.5}$$

$$Z_2 = -\frac{\eta_2 a}{\kappa_2} \left[\frac{(\alpha_2 b + 1) + (\alpha_2 b - 1)e^{2\alpha_2(b-a)}}{(\alpha_2 b + 1)(\alpha_2 a - 1) - (\alpha_2 b - 1)(\alpha_2 a + 1)e^{2\alpha_2(b-a)}} \right] \tag{6.19.6}$$

$$\alpha_j = (i\omega\eta_j/\kappa_j K_{Ej})^{1/2} \tag{6.19.7}$$

$$K_{Ej} = \left[1 - \frac{K_{fj}(1 - K_j/K_{0j})(1 - K_{dryj}/K_{0j})}{\phi K_j(1 - K_{fj}/K_{0j})} \right] K_{Aj} \tag{6.19.8}$$

$$K_{Aj} = \left(\frac{\phi}{K_{fj}} + \frac{1 - \phi}{K_{0j}} - \frac{K_{dryj}}{K_{0j}^2} \right)^{-1} \tag{6.19.9}$$

$$Q_j = \frac{(1 - K_{dryj}/K_{0j})K_{Aj}}{K_j} \tag{6.19.10}$$

Here, $j = 1$ or 2 denotes quantities corresponding to the two different regions; η and K_f are the fluid viscosity and bulk modulus, respectively; ϕ is the porosity; κ is the permeability; and K_0 is the bulk modulus of the solid mineral grains. The saturated bulk and shear moduli, K_j and μ_j, respectively, of region j are obtained from Gassmann's equation using K_{dryj}, μ_{dryj}, and K_{fj}. When the dry-frame moduli are the same in both regions, $\mu_1 = \mu_2 = \mu_{dry}$, because Gassmann predicts no change in the shear modulus upon fluid substitution. At the high-frequency limit, when there is no fluid flow across the fluid interface between regions 1 and 2, the bulk modulus is given by

$$K_\infty(\text{no-flow}) = \frac{K_2(3K_1 + 4\mu_2) + 4\mu_2(K_1 - K_2)S_g}{(3K_1 + 4\mu_2) - 3(K_1 - K_2)S_g} \tag{6.19.11}$$

This assumes that the dry-frame properties are the same in the two regions. The low-frequency limiting bulk modulus is given by Gassmann's relation with an effective fluid modulus equal to the Reuss average of the fluid moduli. In this limit, the fluid pressure is constant and uniform throughout the medium:

$$K(\text{low-frequency}) = \frac{K_2(K_1 - K_{\text{dry}}) + S_g K_{\text{dry}}(K_2 - K_1)}{(K_1 - K_{\text{dry}}) + S_g(K_2 - K_1)} \tag{6.19.12}$$

Note that in White (1975) the expressions for K_E and Q had the P-wave modulus $M_j = K_j + 4\mu_j/3$ in the denominator instead of K_j. As pointed out by Dutta and Seriff (1979), this form does not give the right low-frequency limit. An estimate of the transition frequency separating the relaxed and unrelaxed (no-flow) states is given by

$$f_c \approx \frac{\kappa_2 K_{E2}}{\pi \eta_2 b^2} \tag{6.19.13}$$

which has the length-squared dependence characteristic of diffusive phenomena. When the central sphere is saturated with a very compressible gas (caution: this may not hold at reservoir pressures) R_1, Q_1, $Z_1 \approx 0$. The expression for the effective complex bulk modulus then reduces to

$$K^* = \frac{K_\infty}{1 + 3a^2 R_2 Q_2 K_\infty / (b^3 i\omega Z_2)} \tag{6.19.14}$$

Dutta and Odé obtained more rigorous solutions for the same spherical geometry by solving a boundary-value problem involving Biot's poroelastic field equations. They considered steady-state, time-harmonic solutions for u and w, the displacement of the porous frame and the displacement of the pore fluid relative to the frame, respectively. The solutions in the two regions are given in terms of spherical Bessel and Neumann functions, j_1 and n_1, of order 1. The general solution for w in region 1 for purely radial motion is

$$w(r) = w_c(r) + w_d(r) \tag{6.19.15}$$

$$w_c(r) = C_1 j_1(k_c r) + C_2 n_1(k_c r) \tag{6.19.16}$$

$$w_d(r) = C_3 j_1(k_d r) + C_4 n_1(k_d r) \tag{6.19.17}$$

where C_1, C_2, C_3, and C_4 are integration constants to be determined from the boundary conditions at $r = 0$, $r = a$, and $r = b$. A similar solution holds for region 2, but with integration constants C_5, C_6, C_7, and C_8. The wavenumbers k_c and k_d in each region are given in terms of the moduli and density corresponding to that region:

$$\frac{k_c^2}{\omega^2} = \frac{\rho_f(M\sigma_c + 2\gamma D) - \rho(2\gamma D\sigma_c + 2D)}{(4\gamma^2 D^2 - 2DM)} \tag{6.19.18}$$

$$\frac{k_d^2}{\omega^2} = \frac{\rho_f(M\sigma_d + 2\gamma D) - \rho(2\gamma D\sigma_d + 2D)}{(4\gamma^2 D^2 - 2DM)} \tag{6.19.19}$$

where

$$\rho = (1 - \phi)\rho_0 + \phi\rho_f \tag{6.19.20}$$

$$\gamma = 1 - \frac{K_{\text{dry}}}{K_0} \tag{6.19.21}$$

$$M = K + 4/3\mu \tag{6.19.22}$$

$$D = \frac{K_0}{2}\left[\gamma + \frac{\phi}{K_f}(K_0 - K_f)\right]^{-1} \tag{6.19.23}$$

and σ_c and σ_d are the two complex roots of the quadratic equation

$$(\rho_f M - 2\rho\gamma D)\sigma^2 + \left(Mm - 2\rho D - \frac{i\eta M}{\kappa\omega}\right)\sigma + \left(2m\gamma D - 2\rho_f D - 2i\frac{\eta\gamma D}{\kappa\omega}\right) = 0 \tag{6.19.24}$$

where $m = s\rho_f/\phi$. The tortuosity parameter s (sometimes called the structure factor) is a purely geometrical factor independent of the solid or fluid densities and is never less than 1 (see Section 6.1 on Biot theory). For idealized geometries and uniform flow, s usually lies between 1 and 5. Berryman (1981) obtained the relation

$$s = 1 - \xi\left(1 - \frac{1}{\phi}\right) \tag{6.19.25}$$

where $\xi = \frac{1}{2}$ for spheres and lies between 0 and 1 for other ellipsoids. For uniform cylindrical pores with axes parallel to the pore pressure gradient, s equals 1 (the minimum possible value), whereas for a random system of pores with all possible orientations, $s = 3$ (Stoll, 1977). The solutions for u are given as

$$u(r) = u_c(r) + u_d(r) \tag{6.19.26}$$

$$u_c(r) = \sigma_d w_c(r), \quad u_d(r) = \sigma_c w_d(r) \tag{6.19.27}$$

The integration constants are obtained from the following boundary conditions:

(1) $r \to 0$, $w_1(r) \to 0$ (6.19.28)

(2) $r \to 0$, $u_1(r) \to 0$ (6.19.29)

(3) $r = a$, $u_1(r) = u_2(r)$ (6.19.30)

(4) $r = a$, $w_1(r) = w_2(r)$ (6.19.31)

(5) $r = a$, $\tau_1(r) = \tau_2(r)$ (6.19.32)

(6) $r = a$, $p_1(r) = p_2(r)$ (6.19.33)

(7) $r = b$, $\tau_2(r) = -\tau_0$ (6.19.34)

(8) $r = b$, $w_2(r) = 0$ (6.19.35)

The first two conditions imply no displacements at the center of the sphere because of purely radial flow and finite fluid pressure at the origin. These require $C_2 = C_4 = 0$. Conditions (3)–(6) come from continuity of displacements and stresses at the interface. The bulk radial stress is denoted by τ, and p denotes the fluid pressure. These parameters are obtained from u and w by

$$\tau = M\frac{\partial u}{\partial r} + 2(M - 2\mu)\frac{u}{r} + 2\gamma D\left(\frac{\partial w}{\partial r} + \frac{2}{r}w\right) \tag{6.19.36}$$

$$p = -2\gamma D\left(\frac{\partial u}{\partial r} + \frac{2}{r}u\right) - 2D\left(\frac{\partial w}{\partial r} + \frac{2}{r}w\right) \tag{6.19.37}$$

Condition (7) gives the amplitude τ_0 of the applied stress at the outer boundary, and condition (8) implies that the outer boundary is jacketed. These boundary conditions are not unique and could be replaced by others that may be appropriate for the situation under consideration. The jacketed outer boundary is consistent with non-interacting unit cells. The remaining six integration constants are obtained by solving the linear system of equations given by the boundary conditions. Solving the linear system requires considerable care (Dutta and Odé, 1979). The equations may become ill-conditioned because of the wide range of the arguments of the spherical Bessel and Neumann functions. Once the complete solution for $u(r)$ is obtained, the effective complex bulk modulus for the partially saturated medium is given by

$$K^*(\omega) = -\frac{\tau_0}{\Delta V/V} = -\frac{\tau_0 b}{3u(b)} \tag{6.19.38}$$

The P-wave velocity V_p^* and attenuation coefficient α_p^* are given in terms of the complex P-wave modulus M^* and the effective density p^*

$$\rho^* = S_g[(1 - \phi)\rho_0 + \phi\rho_{f1}] + (1 - S_g)[(1 - \phi)\rho_0 + \phi\rho_{f2}] \tag{6.19.39}$$

$$M^* = (M_r^* + iM_i^*) = (K_r^* + 4\mu^*/3 + iK_i) \tag{6.19.40}$$

$$\mu^* = \mu_1 = \mu_2 = \mu_{dry} \tag{6.19.41}$$

$$V_p^* = (|M^*|/\rho^*)^{1/2}/\cos(\theta_p^*/2) \tag{6.19.42}$$

$$\alpha_p^* = \omega\tan(\theta_p^*/2)/V_p^* \tag{6.19.43}$$

$$\theta_p^* = \tan^{-1}(M_i^*/M_r^*) \tag{6.19.44}$$

Uses

The White and Dutta–Odé models can be used to calculate velocity dispersion and attenuation in porous media with patchy partial saturation.

Assumptions and Limitations

The White and Dutta–Odé models are based on the following assumptions:
- the rock is isotropic;
- all minerals making up the rock are isotropic and linearly elastic;
- the patchy saturation has an idealized geometry consisting of a sphere saturated with one fluid within a spherical shell saturated with another fluid;
- the porosity in each saturated region is uniform, and the pore structure is smaller than the size of the spheres; and
- the patches (spheres) of heterogeneous saturation are much larger than the grain scale but are much smaller than the wavelength.

6.20 Velocity Dispersion, Attenuation, and Dynamic Permeability in Heterogeneous Poroelastic Media

Synopsis

Wave-induced fluid flow in heterogeneous porous media can cause velocity dispersion and attenuation. The scale of the heterogeneities is considered to be much larger than the pore scale but much smaller than the wavelength of the seismic wave. An example would be heterogeneities caused by patchy saturation discussed in Sections 6.18 and 6.19. Müller and Gurevich (2005a, 2005b, 2006) applied the method of statistical smoothing for Biot's poroelasticity equations in a random, heterogeneous, porous medium and derived the effective wavenumbers for the fast and slow P-waves. The parameters of the poroelastic random medium consist of a smooth background component and a zero-mean fluctuating component. The statistical properties of the relative fluctuations of any two parameters X and Y are described by variances and covariances denoted by σ_{XX}^2 and σ_{XY}^2, respectively. The spatial distribution of the random heterogeneities is described by a normalized spatial correlation function $B(\mathbf{r})$. In the following equations, all poroelastic parameters are assumed to have the same normalized spatial correlation function and correlation length, though the general results of Müller and Gurevich (2005a) allow for different correlation lengths associated with each random property.

The effective P-wavenumber $\bar{k}_P$ in three-dimensional random poroelastic media with small fluctuations can be written as (Müller and Gurevich, 2005b)

$$\bar{k}_P = k_P\left(1 + \Delta_2 + \Delta_1 k_{Ps}^2 \int_0^\infty rB(r)\exp(irk_{Ps})\,dr\right) \tag{6.20.1}$$

where k_P is the wavenumber in the homogeneous background and the dimensionless coefficients Δ_1 and Δ_2 are given by

$$\Delta_1 = \frac{\alpha^2 M}{2P_d}\left(\sigma_{HH}^2 - 2\sigma_{HC}^2 + \sigma_{CC}^2 + \frac{32}{15}\frac{\mu^2}{H^2}\sigma_{\mu\mu}^2 - \frac{8}{3}\frac{\mu}{H}\sigma_{H\mu}^2 + \frac{8}{3}\frac{\mu}{H}\sigma_{\mu C}^2\right) \tag{6.20.2}$$

$$\Delta_2 = \Delta_1 + \frac{1}{2}\sigma_{HH}^2 - \frac{4}{3}\frac{\mu}{H}\sigma_{H\mu}^2 + \left(\frac{4\mu}{H} + 1\right)\frac{4}{15}\frac{\mu}{H}\sigma_{\mu\mu}^2 \tag{6.20.3}$$

and the other parameters are: $k_{Ps} = \sqrt{i\omega\eta/\kappa_0 N}$ is the wavenumber of the Biot slow wave in the homogeneous background, ω is the angular frequency, η is the fluid viscosity, κ_0 is the background permeability, μ is the background shear modulus, P_d is the dry (drained) P-wave modulus of the background, H is the saturated P-wave modulus of the background, related to P_d by Gassmann's equation as $H = P_d + \alpha^2 M$, $M = [(\alpha - \phi)/K_0 + \phi/K_f]^{-1}$, ϕ is the background porosity, $\alpha = 1 - K_d/K_0$ is the Biot–Willis coefficient, K_d is the dry bulk modulus, K_0 is the mineral (solid-phase) bulk modulus, K_f is the fluid bulk modulus, $N = MP_d/H$, and $C = \alpha M$

Expressions for velocity dispersion $V(\omega)$ and attenuation or inverse quality factor Q^{-1} are obtained from the real and imaginary parts of the effective wavenumber:

$$V(\omega) = \frac{\omega}{\text{Re}\{\bar{k}_P\}} = V_0\left[1 - \Delta_2 + 2\Delta_1 k_{Psr}^2 \int_0^\infty rB(r)\exp(-rk_{Psr})\sin(rk_{Psr})dr\right] \tag{6.20.4}$$

where $V_0 = \sqrt{H/\rho}=$ constant background P-wave velocity in the saturated porous medium, ρ is the saturated bulk density, and $k_{Psr} = \sqrt{\eta\omega/(2\kappa_0 N)}$ is the real part of the slow P-wave number k_{Ps}

For low-loss media, Q^{-1} can be written as

$$Q^{-1}(\omega) = \frac{2\text{Im}\{\bar{k}_P\}}{\text{Re}\{\bar{k}_P\}} = 4\Delta_1 k_{Psr}^2 \int_0^\infty rB(r)\exp(-rk_{Psr})\cos(rk_{Psr})dr \tag{6.20.5}$$

The term Δ_1 is a measure of the magnitude of attenuation and frequency-dependent velocity dispersion, while Δ_2 produces a frequency-independent velocity shift.

The validity of the approximations for the effective P-wavenumber are expressed mathematically as $\max\{\Delta_1(|k_{Ps}|a)^2, \Delta_2\} \ll 1$.

$$a^2 \gg \frac{\kappa_0 N}{\eta\omega_B} \tag{6.20.6}$$

where a is the correlation length and $\omega_B = \phi\eta/(\kappa_0\rho_f)$ is the Biot characteristic frequency with ρ_f being the fluid density. These conditions arise from the weak-contrast assumption and the low-frequency approximation necessary for the derivation (Müller and Gurevich, 2005b).

The spatial correlation function $B(r)$ and the power spectrum of the fluctuations $\Phi(k)$ are related by a Fourier transform. If the medium is statistically isotropic the relation can be expressed as a three-dimensional Hankel transform:

$$B(r) = \frac{4\pi}{r}\int_0^\infty k\Phi(k)\sin(kr)\,dk \tag{6.20.7}$$

where k is the spatial wavenumber of the fluctuations. In terms of the power spectrum, the expressions for velocity and inverse quality factor are (Müller and Gurevich, 2005b):

$$V(\omega) = V_0 \left[1 - \Delta_2 + 16\pi\Delta_1 \int_0^\infty \frac{k_{\text{Psr}}^4 k^2}{4k_{\text{Psr}}^4 + k^4} \Phi(k)dk \right] \tag{6.20.8}$$

$$Q^{-1}(\omega) = 16\pi\Delta_1 \int_0^\infty \frac{k_{\text{Psr}}^2 k^4}{4k_{\text{Psr}}^4 + k^4} \Phi(k)\, dk \tag{6.20.9}$$

The factor

$$\Theta(k,\ k_{\text{Psr}}) = \frac{k_{\text{Psr}}^2 k^4}{4k_{\text{Psr}}^4 + k^4} \tag{6.20.10}$$

acts as a filter and controls which part of the fluctuation spectrum gives contributions to the velocity dispersion and attenuation.

Results for Specific Correlation Functions

Müller and Gurevich (2005b) give explicit analytic results for specific correlation functions. For an exponential correlation function $B(r) = \exp[-|r|/a]$ and

$$V(\omega) = V_0 \left[1 - \Delta_2 + \Delta_1 \frac{4(ak_{\text{Psr}})^3(1 + k_{\text{Psr}}a)}{(1 + 2k_{\text{Psr}}a + 2k_{\text{Psr}}^2 a^2)^2} \right] \tag{6.20.11}$$

$$Q^{-1}(\omega) = \Delta_1 \frac{4(ak_{\text{Psr}})^2(1 + 2k_{\text{Psr}}a)}{(1 + 2k_{\text{Psr}}a + 2k_{\text{Psr}}^2 a^2)^2} \tag{6.20.12}$$

For a Gaussian correlation function $B(r) = \exp(-r^2/a^2)$, and Muller and Gurevich obtain

$$V(\omega) = V_0 \left\{ 1 - \Delta_2 + \Delta_1 (ak_{\text{Psr}})^2 \frac{\sqrt{\pi}}{4} \sum_{z=z-}^{z=z+} ak_{\text{Psr}}z^* \exp[(ak_{\text{Psr}}z)^2/4]\, \text{erfc}[ak_{\text{Psr}}z/2] \right\} \tag{6.20.13}$$

$$Q^{-1}(\omega) = 2\Delta_1 (ak_{\text{Psr}})^2 \left\{ 1 - \frac{\sqrt{\pi}}{4} \sum_{z=z-}^{z=z+} ak_{\text{Psr}}z \exp[(ak_{\text{Psr}}z)^2/4]\, \text{erfc}[ak_{\text{Psr}}z/2] \right\} \tag{6.20.14}$$

where $z_+ = 1 + i$, $z_- = 1 - i$, and z^* denotes the complex conjugate.

The von Karman correlation function with two parameters, the correlation length a, and the Hurst coefficient v ($0 < v \leq 1$), is expressed as

$$B(r) = 2^{1-v}\Gamma^{-1}(v)\left(\frac{r}{a}\right)^v K_v(r/a) \tag{6.20.15}$$

where Γ is the Gamma function and K_ν is the modified Bessel function of the third kind. When $\nu = \frac{1}{2}$, the von Karman function becomes identical to the exponential correlation function. The equivalent spectral representation for general ν is

$$\Phi(k) = \frac{a^3 \Gamma\left(\nu + \frac{3}{2}\right)}{\pi^{3/2}\Gamma(\nu)(1 + k^2 a^2)^{\nu + 3/2}} \qquad (6.20.16)$$

Using the spectral representation, Müller and Gurevich (2005b) derive expressions for fast P-wave velocity dispersion and inverse quality factor as described here:

$$V(\omega) = V_0\left[1 - \Delta_2 - \frac{1}{2}c_1\Delta_1(\Psi_1 + \Psi_2 + \Psi_3)\right] \qquad (6.20.17)$$

$$Q^{-1}(\omega) = c_1\Delta_1[\Omega_1 + \Omega_2 + \Omega_3] \qquad (6.20.18)$$

with

$$\Psi_1 = -4c_2(1 + \nu)(ak_{Psr})^2 {}_3F_2\left(1, 1 + \frac{\nu}{2}, \frac{3}{2} + \frac{\nu}{2}; \frac{3}{4}, \frac{5}{4}; -4(ak_{Psr})^4\right) \qquad (6.20.19)$$

$$\Psi_2 = \frac{1}{2}\Gamma\left(\nu + \frac{1}{2}\right)(2\nu + 3)c_3 B^{-3/4 - \nu/2}\cos\left[\left(\frac{3}{4} + \frac{\nu}{2}\right)A\right] \qquad (6.20.20)$$

$$\Psi_3 = \Gamma\left(\nu + \frac{3}{2}\right)\left(\frac{3}{2} + \nu\right)B^{-5/4 - \nu/2}[2(ak_{Psr})^2\cos(c_3 A) + \sin(c_3 A)] \qquad (6.20.21)$$

$$\Omega_1 = c_2\,{}_3F_2\left(1, \frac{1}{2} + \frac{\nu}{2}, 1 + \frac{\nu}{2}; \frac{1}{4}, \frac{3}{4}; -4(ak_{Psr})^4\right) \qquad (6.20.22)$$

$$\Omega_2 = -\frac{1}{2}\Gamma\left(\nu + \frac{3}{2}\right)(2\nu + 3)B^{-3/4 - \nu/2}\cos\left[\left(\frac{3}{4} + \frac{\nu}{2}\right)A\right] \qquad (6.20.23)$$

$$\Omega_3 = \Gamma\left(\nu + \frac{5}{2}\right)B^{-5/4 - \nu/2}[2(ak_{Psr})^2\cos(c_3 A) + \sin(c_3 A)] \qquad (6.20.24)$$

$$c_1 = \frac{16\sqrt{\pi}(ak_{Psr})^3}{\Gamma(\nu)(2\nu + 3)} \qquad (6.20.25)$$

$$c_2 = \frac{\Gamma(\nu + 1)(2\nu + 3)}{2\sqrt{\pi}ak_{Psr}} \qquad (6.20.26)$$

$$c_3 = \frac{1}{4} + \frac{\nu}{2} \qquad (6.20.27)$$

$$A = 2\arctan[2(ak_{Psr})^2] \qquad (6.20.28)$$

$$B = 1 + 4(ak_{Psr})^4 \qquad (6.20.29)$$

$${}_pF_q(\xi_1, \ldots, \xi_p; \varsigma_1, \ldots, \varsigma_q; x) = \text{generalized hypergeometric function} \qquad (6.20.30)$$

Low- and High-Frequency Limits

The low-frequency limit when there is enough time for the wave-induced pore pressures to equilibrate is referred to as the "relaxed" or "quasi-static" limit with P-wave phase velocity V_{qs}. The other limiting situation is when frequencies are high enough that there is no time for flow-induced equilibration of pore pressures. This limit is the "unrelaxed" or "no-flow" limit, with phase velocity $V_{nf} > V_{qs}$. Though the no-flow limit might be called the "high-frequency" limit, the wavelengths are still much longer than the scale of the heterogeneities. The limiting velocities for the P-wave are given by (Müller and Gurevich, 2005b)

$$V_{qs} = V_0(1 - \Delta_2) \tag{6.20.31}$$

$$V_{nf} = V_0(1 + \Delta_1 - \Delta_2) \tag{6.20.32}$$

and the relative magnitude of the velocity dispersion effect is

$$\frac{V_{nf} - V_{qs}}{V_0} = \Delta_1 \tag{6.20.33}$$

The limiting velocities do not depend on the spatial correlation functions and are independent of geometry of the heterogeneities. For a porous medium, such that all poroelastic moduli are homogeneous and the only fluctuations are in the bulk modulus of the fluid, the quasi-static limit corresponds to a Reuss average (Wood's average) of the pore-fluid bulk moduli, followed by Gassmann's fluid substitution (see Section 6.3); the no-flow limit corresponds to a Hill's average (Section 4.6) of the saturated P-wave moduli.

In the low-frequency limit, $Q^{-1} \propto \omega$, while for correlation functions that can be expanded in a power series as

$$B(r/a) = 1 - (r/a) + O(r^2/a^2) + \ldots \tag{6.20.34}$$

the high-frequency asymptotic behavior of Q^{-1} is given by $Q^{-1} \propto 1/\sqrt{\omega}$. For a Gaussian correlation function, there is a much faster decrease of attenuation with frequency given by $Q^{-1} \propto 1/\omega$. For all correlation functions, maximum attenuation occurs at the resonance condition when the diffusion wavelength equals the correlation length or $k_{Ps}a \approx 1$. The corresponding frequency is given as $\omega_{max} \approx 2\kappa_0 N/a^2\eta$.

Permeability Fluctuations

The equations in the previous sections for the effective P-wavenumber k_P in three-dimensional random poroelastic media were derived for a medium with a homogeneous permeability. Muller and Gurevich (2006) extended the theory to take into account spatial fluctuations of permeability, and derive expressions for the effective slow P-wavenumber $\bar{k}_{Ps}$. They also identify a dynamic permeability for the heterogeneous poroelastic medium. The dynamic permeability of Müller and Gurevich is different from the dynamic permeability of Johnson et al. (1987), which is frequency-dependent permeability arising from inertial effects at frequencies of the order of Biot's characteristic frequency. The dynamic permeability of Müller and Gurevich is related to wave-induced flow at frequencies much lower than the Biot characteristic frequency.

The relative fluctuation in the property X is denoted by ε_X with zero mean and variance σ_{XX}^2. The fluctuation of permeability, κ, is parameterized in terms of the variance σ_{pp}^2 of the reciprocal permeability $p = 1/\kappa$. The slow P-wave effective wavenumber is given as (Müller et al., 2007):

$$\bar{k}_{Ps}^2 = k_{Ps}^2[1 + \Delta_S \zeta(\omega)] \tag{6.20.35}$$

$$\zeta(\omega) = 1 + k_{Ps}^2 \int_0^\infty rB(r)\exp(ik_{Ps}r)\, dr \tag{6.20.36}$$

$$\Delta_S = \left\langle \left(\frac{\alpha^2 M}{P_d} \varepsilon_\alpha - \varepsilon_{K_f} + \varepsilon_\phi \right)^2 \right\rangle + \frac{\sigma_{pp}^2}{3} \tag{6.20.37}$$

where $\langle \cdot \rangle$ denotes ensemble averages. For a one-dimensional random medium

$$\zeta^{(1D)}(\omega) = 1 + ik_{Ps} \int_0^\infty B(r)\exp(ik_{Ps}r)\, dr \tag{6.20.38}$$

$$\Delta_S^{(1D)} = \left\langle \left(\frac{\alpha^2 M}{P_d} \varepsilon_\alpha - \varepsilon_{K_f} + \varepsilon_\phi \right)^2 \right\rangle + \sigma_{pp}^2 \tag{6.20.39}$$

An effective dynamic permeability κ^* can be derived from the relation $\bar{k}_{Ps} = \sqrt{i\omega\eta/(\kappa^* N)}$ as (Müller et al., 2007):

$$\frac{\kappa^{*(3D)}}{\kappa_0} = 1 - \frac{\sigma_{pp}^2}{3} + \frac{2}{3}\sigma_{pp}^2 k_{Psr}^2 \int_0^\infty rB(r)\exp(-rk_{Psr})\sin(rk_{Psr})\, dr \tag{6.20.40}$$

$$\frac{\kappa^{*(1D)}}{\kappa_0} = 1 - \sigma_{pp}^2 + \sqrt{2}\sigma_{pp}^2 k_{Psr} \int_0^\infty B(r)\exp(-rk_{Psr})\sin(rk_{Psr} + \pi/4)\, dr \tag{6.20.41}$$

In the low-frequency limit,

$$\kappa^{*(3D)}(\omega \to 0) \approx \kappa_0(1 - \sigma_{\kappa\kappa}^2/3) \tag{6.20.42}$$

$$\kappa^{*(1D)}(\omega \to 0) \approx \kappa_0(1 - \sigma_{\kappa\kappa}^2) \tag{6.20.43}$$

The one-dimensional low-frequency limit corresponds to a harmonic average of the permeability for small fluctuations. The high-frequency limit, for both one and three dimensions, is just the arithmetic average of the permeability values. The effective permeability is bounded by the harmonic and arithmetic averages:

$$\left\langle \frac{1}{\kappa} \right\rangle^{-1} = \kappa^{*(1D)}(\omega \to 0) < \kappa^{*(3D)}(\omega \to 0) < \kappa^{*(1D,3D)}(\omega \to \infty) = \langle \kappa \rangle \tag{6.20.44}$$

The one-dimensional result for the frequency-dependent effective permeability can be expressed in a rescaled form as (Müller et al., 2007):

$$\frac{\kappa^{*(1D)}}{\kappa_0} = \left\langle \frac{1}{\kappa} \right\rangle^{-1} \langle \kappa \rangle^{-1} + \sigma_R^2 \sqrt{2} k_{Psr} \int_0^\infty B(r) \exp(-rk_{Psr}) \sin(rk_{Psr} + \pi/4) \, dr \qquad (6.20.45)$$

$$\sigma_R^2 = 1 - \left[\left\langle \frac{1}{\kappa} \right\rangle \langle \kappa \rangle \right]^{-1} = \text{normalized difference of the harmonic and arithmetic averages.}$$

An extended first-order approximation for the effective fast P-wavenumber $\overline{k}_P$ can be obtained by using $\overline{k}_{Ps}$ in place of the background slow P-wavenumber k_{Ps} in the expressions for $\overline{k}_P$. Numerical results by Müller et al. (2007) show that including the permeability fluctuations (with correlation lengths of elastic moduli and permeability fluctuations equal) has the following effects compared with the homogeneous permeability case: the maximum attenuation peak shifts to slightly lower frequencies; the magnitude of maximum attenuation is slightly reduced; there is a broadening of the attenuation peak; and phase velocities are slightly faster.

Uses

The results described in this section can be used to estimate velocity dispersion and attenuation caused by fluid-flow effects in heterogeneous porous media.

Assumptions and Limitations

The equations described in this section apply under the following conditions:
- linear poroelastic media;
- small fluctuations in the material properties of the heterogeneous medium. Numerical tests in one dimension indicate that the model works for relative contrasts of at least 30%;
- isotropic spatial autocorrelation function of the fluctuations; and
- low-frequency approximation. The scale of the heterogeneities is smaller than the wavelength, but larger than the pore scale.

6.21   Waves in a Pure Viscous Fluid

Synopsis

Acoustic waves in a pure viscous fluid are dispersive and attenuate because of the shear component in the wave-induced deformation of an elementary fluid volume. The linearized wave equation in a viscous fluid can be derived from the Navier–Stokes equation (Schlichting, 1951) as

$$\rho \frac{\partial^2 u}{\partial t^2} = c_0^2 \frac{\partial^2 u}{\partial x^2} + \frac{4}{3} \eta \frac{\partial^3 u}{\partial x^2 \partial t} \qquad c_0^2 = \frac{K_f}{\rho} \tag{6.21.1}$$

Then the wavenumber-to-angular-frequency ratio is

$$\frac{k}{\omega} = \sqrt{\frac{c_0^2 + i\gamma\omega}{c_0^4 + \gamma^2\omega^2}} = \frac{e^{i\arctan z/2}}{c_0\sqrt[4]{1+z^2}}$$

$$= \frac{1}{c_0\sqrt[4]{1+z^2}} \left(\sqrt{\frac{\sqrt{1+z^2}+1}{2\sqrt{1+z^2}}} + i\sqrt{\frac{\sqrt{1+z^2}-1}{2\sqrt{1+z^2}}} \right) \tag{6.21.2}$$

$$\gamma = \frac{4\eta}{3\rho}, \quad z = \frac{\gamma\omega}{c_0^2} \tag{6.21.3}$$

and the phase velocity, attenuation coefficient, and inverse quality factor are

$$V = c_0\sqrt{\frac{2(1+z^2)}{\sqrt{1+z^2}+1}} \tag{6.21.4}$$

$$a = \frac{\omega}{c_0}\sqrt{\frac{\sqrt{1+z^2}-1}{2(1+z^2)}} \tag{6.21.5}$$

$$Q^{-1} = 2\sqrt{\frac{\sqrt{1+z^2}-1}{\sqrt{1+z^2}+1}} \tag{6.21.6}$$

If $z \ll 1$ these expressions can be simplified to

$$V = c_0\left[1 + 2\left(\frac{4\pi\eta f}{3\rho c_0^2}\right)^2\right] \qquad a = \frac{8\pi^2\eta f^2}{3\rho c_0^3} \qquad Q^{-1} = \frac{8\pi\eta f}{3\rho c_0^2} \tag{6.21.7}$$

where u is displacement, ρ is density, η is viscosity, K_f is the bulk modulus of the fluid, x is a coordinate, t is time, k is the wavenumber, ω is angular frequency, V is the phase velocity, a is an attenuation coefficient, Q is the quality factor, and f is frequency.

Attenuation is very small at low frequency and for a low-viscosity fluid; however, it may become noticeable in high-viscosity fluids (Figure 6.21.1).

Uses

The equations presented in this section can be used for estimating the frequency dependence of the acoustic velocity in viscous fluids.

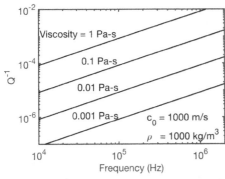

Figure 6.21.1 Inverse quality factor for acoustic waves in viscous fluid

Assumptions and Limitations

The preceding equations assume that the fluid is Newtonian. Many high-viscosity oils are non-Newtonian (i.e., their flow cannot be accurately described by the Navier–Stokes equation).

6.22 Physical Properties of Gases and Fluids

Synopsis

The **bulk modulus** (K) of a fluid or a gas is defined as

$$K = \frac{1}{\beta} = -\frac{dP}{dV/V} = \rho\frac{dP}{d\rho} \tag{6.22.1}$$

where β is compressibility, P is pressure, and V is volume. For small pressure variations (typical for wave propagation), the pressure variation is related to the density variation through the acoustic velocity c_0 (which is 1500 m/s for water at room conditions) as follows:

$$dP = c_0^2\, d\rho. \tag{6.22.2}$$

Therefore,

$$K = \rho c_0^2. \tag{6.22.3}$$

Batzle and Wang (1992) and Han and Batzle (2000) have summarized some important properties of reservoir fluids, which are presented in the following section.

Water and Brine

The **density of brine** ρ_B of salinity S of sodium chloride can be approximated by

$$\rho_B = \rho_W + S\{0.668 + 0.44S + 10^{-6}[300P - 2400P\,S + T(80 + 3T - 3300S - 13P + 47P\,S)]\}$$
$$(6.22.4)$$

where the **density of pure water** (ρ_w) is

$$\begin{aligned}\rho_w = 1 + 10^{-6}(&-80T - 3.3T^2 + 0.00175T^3 \\ &+ 489P - 2TP + 0.016T^2P - 1.3 \times 10^{-5}T^3P \\ &- 0.333P^2 - 0.002TP^2)\end{aligned}$$
$$(6.22.5)$$

In these formulas pressure P is in MPa, temperature T is in degrees Celsius, salinity S is in fractions of one (parts per million divided by 10^6), and density (ρ_B and ρ_w) is in g/cm^3 (Batzle and Wang, 1992). The expression for brine density is intended for NaCl solutions only. Other dissolved salts, even with the same ppm, can have significantly different behavior.

The **acoustic velocity in brine** V_B in m/s is approximated by

$$\begin{aligned}V_B = V_W + S(&1170 - 9.6T + 0.055T^2 - 8.5 \times 10^{-5}T^3 \\ &+ 2.6P - 0.0029TP - 0.0476P^2) \\ &+ S^{3/2}(780 - 10P + 0.16P^2) - 1820S^2\end{aligned}$$
$$(6.22.6)$$

where the **acoustic velocity in pure water** V_w in m/s is

$$V_w = \sum_{i=0}^{4}\sum_{j=0}^{3}\omega_{ij}T^iP^j$$
$$(6.22.7)$$

and coefficients ω_{ij} are

$\omega_{00} = 1402.85$	$\omega_{02} = 3.437 \times 10^{-3}$
$\omega_{10} = 4.871$	$\omega_{12} = 1.739 \times 10^{-4}$
$\omega_{20} = -0.04783$	$\omega_{22} = -2.135 \times 10^{-6}$
$\omega_{30} = 1.487 \times 10^{-4}$	$\omega_{32} = -1.455 \times 10^{-8}$
$\omega_{40} = -2.197 \times 10^{-7}$	$\omega_{42} = 5.230 \times 10^{-11}$
$\omega_{01} = 1.524$	$\omega_{03} = -1.197 \times 10^{-5}$
$\omega_{11} = -0.0111$	$\omega_{13} = -1.628 \times 10^{-6}$
$\omega_{21} = 2.747 \times 10^{-4}$	$\omega_{23} = 1.237 \times 10^{-8}$
$\omega_{31} = -6.503 \times 10^{-7}$	$\omega_{33} = 1.327 \times 10^{-10}$
$\omega_{41} = 7.987 \times 10^{-10}$	$\omega_{43} = -4.614 \times 10^{-13}$

We define the **gas–water ratio** R_G as the ratio of the volume of dissolved gas at standard conditions to the volume of brine. For temperatures below 250 °C, the maximum amount of methane that can go into solution in brine is

$$\log_{10}(R_G) = \log_{10}(0.712P|T - 76.71|^{1.5} + 3676P^{0.64} - 4 - 7.786S(T + 17.78)^{-0.306}$$
$$(6.22.8)$$

If K_B is the bulk modulus of the gas-free brine and K_G is that of brine with gas–water ratio R_G, then

$$\frac{K_B}{K_G} = 1 + 0.0494R_G \tag{6.22.9}$$

(i.e., the bulk modulus decreases linearly with increasing gas content). Experimental data are sparse for the density of brine, but the consensus is that the density is almost independent of the amount of dissolved gas.

The **viscosity of brine** η in cPs for temperatures below 250 °C is

$$\eta = 0.1 + 0.333S + (1.65 + 91.9S^3)\exp\{-[0.42(S^{0.8} - 0.17)^2 + 0.045]T^{0.8}\} \tag{6.22.10}$$

Additional properties of water, brine, and ice can be found in Appendix A7.

Gas

Natural gas is characterized by its gravity G, which is the ratio of gas density to air density at 15.6 °C and atmospheric pressure. The gravity of methane is 0.56. The gravity of heavier gases may be as large as 1.8. Algorithms for calculating the **gas density** and the **bulk modulus** follow.

Step 1: Calculate absolute temperature T_a as

$$T_a = T + 273.15 \tag{6.22.11}$$

where T is in degrees Celsius.

Step 2: Calculate the pseudo-pressure P_r and the pseudo-temperature T_r as

$$P_r = \frac{P}{4.892 - 0.4048G} \tag{6.22.12}$$

$$T_r = \frac{T_a}{94.72 + 170.75G} \tag{6.22.13}$$

where pressure is in MPa.

Step 3: Calculate **density** ρ_G in g/cm^3 as

$$\rho_G \approx \frac{28.8GP}{ZRT_a} \tag{6.22.14}$$

where

$$Z = aP_r + b + E, \quad E = cd \tag{6.22.15}$$

$$d = \exp\left\{-\left[0.45 + 8\left(0.56 - \frac{1}{T_r}\right)^2\right]\frac{P_r^{1.2}}{T_r}\right\} \tag{6.22.16}$$

$$c = 0.109(3.85 - T_r)^2 \; ; \; b = 0.642T_r - 0.007T_r^4 - 0.52 \tag{6.22.17}$$

$$a = 0.03 + 0.00527\,(3.5 - T_r)^3 \tag{6.22.18}$$

$$R = 8.31441 \text{ J}/(\text{g mol deg})(\text{gas constant}) \tag{6.22.19}$$

Step 4: Calculate the adiabatic **bulk modulus** K_G in MPa as

$$K_G \approx \frac{P\gamma}{1 - P_r f/Z}, \qquad \gamma = 0.85 + \frac{5.6}{P_r + 2} + \frac{27.1}{(P_r + 3.5)^2} - 8.7e^{-0.65(P_r + 1)} \tag{6.22.20}$$

$$f = cdm + a, \qquad m = 1.2\left\{ -\left[0.45 + 8\left(0.56 - \frac{1}{T_r} \right)^2 \right] \frac{P_r^{0.2}}{T_r} \right\} \tag{6.22.21}$$

The preceding approximate expressions for ρ_G and K_G are valid as long as P_r and T_r are not both within 0.1 of unity.

Oil

Oil density under room conditions may vary from under 0.5 to 1 g/cm^3, and most produced oils are in the 0.7–0.8 g/cm^3 range. A reference (standard) density that can be used to characterize an oil ρ_0 is measured at 15.6 °C and atmospheric pressure. A widely used classification of crude oil is the American Petroleum Institute's oil gravity (API gravity). It is defined as

$$\text{API} = \frac{141.5}{\rho_0} - 131.5 \tag{6.22.22}$$

where density is in g/cm^3. API gravity may be about 5 for very heavy oils and about 100 for light condensates.

Acoustic velocity V_P in oil may generally vary with temperature T and **molecular weight** M (Wang and Nur, 1986):

$$V_P(T,M) = V_0 - b\Delta T - a_m\left(\frac{1}{M} - \frac{1}{M_0} \right) \tag{6.22.23}$$
$$\frac{1}{b} = 0.306 - \frac{7.6}{M}$$

In equation 6.22.23, V_0 is the velocity in oil of molecular weight M_0 at temperature T_0; a_m is a positive function of temperature, and thus oil velocity increases with molecular weight. When components are mixed, velocity can be approximately calculated as a fractional average of the end components.

For **dead oil** (oil with no dissolved gas), the effects of pressure and temperature on density are largely independent (Batzle and Wang, 1992). The pressure dependence is

$$\rho_P = \rho_0 + (0.00277P - 1.71 \times 10^{-7}P^3)(\rho_0 - 1.15)^2 + 3.49 \times 10^{-4}P \tag{6.22.24}$$

where ρ_P is the density in g/cm³ at pressure P in MPa. The temperature dependence of density at a given pressure P is

$$\rho = \rho_P / [0.972 + 3.81 \times 10^{-4} (T + 17.78)^{1.175}] \tag{6.22.25}$$

where temperature is in degrees Celsius.

The **acoustic velocity** in **dead oil** depends on pressure and temperature as

$$V_P(m/s) = 2096 \left(\frac{\rho_0}{2.6 - \rho_0} \right)^{1/2} - 3.7T + 4.64P \\ + 0.0115 [4.12 (1.08 \rho_0^{-1} - 1)^{1/2} - 1] TP \tag{6.22.26}$$

or, in terms of API gravity as

$$V_P(m/s) = 15462 (77.1 + API)^{-1/2} - 3.7T + 4.64P \\ + 0.0115 \, [0.36 (API + 0.482)^{1/2} - 1] TP. \tag{6.22.27}$$

Live Oil

Large amounts of **gas can be dissolved in oil**. The original fluid *in situ* is usually characterized by **the gas–oil–ratio**, R_G, the volume ratio of liberated gas to remaining oil when the fluid is taken to atmospheric pressure and 15.6 °C. The maximum amount of gas that can be dissolved in an oil is a function of pressure, temperature, and the composition of both the gas and the oil (Standing, 1962) and can be estimated as

$$R_G^{(max)} = 0.02123G \left[(P + 0.176) \exp \left(\frac{4.072}{\rho_0} - 0.00377T \right) \right]^{1.205} \tag{6.22.28}$$

or, in terms of API gravity:

$$R_G^{(max)} = 2.03G [(P + 0.176) \exp(0.02878 \, API - 0.00377T)]^{1.205} \tag{6.22.29}$$

where R_G is in liters/liter (1 L/L = 5.615 ft³/bbl) and G is the gas gravity. Temperature is in degrees Celsius, and pressure is in MPa. (Note that equations 6.22.28 and 6.22.29 are slightly modified from that in Batzle and Wang, in order to conform to Standing, 1962.) The Batzle–Wang expression is

$$R_{G-BW}^{(max)} = 2.03G [P \exp(0.02878 API - 0.00377T)]^{1.205} \tag{6.22.30}$$

At a given GOR, the pressure at which the oil becomes saturated with gas is the **Bubble point pressure**, P_b, which can be derived from equation 6.22.28 and 6.22.29:

$$P_b = \left(\frac{R_G}{0.02123 \, G} \right)^{0.8299} \cdot \exp \left(-\frac{4.072}{\rho_0} + 0.00377 \, T \right) - 0.176, \tag{6.22.31}$$

or in terms of API gravity:

$$P_b = \left(\frac{R_G}{2.03\ G}\right)^{0.8299} \cdot \exp(-0.02878\ API + 0.00377\ T) - 0.176. \tag{6.22.32}$$

The Batzle–Wang expression for bubble point pressure, derived from equation 6.22.30 is

$$P_{b-BW} = \left(\frac{R_G}{2.03\ G}\right)^{.8299} \cdot \exp(-0.02878\ API + 0.00377\ T). \tag{6.22.33}$$

Oil at bubble point is also sometimes referred to as "saturated" oil.

Density of Live Oil

The true density of live oil, ρ_b^{live}, at **bubble point** (saturation) can be derived exactly from mass balance (Standing 1962; Batzle and Wang, 1992; Ahmed, 2001):

$$\rho_b^{live} = \frac{\rho_0 + 0.0012 G R_G}{B_{ob}}\ \text{g/cc} \tag{6.22.34}$$

Parameter B_{ob} is the **solution formation volume factor at the bubble point**, estimated by Standing (1962) as

$$B_{ob} = 0.972 + 0.00038\left[2.495\ R_G\left(\frac{G}{\rho_0}\right)^{1/2} + T + 17.8\right]^{1.175}. \tag{6.22.35}$$

Batzle and Wang (1992) suggest correcting the true density for pressures above bubble point by substituting ρ_b^{live} from equation 6.22.34, in place of ρ_0 in equation 6.22.24. However, this approach has been criticized by Han and Batzle (2000) as not consistent with standard engineering practice. Furthermore, ρ_{BW}^{live} computed this way, does not approach ρ_b^{live} as $P{\to}P_b$, as it should. There has been some confusion in the literature about how to implement equation 6.22.24 to correct the live oil density. One ad hoc approach is to substitute ρ_b^{live} for ρ_0 and substitute $(P - P_b)$ for P in equation 6.22.24 as follows:

$$\rho_{\text{P-ad hoc}} = \rho_b^{live} + (0.00277(P - P_b)$$
$$-1.71 \times 10^{-7}(P - P_b)^3)(\rho_b^{live} - 1.15)^2 + 3.49 \times 10^{-4}(P - P_b) \tag{6.22.36}$$

Another ad hoc approach is to use equation 6.22.24 twice: once to *subtract* the pressure correction to move from P_b to 0, and again to *add* the pressure correction from 0 to $P \geq P_b$:

$$\rho_{\text{P-ad hoc}} = \rho_b^{live} - (0.00277 P_b - 1.71 \times 10^{-7} P_b^3)(\rho_b^{live} - 1.15)^2 + 3.49 \times 10^{-4} P_b)$$
$$+ (0.00277 P - 1.71 \times 10^{-7} P^3)(\rho_b^{live} - 1.15)^2 + 3.49 \times 10^{-4} P). \tag{6.22.37}$$

A more widely used correction (Vasquez and Beggs, 1980; Ahmed, 2001) for density and formation volume factor at pressure *above* bubble point is estimated from the isothermal compressibility as:

$$\rho^{live} = \rho_b^{live} \left(\frac{P}{P_b} \right)^A, \qquad P \geq P_b ,$$

(6.22.38)

and

$$B_o = B_{ob} \left(\frac{P_b}{P} \right)^A, \qquad P \geq P_b ,$$

(6.22.39)

where (Ahmed, p. 108)

$$A = 10^{-5}(28.08 \, R_G + 30.96 \, T - 1180 \, G + 12.61 \, API - 882.6)$$

(6.22.40)

Note that combining equations 6.22.34 and 6.22.38–6.22.49 yields the following expression for density above bubble point:

$$\rho^{live} = \frac{\rho_0 + 0.0012GR_G}{B_o} , P \geq P_b.$$

(6.22.41)

The Pseudo Density of Live Oil

A **pseudo-liquid** is an ideal mixture of oil and fictitious liquid states of ethane and methane at standard conditions (Katz and Standing, 1942; Standing and Katz, 1942a, 1942b, 1962, 1981; McCain, 1990). Katz (1942) gives procedures for estimating the apparent liquid densities of methane and ethane, which vary with the live oil molecular composition. The **apparent liquid density**, ρ_{ga0}, of the combined natural gases at **standard conditions** can be estimated using (Katz, 1942; Han and Batzle, 2000):

$$\rho_{ga0} = 0.61731 \cdot 10^{-0.00326API} + [1.5177 - 0.54349 \log_{10}(API)] \log_{10}(G) .$$

(6.22.42)

The **pseudo liquid density**, ρ_{p0}, of the oil-gas mixture **at standard conditions** is given by (McCain, 1990; Han and Batzle, 2000)

$$\rho_{p0} = \rho_0(1 - v_g) + \rho_{ga0}v_g ,$$

(6.22.43)

where v_g is the apparent volume fraction of the pseudo-liquid **gas at standard conditions**:

$$v_g = \frac{.00123RG}{\rho_{ga0} + .00123RG} .$$

(6.22.44)

Equation 6.21.43 is equivalent to

$$\rho_{p0} = \frac{\rho_0 + .00123GR}{1 + .00123GR/\rho_{ga0}}.$$ (6.22.45)

The Katz–Standing method for adjusting the surface pseudo-liquid density to reservoir temperature and pressure, based on the coefficient of isothermal compressibility and isobaric thermal expansion coefficient, is

$$\rho_p = \rho_{p0} + \Delta\rho_{pP} + \Delta\rho_{pT}$$ (6.22.46)

where

$$\Delta\rho_{pP} = 10^{-3}[0.388 + 37.6 \cdot 10^{-2.65\rho_{p0}}]P - 10^{-6}[1.01 + 886 \cdot 10^{-3.76\rho_{p0}}]P^2$$ (6.22.47)

$$\Delta\rho_T = -10^{-5}(T - 15.6)^{0.951}[8.40 + 82.1(\rho_{p0} + \Delta\rho_P)^{-0.951}]$$
$$+10^{-4}(T - 15.6)^{0.475}[4.58 - 4.94 \cdot 10^{-(\rho_{p0}+\Delta\rho_P)}]$$ (6.22.48)

By comparison, Batzle and Wang (1992) use the following empirical expression for pseudo-density at bubble point:

$$\rho_{pb-BW} = \frac{\rho_0}{B_{ob}(1 + .001R_G)}$$ (6.22.49)

Han and Batzle (2000) use yet another empirical expression used specifically to estimate the sonic velocity, as discussed in the following section.

P-Wave Velocity of Live Oil

The pseudo-liquid density has been used by Batzle and Wang (1992) and Han and Batzle (2000) to estimate the P-wave velocity in oil. Batzle and Wang's expression for **live oil V_P at the bubble point** (i.e., full gas saturation) is estimated empirically by substituting their pseudo density (equation 6.22.49) in place of ρ_o in the expression for dead oil velocity, equation 6.22.26:

$$V_P(\text{m/s}) = 2096\left(\frac{\rho_{pb-BW}}{2.6 - \rho_{pb-BW}}\right)^{1/2} - 3.7T + 4.64P$$

$$+0.0115[4.12(1.08\rho_{pb-BW}^{-1} - 1)^{1/2} - 1]TP$$ (6.22.50)

Han and Batzle (2000) present an updated way to estimate P-wave velocity in live oil. They allow for deviations from ideal mixing by replacing equation 6.22.43 with a "velocity pseudo-density" at standard conditions given by:

$$\rho_{pV0} = \rho_0(1 - v_g) + \varepsilon\rho_{ga0}v_g$$ (6.22.51)

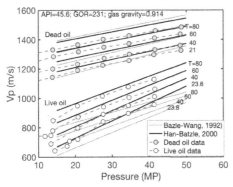

Figure 6.22.1 Comparisons of laboratory data on light 'Reinecke oil' measured by Han and Batzle (2000). The Han–Batzle models fall closer to the data than the Batzle–Wang model.

where $\varepsilon = 0.133$ is an empirical gas coefficient, and ρ_{ga0} is the surface gas apparent density, equation 6.22.42. The Han–Batzle (2000) empirical expression for velocity is given by

$$V_P = [1900.3\rho_{pV0}^{0.6477} - 256.2] - [3.044 + 0.012(141.5/\rho_{pV0} - 131.5)]T$$
$$+ [3 + 0.031(141.5/\rho_{pV0} - 131.5)]P + [0.3356\exp(-4.036\rho_{pV0})]PT$$

$$(6.22.52)$$

The first term in equation 6.22.52 is the velocity at $T = 0^\circ C$ and $P = 0\, MPa$ Equation 6.22.52 is applicable for API of 15–55, GOR up to 250 L/L, pressure up to 55.2 MPa, and temperature up to 100 °C. Han and Batzle's (2000) empirical expression, equation 6.22.52, is based on a wider range of measurements than the original equations of Batzle and Wang.

A comparison of the Batzle–Wang and Han–Batzle models is shown in Figure 6.22.1. Dots show lab data on both dead and live 'Reinecke oil' samples from Han and Batzle (2000). In both cases, API=45.6. For the live oil, GOR=231 L/L and gas gravity=0.914. Both models over-predict the dead oil velocities at all but the coldest temperature. The Han–Batzle model falls much closer to the live oil data than the Batzle–Wang model.

The adiabatic bulk modulus of dead or live oil is found from $K^{live} = \rho^{live} V_P^2$.

Example: Use the Batzle–Wang equations to calculate the density and acoustic velocity of live oil of 30 API gravity at 80 °C and 30 MPa. The gas-oil ratio R_G is 100, and the gas gravity G is 0.6. Compare with the equations of Han and Batzle. Calculate ρ_0 from API as (equation 6.22.22):

$$\rho_0 = \frac{141.5}{API + 131.5} = \frac{141.5}{30 + 131.5} = 0.876\ \mathrm{g/cm^3}$$

Check to make sure that the GOR is within the bubble point limits (equation 6.22.30):

$$R_{\text{G-BW}}^{(\text{max})} = 2.03G[Pexp(0.02878 \text{ API} - 0.00377T)]^{1.205}$$
$$= 2.03 \times 0.6[30 \times \exp(0.02878 \times 30 - 0.00377 \times 80)]^{1.205} = 144 \, L/L$$

which is <1% different than that predicted by equation 6.22.29. The bubble point pressure at GOR=100 is found from equation 6.22.33:

$$P_{b-BW} = \left(\frac{R_G}{2.03 \, G}\right)^{.8299} \cdot \exp(-0.02878 \, API + 0.00377 \, T)$$
$$= \left(\frac{100}{2.03 \times 0.6}\right)^{.8299} \cdot \exp(-0.02878 \times 30 + 0.00377 \times 80) = 22.12 \, MPa$$

Calculate the formation volume factor at bubble point using equation 6.22.35:

$$B_{ob} = 0.972 + 0.00038\left[2.495R_G\left(\frac{G}{\rho_0}\right)^{1/2} + T + 17.8\right]^{1.175}$$
$$= 0.972 + 0.00038\left[2.495 \times 100\left(\frac{0.6}{0.876}\right)^{1/2} + 80 + 17.8\right]^{1.175} = 1.287$$

The corrected formation volume fraction for pressure above bubble point is found using equations 6.22.39 and 6.22.40:

$$A = 10^{-5}(28.08 \, R_G + 30.96 \, T - 1180 \, G + 12.61 \, API - 882.6)$$
$$= 10^{-5}(28.08 \times 100 + 30.96 \times 80 - 1180 \times 0.6$$
$$+ 12.61 \times 30 - 882.6) = 0.0407$$

$$B_o = B_{ob}\left(\frac{P_b}{P}\right)^A$$
$$= 1.287 \times \left(\frac{22.12}{30}\right)^{.0407} = 1.271$$

Calculate the true density at bubble point using equation 6.22.34

$$\rho_b^{live} = \frac{\rho_0 + 0.0012GR_G}{B_{ob}}$$
$$= \frac{0.876 + 0.0012 \times 0.6 \times 100}{1.287} = 0.737 \, g/cm^3$$

The Batzle–Wang density corrected for pressure above bubble point is found by putting ρ_b^{live} in place of ρ_0 in equation 6.22.24:

$$\rho_{\text{P-BW}} = \rho_b^{live} + (0.00277P - 1.71 \times 10^{-7}P^3)(\rho_b^{live} - 1.15)^2 + 3.49 \times 10^{-4}P$$
$$= 0.737 + (0.00277 \times 30 - 1.71 \times 10^{-7} \times 30^3)(0.737 - 1.15)^2$$
$$+ 3.49 \times 10^{-4} \times 30 = 0.761 \text{ g/cm}^3$$

The ad hoc alternative implementation of Batzle–Wang density, equation 6.22.36, yields

$$\rho_{\text{P-BW}} = \rho_b^{live} + (0.00277(\text{P-P}_b) - 1.71 \times 10^{-7}(\text{P-P}_b)^3)(\rho_b^{live} - 1.15)^2$$

$$+ 3.49 \times 10^{-4}(\text{P-P}_b)$$

$$= 0.737 + (0.00277(30 - 22.12) - 1.71 \times 10^{-7}(30 - 22.12)^3)$$

$$\times (0.737 - 1.15)^2 + 3.49 \times 10^{-4}(30 - 22.12)$$

$$= 0.744 g/cm^3$$

In contrast, the corrected density from equation 6.22.38 is

$$\rho^{live} = \rho_b^{live}\left(\frac{P}{P_b}\right)^A$$

$$= 0.737\left(\frac{30}{22.12}\right)^A = 0.746 g/cm^3$$

The Batzle–Wang pseudo-density at bubble point ρ_{pb-BW} is found using equation 6.22.49:

$$\rho_{pb-BW} = \frac{\rho_0}{B_{ob}}(1 + .001R_G)^{-1}$$

$$= \frac{0.876}{1.287}(1 + .001 \times 100)^{-1} = 0.619 \ g/cm^3$$

The Batzle–Wang velocity in live oil is found from equation 6.22.49 in 6.22.26:

$$V_{\text{P-BW}} = 2096\left(\frac{\rho_{pb-BW}}{2.6 - \rho_{pb-BW}}\right)^{1/2} - 3.7T + 4.64P$$

$$+ 0.0115[4.12(1.08/\rho_{pb-BW} - 1)^{1/2} - 1]TP$$

$$= 2096\left(\frac{0.619}{2.6 - 0.619}\right)^{1/2} - 3.7 \times 80 + 4.64 \times 30$$

$$+ 0.0115[4.12(1.08/0.619 - 1)^{1/2} - 1] \times 80 \times 30$$
$$= 1086 \text{ m/s}$$

The Han–Batzle apparent liquid density of gas from equation 6.22.42 is

$$\rho_{ga0} = 0.61731 \cdot 10^{-0.00326API} + [1.5177 - 0.54349\log_{10}(API)]\log_{10}(G)$$
$$= 0.61731 \cdot 10^{-0.00326\times30} + [1.5177 - 0.54349\log_{10}(30)]\log_{10}(0.6)$$
$$= 0.3343 \ g/cm^3$$

and the volume of apparent liquid gas from equation 6.22.44 is

$$v_g = \frac{.00123RG}{\rho_{ga0} + .00123RG}$$
$$= \frac{.00123 \times 100 \times 0.6}{0.3343 + .00123 \times 100 \times 0.6} = 0.181$$

The Han–Batzle velocity pseudo-density from equation 6.22.51 is

$$\rho_{pV0} = \rho_0(1 - v_g) + \varepsilon\rho_{ga0}v_g$$
$$= 0.876(1 - 0.181) + 0.133 \times 0.3343 \times 0.181 = 0.725 \ g/cm^3$$

The Han–Batzle velocity from equation 6.22.52 is

$$V_P = [1900.3\rho_{pV0}^{0.6477} - 256.2] + [3.044 + 0.012(141.5/\rho_{pV0} - 131.5)]P$$
$$- [3 + 0.031(141.5/\rho_{pV0} - 131.5)]T + [0.3356\exp(-4.036\rho_{pV0})]TP$$
$$= [1900.3 \times 0.725^{0.6477} - 256.2] + [3.044 + 0.012(141.5/0.725 - 131.5)]30$$
$$- [3 + 0.031(141.5/0.725 - 131.5)]80 + [0.3356\exp(-4.036 \times 0.725)]80 \times 30$$
$$= 1174 \ m/s$$

The Batzle–Wang bulk modulus is

$$K_{BW} = \rho V_P^2 = 0.761 \times 1000 \times 1086^2 = 0.897 \ GPa$$

The Han–Batzle bulk modulus is

$$K = \rho V_P^2 = 0.746 \times 1000 \times 1174^2 = 1.028 \ GPa$$

The **viscosity of dead oil** $((\eta)$ decreases rapidly with increasing temperature. At room pressure for a gas-free oil, we have

$$\log_{10}(\eta + 1) = 0.505y(17.8 + T)^{-1.163} \tag{6.21.53}$$

$$\log_{10}(y) = 5.693 - 2.863/\rho_0 \tag{6.22.54}$$

Pressure has a smaller influence on viscosity and can be estimated independently of the temperature influence. If oil viscosity is η_0 at a given temperature and room pressure, its viscosity at pressure P and the same temperature is

$$\eta = \eta_0 + 0.145PI \tag{6.22.55}$$

$$\log_{10}(I) = 18.6[0.1 \quad \log_{10}(\eta_0) + (\log_{10}(\eta_0) + 2)^{-0.1} - 0.985] \tag{6.22.56}$$

Viscosity in these formulas is in centipoise, temperature is in degrees Celsius, and pressure is in MPa.

Additional discussion of properties of heavy oil, melt, and magma can be found in Sections 8.8, 8.9, and 8.10.

Condensates

Associated gases are those commonly found with petroleum. The properties of high gravity associated gases are typically rich in ethane through pentane, while **gas condensates** have high concentrations of heptane-plus. The empirical equations presented earlier, for example by Batzle and Wang (1992), are not suitable for these mixtures. Expressions for predicting properties of high gravity gases and condensates are presented here. Most of the discussion is on the *isothermal* bulk modulus, which is usually smaller than the *adiabatic* modulus that Batzle and Wang try to predict.

Isothermal Density and Compressibility

The **principle of corresponding states** says that two substances at the same conditions referenced to their **critical temperature**, T_c, and **pressure**, P_c, will have similar properties (Katz *et al.*, 1959; Sutton, 2005). To take advantage of the principle, the **reduced pressure** P_r and **reduced temperature** T_r are defined:

$$P_r = \frac{P}{P_c} \tag{6.22.57}$$

$$T_r = \frac{T}{T_c} \tag{6.22.58}$$

Many of the published relations for PVT properties of gas and condensates are expressed in terms of P_r and T_r, rather than absolute pressure P and temperature T. As a result, the expressions can describe a fairly broad range of compositions. The reduced properties, P_r and T_r, are dimensionless. The temperatures in equation 6.22.58 must be expressed in absolute units (Kelvin or Rankine). For hydrocarbon gas *mixtures*, the critical properties are replaced with **pseudo critical properties**. Much of the research on PVT properties of gas and condensates has focused on finding empirical relations in terms of P_r and T_r.

When there is mixture of hydrocarbon and non-hydrocarbon molecules in the gas or condensate, the mole fraction of hydrocarbons is computed by subtracting off the mole fractions of the non-hydrocarbons (Standing, 1981), for example:

Table 6.22.1 *Molecular weight and critical properties for air,*
CO_2, H_2S, *and* N_2

	M	P_c (psia)	T_c (°R)
Air	28.964		
CO_2	44.01	1073	547.7
H_2S	34.1	1306.5	672.47
N_2	28.0134	492.3	227.17

$$y_{HC} = 1 - y_{H_2S} - y_{CO_2} - y_{N_2} \tag{6.22.59}$$

where y_x is the mole fraction of each species. The hydrocarbon gas gravity G_{gHC} is then found from the measured gravity of the mixture, G_g, using

$$G_{gHC} = \frac{G_g - (y_{H_2S}M_{H_2S} + y_{CO_2}M_{CO_2} + y_{N_2}M_{N_2})/M_{air}}{y_{HC}} \tag{6.22.60}$$

where G_x is the gravity and M_x is the molecular weight of each species.

Standing's (1981) empirical expressions for pseudo critical pressure and temperature of associated gases with very low concentrations of non-hydrocarbon molecules are given by

$$P_{pcHC} = 667 + 15G_{gHC} - 37.5G_{gHC}^2 \text{ (psia)} \tag{6.22.61}$$

$$T_{pcHC} = 168 + 325G_{gHC} - 12.5G_{gHC}^2 \text{ (°R)} \tag{6.22.62}$$

Sutton (2005) determined slightly updated empirical expressions for pseudo critical properties of associated gases, using measured data base of Z-factors, given by

$$P_{pcHC} = 671.1 + 14G_{gHC} - 34.3G_{gHC}^2 \text{ (psia)} \tag{6.22.63}$$

$$T_{pcHC} = 120.1 + 429G_{gHC} - 62.9G_{gHC}^2 \text{ (°R)} \tag{6.22.64}$$

And from a broader database of gas condensates with only small amounts of non-hydrocarbon gases, Sutton (2005) found (in psia and °R):

$$P_{pcHC} = 744 - 125.4G_{gHC} + 5.9G_{gHC}^2 \text{ (psia)} \tag{6.22.65}$$

$$T_{pcHC} = 164.3 + 357.7G_{gHC} - 67.7G_{gHC}^2 \text{ (°R)} \tag{6.22.66}$$

Elsharkawy *et al.* (2001) give expressions for pseudo-critical properties that are accurate to within 1.58% absolute error over the range $0.6 < G_g < 1.4$;

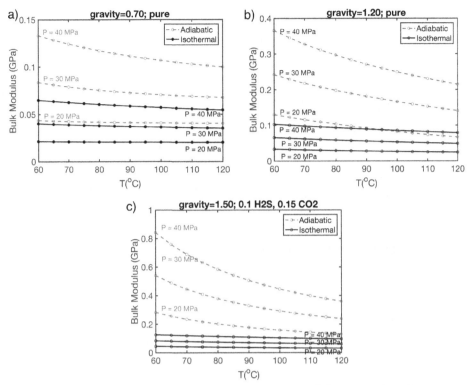

Figure 6.22.2 Comparison of adiabatic and isothermal bulk moduli for gases and condensates. Batzle–Wang predictions for adiabatic moduli are unreasonably large at large gravity, compared with the predicted isothermal moduli. (a) Light hydrocarbon gas, gravity = 0.7. $K_S/K_T \approx 2$. (b) Heavy hydrocarbon gas, gravity = 1.2. $K_S/K_T \approx 3.5$. (c) Condensate gas with mole fraction = 10% of H2S and 15% CO2, gravity = 1.50. $K_S/K_T > 5$.

$$P_{pc} = 787.06 - 147.34G_g + 7.916G_g^2 \tag{6.22.67}$$

$$T_{pc} = 149.18 + 358.14G_g - 66.976G_g^2 \tag{6.22.68}$$

When non-hydrocarbon gases are present (CO_2, H_2S, N_2), the pseudocritical temperature and pressure of high gravity gas and condensate mixtures can be estimated using (Standing, 1981)

$$P_{pc}^* = y_{HC}P_{pcHC} + y_{H_2S}P_{cH_2S} + y_{CO_2}P_{cCO_2} + y_{N_2}P_{cN_2} \tag{6.22.69}$$

$$T_{pc}^* = y_{HC}T_{pcHC} + y_{H_2S}T_{H_2S} - y_{CO_2}T_{cCO_2} + y_{N_2}T_{cN_2} \tag{6.22.70}$$

where subscript c indicates critical point properties and pc indicates pseudocritical properties.

Wichert and Aziz (1972) found that a further correction to the pseudocritical properties is necessary in the presence of H_2S and CO_2 in the gas:

$$\varepsilon = 120[(y_{CO_2} + y_{H_2S})^{0.9} - (y_{CO_2} + y_{H_2S})^{1.6}] + 15(y_{H_2S}^{0.5} - y_{H_2S}^4) \; (°R) \qquad (6.22.71)$$

$$T_{pc} = T_{pc}^* - \varepsilon \; (°R) \qquad (6.22.72)$$

$$P_{pc} = \frac{P_{pc}^*(T_{pc}^* - \varepsilon)}{T_{pc}^* + y_{H_2S}(1 - y_{H_2S})\varepsilon} \; \text{(psia)} \qquad (6.22.73)$$

Sutton (2005) gives a slight update to the Wichert and Aziz correction by fitting to a larger dataset:

$$\varepsilon = 107.6[(y_{CO_2} + y_{H_2S}) - (y_{CO_2} + y_{H_2S})^{2.2}] + 5.9(y_{H_2S}^{0.06} - y_{H_2S}^{0.68}) \; (°R) \qquad (6.22.74)$$

although the results of using equation 6.22.69 and 6.22.74 are very similar.

From the pseudocritical temperature and pressure, equations 6.22.61–6.22.72, the pseudo reduced pressure and temperature can be found:

$$P_{pr} = \frac{P}{P_{pc}} \qquad (6.22.75)$$

$$T_{pr} = \frac{T}{T_{pc}} \qquad (6.22.76)$$

The gas Z-factor can be found using the method of Dranchuk and Abou-Kassem (1975), known as the DAK method, as follows. The pseudo reduced density is approximated as

$$\rho_{pr} = \frac{0.27 P_{pr}}{Z T_{pr}} \qquad (6.22.77)$$

The Z-factor can be estimated from

$$
\begin{aligned}
Z = 1 &+ \left(A_1 + \frac{A_2}{T_{pr}} + \frac{A_3}{T_{pr}^3} + \frac{A_4}{T_{pr}^4} + \frac{A_5}{T_{pr}^3} \right) \rho_{pr} \\
&+ \left(A_6 + \frac{A_7}{T_{pr}} + \frac{A_8}{T_{pr}^2} \right) \rho_{pr}^2 \\
&- A_9 \left(\frac{A_{10}}{T_{pr}} + \frac{A_{11}}{T_{pr}^2} \right) \rho_{pr}^5
\end{aligned}
\qquad (6.22.78)
$$

where $A_1 = 0.3265$, $A_2 = 1.0700$, $A_3 = -0.5339$, $A_4 = 0.01569$, $A_5 = -0.05165$, $A_6 = 0.5475$, $A_7 = -0.7361$, $A_8 = 0.1844$, $A_9 = 0.1056$, $A_{10} = 0.6134$, $A_{11} = 0.7210$. Equations 6.22.77 and 6.22.78 are solved iteratively. Once Z and ρ_{pr} are determined, we can find

$$\left(\frac{\partial Z}{\partial \rho_{pr}}\right)_{T_r} = 1 + \left(A_1 + \frac{A_2}{T_{pr}} + \frac{A_3}{T_{pr}^3} + \frac{A_4}{T_{pr}^4} + \frac{A_5}{T_{pr}^3}\right)$$
$$+ 2\left(A_6 + \frac{A_7}{T_{pr}} + \frac{A_8}{T_{pr}^2}\right)\rho_{pr} - 5A_9\left(\frac{A_7}{T_{pr}} + \frac{A_8}{T_{pr}^2}\right)\rho_{pr}^4 \tag{6.22.79}$$
$$+ \left(2A_{10}\frac{\rho_{pr}}{T_{pr}^3} + 2A_{10}A_{11}\frac{\rho_{pr}^3}{T_{pr}^3} - 2A_{10}A_{11}^2\frac{\rho_{pr}^5}{T_{pr}^3}\right)\exp(-A_{11}\rho_{pr}^2)$$

Finally, the pseudo reduced isothermal compressibility is given by

$$C_{pr} = C_g P_{pc} = \frac{1}{P_{pr}} - \frac{1}{ZT_{pr}}\left(\frac{0.27(\partial Z/\partial \rho_{pr})_{T_{pr}}}{Z + \rho_{pr}(\partial Z/\partial \rho_{pr})_{T_{pr}}}\right) \tag{6.22.80}$$

and the compressibility is given by

$$C_g = C_{pr}/P_{pc} \tag{6.22.81}$$

The absolute gas density is calculated from

$$rho_g = \frac{PM_g}{RZT} \tag{6.22.82}$$

where M_g is the molecular weight of the gas, and R is the gas constant.

The adiabatic bulk modulus, K_S (which we use for seismic analysis), is always larger than the isothermal bulk modulus, K_T. The relation between them is given by

$$\frac{K_S}{K_T} = \frac{C_P}{C_V} = \gamma \tag{6.22.83}$$

where $\gamma = C_P/C_V$, C_P is the heat capacity at constant pressure, and C_V is the heat capacity at constant volume. Empirically expressions for γ are not readily available for heavy gas and condensate mixtures. However, tables for individual pure gases often report γ in the range 1.2–1.4. Figure 6.22.2 compares predictions of adiabatic bulk moduli by Batzle and Wang (1992), suitable for light gas, with the isothermal bulk moduli predicted using the equations in this section. For light gas (Figure 6.22.2a), the predicted $K_S/K_T \approx 2$, which appears plausible. However for heavier gas (Figure 6.22.2b), and gas with contaminants (e.g., H2S and CO2) (Figure 6.22.2c) the ratio is much larger, illustrating the inappropriateness of the Batzle–Wang formulas for these high gravities.

Uses

The equations presented in this section are used to estimate acoustic velocities and densities of pore fluids.

7 Empirical Relations

7.1 Velocity–Porosity Models: Critical Porosity and Modified Upper and Lower Bounds

Synopsis

Nur *et al.* (1991, 1995) and other workers have championed the simple, if not obvious, idea that the P and S velocities of rocks should trend between the velocities of the mineral grains in the limit of low porosity and the values for a mineral–pore-fluid suspension in the limit of high porosity. This idea is based on the observation that for most porous materials there is a **critical porosity**, ϕ_c, that separates their mechanical and acoustic behavior into two distinct domains. For porosities lower than ϕ_c, the mineral grains are load-bearing, whereas for porosities greater than ϕ_c, the rock simply "falls apart" and becomes a suspension, in which the fluid phase is load-bearing. The transition from solid to suspension is implicit in the well-known empirical velocity–porosity relations of Raymer *et al.* (1980) discussed in Section 7.4.

In the **suspension domain**, $\phi > \phi_c$, the effective bulk and shear moduli can be estimated quite accurately using the Reuss (isostress) average:

$$K_R^{-1} = (1 - \phi)K_0^{-1} + \phi K_{fl}^{-1}, \quad \mu_R = 0 \tag{7.1.1}$$

where K_0 and K_{fl} are the bulk moduli of the mineral material and the fluid, respectively. The effective shear modulus of the suspension is zero because the shear modulus of the fluid is zero.

In the **load-bearing** domain, $\phi < \phi_c$, the moduli decrease rapidly from the mineral values at zero porosity to the suspension values at the critical porosity. Nur found that this dependence can often be approximated with a straight line when expressed as ρV^2 versus porosity. Figure 7.1.1 illustrates this behavior with laboratory ultrasonic sandstone data from Han (1986) for samples at 40MPa effective pressure with clay content <10% by volume. In the figure, ρV_P^2 is the P-wave modulus $K + (4/3)\mu$ and ρV_S^2 is the shear modulus.

A geometric interpretation of the mineral-to-critical-porosity trend is simply that if we make the porosity large enough, the grains must lose contact and the rock must lose its rigidity. The geologic interpretation is that, at least for clastics, the weak suspension

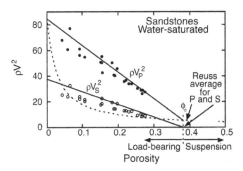

Figure 7.1.1 Critical porosity behavior: ρV_P^2 (P-wave modulus) versus porosity and ρV_S^2 (shear modulus) versus porosity, both trending between the mineral value at zero porosity and the Reuss average at critical porosity

state at critical porosity, ϕ_c, describes the sediment when it is first deposited before compaction and diagenesis. The value of ϕ_c is determined by the grain sorting and angularity at deposition. Subsequent compaction and diagenesis move the sample along an upward trajectory as the porosity is reduced and the elastic stiffness is increased.

Although there is nothing special about a linear trend of ρV^2 versus ϕ, it does describe sandstones fairly well, and it leads to convenient mathematical properties. For dry rocks, the bulk and shear moduli can be expressed as the linear functions

$$K_{dry} = K_0 \left(1 - \frac{\phi}{\phi_c}\right) \tag{7.1.2}$$

$$\mu_{dry} = \mu_0 \left(1 - \frac{\phi}{\phi_c}\right) \tag{7.1.3}$$

where K_0 and μ_0 are the mineral bulk and shear moduli, respectively. Thus, the dry-rock bulk and shear moduli trend linearly between K_0, μ_0 at $\phi = 0$ and $K_{dry} = \mu_{dry} = 0$ at $\phi = \phi_c$

This linear dependence can be thought of as a **modified Voigt average** (see Section 4.1), where one end-member is the mineral and the other end-member is the suspension at the critical porosity, which can be measured or estimated using the Reuss average. Then we can write

$$M_{MV} = (1 - \phi')M_0 + \phi'M_c \tag{7.1.4}$$

where M_0 and M_c are the moduli (bulk or shear) of the mineral material at zero porosity and of the suspension at the critical porosity. The porosity is scaled by the critical porosity, $\phi' = \phi / \phi_c$, and thus ϕ' ranges from 0 to 1 as ϕ ranges from 0 to ϕ_c Note that using the suspension modulus M_c in this form automatically incorporates the effect of pore fluids on the modified Voigt average, which is equivalent to applying Gassmann's relations to the dry-rock-modulus-porosity relations (see Section 6.3 on Gassmann's relations).

The critical porosity value depends on the internal structure of the rock. It may be very small for cracked rocks, large for foam-like rocks, and intermediate for granular rocks. Examples of critical porosity behavior in sandstones, dolomites, pumice, and cracked igneous rocks are shown in Figure 7.1.2 and Table 7.1.1.

Table 7.1.1 *Typical values of critical porosity*

Material	Critical porosity
Natural rocks	
Sandstones	40%
Limestones	60%
Dolomites	40%
Pumice	80%
Chalks	65%
Rock salt	40%
Cracked igneous rocks	5%
Oceanic basalts	20%
Artificial rocks	
Sintered glass beads	40%
Glass foam	90%

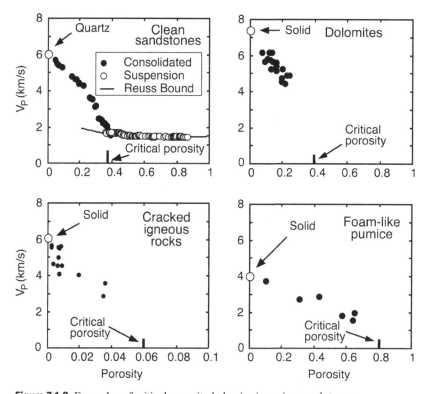

Figure 7.1.2 Examples of critical-porosity behavior in various rock types

Bound Equations as Stiff and Soft "Interpolators"

The **modified Voigt average** illustrated in Figure 7.1.1 can be thought of as interpolation between well constrained end point moduli. The elastic moduli at zero porosity can be reasonably assumed to be those of the mineral making up the rock; the elastic moduli at critical porosity can be estimated using the Reuss average (Section 4.1) mix of minerals and pore fluid, representing the very soft sediment at deposition. Here, the modified Voigt average is nothing more than a heuristic trend line interpolating between the moduli at $\phi = 0$ and $\phi = \phi_c$. Because the Voigt average is also used as an upper bound we can think of it as a "stiff" interpolation.

A more commonly used trend line between mineral and suspension is the Modified Upper Hashin–Shtrikman bound (MUHS), illustrated in Figure 7.1.3. The solid black curve is the MUHS interpolating between the sediment at critical porosity and pure mineral. The solid gray curve is a variation of it, which begins with the contact cement model (Section 5.5) near critical porosity, and then merges with the MUHS. Dashed curves are Modified Lower Hashin–Shtrikman bounds (MLHS) used to represent sorting trends. The lowest MLHS is the suspension line. Each of the other MLHS curves indicates sediments at constant depth but variable porosity due to sorting differences. We refer to the MUHS and MLHS in this context as "stiff" and "soft interpolators," respectively.

> **Caution:**
>
> The use of the Voigt, Reuss, and Hashin–Shrikman to describe depth and sorting trends is heuristic. They should not be interpreted as bounds in this context.

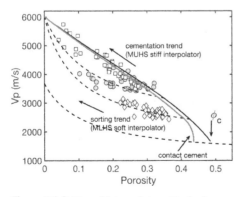

Figure 7.1.3 Use of interpolators. Dashed curves: MLHS sorting trends. Lowest dashed curve is the Reuss average of quartz and water, representing soft sediments at deposition. Solid black curve: MUHS cementation trend between critical porosity and pure mineral. Solid gray curve: variation which blends the contact cement model with MUHS.

Uses

The equations presented in this section can be used for relating velocity and porosity.

Assumptions and Limitations

The model discussed in this section has the following limitations:
- the critical porosity result is empirical; and
- because only the variation with porosity is described, one must apply other corrections to account for parameters such as clay content.

Extensions

Similar descriptions of the failure strength of porous media can be quantified in terms of the critical porosity.

7.2   Velocity–Porosity Models: Wyllie's Time Average and Geertsma's Empirical Relations for Compressibility

Synopsis

Measurements by Wyllie *et al.* (1956, 1958, 1963) revealed that a relatively simple monotonic relation often can be found between velocity and porosity in sedimentary rocks when: (1) they have relatively uniform mineralogy, (2) they are fluid-saturated, and (3) they are at high effective pressure. Wyllie *et al.* approximated these relations with the expression

$$\frac{1}{V_P} = \frac{\phi}{V_{P-fl}} + \frac{1 - \phi}{V_{P-0}} \tag{7.2.1}$$

where V_P, V_{P-0}, and V_{P-fl} are the P-wave velocities of the saturated rocks, of the mineral material making up the rocks, and of the pore fluid, respectively. Some useful values for V_{P-0} are shown in Table 7.2.1.

The interpretation of this expression is that the total transit time is the sum of the transit time in the mineral plus the transit time in the pore fluid. Hence, it is often called the time-average equation.

Table 7.2.1 *Typical mineral P-wave velocities*

	V_{P-0} (m/s)
Sandstones	5480–5950
Limestones	6400–7000
Dolomites	7000–7925

> **Caution:**
>
> The time-average equation is heuristic and cannot be justified theoretically. The argument that the total transit time can be written as the sum of the transit time in each of the phases is a seismic ray theory assumption and can be correct only if: (1) the wavelength is small compared with the typical pore size and grain size; and (2) the pores and grains are arranged as homogeneous layers perpendicular to the ray path. Because neither of these assumptions is even remotely true, the agreement with observations is only fortuitous. Attempts to overinterpret observations in terms of the mineralogy and fluid properties can lead to errors. An illustration of this point is that a form of the time-average equation is sometimes used to interpret shear velocities. To do this, a finite value of shear velocity in the fluid is used, which is clearly nonsense.

Geertsma (1961) suggested the following empirical estimate of bulk modulus in dry rocks with porosities $0 < \phi < 0.3$:

$$\frac{1}{K_{\text{dry}}} = \frac{1}{K_0}(1 + 50\phi) \tag{7.2.2}$$

where K_{dry} is the dry-rock bulk modulus and K_0 is the mineral bulk modulus.

Uses

These equations can be used to relate properties to porosity in rocks, empirically.

Assumptions and Limitations

The use of the time-average equation requires the following serious considerations:
- the rock is isotropic;
- the rock must be fluid saturated;
- the time-average equation works best if rocks are at high enough effective pressure to be at the "terminal velocity," which is usually of the order of 30 MPa. Most rocks show an increase of velocity with pressure owing to the progressive closing of compliant crack-like parts of the pore space, including microcracks, compliant grain boundaries, and narrow tips of otherwise equant-shaped pores. Usually the velocity appears to level off at high pressure, approaching a limiting "terminal" velocity when, presumably, all the crack-like pore space is closed. Because the compliant fraction of the pore space can have a very small porosity and yet have a very large effect on velocity, its presence, at low pressures, can cause a very poor correlation between porosity and velocity; hence, the requirement for high effective pressure. At low pressures or in uncompacted situations, the time-average equation tends to over-predict the velocity and porosity. Log analysts sometimes use

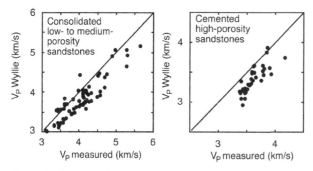

Figure 7.2.1 Comparison of predicted and measured velocity in water-saturated medium-to low-porosity shaly sandstones (40 MPa effective pressure). The velocity in pure quartz is taken at 6.038 km/s, which follows from the bulk modulus, shear modulus, and density being 38 GPa, 44 GPa, and 2.65 g/cm^3, respectively. The velocity in clay is 3.41 km/s, which follows from the bulk modulus, shear modulus, and density being 21 GPa, 7 GPa, and 2.58 g/cm^3, respectively.

a compaction correction, which is an empirical attempt to correct for the effect of compliant porosity. The time-average equation underpredicts velocities in consolidated low-to-medium-porosity rocks and in high-porosity cemented rocks, as shown in Figure 7.2.1.

- the time-average relation assumes a single homogeneous mineralogy. Empirical corrections for mixed mineralogy, such as shaliness, can be attempted to adjust for this; and
- the time-average equation usually works best for intermediate porosities (see the Raymer equations in Section 7.4).

Extension

The time-average equation can be extended in the following ways:
- for mixed mineralogy one can often use an effective average velocity for the mineral material; and
- empirical corrections can sometimes be found for shaliness, compaction, and secondary porosity, but they should be calibrated when possible.

7.3 Vernik–Kachanov Clastics Models

Based on Mori–Tanaka's effective field scheme for effective elastic moduli, combined with empirical parameters obtained by fitting to sandstone data, Vernik (2016) gives a model for dry bulk and shear moduli, K_{dry} and μ_{dry} respectively of consolidated sandstones (arenites):

$$K_{dry} = K_0 \left[1 + \frac{p\phi}{1-\phi} + A(v_0) \frac{\varepsilon_0 \exp(-d\sigma)}{1-\phi} \right]^{-1} \tag{7.3.1}$$

$$\mu_{dry} = \mu_0 \left[1 + \frac{q\phi}{1-\phi} + B(v_0) \frac{\varepsilon_0 \exp(-d\sigma)}{1-\phi} \right]^{-1} \tag{7.3.2}$$

where K_0, μ_0 and v_0 are the mineral matrix bulk and shear moduli, and Poisson's ratio, while ϕ is the porosity. For arenites, Vernik's recommended values for the matrix properties are: $K_0 = 35.6$ GPa, $\mu_0 = 33.0$ GPa, and $v_0 = 0.146$. The pore shape factors p and q can be empirically taken to be equal (for porosities > 0.03) and are related to porosity through an empirical relation: $p = q = 3.6 + b\phi$ with b ranging from 8 to 12. The crack density at zero effective stress, $\varepsilon_0 = 0.3 + 1.6\phi$ and $d = 0.07$ (MPa^{-1}) with σ being the effective stress in MPa. $A(v_0)$ and $B(v_0)$ are shape factors for circular cracks given by:

$$A(v_0) = \frac{16(1-v_0^2)}{9(1-2v_0)} \tag{7.3.3}$$

$$B(v_0) = \frac{32(1-v_0)(5-v_0)}{45(2-v_0)} \tag{7.3.4}$$

For the matrix values previously given for arenites, $A = 2.46$ and $B = 1.59$. With these empirical parameters, Vernik's equations for moduli (in GPa) for consolidated sandstones becomes:

$$K_{dry} = 35.6 \left[1 + \frac{3.6\phi + b\phi^2}{1-\phi} + \frac{(0.738 + 3.936\phi)\exp(-0.07\sigma)}{1-\phi} \right]^{-1} \tag{7.3.5}$$

$$\mu_{dry} = 33.0 \left[1 + \frac{3.6\phi + b\phi^2}{1-\phi} + \frac{(0.477 + 2.544\phi)\exp(-0.07\sigma)}{1-\phi} \right]^{-1} \tag{7.3.6}$$

The fitting parameters are based on core data for arenites with effective pressures ranging from 10 to 70 MPa, and porosities from 0.05 to 0.3.

Vernik and Kachanov (2010) and Vernik (2016) give two models for high-porosity poorly consolidated sands. Vernik's soft sand model 1 describes mechanical consolidation behavior, starting from very high initial porosity ("critical" porosity, ϕ_c) to the consolidation porosity, $\phi_{con} < \phi_c$. The dry P-wave modulus, M_{dry} and shear modulus μ_{dry} are given by:

$$M_{dry} = M_{con} \left(1 - \frac{\phi - \phi_{con}}{\phi_c - \phi_{con}} \right)^n \tag{7.3.7}$$

$$\mu_{dry} = \mu_{con} \left(1 - \frac{\phi - \phi_{con}}{\phi_c - \phi_{con}} \right)^m \tag{7.3.8}$$

where M_{con} and μ_{con} are the moduli from the consolidated sandstone model at $\phi = \phi_{con}$. Empirically, Vernik (2016) recommends $n = 2.00$ and $m = 2.05$ to provide a reasonable fit.

Vernik's soft sand model 2 describes modulus-porosity trends dues to grain-sorting variations with relatively constant state of mechanical diagenesis within limited stratigraphic intervals (Vernik, 2016). The equations for dry P-wave modulus, M_{dry} and shear modulus μ_{dry} are given by:

$$M_{dry} = M_0 \left(1 + p \frac{\phi}{1-\phi} \right)^{-1} \tag{7.3.9}$$

$$\mu_{dry} = \mu_0 \left(1 + q \frac{\phi}{1-\phi} \right)^{-1} \tag{7.3.10}$$

Empirically the shape factors p and q can be taken to be equal for practical purposes, with p varying from about 5 to 45. Vernik's soft sand model 2 trends are very similar to the trends given by the modified lower Hashin–Shtrikman curves. Vernik (2016) also suggests an empirical relation between p and the effective stress σ (in MPa), $p = A\sigma^{-B}$, with A ranging from about 80 to 580 and B from 0.8 to 1.2.

Uses

To model elastic moduli of consolidated sandstones as a function of porosity and effective stress. To obtain moduli for saturated sandstones, the dry moduli should be transformed to saturated moduli using Gassmann's equations with the appropriate fluid properties. To model mechanical compaction and grain sorting trends of elastic moduli versus porosity in high porosity, loosely consolidated sands.

Assumptions and Limitations

The functional form of the Vernik–Kachanov model is based on Mori–Tanaka's effective field approximation for effective elastic moduli. The parameters are obtained by empirical fit to core database. Not applicable for loose, unconsolidated sands.

Portions of the models involve empirical calibrations. The soft sand model 1 for mechanical compaction is only valid for $\phi_{con} \leq \phi < \phi_c$.

7.4 Velocity–Porosity Models: Raymer–Hunt–Gardner Relations

Synopsis

Raymer *et al.* (1980) suggested improvements to Wyllie's empirical velocity-to-travel time relations as follows:

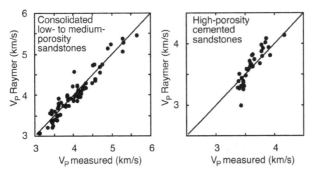

Figure 7.4.1 Comparison of predicted and measured velocities in water-saturated shaly sandstones (40 MPa effective pressure). The velocity in pure quartz is taken at 6.038 km/s, which follows from the bulk modulus, shear modulus, and density being 38 GPa, 44 GPa, and 2.65 g/cm³, respectively. The velocity in clay is 3.41 km/s, which follows from the bulk modulus, shear modulus, and density being 21 GPa, 7 GPa, and 2.58 g/cm³, respectively.

$$V = (1 - \phi)^2 V_0 + \phi V_{fl}, \quad \phi < 37\% \tag{7.4.1}$$

$$\frac{1}{\rho V^2} = \frac{\phi}{\rho_{fl} V_{fl}^2} + \frac{1 - \phi}{\rho_0 V_0^2}, \quad \phi > 47\% \tag{7.4.2}$$

where V, V_{fl}, and V_0 are the velocities in the rock, the pore fluid, and the minerals, respectively. The terms ρ, ρ_{fl}, and ρ_0 are the densities of the rock, the pore fluid, and the minerals, respectively. Note that the second relation is the same as the isostress or Reuss average (see Section 4.1) of the P-wave moduli. A third expression for intermediate porosities is derived as a simple interpolation of these two:

$$\frac{1}{V} = \frac{0.47 - \phi}{0.10} \frac{1}{V_{37}} + \frac{\phi - 0.37}{0.10} \frac{1}{V_{47}} \tag{7.4.3}$$

where V_{37} is calculated from the low-porosity formula at $\phi = 0.37$, and V_{47} is calculated from the high-porosity formula at $\phi = 0.47$.

Figure 7.4.1 shows a comparison of predicted and measured velocities in water-saturated shaly sandstones at 40 MPa effective pressure. Figure 7.4.2 compares the predictions of Raymer *et al.* (1980), Wyllie *et al.* (1956), and Gardner *et al.* (1974) for velocity versus porosity to data for water-saturated clay-free sandstones. None of the equations adequately models the uncemented sands.

Uses

The Raymer–Hunt–Gardner relations have the following uses:
• to estimate the seismic velocities of rocks with a given mineralogy and pore fluid; and

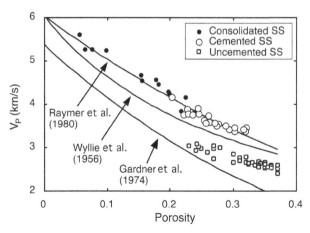

Figure 7.4.2 Velocity versus porosity in water-saturated clay-free sandstones. The Wyllie *et al.* (1956) equation underestimates the consolidated rock values. The Gardner *et al.* (1974) equation underpredicts all of the measured values. None of the equations adequately models the uncemented sands. For the predictions, the mineral is taken as quartz.

- to estimate the porosity from measurements of seismic velocity and knowledge of the rock type and pore-fluid content.

Assumptions and Limitations

The use of the Raymer–Hunt–Gardner relations requires the following considerations:
- the rock is isotropic;
- all minerals making up the rock have the same velocities;
- the rock is fluid-saturated;
- the method is empirical. See also the discussion and limitations of Wyllie's time-average equation (Section 7.2);
- these relations should work best at high enough effective pressure to be at the "terminal velocity," usually of the order of 30 MPa. Most rocks show an increase of velocity with pressure owing to the progressive closing of compliant crack-like parts of the pore space, including microcracks, compliant grain boundaries, and narrow tips of otherwise equant-shaped pores. Usually the velocity appears to level off at high pressure, approaching a limiting "terminal" velocity, when, presumably, all the crack-like pore space is closed. Because the compliant fraction of the pore space can have very small porosity and yet have a very large effect on velocity, its presence, at low pressures, can cause a very poor correlation between porosity and velocity; hence, the requirement for high effective pressure. These relations work well for consolidated low-to-medium porosity and high-porosity cemented sandstones; and
- these relations should not be used for unconsolidated uncemented rocks.

7.5 Velocity–Porosity–Clay Models: Han's Empirical Relations for Shaly Sandstones

Synopsis

Han (1986) found empirical regressions relating ultrasonic (laboratory) velocities to porosity and clay content. These were determined from a set of 80 *well-consolidated* Gulf Coast sandstones with porosities, ϕ, ranging from 3% to 30% and clay volume fractions, C, ranging from 0% to 55%. The study found that clean sandstone velocities at constant effective stress can be related empirically to porosity alone with very high accuracy. When clay is present, the correlation with porosity is relatively poor but becomes very accurate if clay volume is also included in the regression. The regressions are shown in Figure 7.5.1 and Table 7.5.1.

Eberhart–Phillips (1989) used a multivariate analysis to investigate the combined influences of effective pressure, porosity, and clay content on Han's measurements of velocities in water-saturated shaly sandstones. She found that the water-saturated P- and S-wave ultrasonic velocities (in km/s) could be described empirically by

$$V_{\mathrm{P}} = 5.77 - 6.94\phi - 1.73\sqrt{C} + 0.446 \left(P_{\mathrm{e}} - 1.0e^{-16.7P_{\mathrm{e}}} \right) \tag{7.5.1}$$

$$V_{\mathrm{S}} = 3.70 - 4.94\phi - 1.57\sqrt{C} + 0.361 \left(P_{\mathrm{e}} - 1.0e^{-16.7P_{\mathrm{e}}} \right) \tag{7.5.2}$$

where P_{e} is the effective pressure in kilobars. The model accounts for 95% of the variance and has a root mean squared (rms) error of 0.1 km/s.

Uses

These relations can be used to relate velocity, porosity, and clay content empirically in shaly sandstones.

Table 7.5.1 *Han's empirical relations between ultrasonic V_P and V_S in km/s with porosity and clay volume fractions*

Clean sandstones (determined from ten samples)		
Water-saturated		
40 MPa	$V_{\mathrm{P}} = 6.08 - 8.06\phi$	$V_{\mathrm{S}} = 4.06 - 6.28\phi$
Shaly sandstones (determined from 70 samples)		
Water-saturated		
40 MPa	$V_{\mathrm{P}} = 5.59 - 6.93\phi - 2.18C$	$V_{\mathrm{S}} = 3.52 - 4.91\phi - 1.89C$
30 MPa	$V_{\mathrm{P}} = 5.55 - 6.96\phi - 2.18C$	$V_{\mathrm{S}} = 3.47 - 4.84\phi - 1.87C$
20 MPa	$V_{\mathrm{P}} = 5.49 - 6.94\phi - 2.17C$	$V_{\mathrm{S}} = 3.39 - 4.73\phi - 1.81C$
10 MPa	$V_{\mathrm{P}} = 5.39 - 7.08\phi - 2.13C$	$V_{\mathrm{S}} = 3.29 - 4.73\phi - 1.74C$
5 MPa	$V_{\mathrm{P}} = 5.26 - 7.08\phi - 2.02C$	$V_{\mathrm{S}} = 3.16 - 4.77\phi - 1.64C$
Dry		
40 MPa	$V_{\mathrm{P}} = 5.41 - 6.35\phi - 2.87C$	$V_{\mathrm{S}} = 3.57 - 4.57\phi - 1.83C$

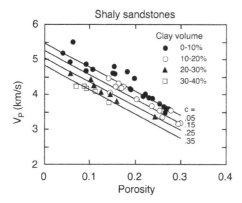

Figure 7.5.1 Han's water-saturated ultrasonic velocity data at 40 MPa compared with his empirical relations evaluated at four different clay fractions

Assumptions and Limitations

The preceding relations have the following limitations:

- These relations are empirical, and thus, strictly speaking, they apply only to the set of rocks studied. However, the result should extend in general to many consolidated sandstones. In any case, the key result is that clay content is an important parameter for quantifying velocity; if possible, the regression coefficients should be recalibrated from cores or logs at the site being studied, but be sure to include clay.

- Han's linear regression coefficients change slightly with confining pressure. They are fairly stable above about 10MPa; below this, they vary more, and the correlation coefficients degrade.

- A common mistake is to try to overinterpret the empirical coefficients by comparing the equations, for example, to Wyllie's time-average equation (see Section 7.2). This can lead to nonsensical interpreted values for the velocities of water and clay. This is not surprising, for Wyllie's equations are only heuristic.

- It is dangerous to extrapolate the results to values of porosity or clay content outside the range of the experiments. Note, for example, that the intercepts of the various equations corresponding to no porosity and no clay do not agree with each other and generally do not agree with the velocities in pure quartz.

7.6   Velocity–Porosity–Clay Models: Tosaya's Empirical Relations for Shaly Sandstones

Synopsis

On the basis of their measurements, Tosaya and Nur (1982) determined empirical regressions relating ultrasonic (laboratory) P- and S-wave velocities to porosity and clay content. For water-saturated rocks at an effective pressure of 40 MPa, they found

$$V_P(\text{km/s}) = 5.8 - 8.6\phi - 2.4C \tag{7.6.1}$$

$$V_S(\text{km/s}) = 3.7 - 6.3\phi - 2.1C \tag{7.6.2}$$

where ϕ the porosity and C is the clay content by volume. See also Han's relation in Section 7.5.

Uses

Tosaya's relations can be used to relate velocity, porosity, and clay content empirically in shaly sandstones.

Assumptions and Limitations

Tosaya's relations have the following limitations:
- These relations are empirical, and thus strictly speaking, they apply only to the set of rocks studied. However, the result should extend in general to many consolidated sandstones. In any case, the key result is that clay content is an important parameter for quantifying velocity; if possible, the regression coefficients should be recalibrated from cores or logs at the site being studied, but be sure to include clay.
- The relation previously given holds only for the high effective pressure value of 40 MPa.
- A common mistake is to try to overinterpret the empirical coefficients by comparing the equations, for example, to Wyllie's time-average equation (see Section 7.2). This can lead to nonsensical interpreted values for the velocities of water and clay. This is not surprising, because Wyllie's equations are heuristic only.
- It is dangerous to extrapolate the results to values of porosity or clay content outside the range of the experiments.

7.7   Velocity–Porosity–Clay Models: Castagna's Empirical Relations for Velocities

Synopsis

On the basis of log measurements, Castagna *et al.* (1985) determined empirical regressions relating velocities with porosity and clay content under water-saturated conditions. See also Section 7.9 on V_P–V_S relations.

For mudrock (clastic silicate rock composed primarily of clay and silt-sized particles), they found the relation between V_P and V_S (in km/s) to be

$$V_P(km/s) = 1.36 + 1.16V_S \tag{7.7.1}$$

where V_P and V_S are the P- and S-wave velocities, respectively.

For shaly sands of the Frio formation they found

$$V_P(km/s) = 5.81 - 9.42\phi - 2.21C \tag{7.7.2}$$

$$V_S(km/s) = 3.89 - 7.07\phi - 2.04C \tag{7.7.3}$$

where ϕ is porosity and C is the clay volume fraction.

Uses

Castagna's relations for velocities can be used to relate velocity, porosity, and clay content empirically in shaly sandstones.

Assumptions and Limitations

Castagna's empirical relations have the following limitations:
- These relations are empirical, and thus strictly speaking they apply only to the set of rocks studied.
- A common mistake is to try to overinterpret the empirical coefficients by comparing the equations, for example, to Wyllie's time-average equation (see Section 7.2). This can lead to nonsensical interpreted values for the velocities of water and clay. This is not surprising because Wyllie's equations are heuristic only; there is no theoretical justification for them, and they do not represent an empirical best fit to any data.

7.8 V_P–V_S–Density Models: Brocher's Compilation

Synopsis

Brocher (2005) compiled data on crustal rocks from the laboratory, wireline logs, VSP (vertical seismic profiling), and field tomography, from which he found empirical relations among V_P, V_S, bulk density (ρ), and Poisson's ratio (v) for crustal rocks. The data represented a broad range of lithologies, including sandstones, shales, mafics, gabbros, calcium-rich rocks (dolomites and anorthosite), and other crystalline rocks. The resulting empirical relations are intended to represent the average behavior of crustal rocks over a large depth range. Brocher compared his results with an excellent compilation of other published empirical relations, which are summarized in this section.

V_P–Density

Brocher (2005) computed the following polynomial fit to the Nafe–Drake curve (Ludwig *et al.,* 1970) relating P-wave velocity and bulk density, previously presented only graphically:

$$\rho(\text{g/cm}^3) = 1.6612 V_P(\text{km/s}) - 0.4721 V_P^2 + 0.0671 V_P^3 - 0.0043 V_P^4 + 0.000106 V_P^5 \quad (7.8.1)$$

Equation 7.8.1 is intended for velocities in the range $1.5 < V_P < 8.5$ km/s and all crustal rocks, except crustal mafic and calcium-rich rocks. Brocher also presented an inverse relation

$$V_P(\text{km/s}) = 39.128 \rho(\text{g/cm}^3) - 63.064 \rho^2 + 37.083 \rho^3 - 9.1819 \rho^4 + 0.8228 \rho^5 \quad (7.8.2)$$

which is accurate in the density range $2.0 < \rho < 3.5$ g/cm³. Equations 7.8.1 and 7.8.2 can be compared with an equation from Gardner *et al.* (1974) for sedimentary rocks:

$$\rho(\text{g/cm}^3) = 1.74 V_P^{0.25} \quad (7.8.3)$$

which is valid for velocities in the range $1.5 < V_P < 6.1$.

Christensen and Mooney (1995) published a relation between P-wave velocity and bulk density in crystalline rocks at 10 km depth (except for volcanic and monomineralic rocks) and in the range $5.5 < V_P < 7.5$ km/s:

$$\rho(\text{g/cm}^3) = 0.541 + 0.360 V_P \quad (7.8.4)$$

Godfrey *et al.* (1997) proposed a linear relation for basalt, diabase, and gabbro, based on data reported by Christensen and Mooney (1995):

$$\rho(\text{g/cm}^3) = 2.4372 + 0.0761 V_P \quad (7.8.5)$$

Equation 7.8.5 is also intended for crustal rocks at 10 km depth in the velocity range $5.9 < V_P < 7.1$ km/s.

Equations 7.8.1 and 7.8.3–7.8.5 are compared in Figure 7.8.1.

V_P–V_S

Brocher (2005) derived the following expression relating V_P and V_S in a broad range of lithologies (excluding calcium-rich and mafic rocks, gabbros, and serpentinites), using data from Ludwig *et al.* (1970):

$$V_S(\text{km/s}) = 0.7858 - 1.2344 V_P + 0.7949 V_P^2 - 0.1238 V_P^3 + 0.0064 V_P^4 \quad (7.8.6)$$

Equation 7.8.6 is accurate for velocities in the range $1.5 < V_P < 8.0$ km/s. The inverse is given as

$$V_P(\text{km/s}) = 0.9409 + 2.0947 V_S - 0.8206 V_S^2 + 0.2683 V_S^3 - 0.0251 V_S^4 \quad (7.8.7)$$

Brocher's relation, intended more specifically for calcium-rich (dolomites and anorthosites), mafic rocks, and gabbros, labeled the "mafic line" in Figure 7.8.2, is given by

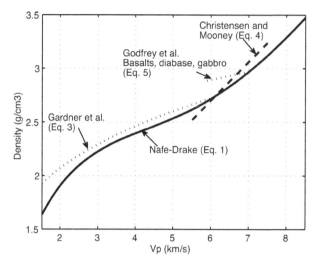

Figure 7.8.1 Brocher's comparison of empirical V_P-bulk density trends for crustal rocks, equations 7.8.1, 7.8.3–7.8.5

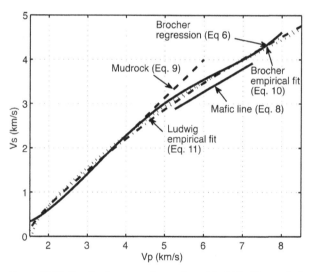

Figure 7.8.2 Brocher's comparison of empirical V_P–V_S trends for crustal rocks, equations 7.8.6, 7.8.8–7.8.11

$$V_S(\text{km/s}) = 2.88 + 0.52(V_P - 5.25), \qquad 5.25 < V_P < 7.25 \tag{7.8.8}$$

Equations 7.8.7 and 7.8.8 can be compared with the well-known mudrock line of Castagna *et al.* (1985):

$$V_S(\text{km/s}) = (V_P - 1.36)/1.16 \tag{7.8.9}$$

Figure 7.8.2 compares equations 7.8.6 and 7.8.8–7.8.11.

V_P–Poisson's Ratio

Brocher (2005) computed the following equation, labeled "Brocher's empirical fit," relating P-wave velocity and Poisson's ratio, valid in the range, $1.5 < V_P < 8.5$ km/s:

$$v = 0.8835 - 0.315 V_P + 0.0491 V_P^2 - 0.0024 V_P^3 \tag{7.8.10}$$

A similar relation, derived by Brocher from data by Ludwig *et al.* (1970), labeled "Ludwig's empirical fit," is

$$v = 0.769 - 0.226 V_P + 0.0316 V_P^2 - 0.0014 V_P^3 \tag{7.8.11}$$

Equations 7.8.8–7.8.11 are compared in Figure 7.8.3.

Uses

These equations can be used to link V_P, V_S, and density in crustal rocks.

Assumptions and Limitations

The equations in this section are empirical. Many of them, particularly those of Brocher, are intended to represent a broad range of crustal lithologies and depths. Dependence of the relations on pore fluids is not shown; it can be assumed that these represent water-saturated rocks.

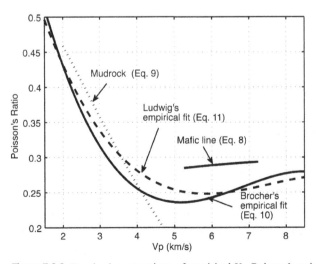

Figure 7.8.3 Brocher's comparison of empirical V_P–Poisson's ratio trends for crustal rocks, equations 7.8.8–7.8.11

7.9 V_P-V_S Relations

Synopsis

V_P-V_S relations are key to the determination of lithology from seismic or sonic log data as well as for direct seismic identification of pore fluids using, for example, AVO analysis. Castagna *et al.* (1993) give an excellent review of the subject.

There is a wide and sometimes confusing variety of published V_P-V_S relations and V_S prediction techniques, which at first appear to be quite distinct. However, most reduce to the same two simple steps.

1. Establish empirical relations among V_P, V_S, and porosity, ϕ, for one reference pore fluid – most often water-saturated or dry.
2. Use Gassmann's (1951) relations to map these empirical relations to other pore-fluid states (see Section 6.3).

Although some of the effective-medium models summarized in Chapter 4 predict both P- and S-velocities on the basis of idealized pore geometries, the fact remains that the most reliable and most often used V_P-V_S relations are empirical fits to laboratory or log data, or both. The most useful role of theoretical methods is extending these empirical relations to different pore fluids or measurement frequencies, which accounts for the two steps previously listed.

We summarize here a few of the popular V_P-V_S relations compared with laboratory and log data sets and illustrate some of the variations that can result from lithology, pore fluids, and measurement frequency.

Some Empirical Relations

Limestones

Figure 7.9.1 shows laboratory ultrasonic V_P-V_S data for water-saturated limestones from Pickett (1963), Milholland *et al.* (1980), and Castagna *et al.* (1993), as compiled by Castagna *et al.* (1993). Superimposed, for comparison, are Pickett's (1963) empirical limestone relation derived from laboratory core data:

$$V_S = V_P/1.9 \ (\text{km/s}) \tag{7.9.1}$$

and a least-squares polynomial fit to the data derived by Castagna *et al.* (1993):

$$V_S = -0.055V_P^2 + 1.017V_P - 1.031 \ (\text{km/s}) \tag{7.9.2}$$

At higher velocities, Pickett's straight line fits the data better, although at lower velocities (higher porosities), the data deviate from a straight line and trend toward the water point, $V_P = 5$ km/s, $V_S = 0$. In fact, this limit is more accurately described as a suspension of grains in water at the critical porosity (see the following discussion), at which the grains lose contact and the shear velocity vanishes.

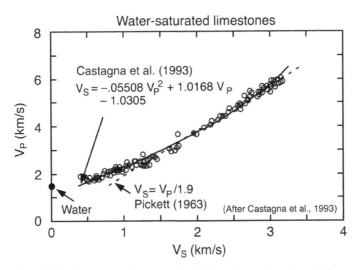

Figure 7.9.1 Laboratory ultrasonic V_P–V_S data for water-saturated limestones

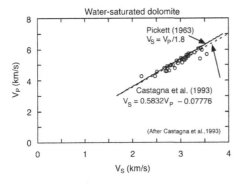

Figure 7.9.2 Laboratory V_P–V_S data for water-saturated dolomites

Dolomite

Figure 7.9.2 shows laboratory V_P–V_S data for water-saturated dolomites from Castagna *et al.* (1993). Superimposed, for comparison, are Pickett's (1963) dolomite (laboratory) relation

$$V_S = V_P/1.8 \text{ (km/s)} \tag{7.9.3}$$

and a least-squares linear fit (Castagna *et al.*, 1993)

$$V_S = 0.583V_P - 0.078 \text{ (km/s)} \tag{7.9.4}$$

For the data shown, the two relations are essentially equivalent. The data range is too limited to speculate about behavior at much lower velocity (higher porosity).

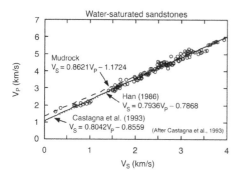

Figure 7.9.3 Laboratory V_P–V_S data for water-saturated sandstones

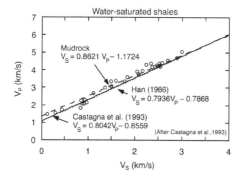

Figure 7.9.4 Laboratory V_P–V_S data for water-saturated shales

Sandstones and Shales

Figures 7.9.3 and 7.9.4 show laboratory V_P–V_S data for water-saturated sandstones and shales from Castagna *et al.* (1985, 1993) and Thomsen (1986), as compiled by Castagna *et al.* (1993).

Superimposed, for comparison, are a least-squares linear fit to these data offered by Castagna *et al.* (1993),

$$V_S = 0.804 V_P - 0.856 \text{ (km/s)} \tag{7.9.5}$$

the famous "mudrock line" of Castagna *et al.* (1985), which was derived from in situ data,

$$V_S = 0.862 V_P - 1.172 \text{ (km/s)} \tag{7.9.6}$$

and the following empirical relation of Han (1986), which is based on laboratory ultrasonic data:

$$V_S = 0.794 V_P - 0.787 \text{ (km/s)} \tag{7.9.7}$$

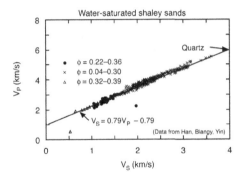

Figure 7.9.5 Laboratory ultrasonic V_P–V_S data for water-saturated shaly sands

Of these three relations, those by Han (1986) and Castagna *et al.* (1985) are essentially the same and give the best overall fit to the sandstones. The mudrock line predicts systematically lower V_S because it is best suited to the most shaly samples, as seen in Figure 7.9.4. Castagna *et al.* (1993) suggest that if the lithology is well known, one can fine tune these relations to slightly lower V_S/V_P for high shale content and higher V_S/V_P in cleaner sands. When the lithology is not well constrained, the Han and the Castagna *et al.* lines give a reasonable average.

Figure 7.9.5 compares laboratory ultrasonic data for a larger set of water-saturated sands. The lowest porosity samples ($\phi = 0.04$–0.30) are from a set of consolidated shaly Gulf Coast sandstones studied by Han (1986). The medium porosities ($\phi = 0.22$–0.36) are poorly consolidated North Sea samples studied by Blangy (1992). The very-high-porosity samples ($\phi = 0.32$–0.39) are unconsolidated clean Ottawa sand studied by Yin (1992). The samples span clay volume fractions from 0% to 55%, porosities from 0.04 to 0.39, and confining pressures from 0 to 40 MPa. In spite of this, there is a remarkably systematic trend well represented by Han's relation as follows:

$$V_S = 0.79V_P - 0.79 \; (\text{km/s}) \tag{7.9.8}$$

Sandstones: More on the Effects of Clay

Figure 7.9.6 shows again the ultrasonic laboratory data for 70 water-saturated shaly sandstone samples from Han (1986). The data are separated by clay volume fractions greater than 25% and less than 25%. Regressions to each part of the data set are shown as follows:

$$V_S = 0.842V_P - 1.099, \; \text{clay} > 25\% \tag{7.9.9}$$

$$V_S = 0.754V_P - 0.657, \; \text{clay} < 25\% \tag{7.9.10}$$

The mudrock line (upper line) is a reasonable fit to the trend but is skewed toward higher clay and lies almost on top of the regression for clay >25%.

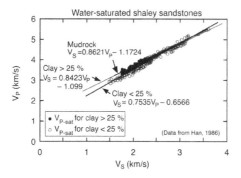

Figure 7.9.6 Laboratory ultrasonic V_P–V_S data for water-saturated shaly sands, differentiated by clay content.

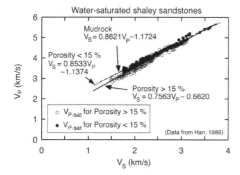

Figure 7.9.7 Laboratory ultrasonic V_P–V_S data for water-saturated shaly sands, differentiated by porosity

Sandstones: Effects of Porosity

Figure 7.9.7 shows the laboratory ultrasonic data for water-saturated shaly sandstones from Han (1986) separated into porosity greater than 15% and less than 15%. Regressions to each part of the data set are shown as follows:

$$V_S = 0.756V_P - 0.662, \text{ porosity} > 15\% \tag{7.9.11}$$

$$V_S = 0.853V_P - 1.137, \text{ porosity} < 15\% \tag{7.9.12}$$

Note that the low-porosity line is very close to the mudrock line, which as we saw previously, fits the high clay values, whereas the high-porosity line is similar to the clean sand (low-clay) regression in Figure 7.9.6.

Sandstones: Effects of Fluids and Frequency

Figure 7.9.8 compares V_P–V_S at several conditions based on the shaly sandstone data of Han (1986). The "dry" and "saturated ultrasonic" points are the measured ultrasonic data. The "saturated low-frequency" points are estimates of low-frequency saturated

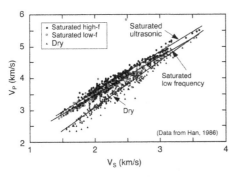

Figure 7.9.8 V_P–V_S for shaly sandstone under several different conditions described in the text

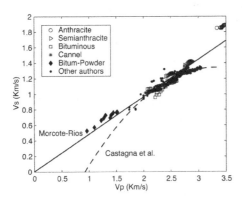

Figure 7.9.9 V_P versus V_S for different coals. The solid line is the best fit through all data. The dashed line is for some bituminous coals.

data computed from the dry measurements using the low-frequency Gassmann's relations (see Section 6.3). It is no surprise that the water-saturated samples have higher V_P/V_S because of the well-known larger effects of pore fluids on P-velocities than on S-velocities. Less often recognized is that the velocity dispersion that almost always occurs in ultrasonic measurements appears to increase V_P/V_S systematically.

Coal

Figure 7.9.9 shows laboratory ultrasonic data for coal (anthracite, semianthracite, bituminous, cannel, and bituminous powder), measured by Morcote *et al.* (2010), plus data reported by Greenhalgh and Emerson (1986), Yu *et al.* (1993), and Castagna *et al.* (1993) on bituminous coals. The Morcote regression

$$V_S = 0.4811 V_P + 0.00382 \ (\text{km/s}) \tag{7.9.13}$$

is for all data in the plot. In addition, a quadratic fit is shown for the data published by Greenhalgh and Emerson (1986) and Castagna *et al.* (1993), given by

$$V_S = -0.232V_P^2 + 1.5421V_P - 1.214 \text{ (km/s)} \tag{7.9.14}$$

Critical Porosity Model

The P- and S-velocities of rocks (as well as their V_P/V_S ratio) generally trend between the velocities of the mineral grains in the limit of low porosity and the values for a mineral-pore-fluid suspension in the limit of high porosity. For most porous materials there is a **critical porosity**, ϕ_c, that separates their mechanical and acoustic behavior into two distinct domains. For porosities lower than ϕ_c the mineral grains are load-bearing, whereas for porosities greater than ϕ_c the rock simply "falls apart" and becomes a suspension in which the fluid phase is load-bearing (see Section 7.1 on critical porosity). The transition from solid to suspension is implicit in the Raymer *et al.* (1980) empirical velocity–porosity relation (see Section 7.4) and the work of Krief *et al.* (1990), which is discussed in this chapter.

A geometric interpretation of the mineral-to-critical-porosity trend is simply that if we make the porosity large enough, the grains must lose contact and their rigidity. The geologic interpretation is that, at least for clastics, the weak suspension state at critical porosity, ϕ_c, describes the sediment when it is first deposited before compaction and diagenesis. The value of ϕ_c is determined by the grain sorting and angularity at deposition. Subsequent compaction and diagenesis move the sample along an upward trajectory as the porosity is reduced and the elastic stiffness is increased.

The value of ϕ_c depends on the rock type. For example $\phi_c \approx 0.4$ for sandstones; $\phi_c \approx 0.7$ for chalks; $\phi_c \approx 0.9$ for pumice and porous glass; and $\phi_c \approx 0.02-0.03$ for granites.

In the **suspension domain**, the effective bulk and shear moduli of the rock K and μ can be estimated quite accurately by using the Reuss (isostress) average (see Section 4.1 on the Voigt–Reuss average and Section 7.1 on critical porosity) as follows:

$$\frac{1}{K} = \frac{\phi}{K_f} + \frac{1 - \phi}{K_0}, \quad \mu = 0 \tag{7.9.15}$$

where K_f and K_0 are the bulk moduli of the fluid and mineral and ϕ is the porosity.

In the **load-bearing domain**, $\phi < \phi_c$, the moduli decrease rapidly from the mineral values at zero porosity to the suspension values at the critical porosity. Nur *et al.* (1995) found that this dependence can often be approximated with a straight line when expressed as modulus versus porosity. Although there is nothing special about a linear trend of modulus versus ϕ, it does describe sandstones fairly well, and it leads to convenient mathematical properties. For dry rocks, the bulk and shear moduli can be expressed as the linear functions

$$K_{dry} = K_0 \left(1 - \frac{\phi}{\phi_c} \right) \tag{7.9.16}$$

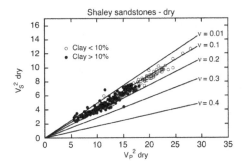

Figure 7.9.10 Velocity data from Han (1986) illustrating that Poisson's ratio is approximately constant for dry sandstones

$$\mu_{dry} = \mu_0 \left(1 - \frac{\phi}{\phi_c} \right) \tag{7.9.17}$$

where K_0 and μ_0 are the mineral bulk and shear moduli, respectively. Thus, the dry-rock bulk and shear moduli trend linearly between K_0, μ_0 at $\phi = 0$, and $K_{dry} = \mu_{dry} = 0$ at $\phi = \phi_c$. At low frequency, changes of pore fluids have little or no effect on the shear modulus. However, it can be shown (see Section 6.3 on Gassmann) that with a change of pore fluids the straight line in the K–ϕ plane remains a straight line, trending between K_0 at $\phi = 0$ and the Reuss average bulk modulus at $\phi = \phi_c$. Thus, the effect of pore fluids on K or $\rho V^2 = K + \frac{4}{3}\mu$ is automatically incorporated by the change of the Reuss average at $\phi = \phi_c$.

The relevance of the critical porosity model to V_P–V_S relations is simply that V_S/V_P should generally trend toward the value for the solid mineral material in the limit of low porosity and toward the value for a fluid suspension as the porosity approaches φ_c (Castagna et al., 1993). Furthermore, if the modulus-porosity relations are linear (or nearly so), it follows that V_S/V_P for a dry rock at any porosity ($0 < \varphi < \varphi_c$) will equal the V_S/V_P of the mineral. The same is true if K_{dry} and μ_{dry} are any other functions of porosity but are proportional to each other $[K_{dry}(\phi) \propto \mu_{dry}(\phi)]$ Equivalently, the Poisson ratio v for the dry rock will equal the Poisson ratio of the mineral grains, as is often observed (Pickett, 1963; Krief et al., 1990).

$$\left(\frac{V_S}{V_P} \right)_{dry\ rock} \approx \left(\frac{V_S}{V_P} \right)_{mineral} \tag{7.9.18}$$

$$\nu_{dry\ rock} \approx \nu_{mineral} \tag{7.9.19}$$

Figure 7.9.10 illustrates the approximately constant dry-rock Poisson ratio observed for a large set of ultrasonic sandstone velocities (from Han, 1986) over a large range of effective pressures ($5 < P_{eff} < 40\text{MPa}$) and clay contents ($0 < C < 55\%$ by volume).

To summarize, the critical porosity model suggests that P- and S-wave velocities trend systematically between their mineral values at zero porosity to fluid-suspension values ($V_S = 0$, $V_P = V_{suspension} \approx V_{fluid}$) at the critical porosity ϕ_c, which is a characteristic of each class of rocks. Expressed in the modulus versus porosity domain, if dry rock ρV_P^2 versus ϕ is proportional to ρV_S^2 versus ϕ (for example, both ρV_P^2 and ρV_S^2 are linear in ϕ), then V_S/V_P of the dry rock will be equal to V_S/V_P of the mineral.

The V_P–V_S relation for different pore fluids is found using Gassmann's relation, which is applied automatically if the trend terminates on the Reuss average at ϕ_c (see a discussion of this in Section 6.3 on Gassmann's relation).

Krief's Relations

Krief *et al.* (1990) suggested a V_P–V_S prediction technique that very much resembles the critical porosity model. The model again combines the same two elements.

1. An empirical V_P–V_S–ϕ relation for water-saturated rocks, which we will show is approximately the same as that predicted by the simple critical porosity model.
2. Gassmann's relation to extend the empirical relation to other pore fluids.

If we model the dry rock as a porous elastic solid, then with great generality, we can write the dry-rock bulk modulus as

$$K_{dry} = K_0(1 - \beta) \tag{7.9.20}$$

where K_{dry} and K_0 are the bulk moduli of the dry rock and mineral and β is Biot's coefficient (see Section 4.7 on compressibilities and Section 2.10 on the deformation of cavities). An equivalent expression is

$$\frac{1}{K_{dry}} = \frac{1}{K_0} + \frac{\phi}{K_\phi} \tag{7.9.21}$$

where K_ϕ is the pore-space stiffness (see Section 2.10) and ϕ is the porosity, so that

$$\frac{1}{K_\phi} = \frac{1}{v_p}\frac{dv_p}{d\sigma}\Big|_{P_p=\text{constant}}, \quad \beta = \frac{dv_p}{dV}\Big|_{P_p=\text{constant}} = \frac{\phi K_{dry}}{K_\phi} \tag{7.9.22}$$

where v_p is the pore volume, V is the bulk volume, σ is confining pressure, and P_p is pore pressure. The parameters β and K_ϕ are two equivalent descriptions of the pore-space stiffness. Ascertaining β versus ϕ or K_ϕ versus ϕ determines the rock bulk modulus K_{dry} versus ϕ.

Krief *et al.* (1990) used the data of Raymer *et al.* (1980) to find a relation for β versus ϕ empirically as follows:

$$(1 - \beta) = (1 - \phi)^{m(\phi)}, \quad \text{where} \quad m(\phi) = 3/(1 - \phi) \tag{7.9.23}$$

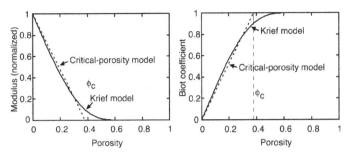

Figure 7.9.11 Left: bulk and shear moduli (same curves when normalized by their mineral values) as predicted by Krief's model and a linear critical porosity model. Right: Biot coefficient predicted by both models.

Next, they used the empirical result shown by Pickett (1963) and others that the dry-rock Poisson ratio is often approximately equal to the mineral Poisson ratio, or $\mu_{dry}/K_{dry} = \mu_0/K_0$. Combining these two empirical results gives

$$K_{dry} = K_0(1 - \phi)^{m(\phi)} \tag{7.9.24}$$

$$\mu_{dry} = \mu_0(1 - \phi)^{m(\phi)}, \quad \text{where} \quad m(\phi) = 3/(1 - \phi) \tag{7.9.25}$$

Plots of K_{dry} versus ϕ, μ_{dry} versus ϕ, and β versus ϕ are shown in Figure 7.9.11.

It is clear from these plots that the effective moduli K_{dry} and μ_{dry} display the critical porosity behavior, for they approach zero at $\phi \approx 0.4$–0.5 (see the previous discussion). This is no surprise because $\beta(\phi)$ is an empirical fit to shaly sand data, which always exhibit this behavior.

Compare these with the linear moduli–porosity relations for dry rocks suggested by Nur *et al.* (1995) for the critical porosity model

$$K_{dry} = K_0\left(1 - \frac{\phi}{\phi_c}\right), \quad 0 \leq \phi \leq \phi_c \tag{7.9.26}$$

$$\mu_{dry} = \mu_0\left(1 - \frac{\phi}{\phi_c}\right) \tag{7.9.27}$$

where K_0 and μ_0 are the mineral moduli and ϕ_c is the critical porosity. These imply a Biot coefficient of

$$\beta = \begin{cases} \phi/\phi_c, & 0 \leq \phi \leq \phi_c \\ 1, & \phi > \phi_c \end{cases} \tag{7.9.28}$$

As shown in Figure 7.9.11, these linear forms of K_{dry}, μ_{dry}, and β are essentially the same as Krief's expressions in the range $0 < \phi < \phi_c$.

The Reuss average values for the moduli of a suspension, $K_{dry} = \mu_{dry} = 0$; $\beta = 1$ are essentially the same as Krief's expressions for $\phi > \phi_c$. Krief's nonlinear form results from trying to fit a single function $\beta(\phi)$ to the two mechanically distinct domains, $\phi < \phi_c$ and $\phi > \phi_c$. The critical porosity model expresses the result with simpler piecewise functions.

Expressions for any other pore fluid are obtained by combining the expression $K_{dry} = K_0(1 - \beta)$ of Krief et al. with Gassmann's equations. Although these are also nonlinear, Krief et al. suggest a simple approximation

$$\frac{V_{P-sat}^2 - V_{fl}^2}{V_{S-sat}^2} = \frac{V_{P0}^2 - V_{fl}^2}{V_{S0}^2} \tag{7.9.29}$$

where V_{P-sat}, V_{P0}, and V_{fl} are the P-wave velocities of the saturated rock, the mineral, and the pore fluid, respectively, and V_{S-sat} and V_{S0} are the S-wave velocities in the saturated rock and mineral. Rewriting this slightly gives

$$V_{P-sat}^2 = V_{fl}^2 + V_{S-sat}^2 \left(\frac{V_{P0}^2 - V_{fl}^2}{V_{S0}^2} \right) \tag{7.9.30}$$

which is a straight line (in velocity-squared) connecting the mineral point (V_{P0}^2, V_{S0}^2) and the fluid point $(V_{fl}^2, 0)$. We suggest that a more accurate (and nearly identical) model is to recognize that velocities tend toward those of a suspension at high porosity rather than toward a fluid, which yields the modified form

$$\frac{V_{P-sat}^2 - V_R^2}{V_{S-sat}^2} = \frac{V_{P0}^2 - V_R^2}{V_{s0}^2} \tag{7.9.31}$$

where V_R is the velocity of a suspension of minerals in a fluid given by the Reuss average (Section 4.1) or Wood's relation (Section 4.4) at the critical porosity.

It is easy to show that this modified form of Krief's expression is exactly equivalent to the linear (modified Voigt) K versus ϕ and μ versus ϕ relations in the critical porosity model with the fluid effects given by Gassmann.

Greenberg and Castagna's Relations

Greenberg and Castagna (1992) have given empirical relations for estimating V_S from V_P in multimineralic, brine-saturated rocks based on empirical, polynomial V_P–V_S relations in pure monomineralic lithologies (Castagna et al., 1993). The shear-wave velocity in brine-saturated composite lithologies is approximated by a simple average of the arithmetic and harmonic means of the constituent pure lithology shear velocities:

$$V_S = \frac{1}{2} \left\{ \left[\sum_{i=1}^{L} X_i \sum_{j=0}^{N_i} a_{ij} V_P^j \right] + \left[\sum_{i=1}^{L} X_i \left(\sum_{j=0}^{N_i} a_{ij} V_P^j \right)^{-1} \right]^{-1} \right\} \tag{7.9.32}$$

Table 7.9.1 *Regression coefficients for pure lithologies*

Lithology	a_{i2}	a_{i1}	a_{i0}	R^2
Sandstone	0	0.804 16	−0.855 88	0.983 52
Limestone	−0.055 08	1.016 77	−1.030 49	0.990 96
Dolomite	0	0.583 21	−0.077 75	0.874 44
Shale	0	0.769 69	−0.867 35	0.979 39

Note:
[a] V_P and V_S in km/s: $V_S = a_{i2} V_P^2 + a_{i1} V_P + a_{i0}$ (Castagna *et al.*, 1993).

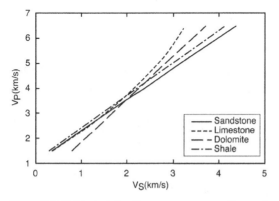

Figure 7.9.12 Typical V_P–V_S curves corresponding to the regression coefficients in Table 7.9.1

$$\sum_{i=1}^{L} X_i = 1 \qquad\qquad (7.9.33)$$

where L is the number of pure monomineralic lithologic constituents, X_i are the volume fractions of lithological constituents, a_{ij} are the empirical regression coefficients, N_i is the order of polynomial for constituent i, and V_P and V_S are the P- and S-wave velocities (km/s), respectively, in composite brine-saturated, multimineralic rock.

Castagna *et al.* (1993) gave representative polynomial regression coefficients for pure monomineralic lithologies as detailed in Table 7.9.1. Note that the preceding relation is for 100% brine-saturated rocks. To estimate V_S from measured V_P for other fluid saturations, Gassmann's equation has to be used in an iterative manner. In the following, the subscript b denotes velocities at 100% brine saturation, and the subscript f denotes velocities at any other fluid saturation (e.g., oil or a mixture of oil, brine, and gas). The method consists of iteratively finding a (V_P, V_S) point on the brine relation that transforms, with Gassmann's relation, to the measured V_P and the unknown V_S for the new fluid saturation. The resulting curves are shown in Figure 7.9.12. The steps are as follows:

1. Start with an initial guess for V_{Pb}.
2. Calculate V_{Sb} corresponding to V_{Pb} from the empirical regression.
3. Perform fluid substitution using V_{Pb} and V_{Sb} in the Gassmann equation to obtain V_{Sf}.
4. With the calculated V_{Sf} and the measured V_{Pf}, use the Gassmann relation to obtain a new estimate of V_{Pb}. Check the result against the previous value of V_{Pb} for convergence. If convergence criterion is met, stop; if not, go back to step 2 and continue.

When the measured P-velocity and desired S-velocity are for 100% brine saturation, then of course. iterations are not required. The desired V_S is obtained from a single application of the empirical regression. This method requires prior knowledge of the lithology, porosity, saturation, and elastic moduli and densities of the constituent minerals and pore fluids.

Example: Estimate, using the Greenberg–Castagna empirical relations, the shear-wave velocity in a brine-saturated shaly sandstone (60% sandstone, 40% shale) with $V_P =3.0$ km/s.

Here $L = 2$ with X_1 (sandstone) = 0.6 and X_2 (shale) = 0.4.
The regressions for pure lithologic constituents give us

$$V_{S\text{-sand}} = 0.804\,16\, V_P - 0.855\,88 = 1.5566 \text{ km/s}$$
$$V_{S\text{-shale}} = 0.769\,69\, V_P - 0.867\,35 = 1.4417 \text{ km/s}$$

The weighted arithmetic and harmonic means are

$$V_{S\text{-arith}} = 0.6 V_{S\text{-sand}} + 0.4 V_{S\text{-shale}} = 1.5106 \text{ km/s}$$
$$V_{S\text{-harm}} = (0.6/V_{S\text{-sand}} + 0.4/V_{S\text{-shale}})^{-1} = 1.5085 \text{ km/s}$$

and finally the estimated V_S is given by

$$V_S = \frac{1}{2}(V_{S\text{-arith}} + V_{S\text{-harm}}) = 1.51 \text{ km/s}$$

Vernik's Relations

The Greenberg–Castagna (1992) empirical relations discussed previously are appropriate for consolidated rocks with P-wave velocities greater than about 2.6 km/s. However, when extrapolated to low velocities, these relations yield unphysical results – specifically $V_S = 0$ at $V_P = 1.06$ km/s, which is slower than the V_P for water. Vernik *et al.* (2002) developed a nonlinear regression for brine-saturated sandstones that honors the observed high-velocity limit for arenites ($V_P \approx 5.48$ km/s, $V_S \approx 3.53$ km/s) and the velocity of a quartz–water suspension ($V_P \approx 1.7$ km/s at $V_S = 0$) at a critical porosity $\phi \approx 0.4$:

$$V_S = (-1.267 + 0.372V_P^2 + 0.00284V_P^4)^{1/2} \; (\text{km/s}) \tag{7.9.34}$$

Their corresponding expression for shale is

$$V_S = (-0.79 + 0.287V_P^2 + 0.00284V_P^4)^{1/2} \; (\text{km/s}) \tag{7.9.35}$$

The V_P/V_S ratio in organic rich rocks tend to be between 1.5–1.8, based on empirical observations from well data (Vernik, 2016). Vernik (2016) and Vernik *et al.* (2017) give an empirical V_P/V_S relation for organic unconventional source rocks that takes into account TOC content:

$$V_S = bV_P + a_0 + (a_{ref} - a_0)\frac{TOC}{TOC_{ref}} \tag{7.9.36}$$

The "zero TOC" line is given by $b = 0.58$ and $a_0 = -0.22$: $V_S = 0.58V_P - 0.22$. Increasing kerogen content tends to reduce the V_P/V_S ratio. A reference TOC line (when TOC = TOC_{ref}) is given by

$$V_S = 0.58V_P + a_{ref} \tag{7.9.37}$$

and this line corresponds to maximum TOC in the area such that all the data are bounded by the two parallel lines, the zero TOC line and the reference TOC lines. The values of a_{ref} and TOC_{ref} were calibrated with well data from 9 different wells in different source rock formations as shown in the following table. The residual error of the fit was about 2.8% (Vernik *et al.*, 2017). This relation requires a TOC log (core-calibrated when possible), in addition to the P-wave sonic log to predict the shear velocity.

Well	TOC_{ref} (%)	a_{ref} (km/s)
Bakken1	13.0	0.1
Bakken2	15.0	0.10
Woodford	8.0	0.22
Avalon	8.0	0.22
Eagle Ford	7.0	0.10
Wolfcamp	6.0	0.18
Cline	6.0	0.10
Marcellus1	9.0	0.30
Marcellus2	14.0	0.30

In the previous table, 1 and 2 refer to two different wells.

Williams's Relation

Williams (1990) used empirical V_P–V_S relations from acoustic logs to differentiate hydrocarbon-bearing sandstones from water-bearing sandstones and shales statistically. His least-squares regressions are:

$$V_P/V_S = 1.182 + 0.00422 \, \Delta t_S \quad \text{(water-bearing sands)} \tag{7.9.38}$$

$$V_P/V_S = 1.276 + 0.003\,74 \, \Delta t_S \quad \text{(shale)} \tag{7.9.39}$$

where Δt_S is the shear-wave slowness in μs/ft. The effect of replacing water with more compressible hydrocarbons is a large decrease in P-wave velocity with little change (slight increase) in S-wave velocity. This causes a large reduction in the V_P/V_S ratio in hydrocarbon sands compared with water-saturated sands having a similar Δt_S. A measured V_P/V_S and Δt_S is classified as either water-bearing or hydrocarbon-bearing by comparing it with the regression and using a statistically determined threshold to make the decision. Williams chose the threshold so that the probability of correctly identifying a water-saturated sandstone is 95%. For this threshold a measured V_P/V_S is classified as water-bearing if

$$V_P/V_S \text{ (measured)} \geq \min[V_P/V_S \text{ (sand)}, \ V_P/V_S \text{ (shale)}] - 0.09$$

and as potentially hydrocarbon-bearing otherwise. Williams found that when $\Delta t_S < 130$ μs/ft (or $\Delta t_P < 75$ μs/ft), the rock is too stiff to give any statistically significant V_P/V_S anomaly upon fluid substitution.

Xu and White's Relation

Xu and White (1995) developed a theoretical model for velocities in shaly sandstones. The formulation uses the Kuster–Toksöz and differential effective-medium theories to estimate the dry rock P- and S-velocities, and the low-frequency saturated velocities are obtained from Gassmann's equation. The sand–clay mixture is modeled with ellipsoidal inclusions of two different aspect ratios. The sand fraction has stiffer pores with aspect ratio $\alpha \approx 0.1 - 0.15$, whereas the clay-related pores are more compliant with $\alpha \approx 0.02 - 0.05$. The velocity model simulates the "V"-shaped velocity–porosity relation of Marion *et al.* (1992) for sand–clay mixtures. The total porosity $\phi = \phi_{\text{sand}} + \phi_{\text{clay}}$, where ϕ_{sand} and ϕ_{clay} are the porosities associated with the sand and clay fractions, respectively. These are approximated by

$$\phi_{\text{sand}} = (1 - \phi - V_{\text{clay}}) \frac{\phi}{1 - \phi} = V_{\text{sand}} \frac{\phi}{1 - \phi} \tag{7.9.40}$$

$$\phi_{\text{clay}} = V_{\text{clay}} \frac{\phi}{1 - \phi} \tag{7.9.41}$$

where V_{sand} and V_{clay} denote the volumetric sand and clay content, respectively. The shale volume from logs may be used as an estimate of V_{clay}. Though the log-derived shale volume includes silts and overestimates clay content, results obtained by Xu and White justify its use. The properties of the solid mineral mixture are estimated by a Wyllie time average of the quartz and clay mineral velocities and arithmetic average of their densities by

$$\frac{1}{V_{P_0}} = \left(\frac{1 - \phi - V_{clay}}{1 - \phi}\right)\frac{1}{V_{Pquartz}} + \frac{V_{clay}}{1 - \phi}\frac{1}{V_{Pclay}} \tag{7.9.42}$$

$$\frac{1}{V_{S_0}} = \left(\frac{1 - \phi - V_{clay}}{1 - \phi}\right)\frac{1}{V_{Squartz}} + \frac{V_{clay}}{1 - \phi}\frac{1}{V_{Sclay}} \tag{7.9.43}$$

$$\rho_0 = \left(\frac{1 - \phi - V_{clay}}{1 - \phi}\right)\rho_{quartz} + \frac{V_{clay}}{1 - \phi}\rho_{clay} \tag{7.9.44}$$

where the subscript 0 denotes the mineral properties. These mineral properties are then used in the Kuster–Toksöz formulation along with the porosity and clay content to calculate dry-rock moduli and velocities. The limitation of small-pore concentration of the Kuster–Toksöz model is handled by incrementally adding the pores in small steps so that the noninteraction criterion is satisfied in each step. Gassmann's equations are used to obtain low-frequency saturated velocities. High-frequency saturated velocities are calculated by using fluid-filled ellipsoidal inclusions in the Kuster–Toksöz model.

The model can be used to predict shear-wave velocities (Xu and White, 1994). Estimates of V_S may be obtained from known mineral matrix properties and measured porosity and clay content or from measured V_P and either porosity or clay content.

Xu and White recommend using measurements of P-wave sonic log because it is more reliable than estimates of shale volume and porosity.

Raymer-Form V_S Prediction

Dvorkin (2007, personal communication) uses the Raymer–Hunt–Gardner functional form $V_P = (1 - \phi)^2 V_{Ps} + \phi V_{Pf}$, where V_{Ps} and V_{Pf} denote the P-wave velocity in the solid and in the pore-fluid phases, respectively, and ϕ is the total porosity, to relate the S-wave velocity in dry rock to porosity and mineralogy as $V_{Sdry} = (1 - \phi)^2 V_{Ss}$, where V_{Ss} is the S-wave velocity in the solid phase. Assuming that the shear modulus of rock does not depend on the pore fluid, V_S in wet rock is $V_{Swet} = V_{Sdry}\sqrt{\rho_{bdry}/\rho_{bwet}}$, where ρ_{bdry} and ρ_{bwet} denote the bulk density of the dry and wet rock, respectively. This equation for V_S prediction reiterates the critical porosity concept of Nur et al. (1995): the V_P/V_S ratio in dry rock equals that in the solid phase. However, the velocity–porosity trend that follows from this equation differs somewhat from the traditional critical porosity trend. This V_S predictor exhibits a velocity–porosity trend essentially identical to that from the stiff-sand (the modified upper Hashin–Shtrikman) model (Figure 7.9.13).

Uses

The relations discussed in this section can be used to relate P- and S-velocity and porosity empirically for use in lithology detection and direct fluid identification.

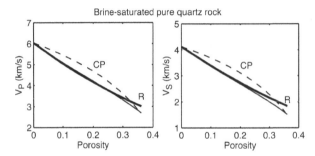

Figure 7.9.13 The stiff-sand (slender curves), critical porosity (dashed curves labeled "CP"), and RHG velocity predictions (heavy curves, labeled "R") for P- (left) and S-wave (right) velocity in pure-quartz, brine-saturated rock

Assumptions and Limitations

Strictly speaking, the empirical relations discussed in this section apply only to the set of rocks studied.

7.10   Velocity–Density Relations

Synopsis

Many seismic modeling and interpretation schemes require, as a minimum, P-wave velocity V_P, S-wave velocity, V_S, and bulk density ρ_b. Laboratory and log measurements can often yield all three together. But there are many applications where only V_P is known, and density or V_S must be estimated empirically from V_P. Section 7.9 summarizes some V_P–V_S relations. Here we summarize some popular and useful V_P–density relations. Castagna *et al.* (1993) give a very good summary of the topic. See also Section 7.8 on Brocher's relations.

Density is a simple volumetric average of the rock constituent densities and is closely related to porosity by

$$\rho_b = (1 - \phi)\rho_0 + \phi\rho_{fl} \tag{7.10.1}$$

where ρ_0 is the density of mineral grains, ρ_{fl} is the density of pore fluids, and ϕ is porosity.

The problem is that velocity is often not very well related to porosity (and therefore to density). Cracks and crack-like flaws and grain boundaries can substantially decrease V_P and V_S, even though the cracks may have near-zero porosity.

Velocity–porosity relations can be improved by fluid saturation and high effective pressures, both of which minimize the effect of thin cracks. Consequently, we also expect velocity–density relations to be more reliable under high effective pressures and fluid saturation.

Gardner *et al.* (1974) suggested a useful empirical relation between P-wave velocity and density that represents an average over many rock types:

$$\rho_b \approx 1.741 V_P^{0.25} \tag{7.10.2}$$

where V_P is in km/s and ρ_b is in g/cm³, or

$$\rho_b \approx 0.23 V_P^{0.25} \tag{7.10.3}$$

where V_P is in ft/s.

More useful predictions can be obtained by using the lithology-specific forms given by Gardner *et al.* (1974). Castagna *et al.* (1993) suggested slight improvements to Gardner's relations and summarized these, as shown in Table 7.10.1, in both polynomial and power-law form. Figures 7.10.1–7.10.4 show Gardner's relations applied to laboratory data.

Assumption and Limitations

Gardner's relations are empirical.

Table 7.10.1 *Polynomial and power-law forms of the Gardner* et al. *(1974) velocity–density relationships presented by Castagna* et al. *(1993). Units are km/s and g/cm3 for velocity and density, respectively.*

Coefficients for the equation $\rho_b = aV_P^2 + bV_P + c$

Lithology	a	b	c	V_P range (km/s)
Shale	−0.0261	0.373	1.458	1.5–5.0
Sandstone	−0.0115	0.261	1.515	1.5–6.0
Limestone	−0.0296	0.461	0.963	3.5–6.4
Dolomite	−0.0235	0.390	1.242	4.5–7.1
Anhydrite	−0.0203	0.321	1.732	4.6–7.4

Coefficients for the equation $\rho_b = dV_P^f$

Lithology	d	f	V_P range (km/s)
Shale	1.75	0.265	1.5–5.0
Sandstone	1.66	0.261	1.5–6.0
Limestone[a]	1.36	0.386	3.5–6.4
Dolomite	1.74	0.252	4.5–7.1
Anhydrite	2.19	0.160	4.6–7.4

Note:
[a] Coefficients for limestone have been revised here to better reflect observed trends

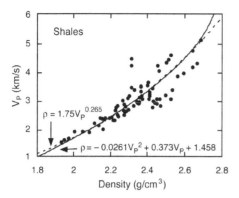

Figure 7.10.1 Both forms of Gardner's relations applied to log and laboratory shale data, as presented by Castagna *et al.* (1993)

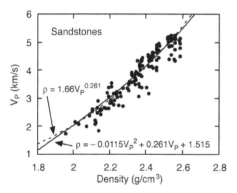

Figure 7.10.2 Both forms of Gardner's relations applied to log and laboratory sandstone data, as presented by Castagna *et al.* (1993)

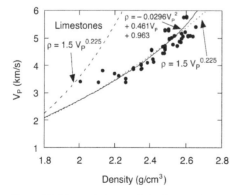

Figure 7.10.3 Both forms of Gardner's relations applied to laboratory limestone data. Note that the published power-law form does not fit as well as the polynomial. We also show a revised power-law form fit to these data, which agrees very well with the polynomial.

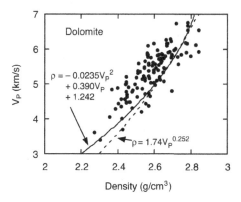

Figure 7.10.4 Both forms of Gardner's relations applied to laboratory dolomite data

7.11 Eaton and Bowers Pore-Pressure Relations

Synopsis

Following Eaton (1975), Gutierrez *et al.* (2006) present the following equation for pore-pressure prediction from measured P-wave velocity:

$$P_{over} - P_p = (P_{over} - P_{hyd})(V_p/V_{pn})^3 \qquad (7.11.1)$$

where P_p is the actual pore pressure, P_{over} is the vertical overburden stress, P_{hyd} is the normal hydrostatic pressure, V_p is the measured velocity, and V_{pn} is the normal-compaction velocity. This equation implies that if $V_p = V_{pn}$, i.e., the rock is normally compacted, $P_p = P_{hyd}$. Conversely, if $V_p < V_{pn}$, i.e., the measured velocity is smaller than the normal velocity, $P_p > P_{hyd}$, which means that the rock is overpressured. The Eaton equation is applicable to overpressure due to undercompaction of shale, which occurs during monotonic overburden stress increase due to burial.

The modified Eaton–Yale pore pressure relation (Yale *et al.*, 2018) extends Eaton's relation to account for age and lithification of shale reservoirs, as well as incorporates corrections to the sonic velocities for lithologic and porosity variations. The modified Eaton–Yale equation is given by:

$$P_{over} - \alpha P_p = C(P_{over} - \alpha P_{hyd})(V_{p-corr}/V_{pn-corr})^{EE} \qquad (7.11.2)$$

where EE is the Eaton exponent, varying between 2 and 3 for young sediments, but can be lower for lithified basin; C is a calibration parameter, and α is Biot's coefficient, which is close to 1 for shallow young sediments but can vary between 0.3 to 0.8 for older lithified unconventional basins. Yale *et al.* (2018) note that inclusion of α is critical in unconventional reservoirs. V_{p-corr} is the measured velocity, corrected for porosity variations, and $V_{pn-corr}$ is the normal compaction velocity, corrected for lithologic or mineralogic variations. The velocity-porosity variations are modeled using a critical porosity model, while mineralogic variations are modeled using a Voigt–Reuss–Hill average of mineral fractions. Like the Eaton method, the modified Eaton–Yale method also makes use of the notion that $V_p < V_{pn}$ is an indication of overpressure, but it also recognizes that in addition to stress, porosity and mineralogy also affect velocities. By correcting for porosity and lithology variations, the method attempts to get a compressional velocity where overpressure is the main controlling factor. Use of this method requires a wide range of petrophysical inputs in addition to sonic logs and density logs, including porosity logs, mineral volume fractions, and well-calibrated petrophysical models, ideally calibrated with core XRD analysis. The method can be sensitive to the petrophysical models used for the corrections.

Bowers (1995) considers overpressure generation not only due to undercompaction but also due to tectonic unloading, such as occurs during uplift of rock. Gutierrez *et al.* (2006) present Bowers's equation as

$$P_{over} - P_p = [(V_p - 5000)/a]^{1/b} \qquad (7.11.3)$$

where pressure is in psi and velocity is in ft/s. The coefficients calibrated to the monotonic-compaction data from the Gulf of Mexico are $a = 9.18448$ and $b = 0.764984$. These coefficients may be different in environments where abnormal pore pressure is generated by unloading, kerogen maturation, or clay mineral transformations.

Assumption and Limitations

The constants and trends used in these equations are highly site-specific and require thorough calibration using real pore-pressure measurements. As a result, we caution against using the constants appropriate for one basin at another location. Moreover, these constants may vary even within the same basin between different fault blocks. The normal compaction trends for the P-wave velocity are sometimes simply not present in basins where all drilled wells encounter overpressure, i.e., the measured velocity is always abnormally slow. In this case, an assumed normal velocity trend has to be adopted.

7.12   Kan and Swan Pore-Pressure Relations

Synopsis

Hottman and Johnson (1965) established an empirical relation between the pore-pressure gradient, $P(z)/z$, and Δt, the departure of sonic interval transit time from a background trend of transit time with depth. The background trend is for normally pressured shale undergoing normal compaction. The relation is given by

Table 7.12.1 *Kan and Swan's (2001) empirical coefficients*

Basin and age	c_1 (Pa/μs)	c_2 (Pa·m/μs²)
Gulf of Mexico Miocene	143	−0.42
Gulf of Mexico Pliocene	67.6	−0.10
Gulf of Mexico Pleistocene	42.7	−0.028
North Sea	13.1	0.168
Alaska	36.5	0.48
Northwest Australia	42.4	0.22
South China Sea	32.8	1.54

$$P(z)/z = R_w + c_1 \Delta t + c_2 (\Delta t)^2 \tag{7.12.1}$$

where R_w is the hydrostatic pore-pressure gradient (in Pa/m), z is the depth in meters, and $\Delta t = t(z) - t_0 \exp(-z/k)$ is the interval transit time departure in μs/m.

The normal compaction shale trend is given by $t_0 \exp(-z/k)$, where t_0 is the transit time of the shale at $z = 0$ and k is determined from log data in normally pressured shale. Kan and Swan (2001) give a compilation of coefficients c_1 and c_2 for specific basins and ages, determined by empirical regression (Table 7.12.1). With the appropriate coefficients, Kan and Swan expect the empirical equation to be reasonably accurate for pressure gradients up to 20 kPa/m as long as the overpressure is caused only by undercompaction.

Assumption and Limitations

- The relation is empirical.
- The relation applies strictly only to overpressure caused by undercompaction and not to other mechanisms such as aquathermal pressure, kerogen maturation, and other late-stage mechanisms.

7.13 Attenuation and Quality Factor Relations

Klimentos and McCann (1990) present a statistical relation between the compressional-wave attenuation coefficient α (measured in dB/cm), porosity ϕ, and clay content C (both measured in volume fraction) obtained in water-saturated sandstone samples. The porosity of the samples ranges from 0.05 to 0.30, and the clay content ranges from zero to 0.25. The differential pressure is 40 MPa and the frequency is 1 MHz. The relation is

$$\alpha = 3.15\phi + 24.1C - 0.132 \tag{7.13.1}$$

with a correlation coefficient of 0.88.

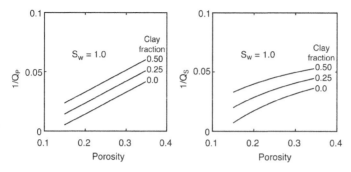

Figure 7.13.1 The P- and S-wave inverse quality factors in fully water-saturated sandstone versus porosity for a clay content of zero, 0.25, and 0.5 (from bottom to top). The effective pressure is 30 MPa, the frequency is 1 MHz, and the permeability is 100 mD. The inverse quality factors increase with increasing clay content.

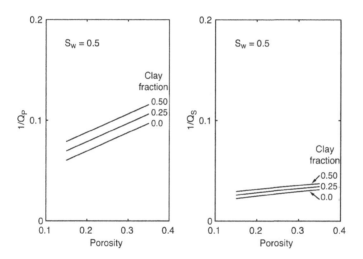

Figure 7.13.2 Same as Figure 7.13.1 but with a water saturation of 0.5. Again, the clay content is zero, 0.25, and 0.5 (from bottom to top). The effective pressure is 30 MPa, frequency is 1 MHz, and permeability is 100 mD. The inverse quality factors increase with increasing clay content.

Koesoemadinata and McMechan (2001) use a large number of experimental results in sandstones to obtain statistical relations between the P- and S-wave inverse quality factors and the effective pressure, porosity, clay content, water saturation, permeability, and frequency. Figures 7.13.1 and 7.13.2 display selected results from these relations.

Uses

These relations can be used to estimate attenuation versus porosity, clay content, saturation, frequency, pressure, and permeability.

Assumptions and Limitations

These relations are strictly empirical and mostly valid at laboratory ultrasonic frequencies. They may not be applicable in the well-log, VSP, and/or seismic frequency ranges.

7.14 Velocity–Porosity–Strength Relations

Synopsis

Chang *et al.* (2004) and Zoback (2007), summarize 30 empirical equations relating physical properties (such as velocity, Young's modulus, and porosity) to unconfined compressive strength (UCS) in sandstone, shale, and limestone and dolomite; see Tables 7.14.1–7.14.3. While some equations work reasonably well (for example, some strength–porosity relationships for sandstone and shale), data on rock strength show considerable scatter around the empirical regressions, emphasizing the importance of local calibration before utilizing any of the relationships presented. Nonetheless, some reasonable correlations can be found between velocity, porosity, and rock strength that can be used for applications related to well-bore stability, where having a lower-bound estimate of rock strength is especially useful.

Table 7.14.1 *Empirical relations for sandstones (Chang* et al.*, 2004; Zoback, 2007)*

UCS =		Equation number
$0.035 V_P - 31.5$	Thuringia, Germany; Freyburg (1972)	(1)
$1200 \exp(-0.036\Delta t)$	Bowen Basin, Australia. Fine-grained consolidated and unconsolidated sandstone; McNally (1987)	(2)
$1.4138 \times 10^7 \Delta t^{-3}$	Gulf Coast, weak unconsolidated sandstone	(3)
$3.3 \times 10^{-20}\rho^2 V_P^4$ $\times [(1+v)/(1-v)]^2$ $\times (1-2v)[1+0.78V_{clay}]$	Gulf Coast sandstones with UCS > 30MPa; Fjaer *et al.* (2008)	(4)
$1.745 \times 10^{-9}\rho V_P^2 - 21$	Cook Inlet, Alaska; coarse-grained sandstone and conglomerates; Moos *et al.* (1999)	(5)
$42.1 \exp(1.9 \times 10^{-11}\,\rho V_P^2)$	Australia; consolidated sandstones with UCS >80 MPa, porosity between 0.05 and 0.12; Chang *et al.* (2004)	(6)
$3.87 \exp(1.14 \times 10^{-10}\rho V_P^2)$	Gulf of Mexico; Chang *et al.* (2004)	(7)
$46.2\exp(0.027E)$	Chang *et al.* (2004)	(8)
$2.28 + 4.1089E$	Worldwide; Bradford *et al.* (1998)	(9)
$254(1 - 2.7\phi)^2$	Worldwide; very clean, well-consolidated sandstones with porosity < 0.3; Vernik *et al.* (1993)	(10)
$277\exp(-10\phi)$	Sandstones with 2 < UCS < 360 MPa, and 0.002 < ϕ < 0.3; Chang *et al.* (2004)	(11)

Table 7.14.2 *Empirical relations for shales (Chang* et al., *2004; Zoback, 2007)*

UCS =		Equation number
$0.77(304.8/\Delta t)^{2.93}$	North Sea, high-porosity Tertiary shales; Horsrud (2001)	(12)
$0.43(304.8/\Delta t)^{3.2}$	Gulf of Mexico, Pliocene and younger shales; Chang *et al.* (2004)	(13)
$1.35(304.8/\Delta t)^{2.6}$	Worldwide; Chang *et al.* (2004)	(14)
$0.5(304.8/\Delta t)^{3}$	Gulf of Mexico; Chang *et al.* (2004)	(15)
$10(304.8/\Delta t - 1)$	North Sea, high-porosity Tertiary shales; Lal (1999)	(16)
$7.97E^{0.91}$	North Sea, high-porosity Tertiary shales; Horsrud (2001)	(17)
$7.22E^{0.712}$	Strong, compacted shales; Chang *et al.* (2004)	(18)
$1.001 \, \phi^{-1.143}$	Low-porosity (<0.1), high-strength (~79MPa) shales; Lashkaripour and Dusseault (1993)	(19)
$2.922 \, \phi^{-0.96}$	North Sea, high-porosity Tertiary shales; Horsrud (2001)	(20)
$0.286 \, \phi^{-1.762}$	High-porosity (>0.27) shales; Chang *et al.* (2004)	(21)

Table 7.14.3 *Empirical relations for limestone and dolomite (Chang* et al., *2004; Zoback, 2007)*

UCS		Equation number
$(7682/\Delta t)^{1.82}/145$	Militzer and Stoll (1973)	(22)
$10^{(2.44 + 109.14/\Delta t)}/145$	Golubev and Rabinovich (1976)	(23)
$13.8E^{0.51}$	Limestone with 10 < UCS < 300 MPa; Chang *et al.* (2004)	(24)
$25.1E^{0.34}$	Dolomite with 60 < UCS < 100 MPa; Chang *et al.* (2004)	(25)
$276(1-3\phi)^{2}$	Korobcheyev deposit, Russia; Rzhevsky and Novik (1971)	(26)
$143.8 \exp(-6.95\phi)$	Middle East, low- to moderate-porosity (0.05–0.2), high UCS (30–150 MPa); Chang *et al.* (2004)	(27)
$135.9 \exp(-4.8\phi)$	low- to moderate-porosity (0.0–0.2), high UCS (10–300 MPa); Chang *et al.* (2004)	(28)

Units and symbols in the tables are:

UCS (MPa):   unconfined compressive strength
Δt (μs/ft):   P-wave sonic transit time
V_P (m/s):   P-wave velocity
ρ (kg/m^3):   bulk density
E (GPa):   Young's modulus
ϕ (fraction):   porosity
V_{clay} (fraction):   clay fraction
GR (API):   gamma-ray log value
v: Poisson's ratio

In addition to UCS, another material parameter of interest in estimating rock strength is the angle of internal friction, $\theta = \tan^{-1}\mu$, where μ is the coefficient of internal friction. Chang *et al.* (2004) give the relations for θ, summarized in Table 7.14.4.

Table 7.14.4 *Angles of internal friction*

θ (deg)=		Equation number
$\sin^{-1}[(V_P - 1000)/(V_P + 1000)]$	Shales; Lal (1999)	(29)
57.8-105ϕ Sandstone	Weingarten and Perkins (1995)	(30)
$\tan^{-1}\left[\dfrac{(GR - GR_{\text{sand}})\mu_{\text{shale}} + (GR_{\text{shale}} - GR)\mu_{\text{sand}}}{GR_{\text{shale}} - GR_{\text{sand}}}\right]$	Shaly sandstones; Chang *et al.* (2004)	(31)

These relations are of doubtful reliability, but show trends that are approximately in agreement with observations (Wong *et al.*, 1993; Horsrud, 2001).

Use of equation (30) in Table 7.14.4 requires the reference values of GR and μ (coefficient of internal friction) for pure shale and pure sand end-members. These have to be assumed or calibrated from well logs.

Uses

These equations can be used to estimate rock strength from properties measurable with geophysical well logs.

Assumptions and Limitations

- The relations described in this section are obtained from empirical fits that show considerable scatter, especially for carbonates. Relations should be locally calibrated when possible.
- Relations for well-consolidated rocks (e.g., equations (10), (18), and (19) in the tables) should not be used for estimating strength of poorly consolidated, weak rocks. Equations (3) and (5) provide reasonable estimates for weak sands.
- Equation (11) for sandstones gives a reasonable fit to UCS for porosities >0.1, with 80% of the data within ±30 MPa.
- Equations (15), and (19)–(21) for shales predict shale strength fairly well. For porosities > 0.1, equations (19)–(21) fit 90% of the data to within ±10MPa.

7.15   Birch's Law

Synopsis

Based on extensive high pressure measurements, Birch (1960, 1961) showed a linear relationship, known as Birch's law, between the compressional wave velocity, V_P, and bulk density, ρ, in *zero-porosity* rocks:

$$V_P = a(\overline{M}) + b\rho \tag{7.15.1}$$

Table 7.15.1 *Mean atomic weight of minerals with ideal compositions (from Birch, 1961, table 13)*

Chrysoberyl	$BeAl_2O_4$	18.14
Spodumene	$LiAlSi_2O_6$	18.60
Brucite	$Mg(OH)_2$	19.44
Serpentine	$Mg_3Si_2O_5(OH)_4$	19.80
Chlorite	$Mg_5Al_2Si_3O_{10}(OH)_8$	19.85
Talc	$Mg_3Si_4O_{10}(OH)_2$	19.96
Anthophyllite	$Mg_7Si_8O_{22}(OH)_2$	20.02
Quartz	SiO_2	20.03
Enstatite	$MgSiO_3$	20.08
Glaucophane	$Na_2Mg_3Al_2Si_8O_{22}(OH)_2$	20.09
Forsterite	Mg_2SiO_4	20.10
Pyrope	$Mg_3Al_2Si_3O_{12}$	20.15
Periclase	MgO	20.16
Albite	$NaAlSi_3O_8$	20.17
Cordierite	$Mg_2Al_4Si_5O_{18}$	20.17
Jadeite	$NaAlSi_2O_6$	20.21
Kyanite	Al_2SiO_5	20.25
Spinel	$MgAl_2O_4$	20.32
Corundum	Al_2O_3	20.39
Phlogopite	$KMg_3AlSi_3O_{10}(OH)_2$	20.86
Muscovite	$KAl_3Si_3O_{10}(OH)_2$	20.96
Orthoclase	$KAlSi_3O_8$	21.40
Anorthite	$CaAl_2Si_2O_8$	21.40
Zoisite	$Ca_2Al_3Si_3O_{12}(OH)$	21.63
Diopside	$CaMgSi_2O_6$	21.65
Grossularite	$Ca_3Al_2Si_3O_{12}$	22.52
Chloritoid	$Fe_2Al_4Si_2O_{10}(OH)_4$	22.90
Wollastonite	$CaSiO_3$	23.23
Spessartite	$Mn_3Al_2Si_3O_{12}$	24.75
Almandite	$Fe_3Al_2Si_3O_{12}$	24.88
Andradite	$Ca_3Fe_2Si_3O_{12}$	25.40
Ferrosilite	$FeSiO_3$	26.38
Rutile	TiO_2	26.63
Fayalite	Fe_2SiO_4	29.11
Ilmenite	$FeTiO_3$	30.35
Hematite	Fe_2O_3	31.94
Magnetite	Fe_3O_4	33.08

where $a(\overline{M})$ is a parameter that depends on the mean atomic weight $\overline{M}$, $a(\overline{M}) = -1.87$ km/s for $\overline{M} = 20 - 22$, and $b = 3.05$ (b is assumed to be independent of $\overline{M}$). The velocity is in km/s and density ρ is in g/cm^3.

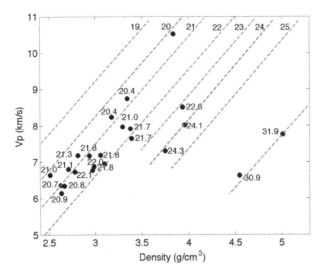

Figure 7.15.1 Velocity at 10 kbars (1 GPa) versus density for silicates and oxides. Numbers are mean atomic weights. Dashed lines suggest the variation of velocity with density and mean atomic weight (from Birch 1961, figure 3).

Birch's work was one of the first rock physics studies to relate rock composition to elastic properties. It provided the key to determining the composition of the deep interior of the Earth from teleseismic travel times.

Birch's law does not hold through structural phase transitions.

7.16   Kerogen Properties

Synopsis

Modeling and interpreting seismic signatures of organic-rich source rocks requires as inputs the properties of kerogen, an important constituent of source rocks. While geochemical properties of kerogen have been studied extensively, their mechanical and elastic properties are less well known. An added complication is that the elastic moduli and density of kerogen depend on the level of maturation and type of kerogen. Following the classification existing in the literature (Tissot *et al.*, 1974; Demaison *et al.*, 1983), kerogen type can be subdivided in four categories:

Type I: this is a kerogen, which when immature is characterized by high atomic H/C (~ 1.5), very high hydrogen Index (HI, > 600 mg HC/g TOC), and low atomic O/C (< 0.1). Type I kerogens are dominated by alginite macerals (very oil prone maceral type), which show low reflectance, high transmittance, and intense fluorescence at low levels of maturity;

Table 7.16.1 *Relations between stages of maturity and vitrinite reflectance*

Stage of Maturity	Ro (%)
Immature	0.2–0.6
Mature	
Early	0.60–0.65
Peak	0.65–0.9
Late	0.90–1.35
Postmature	

Type II: this is a kerogen which when immature shows high atomic H/C (1.2–1.5), high hydrogen index (300–600 mg HC/g TOC) and lower atomic O/C ratio compared to type III and IV. The kerogen is dominated by maceral type from the liptinite group (oil prone), while vitrinite and inertinite can occur in lesser amounts;

Type III: this kerogen is characterized by low atomic H/C (0.7–1.0), low HI (50–200 mg HC/g TOC), and high O/C ratio (up to ~0.3). Maceral types belong to the vitrinite group (gas prone macerals) and show from medium to high reflectance and transmittance, and no fluorescence unless lesser amounts of liptinite are present. Type III organic matter yields less hydrocarbons than type I or II during maturation;

Type IV: type IV kerogen represents *dead* carbon characterized by low H/C and HI ratios (< 0.7 and < 50 mg HC/g TOC, respectively). This type of kerogen is dominated by inertinite, which show high reflectance, no fluorescence, and are opaque in transmitted light. In addition, they generate little or no hydrocarbons during maturation.

During maturation with increase in temperature, solid kerogen is converted to hydrocarbons, often with development of kerogen porosity. Kerogen maturation is traditionally described with the *vitrinite reflectance*, R_0, as shown in Table 7.16.1 (from Peters and Cassa, 1994).

The transformation ratio (TR) is defined as the ratio of converted carbon to the total convertible carbon:

$$TR = \frac{wt\% C_{loss}}{\max(wt\% C_{loss})} \qquad (7.16.1)$$

Modica and Lapierre (2012) suggest an empirical relation between vitrinite reflectance and the transformation ratio for type II kerogen:

$$TR = \frac{100\%}{1 + 20645.5e^{-12.068Ro}} \qquad (7.16.2)$$

Researchers at Institut Français du Pétrole (IFP) have developed a $TR - Ro$ relation based on data from different basins (Ungerer, 1993). The IFP relation, re-derived by Burnham (2016, personal communication) can be given as:

$$Ro = 66.593TR^6 - 160.46TR^5 + 137.65TR^4 - 40.702TR^3 - 2.3354TR^2 \\ + 3.7095TR + 0.3668$$

This relation is more appropriate for type III (North Sea and Mahakam petroleum basins).

A more general procedure to estimate this relationship would be to estimate the transformation ratio from kerogen kinetics, e.g., Behar *et al.* (1997), combined with a vitrinite reflectance estimator such as Easy%Ro (Sweeney and Burnham, 1990), Basin%Ro (Nielsen *et al.*, 2017), and Easy%RoDL (Burnham *et al.*, 2016). The $TR - Ro$ relation is strongly dependent on the kerogen type, but Waples and Marzi (1998) argue that if the composition is properly calibrated, then the $TR - Ro$ relationship is robust over plausible geologic thermal histories. All models for vitrinite reflectance are approximations and require calibration.

The generation of hydrocarbons causes kerogen porosity to form (Jarvie *et al.*, 2007; Loucks *et al.*, 2009; Modica and Lapierre, 2012; Milner *et al.*, 2010). Studies have shown that the amount and shape of nanopores can vary as a function of kerogen type and the thermal history (Loucks *et al.*, 2009; Curtis *et al.*, 2011). Models have been suggested (e.g., Modica and Lapierre, 2012) to estimate the amount of porosity generated from an initial kerogen, depending on its initial richness, lability, and maturity. A linear relationship is sometimes used to model the evolution of pores in kerogen as function of the transformation ratio (TR):

$$\phi_K = kC_C TR = \phi_{max} TR; \quad \text{where } \phi_{max} = kC_c \tag{7.16.3}$$

$$C_C = \frac{1}{TR}\left(1 - \frac{TOC}{iTOC}\right) \tag{7.16.4}$$

$$TR = \frac{S2_i - S2}{S2_i} \tag{7.16.5}$$

$$C_C = \frac{HI_0 \alpha}{1000} \tag{7.16.6}$$

In equation 7.16.3, k is the ratio of kerogen mass to labile mass, ϕ_k is the pore fraction in the kerogen, and C_c is the lability of the organic matter (Modica and Lapierre, 2012; Galford *et al.*, 2013). The value of k typically ranges from 0.68 to 0.95 with 0.80 being a common value (Crain and Holgate, 2014). Lability (C_c) is the fraction of organic carbon that can be converted to hydrocarbon and is typically less than 0.5 (Modica and Lapierre, 2012). Equation 7.16.4 (Daly and Edman, 1987; Modica and Lapierre, 2012) can be used to estimate the lability based on the total organic carbon (TOC), transformation ratio (TR) and initial TOC $(iTOC)$. The transformation ratio can be estimated directly from pyrolysis data using equation 7.16.5

Table 7.16.2 *Porosity parameters for different kerogen types where* k *is the ratio of kerogen mass to labile mass,* C_c *is the lability of the organic matter, and* ϕ_{max} *is the maximum kerogen porosity (equal to* kC_c*) where* k *is assumed to be 1.118 (Modica and Lapierre, 2012).* C_c *is calculated from equation 7.16.6 assuming an* α *value of 0.85 if needed (Jarvie, 2012). Mowry formation initial hydrogen index is back calculated from lability. (Compiled by M. Al Ibrihim, personal communication)*

Kerogen	Formation	HI_0 (mg/g)	C_c	ϕ_{max}	HI_0 Reference
Type I	Typical values	> 600	> 0.510	> 0.57	(Peters *et al.*, 2005)
Type II	Typical values	300 – 600	0.255 – 0.510	0.29 – 0.57	(Peters *et al.*, 2005)
Type II/III	Typical values	200 – 300	0.170 – 0.255	0.19 – 0.29	(Peters *et al.*, 2005)
Type III	Typical values	50 – 200	0.043 – 0.170	0.05 – 0.19	(Peters *et al.*, 2005)
Type IV	Typical values	< 50	< 0.043	< 0.05	(Peters *et al.*, 2005)
	Mowry	424	0.360	0.40	(Modica and Lapierre, 2012)
	Eagle Ford	664	0.564	0.63	(Galford *et al.*, 2013)
	Eagle Ford	411	0.349	0.39	(Jarvie, 2012)
	Haynesville	722	0.614	0.69	(Jarvie, 2012)
	Macellus	507	0.431	0.48	(Jarvie, 2012)
	Bossier	419	0.356	0.40	(Jarvie, 2012)
	Barnett	434	0.369	0.41	(Jarvie, 2012)
	Muskwa	532	0.452	0.51	(Jarvie, 2012)
	Woodford	503	0.428	0.48	(Jarvie, 2012)
	Utica	379	0.322	0.36	(Jarvie, 2012)
	Montney	354	0.301	0.34	(Jarvie, 2012)
	Fayetteville	404	0.343	0.38	(Jarvie, 2012)

(Bordenave *et al.*, 1993) where $S2_i$ is the initial generation potential, i.e., the generation potential of immature samples, and $S2$ is the generation potential. In equation 7.16.6 HI_0 is the initial hydrogen index (in mg/g) and α is a stoichiometric factor typically ranging from 0.82 to 0.88 with 0.85 suitable for marine kerogen (Jarvie, 2012). Table 7.16.2 shows parameters that can be used for the different kerogen types.

As hydrocarbons are generated during thermal maturation, the density of the remaining solid kerogen tends to increase (Okiongbo *et al.*, 2005; Ward, 2010; Ungerer *et al.*, 2014; Rudnicki, 2016; Jagadisan *et al.*, 2017). Some studies suggest using a linear relationship between vitrinite reflectance and kerogen density (Ward, 2010). Others suggest a non-linear relationship (Alfred and Vernik, 2013; Rudnicki, 2016) such as

$$\rho_k = \alpha_p Ro^{\beta_p} \text{ where } 0.4 \leq Ro \leq 1.2 \tag{7.16.7}$$

where α_p and β_p are parameters that depend on the kerogen type (Alfred and Vernik, 2013). A modification to the model can be made so that the density values are kept constant below vitrinite reflectance of 0.4 and above 1.2 following Rudnicki (2016). Table 7.16.3 lists the parameters estimated from different datasets in the literature. In compiling this table, where needed, vitrinite reflectance has been estimated from the pyrolysis maximum temperature T_{max} using the relation

Table 7.16.3 *Density parameters for different kerogen types where α_ρ and β_ρ are parameters in equation 7.16.7. α_ρ and β_ρ are fitted using a sum of squared residuals objective function on the data from the reference papers. Note that Ward's original relation is a linear relation given as $\rho_k = 0.342 Ro + 0.972$ (M. Al Ibrihim, personal communication)*

Kerogen	Formation	α_ρ	β_ρ	Data reference
Type I	Kazhdumi	1.357	0.06	(Kinghorn and Rahman, 1983)
Type II	Kimmeridge Clay	1.293	0.20	(Alfred and Vernik, 2013)
	Marcellus	1.348	0.26	(Ward, 2010)
	Not available	1.184	0.54	(Vandenbroucke and Largeau, 2007)
Type III	Compilation	1.725	0.10	(Kinghorn and Rahman, 1983)

$Ro = 0.0180 T_{max} - 7.16$ (Jarvie *et al.*, 2001). Note that this relation is not suitable for type I kerogen.

Walters et al. (2007) give an empirical relation relating kerogen density (in gm/cc) to the atomic hydrogen to carbon ratio (H/C), and the average molecular weight (wt_{mol}):

$$\rho_k = \frac{1}{0.5129 + 0.298(H/C) + (23.131/wt_{mol})} \approx \frac{1}{0.5129 + 0.298(H/C)} \qquad (7.16.8)$$

Since the molecular weights are large, the term involving the molecular weight can be ignored for practical purposes. With maturity and increasing transformation ratio, (H/C) decreases from its initial value, and the kerogen density correspondingly increases.

Elastic moduli of solid kerogen are difficult to measure accurately, largely due to the tiny size of macerals. Table 7.16.4 summarizes measurements from the literature, most using nano-indentation or atomic force microscopy (AFM). Nano-indentation typically yields the "indentation modulus" E_i

$$\frac{1 - v_g^2}{E_g} = \frac{1}{E_i} - \frac{1 - v_{tip}^2}{E_{tip}} \qquad (7.16.9)$$

where E_g and v_g are the Young's modulus and Poisson's ratio of the sample material (kerogen or mineral), and E_{tip} and v_{tip} are the Young's modulus and Poisson's ratio of the indenter material (usually diamond, which has a Young's modulus of 1141 GPa and a Poisson's ration of 0.07). In order to find the sample bulk modulus, shear modulus, and Young's modulus, an estimate must be made of the sample Poisson's ratio (typically 0.2–0.3).

For comparison, laboratory measurements on coal Morcote *et al.* (2010) suggest the following correlations between grain density (ρ_S) and elastic bulk (K_S) and shear (μ_S) moduli:

$$K_S \approx -5.7 + 0.0086 \rho_S \qquad (7.16.10)$$

Table 7.16.4 *Examples of measured kerogen Young's modulus from the literature (compiled by Alsinan, 2016)*

Formation	Samples	Method	TOC wt%	HI mg/g	T max °C	Ro %	E GPa	Reference
Woodford	⊥ Bedding ‖ Bedding	Modified AFM	22	529	421	-	10.5 ± 1 11.1 ± 0.6	Zeszotarski *et al.* (2004)
Woodford	URDXX95a URDXX95b	Nano-indentation	7.1	- -	436	0.51	8.7 ± 1 2 ± 0.15	Kumar *et al.* (2012)
Kimmeridge	K2	Nano-indentation	49	-	-	0.37–0.5	5.4 ± 0.4	Kumar (2012)
Bazhenov	Sample E	Nano-indentation	2.83	279	-	0.76	4.7–5.9	Ahmadov (2011)
Lockatong	Sample H	Nano-indentation	2.84	2	-	2.58	9.5–11.9	Ahmadov (2011)
Green River	-	Nano-indentation					5–11	Katti *et al.* (2013)
Green River		Nano-indentation					5.39±0.49	Alsinan (2016)
Bakken	7216 7221	Nano-indentation	17.7 16.01	663 636	428.6 431.2	- -	20.4 20.0	Zargari *et al.* (2013)

$$\mu_S \approx -3.3 + 0.0039\rho_S. \qquad (7.16.11)$$

where density is in kg/m^3 and moduli are in GPa.

Similar positive correlations between elastic moduli and density associated with increased molecular packing and crosslinking have also been reported for synthetic polymers (e.g., Hay and Herbert, 2011).

Assumptions and Limitations

Many of these are empirical relations, and should be calibrated to local data when available. Elastic properties of kerogen and their dependence with kerogen maturity is not well known and there exists very few experimental datasets.

Uses

Elastic moduli and density of kerogen can be used as inputs to model elastic properties and seismic velocities of organic rich shales. A typical workflow is to first model the effective elastic moduli of the inorganic constituents (with their associated inorganic porosity) and the organic kerogen (with its associated kerogen porosity, which is a function of maturity). The porous inorganic fraction and the porous organic fractions are modeled using for example an effective medium model such as the differential effective medium model. Then the elastic properties of the inorganic and organic parts are combined using an effective medium model such as the Backus average or the differential effective medium model.

8 Flow and Diffusion

8.1 Darcy's Law

Synopsis

It was established experimentally by Darcy (1856) that the fluid flow rate in a fluid-saturated porous medium is linearly related to the pressure gradient by the following equation:

$$Q_x = -A \frac{\kappa}{\eta} \frac{\partial P}{\partial x} \tag{8.1.1}$$

where Q_x is the volumetric fluid flow rate in the x direction, κ is the permeability of the medium, η is the dynamic viscosity of the fluid, P is the fluid pressure, and A is the cross-sectional area of the sample normal to the pressure gradient.

This can be expressed more generally as

$$\mathbf{Q} = -\frac{\mathbf{\kappa}}{\eta} A \ \text{grad} \ P \tag{8.1.2}$$

where $\mathbf{Q}$ is the vector fluid volumetric velocity field and $\mathbf{\kappa}$ is the permeability tensor.

Permeability, $\mathbf{\kappa}$, has units of area (m^2 in SI units), but the more convenient and traditional unit is the **Darcy**:

$$1 \ \text{Darcy} = 0.986923 \times 10^{-12} \ m^2 \tag{8.1.3}$$

In a water-saturated rock with a permeability of 1 Darcy, a pressure gradient of 1 bar/cm gives a flow velocity of 1 cm/s.

The **hydraulic conductivity** K is related to permeability κ as

$$K = \frac{\kappa \rho g}{\eta} \tag{8.1.4}$$

where $g = 9.81 \ m/s^2$ is the acceleration due to gravity. Hence, the units of hydraulic conductivity are m/s.

For water with density 1 $g/cm^3 = 1000 \ kg/m^3$ and dynamic viscosity 1cPs = 0.001 Pas, the permeability of 1 mD = $10^{-15} \ m^2$ translates into a hydraulic conductivity of $9.81 \times 10^{-9} \ m/s = 9.81 \times 10^{-7} \ cm/s \approx 10^{-6} \ cm/s$. Hydraulic conductivity is often used in ground water flow applications.

If the linear (not volumetric) velocity field V is considered,

$$\mathbf{V} = -\frac{\kappa}{\eta} \text{grad } P \tag{8.1.5}$$

Using this latter notation, Darcy's law for **multiphase flow** of immiscible fluids in porous media (with porosity ϕ) is often stated as

$$V_i = -\frac{\kappa_{ri}\mathbf{K}}{\eta_i} \text{grad } P_i \tag{8.1.6}$$

where the subscript i refers to each phase and κ_{ri} is the relative permeability of phase i. Simultaneous flow of multiphase immiscible fluids is possible only when the saturation of each phase is greater than the irreducible saturation, and each phase is continuous within the porous medium. The relative permeabilities depend on the saturations S_i and show hysteresis, for they depend on the path taken to reach a particular saturation. The pressures P_i in any two phases are related by the capillary pressure P_c, which itself is a function of the saturations. For a two-phase system with fluid 1 as the wetting fluid and fluid 2 as the nonwetting fluid, $P_c = P_2 - P_1$. The presence of a residual nonwetting fluid can interfere considerably with the flow of the wetting phase. Hence the maximum value of κ_{r1} may be substantially less than 1. There are extensions of Darcy's law for multiphase flow (Dullien, 1992) that take into account cross-coupling between the fluid velocity in phase i and the pressure gradient in phase j. The cross coupling becomes important only at very high viscosity ratios $(\eta_i/\eta_j \gg 1)$ because of an apparent lubricating effect.

In a one-dimensional, immiscible displacement of fluid 2 by fluid 1 (e.g., water displacing oil in a water flood), the time history of the saturation $S_1(x, t)$ is governed by the following equation (Marle, 1981):

$$\frac{V}{\phi}\left(\frac{d\xi_1}{dS_1}\frac{\partial S_1}{\partial x} + \frac{\partial}{\partial x}\left(\Psi_1 \frac{\partial S_1}{\partial x}\right)\right) + \frac{\partial S_1}{\partial t} = 0 \tag{8.1.7}$$

where

$$V = V_1 + V_2 \tag{8.1.8}$$

$$\xi_1 = \frac{\eta_2/\kappa_{r2} + \kappa(\rho_1 - \rho_2)g/V}{\eta_1/\kappa_{r1} + \eta_2/\kappa_{r2}} \tag{8.1.9}$$

$$\Psi_1 = \frac{\kappa}{V}\frac{1}{\eta_1/\kappa_{r1} + \eta_2/\kappa_{r2}\frac{dP_c}{dS_1}} \tag{8.1.10}$$

where V_1 and V_2 are the Darcy fluid velocities in phases 1 and 2, respectively, and g is the acceleration due to gravity. The requirement that the two phases completely fill the pore space implies $S_1 + S_2 = 1$. Neglecting the effects of capillary pressure gives the **Buckley–Leverett** equation for immiscible displacement:

$$\left(\frac{V}{\phi}\frac{d\xi_1}{dS_1}\right)\frac{\partial S_1}{\partial x} + \frac{\partial S_1}{\partial t} = 0 \tag{8.1.11}$$

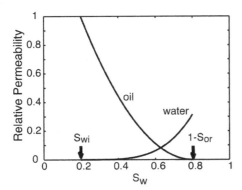

Figure 8.1.1 Relative permeability for water (the rising curve) and oil (falling curve) according to equations of Desbrandes (1985) for an irreducible water saturation of 0.2 and a residual oil saturation of 0.2.

This represents a saturation wavefront traveling with velocity $(V/\phi)(d\xi_1/dS_1)$.

Desbrandes (1985) presented the following empirical equations that link the relative permeability for water (κ_{rw}) and oil (κ_{ro}) to water saturation (S_w), irreducible water saturation (S_{wi}), and residual oil saturation (S_{or}):

$$\kappa_{rw} = \left(\frac{S_w - S_{wi}}{1 - S_{wi}}\right)^4 \tag{8.1.12}$$

$$\kappa_{ro} = \left(\frac{1 - S_{or} - S_w}{1 - S_{or} - S_{wi}}\right)^2 \tag{8.1.13}$$

where saturation can vary from 0 to 1. These relative permeability curves are plotted in Figure 8.1.1.

Single Phase Gas Flow

Because gas is highly compressible, its density can vary substantially with pressure. Therefore, in steady flow, the volume flow rate will vary between the input and output flow surfaces of a porous sample. Permeability can be computed by assuming isothermal conditions and that the product of flow velocity and density is constant (i.e., uniform mass flux). Thus, the following form of Darcy's law can be used (Dullien, 1979):

$$V_2 = \frac{\kappa}{\eta}\frac{(P_1^2 - P_2^2)}{2P_2 L} = \frac{\kappa}{\eta}\frac{P_m}{P_2}\frac{\Delta P}{L}, \tag{8.1.14}$$

where L is the sample length, V_2 is the outlet flow velocity, P_1 and P_2 are the inlet and outlet pressures, $\Delta P = P_1 - P_2$, and $P_m = (P_1 + P_2)/2$.

Diffusivity

If the fluid and the matrix containing it are compressible and elastic, the saturated system can take on the behavior of the diffusion equation. If we combine Darcy's law with the equation of mass conservation given by

$$\nabla \cdot (\rho \mathbf{V}) + \frac{\partial(\rho\phi)}{\partial t} = 0 \tag{8.1.15}$$

plus Hooke's law expressing the compressibility of the fluid, β_{fl}, and of the pore volume, β_{pv}, and drop nonlinear terms in pressure, we obtain the classical diffusion equation:

$$\nabla^2 P = \frac{1}{D}\frac{\partial P}{\partial t} \tag{8.1.16}$$

where D is the diffusivity:

$$D = \frac{\kappa}{\eta\phi(\beta_{fl} + \beta_{pv})} \tag{8.1.17}$$

One-Dimensional Diffusion

Consider the one-dimensional diffusion that follows an initial fluid pressure pulse:

$$P = P_0\,\delta(x) \tag{8.1.18}$$

We obtain the standard result, illustrated in Figure 8.1.2, that

$$P(x,\,t) = \frac{P_0}{\sqrt{4\pi Dt}}e^{-x^2/4Dt} = \frac{P_0}{\sqrt{4\pi Dt}}e^{-\tau/t} \tag{8.1.19}$$

where the characteristic time depends on the length scale x and the diffusivity:

$$\tau = \frac{x^2}{4D} \tag{8.1.20}$$

Next, consider a semi-infinite ($z \geq 0$) homogeneous medium, with initial fluid pressure $P = 0$. The surface pressure is $P(z = 0) = P_0 H(t)$ where $H(t)$ is the Heaviside function. The resulting solution for one-dimensional diffusion is

$$P = P_0 erfc\left(\frac{z}{2\sqrt{Dt}}\right) \tag{8.1.21}$$

If, instead, the surface pressure of the fluid is a harmonic function in time, $P(z = 0) = P_0 \cos(\omega t - \varepsilon)$, then the resulting solution for one-dimensional diffusion is

$$P(z,t) = P_0 e^{-z\sqrt{\omega/2D}}\cos(\omega t - z\sqrt{\omega/2D} - \varepsilon)\,. \tag{8.1.22}$$

The harmonic solution describes a damped pressure wave propagating with wavelength $\lambda = \sqrt{8\pi^2 D/\omega}$. The amplitude falls off with distance like $e^{-2\pi z/\lambda}$. The phase velocity (the propagation of peaks and troughs) is $\sqrt{2D\omega}$.

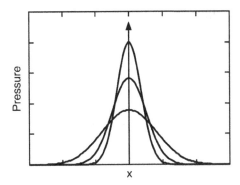

Figure 8.1.2 Pressure curves for one-dimensional diffusion

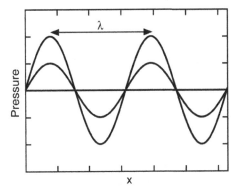

Figure 8.1.3 Decay of an instantaneous sinusoidal pore-pressure disturbance in a saturated system

Consider an instantaneous spatially sinusoidal pore pressure disturbance in the saturated system, as shown in Figure 8.1.3. The disturbance will decay approximately as $e^{-t/\tau}$, where the diffusion time is again related to the length and diffusivity by

$$\tau_{\mathrm{d}} = \frac{\lambda^2}{4D} \qquad (8.1.23)$$

It is interesting to ask, when is the diffusion time equal to the period of the seismic wave causing such a disturbance? The seismic period is

$$\tau_{\mathrm{s}} = \frac{\lambda}{V_{\mathrm{p}}} \qquad (8.1.24)$$

Equating τ_{d} to τ_{s} gives

$$\tau_{\mathrm{d}} = \tau_{\mathrm{s}} \Rightarrow \frac{\lambda^2}{4D} = \frac{\lambda}{V_{\mathrm{p}}} \qquad (8.1.25)$$

which finally gives

$$\tau = \frac{4D}{V_{\mathrm{p}}} \qquad (8.1.26)$$

For a rock with a permeability of 1 milliDarcy (1 mD), the critical frequency $(1/\tau)$ is 10 MHz.

Saturated Porous Sphere

A spherical shell of porous material with internal radius R_1 held at fluid pressure P_1 and outer radius R_2 held at fluid pressure P_2 has the steady radial fluid pressure field

$$P = \frac{R_1 P_1 (R_2 - r) + R_2 P_2 (r - R_1)}{r(R_2 - R_1)} \tag{8.1.27}$$

where r is the radial distance.

A solid porous sphere with radius R, initial fluid pressure $P = 0$, and surface fluid pressure $P_0 H(t)$ has the solution

$$P(r,t) = \frac{RP_0}{r} \sum_{n=0}^{\infty} \left\{ erfc \frac{(2n+1)R - r}{2\sqrt{Dt}} - erfc \frac{(2n+1)R + r}{2\sqrt{Dt}} \right\}, \ r \leq R. \tag{8.1.28}$$

An infinite medium with a spherical cavity with radius R, initial pressure $P = 0$, and surface pressure $P_0 H(t)$ has the solution

$$P(r,t) = \frac{RP_0}{r} erfc \left(\frac{r - R}{2\sqrt{Dt}} \right), \ r \geq R \tag{8.1.29}$$

Klinkenberg Effect and Knudsen Flow

When the mean free path of gas molecules is comparable to or larger than the dimensions of the pore space, the continuum description of gas flow becomes invalid. In these cases, measured permeability to gas is larger than the permeability to liquid. This is sometimes thought of as the increase in apparent gas permeability caused by slip at the gas-mineral interface. This is known as the Klinkenberg effect (Bear, 1972). The Klinkenberg correction is given by

$$\kappa_g = \kappa_1 \left(1 + \frac{4c\lambda}{r} \right) = \kappa_1 \left(1 + \frac{b}{P} \right) \tag{8.1.30}$$

where κ_g is the gas permeability, κ_1 is the liquid permeability, λ is the mean free path of gas molecules at the pressure P at which κ_g is measured, $c \approx 1$ is a proportionality factor, b is an empirical parameter that is best determined by measuring κ_g at several pressures, and r is the radius of the capillary. The mean free path is the average distance traveled by molecules between collisions and is given by

$$\lambda = \frac{K_B T}{\sqrt{2}\,\pi\sigma^2 P} \tag{8.1.31}$$

where T is the Kelvin temperature, P is the pressure (Pa), K_B is Boltzmann's constant ($1.3805 \ 10^{-23}$ J/K), and σ is the collision diameter (m). Some values of σ are 0.42 nm for methane, 0.47 nm for ethane, and 0.44 nm for carbon dioxide.

The Knudsen number, Kn is defined as

$$Kn = \frac{\lambda}{r},$$ (8.1.32)

and can be used to define flow regimes (Ziarani and Aguilera, 2012; Karniadakis *et al.*, 2005). When $Kn < 0.01$, the flow is in the **viscous flow** domain, and the equations of continuum flow apply. When $0.01 < Kn < 0.1$, we are in the **slip flow** domain, where the flow can still be described as a continuum but with slipping boundary conditions as described by the Klinkenberg equation. When $0.1 < Kn < 10$, we are in the **transition flow** domain, where both continuum and diffusive effects occur. In this domain, the Klinkenberg correction begins to fail. When $Kn > 10$ we are in the **Knudsen flow** domain, where the continuum description completely fails and transport requires a diffusive description.

Knudsen's flow correction (Ziarani and Aguilera, 2012; Beskok and Karniadakis, 1999) is given by

$$k_g = f_c k_l$$ (8.1.33)

where

$$f_c = [1 + \alpha(Kn) \ Kn] \left[1 + \frac{4Kn}{1 + Kn}\right].$$ (8.1.34)

The factor α is a dimensionless rarefaction coefficient. An estimate of α is proposed by Civan (2010):

$$\alpha = \frac{1.358}{1 + 0.170 \ Kn^{-0.4348}}$$ (8.1.35)

Knudsen's flow correction is reported to be more accurate than the Klinkenberg correction (Ziarani and Aguilera, 2012), especially at Knudsen number greater than 0.1. Klinkenberg's correction factor in terms of the Knudsen number is

$$f_K \approx 1 + 4Kn .$$ (8.1.36)

Figure 8.1.4 Compares the Knudsen and Klinkenberg correction factors vs. the Knudsen number

Assumptions and Limitations

The following considerations apply to the use of Darcy's law:
- Darcy's law applies to a representative elementary volume much larger than the grain or pore scale.
- Darcy's law is applicable when inertial forces are negligible in comparison to pressure gradient and viscous forces, and the Reynolds number Re is small ($Re \approx 1-10$). The Reynolds number for porous media is given by $Re = \rho Vl/\eta$, where ρ is the fluid

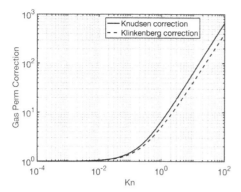

Figure 8.1.4 Comparison of Knudsen and Klinkengerg gas permeability correction factors f_c and f_K vs. Knudsen number

density, η is the fluid viscosity, V is the fluid velocity, and l is a characteristic length of fluid flow determined by pore dimensions. At high Re, inertial forces can no longer be neglected in comparison with viscous forces, and Darcy's law breaks down.

- Some authors mention a minimum threshold pressure gradient below which there is very little flow (Bear, 1972). Non-Darcy behavior below the threshold pressure gradient has been attributed to streaming potentials in fine-grained soils, immobile adsorbed water layers, and clay-water interaction, giving rise to non-Newtonian fluid viscosity.

Extensions

When inertial forces are not negligible (large Reynolds number), Forchheimer suggested a nonlinear relation between the fluid flux and the pressure gradient (Bear, 1972) as follows:

$$\frac{dP}{dx} = aV + bV^2 \tag{8.1.37}$$

where a and b are constants.

Another extension to Darcy's law was proposed by Brinkman by adding a Laplacian term to account for solid-fluid interface interactions (Sahimi, 1995). The modified equation, sometimes termed the Darcy–Brinkman equation is:

$$\nabla P = -\frac{\eta'}{K} V + \eta' \nabla^2 V \tag{8.1.38}$$

where the viscosity η' can, in principle, be different from the pure fluid viscosity, η, though in Brinkman's original paper and in many applications they are taken to be the same (Sahimi, 1995). The Darcy–Brinkman equation is useful for modeling flow in very high-porosity media, and also in transition zones between porous media flow and open-channel flow.

8.2   Viscous Flow

Synopsis

In a Newtonian, viscous, incompressible fluid, stresses and velocities are related by Stokes' law (Segel, 1987):

$$\sigma_{ij} = -P\delta_{ij} + 2\eta D_{ij} \tag{8.2.1}$$

$$D_{ij} = \frac{1}{2}\left(\frac{\partial V_i}{\partial x_j} + \frac{\partial V_j}{\partial x_i}\right) \tag{8.2.2}$$

where σ_{ij} denotes the elements of the stress tensor, V_i represents the components of the velocity vector, P is pressure, and η is dynamic viscosity.

For a simple shear flow between two walls, this law is called the Newton friction law and is expressed as

$$\tau = \eta\frac{dV}{dy}, \tag{8.2.3}$$

where τ is the shear stress along the flow, V is velocity along the flow, and y is the coordinate perpendicular to the flow.

The Navier–Stokes equation for a Newtonian viscous incompressible flow is (e.g., Segel, 1987)

$$\rho\left(\frac{\partial \mathbf{V}}{\partial t} + \mathbf{V}\cdot\,\text{grad }\mathbf{V}\right) = -\text{grad }P + \eta\Delta\mathbf{V} \tag{8.2.4}$$

where ρ is density, t is time, V is vector velocity, and Δ is the Laplace operator.

Useful Examples of Viscous Flow

(a) Steady two-dimensional laminar flow between two walls (Lamb, 1945):

$$V(y) = -\frac{1}{2\eta}\frac{dP}{dx}(R^2 - y^2), \quad Q = -\frac{2}{3\eta}\frac{dP}{dx}R^3 \tag{8.2.5}$$

where $2R$ is the distance between the walls and Q is the volumetric flow rate per unit width of the slit (Figure 8.2.1).

(b) Steady two-dimensional laminar flow in a circular pipe:

$$V(y) = -\frac{1}{4\eta}\frac{dp}{dx}(R^2 - y^2), \quad Q = -\frac{\pi}{8\eta}\frac{dP}{dx}R^4 \tag{8.2.6}$$

where R is the radius of the pipe (see Figure 8.2.1).

(c) Steady laminar flow in a pipe of elliptical cross-section (Lamb, 1945):

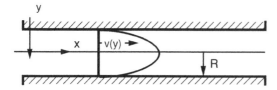

Figure 8.2.1 Steady laminar flow in a two-dimensional slot, or in a pipe of circular cross-section

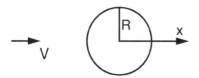

Figure 8.2.2 Steady laminar flow past a sphere

$$Q = -\frac{\pi}{4\eta}\frac{dP}{dx}\frac{a^3 b^3}{a^2 + b^2} \tag{8.2.7}$$

where a and b are the semiaxes of the cross-section.

(d) Steady laminar flow in a pipe of rectangular cross-section:

$$Q = -\frac{1}{24\eta}\frac{dP}{dx}ab(a^2 + b^2) + \frac{8}{\pi^5\eta}\frac{dP}{dx}\sum_{n=1}^{\infty}\frac{1}{(2n-1)^5}$$
$$\times \left[a^4\tanh\left(\pi b\frac{2n-1}{2a}\right) + b^4\tanh\left(\pi a\frac{2n-1}{2b}\right)\right] \tag{8.2.8}$$

where a and b are the sides of the rectangle. For a square this equation yields

$$Q = -0.035144\frac{a^4}{\eta}\frac{dP}{dx} \tag{8.2.9}$$

(e) Steady laminar flow in a pipe of equilateral triangular cross-section with the length of a side b:

$$Q = -\frac{\sqrt{3}b^4}{320\eta}\frac{dP}{dx} \tag{8.2.10}$$

(f) Steady laminar flow past a sphere (pressure on the surface of the sphere depends on the x coordinate only):

$$P = -\frac{3}{2}\eta\frac{x}{R^2}V \tag{8.2.11}$$

where V is the undisturbed velocity of the viscous flow (velocity at infinity), and the origin of the x-axis is in the center of the sphere of radius R (Figure 8.2.2). The total resistance force is $6\pi\eta VR$. The combination

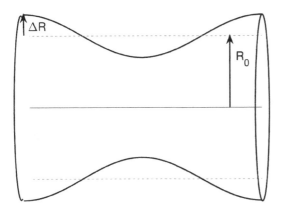

Figure 8.2.3 Circular pipe with radius varying sinusoidally along the pipe axis

$$Re = \frac{\rho V R}{\eta} \tag{8.2.12}$$

where R is the characteristic length of a flow (e.g., the pipe radius) and Re is the **Reynolds number**. Flows where $Re < 1$ are called **creeping flows**. (Viscous forces are the dominant factors in such flows.) A flow becomes turbulent (i.e., nonlaminar) if $Re > 2000$.

(g) Steady laminar flow in a circular pipe with sinusoidally varying radius:
Sisavath *et al.* (2001) give the solution for creeping flow in a pipe with sinusoidally varying radius, R (Figure 8.2.3):

$$R(z) = R_0 \left[1 + \frac{\Delta R}{R_0} \sin\left(\frac{2\pi z}{\lambda} \right) \right] \tag{8.2.13}$$

The normalized fluctuation of the radius about the mean is $\bar{\delta} = \Delta R / R_0$, and the normalized "wavelength" of the fluctuation is $\bar{\lambda} = \lambda / R_0$.

If the radius is slowly varying ($\delta/\lambda \ll 1$), the permeability is approximately

$$\kappa = \kappa_0 \left(\frac{2(1 - \bar{\delta}^2)^{7/2}}{2 + 3\bar{\delta}^2} \right) \left(1 + \frac{16\pi^2}{3} \frac{\bar{\delta}^2}{\bar{\lambda}^2} \frac{(1 - \bar{\delta}^2)}{2 + 3\bar{\delta}^2} \right)^{-1} \tag{8.2.14}$$

where κ_0 is the permeability of a straight pipe with the same radius R_0. If $\bar{\delta} = 0.3$, the fluctuations lower the permeability by a factor of 0.15.

Fanning Friction Factor

A common way to describe the flow capacity of a straight tube or duct is the *friction factor*, which is related to the average shear stress $\bar{\tau}$ applied to the tube walls during flow. The **Fanning friction factor** f is the dimensionless number

$$f = \frac{2\bar{\tau}}{\rho V^2} , \tag{8.2.15}$$

where ρ is the fluid density, and V is the average velocity of the fluid. The **Darcy friction factor** is $f_D = 4f$. For steady flow, the shear stress integrated over the perimeter P must be related to the pressure gradient $\partial p / \partial z$ along the axis of the pipe, multiplied by the tube cross-sectional area A:

$$\bar{\tau} = \frac{A}{P} \frac{\partial p}{\partial z} \tag{8.2.16}$$

Therefore, the friction factor becomes

$$f = \frac{2}{\rho V^2} \frac{A}{P} \frac{\partial p}{\partial z} \tag{8.2.17}$$

The friction factor is often combined with the Reynolds number $\mathrm{Re} = \rho V L / \eta$, where η is the fluid viscosity, and L is a characteristic width dimension of the pipe:

$$f \, \mathrm{Re} = \frac{2L}{V\eta} \left(\frac{A}{P} \right) \frac{\partial p}{\partial z} . \tag{8.2.18}$$

One common definition of length scale is $L_H = 4A/P$; another definition is $L_{\sqrt{A}} = \sqrt{A}$. These lead to slightly different Reynolds numbers, Re_H and $\mathrm{Re}_{\sqrt{A}}$. The fluid flux Q in the pipe can be written in terms of the friction factor. The two forms resulting from Re_H and $\mathrm{Re}_{\sqrt{A}}$ are

$$Q = AV = \frac{2}{f \, \mathrm{Re}_{\sqrt{A}}} \eta \left(\frac{A^{5/2}}{P} \right) \frac{\partial p}{\partial z}, \tag{8.12.19}$$

$$Q = AV = \frac{8}{f \, \mathrm{Re}_H \eta} \left(\frac{A^3}{P^2} \right) \frac{\partial p}{\partial z} \tag{8.2.20}$$

Hence, for laminar flow the permeability, assuming a straight pipe, can be written as

$$\kappa = \frac{2}{f \, \mathrm{Re}_{\sqrt{A}} A_{sample}} \left(\frac{A^{5/2}}{P} \right)$$
$$= \frac{8}{f \, \mathrm{Re}_H A_{sample}} \left(\frac{A^3}{P^2} \right) \tag{8.2.21}$$

where A_{sample} is the cross sectional area of the material sample that contains the pipe. Note that the relation between the two quantities:

$$\frac{f \, \mathrm{Re}_H}{f \, \mathrm{Re}_{\sqrt{A}}} = \frac{4\sqrt{A}}{P} . \tag{8.2.22}$$

Table 8.2.1 lists straight-pipe friction factors for a number of pipe cross-sectional shapes. Table 8.2.2 lists straight-pipe friction factors for regular polygons.

Table 8.2.1 *Quantities* $f\mathrm{Re}_H$ *and* $f\mathrm{Re}_{\sqrt{A}}$ *for straight pipes with various cross-sections (Bahrami et al., 2005)*

	$f\,\mathrm{Re}_{\sqrt{A}}$	$f\,\mathrm{Re}_H$
	14.18	16
	$\dfrac{2\pi^{3/2}(1+\alpha^2)^{[1]}}{\sqrt{\alpha}E(\sqrt{1-\alpha^2})}$	$\dfrac{2\pi^2(1+\alpha^2)^{[1]}}{E(\sqrt{1-\alpha^2})^2}$
	14.23	14.23
	$\dfrac{12}{\left[1-\dfrac{192}{\pi^5}\alpha\tanh\left(\dfrac{\pi}{2\alpha}\right)\right](1+\alpha)\sqrt{\alpha}}$ [1]	$\dfrac{24}{\left[1-\dfrac{192}{\pi^5}\alpha\tanh\left(\dfrac{\pi}{2\alpha}\right)\right](1+\alpha)^2}$ [1]
	15.20	13.33
	$\dfrac{\phi\sqrt{\phi}}{(1+\phi)g(\phi)^{[2]}}$	$\dfrac{2\phi^2}{(1+\phi)^2 g(\phi)^{[2]}}$
	14.92	14.17
	16.17	15.77

[1] $\alpha = c/b$; $E(.)$ is the complete elliptic integral of the second kind.

[2] $g(\phi) = \dfrac{\tan(2\phi)-2\phi}{16\phi} - \dfrac{128\phi^3}{\pi^5}\sum_{n=1}^{\infty}\left[\dfrac{1}{(2n-1)^2(2n-1+4\phi/\pi)^2(2n-1-4\phi/\pi)}\right]$

Table 8.2.2 *Regular Polygons with number of sides N. (Bahrami et al., 2005)*

N	$f\,\mathrm{Re}_{\sqrt{A}}$	$f\,\mathrm{Re}_H$
3	15.19	13.33
4	14.23	14.23
5	14.04	14.73
6	14.01	15.05
8	14.03	15.41
∞	14.18	16

Uses

The equations presented in this section are used to describe viscous flow in pores and ducts.

Assumptions and Limitations

The equations presented in this section assume that the fluid is incompressible and Newtonian.

Table 8.3.1 *Typical values for surface tension (at 20 °C, unless otherwise noted)*

Interface	γ in mN/m
Water–air	72.80
Octane–air	21.62
Heptane–air	20.14
Mercury–air	486.5
Water–mercury	415
Ethylene glycol–air (25 °C)	47.3
Benzene–air	28.88
Toluene–air	28.52
Ethanol–air	22.39

Figure 8.3.1 Surface tensional forces at a liquid–solid or liquid–liquid interface

8.3   Capillary Forces

Synopsis

A **surface tension**, γ, exists at the interface between two immiscible fluids or between a fluid and a solid (Figure 8.3.1). The surface tension acts tangentially to the interface surface. If τ is a force acting on length l of the surface, then the surface tension is defined as $\gamma = \tau/l$; hence, the unit of surface tension is force per unit length (N/m). A surface tension may be different at interfaces between different materials. For example, γ_{ow} (oil–water) differs from γ_{og} (oil–gas) and from γ_{wg} (water–gas). Some typical values of fluid–fluid surface tensions are shown in Table 8.3.1.

The force equilibrium condition at a triple point of contact between a solid, a liquid, and a gas is given by (Figure 8.3.2) Young's relation:

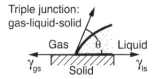

Triple junction:
gas-liquid-solid

Gas   θ   Liquid

γ_{gs} Solid γ_{ls}

Wetting fluid   Non-wetting fluid

Figure 8.3.2 Triple junction at the interface between a solid, a liquid, and a gas. Top: the contact angle at the triple junction. Bottom: schematic of wetting and nonwetting fluids.

$$\gamma_{lg} \cos(\theta) + \gamma_{ls} = \gamma_{gs}, \quad \cos(\theta) = \frac{\gamma_{gs} - \gamma_{ls}}{\gamma_{lg}} \tag{8.3.1}$$

where γ_{lg}, γ_{ls}, and γ_{gs} are the liquid–gas, liquid–solid, and gas–solid interfacial tensions, respectively. The angle θ is the **contact angle**. The liquid within the droplet is **wetting** if $\theta < 90°$ and **nonwetting** if $\theta > 90°$. The equilibrium will not exist if

$$|(\gamma_{gs} - \gamma_{ls})/\gamma_{lg}| > 1 \tag{8.3.2}$$

A similar situation can occur at the triple interface of a solid and two immiscible liquids.

The surface tension between a solid and liquid is difficult to measure. Some techniques are based on directly measuring the contact angle of a drop (goniometry), while others are based on measurements of the force of interaction between the solid and a test liquid (tensiometry). The observed contact angle can be affected by the liquid and solid compositions, the surface roughness, temperature, pressure, heterogeneity along the surface, and the amount of time elapsed since the liquid and solid came into contact. The contact angle for a fluid contact advancing (or recently advanced) along the solid surface is usually larger than the contact angle on a receding (or recently receded) contact.

If a sphere of oil (radius R) is floating inside water, the oil–water surface tension, γ, has the effect of contracting the sphere, causing the pressure inside the sphere, P_o, to be greater than pressure in the surrounding water P_w. The difference between these two pressures is called **capillary pressure**: $P_c = P_o - P_w$ and is given by the Laplace equation:

$$P_c = 2\gamma/R \tag{8.3.3}$$

In general, capillary pressure inside a surface of two principal radii of curvature R_1 and R_2 is

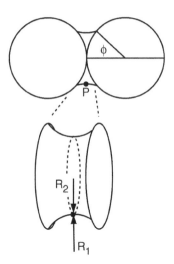

Figure 8.3.3 A pendular ring of fluid between two grains, showing the two radii of curvature at point P

$$P_c = \gamma \left(\frac{1}{R_1} \pm \frac{1}{R_2} \right) \tag{8.3.4}$$

where the plus sign corresponds to the case in which the centers of curvature are located on the same side of the interface (e.g., a sphere), and the minus sign corresponds to the case in which the centers of the curvature are located on opposite sides (e.g., a torus).

If the pore fluid is wetting, a pendular ring may exist at the point of contact of two grains (see Figure 8.3.3). The capillary pressure at any point P on the surface of the ring depends on the contact angle θ of the liquid with the grains and on the radii of curvature, R_1 and R_2, at that point, as follows:

$$P_c = \gamma \left(\frac{1}{R_1} - \frac{1}{R_2} \right) \cos(\theta) \tag{8.3.5}$$

If $\theta = 0$, we have the following expression at the contact point of two identical spherical grains of radius R (Gvirtzman and Roberts, 1991):

$$R_1 = R \frac{\sin(\phi) + \cos(\phi) - 1}{\cos(\phi)}, \qquad R_2 = R \frac{1 - \cos(\phi)}{\cos(\phi)} \tag{8.3.6}$$

where ϕ is the angle between the line connecting the centers of the spheres and the line from the center of the sphere to the edge of the pendular ring. The ring will exist as long as its capillary pressure is smaller than the external pressure. The preceding formulas show that this condition is valid for $\phi < 53°$. The maximum volume of this ring is about 0.09 times the volume of an individual grain. This maximum volume tends to decrease with increasing angle θ (e.g., it is 0.04 times the grain volume for $\theta = 32°$).

When a narrow tube is brought into contact with a wetting liquid, the liquid rises inside the tube (see Figure 8.3.4). The weight of the fluid column is supported by the capillary forces. The height of this rise (h) is

$$h = \frac{2\gamma \cos\theta}{\Delta\rho g r} \tag{8.3.7}$$

where γ is the surface tension between the two fluids in the tube; θ is the contact angle of the wetting liquid on the tube surface; $\Delta\rho$ is the density contrast of the fluids (approximately 1000 kg/m³ for a water–air system); g is the acceleration due to gravity (9.8 m/s²); and r is the radius of the tube. For a tube containing water and air with surface tension at room conditions of about 0.07 N/m, and assuming a contact angle of 30°, we plot the capillary equilibrium height versus the tube radius in Figure 8.3.5.

Imbibition refers to the increase of the wetting-fluid saturation in a rock, while **drainage** refers to the decrease of the wetting-fluid saturation. Imagine a rock fully saturated with water, the wetting phase. If the rock is put in contact with oil (the nonwetting phase), the oil will not enter the pore space because of the capillary forces holding the water in place. If the oil is pressurized to the capillary pressure of the largest pore throats, oil will begin to enter (drainage). Further increases in oil pressure

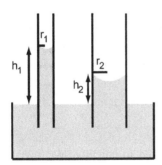

Figure 8.3.4 Capillary rise of a wetting fluid inside tubes of different radii

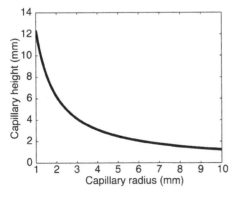

Figure 8.3.5 Capillary rise of a wetting fluid as a function of tube radius, for an air–water system with a contact angle of 30°

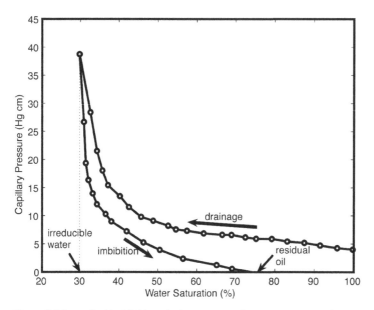

Figure 8.3.6 Typical imbibition–drainage curves for a water-wet rock

will allow the oil to be forced into successively smaller and smaller throats, until eventually a limit is reached, corresponding to the **irreducible water saturation** (see Figure 8.3.6). Dropping the oil pressure causes the water to re-enter the rock, displacing oil. The saturation-pressure curve for imbibition will generally be different than the curve for drainage (see Figure 8.3.6). At zero oil pressure, some **residual oil** will remain in the rock, trapped in bypassed pores.

Uses

The formulas presented in this section can be used to calculate wetting pore-fluid geometry in granular rocks.

Assumptions and Limitations

Some of the formulas in this section are based on idealized geometries. For example, the description of pendular rings is for spherical grains. The equation for capillary rise is for tubes of circular cross-section

8.4   Kozeny–Carman Relation for Flow

Synopsis

The Kozeny–Carman (Carman, 1961) relation provides a way to estimate the permeability of a porous medium in terms of generalized parameters such as porosity, surface

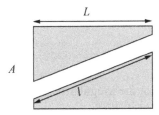

Figure 8.4.1 Solid block with a pipe used for Kozeny–Carman derivations. The notation is explained in the text.

area, and particle size. The derivation is based on viscous fluid flow through a pipe having a circular cross-section with radius R. The flux in the pipe of length l can be written as

$$Q = -\frac{\pi R^4}{8\eta}\frac{\Delta P}{l} \tag{8.4.1}$$

where η is the dynamic viscosity, and ΔP is the pressure drop over the pipe length l. Assume that the direction of this pipe is at an angle to the pressure gradient, which is applied to a permeable block of length $L(L \le l)$ and cross-sectional area A (Figure 8.4.1). Comparison with Darcy's law,

$$Q = -\kappa\frac{A}{\eta}\frac{\Delta P}{L} \tag{8.4.2}$$

where κ is the permeability, gives an effective permeability for the block expressed as

$$\kappa = \frac{\pi R^4}{8A}\frac{L}{l} = \frac{\pi R^4}{8A\tau} \tag{8.4.3}$$

where $\tau = l/L$ is the tortuosity (defined as the ratio of total flow-path length to length of the sample).

The porosity, ϕ, and the specific surface area, S (defined as the pore surface area divided by the sample volume), can be expressed in terms of the properties of the pipe by the following relations:

$$\phi = \frac{\pi R^2 l}{AL} = \frac{\pi R^2}{A}\tau \tag{8.4.4}$$

$$S = \frac{2\pi R l}{AL} = \frac{2\pi R\tau}{A} = \frac{\pi R^2\tau}{A}\frac{2}{R} = \frac{2\phi}{R} \tag{8.4.5}$$

Finally, we can express the permeability of the block with the pipe in terms of the more general properties, ϕ and S, to obtain the Kozeny–Carman relation:

$$\kappa = \frac{1}{2}\frac{\phi^3}{S^2\tau^2} = \frac{\phi}{8\tau^2}R^2 \tag{8.4.6}$$

This relation, which is exact for an ideal circular pipe geometry, is often presented in general terms as

$$\kappa = B \frac{\phi^3}{S^2 \tau^2} \tag{8.4.7}$$

where B is a geometric factor that partly accounts for the irregularities of pore shapes.

The exact expressions for viscous flow through pipes of elliptical, square, and triangular cross-sections allow us to derive the Kozeny–Carman relations for blocks containing these tube shapes. Specifically, for the elliptical cross-section with semiaxes a and b,

$$\kappa = \frac{\pi}{4} \frac{a^3 b^3}{a^2 + b^2} \frac{1}{A\tau} \tag{8.4.8}$$

$$\phi = \frac{\pi a b}{A} \tau \tag{8.4.9}$$

$$S = \frac{\pi \sqrt{2(a^2 + b^2)} l}{AL} \tag{8.4.10}$$

Then,

$$\kappa = \frac{1}{4} \frac{b^2}{1 + (b/a)^2} \frac{\phi}{\tau} = \frac{1}{2} \frac{\phi^3}{S^2 \tau^2}, \quad B = 0.5 \tag{8.4.11}$$

which implies the universality of the latter expression for pores of elliptical cross-section.

Similarly, for a square pipe whose side is a,

$$\kappa = 0.035144 a^2 \frac{\phi}{\tau^2} = 0.562 \frac{\phi^3}{S^2 \tau^2}, \quad B = 0.562 \tag{8.4.12}$$

and for a pipe of equilateral triangular cross-section,

$$\kappa = \frac{\phi a^2}{80 \tau^2} = 0.6 \frac{\phi^3}{S^2 \tau^2}, \quad B = 0.6 \tag{8.4.13}$$

Notice that the B-factors of similar magnitude (0.5, 0.562, and 0.6) relate permeability to $\phi^3 / S^2 \tau^2$ in these three pipe geometries. (See also the discussion of the Fanning friction factor in Section 8.2.)

Srisutthiyakorn (2018) suggests predicting flow in non-circular pipes using the Kozeny–Carman form in equation 8.4.3 with R is replaced by the *apparent radius* $R_A = \sqrt{R_H R_{cir}}$, the geometric mean of the hydraulic radius $R_H = 2\phi/S$ and the radius of a pipe that has the same porosity as the pipe under consideration, R_{cir}. Table 8.4.1 compares estimates of permeability for tubes with six different cross-sectional shapes, computed using equation 8.4.3. For non-circular shapes, R_A yields better estimates than either R_H or R_{cir}.

Table 8.4.1 *Permeability estimates for tubes with various cross-sectional shapes, computed using equation 8.4.3. Values are normalized by the "correct" permeability computed for each case using Lattice Boltzman simulations (Srisutthiyakorn, 2018).*

$\kappa = \dfrac{\pi R_H^4}{8A\tau}$	1	0.8	0.70	0.50	0.31	0.28
$\kappa = \dfrac{\pi R_{Cir}^4}{8A\tau}$	1	1.25	1.13	1.38	1.85	2.18
$\kappa = \dfrac{\pi R_A^4}{8A\tau}$	1	1	0.89	0.83	0.75	0.78

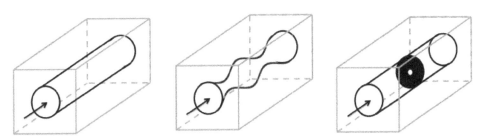

Figure 8.4.2 Straight pipes with circular cross sections: (a) constant radius; (b) radius varying sinusoidally along the pipe axis; (c) pipe with baffle or choke.

Pore Throat Effect

Efforts to extend the usefulness of the Kozeny–Carman equation have included corrections for flow-path cross-sectional shape, tortuosity, scale, surface area, and porosity. An important parameter, missing from the Kozeny–Carman description is the variability of channel diameter along any given flow path. Figure 8.4.2 compares blocks containing three straight pipes with circular cross sections. The first is a simple cylinder, the model from which the Kozeny–Carman originated (equations 8.4.3 and 8.4.7). The second pipe has radius that varies sinusoidally along the pipe axis (equation 8.2.14, from Sisavath *et al.*, 2014) and has less permeability than a cylinder with the same mean radius. The decrease in permeability is larger than can be accounted for by the increase in specific surface area. The third pipe is a cylinder with a baffle or choke in the center – an exaggerated example of a small pore throat. Srisutthiyakorn (2018) simulated flow in the latter model using the Lattice–Boltzmann method and found that the permeability in a pipe 200 times longer than the baffle length could be estimated using an apparent radius $R_A \approx (0.25/R_B + 0.75/R_{cir})^{-1}$, where R_B is the radius of the hole in the baffle, and R_{cir} is the radius of the straight pipe.

Sphere Packs

There is no known exact expression for the permeability of a random sphere pack. However, a common strategy is to extrapolate the Kozeny–Carman equation 8.4.7 from tubes to a packing of identical spheres with diameter d, via the porosity and specific surface area, $S = 6(1 - \phi)/d$. For a random pack of identical spheres Carman (1937) proposed tortuosity $\tau = \sqrt{2}$. (In general, τ is a function of porosity and the density of packing.) For a *monomodal* sphere pack, with $1/B = 2.5$, a common estimate of permeability, based on the Kozeny–Carman equation, is

$$\kappa = \frac{d^2}{180} \frac{\phi^3}{(1 - \phi)^2} ,$$
(8.4.14)

where d is the particle size. Permeability is in the same units as d^2.

Rumpf and Gupta (1971) suggested an alternative expression for the permeability of monodisperse sphere packs:

$$\kappa = \frac{\phi^{5.5} d^2}{5.6} .$$
(8.4.15)

Howells (1974) and Hinch (1977) derived the following expansion for non-dilute sphere packs

$$\frac{\kappa_S}{\kappa} = 1 + \frac{3}{\sqrt{2}}(1 - \phi)^{1/2} + \frac{135}{64}(1 - \phi)\ln(1 - \phi) + 16.456(1 - \phi) + O(1 - \phi) ,$$
(8.4.16)

where κ_S is the Stokes permeability (Torquato, 2002), written in terms of the sphere radius R:

$$\kappa_s = \frac{2R^2}{9(1 - \phi)} ,$$
(8.4.17)

and ϕ is the porosity. Note that the Howells–Hinch result resembles a self-consistent expansion of the Stokes permeability (Torquato, 2002):

$$\kappa = \kappa_S \left(1 + \frac{R}{\sqrt{\kappa}} + \frac{R^2}{3\kappa} \right)^{-1} ,$$
(8.4.18)

or

$$\frac{\kappa_S}{\kappa} = 1 + \frac{3}{\sqrt{2}}(1 - \phi)^{1/2} + \dots .$$
(8.4.19)

Ellipsoidal Particles

Table 8.4.2, from Coelho *et al.* (1997), shows the results of numerically simulated random packs of spheroidal particles with semi-axes l_1 and $l_2 = l_3$. The particle aspect ratio is l_1/l_2, so that particles with aspect ratios <1 are oblate and those with aspect ratios >1 are prolate.

Table 8.4.2 *Simulation results for packing of ellipsoidal particles (Coelho et al., 1997). Column parameters are aspect ratio, ratio of sample dimension to particle dimension (w/R_v), porosity, number of particles in the simulation (N_u), horizontal electrical conductivity $(\bar{\sigma}_{xy}/\sigma_{fluid})$ and horizontal permeability normalized by particle dimension squared (σ_{xy}/R_v^2).*

Aspect ratio	w/R_v	ϕ	N_u	$\bar{\sigma}_{xy}/\sigma_{fluid}$	$10^3 \kappa_{xy}/R_v^2$
0.1	7.43	0.395	592	0.212	1.59
0.2	9.36	0.428	559	0.230	2.29
0.5	12.70	0.412	575	0.205	2.02
1.0	12.50	0.403	279	0.177	2.16
2.0	10.08	0.415	572	0.188	1.77
5.0	5.47	0.567	423	0.325	7.39
10.0	3.45	0.688	305	0.464	19.62

The resulting packing porosity is ϕ. The packing algorithm has a gravitational pull in the z-direction, so that there is an anisotropy to the arrangement. The numerically simulated (finite difference) horizontal electrical conductivity (assuming conducting pore fluid and insulating particles) and the horizontal permeability are denoted by $\bar{\sigma}_{xy}$ and κ_{xy}, respectively. Coelho *et al.* find that the transport properties are controlled primarily by the porosity and are fairly independent of the particle shape (over the range of shapes explored). Permeability is also affected by the characteristic particle dimension, R_v, which is the radius of a sphere having the same volume as the particle.

Mixed Spherical Particle Sizes

Extensive tests (Rumpf and Gupte, 1971; Dullien, 1991) on laboratory data for granular media with mixed particle sizes (poor sorting) suggest that the Kozeny–Carman relation can still be applied using an effective or average particle size $\bar{D}$, defined by

$$\frac{1}{\bar{D}} = \frac{\int D^2 n(D)dD}{\int D^3 n(D)dD} \tag{8.4.20}$$

where $n(D)$ is the number distribution of each size particle. This can be written in terms of a discrete size distribution as follows:

$$\frac{1}{\bar{D}} = \frac{\sum_i D_i^2 n_i}{\sum_i D_i^3 n_i} \tag{8.4.21}$$

This can be converted to a mass distribution by noting that

$$m_i = n_i \frac{4}{3}\pi r_i^3 \rho_i = n_i \left(\frac{1}{6}\pi\rho_i\right) D_i^3 \tag{8.4.22}$$

where m_i, r_i, and ρ_i, are the mass, radius, and density of the ith particle size. Then, one can write

$$\frac{1}{\overline{D}} = \frac{\sum_i (m_i/\rho_i D_i)}{\sum_i (m_i/\rho_i)} \tag{8.4.23}$$

If the densities of all particles are the same, then

$$\frac{1}{\overline{D}} = \frac{\sum_i (m_i/D_i)}{\sum_i m_i} = \sum_i \frac{f_i}{D_i} \tag{8.4.24}$$

where f_i, is either the mass fraction or the volume fraction of each particle size.

An equivalent description of the particle mixture is in terms of specific surface areas. The specific surface area (pore surface area divided by bulk sample volume) for a mixture of spherical particles is

$$\overline{S} = 3(1 - \phi) \sum_i \frac{f_i}{r_i} \tag{8.4.25}$$

Permeability Bounds

The two-point void upper bound on isotropic permeability κ is given by (Rubinstein and Torquato, 1989; Torquato and Rubinstein, 1989; Torquato, 2001; Torquato and Pham, 2004):

$$\kappa \le \frac{2}{3} \frac{l_P^2}{\phi_S^2} \tag{8.4.26}$$

where $\phi_P = 1 - \phi_S$ is the porosity, and l_P is a pore space length scale, defined by

$$l_P^2 = \int_0^\infty [S_2(\mathbf{r}) - \phi_P^2] \mathbf{r} \, d\mathbf{r} \tag{8.4.27}$$

In equation 8.4.27, $S_2(\mathbf{r}) = \langle I^{(P)}(\mathbf{x}) I^{(P)}(\mathbf{x} + \mathbf{r}) \rangle$ is a two-point correlation function, where $I^{(P)}(\mathbf{x})$ is the pore space indicator function. The void bound on permeability is optimal among all microstructures having the same porosity ϕ_P and length scale l_P as defined in equation 8.4.27 (Torquato and Pham, 2004).

The coated spheres model consists of a random pack of spheres that are composed of a spherical inclusion of phase 1 with radius R_1, surrounded by a shell of phase 2 with radius R_M, such that the inclusion volume fraction $\phi_1 = (R_1/R_M)^3$ is fixed for each coated sphere. The model is space-filling, so that there are an infinite number of sizes ranging to infinitesimally small.

Let the number density of spheres with inner radius R_{Ik} be ρ_k. Then the nth moment of the inclusion radius distribution R_I is defined as

$$\langle R_I^n \rangle \equiv \frac{1}{\rho} \sum_{k=1}^{\infty} \rho_k R_{Ik}^n \tag{8.4.28}$$

where ρ is a normalization constant, with units of inverse length cubed. It can be shown that for the coated spheres model, the characteristic scale is (Torquato and Pham, 2004).

$$l_P^2 = \frac{\phi_P \, \phi_S^2 \langle R_I^5 \rangle}{15 \langle R_I^3 \rangle} \tag{8.4.29}$$

For the 2D coated cylinder, model the characteristic scale is

$$l_P^2 = \frac{\phi_P \, \phi_S^2 \langle R_I^4 \rangle}{8 \langle R_I^2 \rangle} \tag{8.4.30}$$

In three dimensions, the permeability is zero if the pore space consists of the isolated inclusions in the coated spheres model. If the pore space surrounds the solid spherical inclusions, then $\phi_I = \phi_S$, and the (not necessarily optimal) upper bound is (Torquato and Pham, 2004)

$$\kappa_{3D} \le \kappa_{S-3D} \left[1 + \frac{1}{5} \phi_S^{1/3} - \phi_S - \frac{1}{5} \phi_S^2 \right], \quad \text{where } \kappa_{S-3D} = \frac{1}{3\phi_S} \frac{\langle R_I^5 \rangle}{\langle R_I^3 \rangle}. \tag{8.4.31}$$

In **two dimensions**, in which there is axial flow inside of parallel cylindrical tubes with porosity $\phi_P = \phi_I$, the **optimum** upper bound on permeability is realized by coated cylinders and is given by

$$\kappa_{2D} \le \frac{\phi_P}{8} \frac{\langle R_I^4 \rangle_{2D}}{\langle R_I^2 \rangle_{2D}} \tag{8.4.32}$$

The upper bound for flow in the outer phase of a coated cylinder model is

$$\kappa_{2D} \le \kappa_{S2D} \left[-\frac{3}{2} + 2\phi_S - \frac{1}{2} \phi_S^2 - \ln\phi_S \right], \quad \text{where } \kappa_{S-2D} = \frac{1}{4\phi_S} \frac{\langle R_I^4 \rangle}{\langle R_I^2 \rangle}. \tag{8.4.33}$$

Percolation

Bourbié *et al.* (1987) discuss a more general form of the Kozeny–Carman equation:

$$\frac{\kappa}{d^2} \propto \phi^n \tag{8.4.34}$$

in which n has been observed experimentally to vary with porosity from $n \ge 7$ ($\phi < 5\%$) to $n \le 2$ ($\phi > 30\%$). The Kozeny–Carman value of $n = 3$ appears to be appropriate for very clean materials such as Fontainebleau sandstone and sintered glass, whereas $n = 4$ or 5 is probably more appropriate for more general natural materials (Bourbié *et al.*, 1987).

Mavko and Nur (1997) suggest that one explanation for the apparent dependence of the coefficient, n, on porosity is the existence of a percolation porosity, ϕ_c below which the remaining porosity is disconnected and does not contribute to flow.

Experiments suggest that this percolation porosity is of the order of 1%–3%, although it depends on the mechanism of porosity reduction. The percolation effect can be incorporated into the Kozeny–Carman relations simply by replacing ϕ by $(\phi - \phi_c)$. The idea is that it is only the porosity *in excess* of the percolation porosity that determines the permeability. Substituting this into the Kozeny–Carman equation gives

$$\frac{\kappa}{d^2} = \frac{1}{72} \frac{(\phi - \phi_c)^3}{[1 - (\phi - \phi_c)]^2 \tau^2} \tag{8.4.35}$$

or

$$\kappa = B \frac{(\phi - \phi_c)^3}{(1 + \phi_c - \phi)^2} d^2 \tag{8.4.36}$$

$$\approx B(\phi - \phi_c)^3 d^2$$

The result is that the derived $n = 3$ behavior can be retained while fitting the permeability behavior over a large range in porosity.

Remember that in all of the preceding formulations of the Kozeny–Carman relation, the unit of permeability is length squared (e.g., m^2), where the length unit is that used for the pipe radius or particle size (e.g., m). The porosity is expressed as a unitless volume fraction, and the unit of the specific surface area is the inverse length (e.g., m^{-1}).

The same equation for permeability in milliDarcy, particle size in mm, and porosity in fraction of unity reads as

$$\frac{\kappa}{d^2} = \frac{10^9}{72} \frac{(\phi - \phi_c)^3}{[1 - (\phi - \phi_c)]^2 \tau^2} \tag{8.4.37}$$

Figure 8.4.3 Compares laboratory permeability data of Fontainebleau sandstone with permeability calculations according to the preceding Kozeny–Carman equation. The percolation porosity is 0.02, and tortuosity varies from 1 (upper curve) to 5 (lower curve) with constant increment 0.5. The grain size is 0.25 mm. The apparent tortuosity values appropriate for this dataset are between 1.5 and 3.

Example: Calculate the permeability of a sandstone sample which has porosity 0.32 and an average grain size of 100 μm. Use the Kozeny–Carman relation modified for a percolation porosity:

$$\kappa = B\frac{(\phi - \phi_c)^3}{(1 + \phi_c - \phi)^2}d^2$$

Assume that $B = 15$ and $\phi_c = 0.035$. The units of B in this equation are such that expressing d in microns gives permeability in milliDarcy. Then

$$\kappa = B\frac{(\phi - \phi_c)^3}{(1 + \phi_c - \phi)^2}d^2 = 15\frac{(0.32 - 0.035)^3}{(1 + 0.035 - 0.32)^2}100^2$$

$$= 6792.3 \text{ milliDarcy} = 6.79 \text{ Darcy}$$

Compare the permeabilities κ_1 and κ_2 of two sandstones that have the same porosity and pore microstructure, but different average grain sizes, $d_1 = 80$ μm and $d_2 = 240$ μm.

Assuming that B and ϕ_c are the same for both sandstones since they have the same pore microstructure, we can express the ratio of their permeabilities as

$$\frac{\kappa_1}{\kappa_2} = \frac{d_1^2}{d_2^2} = \frac{80^2}{240^2} = \frac{1}{9}$$

The sandstone with larger average grain size has a higher permeability (by a factor of 9), even though both have the same total porosity.

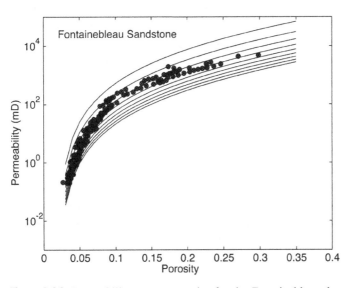

Figure 8.4.3 Permeability versus porosity for the Fontainebleau dataset with theoretical curves superimposed according to the Kozeny–Carman equation. The threshold porosity is 0.02, and the tortuosity varies from 1 (upper curve) to 5 (lower curve) with a constant increment of 0.5.

Comments on Tortuosity

Tortuosity means different things to different people. Some definitions of tortuosity attempt to describe geometric heterogeneity, as might be observed in 2D or 3D rock images. Some definitions of tortuosity attempt to describe factors that retard flow of fluid, current, or diffusing species. And often, tortuosity is treated as a fudge factor to force a model to fit data. Excellent discussions of tortuosity can be found for example in Clennell (1998), Duda *et al.* (2011), and Ghanbarian *et al.* (2013).

The geometric tortuosity, τ_g, can be defined as an average ratio of "shortest" paths through the pore space, L_g, to sample length L_s:

$$\tau_g = \frac{\langle L_g \rangle}{L_s}. \tag{8.4.38}$$

The hydraulic tortuosity, τ_h, is defined in terms of the flux-weighted average streamline path length, L_h, in laminar hydraulic flow:

$$\tau_h = \frac{\int L_h dq}{L_s \int dq} \tag{8.4.39}$$

where q is flux.

The term, "hydraulic tortuosity factor," T_h, is usually defined as the square of the flux-weighted average flow path to the sample length (Clennell, 1997; Dullien, 1992), where $T_h \geq 1$. Both $\sqrt{T_h}$ and T_h have been called the "tortuosity" in the literature. Bear (1972) described the hydraulic "tortuosity coefficient" as $1/T_h$.

It has been argued (Dullien, 1992; Clennell, 1997) that tortuosity is not a physical attribute of the pore space but a property of 1D models chosen to describe hydraulic, electrical, or diffusional transport in porous media. For example, the Kozeny–Carman (Kozeny, 1927; Carman, 1937) equation for absolute permeability, κ, is

$$\kappa = \frac{\phi^3}{\beta \tau_{hK}^2 (1 - \phi^2) S_0^2}, \tag{8.4.40}$$

where ϕ is the porosity, β is a geometric constant, S_0 is the specific surface area, and τ_{hk} is the tortuosity in an "equivalent channel" or capillary model. The electrical tortuosity, τ_e, is defined in terms of the average electrical current flux path length, L_e:

$$\tau_e = \frac{\langle L_e \rangle}{L_s}. \tag{8.4.41}$$

Similarly an "electrical tortuosity *factor*" can be fined as $T_e = \tau_e^2$. The electrical formation factor is $F = R_0/R_w$, where R_0 is the brine-saturated rock resistivity, and R_w

is the brine resistivity. The modified form of Archie's (empirical) law (Archie, 1942; Winsauer *et al.*, 1952) states that

$$F = \frac{a}{\phi^m} \tag{8.4.42}$$

where m is the "cementation exponent," and a is the "lithology factor" or "tortuosity factor." It is common (Wyllie, 1957; Schopper, 1966; Dullien, 1992; Walsh and Brace, 1984; Coleman and Vassilicos, 2008) to infer an electrical tortuosity from measured formation factor using

$$\tau_e^2 = \phi F . \tag{8.4.43}$$

Some authors (Walsh and Brace, 1984; Berryman and Blair, 1986) have equated electrical and hydraulic tortuosities under restrictive assumptions of pore geometry. For example, assuming the capillary model, equations 8.4.40 and 8.4.43 might be combined to yield a relation between permeability and formation factor:

$$\kappa = \frac{\phi^2}{\beta F (1 - \phi)^2 S_0^2} \tag{8.4.44}$$

In general, electrical and hydraulic tortuosities are not the same, which is not surprising since the equations governing each type of flow are fundamentally different. Fluid flow lines tend to be more clustered along wider channels than are electrical flux lines. Hence, in general, the hydraulic tortuosity is larger than the electrical tortuosity. Zhang and Knackstedt (1995) simulated both electrical and hydraulic flow using the lattice gas automaton method and found that the hydraulic and electrical tortuosity factors varied from a few percent to a factor of ten. Using finite element analysis, Saomoto and Katagiri (2015) found that hydraulic tortuosity is on average 15% larger than electrical tortuosity.

Tortuosity has been related to Biot's inertial coupling "structure factor," α (Brown, 1980; Johnson, 1980; Berryman, 1980, 1981, 2003; Johnson *et al.*, 1982). When an object with density ρ and volume V is accelerated while entrained in a fluid with density ρ_f, the effective inertial mass is $(\rho + r\rho_f)V$ where r is a measure of geometry, and $r\rho_f V$ is the "induced mass." This leads to Biot's structure factor $\alpha = r(1 + \phi^{-1})$, which has been equated to tortuosity, $\alpha = \tau^2$ (Berryman, 1980, 1981). Another interpretation (Johnson, 1980; Johnson *et al.*, 1982; Berryman 2003) is in terms of the Biot slow wave velocity $V_{slow} = \sqrt{K_f / \alpha \rho_f}$, where K_f is the fluid bulk modulus. The speed of sound in a pure fluid is $V_0 = \sqrt{K_f / \rho_f} = V_{slow}\sqrt{\alpha}$. Therefore, we might interpret $\tau = \sqrt{\alpha}$ as a measure of the longer wave propagation path, or tortuosity. Brown (1980) showed that $\alpha = F\phi$. For well-separated spheres, the geometric factor in Biot's coupling factor is $r = 1/2$, and this value is often adopted for rocks. However,

$\tau^2 = \alpha = (1 + \phi^{-1})/2$ is *not* the exact tortuosity in a close pack of spheres, as is sometimes assumed.

Berryman and Milton (1988) considered variational bounds on electrical conductivity, σ, of a two-phase isotropic composite. If one of the phases (e.g., the mineral frame) is an insulator, then the Hashin–Shtrikman bounds on conductivity are

$$0 \leq \sigma \leq \frac{2\sigma_w \phi}{3 - \phi} \tag{8.4.45}$$

where σ_w is the conductivity of the fluid phase. Noting that $\tau_e^2 = \phi F$, the lower Hashin–Shtrikman bound on tortuosity is given by (Berryman and Milton, 1988)

$$\tau_{e(HS-)}^2 = 1 + (1 - \phi)/2 \leq \tau_e^2 \tag{8.4.46}$$

Table 8.4.2 (Srisutthiyakorn, 2018) compares tortuosities computed directly from stream lines determined with Lattice–Boltzmann flow simulations. Pore space geometries were taken from sphere packs, Fontainebleau sandstone, bitumen sand, and Grosmont carbonate. Flux-weighted tortuosities are observed to be <1.5. In contrast, tortuosities determined as fitting parameters in the Kozeny–Carman equation are systematically larger.

Having stated many of the caveats, we nevertheless summarize a few published empirical relations of tortuosity to porosity. Carman (1956) observed dye flowing through a glass sphere pack and estimated tortuosity to be approximately $\sqrt{2}$. A commonly cited empirical model for both hydraulic and electrical tortuosity is

$$\tau = 1 - P\log(\phi), \tag{8.4.47}$$

where P depends strongly on the microstructure of the porous material, ranging from $P \approx 1.6$ for wood chips (Pech, 1984) to $P \approx 1.9$ (Mauret and Renaud, 1997) for high porosity beds of spheres and fibers. An alternative empirical model from Mota *et al.* (2001) for binary mixtures of spherical particles is

$$\tau_h = \phi^{-\beta}, \tag{8.4.48}$$

with $\beta \approx 0.4$. Du Plessis and Masliyah (1991) derived an analytical model for isotropic granular media:

$$\tau_h = \frac{\phi}{1 - (1 - \phi)^{2/3}}. \tag{8.4.49}$$

Ahmadi *et al.* (2011) proposed an analytical model for equal–sized spheres:

$$\tau_h = \sqrt{\frac{2\phi}{3[1 - B(1 - \phi)^{2/3}]} + \frac{1}{3}}, \tag{8.4.50}$$

Table 8.4.3 *Average of flux-weighted tortuosity, mean tortuosity, minimum tortuosity, and maximum tortuosity, all determined from Lattice–Boltzmann simulations of streamlines. τ_{KC} is the tortuosity inverted from equation 8.4.7 that makes the Kozeny–Carman permeability match the LB simulations. (Srisutthiyakorn, 2018)*

	Simple cubic pack (SCP)*	Face-centered cubic pack (FCP)*	Finney packs (LX = 3)*	Finney packs (LX = 6)*	Finney packs (LX = 9)*	Finney packs (LX = 12)*	Fontainebleau sandstone	Bituminous sand	Berea sandstone	Grosmont carbonate
$\tau_{FluxWeighted}$	1.01	1.35	1.20	1.25	1.23	1.24	1.46	1.23	1.38	1.38
τ_{Mean}	1.02	1.35	1.21	1.26	1.23	1.25	1.46	1.24	1.37	1.37
τ_{Min}	1.00	1.20	1.06	1.07	1.08	1.10	1.27	1.09	1.13	1.17
τ_{Max}	1.63	2.00	1.75	1.96	1.85	1.86	1.89	2.03	2.25	2.20
τ_{KC}	1.56	1.82	2.54	1.78	1.77	1.83	2.90	1.78	2.55	7.85

* denotes regular and random sphere packs. LX indicates field of view, with smaller numbers corresponding to sample dimensions comparable to the sphere diameter and larger numbers corresponding to sample dimensions much larger than the sphere diameter.

where $B \approx 1.21$ for cubic packings and $B \approx 1.11$ for tetrahedral packings. Koponen *et al.* (1997) proposed

$$\tau_h = 1 + 0.65 \frac{1 - \phi}{(\phi - \phi_t)^{0.19}}. \tag{8.4.51}$$

Lanfrey *et al.* (2010) developed a theoretical expression for geometric tortuosity in a bed of randomly packed identical particles:

$$\tau_g = 1.23 \frac{(1 - \phi)^{4/3}}{\zeta^2 \phi}, \tag{8.4.52}$$

where ζ is the sphericity, equal to 1 for spheres and <1 otherwise. Li and Yu (2011) used a solid fractal model and arrived at

$$\tau_g = (19/18)^{\ln(\phi)/\ln(8/9)}. \tag{8.4.53}$$

Uses

The Kozeny–Carman relation can be used to estimate the permeability from geometric properties of a rock.

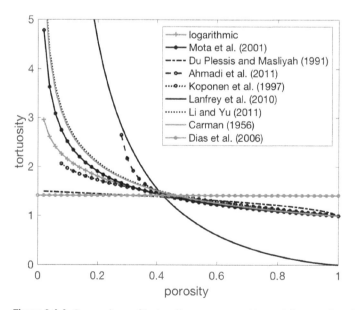

Figure 8.4.4 Comparison of tortuosity versus porosity models, equations 8.4.47–8.4.53

Assumptions and Limitations

The following assumptions and limitations apply to the Kozeny–Carman relation:
- The derivation is heuristic. Strictly speaking, it should hold only for rocks with porosity in the form of circular pipes. Nevertheless, in practice it often gives reasonable results. When possible it should be tested and calibrated for the rocks of interest.
- The rock is isotropic.
- Fluid-bearing rock is completely saturated.

8.5 Permeability Relations with S_{wi}

Synopsis

The following family of empirical equations relate permeability to porosity and irreducible water saturation (Schlumberger, 1991):

the Tixier equation

$$\kappa = 62500 \; \phi^6 / S_{wi}^2 \tag{8.5.1}$$

the Timur equation

$$\kappa = 10000\phi^{4.5} / S_{wi}^2 \quad \text{or} \quad \kappa = 8581\phi^{4.4} / S_{wi}^2 \tag{8.5.2}$$

the Coates–Dumanoir equation

$$\kappa = (90000/m^4)\phi^{2m} / S_{wi}^2 \tag{8.5.3}$$

where m is the cementation exponent (see Archie's law, Section 9.4); and the Coates equation

$$\kappa = 4900\phi^4 (1 - S_{wi})^2 / S_{wi}^2 \tag{8.5.4}$$

where the permeability is in milliDarcy and porosity and irreducible water saturation are unitless volume fractions.

The irreducible water saturation is by definition the lowest water saturation that can be achieved in a core plug by displacing the water by oil or gas. The state is usually achieved by flowing oil or gas through a water-saturated sample, or by spinning the sample in a centrifuge to displace the water with oil or gas. The term is somewhat imprecise, because the irreducible water saturation is dependent on the final drive pressure (when flowing oil or gas) or the maximum speed of rotation (in a centrifuge). The related term, connate water saturation, is the lowest water saturation found *in situ*. The smaller the grain size, the larger the irreducible water saturation, because the capillary forces that retain water are stronger in thinner capillaries.

Figure 8.5.1 compares laboratory permeability data for Fontainebleau sandstone with permeability calculations according to the preceding equations. Fontainebleau is

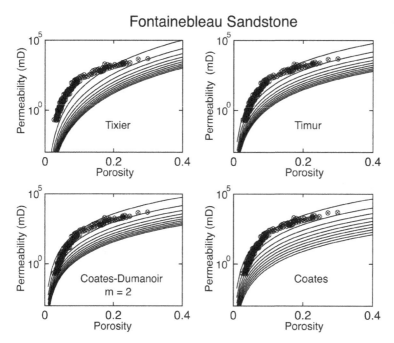

Figure 8.5.1 Permeability versus porosity for the Fontainebleau dataset, with theoretical curves superimposed according to the previous equations. The irreducible water saturation varies from 0.05 for the upper curve to 0.50 for the lower curve, with a constant increment of 0.05. The cementation exponent in the Coates–Dumanoir equation is 2 in this example.

an extremely clean and well-sorted sandstone with large grain size (about 0.25 mm). As a result, we expect very low irreducible water saturation. This is consistent with Figure 8.5.1, where these data are matched by the curves with S_{wi} between 0.05 and 0.10.

Figure 8.5.2 compares permeability calculations according to these equations with the Kozeny–Carman formula. All equations provide essentially identical results, except for the Tixier equation, which gives smaller permeability in the low-to-medium porosity range.

Combining the Kozeny–Carman equation for permeability (κ_{KC}):

$$\frac{\kappa_{KC}}{d^2} = \frac{10^9}{72} \frac{(\phi - \phi_c)^3}{[1 - (\phi - \phi_c)]^2 \tau^2} \tag{8.5.5}$$

where κ_{KC} is in mD, d is in mm, and porosity (ϕ and ϕ_c) varies from 0 to 1, with the Timur equation:

$$\kappa_{TIM} = 8581\phi^{4.4}/S_{wi}^2 \tag{8.5.6}$$

yields the following relation between S_{wi} and d:

$$S_{wi} = \frac{0.025}{d} \frac{\phi^{2.2}[1 - (\phi - \phi_c)]\tau}{(\phi - \phi_c)^{1.5}} \tag{8.5.7}$$

where d is in mm. For $\phi_c = 0$ this equation becomes

$$S_{wi} = \frac{0.025\phi^{0.7}(1-\phi)\tau}{d} \tag{8.5.8}$$

Assume $\tau = 2$. Then the curves of S_{wi} versus ϕ according to the last equation are plotted in Figure 8.5.3.

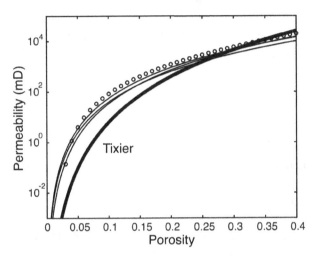

Figure 8.5.2 Permeability versus porosity according to the Tixier equation (bold curve), Timur, Coates–Dumanoir, and Coates equations (fine curves); and the Kozeny–Carman equation (open symbols). In these calculations we used a tortuosity of 2.5; an irreducible water saturation of 0.1; a threshold porosity of 0.02; and a grain size of 0.25 mm.

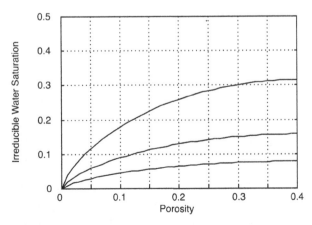

Figure 8.5.3 Irreducible water saturation versus porosity according to the last equation in the preceding text. The curves from top to bottom are for grain sizes of 0.05, 0.1, and 0.2 mm, respectively.

The result shown in Figure 8.5.3 is somewhat counterintuitive: the irreducible water saturation increases with increasing porosity. This is due to the fact that in Timur's equation, and at fixed permeability, S_{wi} increases with increasing ϕ. An explanation may be that to maintain fixed permeability at increasing porosity, the pore size needs to reduce, thus resulting in increasing S_{wi}.

Caution:

This S_{wi} equation is obtained by the ad hoc equating of two different permeability equations, one of which is based on a crude idealization of the pore-space geometry, while the other is purely empirical. Therefore, the resulting S_{wi} may deviate from factual experimental values.

Uses

The permeability relations with irreducible water saturation can be used to estimate the permeability from this parameter often available from laboratory experiments.

Assumptions and Limitations

The following assumptions and limitations apply to the permeability relations with irreducible water saturation:
- These equations are strictly empirical, although they implicitly use the physics-based relation between the grain size and irreducible water saturation. The functional forms used in these equations have to be calibrated, whenever possible, to site-specific data.
- The rock is isotropic.
- Fluid-bearing rock is completely saturated.

8.6   Permeability of Fractured Formations

Synopsis

Many models for the permeability of fractured formations are based on the expression for viscous flow rate between two parallel walls:

$$Q = -\frac{h^3}{12\eta}\frac{\Delta P}{l} \tag{8.6.1}$$

where h is the aperture of the fracture, Q is the volume flow rate per unit length of fracture (normal to the direction of flow), η is the fluid viscosity, and ΔP is the pressure drop over length l.

For a set of parallel fractures with spacing D the fracture porosity is

$$\phi = h/D \tag{8.6.2}$$

A tortuosity can be defined in the similar way as in the Kozeny–Carman formalism (Figure 8.4.1) or, through the angle α of the fracture plane to the pressure gradient vector:

$$\tau = 1/\cos\alpha \tag{8.6.3}$$

Then, using Darcy's definition of permeability, the permeability of a formation with parallel fracture planes at angle α relative to the pressure gradient is

$$\kappa = \frac{\phi h^2}{12\tau^2} = \frac{\phi h^2}{12} \cos^2\alpha \tag{8.6.4}$$

If the matrix of the rock (the material between the fractures) has permeability κ_m then the combined total permeability κ_t will simply be the sum of the two, the matrix permeability and the permeability of the fractures:

$$\kappa_t = \kappa_m + \kappa = \kappa_m + \frac{\phi h^2}{12} \cos^2\alpha \tag{8.6.5}$$

Since real fracture surfaces are irregular, the simple model for parallel plates must be modified (see Jaeger *et al.*, 2007, for an excellent summary). A common approach is to approximate the irregular channel with a smooth parallel channel having an effective **hydraulic aperture** h_H It was shown (Beran, 1968) that the hydraulic aperture is bounded as

$$\langle h^{-3} \rangle^{-1} \le h_H^3 \le \langle h^3 \rangle \tag{8.6.6}$$

where the operator $\langle \cdot \rangle$ is the average over the plane of the fracture. The lower bound $\langle h^{-3} \rangle^{-1/3}$ corresponds to the case where the aperture fluctuates only in the direction of flow, while the upper bound $\langle h^3 \rangle^{1/3}$ corresponds to the case where the aperture varies only in the direction normal to flow.

Renshaw (1995) gave an expression for hydraulic aperture

$$h_H^3 = \langle h \rangle^3 [1 + \sigma_h^2/\langle h \rangle^2]^{-3/2} \tag{8.6.7}$$

where σ_h^2 is the variance of h. Other analyses have been presented by Elrod (1979), Dagan (1993), and Jaeger *et al.* (2007).

Uses

The permeability relations for a fractured formation can be used in reservoirs, such as chalks, carbonates, and shales, where fluid transport is dominated by natural or artificial fractures.

Assumptions and Limitations

The following assumptions and limitations apply to the permeability relations in fractured formations:

- these equations are derived for an idealized pore geometry and thus may fail in an environment without adequate hydraulic communication between fractures and/or complex fracture geometry;
- fluid-bearing rock is completely saturated.

8.7   Diffusion and Filtration: Special Cases

Synopsis

Nonlinear Diffusion

Some rocks, such as coals, exhibit strong sensitivity of permeability (κ) and porosity (ϕ) to net pressure changes. During an injection test in a well, apparent permeability may be 10–20 times larger than that registered during a production test in the same well. In this situation, the assumption of constant permeability, which leads to the linear diffusion equation, is not valid. The following nonlinear diffusion equation must be used (Walls $et\ al.$, 1991):

$$\frac{\eta\phi(\beta_{\text{fl}}+\beta_{\text{pv}})}{\kappa}\frac{\partial P}{\partial t} = \Delta P + (\beta_{\text{fl}}+\gamma)(\nabla P)^2 \tag{8.7.1}$$

where fluid compressibility β_{fl} and pore-volume compressibility β_{pv} are defined as

$$\beta_{\text{fl}} = \frac{1}{\rho}\frac{\partial\rho}{\partial P}, \quad \beta_{\text{pv}} = \frac{1}{\upsilon_{\text{p}}}\frac{\partial\upsilon_{\text{p}}}{\partial P} \tag{8.7.2}$$

where υ_{p} is the pore volume and ρ is the fluid density. The permeability-pressure parameter γ is defined as

$$\gamma = \frac{1}{\kappa}\frac{\partial\kappa}{\partial P} \tag{8.7.3}$$

For one-dimensional plane filtration, the diffusion equation previously given is

$$\frac{\eta\phi(\beta_{\text{fl}}+\beta_{\text{pv}})}{\kappa}\frac{\partial P}{\partial t} = \frac{\partial^2 P}{\partial x^2} + (\beta_{\text{fl}}+\gamma)\left(\frac{\partial P}{\partial x}\right)^2 \tag{8.7.4}$$

For one-dimensional radial filtration, it is

$$\frac{\eta\phi(\beta_{\text{fl}}+\beta_{\text{pv}})}{\kappa}\frac{\partial P}{\partial t} = \frac{\partial^2 P}{\partial r^2} + \frac{1}{r}\frac{\partial P}{\partial r} + (\beta_{\text{fl}}+\gamma)\left(\frac{\partial P}{\partial r}\right)^2 \tag{8.7.5}$$

Hyperbolic Equation of Filtration (Diffusion)

The diffusion equations previously presented imply that changes in pore pressure propagate through a reservoir with infinitely high velocity. This artifact results from using the original Darcy's law, which states that volumetric fluid flow rate and pressure gradient are linearly related. In fact, according to Newton's second law, pressure gradient (or acting force) should also be proportional to acceleration (time derivative of the fluid flow rate). This modified Darcy's law was first used by Biot (1956) in his theory of dynamic poroelasticity:

$$-\frac{\partial P}{\partial x} = \frac{\eta}{\kappa} V_x + \frac{\tau \rho}{\phi} \frac{\partial V_x}{\partial t} \tag{8.7.6}$$

where τ is tortuosity. The latter equation, if used instead of the traditional Darcy's law,

$$-\frac{\partial P}{\partial x} = \frac{\eta}{\kappa} V_x \tag{8.7.7}$$

yields the following hyperbolic equation that governs plane one-dimensional filtration:

$$\frac{\partial^2 P}{\partial x^2} = \tau \rho (\beta_{\text{fl}} + \beta_{\text{pv}}) \frac{\partial^2 P}{\partial t^2} + \frac{\eta \phi (\beta_{\text{fl}} + \beta_{\text{pv}})}{\kappa} \frac{\partial P}{\partial t} \tag{8.7.8}$$

This equation differs from the classical diffusion equation because of the inertia term,

$$\tau \rho (\beta_{\text{fl}} + \beta_{\text{pv}}) \frac{\partial^2 P}{\partial t^2} \tag{8.7.9}$$

Changes in pore pressure propagate through a reservoir with a finite velocity, c:

$$c = \sqrt{\frac{1}{\tau \rho (\beta_{\text{fl}} + \beta_{\text{pv}})}} \tag{8.7.10}$$

This is the velocity of the slow Biot wave in a very rigid rock.

Uses

The equations presented in this section can be used to calculate fluid filtration and pore-pressure pulse propagation in rocks.

Assumptions and Limitations

The equations presented in this section assume that the fluid is Newtonian and the flow is isothermal.

8.8   Heavy Oil Viscosity and Shear Modulus

Synopsis

Heavy oil and extra heavy oil usually consist of large fractions of high molecular weight, non-parafinic compounds such as resins and asphaltenes and a low fraction of volatile, low molecular weight compounds such as alkanes. Biodegraded oils have a deficiency of hydrogen and a high proportion of nitrogen, sulfur, oxygen, and asphaltic molecules in the carbon network (Rojas, 2010).

Heavy oil and extra heavy oil generally have very high viscosity. The viscosity is usually very sensitive to temperature, and less sensitive to pressure. The viscosity of heavy oil is, in general, poorly correlated to API gravity (Hinkle, 2007; Gray, 2010), but the sensitivity of heavy oil to dissolved gas can be more systematic.

Heavy oil viscosity and moduli tend to decrease with increasing temperature. Extensive studies of this behavior have been done for polymers, as well as for oil, heavy oil, and bitumen. Figure 8.8.1 shows heavy oil viscosity vs. temperature trends, compiled by Dusseault (2006), compared with predictions by Puttagunta *et al.* (1988) and Beggs and Robinson (1975).

Published Empirical Correlations

There are many published correlations of viscosity with temperature, pressure, and GOR. We summarize here a few that we found to be often-quoted and/or useful. While these can be useful to estimate plausible viscosity values, the best practice is to always calibrate to lab measurements of the oil of interest. Useful published summaries of the various models can be found in Bennison (1998) and Rojas (2010).

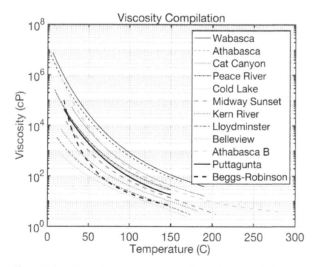

Figure 8.8.1 Viscosity vs. temperature for various North American heavy oils, compiled by Dusseault, 2006), and compared with predictions of Puttagunta *et al.* (1988) and Beggs–Robinson (1975)

Beggs and Robinson (1975) – Correlation for Viscosity of Dead and Gas Saturated Oils

For dead heavy oils at $T > 0^\circ C$:

$$\ln(\eta_{od} + 1) = 0.505y(17.8 + T)^{-1.163}$$
$$\ln(y) = 5.693 - 2.863/\rho_0$$

where η_{od} is the dead oil viscosity in cP, $T > 0$ is temperature in $^\circ C$, and ρ_0 is the oil density at standard conditions.

For saturated (at bubble point) oils:

$$\eta_{ob} = a\eta_{od}^b$$
$$a = 10.715(R_S + 100)\big)^{-0.515}$$
$$b = 5.44(R_S + 150)^{-0.338}$$

where η_{ob} is the oil viscosity at bubble point and R_S is the gas-oil ratio (Scf/Sbl) at bubble point.

DeGuetto *et al.* (1995) – Correlation for Viscosity of Dead Heavy Oils

For heavy oil ($10 < API < 22$):

$$\log_{10}(\eta_{od} + 1) = y$$
$$\log_{10}(y) = 2.064922 - 0.01793\ API - 0.7022\log_{10}(T)$$

For extra heavy oil: ($6 < API < 10$):

$$\log_{10}(\eta_{od} + 1) = y$$
$$\log_{10}(y) = 1.90296 - 0.012619\ API - 0.61748\log_{10}(T)$$

where η_{od} is viscosity in cP, T is temperature in $^\circ F$, and API is the oil gravity.

Bennison (1998) – Correlation for Heavy Dead Oils from N. Sea Fields

For heavy oils limited to $T < 121^\circ C\ (250^\circ F)$, API < 20

$$\eta_{od} = 10^x T^y$$
$$x = 0.052\ API^2 - 2.2704\ API - 5.7567$$
$$y = -0.0222\ API^2 + 0.9415 API - 12.839$$

where η_{od} is viscosity in cP, T is temperature in $^\circ F$ and API is the oil gravity.

Puttagunta *et al.* (1988) – Correlations for Dead and Live Oils, Ranging from Medium Weight to Bitumen

$$\eta = \exp(x)$$
$$x = 2.30259\left[\frac{b}{(1 + (T - 30)/303.15)^S} + C\right] + B_0 P\exp(d\,T) - C_t$$

where η is viscosity in $Pa - s$, and T is temperature in $°C$, and P is pressure in MPa. The various constants are:

$$C = -3.0020$$
$$b = \log_{10}(\eta_{30}) - C$$
$$S = 0.0066940\, b + 3.53641$$
$$B_0 = 0.0047424\, b + 0.0081709$$
$$d = -\ 0.0015646\, b + 0.0061814$$

where η_{30} is the viscosity at 30 °C and 101.3 kPa pressure.

The last term in the equation is a correction for all gases dissolved in the oil:

$$C_t = CO_{CH4}X_{CH4}\exp(-E_{CH4}T) + CO_{CO2}X_{CO2}\exp(-E_{CO2}T)$$
$$+ CO_{C2H6}X_{C2H6} \cdot \exp(-E_{C2H6}\ T)$$
$$CO_{CH4} = 0.027659\, b - 0.014549$$
$$E_{CH4} = 0.00047360\, b + 0.0072533$$
$$CO_{CO2} = 0.029016\, b - 0.012626$$
$$E_{CO2} = 0.0028338\, b - 0.0023982$$
$$CO_{C2H6} = 0.028971\, b - 0.0053846$$
$$E_{C2H6} = 0.0027408\, b - 0.0029816$$

where η_{30} is a reference viscosity at $T = 30°C$ and atmospheric pressure, and X_{CH4}, X_{CO2}, and X_{C2H6} are the gas concentrations in mole%. Setting concentrations to zero yields the dead-oil viscosity.

Hossain *et al.* (2005) – Correlations for Dead and Gas Saturated Heavy Oils

For dead heavy oil:

$$\eta_{od} = 10^{-0.71523\ API + 22.13766}\, T^{0.269024\ API - 8.268047}$$

For gas-saturated heavy oil:

$$\eta_{ob} = A\eta_{od}^{B}$$
$$A = 1 - 0.00171883\ R_S + 1.58081\ \cdot 10^{-6}\ R_S^2$$
$$B = 1 - 0.00205246\ R_S + 3.47559\ \cdot 10^{-6}\ R_S^2$$

where η_{od} is the dead-oil viscosity in cP, η_{ob} is saturated-oil viscosity in cP, T is temperature in $°F$, API is the oil gravity, and R_S is the (saturated) solution gas ratio (Scf/Sbl).

Rojas (2010) – Correlations for Complex Viscosity in Heavy Oils Spanning Temperatures from the "Glass" Phase through the "Liquid" Phase

$$\eta^* = K\omega^{n-1}\exp\left[\frac{E}{R}\left(\frac{1}{T_G - T} + \frac{1}{T_L - T}\right)\right]$$

E=activation energy

ω frequency

T_G glass point temperature

T_L liquid point temperature

K constant

n power-law constant

where η is the dead-oil viscosity in $Pa - s$, T is temperature in °C, T_G is the glass point temperature, and is T_L the liquid point temperature. The constants E, T_G, T_L, and K are found empirically for each oil.

Javanbakhti *et al.* (2018) – empirical relations for bitumen shear modulus μ_a as a function of temperature, T, angular frequency $\omega = 2\pi f$, and viscosity, η, with the viscosity itself being another empirical function of temperature and API. Their empirical relation is given as:

$$\mu_a = \mu_{max}\exp[-B\exp(-C\ln(\tau\omega))]$$

where $\tau = \eta/\mu_{max}$ is the relaxation time, with $\mu_{max} = 0.826$ GPa

$$B = 0.01859T + 0.1922\log_{10}(\eta) + 0.56221$$

$$C = 0.00835\log_{10}(\eta) + 0.07514$$

$$\log_{10}(\eta) = -1.23667API + 29.6079 + (0.465549API - 11.09838)\log_{10}T$$

where the temperature is in Fahrenheit and the viscosity is in cP. The relation for viscosity is based on measurements on extracted oil samples from over 100 core plugs measured at temperatures from 35 °C to 75 °C and API ranging from 5.8 to 10 API.

8.9   Particles and Bubbles in a Viscoelastic Background

A suspension of particles, droplets, or bubbles in a Newtonian fluid will generally exhibit a macro-scale viscosity that is different from that of the pure fluid. The relative viscosity H is defined as the ratio of the effective macro-scale viscosity η_{eff} to the viscosity of the background liquid phase η_b:

$$\eta_{eff} = \eta_b H . \qquad (8.9.1)$$

The presence of the suspended phase can cause either an increase or decrease in viscosity. Factors controlling the effective viscosity include: the volume fraction of the suspended phase, the ratio of inclusion to supporting fluid viscosities, the amount of inclusion deformation under shear flow, the shear strain rate, the inter-fluid surface tension, the magnitude of the overall strain, and whether the motion is steady or transient (Manga et al., 1998; Rust and Manga, 2002; Manga and Loewenberg, 2001; Murai and Oiwa, 2006).

Einstein (1906) showed that the effective viscosity of a Newtonian liquid containing a dilute suspension of rigid spherical particles can be written as

$$\eta_{eff} = \eta_b \cdot H; \quad H = 1 + C\varphi, \tag{8.9.2}$$

where φ is the volume fraction of particles, and $C = 2.5$ is a the shape parameter. A number of authors have pushed the analysis to higher particle concentrations (Guth and Simha, 1936; Thomas, 1965; Farris, 1968; Kitano et al., 1981; Brouwers, 2006). Brouwers (2010) successfully extended Einstein's theory (equation 8.9.2) to the limit of particle jamming. His expression for equal-sized rigid spherical particles is

$$H(\varphi) = \left(\frac{1 - \varphi}{1 - \varphi/\varphi_{jam}} \right)^{C\varphi_{jam}//(1-\varphi_{jam})}, \tag{8.9.3}$$

where and $C = 2.5$ for spheres, and $\varphi_{jam} \approx 0.64$ is the particle solid fraction at a jammed random close packing. Equation 8.9.3 predicts $H(\varphi) = 1 + C\varphi$ (for $\varphi \to 0$), and $H \to \infty$ (for $\varphi \to \varphi_{jam}$). Brouwers (2010) also extended equation 8.9.3 to the case of a bimodal mixture of spherical particles.

Taylor (1932, 1934) generalized Einstein's result for a dilute distribution of viscous spherical inclusions, which might represent bubbly magma or an oil-water emulsion:

$$H(\varphi) = 1 + \frac{5}{2}\varphi \left(\frac{\eta_{inc} + (2/5)\eta_b}{\eta_{inc} + \eta_b} \right), \tag{8.9.4}$$

where η_{incl} is the viscosity of the inclusion phase. Taylor's expression is valid when the fluid flow has become steady (Manga and Loewenberg, 2001) and the strain rate is low, so that the surface tension can maintain the inclusion's spherical shape. Taylor's expression becomes the same as Einstein's in the rigid limit, $\eta_{inc}/\eta_b \to \infty$. In the limit $\eta_{inc}/\eta_b \to 0$ when the inclusions are inviscid gas bubbles, Taylor's equation reduces to $H = 1 + \varphi$. The difference between the expressions for rigid and inviscid spheres results from the different boundary conditions—no slip for the rigid solid particle and free slip and large inclusion currents for the bubble. For both extremes, $H > 1$, due to the deformed flow and increased shear dissipation in the background fluid around the spherical inclusions.

The results change when the bubbles deform, resulting from the competing effects of viscous shear stress, which acts to deform the bubbles, and surface tension, which acts to maintain the spherical bubble shape (Palierne, 1990; Pal, 2003). The ratio of viscous shear stress to surface tension acting across a fluid-fluid boundary is described by the capillary number, $Ca = \eta_b V/\gamma$, where V is a characteristic velocity and γ is the surface tension. When describing bubbles and fluid inclusions the capillary number is often expressed as $Ca = r\dot{\varepsilon} \, \eta_b/\gamma$, where r is the bubble radius, and $\dot{\varepsilon}$ is the shear strain rate. When the viscous forces are very small ($Ca \to 0$) the bubble remains spherical and Taylor's expression applies. Frankel and Acrivos (1970) proposed an expression for relative viscosity with dilute concentrations of only slightly deformed bubbles. When the viscosity of the bubble phase is much smaller than the viscosity of the fluid phase, their expression is

$$H(\varphi) = 1 + \frac{1 - (12/5)Ca^2}{1 + ((6/5)Ca)^2}\varphi. \tag{8.9.5}$$

Equation 8.9.5 works well for $Ca < 1$ (Rust and Manga, 2002), i.e., low shear rate and high surface tension. Note that the coefficient on φ becomes negative when $Ca > 0.65$. As $Ca \to \infty$ the Frankel and Acrivos equation goes to

$$H = 1 - \frac{5}{3}\varphi, \tag{8.9.6}$$

which is the expression predicted for spherical inviscid inclusions with no surface tension (Dewey, 1947; Rust and Manga, 2002).

Heuristically, we might combine the equations of Taylor, Frankel, and Acrivos to get the following expression for dilute concentrations:

$$H \approx 1 + \varphi \left(\frac{1 + \dfrac{5}{2}\dfrac{\eta_{inc}}{\eta_b} - \dfrac{12}{5}Ca^2}{1 + \dfrac{\eta_{inc}}{\eta_b} + \dfrac{36}{25}Ca^2} \right) \tag{8.9.7}$$

This equation goes to the limits previously expressed: equation 8.9.4 when $Ca \to 0$; equation 8.9.6 when $Ca \to \infty$; equation 8.9.5 when $\eta_{inc}/\eta_b \to 0$; and equation 8.9.2 when $\eta_{inc}/\eta_b \to \infty$.

Pal (2002) suggested a series of models for the complex shear modulus μ_{eff}^* of a general linear viscoelastic background material containing viscoelastic spherical particles, with small surface tension:

$$\left(\frac{\mu_{eff}^*}{\mu_b^*} \right) \left(\frac{\mu_{eff}^* - \mu_{inc}^*}{\mu_b^* - \mu_{inc}^*} \right)^{-2.5} = H, \tag{8.9.8}$$

where H is a function of particle concentration, μ_b^* is the complex shear modulus of the continuous background phase, and μ_{inc}^* is the complex shear modulus of the included particles. Equation 8.9.8 is valid to high concentrations of inclusions and was derived recursively from expressions for lower concentrations, using essentially the same scheme as the Differential Effective Medium (DEM) approximation, which was previously used to describe effective elastic moduli of porous rocks (Norris, 1985; Zimmerman, 1991). For dilute concentrations, H might be approximated using equation 8.9.4. In order to approximate effects of jamming, Pal considered several ad hoc expressions for which H would become infinite when $\varphi = \varphi_{jam}$:

$$H \approx \left(1 - \frac{\varphi}{\varphi_{jam}}\right)^{-2.5\varphi_{jam}} \quad \text{or } H \approx \exp\left(\frac{2.5\varphi}{1 - \varphi/\varphi_{jam}}\right). \tag{8.9.9}$$

As an alternative to equation 8.9.9, one could use Brouwer's expression for H (equation 8.9.3), which yields the following expression for effective complex modulus at high capillary number:

$$\left(\frac{\mu_{eff}^*}{\mu_b^*}\right)\left(\frac{\mu_{eff}^* - \mu_{inc}^*}{\mu_b^* - \mu_{inc}^*}\right)^{-2.5} = \left(\frac{1 - \varphi}{1 - \varphi/\varphi_{jam}}\right)^{2.5\varphi_{jam}/(1-\varphi_{jam})}. \tag{8.9.10}$$

Equation 8.9.10 reduces to the following limiting forms:

If $|\mu_{inc}^*/\mu_b^*| \ll 1$: $\qquad \mu_{eff}^* = \mu_b^* H^{-2/3}$ (8.9.11)

If $|\mu_b^*/\mu_{inc}^*| \ll 1$; $\qquad \mu_{eff}^* = \mu_b^* H$ (8.9.12)

If $\varphi \to \varphi_{jam}$: $\qquad \mu_{eff}^* = \mu_{inc}^*$ (8.9.13)

For small φ : $\qquad H = 1 + (5/2)\varphi$ (8.9.14)

At jamming, our modified form of Pal's effective modulus (equation 8.9.10) becomes equal to the particle modulus; perhaps a more realistic limit at jamming would be the modulus of a porous random pack of spheres, which could be simulated by interpreting the inclusions as isolated fragments of the background material at jamming (Mukerji *et al.*, 1995). The complex shear modulus in equation 8.9.10 can be expressed as complex viscosity with

$$\eta_{eff}^* = \eta_R + i\eta_I = \mu_{eff}^*(\omega)/i\omega. \tag{8.9.15}$$

The differential form of Pal's (2002) model is

$$d\mu_{eff}^* = \mu_{eff}^* \frac{5(\mu_{inc}^* - \mu_{eff}^*)}{(2\mu_{inc}^* + 3\mu_{eff}^*)} \frac{d\varphi}{(1 - \varphi/\varphi_{jam})} \tag{8.9.16}$$

which can be integrated from $\varphi = 0$ to $\varphi = \varphi_{jam}$ with the initial condition $\mu_{eff}^* = \mu_b^*$ at $\varphi = 0$, to obtain equation 8.9.8. For Brouwers' modification, equation 8.9.10, the differential form is only slightly different:

$$d\mu_{eff}^* = \mu_{eff}^* \frac{5(\mu_{inc}^* - \mu_{eff}^*)}{(2\mu_{inc}^* + 3\mu_{eff}^*)} \frac{d\varphi}{(1 - \varphi/\varphi_{jam})(1 - \varphi)} \tag{8.9.17}$$

Pal (2003) extended his theory to the case of arbitrary capillary number. In the differential form, we have:

$$d\mu_{eff}^* = 5\mu_{eff}^* \frac{[4(2\mu_{eff}^* + 5\mu_{inc}^*) + (Ca/\mu_{eff}^*)(\mu_{inc}^* - \mu_{eff}^*)(16\mu_{eff}^* + 19\mu_{inc}^*)]}{[40(\mu_{eff}^* + \mu_{inc}^*) + (Ca/\mu_{eff}^*)(2\mu_{inc}^* + 3\mu_{eff}^*)(16\mu_{eff}^* + 19\mu_{inc}^*)]} \frac{d\varphi}{(1 - \varphi/\varphi_{jam})}$$

$$(8.9.18)$$

For the Brouwers modification, this is replaced by

$$d\mu_{eff}^* = 5\mu_{eff}^* \frac{[4(2\mu_{eff}^* + 5\mu_{inc}^*) + (Ca/\mu_{eff}^*)(\mu_{inc}^* - \mu_{eff}^*)(16\mu_{eff}^* + 19\mu_{inc}^*)]}{[40(\mu_{eff}^* + \mu_{inc}^*) + (Ca/\mu_{eff}^*)(2\mu_{inc}^* + 3\mu_{eff}^*)(16\mu_{eff}^* + 19\mu_{inc}^*)]} \frac{d\varphi}{(1 - \varphi/\varphi_{jam})(1 - \varphi)}$$

$$(8.9.19)$$

8.10 Viscosity of Silicate Melts and Magma

Synopsis

Melt Viscosity

Giordano *et al.* (2008) found an empirical non-Arrhenian model for Newtonian viscosity of a broad range of silicate melts. The model is based on more than 1770 measurements, and predicts viscosity as a function of composition and temperature (but not pressure). The supporting data span melts from basic to silicic, from subalkaline to peralkaline, and from metaluminous to peraluminous. The expression is a Vogel–Fulcher–Tammann (VFT) (Vogel, 1921; Fulcher, 1925; Tammann and Hesse, 1926) equation:

$$\log\eta = A + \frac{B}{T_K - C} \tag{8.10.1}$$

where viscosity η is in Pa-s, temperature T_K is in Kelvin, and A, B, and C are empirical constants. Parameter A is interpreted as $A = \log\eta_\infty$ where η_∞ is the limiting low viscosity at very high temperature. Several authors have found support for taking a fixed limiting viscosity for a wide range of melt compositions. Giordano *et al.* take $A \approx -4.55$.

Parameters B and C depend on composition. The M and N below are oxide mole% linked to each coefficient b and c in Table 8.10.1. Double subscript indicates product of mole% of the two oxide groups shown in those table entries:

$$B = \sum_{i=1}^{7}[b_i M_i] + \sum_{j=1}^{3}[b_{1j}(M1_{1j} \cdot M2_{1j})] \tag{8.10.2}$$

$$C = \sum_{i=1}^{6}[c_i N_i] + [c_{11}(N1_{11} \cdot N2_{11})] \tag{8.10.3}$$

Table 8.10.1 *Coefficient for calculation of viscosity parameters B and C (A=−4.55) from melt compositions expressed as mol% oxides. From the model by Giordano et al. (2007)*

	Oxides	Values		Oxides	Values
b_1	$SiO_2 + TiO_2$	159.6	c_1	SiO_2	2.75
b_2	Al_2O_3	-173.3	c_2	$TA^{(c)}$	15.7
b_3	$FeO(T) + MnO + P_2O_5$	72.1	c_3	$FM^{(d)}$	8.3
b_4	MgO	75.7	c_4	CaO	10.2
b_5	CaO	-39.0	c_5	$NK^{(e)}$	-12.3
b_6	$Na_2O + V^{(b)}$	-84.1	c_6	$\ln\left(1 + V^{(b)}\right)$	-99.5
b_7	$V^{(b)} + \ln(1 + H_2O)$	141.5	c_{11}	$\left(Al_2O_3 + FM^{(d)} + CaO - P_2O_5\right)*$	0.30
				$\left(NK^{(e)} + V^{(b)}\right)$	
b_{11}	$(SiO_2 + TiO_2) * \left(FM^{(d)}\right)$	-2.43			
b_{12}	$(SiO_2 + TA + P_2O_5)*$	-0.91			
	$\left(NK^{(e)} + H_2O\right)$				
b_{13}	$(Al_2O_3) * \left(NK^{(e)}\right)$	17.6			

[b] Sum of $H_2O + F_2O_{-1}$
[c] Sum of $TiO_2 + Al_2O_3$
[d] Sum of $FeO(T) + MnO + MgO$
[e] Sum of $Na_2O + K_2O$

Magma Viscosity

Magma is a multiphase mixture of melt, solid phenocrysts, and gas bubbles. Phenocrysts increase the effective viscosity relative to the viscosity of the melt phase. (A discussion of models predicting the effect of particles on viscosity can be found in Section 8.9.) At small strain rates and phenocryst contents, magma behavior is approximately Newtonian. At larger phenocryst contents magmas develop a yield strength and a strain-rate dependent viscosity (Pinkerton and Stevenson, 1992).

Bubbles can either increase or decrease the effective viscosity, depending on capillary number, $Ca = \eta_m \dot{\varepsilon} r / \gamma$, where η_m is the melt viscosity, $\dot{\varepsilon}$ is the shear strain rate, r is the bubble size, and γ is the surface tension. (See Section 8.9.) When $Ca \ll 1$, capillary forces maintain spherical bubble shape, leading to increased effective viscosity; when $Ca \gg 1$, viscous shear forces dominate, the bubbles can elongate, and the effective viscosity decreases.

Takeuchi (2011) compiled melt compositions (basic to rhyolitic), water contents, temperatures, and phenocryst contents on 83 erupted magmas; he then computed viscosities under pre-eruptive conditions using the model of Giordano *et al.* (2008), which was previously discussed. A subset of his results can be found in Appendix. Melt phase viscosities tend to decrease with increasing temperature (Figure 8.10.1) and decrease with increasing SiO_2 content (though the Giordano *et al.* model indicates a much more complicated dependence on composition). Magma viscosities show a weaker correlation with SiO_2 content than melt (Figure 8.10.2), because phenocrysts increase the resistance to flow and increase the effective viscosity.

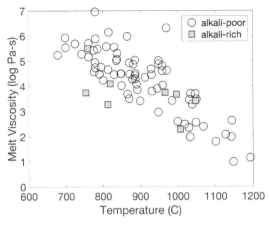

Figure 8.10.1 Melt viscosity versus temperature computed for 83 magmas spanning from basic to rhyolitic compositions. Redrawn from a figure in Takeuchi (2011).

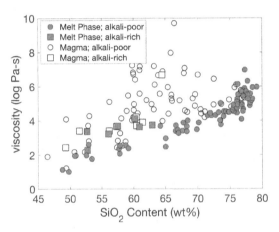

Figure 8.10.2 Magma and melt viscosity versus SiO_2 content (redrawn from a figure in Takeuchi, 2011). Gray symbols are for the melt phase. Open circles are for magma containing phenocrysts. Melt viscosity tends to increase with increasing SiO_2 content. Phenocrysts always increase magma viscosity.

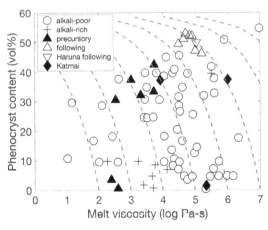

Figure 8.10.3 Melt phase viscosity and magma viscosity for different contents of phenocrysts (after Takeuchi, 2011). Dashed lines are contours of constant log viscosity as predicted by the Einstein model (see Section 8.9). "Precursory" erupted magmas often have lower viscosity than the following magmas.

Table 8.10.2 *Melt compositions (basic to rhyolitic), water contents, temperatures, and phenocryst contents on 83 erupted magmas, compiled by Takeuchi (2011); he then computed viscosities under pre-eruptive conditions using the model of Giordano et al. (2008).*

Volcano	Main Eruption Style	Bulk SiO$_2$ Content (wt%)	Melt SiO$_2$ Content (wt%)	Melt water Content (wt%)	Temperature (°C)	Phenocryst Content (vol%)	Melt Viscosity (log Pa-s)	Magma Viscosity (log Pa-s)	Reference
Laacher See	Plinian	57.4	57.4	5.5	755	1	3.7	3.7	19,49
Campi Flegrei	Caldera forming	60.5	60.8	2.0	1000	5	3.6	3.7	31
Vesuvius	Plinian	56.0	56.0	6.2	815	10	3.2	3.4	6,48
Stromboli	Eruptive paroxysm	49.4	49.8	2.6	1150	11	1.0	1.2	10
Etna	Plinian	49.3	52.7	3.0	1010	10	2.3	2.5	7,17
Santorini	Caldera forming	69.1	73.6	2.8	885	15	5.1	5.4	9
Fantale	Postcaldera Lava flow	70.0	73.9	4.8	800	14	4.3	4.6	15,65
Taupo	Caldera forming	76.2	76.2	4.3	839	5	5.1	5.2	12
Taupo	Caldera forming	75.7	76.9	4.5	760	10	5.7	5.9	67
Krakatau	Caldera forming	68.3	72.0	4.0	885	11	4.3	4.6	30
Agung	Lava flow Sub-Plinian, Vulcanian	57.6	59.1	3.0	1070	19	2.4	2.8	50
Pinatubo	Caldera forming	64.0	78.5	6.3	760	47	5.2	6.9	66,39,47
		59.2	64.8	5.0	950	38	3.0	4.1	40,43,55
Kikai	Lava dome	70.6	77.8	1.0	970	18	6.3	6.7	38,45
Kikai	Explosive	69.3	76.9	3.0	960	17	4.6	5.0	
Kikai	Scoria cone building	53.0	57.8	2.0	1125	29	2.1	2.8	38,45,25
Kikai	Caldera forming	71.0	73.4	3.7	960	10	4.0	4.2	38,45
Sakurajima	Vulcanian	60.0	67.6	3.0	1030	31	3.4	4.2	37,69
Aira	Caldera forming	75.0	76.3	6.0	780	13	5.0	5.2	2,59,32
Unzen	Lava dome	64.6	68.2	6.0	900	25	3.4	4.0	35,44,61,26
Aso	Caldera forming	51.0	66.8	5.9	870	25	3.5	4.1	24
		54.0	68.2	5.7	865	25	3.7	4.3	24
		68.0	72.0	4.9	830	7.5	4.5	4.6	24
Asama	Vulcanian	61.5	69.8	3.0	1050	35	3.5	4.5	51,33
Haruna	Plinian	61.0	78.1	5.3	835	50	5.0	7.0	54
Miyakejima	Phreatomagmatic	50.6	53.4	1.8	1100	17	1.8	2.2	46

Hokkaido Komagatake	Plinian	60.0	76.8	3.0	950	48	5.0	6.8	55,57
		58.0	67.0	3.0	1040	33	3.3	4.2	55,57
Hokkaido Komagatake	Plinian	61.0	76.7	3.0	960	53	4.8	7.2	55,56
		58.3	57.9	2.0*	1050**	1	2.6	2.6	55,56
Hokkaido Komagatake	Plinian	59.8	76.9	3.0	950	53	4.9	7.2	55,56
		58.9	64.1	1.5	1035	34	3.7	4.6	55,56
Hokkaido Korragatake	Plinian	59.9	75.8	3.0	970	52	4.6	6.8	55,56
		57.4	57.6	3.0	1030	4	2.4	2.4	55,56
Usu	Plinian	75.0	76.1	6.4	780	4	4.7	4.8	58
Tarumai	Plinian	60.3	75.9	4.0	930	50	4.5	6.4	55,56
		56.2	66.4	2.0	1045	43	3.7	5.1	55,56
Karymsky	Caldera forming	69.6	75.6	3.6	871	26	4.9	5.6	23
Karymsky	Caldera forming	69.5	74.3	4.3	883	21	4.4	4.9	23
Shiveluch	Lava dome Vulcanian	62.0	68.6	5.1	850	39	4.0	5.2	11
Aniakchak	Caldera forming	58.7	58.6	3.5	1000	10	2.6	2.8	26
		70.6	71.5	4.0	885	9	4.3	4.5	26
Katmai	Caldera forming	60.0	67.6	3.3	945	37.5	3.9	5.0	18
		67.2	79.1	2.5	865	37.5	6.0	7.1	18,27
		77.2	78.6	4.0	838	2	5.3	5.3	27,8
Rainier	Plinian	64.0	75.4	3.0	930	54	4.7	7.2	60
		58.5	64.7	5.0	1020	31	2.5	3.3	60
St. Helens	Plinian	62.8	73.0	4.6	930	40	3.8	5.0	42
St. Helens	Plinian	63.2	70.2	4.6	893	31	3.9	4.7	14
St. Helens	Plinian	67.2	74.8	4.8	847	27	4.5	5.2	14
St. Helens	Plinian	64.1	72.8	3.7	913	41	4.4	5.6	14
St. Helens	Plinian	63.3	74.3	4.3	870	46	4.5	6.1	14
St. Helens	Plinian	65.5	75.6	5.6	795	40	4.8	6.0	14
St. Helens	Plinian	65.9	74.8	6.3	791	32	4.5	5.3	14
St. Helens	Plinian	65.8	75.5	6.5	777	35	4.6	5.6	14

Table 8.10.2 *(cont.)*

Volcano	Main Eruption Style	Bulk SiO$_2$ Content (wt%)	Melt SiO2 Content (wt%)	Melt water Content (wt%)	Temperature (°C)	Phenocryst Content (vol%)	Melt Viscosity (log Pa-s)	Magma Viscosity (log Pa-s)	Reference
Crater Lake	Caldera forming	70.4	73.2	3.9	885	10	4.5	4.7	3,4
Inyo	Lava dome	72.6	78.1	4.1	880	5	4.9	5.0	20,62,16
Long Valley	Caldera forming	76.8	77.4	5.0	725	5	5.7	5.8	5,1,21
Valles Caldera	Plinian		77.8	6.0	697	7.5	5.6	5.7	52,53
Valles Caldera	Caldera forming		77.5	5.0	697	5	6.0	6.1	63,52,53
Valles Caldera	Plinian		79.3	5.0	777	5	5.5	5.5	52,53
Valles Caldera	Plinian		78.7	6.0	813	5	4.7	4.8	52,53
Valles Caldera	Plinian		79.0	3.0	813	5	6.2	6.2	52,53
Kilauea	Hawaiian	50.8	50.8	0.4	1140	0	2.0	2.0	13,36,63
Kilauea	Hawaiian	46.4	48.8	0.7	1190	30	1.2	1.9	29,34,41,63
Volcan Colima	Lava dome vulcanian	60.0	66.6	3.5	985	40	3.4	4.6	28
Masaya	Plinian	52.4	52.6	3.0	1035	10	2.0	2.2	67,17

1 Anderson et al. [2000]; 2 Aramaki [1984]; 3 Bacon and Druitt [1988]; 4 Bacon et al. [1992]; 5 Bailey et al. [1976]; 6 Cioni et al. [1995]; 7 Coltelli et al. [1998]; 8 Coombs and Gardner [2001]; 9 Cottrell et al. [1999]; 10 Di Carlo et al. [2006]; 11 Dirksen et al. [2006]; 12 Dunbar et al. [1989a, 1989b]; 13 Garcia and Wolfe [1988]; 14 Gardner et al. [1995]; 15 Gibson [1970]; 16 Ghiorso and Sack [1991]; 17 Goepfert and Gardner [2010]; 18 Hammer et al. [2002]; 19 Harms et al. [2004]; 20 Hervig et al. [1989]; 21 Hildreth and Wilson [2007]; 22 Holtz et al. [2004]; 23 Izbekov et al. [2004]; 24 Kaneko et al. [2007]; 25 Kawanabe and Saito [2002]; 26 Larsen [2006]; 27 Lowenstern [1993]; 28 Luhr [2002]; 29 Macdonald and Katsura [1961]; 30 Mandeville et al. [1996]; 31 Marianelli et al. [2006]; 32 Miyagi and Yurimoto [1995]; 33 Miyake et al. [2005]; 34 Murata and Richter [1966]; 35 Nakada and Motomura [1999]; 36 Neal et al. [1988]; 37 Okumura et al. [2004]; 38 Ono et al. [1982]; 39 Pallister et al. [1992]; 40 Pallister et al. [1996]; 41 Richter and Murata [1966]; 42 Rutherford et al. [1985]; 43 Rutherford and Devine [1996]; 44 Sato et al. [1999]; 45 Saito et al. [2001, 2002]; 46 Saito et al. [2005]; 47 Scaillet and Evans [1999]; 48 Scaillet et al. [2008]; 49 Schmincke et al. [1999]; 50 Self and King [1996]; 51 Shimano et al. [2005]; 52 Stix and Gorton [1990];53 Stix and Layne [1996]; 54 Suzuki and Nakada [2007]; 55 Takeuchi [2002]; 56 Takeuchi [2004]; 57 Takeuchi and Nakamura [2001]; 58 Tomiya et al. [2010]; 59 Tsukui and Aramaki [1990]; 60 Venezky and Rutherford [1997]; 61 Venezky and Rutherford [1999]; 62 Vogel et al. [1989]; 63 Wallace and Anderson [1998]; 64 Warshaw and Smith [1988]; 65 Webster et al. [1993]; 66 Westrich and Gerlach [1992]; 67 Williams [1983]; 68 Wilson et al. [2005]; 69 Yamanoi et al. [2008]

9 Electrical Properties

9.1 Bounds and Effective Medium Models

Synopsis

If we wish to predict the effective dielectric permittivity ε of a mixture of phases theoretically, we generally need to specify: (1) the volume fractions of the various phases, (2) the dielectric permittivity of the various phases, and (3) the geometric details of how the phases are arranged relative to each other. If we specify only the volume fractions and the constituent dielectric permittivities, then the best we can do is to predict the upper and lower bounds.

Bounds: The best bounds, defined as giving the narrowest possible range without specifying anything about the geometries of the constituents, are the **Hashin–Shtrikman (HS)** (Hashin and Shtrikman, 1962) bounds. For a two-phase composite, the Hashin–Shtrikman bounds for dielectric permittivity are given by

$$\varepsilon^{\pm} = \varepsilon_1 + \frac{f_2}{(\varepsilon_2 - \varepsilon_1)^{-1} + f_1/(3\varepsilon_1)} \tag{9.1.1}$$

where ε_1 and ε_2 are the **dielectric permittivities** of individual phases, and f_1 and f_2 are volume fractions of individual phases.

Upper and lower bounds are computed by interchanging which material is termed 1 and which is termed 2. The expressions give the upper bound when the material with higher permittivity is termed 1 and the lower bound when the lower permittivity material is termed 1.

A more general form of the bounds, which can be applied to more than two phases (Berryman, 1995), can be written as

$$\varepsilon^{\text{HS}+} = \Pi(\varepsilon_{\max}), \quad \varepsilon^{\text{HS}-} = \Pi(\varepsilon_{\min}) \tag{9.1.2}$$

$$\Pi(z) = \left\langle \frac{1}{\varepsilon(r) + 2z} \right\rangle^{-1} - 2z \tag{9.1.3}$$

where z is just the argument of the function $\Pi(\cdot)$, and r is the spatial position. The brackets $\langle \cdot \rangle$ indicate an average over the medium, which is the same as an average over the constituents weighted by their volume fractions.

Spherical inclusions: Estimates of the effective dielectric permittivity, ε^*, of a composite may be obtained by using various approximations, both self-consistent and non-self-consistent. The **Clausius–Mossotti (CM) formula** for a two-component material with spherical inclusions of material 2 in a host of material 1 is given by

$$\frac{\varepsilon_{CM}^* - \varepsilon_1}{\varepsilon_{CM}^* + 2\varepsilon_1} = f_2 \frac{\varepsilon_2 - \varepsilon_1}{\varepsilon_2 + 2\varepsilon_1} \tag{9.1.4}$$

or equivalently

$$\varepsilon_{CM}^* = \Pi(\varepsilon_1) \tag{9.1.5}$$

This non-self-consistent estimate, also known as the **Lorentz–Lorenz** or **Maxwell–Garnett (MG) equation**, actually coincides with the Hashin–Shtrikman bounds. The two bounds are obtained by interchanging the role of spherical inclusions and host material.

For multiphase composites, with N-1 types of spherical inclusions, each with volume fraction f_i, the Maxwell–Garnett equation generalizes to

$$\frac{\varepsilon_{MG}^* - \varepsilon_1}{\varepsilon_{MG}^* + 2\varepsilon_1} = \sum_{i=2}^{N} f_i \frac{\varepsilon_i - \varepsilon_1}{\varepsilon_i + 2\varepsilon_1} \tag{9.1.6}$$

The self-consistent (SC) or **coherent potential approximation (CPA)** (Bruggeman, 1935; Landauer, 1952; Berryman, 1995) for the effective dielectric permittivity ε_{SC}^* of a composite made up of spherical inclusions of N phases may be written as

$$\sum_{i=1}^{N} f_i \frac{\varepsilon_i - \varepsilon_{SC}^*}{\varepsilon_i + 2\varepsilon_{SC}^*} = 0 \tag{9.1.7}$$

or

$$\varepsilon_{SC}^* = \Pi(\varepsilon_{SC}^*) \tag{9.1.8}$$

The solution, which is a fixed point of the function $\Pi(\varepsilon)$, is obtained by iteration. In this approximation, all N components are treated symmetrically with no preferred host material.

In the **Differential Effective Medium (DEM)** approach (Bruggeman, 1935; Sen et al., 1981), infinitesimal increments of inclusions are added to the host material until the desired volume fractions are reached. For a two-component composite with material 1 as the host containing spherical inclusions of material 2, the effective dielectric permittivity ε_{DEM}^* is obtained by solving the differential equation

$$(1-y)\frac{d}{dy}[\varepsilon^*_{DEM}(y)] = \frac{\varepsilon_2 - \varepsilon^*_{DEM}(y)}{\varepsilon_2 + 2\varepsilon^*_{DEM}(y)}[3\varepsilon^*_{DEM}(y)] \tag{9.1.9}$$

where $y = f_2$ is the volume fraction of spherical inclusions. The analytic solution with the initial condition $\varepsilon^*_{DEM}(y=0) = \varepsilon_1$ is (Berryman, 1995)

$$\left[\frac{\varepsilon_2 - \varepsilon^*_{DEM}(y)}{\varepsilon_2 - \varepsilon_1}\right]\left[\frac{\varepsilon_1}{\varepsilon^*_{DEM}(y)}\right]^{1/3} = 1 - y \tag{9.1.10}$$

The DEM results are path–dependent and depend on which material is chosen as the host. The **Hanai–Bruggeman** model (Bruggeman, 1935; Hanai, 1968) starts with the rock as the host into which infinitesimal amounts of spherical inclusions of water are added. This results in a rock with zero direct current (DC) conductivity because at each stage the fluid inclusions are isolated and there is no conducting path (usually the rock mineral itself does not contribute to the DC electrical conductivity). Sen *et al.* (1981) in their self-similar model of coated spheres start with water as the initial host and incrementally add spherical inclusions of mineral material. This leads to a composite rock with a finite DC conductivity because a conducting path always exists through the fluid. Both the Hanai–Bruggeman and the Sen *et al.* formulas are obtained from the DEM result with the appropriate choice of host and inclusion.

Equivalence statement: Bounds and estimates for electrical conductivity, σ, can be obtained from the preceding equations by replacing ε everywhere with σ. This is because the governing relations for dielectric permittivity and electrical conductivity (and other properties such as magnetic permeability and thermal conductivity) are mathematically equivalent (Berryman, 1995). The relationship between the dielectric permittivity, the electrical field, $\mathbf{E}$, and the displacement field, $\mathbf{D}$, is $\mathbf{D} = \varepsilon\mathbf{E}$. In the absence of charges $\nabla \cdot \mathbf{D} = 0$, and $\nabla \times \mathbf{E} = 0$, because the electric field is the gradient of a potential. Similarly, for:

- electrical conductivity, σ, $\mathbf{J} = \sigma\mathbf{E}$ from Ohm's law, where $\mathbf{J}$ is the current density, $\nabla \cdot \mathbf{J} = 0$ in the absence of current source and sinks, and $\nabla \times \mathbf{E} = 0$;
- magnetic permeability, μ, $\mathbf{B} = \mu\mathbf{H}$, where $\mathbf{B}$ is the magnetic induction, $\mathbf{H}$ is the magnetic field, $\nabla \cdot \mathbf{B} = 0$, and in the absence of currents $\nabla \times \mathbf{H} = 0$; and
- thermal conductivity, κ, $\mathbf{q} = -\kappa\nabla\theta$ from Fourier's law for heat flux, $\mathbf{q}$, and temperature, θ, $\nabla \cdot \mathbf{q} = 0$ when heat is conserved, and $\nabla \times \nabla\theta = 0$.

Ellipsoidal inclusions: Estimates for the effective dielectric permittivity of composites with nonspherical, ellipsoidal inclusions require the use of depolarizing factors L_a, L_b, L_c along the principal directions a, b, c of the ellipsoid. The generalization of the

Maxwell–Garnett relation for randomly oriented identical ellipsoidal inclusions in an isotropic composite is (Sihvola, 1999; Torquato, 2002)

$$\varepsilon_{MG}^* = \varepsilon_m + \frac{\dfrac{\phi\varepsilon_m}{3}\displaystyle\sum_{j=1}^{3}\frac{(\varepsilon_i - \varepsilon_m)}{\varepsilon_m + L_j(\varepsilon_i - \varepsilon_m)}}{1 - \dfrac{\phi}{3}\displaystyle\sum_{j=1}^{3}\frac{L_j(\varepsilon_i - \varepsilon_m)}{\varepsilon_m + L_j(\varepsilon_i - \varepsilon_m)}} \tag{9.1.11}$$

The superscripts m and i refer to the host matrix phase and the inclusion phase

The self-consistent (SC) estimate for randomly oriented ellipsoidal inclusions in an isotropic composite is (Torquato, 2002):

$$\sum_{i=1}^{N} f_i(\varepsilon_i - \varepsilon_{SC}^*)R^{*i} = 0 \tag{9.1.12}$$

$$R^{mi} = \frac{\varepsilon_m}{3}\sum_{j=a,b,c}\frac{1}{\varepsilon_m + L_j(\varepsilon_i - \varepsilon_m)} \tag{9.1.13}$$

where R^{mi} is the **electric field concentration factor**, and is a function of the **depolarizing factors** L_a, L_b, L_c along each of the principal directions. Note that the field concentration factor R^{mi} in equation 9.1.13 is slightly different than the notation used in Berryman (1995), scaled by a factor of $3\varepsilon_m$. In the self-consistent formula, the superscript * on R indicates that ε_m should be replaced by ε_{SC}^* in the expression for R. The differential effective medium (DEM) equation for identical randomly oriented ellipsoids is

$$(1 - y)\frac{d}{dy}[\varepsilon_{DEM}^*(y)] = \frac{\varepsilon_2 - \varepsilon_{DEM}^*(y)}{\varepsilon_2 + 2\varepsilon_{DEM}^*(y)}R^{m*} \tag{9.1.14}$$

In the DEM formula, the superscript * on R indicates that ε_m should be replaced by ε_{DEM}^* in the expression for R.

For spheroids ($a = b$) of arbitrary aspect ratio $\alpha = c/a$, the depolarization factors can be found from

$$L_c = \begin{cases} \dfrac{\gamma^2}{\gamma^2 - 1}\left[1 - (arctan\sqrt{\gamma^2 - 1})/\sqrt{\gamma^2 - 1}\right] \; ; \; \gamma = \dfrac{1}{\alpha} > 1, \; \textit{oblate spheroid} \\[4mm] \dfrac{\gamma^2}{\gamma^2 - 1}\left[1 - \dfrac{1}{2}ln\left(\dfrac{1 + \sqrt{1 - \gamma^2}}{1 - \sqrt{1 - \gamma^2}}\right)/\sqrt{1 - \gamma^2}\right] \; ; \; \gamma = \dfrac{1}{\alpha} < 1, \; \textit{prolate spheroid} \end{cases}$$
$$\tag{9.1.15}$$

$$L_a = L_b = (1 - L_c)/2 \tag{9.1.16}$$

Depolarizing factors and the coefficient R for some specific spheroidal shapes are given in the Table 9.1.1. Depolarizing factors for more general ellipsoidal shapes are tabulated by Osborn (1945) and Stoner (1945).

Table 9.1.1 *Field concentration factor R and depolarizing factors L_j for some specific shapes. The subscripts m and i refer to the background and inclusion materials (from Torquato, 2002).*

Inclusion shape	L_a, L_b, L_c	R^{mi}
Spheres	$\dfrac{1}{3}, \dfrac{1}{3}, \dfrac{1}{3}$	$\dfrac{3\varepsilon_m}{\varepsilon_i + 2\varepsilon_m}$
Needles	$\dfrac{1}{2}, \dfrac{1}{2}, 0$	$\dfrac{\varepsilon_m}{3}\left(\dfrac{1}{\varepsilon_m} + \dfrac{4}{\varepsilon_i + \varepsilon_m}\right)$
Disks	$0, 0, 1$	$\dfrac{\varepsilon_m}{3}\left(\dfrac{1}{\varepsilon_i} + \dfrac{2}{\varepsilon_m}\right)$

Ellipsoidal inclusion models have been used to model the effects of pore-scale fluid distributions on the effective dielectric properties of partially saturated rocks theoretically (Knight and Nur, 1987; Endres and Knight, 1992).

Layered media: Exact results for the long-wavelength effective dielectric permittivity of a layered medium (layer thicknesses much smaller than the wavelength) are given by (Sen *et al.*, 1981)

$$\varepsilon_{\parallel}^* = \langle \varepsilon_i \rangle \tag{9.1.17}$$

and

$$\frac{1}{\varepsilon_{\perp}^*} = \left\langle \frac{1}{\varepsilon_i} \right\rangle \tag{9.1.18}$$

for fields parallel to the layer interfaces and perpendicular to the interfaces, respectively, where ε_i, is the dielectric permittivity of each constituent layer. The direction of wave propagation is perpendicular to the field direction.

Uses

The equations presented in this section can be used for the following purposes:
- to estimate the range of the average mineral dielectric permittivity for a mixture of mineral grains; and
- to compute the upper and lower bounds for a mixture of mineral and pore fluid.

Assumptions and Limitations

The following assumption and limitation apply to the equations in this section:
- most inclusion models assume the rock is isotropic; and
- effective medium theories are valid when wavelengths are much longer than the scale of the heterogeneities.

9.2   Velocity Dispersion and Attenuation

Synopsis

The complex wavenumber associated with propagation of electromagnetic waves of angular frequency ω is given by $k = \omega\sqrt{\varepsilon\mu}$, where both ε, the dielectric permittivity, and μ, the magnetic permeability, are in general frequency-dependent, complex quantities denoted by

$$\varepsilon = \varepsilon' - i\left(\frac{\sigma}{\omega} + \varepsilon''\right) \tag{9.2.1}$$

$$\mu = \mu' - i\mu'' \tag{9.2.2}$$

where σ is the electrical conductivity. For most nonmagnetic earth materials, the **magnetic permeability** equals μ_0, the magnetic permeability of free space. The dielectric permittivity normalized by ε_0, the dielectric permittivity of free space, is often termed the **relative dielectric permittivity**, or **dielectric constant**, κ, which is a dimensionless measure of the dielectric behavior. The dielectric susceptibility $\chi = \kappa - 1$.

The real part, k_R, of the complex wavenumber describes the propagation of an electromagnetic wavefield, whereas the imaginary part, k_I, governs the decay in field amplitude with propagation distance. A plane wave of amplitude E_0 propagating along the z direction and polarized along the x direction may be described by

$$E_x = E_0 e^{i(\omega t - kz)} = E_0 e^{-k_I z} e^{i(\omega t - k_R z)} \tag{9.2.3}$$

The **skin depth**, the distance over which the field amplitude falls to $1/e$ of its initial value, is equal to $1/k_I$. The dissipation may also be characterized by the loss tangent, the ratio of the imaginary part of the dielectric permittivity to the real part:

$$\tan\delta = \frac{\sigma}{\omega\varepsilon'} + \frac{\varepsilon''}{\varepsilon'} \tag{9.2.4}$$

Relations between the various parameters may be easily derived (e.g., Guéguen and Palciauskas, 1994). With $\mu = \mu_0$:

$$k_R = \omega\sqrt{\mu_0\varepsilon'}\sqrt{\frac{1 + \cos\delta}{2\cos\delta}} \tag{9.2.5}$$

$$k_I = \omega\sqrt{\mu_0\varepsilon'}\sqrt{\frac{1 - \cos\delta}{2\cos\delta}} \tag{9.2.6}$$

$$\varepsilon' = \frac{k_R^2 - k_I^2}{\mu_0\omega^2}, \quad \frac{\sigma}{\omega} + \varepsilon'' = \frac{2k_R k_I}{\mu_0\omega^2} \tag{9.2.7}$$

$$\tan \delta = \frac{2k_R k_I}{k_R^2 - k_I^2} \tag{9.2.8}$$

$$V = \frac{\omega}{k_R} = \frac{1}{\sqrt{\mu_0 \varepsilon'}} \sqrt{\frac{2 \cos \delta}{1 + \cos \delta}} \tag{9.2.9}$$

where V is the phase velocity. When there is no attenuation $\delta = 0$ and

$$V = \frac{1}{\sqrt{\mu_0 \varepsilon'}} = \frac{c}{\sqrt{\kappa'}} \tag{9.2.10}$$

where $c = 1/\sqrt{\mu_0 \varepsilon_0}$ is the speed of light in a vacuum and κ' is the real part of the dielectric constant.

In the high-frequency *propagation* regime ($\omega \gg \sigma/\varepsilon'$), displacement currents dominate, whereas conduction currents are negligible. Electromagnetic waves propagate with little attenuation and dispersion. In this high-frequency limit the wavenumber is

$$k_{\text{hif}} = \omega \sqrt{\mu_0 \varepsilon'} \sqrt{1 - i \frac{\varepsilon''}{\varepsilon'}} \tag{9.2.11}$$

In the low-frequency *diffusion* regime ($\omega \ll \sigma/\varepsilon'$), conduction currents dominate, and an electromagnetic pulse tends to spread out with a $\sqrt{t}$ time dependence characteristic of diffusive processes. The wavenumber in this low-frequency limit is

$$k_{\text{low}\,f} = (1 - i) \sqrt{\frac{\omega \sigma \mu_0}{2}} \tag{9.2.12}$$

For typical crustal rocks the diffusion region falls below about 100 kHz.

The **Debye** (1945) and the **Cole–Cole** (1941) models are two common phenomenological models that describe the frequency dependence of the complex dielectric constant. In the Debye model, which is identical to the standard linear solid model for viscoelasticity (see Section 3.8), the dielectric constant as a function of frequency is given by

$$\kappa(\omega) = \kappa' - i\kappa'' = \kappa_\infty + \frac{\kappa_0 - \kappa_\infty}{1 + i\omega\tau} \tag{9.2.13}$$

$$\kappa' = \kappa_\infty + \frac{\kappa_0 - \kappa_\infty}{1 + (\omega\tau)^2} \tag{9.2.14}$$

$$\kappa'' = (\kappa_0 - \kappa_\infty) \frac{\omega\tau}{1 + (\omega\tau)^2} \tag{9.2.15}$$

where τ is the characteristic relaxation time, κ_0 is the low-frequency limit, and κ_∞ is the high-frequency limit.

The Cole–Cole model is given by

$$\kappa(\omega) = \kappa_\infty + \frac{\kappa_0 - \kappa_\infty}{1 + (i\omega\tau)^{1-\alpha}}, \quad 0 \le \alpha \le 1 \tag{9.2.16}$$

When the parameter α equals 0, the Cole–Cole model reduces to the Debye model. Sherman (1988) described the frequency-dependent dielectric constant of brine-saturated rocks as a sum of two Debye models – one for interfacial polarization (below 1 GHz) and another for dipole polarization in brine (above 1 GHz).

The amplitude reflection and transmission coefficients for uniform, linearly polarized, homogeneous plane waves are given by **Fresnel's equations** (e.g., Jackson, 1975). For a plane wave incident from medium 1 onto an interface between two isotropic, homogeneous half-spaces, medium 1 and 2, the equations are as follows.

• Electric field transverse to the plane of incidence (TE mode):

$$\frac{E_r^\perp}{E_i^\perp} = \frac{\sqrt{\mu_1\varepsilon_1}\,\cos\theta_i - (\mu_1/\mu_2)\sqrt{\mu_2\varepsilon_2 - \mu_1\varepsilon_1\,\sin^2\theta_i}}{\sqrt{\mu_1\varepsilon_1}\,\cos\theta_i + (\mu_1/\mu_2)\sqrt{\mu_2\varepsilon_2 - \mu_1\varepsilon_1\,\sin^2\theta_i}} \tag{9.2.17}$$

$$\frac{E_t^\perp}{E_i^\perp} = \frac{2\sqrt{\mu_1\varepsilon_1}\,\cos\theta_i}{\sqrt{\mu_1\varepsilon_1}\,\cos\theta_i + (\mu_1/\mu_2)\sqrt{\mu_2\varepsilon_2 - \mu_1\varepsilon_1\,\sin^2\theta_i}} \tag{9.2.18}$$

• Electric field in the plane of incidence (TM mode):

$$\frac{E_r^\|}{E_i^\|} = \frac{\sqrt{\mu_1\varepsilon_1}\sqrt{\mu_2\varepsilon_2 - \mu_1\varepsilon_1\,\sin^2\theta_i} - \mu_1\varepsilon_2\,\cos\theta_i}{\sqrt{\mu_1\varepsilon_1}\sqrt{\mu_2\varepsilon_2 - \mu_1\varepsilon_1\,\sin^2\theta_i} + \mu_1\varepsilon_2\,\cos\theta_i} \tag{9.2.19}$$

$$\frac{E_t^\|}{E_i^\|} = \frac{2\sqrt{\mu_1\varepsilon_1\mu_2\varepsilon_2}\,\cos\theta_i}{\sqrt{\mu_1\varepsilon_1}\sqrt{\mu_2\varepsilon_2 - \mu_1\varepsilon_1\,\sin^2\theta_i} + \mu_1\varepsilon_2\,\cos\theta_i} \tag{9.2.20}$$

where θ_i is the angle of incidence; $E_i^\perp$, $E_r^\perp$, $E_t^\perp$ are the incident, reflected, and transmitted transverse electric field amplitudes, respectively; $E_i^\|$, $E_r^\|$, $E_t^\|$ are the incident, reflected, and transmitted parallel electric field amplitudes, respectively; μ_1, μ_2 are the magnetic permeability of medium 1 and 2; and ε_1, ε_2 are the dielectric permittivity of medium 1 and 2.

The positive direction of polarization is taken to be the same for incident, reflected, and transmitted electric fields. In terms of the electromagnetic plane-wave impedance $Z = \sqrt{\mu/\varepsilon} = \omega\mu/k$, the amplitude reflection coefficients may be expressed as

$$R_{TE}^\perp = \frac{E_r^\perp}{E_i^\perp} = \frac{Z_2\,\cos\theta_i - Z_1\sqrt{1 - (Z_1\varepsilon_1/Z_2\varepsilon_2)^2\,\sin^2\theta_i}}{Z_2\,\cos\theta_i + Z_1\sqrt{1 - (Z_1\varepsilon_1/Z_2\varepsilon_2)^2\,\sin^2\theta_i}} \tag{9.2.21}$$

$$R_{\text{TM}}^{\parallel} = \frac{E_{\text{r}}^{\parallel}}{E_{\text{i}}^{\parallel}} = \frac{Z_2\sqrt{1 - (Z_1\varepsilon_1/Z_2\varepsilon_2)^2 \sin^2\theta_{\text{i}}} - Z_1 \cos\theta_{\text{i}}}{Z_2\sqrt{1 - (Z_1\varepsilon_1/Z_2\varepsilon_2)^2 \sin^2\theta_{\text{i}}} + Z_1 \cos\theta_{\text{i}}} \qquad (9.2.22)$$

For normal incidence, the incident plane is no longer uniquely defined, and the difference between transverse and parallel modes disappears. The reflection coefficient is then

$$R_{\text{TE}}^{\perp} = \frac{Z_2 - Z_1}{Z_2 + Z_1} = R_{\text{TM}}^{\parallel} \qquad (9.2.23)$$

Electromagnetic wave propagation in layered media can be calculated using propagator matrices (Ward and Hohmann, 1987). The electric and magnetic fields at the top and bottom of the stack of layers are related by a product of propagator matrices, one for each layer. The calculations are done in the frequency domain and include the effects of all multiples. For waves traveling perpendicularly to the layers with layer impedances and thicknesses Z_j and d_j, respectively,

$$\begin{bmatrix} E_y \\ H_x \end{bmatrix}_{j-1} = \mathbf{A}_j \begin{bmatrix} E_y \\ H_x \end{bmatrix}_j \qquad (9.2.24)$$

Each layer matrix $\mathbf{A}_j$, has the form

$$\mathbf{A}_j = \begin{bmatrix} \cosh(ik_j d_j) & -Z_j \sinh(ik_j d_j) \\ -\dfrac{1}{Z_j}\sinh(ik_j d_j) & \cosh(ik_j d_j) \end{bmatrix} \qquad (9.2.25)$$

Uses

The results described in this section can be used for computing electromagnetic wave propagation, velocity dispersion, and attenuation.

Assumptions and Limitations

The results described in this section are based on the following assumptions:
- isotropic homogeneous media, except for layered media; and
- plane-wave propagation.

9.3 Empirical Relations for Composites

Synopsis

The **Lichtnecker–Rother** empirical formula for the effective dielectric constant κ^* of a mixture of N constituents is given by a simple volumetric power-law average of the dielectric constants of the constituents (Sherman, 1986; Guéguen and Palciauskas, 1994):

$$\kappa^* = \left[\sum_{i=1}^{N} f_i(\kappa_i)^{\gamma} \right]^{1/\gamma}, \quad -1 \le \gamma \le 1 \tag{9.3.1}$$

where κ_i is the dielectric constant of individual phases and f_i are the volume fractions of individual phases.

For $\gamma = \dfrac{1}{2}$ this is equivalent to the **complex refractive index method (CRIM)** formula:

$$\sqrt{\kappa^*} = \sum_{i=1}^{N} f_i \sqrt{\kappa_i} \tag{9.3.2}$$

The CRIM equation (Meador and Cox, 1975; Endres and Knight, 1992) is analogous to the time-average equation of Wyllie (see Section 7.2) because the velocity of electromagnetic wave propagation is inversely proportional to $\sqrt{\kappa}$. The CRIM empirical relation has been found to give reasonable results at high frequencies (above ~ 0.5 GHz).

The **Odelevskii** formula for two phases is (Shen, 1985)

$$\varepsilon^* = B + \left(B^2 + \frac{1}{2} \varepsilon_1 \varepsilon_2 \right)^{1/2} \tag{9.3.3}$$

$$B = \frac{1}{4}[(3f_1 - 1)\varepsilon_1 + (3f_2 - 1)\varepsilon_2] \tag{9.3.4}$$

Typically, only the real part of the dielectric permittivity is used in this empirical formula.

Topp's relation (Topp *et al.*, 1980), based on measurements on a variety of soil samples at frequencies of 20 MHz to 1 GHz, is used widely in interpretation of time domain reflectometry (TDR) measurements for volumetric soil water content. The empirical relations are

$$\theta_v = -5.3 \times 10^{-2} + 2.92 \times 10^{-2}\,\kappa_a - 5.5 \times 10^{-4}\,\kappa_a^2 + 4.3 \times 10^{-6}\,\kappa_a^3 \tag{9.3.5}$$

$$\kappa_a = 3.03 + 9.3\theta_v + 146.0\theta_v^2 - 76.7\theta_v^3 \tag{9.3.6}$$

where κ_a is the apparent dielectric constant as measured by pulse transmission (such as in TDR or coaxial transmission lines) with dielectric losses excluded, and θ_v is the volumetric soil water content, the ratio of volume of water to the total volume of the sample. The estimation error for the data of Topp *et al.* (1980) was about 1.3%. The relations do not violate the Hashin–Shtrikman bounds for most materials. They do not give good estimates for soils with high clay content or organic matter and should be

Table 9.3.1 *Regression coefficients and coefficient of determination* R^2 *for empirical relations between volumetric water content and dielectric constant (real part) at different portable dielectric probe frequencies (Brisco et al., 1992)*

$\theta_v = a + b\kappa' + c\kappa'^2 + d\kappa'^3$

Frequency (GHz)	a	b	c	d	R^2
9.3 (X-band)	-3.58×10^{-2}	4.23×10^{-2}	-0.153×10^{-4}	17.7×10^{-6}	0.86
5.3 (C-band)	-1.01×10^{-2}	2.62×10^{-2}	-4.71×10^{-4}	4.12×10^{-6}	0.91
1.25 (L-band)	-2.78×10^{-2}	2.80×10^{-2}	-5.86×10^{-4}	5.03×10^{-6}	0.95
0.45 (P-band)	-1.88×10^{-2}	2.46×10^{-2}	-4.34×10^{-4}	3.61×10^{-6}	0.95

recalibrated for such material. Brisco *et al.* (1992) published empirical relations between volumetric soil water content and the real part of the dielectric constant measured by portable dielectric probes (PDP) utilizing frequencies from 0.45 to 9.3 GHz. The PDP measure both the real and imaginary components of the dielectric constant. In general TDR can sample soil layers 0–5 cm or deeper, whereas the PDP sample layers of about 1 cm thickness. Empirical relations in different frequency bands from Brisco *et al.* are summarized in Table 9.3.1.

Olhoeft (1979) obtained the following empirical relation between the measured effective dielectric constant and density for dry rocks:

$$\kappa' = \left((\kappa'_0)^{1/\rho_0} \right)^{\rho} = 1.91^{\rho} \tag{9.3.7}$$

where κ'_0, ρ_0 are the mineral dielectric constant and the mineral density (g/cm^3), and κ', ρ are the dry-rock dielectric constant and dry bulk density (g/cm^3).

The coefficient 1.91 was obtained from a best fit to data on a variety of terrestrial and lunar rock samples. The relation becomes poor for rocks with water-containing clays and conducting minerals such as sulfides and magnetite.

Knight and Nur (1987) measured the complex dielectric constant of eight different sandstones at different saturations and frequencies. They obtained a power-law dependence of κ' (the real part of the complex dielectric constant) on frequency ω expressed by

$$\kappa' = A\omega^{-\alpha} \tag{9.3.8}$$

where A and α are empirical parameters determined by fitting to the data and depend on saturation and rock type. For the different samples measured by Knight and Nur, α ranged from 0.08 to 0.266 at a saturation of 0.36 by deionized water. At the same saturation log A ranged from about 1.1 to 1.8.

Mazáč *et al.* (1990) correlated aquifer hydraulic conductivity K determined from pumping tests with electrical resistivities ρ interpreted from vertical electrical sounding. Their relation is

$$K(10^{-5}\text{m/s}) = \frac{\rho^{1.195}(\text{ohm m})}{97.5} \tag{9.3.9}$$

The correlation coefficient was 0.871.

Koesoemadinata and McMechan (2003) have given empirical relations between the dielectric constant (κ), and porosity (ϕ), clay content (C), density (ρ), and permeability (k). The regressions are based on data from about 30 samples of Ferron sandstone, a fine- to medium-grained sandstone with porosity ranging from 10% to 24% and clay content ranging from 10% to 22%. The κ data were measured on dry (<1% water saturation by volume) samples at 50 MHz.

$$\kappa = -2.90299\,\ln\phi + 12.8161 \qquad R^2 = 0.66 \tag{9.3.10}$$

$$\kappa = -0.35978\,\ln k + 5.35307 \qquad R^2 = 0.78 \tag{9.3.11}$$

$$\kappa = -0.2572\,\ln k - 0.07298\,\phi + 6.41999 \quad R^2 = 0.82 \tag{9.3.12}$$

$$\kappa = 7.1806\rho - 11.0397 \qquad R^2 = 0.65 \tag{9.3.13}$$

$$\kappa = 6.6611\rho + 0.0818C - 11.1742 \qquad R^2 = 0.79 \tag{9.3.14}$$

In the previous empirical regressions, permeability is in mD, density is in g/cm^3, and porosity and clay content are given as a percentage, while κ is dimensionless.

Uses

The equations presented in this section can be used to relate rock and soil properties such as porosity, saturation, soil moisture content, and hydraulic conductivity to electrical measurements.

Assumptions and Limitations

The relations are empirical and therefore strictly valid only for the data set from which they were derived. The relations may need to be recalibrated to specific locations or rock and soil types.

9.4 Electrical Conductivity in Porous Rocks

Synopsis

Most crustal rocks are made up of minerals that are semiconductors or insulators (silicates and oxides). Conducting currents in fluid-saturated rocks caused by an applied DC voltage arise primarily from the flow of ions within the pore fluids. The ratio of the conductivity of the pore fluid to the bulk conductivity of the fully saturated rock is known as the **formation factor,** F (Archie, 1942):

$$F = \frac{\sigma_w}{\sigma} = \frac{R_0}{R_w} \tag{9.4.1}$$

where σ_w, R_w are the conductivity and the resistivity of the pore fluid, and σ, R_0 are the conductivity and the resistivity of rock fully saturated with formation water.

The Hashin–Shtrikman lower bound on F for a rock with porosity ϕ and non-conducting mineral is (Berryman, 1995)

$$F^{HS-} = 1 + \frac{3}{2}\frac{1-\phi}{\phi} \tag{9.4.2}$$

which corresponds to the Hashin–Shtrikman upper bound on conductivity

$$\sigma^{HS+} = \frac{2\phi}{3-\phi}\sigma_w \tag{9.4.3}$$

where σ_w is the conductivity of the fluid and the mineral is non-conducting.

The DEM theory of Sen et al. (1981) predicts (as $\omega \rightarrow 0$ where ω is the frequency)

$$\sigma = \sigma_w \phi^{3/2} \tag{9.4.4}$$

or

$$F_{DEM} = \phi^{-3/2} \tag{9.4.5}$$

In this version of the DEM model, spheres of nonconducting mineral grains are embedded in the conducting fluid host so that a conducting path always exists through the fluid for all porosities (see Section 9.1). The exponent depends on the shape of the inclusions and is greater than 1.5 for plate- or needle-like inclusions.

Archie's law (1942), which forms the basis for resistivity log interpretation, is an empirical relation relating the formation factor to the porosity in *brine*-saturated *clean* (no shale) reservoir rocks:

$$F = \phi^{-m} \tag{9.4.6}$$

The exponent m (sometimes termed the cementation exponent) varies between approximately 1.3 and 2.5 for most sedimentary rocks and is close to 2 for sandstones. For natural and artificial unconsolidated sands and glass beads, m is close to 1.3 for spherical grains and increases to 1.9 for thin disk-like grains (Wyllie and Gregory, 1953; Jackson et al., 1978). Carbonates show a much wider range of variation and have m values as high as 5 (Focke and Munn, 1987).

The lower bound on m, corresponding to the upper Hashin–Shtrikman bound on conductivity, can be found from $\phi^m = \sigma^{HS+}/\sigma_w$ and is a function of porosity:

$$m^{HS-} = \log(2\phi/(3-\phi))/\log(\phi) \tag{9.4.7}$$

From equation 9.4.7, we find $\lim\limits_{\phi \to 0} m(\phi) = 1$ and $\lim\limits_{\phi \to 1} m(\phi) = 3/2$.

Archie's law is sometimes written as

$$F = (\phi - \phi_0)^{-m} \tag{9.4.8}$$

or

$$F = a\phi^{-m} \tag{9.4.9}$$

where ϕ_0 is a percolation porosity below which there are no conducting pathways and the rock conductivity is zero, and a is an empirical constant close to 1. A value different from 1 (usually greater than 1) results from trying to fit an Archie-like model to rocks that do not follow Archie behavior. Clean, well-sorted sands with electrical conduction occurring only by diffusion of ions in the pore fluid are best described by Archie's law. Shaly sands, rocks with moldic secondary porosity, and rocks with isolated microporous grains are examples of non-Archie rocks (Herrick, 1988). Archie's law typically applies when the pore fluid is the only conducting phase and the minerals are all non-conducting. Glover *et al.* (2000) and Glover (2010) gave generalized forms of Archie's equations when more than one phase is conducting. For two conducting phases with conductivities σ_1 and σ_2 and volume fractions $(1 - \phi_2)$ and ϕ_2 respectively, the effective conductivity is given by (Glover *et al.*, 2000):

$$\sigma_{eff} = \sigma_1(1 - \phi_2)^p + \sigma_2\phi_2^m \tag{9.4.10}$$

$$p = \frac{\log(1 - \phi_2^m)}{\log(1 - \phi_2)} \tag{9.4.11}$$

For n conducting phases the effective conductivity is (Glover, 2010):

$$\sigma_{eff} = \sum_i^n \sigma_i\phi_i^{m_i} \tag{9.4.12}$$

$$m_j = \log(1 - \sum_{i \neq j} \phi_i^{m_i})/\log(1 - \sum_{i \neq j} \phi_i) \tag{9.4.13}$$

where σ_i and ϕ_i are the conductivities and volume fractions of the constituent phases. The equation for the exponent m_i has to be solved by numerical iteration. Glover (2010) gives a first order approximation as:

$$m_j = \sum_{i \neq j} \phi_i^{m_i}/\sum_{i \neq j} \phi_i \tag{9.4.14}$$

Archie's second law for saturation relates the DC resistivity, R_t, of a partially saturated rock to the brine saturation, S_w, and the porosity by

$$S_w^{-n} = \frac{R_t}{R_0} = \phi^m \frac{R_t}{R_w} \tag{9.4.15}$$

where R_0 is the DC resistivity of the same rock at $S_w = 1$, and the saturation exponent, n, derived empirically, is around 2. The value of n depends on the type of the pore fluid and is different for gas-brine saturation versus oil-brine saturation. Experimentally, saturation exponents for oil-wet porous media have been found to be substantially higher ($n \approx 2.5 - 9.5$) than for water-wet media (Sharma *et al.*, 1991). In terms of conductivity Archie's second law may be expressed as

$$\sigma_t = (S_w^n \phi^m)\sigma_w \tag{9.4.16}$$

where $\sigma_t = 1/R_t$ is the conductivity of the partially saturated rock. Archie's empirical relations have been found to be applicable to a remarkably wide range of rocks (Ransom, 1984).

Shaly sands: Electrical conductivity in shaly sands is complicated by the presence of clays. Excess ions in a diffuse double layer around clay particles provide current conduction pathways along the clay surface in addition to the current flow by ions diffusing through the bulk pore fluid. Minerals like silica also develop an electrical double layer when in contact with brine. The electrical double layer consists of the Stern layer formed by ions sorbed directly on the mineral surface and the diffuse layer, further away from the mineral surface consisting of an excess of counter-ions. The electrical double layer is responsible for excess surface electrical conductivity with respect to the pore water conductivity. (Revil and Glover, 1997, 1998; Revil and Florsch, 2010; Revil, 2012; Revil and Jardani, 2013). The conductivity of this surface layer depends on the brine conductivity, and hence the overall bulk conductivity of the saturated rock is a nonlinear function of the brine conductivity. A wide variety of formulations have been used to model conductivity in shaly sands, and Worthington (1985) describes over 30 shaly sand models used in well log interpretation. Almost all of the models try to modify Archie's relation and account for the excess conductivity by introducing a shale conductivity term X:

$$\sigma = \frac{1}{F}\sigma_w, \text{ clean sands, Archie} \tag{9.4.17}$$

$$\sigma = \frac{1}{F}\sigma_w + X, \text{ shaly sands} \tag{9.4.18}$$

The various models differ in their choice of X. Some of the earlier models described X in terms of the volume of shale V_{sh}, as determined from logs:

Simandoux (1963):

$$\sigma = \frac{1}{F}\sigma_w + V_{sh}\sigma_{sh} \tag{9.4.19}$$

Poupon and Leveaux (1971), *"Indonesia formula"*:

$$\sqrt{\sigma} = \sqrt{\frac{1}{F}\sigma_w} + V_{sh}^\alpha \sqrt{\sigma_{sh}}, \quad \alpha = 1 - \frac{V_{sh}}{2} \tag{9.4.20}$$

where σ_{sh} is the conductivity of fully brine-saturated shale. Although these equations are applicable to log interpretation and may be used without calibration with core data, they do not have much physical basis and do not allow a complete representation of conductivity behavior for all ranges of σ_w. More recent models attempt to capture the physics of the diffuse ion double layer surrounding clay particles. Of these, the **Waxman–Smits** model (Waxman and Smits, 1968) and its various modifications such as the dual-water model (Clavier *et al.*, 1984) and the Waxman–Smits–Juhász model (Juhász, 1981) are the most widely accepted. The Waxman–Smits formula is

$$\sigma = \frac{1}{F}(\sigma_w + BQ_v) \tag{9.4.21}$$

$$B = 4.6(1 - 0.6e^{-\sigma_w/1.3}) \tag{9.4.22}$$

$$Q_v = \frac{CEC(1 - \phi)\rho_0}{\phi} \tag{9.4.23}$$

where CEC is the cation exchange capacity and ρ_0 is the mineral grain density.

Note that here and in the following we use shaly sand equations $F = a\phi^{-m}$ and not σ_w/σ. The **cation exchange capacity** (CEC) is a measure of the excess charges on the mineral surface and is expressed in Coulombs per unit mass of solid mineral, while Q_v is the excess charge per unit pore volume (Coulombs per m^3). CEC is commonly reported in units of meq g^{-1} (meq = milli mole equivalent charge = $1 \times 10^{-3} eN$) where $e = 1.6 \times 10^{-19}$ C is the elementary electron charge and $N = 6.023 \times 10^{23}$ mol^{-1} is Avogadro's constant. In SI units 1 meq g^{-1} = 96320 C kg^{-1} (Revil and Jardani, 2013). Clays often have an excess negative electrical charge within the sheet-like particles. This is compensated by positive counterions clinging to the outside surface of the dry clay sheets. The resulting positive surface charge is a property of the dry clay mineral and is called the cation exchange capacity (Clavier *et al.*, 1984). In the presence of an electrolytic solution such as brine, the electrical forces holding the positive counterions at the clay surface are reduced. The counterions can move along the surface contributing to the electrical conductivity. The average mobility of the ions is described by B. The parameter B is a source of uncertainty, and several expressions for it have been developed since the original paper. Juhász (1981) gives the following expressions for B:

$$B = \frac{-5.41 + 0.133T - 1.253 \times 10^{-4}T^2}{1 + R_w^{1.23}(0.025T - 1.07)} \tag{9.4.24}$$

where T is the temperature in degrees Fahrenheit or

$$B = \frac{-1.28 + 0.225T - 4.059 \times 10^{-4}T^2}{1 + R_w^{1.23}(0.045T - 0.27)} \tag{9.4.25}$$

for temperature in degrees Celsius. Application of the Waxman–Smits equation requires calibration with core CEC measurements. CEC is linearly related to the specific surface area (surface area per unit mass) S_{sp} through the surface charge density Q_s of the mineral surfaces: $Q_s = CEC/S_{sp}$. Revil and Jardani (2013) present compilation of data on CEC and S_{sp} for various clays. The surface charge density for most clays at near neutral pH is approximately constant equal to 1 to 3 electron charges ($e = 1.6 \times 10^{-19}$ C) per nm^2. Kaolinite and Chlorite have CEC values typically ranging from 0.02 to 0.1 meq g^{-1} with specific surface area of 10–50 $m^2\,g^{-1}$; for Illite CEC is around 0.2 to 0.41 meq g^{-1} with specific surface area of around 100 $m^2\,g^{-1}$; Smectite has a higher CEC of 1 to 2 meq g^{-1} with specific surface area ranging from 500 to 1000 $m^2\,g^{-1}$.

The normalized Waxman–Smits or Waxman–Smits–Juhász model (Juhász, 1981) does not require CEC data because it uses V_{sh} derived from logs to estimate Q_v by normalizing it to the shale response. In this model,

$$BQ_v = Q_{vn}(\sigma_{wsh} - \sigma_w) \tag{9.4.26}$$

$$Q_{vn} = \frac{Q_v}{Q_{vsh}} = \frac{V_{sh}\phi_{sh}}{\phi} \tag{9.4.27}$$

where ϕ is the total porosity (density porosity), ϕ_{sh} is the total shale porosity, and σ_{wsh} is the shale water conductivity obtained from $\sigma_{wsh} = F\sigma_{sh}$, where σ_{sh} is the conductivity of 100% brine-saturated shale. The normalized Q_v ranges from 0 in clean sands to 1 in shales. Brine saturation S_w can be obtained from these models by solving (Bilodeaux, 1997)

$$S_w = \left[\frac{FR_w}{R_t(1 + R_w BQ_v/S_w)}\right]^{1/n} \tag{9.4.28}$$

For $n = 2$ the explicit solution (ignoring the negative root) is

$$S_w = \sqrt{\frac{FR_w}{R_t} + \left(\frac{BQ_v R_w}{2}\right)^2} - \frac{BQ_v R_w}{2} \tag{9.4.29}$$

The dual-water model divides the total water content into the bound clay water, for which the conductivity depends only on the clay counterions, and the far water, away from the clay, for which the conductivity corresponds to the ions in the bulk formation water (Clavier et al., 1984). The bound water reduces the water conductivity σ_w by a factor of $(1 - \alpha v_Q Q_v)$. The dual-water model formula is (Clavier et al., 1984; Sen and Goode, 1988)

$$\sigma = \phi^m[\sigma_w(1 - \alpha v_Q Q_v) + \beta Q_v] \tag{9.4.30}$$

where v_Q is the amount of clay water associated with 1 milliequivalent of clay counterions, β is the counterion mobility in the clay double layer, and α is the ratio of the diffuse double-layer thickness to the bound water layer thickness. At high salinities (salt concentration exceeding 0.35 mol/ml) $\alpha = 1$. At low salinities it is a function of σ_w, and is given by

$$\alpha = \sqrt{\frac{\gamma_1 \langle n_1 \rangle}{\gamma \langle n \rangle}} \tag{9.4.31}$$

where $\langle n \rangle$ is the salt concentration in bulk water at 25 °C in mol/ml, γ is the NaCl activity coefficient at that concentration, $\langle n_1 \rangle = 0.35$ mol/ml, and $\gamma_1 = 0.71$, i.e., the corresponding NaCl activity coefficient.

Although v_Q and β have a temperature and salinity dependence, Clavier *et al.* (1984) recommend the following values for v_Q and β:

$$v_Q = 0.28 \text{ ml / meq} \tag{9.4.32}$$

$$\beta = 2.05 \text{ (S/m)}/(\text{meq}/\text{cm}^3) \tag{9.4.33}$$

These values are based on analysis of CEC data for clays and conductivity data on core samples. At low salinities v_Q varies with $\sqrt{T}$ and increases by about 26% from 25 °C to 200 °C.

Generalizing from theoretical solutions for electrolytic conduction past charged spheres in the presence of double layers, Sen and Goode (1988) proposed the following shaly-sand equation:

$$\sigma = \frac{1}{F}\left(\sigma_w + \frac{AQ_v}{1 + CQ_v/\sigma_w}\right) + EQ_v \tag{9.4.34}$$

The constants A and C depend on pore geometry and ion mobility, and the term EQ_v accounts for conductivity by surface counterions even when water conductivity is zero. Sen and Goode were able to express the relation in terms of Archie's exponent m by fitting to core data (about 140 cores)

$$\sigma = \phi^m\left(\sigma_w + \frac{m1.93Q_v}{1 + 0.7/\sigma_w}\right) + 1.3\phi^m Q_v \tag{9.4.35}$$

where conductivities are in mho/m and Q_v is in meq/ml.

In the limit of no clay ($Q_v = 0$) or no counterion mobility ($A = C = 0$) the expression reduces to Archie's equation. In the limits of high and low brine conductivity, with nonzero Q_v, the expression becomes

$$\sigma = \frac{1}{F}(\sigma_w + AQ_v) + EQ_v, \quad \text{high}-\sigma_w\text{limit} \tag{9.4.36}$$

$$\sigma = \frac{1}{F}\left(1 + \frac{A}{C}\right)\sigma_w + EQ_v, \quad \text{low}-\sigma_w\text{limit} \tag{9.4.37}$$

At low σ_w the σ versus σ_w curve has a higher slope than at the high-σ_w limit. At high σ_w the electric current is more concentrated in the pore-space bulk fluid than in the clay double layer, whereas for low σ_w the currents are mostly concentrated within the double layer. This gives rise to the curvature in the σ versus σ_w behavior.

Skold *et al.* (2011) and Revil (2012) developed a model for conductivity of fully saturated shaly sands based on differential effective medium theory for grains with an electrical double layer (Stern layer and the diffuse layer) in an electrolyte. The complex conductivity is given as (Revil and Jardani, 2013)

$$\sigma^* = \frac{\sigma_W}{F} + \sigma_S^* \tag{9.4.38}$$

The complex surface conductivity σ_S^* depends on the CEC and the mobility of counterions in the diffuse layer and the Stern layer.

$$\sigma_S^* = \frac{2}{3}\rho_0 CEC[(1-f)\beta_{(+)} - i\beta_{(+)}^S f] \tag{9.4.39}$$

where f is the fraction of counterions in the Stern layer, $\beta_{(+)}$ and $\beta_{(+)}^S$ are the mobility of the counterions in the diffuse layer and the Stern layer respectively. The mobility in the diffuse layer is the same as the mobility in the bulk pore water, but the mobility in the Stern layer is much smaller. For Sodium ions $\beta_{(+)}(Na^+, 25\ °C) = 5.2\ x\ 10^{-8}\ m^2\ s^{-1}\ V^{-1}$ and $\beta_{(+)}^S(Na^+, 25\ °C) = 1.5\ x\ 10^{-10}\ m^2\ s^{-1}\ V^{-1}$ and $f \approx 0.9$ for illite and smectite (Revil and Jardani, 2013).

The total (quasi-static) electrical current flow in a porous media saturated with an electrolyte consists of the current flow due to an applied electrical field **E** as well as a contribution from the streaming current due to the drag of excess charges in the electrical diffuse layer by fluid flow. The total current density for quasi-static conditions is (Revil *et al.*, 2011)

$$\mathbf{J} = \sigma\mathbf{E} + \hat{Q}_V\mathbf{U} \tag{9.4.40}$$

where **U** is the Darcy velocity of fluid flow in the porous media, and $\hat{Q}_V$ is the *effective* excess moveable charge. When the thickness of the electrical diffuse layer is much smaller than the effective pore radius, $\hat{Q}_V$ can be much smaller than the excess charge density in the diffuse layer. Jardani *et al.* (2007) give an empirical relation relating $\hat{Q}_V$(in C m^{-3}) to permeability k (in m^2):

$$\log_{10}\hat{Q}_V = -9.235 - 0.822\log_{10}k \tag{9.4.41}$$

The empirical relation is based on compilation of data for many different porous materials including sandstones, limestones, clayrock, glass beads, sand and clayey soils and fluids with pH ranging from 6 to 8.5.

Uses

The equations presented in this section can be used to interpret resistivity logs and in the forward model component of workflows for inverting electrical resistivity data to estimate soil and rock properties.

Assumptions and Limitations

Models for log interpretation involve much empiricism, and empirical relations should be calibrated to specific locations and formations.

9.5 Cross-Property Bounds and Relations between Elastic and Electrical Parameters

As discussed throughout this book, most effective properties of rocks are controlled by microgeometry. It's not surprising, therefore, that knowing one effective property of the medium (e.g., electrical conductivity) might help to constrain another effective property of the medium (e.g., bulk modulus), even though the two properties may be governed by very different constitutive relations (Torquato, 2002). Cross-property relations have several uses: predicting a missing property (e.g., seismic V_P) from a measured property (e.g., resistivity), yielding improved rock or fluid description from multiple measurement types, and providing a check on the consistency of sets of measurements.

Electrical–Elastic Cross Bounds

Formal *bounds* on cross-property have been widely researched (Beran, 1965, 1968; Prager, 1969; Kantor and Bergman, 1984; Milton, 1984; Berryman and Milton, 1988; Gibiansky and Torquato, 1996, 1998). The simplest, but not particularly useful, cross bounds are the Hashin–Shtrikman bounds. Consider, for example, a two-phase composite with phase electrical conductivities (σ_1, σ_2), bulk moduli (K_1, K_2), shear moduli (μ_1, μ_2), and volume fractions (f_1, f_2). The Hashin–Shtrikman bounds on effective bulk modulus and electrical conductivity are, in Berryman's canonical form (Section 4.3)

$$K^{HS\pm} = \left[\frac{f_1}{K_1 + \frac{4}{3}\mu_\pm} + \frac{f_2}{K_2 + \frac{4}{3}\mu_\pm} \right]^{-1} - \frac{4}{3}\mu_\pm = \left[\frac{f_1}{K_1 + y_K^\pm} + \frac{f_2}{K_2 + y_K^\pm} \right]^{-1} - y_K^\pm .$$

$$(9.5.1)$$

$$\sigma^{HS\pm} = \left[\frac{f_1}{\sigma_1 + 2\sigma_\pm} + \frac{f_2}{\sigma_2 + 2\sigma_\pm} \right]^{-1} - 2\sigma_\pm = \left[\frac{f_1}{\sigma_1 + y_\sigma^\pm} + \frac{f_2}{\sigma_2 + y_\sigma^\pm} \right]^{-1} - y_\sigma^\pm \quad (9.5.2)$$

where $y_K^\pm = \frac{4}{3}\mu_\pm = \frac{4}{3}\left\{ \begin{array}{l} \max(\mu_1,\mu_2) \\ \min(\mu_1,\mu_2) \end{array} \right\}$ and $y_\sigma^\pm = 2\sigma_\pm = 2\left\{ \begin{array}{l} \max(\sigma_1,\sigma_2) \\ \min(\sigma_1,\sigma_2) \end{array} \right\}$ are the

Y-transforms of K_{eff} and σ_{eff}, respectively:

$$y_K(K_{eff}) = -f_2 K_1 - f_1 K_2 + \frac{f_1 f_2 (K_1 - K_2)^2}{f_1 K_1 + f_2 K_2 - K_{eff}} \tag{9.5.3}$$

$$y_\sigma(\sigma_{eff}) = -f_2 \sigma_1 - f_1 \sigma_2 + \frac{f_1 f_2 (\sigma_1 - \sigma_2)^2}{f_1 \sigma_1 + f_2 \sigma_2 - \sigma_{eff}} \tag{9.5.4}$$

Since the Hashin–Shtrikman bounds are independent of geometry, they don't provide a means to constrain one of the effective properties by another. The exception would be if one of the effective properties falls on a Hashin–Shtrikman bound: a rock whose bulk modulus falls on the *upper* Hashin–Shtrikman bound will have electrical conductivity on the *lower* Hashin–Shtrikman bound (and vice versa), if the pore-filling phase is electrically more conductive than the mineral.

Berryman and Milton (1988) introduced electrical-elastic cross bounds that are narrower than the Hashin–Shtrikman bounds, based on the three-point geometric parameters ζ_1 and $\zeta_2 = 1 - \zeta_1 (0 \le \zeta_2 \le 1)$ that were discussed in Section 4.3. If the effective bulk modulus K_{eff} is known, then the bounds of Beran and Molyneux (1966) can be inverted to put the following limits on ζ_2 (if $\mu_1 \ge \mu_2$):

$$\zeta_2^-(y_K) \equiv \frac{4/(3y_K) - 1/\mu_1}{1/\mu_2 - 1/\mu_1} \le \zeta_2 \le \frac{\mu_1 - 3y_K/4}{\mu_1 - \mu_2} \equiv \zeta_2^+(y_K) \tag{9.5.5}$$

The corresponding bounds on electrical conductivity are then found from the range $\zeta_2^- \le \zeta_2 \le \zeta_2^+$:

$$\left[\frac{f_1}{\sigma_1 + y_\sigma^-} + \frac{f_2}{\sigma_2 + y_\sigma^-} \right]^{-1} - y_\sigma^- \le \sigma_{eff} \le \left[\frac{f_1}{\sigma_1 + y_\sigma^+} + \frac{f_2}{\sigma_2 + y_\sigma^+} \right]^{-1} - y_\sigma^+ \tag{9.5.6}$$

where

$$y_\sigma^+ = 2\langle\sigma\rangle_\zeta \equiv 2(\zeta_1^+ \sigma_1 + \zeta_2^+ \sigma_2), \quad y_\sigma^- = 2\langle\sigma^{-1}\rangle_\zeta^{-1} \equiv 2(\zeta_1^- \sigma_1^{-1} + \zeta_2^- \sigma_2^{-1})^{-1} \tag{9.5.7}$$

If instead, we know the effective conductivity, σ_{eff}, then we can solve for the limits on ζ_2 (if $\sigma_1 > \sigma_2$):

$$\zeta_2^-(y_\sigma) \equiv \frac{2/y_\sigma - 1/\sigma_1}{1/\sigma_2 - 1/\sigma_1} \le \zeta_2 \le \frac{\sigma_1 - y_\sigma/2}{\sigma_1 - \sigma_2} \equiv \zeta_2^+(y_\sigma) . \tag{9.5.8}$$

The range of ζ_2 determines the maximum and minimum bulk moduli that are possible for the given σ:

$$\left[\frac{f_1}{K_1 + y_K^-} + \frac{f_2}{K_2 + y_K^-} \right]^{-1} - y_K^- \le K_{eff} \le \left[\frac{f_1}{K_1 + y_K^+} + \frac{f_2}{K_2 + y_K^+} \right]^{-1} - y_K^+ \tag{9.5.9}$$

where

$$y_K^+ = \frac{4}{3}\langle\mu\rangle_\zeta \equiv \frac{4}{3}(\zeta_1^+\mu_1 + \zeta_2^+\mu_2) \qquad y_\sigma^- = \frac{4}{3}\langle\mu^{-1}\rangle_\zeta^{-1} \equiv \frac{4}{3}(\zeta_1^-\mu_1^{-1} + \zeta_2^-\mu_2^{-1})^{-1} \qquad (9.5.10)$$

Gibiansky and Torquato (1996) derived cross bounds (GT) between bulk modulus and electrical conductivity of two-phase materials that are narrower than the Hashin–Shtrikman and Berryman–Milton bounds. Their method uses five hyperbolic curves defined in the $K - \sigma$ plane:

$$GT1 : Hyp[(\sigma_{1^*}, K_{1^*}), (\sigma_{2^*}, K_{2^*}), (\sigma_V, K_V)] \qquad (9.5.11)$$

$$GT2 : Hyp[(\sigma_{1^*}, K_{1^*}), (\sigma_{2^*}, K_{2^*}), (\sigma_1, K_1)] \qquad (9.5.12)$$

$$GT3 : Hyp[(\sigma_{1^*}, K_{1^*}), (\sigma_{2^*}, K_{2^*}), (\sigma_2, K_2)] \qquad (9.5.13)$$

$$GT4 : Hyp[(\sigma_{1^*}, K_{1^*}), (\sigma_{2^*}, K_{2^*}), (\sigma_{1^{**}}, K_R)] \qquad (9.5.14)$$

$$GT5 : Hyp[(\sigma_{1^*}, K_{1^*}), (\sigma_{2^*}, K_{2^*}), (\sigma_{2^{**}}, K_R)] \qquad (9.5.15)$$

The $Hyp[(x_1, y_1), (x_2, y_2), (x_3, y_3)]$ function defines a hyperbola passing through points
(x_1, y_1), (x_2, y_2), and (x_3, y_3). It can be defined parametrically by

$$x = \gamma x_1 + (1 - \gamma)x_2 - \frac{\gamma(1 - \gamma)(x_1 - x_2)^2}{(1 - \gamma)x_1 + \gamma x_2 - x_3} \qquad (9.5.16)$$

$$y = \gamma y_1 + (1 - \gamma)y_2 - \frac{\gamma(1 - \gamma)(y_1 - y_2)^2}{(1 - \gamma)y_1 + \gamma y_2 - y_3} \qquad (9.5.17)$$

where $\gamma \in [0, 1]$. Arguments to the five Hyp functions are defined as follows: σ_{1^*} and σ_{2^*} are the Hashin–Shtrikman bounds on conductivity; K_{1^*} and K_{2^*} are the Hashin–Shtrikman bounds on bulk modulus; σ_V and σ_R are the Voigt and Reuss bounds on conductivity; K_V and K_R are the Voigt and Reuss bounds on bulk modulus; and $\sigma_{1^{**}}$ and $\sigma_{2^{**}}$ are the Hashin–Shtrikman bounds evaluated with (unphysical) values $-\sigma_1$ and $-\sigma_2$ assigned as the phase conductivities. Of the five curves, the outermost pair define the desired bounds. Curves GT2 and GT3 are realizable and correspond to doubly coated spheres, which are also the limiting cases of the embedded bound method of Mavko and Saxena (2013) (Section 6.8). Curve GT1 is known to be realizable at three discrete points that correspond to very special and rare polycrystalline arrangement of the two phases (Milton, 1981; Gibiansky and Milton, 1993; Vinogradov and Milton, 2005); it is not known if the remainder of the GT1 curve is realizable. It is also not known if curves GT4 and GT5 are realizable.

Figure 9.5.1 compares the Gibiansky and Torquato bounds, the embedded bounds, and the Hashin–Shtrikman bounds, for a quartz-rich rock with a conductive pore fluid, and porosity 0.06. For this composition, all pairs of K_{eff} and σ_{eff} must lie within the envelope defined by curves GT1 and GT4. We can see that is it impossible for a pair

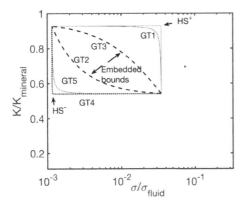

Figure 9.5.1 Comparison of the Hashin–Shtrikman and Gibiansky and Torquato cross bounds for a quartz-rich rock saturated with brine (the fluid is 1000 times more conductive than the mineral). The porosity is 0.06. The effective bulk modulus K is normalized by the bulk modulus of the mineral; the effective conductivity σ is normalized by the fluid conductivity.

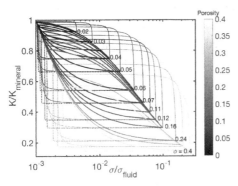

Figure 9.5.2 Similar to Figure 9.5.1. Comparison of the Hashin–Shtrikman and Gibiansky and Torquato cross bounds for a quartz-rich rock saturated with brine (the fluid is 1000 times more conductive than the mineral). Labeled envelopes for porosities ranging from 0 to 0.4 are superimposed. The effective bulk modulus K is normalized by the bulk modulus of the mineral; the effective conductivity σ is normalized by the fluid conductivity.

(K_{eff}, σ_{eff}) to be both on the upper Hashin–Shtrikman bounds or both on the lower Hashin–Shtrikman bounds. It is also observed that in general the embedded bounds (GT2 and GT3) lie inside of the Gibiansky and Torquato bounds.

When the volume fractions are not known, then the bounded region becomes the union of GT envelopes over all possible porosities. Figure 9.5.2 is similar to Figure 9.5.1, but now with stacked curves representing porosities ranging between 0 and 0.4. Constraints between electrical and elastic properties exist even if the porosity is not known; however, knowing porosity substantially narrows the range.

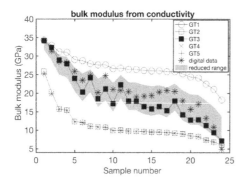

Figure 9.5.3 Digital data points against the corresponding GT bounds for 24 sandstone samples, plotted in order of increasing sample porosity. The GT curves are estimated using the digital estimate of electrical conductivity combined with information on porosity and phase property end-points. The digital sandstone samples, spanning a wide range of porosity values, lie within a narrow bulk modulus band around the embedded curve 'GT3', shaded and labeled as 'reduced range'.

Figure 9.5.3 shows digitally computed effective bulk modulus for sandstones, compared with the GT bounds (Dutta, 2018). For each sample, the corresponding digital estimate of electrical conductivity is combined with its porosity to obtain the five GT curves for bulk modulus (labeled 'GT1'–'GT5'). The black stars are the FEM computed digital bulk modulus for each sample. Most of the digital bulk modulus values lie between the embedded bound, GT3, and the partially realizable bound, GT1, on average lying closer to GT3 than any other GT curve. Three samples lie within the embedded bounds. Dutta (2018) suggests an empirical 'reduced range' shown by the shaded band, and described by linear combinations of GT1 and GT3:

$$K_{upper} \approx 0.4\, K_{GT3} + 0.6\, K_{GT1} \tag{9.5.18}$$

$$K_{lower} \approx 0.7\, K_{GT3} + 0.3\, K_{GT2} \tag{9.5.19}$$

In a similar exercise, Dutta found that digital values of conductivity predicted from bulk modulus and porosity fell in the reduced range

$$\sigma_{upper} \approx 0.6\, \sigma_{GT3} + 0.4\, \sigma_{GT1} \tag{9.5.20}$$

$$\sigma_{lower} \approx 0.7\, \sigma_{GT3} + 0.3\, K_{GT2} \tag{9.5.21}$$

Electrical–Electrical Cross Bounds

Rigorous cross-bounds of two-phase isotropic composites also exist among effective properties electrical conductivity, dielectric constant, thermal conductivity, and magnetic permeability. Consider a composite with volume fractions f_1 and f_2. Phase properties λ_1 and λ_2 result in effective property λ_e, and phase properties σ_1 and σ_2 result in effective property σ_e.

If λ_e is known, then bounds on σ_e are given by the equation (Bergman, 1978)

$$\frac{\lambda_2 - \lambda_1}{\lambda_e - f_1\lambda_1 - f_2\lambda_2} - \frac{\sigma_2 - \sigma_1}{\sigma_{e\pm} - f\sigma_1 - f_2\sigma_2} = \frac{3(\lambda_2\sigma_1 - \lambda_1\sigma_2)}{f_1f_2(\lambda_2 - \lambda_1)(\sigma_2 - \sigma_1)} \quad (9.5.22)$$

or more explicitly:

$$\sigma_{e\pm} = (f_1\sigma_1 + f_2\sigma_2) + (\sigma_2 - \sigma_1) \left[\frac{3(\lambda_1\sigma_2 - \lambda_2\sigma_1)}{f_1f_2(\lambda_2 - \lambda_1)(\sigma_2 - \sigma_1)} + \frac{(\lambda_2 - \lambda_1)}{\lambda_e - (f_1\lambda_1 + f_2\lambda_2)} \right]^{-1} \quad (9.5.23)$$

In equations 9.5.22 and 9.5.23, $\sigma_{e\pm}$ is the upper (lower) bound when the sign of $(\lambda_2\sigma_1 - \lambda_1\sigma_2)/(\sigma_1 - \sigma_2)$ is positive (negative).

Avellaneda *et al.* (1988) showed that when $\sigma_1/\sigma_2 \leq \lambda_2/\lambda_1$ another bound is

$$\frac{(\sigma_1 + 2\sigma_2)}{(\sigma_1 - \sigma_2)} \left[\frac{(\sigma_2 + 2\sigma_1)}{(\sigma_1 - \sigma_2)} - \phi_2 \frac{(\sigma_{e\mp} + 2\sigma_1)}{(\sigma_1 - \sigma_e)} \right] = \frac{(\lambda_1 + 2\lambda_2)}{(\lambda_1 - \lambda_2)} \left[\frac{(\lambda_2 + 2\lambda_1)}{(\lambda_1 - \lambda_2)} - \phi_2 \frac{(\lambda_e + 2\lambda_1)}{(\lambda_1 - \lambda_e)} \right] \quad (9.5.24)$$

or more explicitly:

$$\sigma_{e\mp} = \frac{(B + 2)\sigma_1}{(B - 1)}; \quad B = \frac{-\Delta\sigma}{f_2\sigma_{12}} \left[\frac{\lambda_{12}\lambda_{21}}{\Delta\lambda^2} - \frac{\sigma_{12}\sigma_{21}}{\Delta\sigma^2} - \frac{f_2\lambda_{12}}{\Delta\lambda} \frac{(\lambda_e + 2\lambda_1)}{(\lambda_e - \lambda_1)} \right] \quad (9.5.25)$$

where

$$\Delta\lambda = \lambda_2 - \lambda_1; \quad \Delta\sigma = \sigma_2 - \sigma_1; \quad \lambda_{12} = \lambda_1 + 2\lambda_2; \quad \lambda_{21} = \lambda_2 + 2\lambda_1;$$

$$\sigma_{12} = \sigma_1 + 2\sigma_2; \quad \sigma_{21} = \sigma_2 + 2\sigma_1 \quad (9.5.26)$$

when $\sigma_1/\sigma_2 > \lambda_2/\lambda_1$, the subscripts in equations 9.5.24–9.5.26 must be interchanged. In equations 9.5.24–9.5.26, $\sigma_{e\mp}$ is the lower (upper) bound when $(\lambda_2\sigma_1 - \lambda_1\sigma_2)/(\sigma_1 - \sigma_2)$ is positive (negative).

Hence, equations 9.5.22–9.5.26 provide the best-known upper and lower cross-bounds between properties λ_1 and σ_1. It is useful to note that *substitution bounds* can also be found by setting parameters $\lambda_i \rightarrow \lambda_i^{(A)}$ and $\sigma_i \rightarrow \lambda_i^{(B)}$, where $\lambda_i^{(A)}$ and $\lambda_i^{(B)}$ are both the same type of property, with a change in phase property values. For example, we can find the substitution bounds between effective dielectric constant ε_e^A with initial phases quartz and oil ($\varepsilon_1^A = 4.6$ and $\varepsilon_2^A = 2$) and effective dielectric constant ε_e^B with substituted phases quartz and water ($\varepsilon_1^B = 4.6$ and $\varepsilon_2^B = 80$).

Empirical Relations

Faust (1953) presents an empirical relation between the measured resistivity R_0 of a water-saturated formation and V_P:

$$V_P = 2.2888 \left(Z \frac{R_0}{R_w} \right)^{1/6} \quad (9.5.27)$$

where R_w is the resistivity of formation water, Z is depth in km, and V_P is in km/s.

Faust's original form of the empirical relation is

$$V(\text{ft/s}) = \gamma (Z[R_t])^{1/6} \tag{9.5.28}$$

$$\gamma = 1948 \tag{9.5.29}$$

where depth is in feet and $[R_t]$ is a dimensionless ratio of the average formation resistivity to the average formation water resistivity.

Hacikoylu et al. (2006) show that this equation is only applicable for consolidated sandstone with small clay content. Hacikoylu et al. (2006) use theoretical velocity–porosity relations for soft sediment in combination with equations for the formation factor to obtain a velocity–resistivity relation appropriate for soft sediment. They show that this theoretical dependence can be approximated by the following equation:

$$V_P = \frac{R_0/R_w}{0.9 + c(R_0/R_w)} \tag{9.5.30}$$

where V_P is in km/s and the coefficient c varies between 0.27 and 0.32 for Gulf of Mexico shale data.

Figure 9.5.4 displays V_P versus the formation factor $F = R_0/R_w$ for both equations, using $Z = 2$ km and $c = 0.3$.

Koesoemadinata and McMechan (2003) have given empirical relations between the ultrasonic P-wave velocity (V_P), dielectric constant (κ), porosity (ϕ), density (ρ), and permeability (k). The regressions are based on data from about 30 samples of Ferron sandstone, a fine- to medium-grained sandstone with porosity ranging from 10% to 24% and clay content ranging from 10% to 22%. The V_P and κ data were measured on dry (< 1% water saturation by volume) samples at 125 kHz and 50 MHz, respectively.

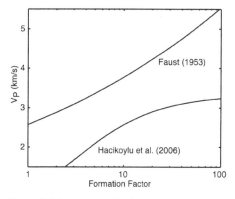

Figure 9.5.4 P-wave velocity versus formation factor according to the Faust (1953) and Hacikoylu et al. (2006) equations

$$\kappa = 0.00119\, V_P + 1.41383 \qquad\qquad R^2 = 0.55 \qquad\qquad (9.5.31)$$

$$\kappa = 0.00058\, V_P - 0.16294\, \ln k + 3.3758 \quad R^2 = 0.64 \qquad\qquad (9.5.32)$$

$$\kappa = 0.00072\, V_P - 0.06851\phi + 3.8133 \quad R^2 = 0.65 \qquad\qquad (9.5.33)$$

$$\kappa = 0.00072\, V_P + 2.3032\rho - 2.4137 \quad R^2 = 0.68 \qquad\qquad (9.5.34)$$

$$V_P = 479.6734\kappa + 380.980 \qquad\qquad R^2 = 0.55 \qquad\qquad (9.5.35)$$

$$V_P = 229.35033\kappa - 104.8169\, \ln k + 1756.658 \; R^2 = 0.64 \qquad (9.5.36)$$

In the previous empirical regressions V_P is in m/s, permeability is in mD, density is in g/cm^3, porosity is given as a percentage, and κ is dimensionless. R^2 is the coefficient of determination for the regression.

Carcione *et al.* (2007) have compiled and derived cross-property relations and bounds relating electrical conductivity to elastic moduli and velocities of rocks. The cross-property relations are based on existing empirical and theoretical relations between electrical conductivity and porosity and between elastic moduli and porosity. The basic approach is as follows. If the relation between porosity, ϕ, and conductivity, σ, is described by $\sigma = f(\phi)$, while the relation between elastic velocity V and porosity is given by $V = g(\phi)$, then the cross-property relation can be obtained by eliminating ϕ to give $\sigma = f(g^{-1}(V))$.

Dry Rocks

For a dry rock with randomly oriented penny-shaped cracks of zero conductivity in an isotropic elastic medium, Bristow (1960) gave the following relations between dry-rock elastic moduli and electrical conductivity:

Bristow

$$\frac{K_s - K_m}{K_m} = \frac{2(1 - v_s^2)}{1 - 2v_s}\frac{\sigma_s - \sigma_m}{\sigma_m} \qquad\qquad (9.5.37)$$

$$\frac{\mu_s - \mu_m}{\mu_m} = \frac{4}{5}(1 - v_s)(5 - v_s)\left(\frac{\sigma_s - \sigma_m}{\sigma_m}\right) \qquad (9.5.38)$$

where K_m, μ_m are the dry-rock bulk and shear moduli, σ_m is the dry-rock electrical conductivity, K_s, μ_s, v_s are bulk modulus, shear modulus, and Poisson ratio for the solid mineral, and σ_s is the electrical conductivity of the solid mineral.

Berryman and Milton (1988) consider the case of an insulating solid mineral ($\sigma_s = 0$) with porosity ϕ filled with an electrically conductive pore fluid of conductivity $\sigma_f \neq 0$ that has either zero fluid bulk modulus or is under drained conditions. They obtained the following bounds relating the effective bulk and shear moduli (K, μ) to the effective electrical conductivity, σ, of the rock:

Berryman–Milton bounds

$$\frac{1}{2}\frac{(1-\phi)\sigma}{(\phi\sigma_f - \sigma)} \leq 1 - \frac{3\phi K}{4\mu_s(1-\phi-K/K_s)} \tag{9.5.39}$$

$$\frac{1}{2}\frac{(1-\phi)\sigma}{(\phi\sigma_f - \sigma)} \leq \frac{21}{5-21B}\left[1 + B - \frac{6A\phi\mu}{(1-\phi)\mu_s - \mu}\right] \tag{9.5.40}$$

$$A = \frac{6(K_s + 2\mu_s)^2}{(3K_s + \mu_s)^2} \tag{9.5.41}$$

$$B = \frac{5\mu_s(4K_s + 3\mu_s)}{(3K_s + \mu_s)^2} \tag{9.5.42}$$

Gibiansky and Torquato (1996) obtained the following cross-property bound for a dry, cracked rock with a finitely conducting solid mineral:

Gibiansky–Torquato bound

$$\frac{1}{K_m} - \frac{1}{K_s} \geq \frac{3\sigma_s}{2\mu_s}\frac{1-v_s}{1+v_s}\left(\frac{1}{\sigma_m} - \frac{1}{\sigma_s}\right) \tag{9.5.43}$$

Wet Rocks

For fluid-saturated rocks with pore-fluid bulk modulus K_f and pore-fluid conductivity σ_f, such that $K_f/K_s \leq \sigma_f/\sigma_s$, Milton (1981) obtained the following inequalities for the effective elastic moduli (K, μ) and the effective electrical conductivity, σ:

Milton Bounds

$$\frac{K}{K_s} \leq \frac{\sigma}{\sigma_s}, \qquad \frac{\mu}{K_s} \leq \frac{3\sigma}{2\sigma_s} \tag{9.5.44}$$

These inequalities remain valid when both $K_f = 0$ and $\sigma_f = 0$.

Combining Archie's relation (see Section 9.4)

$$\sigma = \sigma_f\phi^m \tag{9.5.45}$$

where m is Archie's cementation factor, with the Wyllie time-average relation (see Section 7.2)

$$\phi = \frac{1/V_P - 1/V_0}{1/V_f - 1/V_0} \tag{9.5.46}$$

where V_P is the P-wave velocity in the porous rock, V_0 is the P-wave velocity in the solid mineral grain, and V_f is the wave velocity in the pore fluid gives the following relation between bulk rock conductivity and elastic-wave velocity:

Archie/Time-Average

$$\sigma = \sigma_f \left(\frac{V_0/V_P - 1}{V_0/V_f - 1} \right)^m \tag{9.5.47}$$

Substituting porosity from Archie's equation into the Raymer–Hunt–Gardner velocity–porosity relation (see Section 7.4) gives:

Archie/Raymer

$$V_P = \left[1 - \left(\frac{\sigma}{\sigma_f} \right)^{1/m} \right]^2 V_0 + \left(\frac{\sigma}{\sigma_f} \right)^{1/m} V_f \tag{9.5.48}$$

Similarly, solving for porosity from the Wyllie time-average equation and substituting into various conductivity–porosity relations gives the following cross-property expressions (Carcione *et al.*, 2007):

Glover et al. (2000)/Time-Average

$$\sigma = (1 - \phi)^p \sigma_s + \sigma_f \phi^m, \quad \phi = \frac{1/V_P - 1/V_0}{1/V_f - 1/V_0} \tag{9.5.49}$$

Hermance (1979)/Time-Average

$$\sigma = (\sigma_f - \sigma_s)\phi^m + \sigma_s, \quad \phi = \frac{1/V_P - 1/V_0}{1/V_f - 1/V_0} \tag{9.5.50}$$

Self-Similar (Sen et al., 1981)/Time-Average

$$V_P = \left[\left(\frac{1}{V_f} - \frac{1}{V_0} \right) \left(\frac{\sigma_s - \sigma}{\sigma_s - \sigma_f} \right) \left(\frac{\sigma_f}{\sigma} \right)^{1-1/m} + \frac{1}{V_0} \right]^{-1} \tag{9.5.51}$$

Brito Dos Santos *et al.* (1988) obtained the preceding relation combining the self-similar conductivity model and the Wyllie time-average relation for elastic velocity, for a porous medium with a conducting matrix. It should be noted that the results from the Hermance (1979) and Glover *et al.* (2000) conductivity–porosity relations lie outside the Hashin–Shtrikman bounds.

Raymer–Hunt–Gardner's velocity–porosity relation combined with the Hashin–Shtrikman lower bound on conductivity gives the following relation (Hacikoylu *et al.*, 2006):

HS/Raymer

$$V_P = (1 - \phi + \phi_p)^2 V_0 + \left(\phi - \phi_p\right) V_f, \quad \phi = \frac{3\sigma}{\sigma + 2\sigma_f} \tag{9.5.52}$$

where ϕ_p is the critical porosity (see Section 7.1), taken to be 0.4 in Hacikoylu *et al.* (2006).

HS Models

The Hashin–Shtrikman upper and lower bounds for electrical conductivity (see Section 9.1) can be solved for porosity, which can then be inserted in the corresponding Hashin–Shtrikman bounds for elastic moduli to obtain relations between elastic and electrical properties. The porosity of a two-constituent composite, in terms of the upper and lower Hashin–Shtrikman bounds on the electrical conductivity, is given as follows (Carcione *et al.*, 2007):

$$\phi = \left(\frac{\sigma_s - \sigma_{HS}^-}{\sigma_s - \sigma_f}\right)\left(\frac{\sigma_f + 2\sigma_s}{\sigma_{HS}^- + 2\sigma_s}\right) = \left(\frac{\sigma_s - \sigma_{HS}^+}{\sigma_s - \sigma_f}\right)\left(\frac{3\sigma_f}{\sigma_{HS}^+ + 2\sigma_f}\right) \tag{9.5.53}$$

where σ_{HS}^+ and σ_{HS}^- are the upper and lower Hashin–Shtrikman bounds for electrical conductivity, respectively.

Carcione *et al.* substitute this porosity expression into the expression for the Hashin–Shtrikman lower bound for the elastic bulk modulus, K_{HS}^-, giving:

$$K_{HS}^- = \left[\left(\frac{\sigma_s - \sigma_{HS}^-}{\sigma_s - \sigma_f}\right)\left(\frac{\sigma_f + 2\sigma_s}{\sigma_{HS}^- + 2\sigma_s}\right)\left(\frac{1}{K_f} - \frac{1}{K_s}\right) + \frac{1}{K_s}\right]^{-1} \tag{9.5.54}$$

Similar expressions can be obtained using the upper bound.

> **Caution:**
>
> The electrical-elastic cross bounds, discussed earlier in this section, prohibit certain cases of equation 9.5.49. For example, if the fluid conductivity is larger than the mineral conductivity, then a rock cannot simultaneously realize both the electrical upper Hashin–Shtrikman bound and the elastic upper Hashin–Shtrikman bound. Similarly, a rock cannot simultaneously realize both the electrical lower Hashin–Shtrikman bound and the elastic lower Hashin–Shtrikman bound.

Gassmann-Based Relations

Carcione *et al.* (2007) suggest using Gassmann's relation to derive relations between elastic and electrical properties. Gassmann's relation (see Section 6.3) relates the elastic bulk modulus K_m of a dry rock with mineral modulus K_s to the bulk modulus

K_{sat} of the same rock fully saturated with a fluid of bulk modulus K_f. One form of Gassmann's relation is

$$K_{sat} = \frac{K_s - K_m(\phi) + \phi K_m(\phi)(K_s/K_f - 1)}{1 - \phi - K_m(\phi)/K_s + \phi K_s/K_f} \tag{9.5.55}$$

For the dry bulk modulus, Carcione et al. suggest using a dry modulus–porosity relation based on Krief's equation (see Section 7.9):

$$K_m = K_s(1 - \phi)^{(1-\phi+A)/(1-\phi)} \tag{9.5.56}$$

$$\mu_m = \left(\frac{\mu_s}{K_s}\right) K_m \tag{9.5.57}$$

Other $K_m - \phi$ models, such as the soft-sand or cemented-sand models described in Section 5.5, can also be used. The porosity is then replaced by a porosity derived from one of the many $\sigma - \phi$ relations such as the following:

$$\phi = \left(\frac{\sigma}{\sigma_f}\right)^{1/m} \qquad\qquad \text{Archie} \tag{9.5.58}$$

$$\phi = \left(\frac{\sigma - \sigma_s}{\sigma_f - \sigma_s}\right)^{1/m} \qquad\qquad \text{Hermance} \tag{9.5.59}$$

$$\phi = \left(\frac{\sigma^{1/\gamma} - \sigma_s^{1/\gamma}}{\sigma_f^{1/\gamma} - \sigma_s^{1/\gamma}}\right) \qquad \gamma = 2 \quad \text{CRIM} \tag{9.5.60}$$

$$\phi = \left(\frac{\sigma - \sigma_s}{\sigma_f - \sigma_s}\right)\left(\frac{\sigma_f}{\sigma}\right)^{1-1/m} \qquad \text{self–similar} \tag{9.5.61}$$

$$\phi = \left(\frac{\sigma_s - \sigma}{\sigma_s - \sigma_f}\right)\left(\frac{\sigma_f + 2\sigma_s}{\sigma + 2\sigma_s}\right) \qquad \text{HS lower bound} \tag{9.5.62}$$

$$\phi = \left(\frac{\sigma_s - \sigma}{\sigma_s - \sigma_f}\right)\left(\frac{3\sigma_f}{\sigma + 2\sigma_f}\right) \qquad \text{HS upper bound} \tag{9.5.63}$$

Layered Media

By combining the Backus-averaged elastic modulus (see Section 4.16) and electrical conductivity for a two-constituent layered medium with layering perpendicular to the 3-axis, Carcione et al. (2007) give the following relations:

$$\frac{\sigma_{11} - \sigma_2}{\sigma_1 - \sigma_2} = \frac{c_{66} - \mu_2}{\mu_1 - \mu_2} \tag{9.5.64}$$

$$\frac{\sigma_{33}^{-1} - \sigma_2^{-1}}{\sigma_1^{-1} - \sigma_2^{-1}} = \frac{c_{44}^{-1} - \mu_2^{-1}}{\mu_1^{-1} - \mu_2^{-1}} \tag{9.5.65}$$

Subscripts 1 and 2 denote the properties of the constituent layers, and the anisotropic effective averaged quantities are denoted by the usual two-index notation.

Uses

These relations can be used to estimate seismic velocity from measured electrical properties.

Assumptions and Limitations

The empirical equations may not be valid for rock types and environments that are different from the ones used to establish the relations. The Faust (1953) equation may not be valid in soft, shaley sediment. The coefficient c in the Hacikoylu *et al.* (2006) equation may have to be adjusted for environments other than the Gulf of Mexico. When combining two relations for elastic and electric properties to obtain a cross-property relation, the resulting equation will be a good description of the cross-relation only if the original equations for elastic and electrical properties themselves are good descriptors. These cross-relations have not been tested widely with controlled data. Two of the empirical relations (Glover *et al.* and Hermance) violate the theoretical Hashin–Shtrikman bounds.

9.6   Brine Resistivity

The electrical resistivity of natural formation waters depends strongly on temperature and ionic composition (Table 9.6.1), and weakly on pressure. The temperature variation of brine resistivity R is often predicted using Arps' (1953) equation (Schöen, 1996):

Table 9.6.1 *Example water chemistries in weight percent*. Compiled by Ucok* et al. *(1980)*

Ion	Normal Groundwater	Seawater	Salton Sea Brine	Cerro Prieto Brine (N. Mexico)
Na^+	0.061	1.5	5.04	0.6
K^+	0.061	0.038	1.75	0.17
Ca^{++}	0.0037	0.04	2.8	0.034
Cl^-	0.0082	1.9	15.5	1.1
Mg^{++}	0.00024	0.135	00.0054	0.0016
SO_4^{--}	0.1	0.27		
$(HCO_3)^-$	0.0429	0.014	00.01	0.0011
Total dissolved solids	0.198	3.45	25.8	2.0

* 1 weight percent = 10,000 ppm = 9.98859 g/l

$$R(T) = R_0 \frac{(T_0 + 21.5)}{(T + 21.5)} \, , \quad (T \text{ and } T_0 \text{ in Celsius}) \tag{9.6.1}$$

$$R(T) = R_0 \frac{(T_0 + 6.77)}{(T + 6.77)} \, , \quad (T \text{ and } T_0 \text{ in Fahrenheit}) \tag{9.6.2}$$

where R_0 is the resistivity at a reference temperature T_0 (Choose the correct form for Fahrenheit versus Celsius). Arp's equation is useful for temperatures in the range 10–200 °C (Ucok et al., 1980).

The variation of brine resistivity with salinity can be estimated using the following equation found by empirically fitting measured values at 40 °C from the Schlumberger log interpretation chart (Schlumberger, 2009):

$$\log_{10}(R_0) = 0.0425(\log_{10}(c_{NaCl}))^2 - 1.246\log_{10}(c_{NaCl}) + 3.921 \tag{9.6.3}$$

(where c_{NaCl} is the NaCl concentration in ppm) and then extrapolating to different temperatures using the Arps equation (equation 9.6.1). The Schlumberger table is the most often quoted source for petrophysical and petroleum reservoir work; it is valid in the range 200–280,000 ppm and 10–200°C. Examples of the empirical prediction relative to Schlumberger values at 65°C and 120°C are shown in Figure 9.6.1.

Bigelow (1992) gave the following empirical formula for brine resistivity

$$R = \left(0.0123 + \frac{3647.5}{c_{NaCl}^{0.995}}\right) \frac{81.77}{T_F + 6.77} \tag{9.6.4}$$

$$R = \left(0.0123 + \frac{3647.5}{c_{NaCl}^{0.995}}\right) \frac{45.42}{(T_C + 21.5)} \tag{9.6.5}$$

Ions other than Na^+ and Cl^- also affect the electrical resistivity of brine. A common approach is to convert each ion type to the electrically equivalent concentration of Na^+ and Cl^-, using the formula

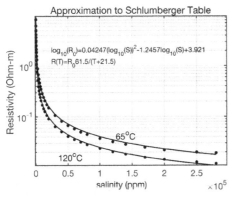

Figure 9.6.1 Comparison of empirical equation (solid curves) with Schlumberger values (dots) at two different temperatures

$$c_{equiv} = \sum_{i=1}^{N} M_i c_i \qquad (9.6.6)$$

where c_{equiv} is the electrically equivalent NaCl concentration (ppm), M_i are the multipliers for each ion type, and c_i are the actual ion concentrations (ppm). Values of c_i for Mg^{++}, Ca^{++}, CO_3^-, K^+, and SO_4^{--} are shown in Figure 9.6.2. Multipliers for Na^+ and Cl^- are 1.

Table 9.6.2 *Selected equivalent Na multipliers (from Ucok* et al., *1980)*

	Multiplier for K^+		Multipliers for Ca^{++}	
ppm	Multiplier 25 °C	Multiplier 100 °C	Multiplier 25 °C	Multiplier 100 °C
10,000	0.331	−0.052	0.723	−0.247
20,000	0.474	0.140	–	–
30,000	0.592	0.331	0.723	0.160
40,000	0.665	0.474	–	–
50,000	0.770	0.675	0.680	0.324
60,000	0.873	0.873	–	–
70,000	0.945	–	0.644	0.505
80,000	1.036	1.215	–	–
90,000	1.127	1.360	0.637	0.576
100,000	1.210	1.497	0.634	0.612
120,000	1.420	1.574	0.587	0.630
150,000	1.560	1.480	0.446	0.631
170,000	1.64	1.380	0.397	0.576
200,000	–	1.220	0.280	0.446
250,000	–	–	0.0255	0.225

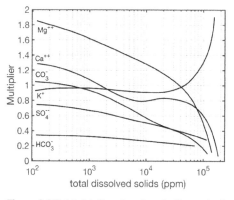

Figure 9.6.2 Multipliers for electrically equivalent ion concentrations of Na^+ (after Schlumberger, 2009)

9.7   Dielectric Constants

Dielectric constants for selected rocks, minerals, water, and ice are listed in Tables 9.7.1–9.7.3.

Table 9.7.1 *Dielectric constants of minerals and rocks*

	Dielectric constant
Quartz (bulk)	4.5–4.7
Calcite (bulk)	7–8
Shale	13–15
Gas	1
Oil	2.2
Anhydrite	6.35
Gypsum	4.16
Sandstone	4.65
Dolostone	6.8
Limestone	7.5 – 9.2

Table 9.7.2 *Dielectric constant of water*

T(°C)	Dielectric constant	T(°C)	Dielectric constant
0	88.00	40	73.28
5	86.40	45	71.59
10	84.11	50	69.94
15	82.22	60	66.74
20	80.36	70	63.68
25	78.54	80	60.78
30	76.75	90	57.98
35	75.00	100	55.33

Table 9.7.3 *Dielectric constants of ice (Hobbs* et al., *1966)*

	T K	Dielectric constant	Density (g/cm³)
Ice I			0.917
	273	91.5	
	262.3	95.0	
	252.2	97.4	
	241	100	
	228.4	104	
	216.3	114	
Ice III			1.155
	243	117	
	253		
Ice V			1.258
	263		
	243	144	
	223		
Ice VI			1.350
	273		
	242	193	
	224		

Appendices

A.1 Typical Rock Properties

Sample of Published Laboratory Data

Shaly Sandstones. Dry and water-saturated ultrasonic velocities in consolidated sandstones, plotted versus porosity, clay volume fraction, and effective pressure. De-Hua Han (1986).

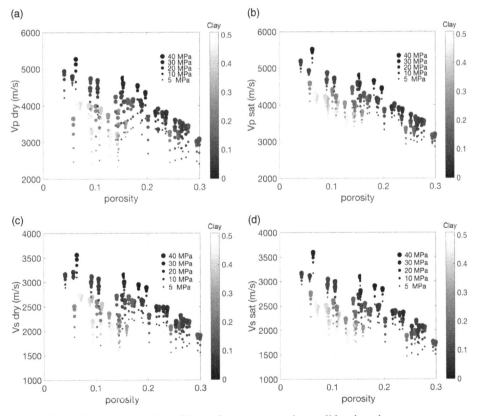

Figure A.1.1 Ultrasonic velocities of dry and water-saturated consolidated sandstones.

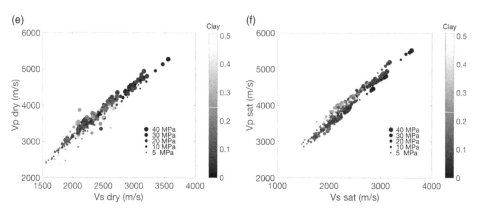

Figure A.1.1 (cont.)

Shaly Sandstones. Porosity, permeability, electrical conductivity, and brine-saturated ultrasonic velocities in shaly sandstones, plotted versus porosity, clay volume fraction, and effective pressure. Saturated with sodium chloride 35 g/l. Conductivity measured at 440 Hz. Tongchen Han (2010).

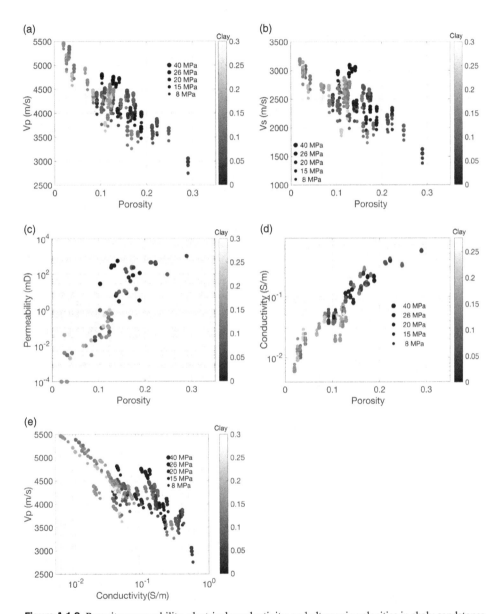

Figure A.1.2 Porosity, permeability, electrical conductivity, and ultrasonic velocities in shaly sandstones.

Shaly Sandstones from Oseberg. Porosity and dry and water-saturated ultrasonic V_P and V_S. Sverre Strandenes, Rock Physics Analysis of the Brent Group Reservoir in the Oseberg Field, SRB Special Report, Stanford University.

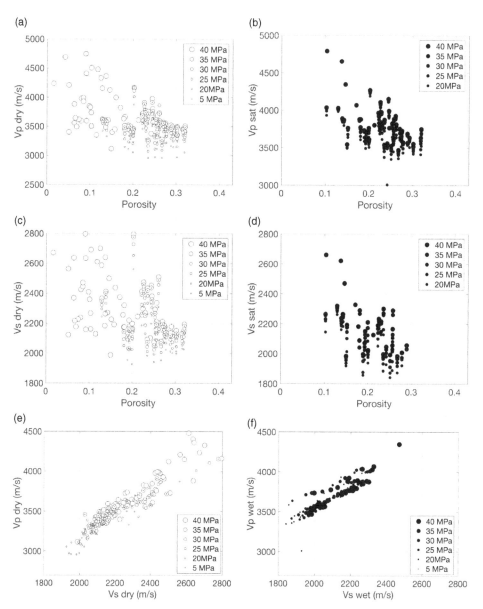

Figure A.1.3 Ultrasonic velocities of dry and water-saturated shaly sandstones from the Oseberg formation.

Weakly Consolidated Sandstones from the Troll Field. Ultrasonic velocities plotted versus porosity and effective pressure for dry and saturated rocks. J.P. Blangy (1992).

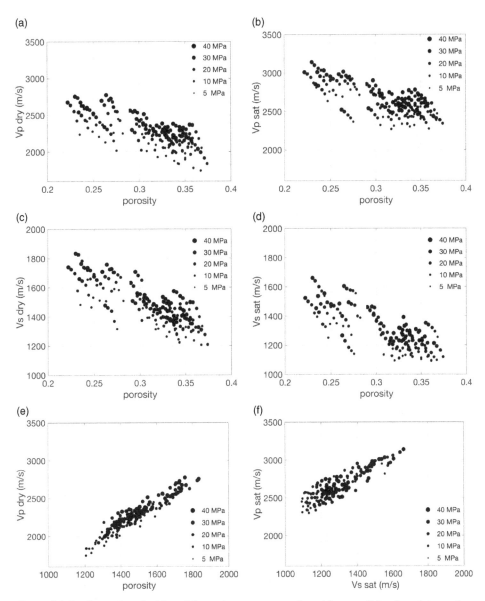

Figure A.1.4 Ultrasonic velocities of dry and water-saturated weakly consolidated sandstones from the Troll field.

Shaly Sandstones. Ultrasonic V_P and Q_P versus porosity and clay. Water-saturated at 40 MPa effective pressure. T. Klimentos and C. McCann (1990).

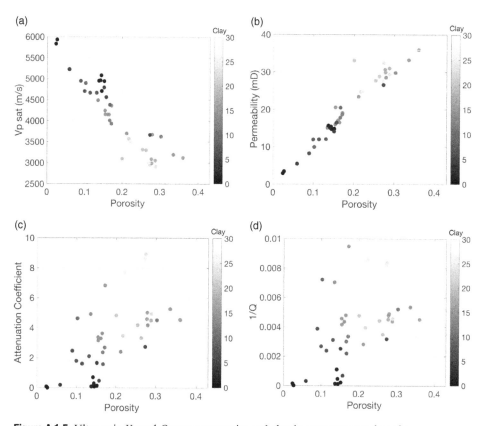

Figure A.1.5 Ultrasonic Vp and Qp versus porosity and clay in water-saturated sandstones.

Sand-Clay Mixtures. Porosity, permeability and dry and saturated ultrasonic V_P and V_S for water-saturated mixtures of clean Ottawa sand and kaolinite powder while uploading pressure. Labels on curves refer to effective pressure in MPa. Gas permeability is taken at nominally zero effective pressure. Yin (1992).

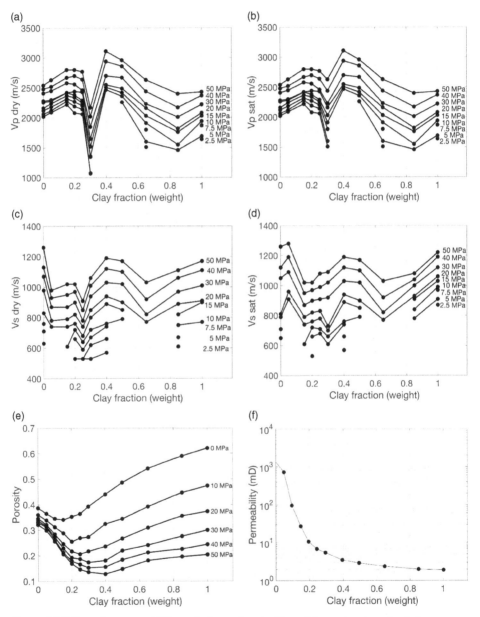

Figure A.1.6 Porosity, permeability, and ultrasonic velocities of dry and water-saturated unconsolidated mixtures of Ottawa sand and kaolinite powder.

Unconsolidated Pomponio Beach Sand. Ultrasonic velocities of dry and water-saturated unconsolidated sand. Arrows indicate loading and unloading cycles of effective pressure. Zimmer (2003).

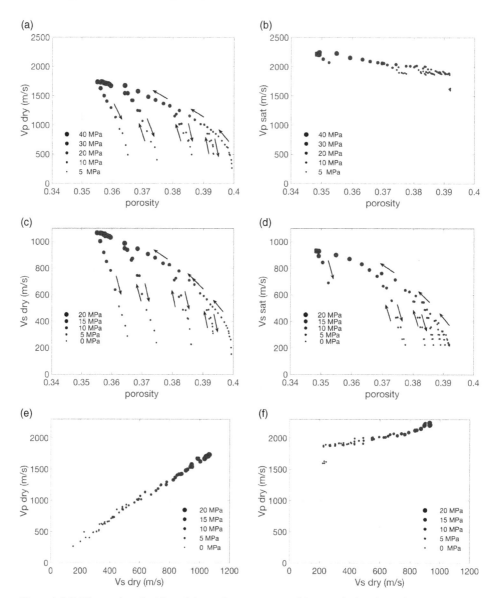

Figure A.1.7 Ultrasonic velocities of dry and water-saturated Pomponio beach sand.

Carbonates. Porosity, permeability, and ultrasonic velocity in water-saturated carbonates. Pore types according to the Choquette and Pray classification system are noted by the gray scale. IP: interparticle; WP: intraparticle; IX: intercrystal; VUG: vuggy; MO: Moldic. Dominant mineralogy (calcite or dolomite) is indicated by dot size. Weger (2006).

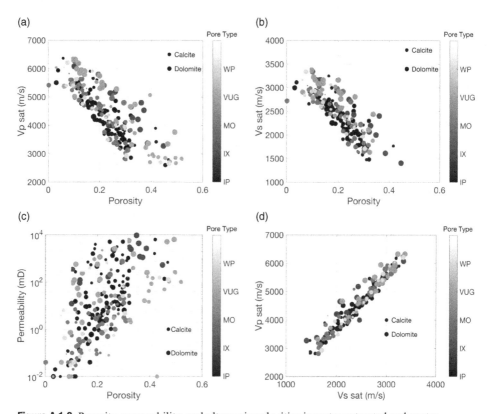

Figure A.1.8 Porosity, permeability, and ultrasonic velocities in water-saturated carbonates

Electrical Conductivity of Shaly Sandstones. Brine-saturated electrical conductivity. Vertical axis is effective electrical conductivity. Labels on curves indicate conductivity of different pore fluids. F is formation factor. Waxman and Smits (1968).

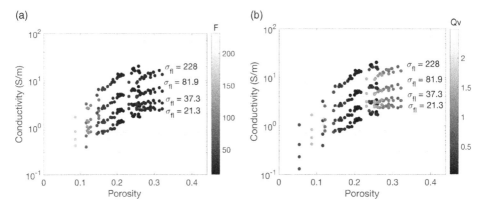

Figure A.1.9 Brine-saturated electrical conductivity in shaly sandstones.

Ultrasonic Measurements on Room-Dry Niobrara Shales. Ro is vitrinite reflectance. Epsilon, gamma, and delta are the Thomsen (1986) parameters. TOC is total organic content. Data compiled from: Vernik (1993, 1994, 2016) and Vernik and Landis (1996).

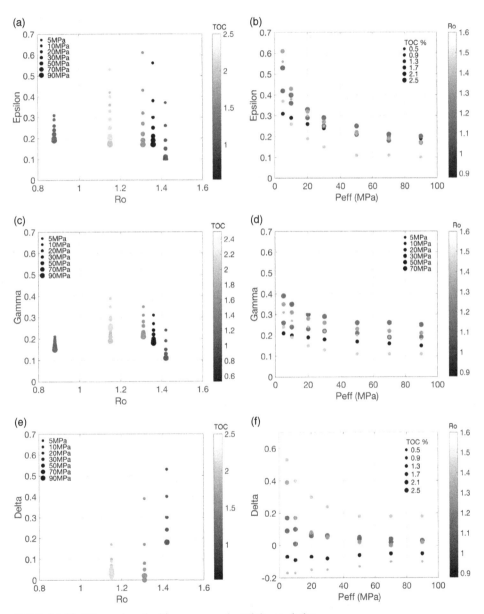

Figure A.1.10 Ultrasonic velocities on room-dry Niobrara shales.

Ultrasonic Measurements on Room-Dry Bakken Shales. Ro is vitrinite reflectance. Epsilon, gamma, and delta are the Thomsen (1986) parameters. TOC is total organic content. Data compiled from: Vernik (1993, 1994, 2016), Vernik and Landis (1996).

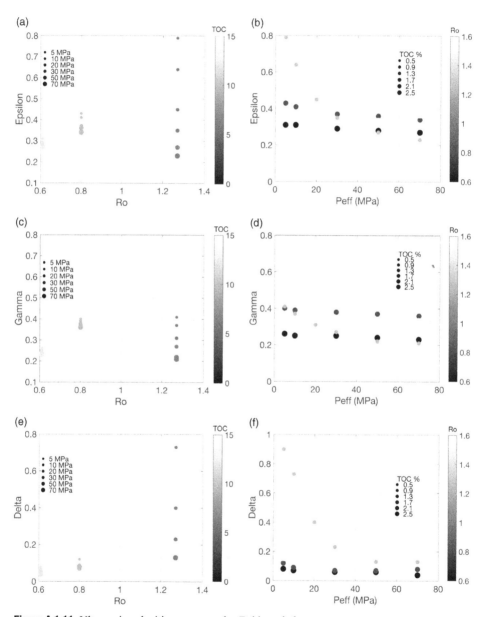

Figure A.1.11 Ultrasonic velocities on room-dry Bakken shales.

Various Mudrocks. Ultrasonic velocity measurements on room-dry mudrocks at high effective pressure (50–70 MPa). Kerogen is in volume fraction. Density is bulk density (g/cm^3). Epsilon, gamma, and delta are the Thomsen (1986) parameters. Data compiled by Vernik (2016) from sources: Vernik and Liu (1997), Sondergeld *et al.* (2000), Lo *et al.* (1986), Hornby (1998), Tosaya (1982), and Johnston and Christensen (1995).

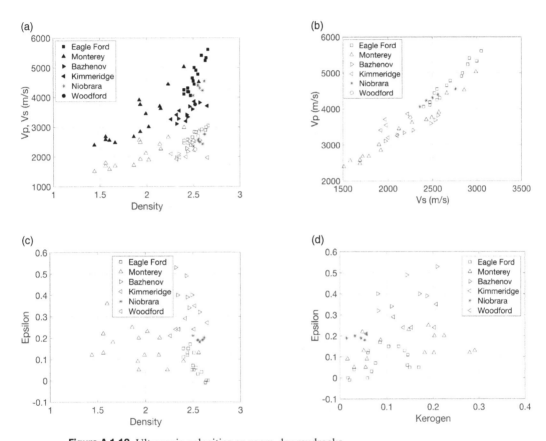

Figure A.1.12 Ultrasonic velocities on room-dry mudrocks.

(e)

(f)

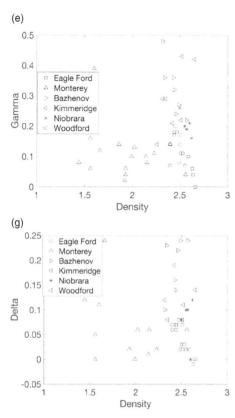

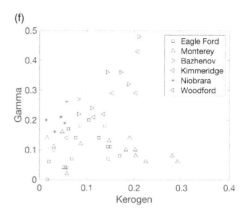

(g)

(h)

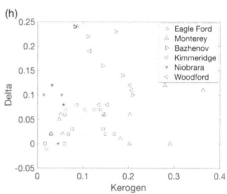

Figure A.1.12 (cont.)

Selected Wireline Data

Oil Well in Alaska. Sandstone/Shale. Top: Wireline curves. From left to right: gamma-ray (GR); water saturation (S_w); the total porosity as computed from the bulk density and pore-fluid properties; V_P (black for in-situ and blue for 100% wet conditions obtained using Gassmann's fluid substitution); and V_S (black for in-situ and blue for 100% wet conditions). Middle: Velocity versus porosity at in-situ conditions (top row) and 100% wet conditions (bottom row). Bottom: Porosity versus GR showing dispersed clay (V-shape) behavior. Color is water saturation (S_w). After Dvorkin, Gutierrez, and Grana (2014).

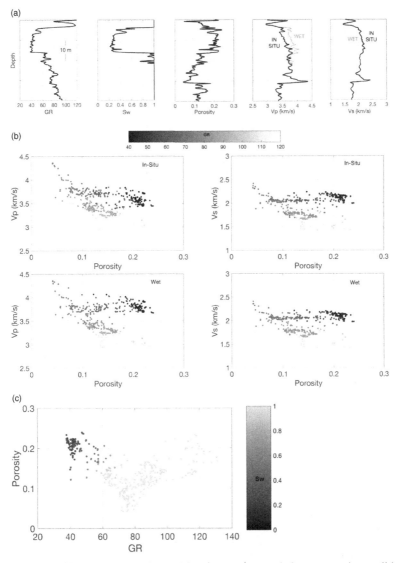

Figure A.1.13 Wireline data from oil-bearing sandstone-shale sequence in a well in Alaska.

Sandstone/Shale. Top: Gas well with residual gas saturation at the top of the interval. In the velocity tracks black curves are for the in-situ conditions while blue curves are for 100% wet conditions as computed using Gassmann's fluid substitution. Middle: Velocity versus porosity color-coded by GR. Top row for in-situ conditions. Bottom row for 100% wet conditions. Bottom: Effect of fluid substitution on the location of gas sand in the velocity versus porosity plane. Left: In-situ conditions. Right: 100% wet conditions. After Dvorkin, Gutierrez, and Grana (2014).

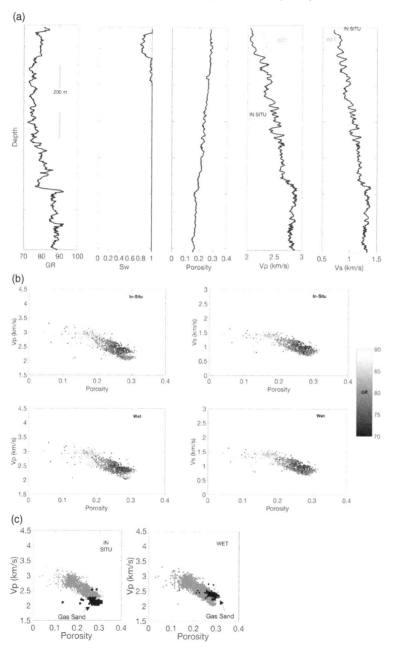

Figure A.1.14 Wireline data from a gas well.

Offshore Well from Clastic (Sandstone/Shale) Environment. Combined effect of lithology and compaction, wireline data. Velocity versus porosity color-coded by depth (left) and GR (right).

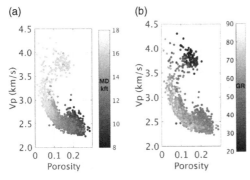

Figure A.1.15 Wireline data from an offshore well in clastic environment.

Gas On-Shore Well (Sandstone/Shale). Effect of lithology and porosity on velocity. The colorbar refers to both plots.

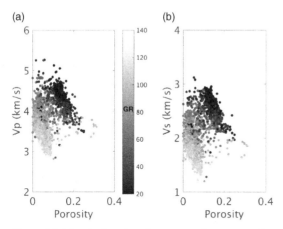

Figure A.1.16 Wireline data from an on-shore gas well.

Three tight gas sand on-shore wells. Effect of lithology and porosity on velocity. The colorbar refers to all plots. The datapoints in the plots are for 100% wet conditions obtained from in-situ measurements via fluid substitution.

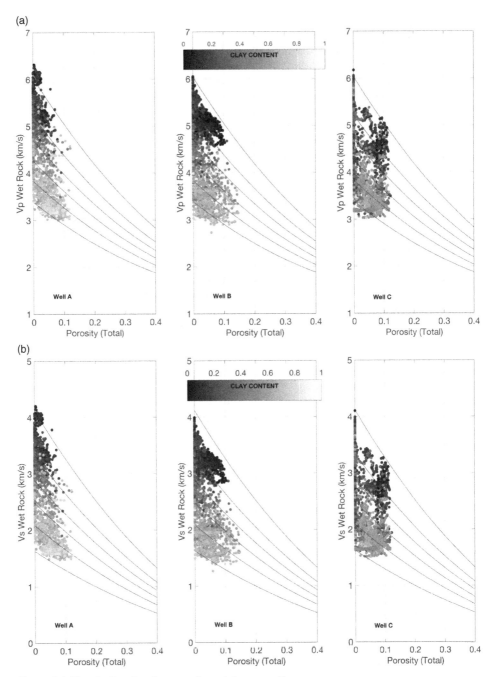

Figure A.1.17 Wireline data from on-shore tight gas wells.

Offshore sandstone/shale gas well. Effect of lithology and porosity on sonic velocity. The data are for 100% wet condition obtained via fluid substation on measured data. The difference between the neutron and density-derived porosity serves as clay indicator.

Figure A.1.18 Wireline data from an offshore gas well.

Barents Sea, sandstone/shale. Effect of porosity and mineralogy on V_P and V_S. The difference between the neutron and density-derived porosity serves as clay indicator. Top: in-situ conditions. Bottom: 100% wet conditions computed from the measured data via fluid substitution.

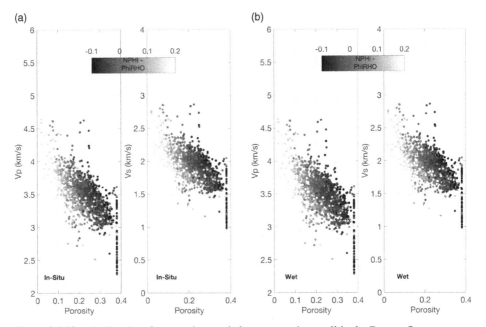

Figure A.1.19 Wireline data from sandstone-shale sequence in a well in the Barents Sea.

A.2   Conversions

Mass/Weight

1 g	$= 10^{-3}$ kg
1 kg	= 2.204623 lb
1 lb	= 0.4535924 kg
1 ton (USA)	= 2000 lb
	= 907.2 kg
1 ton (imperial)	= 2240 lb
	= 1016 kg
1 ton (metric)	= 1000 kg
	= 2204.622 lb
1 oz (avdp.)	= 28.3495 g
1 oz (troy)	= 31.03 48 g

Length

1 m	= 39.37 in
	= 3.2808399 ft
1 cm	= 0.3937 in
	= 0.032808399 ft
	= 0.01 m
1 in	= 2.540005 cm
	= 0.02540005 m
1 ft	= 30.48006 cm
	= 0.3048006 m
1 km	= 0.62137 mile
1 mile	= 1.60935 km
1 nautical mile	= 1.15077 miles
	= 1.852 km
1 μm	$= 10^{-6}$ m
	$= 10^{-4}$ cm
	$= 3.937 \times 10^{-5}$ in
1 Å	$= 10^{-10}$ m
	$= 10^{-8}$ cm
	$= 3.937 \times 10^{-9}$ in
1 light year	$= 9.4543 \times 10^{25}$ Å
	$= 9.454 \times 10^{12}$ km
1 astronomical unit	$= 1.4960 \times 10^{8}$ km
1 hair's breadth	= 0.01 cm
1 arm's length	= 70 cm
1 cubit (Roman)	= 44.4 cm
1 fathom	= 182.88 cm
1 gnat's eye	= 0.0125 cm
1 klick	= 1 km
1 Planck length	$= 1.616 \times 10^{-35}$ m

Velocity/Slowness/Transit Time

1 m/s	$= 3.2808$ ft/s
	$= 0.3048 \times 10^6/(1 \ \mu s/ft)$
1 ft/s	$= 0.3048$ m/s
1 mile/h	$= 1.609$ km/h
	$= 0.447$ m/s
	$= 1.47$ ft/s

Area

1 m^2	$= 10^4$ cm^2
	$= 10.764$ ft^2
	$= 1550.0$ in^2
	$= 2.47104 \times 10^{-4}$ acres
	$= 10^{-4}$ hectares
1 ft^2	$= 929.030$ cm^2
	$= 0.092903$ m^2
	$= 144$ in^2
	$= 2.2957 \times 10^{-5}$ acres
	$= 9.29 \times 10^{-6}$ hectares
1 acre	$= 4.0469 \times 10^7$ cm^2
	$= 4.0469 \times 10^3$ m^2
	$= 43 \ 560$ ft^2
	$= 0.0015625$ mile^2
	$= 0.40469$ hectares
1 hectare	$= 10^8$ cm^2
	$= 10^4$ m^2
	$= 1.0764 \times 10^5$ ft^2
	$= 3.8610 \times 10^{-3}$ mile^2
	$= 2.4711$ acre
1 km^2	$= 10^{10}$ cm^2
	$= 10^6$ m^2
	$= 100$ hectare
	$= 247.105$ acre
	$= 0.386 \ 10$ mile^2

Density

1 g/cm^3	$= 0.036127$ lb/in^3
	$= 62.42797$ lb/ft^3
	$= 1000$ kg/m^3
1 lb/in^3	$= 27.6799$ g/cm^3
	$= 27679.9$ kg/m^3
1 lb/ft^3	$= 0.016018$ g/cm^3

Volume

1 cm^3	= 0.0610238 in^3
1 in^3	= 16.38706 cm^3
1 liter	= 0.264172 gallons
	= 0.035315 ft^3
	= 1.056688 qt
	= 1000 cm^3
1 bbl	= 0.158987 m^3
	= 42 gallons
	= 5.6146 ft^3
1 m^3	= 6.2898106 bbl
1 ft^3	= 7.481 gallon (US)
	= 0.1781 bbl
1 gallon (US)	= 0.1337 ft^3
1 stere	=1 m^3

Force

1 N	= 1 kg m/s^2
1 lb	= 4.4482 N
1 dyn	= 10^{-5} N
1 kg force	= 9.80665 N
	= 9.80665 × 10^5 dyne

Pressure

1 atm (76 cm Hg)	= 1.01325 bar
	= 1.033227 kg force/cm^2
	= 14.695949 psi
1 bar	= 10^6 dyne/cm^2
	= 10^5 N/m^2
	= 0.1 MPa
1 kg force/cm^2	= 9.80665 × 10^5 dyne/cm^2
	= 0.96784 atm
1 psi	= 0.070307 kg-force/cm^2
	= 0.006895 MPa
	= 0.06895 bar
1 kpsi	= 70.307 kg-force/cm^2
	= 6.895 MPa
	= 68.95 bar
1 Pa	= 1 N/m^2
	= 1.4504 × 10^{-4} psi
1 MPa	= 10^6 Pa
	= 145.0378 psi
	= 10 bar
1 kbar	= 100 MPa

Pressure Gradients (or Mud Weight to Pressure Gradient)

1 psi/ft	$= 144$ lb/ft^3
	$= 19.24$ lb/gallons
	≈ 0.0225 MPa/m
	$= 22.5$ kPa/m
lb/gallon	$= 0.052$ psi/ft

Mud Density to Pressure Gradient

1 psi/ft$\Leftrightarrow$2.31 g/cm^3

Viscosity

1 Poise	$= 1$ dyne s/cm^2
1 cP	$= 0.01$ Poise
1 Pa s	$= 10$ Poise
	$= 1000$ cPoise

Permeability

1 Darcy	$= 0.986923 \times 10^{-12}$ m^2
	$= 0.986923$ μm^2
	$= 0.986923 \times 10^{-8}$ cm^2
	$= 1.06 \times 10^{-11}$ ft^2

Gas–Oil Ratio

1 liter/liter $= 5.615$ ft^3/bbl

Units of Electromagnetism

electric charge	coulomb, $A \cdot s$	Q
electric current	ampere, A	I
electric current density	ampere/meter^2, A/m^2	J
resistance, impedance, reactance	ohm, Ω, $kg \cdot m^2 \cdot s^{-3} \cdot A^{-2}$	R, Z, X
resistivity	ohm-meter, $\Omega \cdot m$, $kg \cdot m^3 \cdot s^{-3} \cdot A^{-2}$	ρ
electric capacitance	farad, $kg^{-1} \cdot m^{-2} \cdot s^4 \cdot A^2$	C
electric field strength	volt/meter, $kg \cdot m \cdot A^{-1} \cdot s^{-3}$	**E**
electric displacement field	coulomb/meter^2, $m^{-2} \cdot s \cdot A$	**D**
permittivity	farad/meter, $kg^{-1} \cdot m^{-3} \cdot A^2 \cdot s^4$	ε
electric susceptibility	dimensionless	X_e

conductance, admittance	siemens, Ω^{-1}, $kg^{-1} \cdot m^{-2} \cdot A^2 \cdot s^3$	G, Y
electric conductivity	siemens/meter $kg^{-1} \cdot m^{-3} \cdot A^2 \cdot s^3$	σ, κ
magnetic flux density	tesla, $kg \cdot s^{-2} \cdot A^{-1}$	**B**
magnetic flux	weber, $kg \cdot m^2 \cdot s^{-2} \cdot A^{-1}$	ϕ
magnetic field strength	ampere/meter, $A \cdot m^{-1}$	**H**
inductance	henry, $kg \cdot m^2 \cdot A^{-2} \cdot s^{-2}$	L
magnetic permeability	henry/meter, $kg \cdot m \cdot A^{-2} \cdot s^{-2}$	μ
magnetic susceptibility	dimensionless	χ

Other Conversions

1 year	$= 3.1536 \times 10^7$ s
1 flick	$= 1/705{,}600{,}000$ s
1 erg	$= 10^{-7}$ m^2 kg/s^2
1 joule (J)	$= 10^7$ erg
	$= 1$ m^2 kg/s^2
1 calorie (cal)	$= 4.186$ J
	$= 4.186 \times 10^7$ erg
1 watt	$= 1$ m^2 kg/s^3
1 kilocalorie	$= 4186.8$ m^2 kg/s^2
1 BTU	$= 252$ cal
	$= 1054$ J
1 eV	$= 1.60 \times 10^{-19}$ J
1 weber/m^2	$= 1$ tesla
	$= 10^4$ gauss
1 siemen (S)	$= \Omega^{-1} = 1$ mho $= 1$ amp/volt
1 radian (rad)	$= 57.30°$
1 second of arc	$= 4.848 \times 10^{-6}$ rad
1 ppm	$= 1$ mg per liter of water

A.3   Physical Constants

Astronomical Constants

Mass of the Sun	1.989×10^{33} g
Mass of the Earth	5.976×10^{27} g
Equatorial radius of the Earth	$6.378\ 16 \times 10^8$ cm
Polar radius of the Earth	$6.356\ 78 \times 10^8$ cm
Gravitational acceleration at equator	978.03 cm s^{-2}
Gravitational acceleration at pole	983.20 cm s^{-2}

Fundamental Constants

Gravitational constant, G	6.670×10^{-8} dyne cm^2 g^{-2}
Speed of light, c (in vacuum)	2.998×10^{10} cm s^{-1}
Planck length	1.616×10^{-35} m
Planck time	5.391×10^{-44} s
Gas constant, R	8.314×10^{7} erg K^{-1} mol^{-1}
	0.082057 L atm K^{-1} mol^{-1}
Avogadro's number, N	6.022×10^{23} mol^{-1}
Planck's constant, h	6.625×10^{-27} erg K^{-1}
Boltzmann's constant	1.3805×10^{-16} erg K^{-1}
Mass of an electron	9.110×10^{-28} g
Mass of a proton	1.673×10^{-24} g
Charge of an electron	1.602×10^{-19} C
Electron gyromagnetic ratio, γ_e	$1.760859770 \times 10^{11}$ s^{-1} T^{-1}
Nucleus gyromagnetic ratio, γ	
^{1}H gyromagnetic ratio, γ	$2\pi \times 42.576 \times 10^{6}$ s^{-1} T^{-1}
^{3}He gyromagnetic ratio, γ	$-2\pi \times 32.434 \times 10^{6}$ s^{-1} T^{-1}
^{14}N gyromagnetic ratio, γ	$2\pi \times 3.0766 \times 10^{6}$ s^{-1} T^{-1}
^{17}O gyromagnetic ratio, γ	$-2\pi \times 5.7716 \times 10^{6}$ s^{-1} T^{-1}
permittivity of free space	$8.8542 \cdot 10^{-12}$ farad/m
magnetic permeability of free space	$4\pi \cdot 10^{-7}$ henry/m

Dimensionless Numbers

Bagnold number: Dimensionless ratio of grain collision stresses to viscous fluid stress in flow of granular media. $Ba = \rho d^2 \lambda^{1/2} \dot{\gamma}/\eta$, where ρ is the particle density, d is the particle diameter, λ is the linear concentration, $\dot{\gamma}$ is the shear rate, η is the dynamic viscosity of inter-grain fluid.

Blake number: Dimensionless ratio of inertial to viscous forces in fluid flow through granular media. $B = V\rho D_h/\eta(1 - \phi)$, where V = flow velocity; ρ = fluid density; η = dynamic viscosity; ϕ = porosity; D_h = hydrodynamic diameter.

Bond number: Dimensionless ratio of buoyant forces to capillary forces for fluids in porous media. $Bo = \Delta\rho g L^2/\gamma$, where $\Delta\rho$ = difference in density of the two fluid phases, g= gravitational acceleration; L = characteristic length; γ = surface tension.

Capillary number: Dimensionless number representing the relative influence of viscous forces and surface tension at an interface between two fluids: $Ca = \eta V/\gamma$, where η = dynamic viscosity; V = characteristic velocity; γ = surface tension.

Damköhler number: Dimensionless number often used to describe rates in reactive transport: $Da = t_r/t_c$, where t_r is the reaction rate, and t_c is the convective mass transfer rate.

Deborah number: Dimensionless number often used in rheology. $De = t_c/t_p$ where t_c is the relaxation time of a material to applied stress or displacement, and t_p is the time of the experiment or observation.

Dukhin number: Dimensionless ratio of electric surface conductivity to the electric bulk conductivity in heterogeneous systems. $Du = \kappa^\sigma / K_m d$, where κ^σ is the surface conductivity, K_m is the bulk electrical conductivity, and d is the particle size.

Ekman number: Dimensionless ratio of viscous and Coriolis forces. $Ek = v/2D^2\Omega \sin\varphi$, where D is a characteristic length scale, v is the kinematic viscosity, and $\Omega \sin\varphi$ is the Coriolis frequency.

Euler number: Dimensionless ratio of stream pressure to inertial forces. $Eu = \Delta p/\rho V^2$, where Δp is the pressure drop, ρ is the fluid density, and V is the characteristic flow velocity.

Froude number: Dimensionless ratio of a body's inertial to gravitational forces. $Fr = V/\sqrt{gl}$, where V is a characteristic velocity, g is the acceleration of gravity, and l is a characteristic length.

Galilei number: Ratio of gravitational to viscous forces. $Ga = gL^3/v^2$, where g is the gravitational acceleration, L is a characteristic length, and v is the kinematic viscosity.

Knudsen number: Dimensionless ratio of mean free path to characteristic dimension in gas dynamics. $Kn = \lambda/L$, where $\lambda =$ mean free path; $L =$ characteristic length.

Nusselt number: Dimensionless ratio of convective to conductive heat transfer. $Nu = hd/k$, where h is the convective heat transfer coefficient, d is a characteristic length, and k is the thermal conductivity of the fluid.

Péclet number: Dimensionless number relevant to transport in continuous media: $Pe = t_a/t_m$, where t_a is the advective mass transport rate, and t_m is the diffusive mass transport rate.

Prandtl number: Dimensionless ratio of viscous diffusion rate to thermal diffusion rate. $Pr = c_p\eta/k$, where c_p is the specific heat, η is the dynamic viscosity, and k is the thermal conductivity.

Rayleigh number: Dimensionless number associated with buoyancy-driven flow, also known as free convection. Ra is the ratio (timescale for thermal transport via diffusion)/ (timescale for thermal transport via flow at speed u). $Ra = \rho g \beta \Delta T \; l^3/\eta\alpha$, where g is the acceleration of gravity, ρ is the density, β is the thermal expansion coefficient, l is the characteristic distance, ΔT is the temperature difference across the characteristic length, α is the thermal diffusivity, and η is the viscosity.

Reynolds number: Dimensionless number comparing the relative influence of inertial and viscous forces in fluid flow: $Re = \rho VL/\eta$, where $\rho =$ density; $\eta =$ dynamic viscosity; $V =$ characteristic velocity relative to the object; $L =$ characteristic length.

Rossby number: Ratio of inertial force to Coriolis force. $Ro = U/Lf$, where U is the characteristic velocity, L is the characteristic length scale, and $f = 2\Omega \sin\phi$ is the Coriolis frequency, Ω is the angular frequency of planetary rotation, and ϕ is the latitude.

Weber number: Dimensionless ratio of inertia to surface tension. $We = \rho V^2 l / \sigma$, where ρ is the fluid density, V is its velocity, l is a characteristic length, and σ is the surface tension.

A.4   Moduli and Density of Common Minerals

Table A.4.1 summarizes isotropic moduli, densities, and velocities for many common minerals. The data have been taken from a variety of sources, but we drew heavily from extensive compilations by Ellis *et al.* (1988), Blangy (1992), Castagna *et al.* (1993), and Carmichael (1989).

Caution: since all of the individual mineral crystals are anisotropic, the values reported here for bulk and shear moduli, V_P and V_S are estimates for an isotropic

Table A.4.1 *Moduli, densities, and velocities of common minerals*

Mineral	Bulk modulus (GPa)	Shear modulus (GPa)	Density (g/cm^3)	V_P (km/s)	V_S (km/s)	Poisson's ratio	References
Olivines							
Forsterite	129.8	84.4	3.32	8.54	5.04	0.23	[1–3]
"Olivine"	130	80	3.32	8.45	4.91	0.24	[54]
Garnets							
Almandine	176.3	95.2	4.18	8.51	4.77	0.27	[1]
Zircon	19.8	19.7	4.56	3.18	2.08	0.13	[4,7]
Epidotes							
Epidote	106.5	61.1	3.40	7.43	4.24	0.26	[9]
Dravite	102.1	78.7	3.05	8.24	5.08	0.19	[4–6]
Pyroxenes							
Diopside	111.2	63.7	3.31	7.70	4.39	0.26	[8,9]
Augite	94.1	57.0	3.26	7.22	4.18	0.25	[9]
	13.5	24.1	3.26	3.74	2.72	0.06	[10]
Sheet silicates							
Muscovite	61.5	41.1	2.79	6.46	3.84	0.23	[11]
	42.9	22.2	2.79	5.10	2.82	0.28	[55]
	52.0	30.9	2.79	5.78	3.33	0.25	[24]
Phlogopite	58.5	40.1	2.80	6.33	3.79	0.22	[11]
	40.4	13.4	2.80	4.56	2.19	0.35	[55]
Biotite	59.7	42.3	3.05	6.17	3.73	0.21	[11]
	41.1	12.4	3.05	4.35	2.02	0.36	[55]
Clays							
Kaolinite	1.5	1.4	1.58	1.44	0.93	0.14	[10]
"Gulf clays" (Han)[a]	25	9	2.55	3.81	1.88	0.34	[50,53]
"Gulf clays" (Tosaya)[a]	21	7	2.6	3.41	1.64	0.35	[49,53]
Mixed clays[a]				3.40	1.60		[49]

Table A.4.1 (*cont.*)

Mineral	Bulk modulus (GPa)	Shear modulus (GPa)	Density (g/cm³)	V_P (km/s)	V_S (km/s)	Poisson's ratio	References
				3.41	1.63		[50]
Montmorillonite–illite mixture[a]				3.60	1.85		[51]
Illite[a]				4.32	2.54		[52]
Framework silicates							
Perthite	46.7	23.63	2.54	5.55	3.05	0.28	[54]
Plagioclase feldspar (Albite)	75.6	25.6	2.63	6.46	3.12	0.35	[10]
"Average" feldspar	37.5	15.0	2.62	4.68	2.39	0.32	
Quartz	37	44.0	2.65	6.05	4.09	0.08	[54]
	36.6	45.0	2.65	6.04	4.12	0.06	[14–16]
	36.5	45.6	2.65	6.06	4.15	0.06	[44]
	37.9	44.3	2.65	6.05	4.09	0.08	[47]
Quartz with clay (Han)	39	33.0	2.65	5.59	3.52	0.17	[50,53]
Oxides							
Corundum	252.9	162.1	3.99	10.84	6.37	0.24	[17,18]
Hematite	100.2	95.2	5.24	6.58	3.51	0.14	[19,20]
	154.1	77.4	5.24	7.01	3.84	0.28	[10,12]
Rutile	217.1	108.1	4.26	9.21	5.04	0.29	[21,22]
Spinel	203.1	116.1	3.63	9.93	5.65	0.26	[1]
Magnetite	161.4	91.4	5.20	7.38	4.19	0.26	[4,23,24]
	59.2	18.7	4.81	4.18	1.97	0.36	[10]
Hydroxides							
Limonite	60.1	31.3	3.55	5.36	2.97	0.28	[10]
Sulfides							
Pyrite	147.4	132.5	4.93	8.10	5.18	0.15	[25]
	138.6	109.8	4.81	7.70	4.78	0.19	[10]
Pyrrhotite	53.8	34.7	4.55	4.69	2.76	0.23	[10]
Sphalerite	75.2	32.3	4.08	5.38	2.81	0.31	[26,27]
Sulfates							
Barite	54.5	23.8	4.51	4.37	2.30	0.31	[14]
	58.9	22.8	4.43	4.49	2.27	0.33	[28]
	53.0	22.3	4.50	4.29	2.22	0.32	[7]
Celestite	81.9	21.4	3.96	5.28	2.33	0.38	[4]
	82.5	12.9	3.95	5.02	1.81	0.43	[28]
Anhydrite	56.1	29.1	2.98	5.64	3.13	0.28	[30]
	62.1	33.6	2.96	6.01	3.37	0.27	[48]
Gypsum	42.5	15.7	2.35	5.80			[29,56,57]
Polyhalite			2.78	5.30			[31]
Carbonates							
Calcite	76.8	32.0	2.71	6.64	3.44	0.32	[14]
	63.7	31.7	2.70	6.26	3.42	0.29	[32]
	70.2	29.0	2.71	6.34	3.27	0.32	[33]

Table A.4.1 (*cont.*)

Mineral	Bulk modulus (GPa)	Shear modulus (GPa)	Density (g/cm³)	V_P (km/s)	V_S (km/s)	Poisson's ratio	References
	74.8	30.6	2.71	6.53	3.36	0.32	[43]
	68.3	28.4	2.71	6.26	3.24	0.32	[44]
Siderite	123.7	51.0	3.96	6.96	3.59	0.32	[34]
Dolomite	94.9	45.0	2.87	7.34	3.96	0.30	[35]
	69.4	51.6	2.88	6.93	4.23	0.20	[13]
	76.4	49.7	2.87	7.05	4.16	0.23	[45]
Aragonite	44.8	38.8	2.92	5.75	3.64	0.16	[19,20,36]
Natronite	52.6	31.6	2.54	6.11	3.53	0.26	[53, 54]
Phosphates							
Hydroxyapatite	83.9	60.7	3.22	7.15	4.34	0.21	[4]
Fluorapatite	86.5	46.6	3.21	6.80	3.81	0.27	[37]
Fluorite	86.4	41.8	3.18	6.68	3.62	0.29	[38,39]
Halite	24.8	14.9	2.16	4.55	2.63	0.25	[14,40–42]
			2.16	4.50	2.59		[46]
Sylvite	17.4	9.4	1.99	3.88	2.18	0.27	[40]
Organic							
Kerogen	2.9	2.7	1.3	2.25	1.45	0.14	[53,54]
Zeolites							
Narolite	46.6	28.0	2.25	6.11	3.53	0.25	[53,54]

Note:

[a] Clay velocities were interpreted by extrapolating empirical relations for mixed lifhologies to 100% clay (Castagna *et al.,* 1993).

References: [1] Verma (1960); [2] Graham and Barsch (1969); [3] Kumazawa and Anderson (1969); [4] Hearmon (1956); [5] Mason (1950); [6] Voigt (1890); [7] Huntington (1958); [8] Ryzhova *et al.* (1966); [9] Alexandrov *et al.* (1964); [10] Woeber *et al.* (1963); [11] Alexandrov and Ryzhova (1961a); [12] Wyllie *et al.* (1956); [13] *Log Interpretation Charts* (1984); [14] Simmons (1965); [15] Mason (1943); [16] Koga *et al.* (1958); [17] Wachtman *et al.* (1960); [18] Bernstein (1963); [19] Hearmon (1946); [20] Voigt (1907); [21] Birch (1960a); [22] Joshi and Mitra (1960); [23] Doraiswami (1947); [24] Alexandrov and Ryzhova (1961b); [25] Simmons and Birch (1963); [26] Einspruch and Manning (1963); [27] Berlincourt *et al.* (1963); [28] Seshagiri Rao (1951); [29] Tixier and Alger (1967); [30] Schwerdtner *et al.* (1965); [31] *Formation Evaluation Data Handbook* (1982); [32] Bhimasenacher (1945); [33] Peselmck and Robie (1963); [34] Christensen (1972); [35] Humbert and Plicque (1972); [36] Birch (1960b); [37] Yoon and Newnham (1969); [38] Bergmann (1954); [39] Huffman and Norwood (1960); [40] Spangenburg and Haussuhl (1957); [41] Lazarus (1949); [42] Papadakis (1963); [43] Dandekar (1968); [44] Anderson and Liebermann (1966); [45] Nur and Simmonds (1969b); [46] Birch (1966); [47] McSkimin *et al.* (1965); [48] Rafavich *et al.* (1984); [49] Tosaya (1982); [50] Han *et al.* (1986); [51] Castagna *et al.* (1985); [52] Eastwood and Castagna (1986); [53] Blangy (1992); [54] Carmichael (1989); [55] Ellis *et al.* (1988); [56] Choy *et al.* (1979); [57] Bhalla *et al.* (1984).

polycrystalline assemblage (see Section 4.21). The effective isotropic properties depend also on the crystal sizes and shapes and are therefore not unique. The values typically reported are Voigt–Reuss–Hill effective properties.

Tables A.4.2a–A.4.2g summarize densities and anisotropic elastic constants for many common minerals. The data have been taken from an extensive compilation by Bass (1995).

Table A.4.2a *Elastic moduli of cubic crystals at room pressure and temperature*

Material	ρ (Mg/m^3)	Subscript on c_{ij} (GPa)			K_s (GPa)	G (GPa)	Source
		11	44	12			
Elements, metallic compounds							
Au, gold	19.283	191	42.4	162	171.7	27.6	e
Ag, silver	10.500	122	45.4	92	102.0	29.2	e
C, diamond	3.512	1079	578	124	443.0	535.7	h
Cu, copper	8.932	169	75.3	122	137.3	46.9	d
Fe, α-iron	7.874	230	117	135	166.7	81.5	e
Binary oxides							
CaO, lime	3.346	224	80.6	60	114.7	81.2	d, 1
Fe$_{0.92}$O, wustite	5.681	245.7	44.7	149.3	181.4	46.1	m
MgO, periclase	3.584	294	155	93	160.0	130.3	d
Spinel-structured oxides							
Fe$_3$O$_4$, magnetite	5.206	270	98.7	108	162.0	91.2	e
MgAl$_2$O$_4$, spinel	3.578	282.9	154.8	155.4	197.9	108.5	u
Sulfides							
FeS$_2$, pyrite	5.016	361	105.2	33.6	142.7	125.7	k
PbS, galena	7.597	127	23	24.4	58.6	31.9	e
ZnS, spalerite	4.088	102	44.6	64.6	77.1	31.5	e
Binary Halides							
CaF$_2$, fluorite	3.181	165	33.9	47	86.3	42.4	d
NaCl, halite	2.163	49.1	12.8	12.8	24.9	14.7	d
KC1, sylvite	1.987	40.5	6.27	6.9	18.1	9.4	d

Table A.4.2b *Elastic moduli of hexagonal crystals at room pressure and temperature (Bass, 1995)*

Material	ρ (g/cm^3)	Subscript on c_{ij} (GPa)					K_s (GPa)	G (GPa)	Source
		11	33	44	12	13			
Be$_3$Al$_2$Si$_6$O$_{18}$, beryl	2.698	308.5	283.4	66.1	128.9	118.5	181	79.2	v
C, graphite	2.26	1060	36.5	0.3	180	15	161	109.3	b
Ice (257 K)	–	13.5	14.9	3.09	6.5	5.9	8.72	3.48	d
Ice (270 K)	0.9175	13.7	14.7	2.96	6.97	5.63	8.73	3.40	c

Table A.4.2c *Elastic moduli of trigonal crystals (six moduli) at room pressure and temperature (Bass, 1995)*

Material	ρ (g/cm^3)	Subscript on c_{ij} (GPa)						K_s (GPa)	G (GPa)	Source
		11	33	44	12	13	14			
Al$_2$O$_3$, corundum	3.982	497	501	146.8	162	116	−21.9	253.5	163.2	i
CaCO$_3$, calcite	2.712	144	84.0	33.5	53.9	51.1	−20.5	73.3	32.0	d
Fe$_2$O$_3$, hematite	5.254							206.6	91.0	g
SiO$_2$,s α-quartz	2.648	86.6	106.1	57.8	6.7	12.6	−17.8	37.8	44.3	d
Tourmaline	3.100	305.0	176.4	64.8	108	51	−6	127.2	81.5	j

Table A.4.2d *Elastic moduli of trigonal crystals (seven moduli) at room pressure and temperature (Bass, 1995)*

Material	ρ (g/cm^3)	Subscript on c_{ij} (GPa)							K_s (GPa)	G (GPa)	Source
		11	33	44	12	13	14	15			
Dolomite, CaMg(CQ$_3$)$_2$	3.795	205	113	39.8	71.0	57.4	−19.5	13.7	94.9	45.7	d, f

Table A.4.2e *Elastic moduli of tetragonal crystals (six moduli) at room pressure and temperature (Bass, 1995)*

Material	ρ (Mg/m^3)	Subscript on c_{ij} (GPa) 11	33	44	66	12	13	K_s (GPa)	G (GPa)	Source
SiO$_2$, stishovite	4.290	453	776	252	302	211	203	316	220	r
SiO$_2$, α-cristobalite	2.335	59.4	42.4	67.2	25.7	3.8	−4.4	16.4	39.1	t

Table A.4.2f *Elastic moduli of orthorhombic crystals (nine moduli) at room pressure and temperature (Bass, 1995)*

Material	ρ (g/cm^3)	Subscript on c_{ij} (GPa) 11	22	33	44	55	66	12	13	23	K_s (GPa)	G (GPa)	Source
				Perovskites									
MgSiO$_3$	4.108	515	525	435	179	202		175	117	117	139	246.4	184.2	s
NaMgF$_3$	3.058	126	147	143	47	45		50	50	45	43	75.7	46.7	w
				Pyroxenes									
Enstatite, MgSiO$_3$	3.198	225	178	214	78	76		82	72	54	53	107.8	75.7	q
Ferosilite, FeSiO$_3$	4.002	198	136	175	59	58		49	84	72	55	101	52	a
				Other silicates									
Al$_2$SiO$_5$, andalusite	3.145	233	289	380	100	88	112	98	116	81	162	99.1	o
Al$_2$SiO$_5$, silimanite	3.241	287	232	388	122	81	89	159	83	95	170.8	91.5	o
				Sulfates, carbonates									
Sulfur	2.065	24	21	48	4.3	8.7	7.6	13.3	17.1	15.9	19.1	6.7	d
BaSO$_4$, barite	4.473	89	81	107	12	28	27	48	32	30	55.0	22.8	d
CaSO$_4$, anhydrite	2.963	93.8	185	112	33	27	9.3	17	15	32	54.9	29.3	d
CaCO$_3$, aragonite	2.930	160	87	85	41	26	43	37	1.7	16	46.9	38.5	d

Table A.4.2g *Elastic moduli of monoclinic crystals at room pressure and temperature (Bass, 1995)*

Material	P(g/cm3)	Subscript on c_{ij} (GPa)													K_s (GPa)	G (GPa)	Source
		11	22	33	44	55	66	12	13	23	15	25	35	46			
		Feldspars															
NaAlSi$_3$O$_8$, albite		74	131	128	17	30	32	36	39	31	−6.6	−13	−20	−2.5	56.9	28.6	d
CaAl$_2$Si$_2$O$_8$, anorthite		124	205	156	24	40	42	66	50	42	−19	−7	−18	−1	84.2	39.9	e
labradorite		99	158	150	22	35	37	63	49	27	−2.5	−11	−12	−5.4	74.5	33.7	e
KAlSi$_3$O$_8$, microcline		67	169	118	14	24	36	45	27	20	−0.2	−12	−15	−1.9	55.4	28.1	e
KAlSi$_3$O$_8$, oligoclase		81	163	124	19	27	36	38	53	33	−16	−24	−6	−0.9	62.0	29.3	e
		Silicates															
SiO$_2$, ccesite	2.911	161	230	232	68	73	59	82	103	36	−36	2.6	−3.9	9.9	113.7	61.6	p
KAl$_3$Si$_3$O$_{10}$(OH)$_2$, muscovite	2.844	184	178	59	16	18	72	48	24	22	−2	3.9	1.2	0.5	58.2	35.3	n
		Sulfides, sulfates															
CaSO$_4$, gypsum	2.317	94.5	65.2	50.2	8.6	32	11	38	28	32	−11	6.9	−7.5	−1.1	42.5	15.7	d

References: a, Bass (1984); b, Blakslee *et al.* (1970); c, Gammon *et al.* (1980); d, Hearmon (1979); e, Hearmon (1984); f, Humbert and Plique (1972); g, Liebermann and Schreiber (1968); h, McSkimin and Bond (1972); i, Ohno *et al.* (1986); j, Ozkan and Jamieson (1978); k, Simmons and Birch (1963); 1, Soga (1967); m, Sumino *et al.* (1980); n, Vaughan and Guggenheim (1986); o, Verma (1960); p, Weidner and Carleton (1977); q, Weidner and Hamaya (1983); r, Weidner *et al.* (1982); s, Yeganeh-Haeri *et al.* (1989); t, Yeganeh-Haeri *et al.* (1992); u, Yoneda (1990); v, Yoon and Newham (1973); w, Zhao and Weidner (1993).

Polycrystalline Rock Salt

Rock salt is dominantly halite, but other common minerals can be present such as calcite, dolomite, anhydrite, gypsum, and potash. Gulf coast salt domes are generally more than 97% halite, with the major impurity being anhydrite. Average seawater contains approximately 38,450 ppm dissolved solids, of which about ¾ will precipitate as halite (Lorenz *et al.*, 1981).

Table A.4.3 *In situ measurements of dynamic properties of salt (Hume and Shakoor, 1981)*

	Winnfield Salt Dome, La (salt dome)	Gnome Drift, NM (bedded salt)
V_P	4370 m/s	4085 m/s
V_S	2550 m/s	2150 m/s
Poisson's Ratio	0.241	0.31
Shear Modulus	14.4 GPa	9.8 GPa
Bulk Modulus	23.0 GPa	21.7 GPa
Density	2160 kg/m^3	2020 kg/m^3

Table A.4.4 *Lab measurements of Elastic Modulus vs. pressure (Hume and Shakoor, 1981) Compacted fine pure NaCl powder 298 K. Compiled from Voronov, F.F. and S.B. Grigor'ev (1976).*

P (GPa)	E (GPa)	Shear modulus (GPa)	Bulk Modulus (GPa)	Density kg/m^3
0.0	36.91	14.69	25.23	2614
1.0	40.72	15.97	30.18	2248
2.0	44.02	17.05	35.07	2322
3.0	46.85	17.97	39.79	2388
4.0	49.30	18.75	44.24	2449
5.0	51.44	19.44	48.36	2506
6.0	53.39	20.08	52.08	2559
7.0	55.25	20.72	55.34	2610
8.0	57.16	21.39	58.10	2658

Table A.4.5 *Ultrasonic velocities reagent grade polycrystalline powder NaCl 300 K Bridgeman Anvil Device (Hume and Shakoor, 1981). Data from Frenkel et al. (1976).*

P (GPa)	Vp (m/s)	Vs (m/s)
2.5	5007	2679
3.0	5088	2696
3.5	5165	2710
4.0	5238	2722
4.5	5307	2732
5.0	5380	2750
5.5	5450	2768
6.0	5518	2785
6.5	5585	2804
7.0	5648	2820
7.5	5710	2835
8.0	5769	2849
9.0	5877	2867
10.0	5986	2894
11.0	6092	2923
12.0	6193	2950
13.0	6290	2976
15.0	6478	3033
17.0	6647	3076
19.0	6785	3083
21.0	6915	3088
23.0	7056	3123
25.0	7188	3154
27.0	7304	3169

Table A.4.6 *Ultrasonic velocities on rock salt (Hume and Shakoor, 1981). Data from Heard et al. (1975)*

P (GPa)	Vp m/s	Vs m/s	density
0.000	4078	2419	2.140
0.010	4310	2509	
0.020	4371	2591	2.143

Table A.4.6 (*cont.*)

P (GPa)	Vp m/s	Vs m/s	density
0.050	4419	2629	2.145
0.075	4430	2636	2.149
0.100	4453	2659	2.151
0.150	4463	2657	2.157
0.200	4493	2666	2.164
0.250	4509	2668	2.173
0.300	4538	2676	2.181
0.350	4530	2685	2.186
0.400	4549	2591	2.192

Table A.4.7 *Ultrasonic velocities (km/s) of polycrystalline Halite. Napoleonville Salt Dome, LA. 300 m depth. 5–15 mm crystals. Bulk density 2.165 g/cm3. Assumed porosity~0. (Yan et al., 2016).*

			Confining Pressure (MPa)					
		Wave mode 5		10	20	30	40	50
	23	P	4.637	4.643	4.644	4.647	4.651	4.654
	23	S	2.561	2.567	2.569	2.570	2.572	2.571
	53	P	4.595	4.599	4.603	4.607	4.610	4.612
	53	S	2.531	2.539	2.544	2.546	2.548	2.551
°C	83	P	4.556	4.562	4.566	4.569	4.572	4.575
	83	S	2.518	2.527	2.530	2530	2.534	2.535
	113	P	4.514	4.522	4.526	4.529	4.534	4.536
	113	S	2.491	2.502	2.509	2509	2.512	2.510
	143	P	4.480	4.488	4.493	4.497	4.501	4.504
	143	S	2.466	2.473	2.478	2.478	2.480	2.481

Empirical relation representing table values

$V_P = 4.6910 - 0.01918e^{-0.05164P} + 1.3265 \cdot 10^{-6}PT - 0.001707T + 2.3893 \cdot 10^{-6}T^2 -$

$V_S = 2.5830 - 0.03440e^{-0.2081P} + 0.9889 \cdot 10^{-6}PT - 0.0006058T + 0.96832 \cdot 10^{-6}T^2$

where P is in MPa, and T is in °C.

A.5   Properties of Mantle Minerals

Table A.5.1 shows density and elastic moduli of mantle minerals along with pressure and temperature derivatives. Table A.5.2 shows elastic moduli of olivine phases, along with pressure and temperature derivatives.

Table A.5.1 *Density and moduli of mantle minerals. From Duffy and Anderson (1989).*

Formula (name)	Density[a] g/cm^3	K_S(GPa)	G (GPa)	K'_S	G'_S	$\dot{K}_S$ GPa/K	$\dot{G}$ GPa/K
$(Mg,Fe)_2SiO_4$ (olivine)	$3.222+1.182\,X_{Fe}$	129[b]	$82-31\,X^b_{Fe}$	5.1[c]	1.8[c]	0.016[d]	0.013[e]
$(Mg,Fe)_2SiO_4$ (β-spinel)	$3.472+1.24\,X_{Fe}$	174[f]	$114-41\,X^g_{Fe}$	4.9	1.8	0.018	0.014
$(Mg,Fe)_2SiO_4$ (γ-spinel)	$3.548+1.30\,X_{Fe}$	184[h]	$119-41\,X^h_{Fe}$	4.8	1.8	0.017	0.014
$(Mg,Fe)SiO_3$ (orthopyroxene)	$3.204+0.799\,X_{Fe}$	104[i]	$77-24\,X^i_{Fe}$	5.0	2.0	0.012	0.011
$Ca(Mg,Fe)Si_2O_6$ (clinopyroxene)	$3.277^j+0.380\,X_{Fe}$	$113+7\,X^k_{Fe}$	$67-6\,X^k_{Fe}$	4.5[l]	1.7	0.013	0.010
$NaAlSi_2O_6$ (jadeite)	3.32^j	143[m]	84[m]	4.5	1.7	0.016	0.013
$(Mg,Fe)O$ (magnesiowustite)	$3.583+2.28\,X_{Fe}$	$163-8\,X^n_{Fe}$	$131-77\,X^n_{Fe}$	4.2[k]	2.5[k]	0.016[k]	0.024[k]
Al_2O_3 (corundum)	3.988^j	251[k]	162[k]	4.3[k]	1.8[k]	0.014[k]	0.019[k]
SiO_2 (stishovite)	4.289	316[k]	220[k]	4.0	1.8	0.027	0.018
$(Mg,Fe)_3Al_2Si_3O_{12}$ (garnet)	$3.562+0.758\,X_{Fe}$	$175+1\,X^k_{Fe}$	$90+8\,X^k_{Fe}$	4.9[k]	1.4[k]	0.021[k]	0.010[k]
$Ca_3(Al,Fe)_2Si_3O_{12}$ (garnet)	$3.595+0.265\,X^j_{Fe}$	$169-11\,X^o_{Fe}$	$104-14\,X^o_{Fe}$	4.9	1.6	0.016[p]	0.015[p]
$(Mg,Fe)SiO_3$ (ilmenite)	$3.810+1.10\,X_{Fe}$	212[q]	$132-41\,X^q_{Fe}$	4.3	1.7	0.017	0.017
$(Mg,Fe)SiO_3$ (perovskite)	$4.104+107\,X_{Fe}$	266[r]	153	3.9[r]	2.0	0.031	0.028
$CaSiO_3$ (perovskite)	4.13^s	227	125	3.9	1.9	0.027	0.023
$(Mg,Fe)_4Si_4O_{12}$ (majorite)	$3.518+0.973\,X_{Fe}$	$175+1\,X^t_{Fe}$	$90+8\,X^t_{Fe}$	4.9	1.4	0.021	0.010
$Ca_2Mg_2Si_4O_{12}$ (majorite)	3.53^u	165	104	4.9	1.6	0.016	0.015
$Na_2Al_2Si_4O_{12}$ (majorite)	4.00^v	200	127	4.9	1.6	0.016	0.015

K_S adiabatic bulk modulus

G shear modulus

K'_S pressure derivative of adiabatic bulk modulus

G' pressure derivative of shear modulus

$\dot{K}_S$ absolute value of temperature derivative of adiabatic bulk modulus

$\dot{G}$ absolute value of temperature derivative of shear modulus

X_{Fe} mole fraction of iron

[a] All densities from Jeanloz and Thompson (1983) except where indicated.

[b] Graham *et al.* (1982), Schwab and Graham (1983), Suzuki *et al.* (1983). Sumino and Anderson (1984), and Yeganeh-Haeri and Vaughan (1984).

[c] Schwab and Graham (1983) and Sumino and Anderson (1984).

[d] Suzuki et al. (1983) and Sumino and Anderson (1984).

[e] Schwab and Graham (1983), Suzuki et al. (1983), and Sumino and Anderson (1984).

[f] Sawamoto et al. (1984).

[g] Sawamoto et al. (1984). The effect of Fe on the modulus is from Weidner et al. (1984).

[h] Weidner et al. (1984).

[i] Sumino and Anderson (1984), Bass and Weidner (1984), and Duffy and Vaughan (1988).

[j] Robie et al. (1966).

[k] Sumino and Anderson (1984).

[l] Levien and Prewitt (1981).

[m] Kandelin and Weidner (1988a, 1988b).

[n] Sumino and Anderson (1984). The variation with Fe content is unclear. See Graham and Kim (1986) for a review.

[o] Halleck (1973) and Bass (1986). Linear variation of moduli between end members is assumed (Haniford and Weidner, 1985).

[p] Isaak and Anderson (1987).

[q] Weidner and Ito (1985). The variation with iron content is modeled by analogy with orthopyroxene.

[r] Knittle and Jeanloz (1987).

[s] Bass (1984).

[t] Inferred from Weidner et al. (1987).

[u] Akaogi et al. (1987).

[v] Estimated by Bass and Anderson (1984).

Table A.5.2 *Elastic properties of olivine phases (compiled by Duffy and Anderson, 1988)*

K_S (GPa)	G (GPa)	K'_S	G'_S	$\dot{K}_S$ GPa/K	$\dot{G}$ GPa/K	reference
			α phase			
129	82	5.1	1.8	0.016	0.013	Duffy and Anderson (1989)
129	81	5.2	1.8	0.014	0.014	Bina and Wood (1987)
129	81	5.2	1.79	0.014	0.013	Irifune (1987)
129	80	4.7	1.6	0.016	0.014	Weidner (1986)
			β phase			
174	114	4.9	1.8	0.018	0.014	Duffy and Anderson (1989)
		4.8	2.0	0.017	0.015	Anderson (1988)
174	114	4.8	1.8	0.016	0.023	Bina and Wood (1987)
174	114	4	1.1	0.011	0.011	Irifune (1987)
174	114	4.3	0.9	0.020	0.014	Weidner (1986)
			γ phase			
184	119	4.8	1.8	0.017	0.014	Duffy and Anderson (1989)
		5.0	2.0	0.017	0.015	Anderson (1988)
184	119	4.8	1.8	0.015	0.024	Bina and Wood (1987)
184	119	4.8	1.1	0.012	0.011	Irifune (1987)
184	119	4.3	0.9	0.020	0.014	Weidner (1986)
			Al_2MgO_4 Spinel			
197	108	4.9	0.5	0.016	0.009	Sumino and Anderson (1984)

A.6 Properties of Melts, Magma, and Igneous Rocks

Table A.6.1 shows data for velocities, densities, and bulk moduli of melts. Additional discussion of viscosity of melt and magma can be found in Section 8.10.

Table A.6.1 *Velocities, densities, and adiabatic bulk moduli of melts. Compiled by Bass (1995). Measurements with frequency shown were ultrasonic; others were shock wave.*

Composition	T K	Rho g/cc	$K_{S,\infty}$ GPa	V_P m/s	Freq MHz	Reference
Fe	2490	6.54	94.8	3808		A
	3950	5.54	52.4	3075		A
$CaAl_2Si_2O_8$ (An)	1833	2.56	20.6	2850	3.529	B
	1893		20.4		3.0	C
An^a	1923	2.55	17.9			D
$An_{36}Di_{64}$	1677		23.0			C
$An_{36}Di_{64}{}^a$	1673	2.61	24.2			E
$An_{50}Di_{50}$	1673	2.60	21.6	2885	3.635	B
	1573	2.61	22.1	2910	3.922	B
$An_{50}Ab_{50}$	1753	2.44	17.8	2850	3.858	B
$Ab_{50}Di_{50}$	1698	2.45	18.2	2735	3.662	B
	1598	2.46	19.3	2830	3.943	B
$Ab_{75}Di_{25}$	1753	2.39	16.4	2800	3.565	B
	1648	2.40	16.7	3400	3.833	B
$Ab_{33}An_{33}Di_{33}$	1698	2.49	19.5	2805	3.803	B
	1583	2.50	19.8	2880	3.944	B
$BaSi_2O_5$	1793	3.44	19.5	2390	3.906	B
	1693	3.47	20.2	2410	3.652	B
$CaSiO_3$	1836	2.65	27.1	3120	3.484	B
$CaTiSiO_5$	1753	2.96	19.9	2590	4.014	B
	1653	3.01	20.0	2580	4.013	B
$Cs_2Si_2O_5$	1693	3.14	6.4	1450	3.854	B
	1208	3.34	8.8	2345	4.023	B
$CaMgSi_2O_6$ (Di)^a	1773	2.61	22.4			D
Di	1758	2.60	24.2	3040	3.842	B
	1698	2.61	24.1	3020	3.83	B
$Fe_{1.22}Si_{0.89}O_3$	1693	3.48	19.2	2345	3.665	B
	1598	3.51	20.6	2450	3.680	B
Fe_2SiO_4	1653	3.71	21.4	2400	7.65	B
	1503	3.76	22.6	2450	8.67	B
$K_2Si_2O_5$	1693	2.16	10.3	2190	3.955	B
	1408	2.22	11.9	2600	3.951	B
K_2SiO_3	1698	2.10	7.5	1890	4.909	B
	1498	2.17	8.5	1970	5.242	B
$Li_2Si_2O_5$	1693	2.12	15.0	2670	4.100	B
	1411	2.17	16.3	2740	3.852	B
Li_2SiO_3	1543	2.08	20.7	3160	3.712	B

Table A.6.1 (*cont.*)

Composition	T K	Rho g/cc	$K_{S,\infty}$ GPa	V_P m/s	Freq MHz	Reference
MgSiO$_3$	1913	2.52	20.6	2860	4.040	B
NaCl	1094			1727	8.61	F
	1322			1540	8.61	F
(Na$_2$O)$_{33}$(Al$_2$O$_3$)$_6$(SiO$_2$)$_{61}$	1684		15.8[a]	2653	3.707	F
	1599		16.4[a]	2695	3.764	F
(Na$_2$O)$_{32}$(Al$_2$O$_3$)$_{15}$(SiO$_2$)$_{52}$	1690		18.6[a]	2835	5.558	F
Na$_2$Si$_2$O$_5$	1693	2.20	14.0	2525	3.934	B
	1408	2.26	16.2	2680	3.990	B
Na$_2$SiO$_3$	1573	2.22	15.7	2663	10.1	B
	1458	2.25	17.0	2752	8.4	B
Or$_{78}$An$_{22}$	1783	2.33	13.8	4300	3.836	B
	1598	2.35	14.1	5200	3.923	B
Or$_{61}$Di$_{39}$	1768	2.38	16.0	2795	3.656	B
	1578	2.40	16.5	3470	3.673	B
Rb$_2$Si$_2$O$_5$	1693	2.78	7.8	1678	3.945	B
	1408	2.88	9.9	2130	3.974	B
SrSi$_2$O$_5$	1758	3.02	19.6	2550	3.690	B
	1653	3.04	20.1	2570	3.833	B
Tholeitic Basalt	1708	2.65	17.9	2600	3.839	B
	1505	2.68	18.3	2610	3.909	B
Basalt-Andesite	1803	2.55	18.6	2700	3.790	B
	1503	2.59	19.4	2980	3.863	B
Andesite	1783	2.44	16.1	2775	3.827	B
	1553	2.46	16.6	3850	3.889	B
Ryolite	1803	2.29	13.0	4350	3.664	B
	1553	2.31	13.5	5280	3.723	B

Abbreviations: An: CaAl$_2$Si$_2$O$_8$; Di: CaMgSi$_2$O$_6$, Or: KAISi$_3$O$_8$; Ab: NaAISi$_3$O$_8$.
[a] From shock wave experiments.

A. Hixson, R.S., Winkler, M.A. and Hodgdon, M. L., 1990. Sound speed and thermophysical properties of liquid iron and nickel. *Physical Review B*, **42**, 6485–6491.
B. Rivers, M.L. and Carmichael, I.S.E., 1987. Ultrasonic studies of silicate melts. *Journal of Geophysical Research*, **92**, 9247–9270.
C. Secco, R.A., Manghnani, M.H. and Liu, T.C., 1991. The bulk modulus-attenuation-viscosity systematics of diopside-anorthite melts. *Geophysical Research Letters*, **18**, 93–96.
D. Rigden, S.M., Ahrens, T.J. and Stolper, E.M., 1989. High pressure equation of state of molten anorthite and diopside. *Journal of Geophysical Research*, **94**, 9508–9522.
E. Rigden, S.M., Ahrens, T.J. and Stolper, E.M., 1988. Shock compression of molten silicate: results for a model basaltic composition. *Journal of Geophysical Research*, **93**, 367–382.
F. Kress, V.C., Williams, Q. and Carmichael, I.S.E., 1988. Ultrasonic investigation of melts in the system Na$_2$O-Al$_2$O$_3$-SiO$_2$. *Geochimica et Cosmochimica Acta*, **52**, 283–293.

Table A.6.2 *Elastic Properties of Hawaiian Basalts (Manghnani and Woolard, 1965).*
Multiple values of V_P for a single sample indicate measurements in three mutual directions.

	V_P km/s	V_S km/s	Density g/cm^3	Bulk modulus GPa	Shear modulus GPa	Young's modulus GPa	Poisson's ratio
Olivine basalt	4.95	2.56	2.0	31.5	13.1	34.5	0.317
	4.63						
	4.82						
Olivine basalt	4.65	2.50	2.30	30.5	14.7	38.0	0.292
Olivine basalt	5.65						
	4.38	3.10	2.36	40.3	22.7	52.3	0.264
	5.47						
Olivine basalt	5.08	3.02	2.40	46.	21.8	55.6	0.296
(Ankaramite)							
Olivine basalt	5.52	2.76	2.60	52.7	19.8	52.7	0.330
Eclogite	6.06						
	5.82	2.94	2.81	62.9	24.3	64.5	0.328
	5.86						
Amphibolite	6.90						
	6.75	3.53	2.95	85.	36.7	96.3	0.312
	6.76						
Hawaiite	4.20	2.51	2.59	24.	16.3	40.	0.224
Trachyte	5.18	2.83	2.60	42.2	20.8	54.	0.298

Table A.6.3 *Compressional and shear wave velocities (from Christensen et al., 1980)*

			Velocity (km/s) at Various Pressures (kbar)							
	Wet Bulk density g/cm^3	Mode	0.2	0.4	0.6	0.8	1.0	2.0	4.0	6.0
Flow	2.735	P	5.58	5.66	5.70	5.73	5.76	5.88	5.99	6.04
basalt	2.735	S	2.92	2.95	2.97	2.99	3.01	3.08	3.20	3.25
Flow	2.831	P	6.00	6.09	6.14	6.18	6.22	6.33	6.45	6.56
basalt	2.831	S	3.26	3.29	3.31	3.33	3.34	3.36	3.37	3.37
Flow	2.684	P	5.05	5.15	5.20	5.24	5.28	5.40	5.58	5.71
basalt	2.684	S	2.58	2.63	2.66	2.69	2.71	2.78	2.86	2.91
breccia	2.506	P	4.31	4.41	4.50	4.57	4.62	4.84	5.13	5.32
	2.506	S	2.51	2.56	2.60	2.64	2.67	2.77	2.90	2.97
Vesicular	2.229	P	3.70	3.75	3.77	—	—	—	—	—
basalt	2.229	S	1.84	1.92	1.96	—	—	—	—	—
Vesicular	2.196	P	3.84	3.88	3.90	—	—	—	—	—
basalt	2.196	S	1.93	1.96	1.98	—	—	—	—	—
Vesicular	2.230	P	3.60	3.64	3.66	—	—	—	—	—
basalt	2.230	S	1.81	1.84	1.86	—	—	—	—	—
Breccia	2.087	P	3.25	3.32	3.37	3.41	3.44	3.57	3.75	3.87

Table A.6.3 (*cont.*)

	Wet Bulk density g/cm³	Mode	0.2	0.4	0.6	0.8	1.0	2.0	4.0	6.0
			Velocity (km/s) at Various Pressures (kbar)							
Breccia	2.076	P	3.18	3.22	3.25	3.27	3.29	3.39	3.57	3.72
Breccia	2.064	P	3.65	3.69	3.73	3.75	3.76	3.81	3.90	4.12
breccia	1.980a	P	2.83	2.86	2.89	2.89	2.92	—	—	—
	1.980a	S	1.38	1.42	1.46	1.51	1.55	—	—	—
Flow basalt	2.597	P	4.40	4.45	4.49	4.52	4.53	4.60	4.73	4.84
	2.597	S	2.47	2.50	2.52	2.53	2.54	2.56	2.58	2.59

Table A.6.4 *P-wave velocities in basic rocks (Christensen, 1968)*

	Density g/cm3	0.1	0.5	1.0	2.0	4.0	6.0	8.0	10.0
		Pressure (kb)							
Basalt 1	2.91	5.8	6.03	6.08	6.13	6.21	6.25	6.28	6.33
	2.91	5.9	6.05	6.11	6.15	6.23	6.28	6.32	6.37
	2.88	5.6	5.76	5.86	5.97	6.03	6.10	6.16	6.20
Mean	2.90	5.8	5.95	6.02	6.08	6.16	6.21	6.25	6.30
Basalt 2	2.92	5.8	6.00	6.03	6.06	6.11	6.16	6.20	6.25
	2.91	5.8	6.04	6.07	6.09	6.14	6.20	6.24	6.28
	2.91	5.9	6.08	6.14	6.19	6.22	6.27	6.36	6.35
Mean	2.91	5.8	6.04	6.08	6.11	6.16	6.21	6.27	6.29
Basalt 3	2.95	6.0	6.14	6.19	6.25	6.34	6.36	6.42	6.46
	2.92	5.9	6.09	6.16	6.21	6.27	6.34	6.36	6.39
	2.94	6.0	6.11	6.17	6.25	6.29	6.34	6.38	6.42
Mean	2.94	6.0	6.11	6.17	6.24	6.30	6.35	6.39	6.42

Table A.6.5 *Modal analysis of rock in previous table (Christensen, 1968)*

	Plagioclase	Pyroxene	Magnetite	Calcite	Sericite+Chlorite
Basalt 1	53.4	26.5	7.2	3.2	9.7
Basalt 2	54.2	25.3	6.9	3.0	7.6
Basalt 3	55.8	30.2	6.9	2.6	4.5

Table A.6.6 *Ultrasonic P-wave velocities in dry igneous rocks (from Hughes and Jones, 1950)*

		Pressure in kg/cm2						
	Temp°C	35	176	352	527	703	879	1055
Granite	30°	5519	5798	5999	6109	6163	6205	6232
2.601 g/cm³	100°	——	5606	5801	6052	6081	6127	6183
porosity = .011	154°	——	5319	5515	5667	5863	5974	6077
Andesite	32°	5230	5244	5255	5266	5281	5284	5310
2.618 g/cm³	107°	5133	5176	5208	5226	5230	5251	5273
porosity = .011	158°	5119	5154	5186	5208	5230	5233	5251
Quartz monzonite	30°	5259	5621	5818	5900	5946	5976	5998
2.628 g/cm³	101°	4830	5293	5675	5818	5892	5955	5981
porosity = .014	150°	4484	4959	5341	5591	5746	5830	5892
Diorite	33°	5783	5970	6114	6196	6243	6275	6301
3.026 g/cm³	102°	5434	5733	5980	6088	6170	6196	6232
porosity = .021	151°	5271	5590	5913	6014	6093	6170	6196
Norite	32°	6181	6495	6649	6758	6811	6851	6877
3.057 g/cm³	100°	5860	6303	6550	6675	6771	6824	6851
porosity = .019	152°	—	6143	6483	6656	6745	6811	6831

Table A.6.7 *Ultrasonic S-wave velocities in dry igneous rocks (from Hughes and Jones, 1950)*

		Pressure in kg/cm2						
	Temp °C	35	176	352	527	703	879	1055
Granite	30°	3043	3234	3301	3398	3426	3454	3474
2.601 g/cm³	100°	—	3187	3232	3320	3325	3314	3327
porosity = .011								
Andesite	32°	2734	2784	2815	2832	2840	2846	2844
2.618 g/cm³	107°	2698	2734	2776	2800	2811	2814	2825
porosity = .011	158°	2629	2718	2753	2764	2783	2808	2818
Quartz monzonite	30°	2894	3091	3146	3196	3203	3212	3229
2.628 g/cm³	101°	2811	2894	3107	3183	3202	3203	3226
porosity = .014	150°	2670	2862	2982	3106	3148	3178	3181
Diorite	33°	3056	3173	3254	3302	3310	3318	3322
3.026 g/cm³	102°	2871	3090	3171	3246	3291	3299	3312
porosity = .021	151°	—	3017	3114	3186	3226	3262	3277
Norite	32°	3239	3377	3516	3575	3582	3606	3632
3.057 g/cm³	100°	2916	3243	3392	3507	3546	3550	3558
porosity = .019	152°	—	3231	3306	3435	—	—	3529

Table A.6.8. *Vp vs. pressure in metamorphic rocks (from Christensen 1965)*

Gneiss 1	X	2.643	4.7	5.0	5.5	5.7	5.8	5.92	6.12	6.23	6.29	6.34	6.37
	Y	2.621	5.0	5.4	5.8	5.9	6.01	6.07	6.18	6.29	6.32	6.37	6.41
	Z	2.665	4.6	4.9	5.5	5.7	5.84	5.91	6.05	6.15	6.21	6.24	6.29
Gneiss 2	X	2.661	4.5	4.8	5.3	5.5	5.7	5.79	6.04	6.22	6.29	6.35	6.39
	Y	2.651	4.5	4.9	5.4	5.7	5.8	5.91	6.09	6.22	6.27	6.31	6.35
	Z	2.650	4.6	5.1	5.5	5.7	5.8	5.85	6.04	6.09	6.15	6.21	6.25
Gneiss3	X	2.742	5.1	5.3	5.7	5.9	6.05	6.19	6.32	6.44	6.48	6.54	6.58
	Y	2.745	4.7	5.0	5.5	5.8	5.97	6.06	6.33	6.42	6.48	6.53	6.57
	Z	2.777	5.4	5.5	5.9	6.1	6.14	6.19	6.32	6.43	6.50	6.53	6.56
Gneiss 4	X	2.819	4.6	5.0	5.4	5.7	5.9	6.02	6.32	6.52	6.60	6.68	6.72
	Y	2.877	4.7	5.1	5.5	5.8	6.0	6.08	6.28	6.45	6.52	6.57	6.63
	Z	2.776	5.1	5.3	5.7	5.8	5.92	5.98	6.14	6.23	6.29	6.34	6.38
Gneiss 5	X	2.845	5.5	5.8	5.9	6.1	6.24	6.29	6.35	6.47	6.54	6.58	6.63
	Y	2.850	5.7	5.7	5.8	5.81	5.85	5.88	5.99	6.16	6.36	6.49	6.58
	Z	2.848	5.2	5.4	5.7	5.9	5.94	5.98	6.09	6.18	6.24	6.29	6.33
metagabbro	X	2.98		5.9	5.99	6.07	6.15	6.18	6.32	6.51	6.63	6.73	6.80
	Y	3.00	6.3	6.31	6.36	6.38	6.43	6.46	6.58	6.72	6.79	6.86	6.92
	Z	2.98	5.4	5.68	6.02	6.22	6.33	6.40	6.57	6.69	6.75	6.82	6.84
Epidote amphibolite 1	X	3.11	6.8	6.9	7.1	7.2	7.2	7.29	7.45	7.61	7.73	7.77	7.82
	Y	3.15	6.0	6.3	6.7	6.9	7.0	7.10	7.32	7.52	7.60	7.65	7.69
Epidote amphibolite 2	X	3.25	6.4	6.5	6.8	6.9	7.1	7.15	7.40	7.66	7.75	7.81	7.83
	Y	3.30	6.4	6.6	6.9	7.1	7.2	7.25	7.49	7.65	7.72	7.77	7.80
	Z	3.24	5.8	6.1	6.4	6.7	6.8	6.87	7.07	7.24	7.32	7.36	7.39
Amphibolite 1	X	3.05	7.0	7.1	7.1	7.2	7.18	7.21	7.26	7.37	7.42	7.46	7.49
	Y	3.04	6.9	7.0	7.04	7.09	7.10	7.13	7.20	7.28	7.33	7.38	7.42
	Z	3.04	5.6	5.8	6.05	6.14	6.23	6.29	6.45	6.60	6.68	6.73	6.76
Amphibolite 2	X	3.03	5.8	6.1	6.5	6.7	6.89	6.97	7.25	7.47	7.52	7.58	7.61
	Y	3.02	6.2	6.4	6.6	6.8	6.85	6.92	7.08	7.22	7.26	7.29	7.33
	Z	3.03	4.5	4.9	5.4	5.7	5.9	6.01	6.27	6.44	6.51	6.55	6.59
Garnet Schist	X	2.75	5.4	5.7	6.1	6.3	6.37	6.44	6.58	6.70	6.74	6.77	6.82
	Y	2.76	5.3	5.6	6.0	6.2	6.29	6.44	6.54	6.64	6.71	6.76	6.77
	Z	2.76	5.0	5.3	5.5	5.61	5.67	5.71	5.84	5.95	6.05	6.11	6.17
Kyanite schist 1	X	2.99	5.0	5.4	5.9	6.2	6.34	6.53	6.72	7.07	7.27	7.41	7.50
	Y	2.95	5.6	6.1	6.4	6.55	6.72	6.78	7.01	7.17	7.29	7.35	7.48
	Z	3.07	4.6	5.1	5.5	5.8	5.95	6.12	6.61	6.92	7.03	7.19	7.26
Kyanite schist 2	X	2.73	5.8	6.1	6.4	6.53	6.68	6.73	6.92	7.04	7.17	7.23	7.27
	Y	2.84	5.0	5.3	5.8	6.05	6.15	6.26	6.54	6.75	6.87	6.92	7.00
	Z	2.73	4.6	5.0	5.4	5.57	5.67	5.74	5.87	6.02	6.02	6.12	6.20
Staurolite-garnet schist	X	2.75	5.9	6.4	6.63	6.74	6.78	6.87	6.97	7.16	7.20	7.23	7.25
	Y	2.76	5.6	6.0	6.2	6.41	6.49	6.54	6.66	6.77	6.85	6.89	6.93
	Z	2.75		4.8	5.1	5.25	5.34	5.41	5.54	5.63	5.70	5.78	5.85
Gneiss 6	X	2.76	5.4	5.7	5.9	6.0	6.10	6.17	6.34	6.49	6.55	6.60	6.65
	Y	2.75	5.2	5.5	5.7	5.9	5.98	6.05	6.24	6.44	6.52	6.57	6.63
	Z	2.76	4.1	4.4	4.7	4.9	5.07	5.16	5.43	5.66	5.81	5.90	5.99
Quartzite	X	2.63	5.2	5.5	5.7	5.83	5.91	5.98	6.07	6.16	6.21	6.24	6.28
	Y	2.62	5.6	5.8	5.91	5.97	6.01	6.05	6.12	6.17	6.21	6.26	6.28
	Z	2.63	5.6	5.79	5.95	6.01	6.07	6.11	6.18	6.25	6.29	6.31	6.33

Table A.6.8. (*cont.*)

Feldspathic mica	X	2.68	5.4	5.7	6.1	6.17	6.23	6.27	6.35	6.43	6.49	6.51	6.55
quartzite	Y	2.66	5.2	5.5	5.9	6.06	6.11	6.16	6.23	6.29	6.35	6.39	6.43
	Z	2.68	5.1	5.4	5.69	5.78	5.83	5.86	5.93	6.01	6.07	6.12	6.17
Slate	X	2.77	6.28	6.31	6.34	6.35	6.36	6.37	6.41	6.48	6.54	6.59	6.66
	Y	2.75	6.29	6.31	6.32	6.33	6.34	6.36	6.40	6.46	6.51	6.57	6.63
	Z	2.77	4.94	4.97	5.02	5.04	5.06	5.09	5.15	5.29	5.41	5.50	5.59

A.7 Velocities and Moduli of Ice, Methane Hydrate, and Sea Water

Tables A.7.1 and A.7.2, from Helgerud (2001) summarize ultrasonic velocities, Poisson's ratio, and dynamic elastic moduli for compacted polycrystalline ice and methane hydrate, as a function of temperature (in degrees Celsius) and pressure (in psi).

Table A.7.1 *Regressions of velocities, Poisson s ratio and dynamic elastic moduli versus temperature (-20 to $-5°C$) and piston pressure (3250–4750 psi) in compacted, polycrystalline ice. $F(T, P) = aT + bP + c$ (Helgerud, 2001).*

$F(T, P)$	a	b	c	std
V_P (m/s)	-2.80 ± 0.01	$(1.98 \pm 0.11) \times 10^{-3}$	3870.1 ± 0.5	0.9
V_S (m/s)	-1.31 ± 0.01	$-(1.83 \pm 0.07) \times 10^{-3}$	1949.3 ± 0.3	0.6
v	$-(2.0 \pm 0.2) \times 10^{-5}$	$(6.4 \pm 0.2) \times 10^{-7}$	0.3301 ± 0.0001	0.0002
M (GPa)	$-(2.01 \pm 0.01) \times 10^{-2}$	$(1.41 \pm 0.08) \times 10^{-5}$	13.748 ± 0.0003	0.007
G (GPa)	$-(4.72 \pm 0.02) \times 10^{-3}$	$-(6.6 \pm 0.3) \times 10^{-6}$	3.488 ± 0.001	0.002
K (GPa)	$-(1.38 \pm 0.01) \times 10^{-2}$	$(2.30 \pm 0.09) \times 10^{-5}$	9.097 ± 0.004	0.008

Table A.7.2 *Regressions of velocities, Poisson s ratio and dynamic elastic moduli versus temperature (-15 to $15 °C$) and piston pressure (4000–9000psi) in compacted, polycrystalline methane hydrate. $F(T, P) = aT + bP + c$ (Helgerud, 2001).*

$F(T, P)$	a	b	c	std
V_P (m/s)	-2.27 ± 0.01	$(2.52 \pm 0.07) \times 10^{-3}$	3778.0 ± 0.6	2
V_S (m/s)	-1.04 ± 0.01	$-(1.08 \pm 0.03) \times 10^{-3}$	1963.6 ± 0.3	1
v	0	$(6.83 \pm 0.06) \times 10^{-7}$	0.31403 ± 0.00005	0.0003
M (GPa)	$-(1.47 \pm 0.01) \times 10^{-2}$	$(1.99 \pm 0.05) \times 10^{-5}$	13.527 ± 0.004	0.02
G (GPa)	$-(3.77 \pm 0.03) \times 10^{-3}$	$-(3.9 \pm 0.01) \times 10^{-6}$	3.574 ± 0.001	0.004
K (GPa)	$-(9.71 \pm 0.06) \times 10^{-3}$	$(2.51 \pm 0.03) \times 10^{-5}$	8.762 ± 0.003	0.01

Table A.7.3 *Adiabatic elastic constants of ice, corrected to* $-16°C$. *Compiled by Gammon* et al. *(1983). Samples are (nearly) free of bubbles and impurities.*

C_{11}	C_{12}	C_{13}	C_{33}	C_{44}	reference
			Artificial Ice		
139.61	71.53	57.65	150.13	30.21	a
138.5	70.7	58.1	149.9	31.9	b
133	63	49	142	30.6	c
132.1	67.9	58.0	144.3	28.9	d
137.5	67.7	52.5	149.9	30.7	e
139.1	68.6	53.9	149.8	20.8	f
			Glacier Ice		
139.13	70.26	58.01	150.59	30.11	a
			Lake Ice		
138.76	69.79	56.57	150.71	30.24	a
133.9	68.0	77.9	168.6	37.7	g
			Sea Ice		
142.7	73.2	59.5	147.4	29.8	a

[a] Gammon *et al.* (1983)
[b] Jona and Scherrer (1952)
[c] Bass *et al.* (1957)
[d] Dantl (1969)
[e] Brockkamp and Querfurth (1964)
[f] Proctor (1966)
[g] Bogorskiy (1964)

Gammon *et al.* (1983) measured properties of single crystal manmade ice, lake ice, sea ice, and glacier ice, using Brillouin scattering. Their compiled lab measurements are shown in Table A.7.3. The measured temperature-dependence of density ρ (Butkovich, 1955; Gammon *et al.*, 1983) can be approximated empirically as

$$\rho^{-1} = \rho_0^{-1}[1 + 1.576 \cdot 10^{-4}T - 2.778 \cdot 10^{-7}T^2 + 8.850 \cdot 10^{-9}T^3 - 1.778 \cdot 10^{-10}T^4]$$

where $\rho_0 = 916.7$ kg/m^3 is the density at $0°C$, and T is the temperature in $°C$.

Gammon *et al.* give the following empirical equation for the temperature dependence of elastic constants:

$$C_{ij}(T) = C_{ij}(T_m) \frac{1 - 0.001418T}{1 - 0.001418T_m}$$

where C_{ij} is any one of the elastic constants, temperature T is in $°C$ and $C_{ij}(T_m)$ is the measured elastic constant at temperature T_m. The P-wave velocity is given empirically by

$$V_P(T) = V_P(T_m) \frac{1 - 0.0006196T}{1 - 0.0006196T_m}$$

Table A.7.4 *Young's modulus floating sea ice, computed from measured sonic velocities on cores samples, and assuming Poisson's ratio = 0.295. T = − 3 to − 13 °C. From Pounder and Langlebein (1964).*

Number of samples	Porosity	E_{obs} (GPa)	E_{calc} [*] (GPa)
		Polar ice – vertical cores	
9	.00714	9.775	9.975
9	.02461	9.443	9.914
10	.03656	9.434	9.872
10	.04157	9.467	9.854
10	.04624	9.442	9.838
10	.05065	9.461	9.822
10	.05587	9.624	9·804
10	.06192	9.469	9.783
10	.07130	9.265	9.750
10	.1098	9.270	9.615
		Biennial ice – vertical cores	
10	.01707	9.284	9.940
10	.02490	9.298	9.913
10	.02802	9.470	9.902
10	.0311	9.602	9.891
10	.03413	9.658	9.880
10	.03731	9.529	9.869
10	.04487	9.578	9.843
10	.05468	9.621	9.808
10	.06254	9.425	9.780
9	.08873	9.541	9.689
		Biennial ice – horizontal cores	
11	.02914	9.372	9.898
10	.02873	9.494	9.899

[*] modulus predicted by Langlebein (1962) for sea ice: $E = 10.00 - 3.5\phi$ where ϕ is brine content (saturated porosity).

Gammon *et al.* suggest that the concentration of dissolved solids in water at the time of freezing does not significantly affect the elastic properties of the resulting mono-crystals.

Elastic constants of sea ice were estimated by Pounder and Langlebein (1964) and summarized in Table A.7.4. Data from a floating ice sheet near Ellef Ringnes Island, N. W. T., Canada. They observe Poisson's ratio to be fairly independent of salinity, temperature, and type of sea ice. Polar ice density: 894 kg/m^3.

Table A.7.5 *Phase relations for typical sea water. Amount (in g/kg) of ions and H_2O in brine, and salts and ice under equilibrium at different temperatures. Salinity (total salts) $S = 34.325$ parts per thousand. From Assur (1958).*

T °C	Dissolved ions in brine							Precipitated salts						
	K^+	Ca^+	Mg^{++}	$SO_4^=$	Na^+	Cl^-	H_2O	$MgCl\cdot 12H_2O$	KCl	$NaCl\cdot 2H_2O$	$MgCl\cdot 8H_2O$	$NaSO_4\cdot 10H_2O$	$CaCO_3\cdot 6H_2O$	Ice
0	.380	.400	1.272	2.649	10.556	18.980	965.675	0	0	0	0	0	0	0
-2	.380	.400	1.272	2.649	10.556	18.980	878.905	0	0	0	0	0	0	86.770
-4	.380	.386	1.272	2.649	10.556	18.980	451.614	0	0	0	0	0	.072	515.008
-6	.380	.374	1.272	2.649	10.556	18.980	309.301	0	0	0	0	0	.133	656.227
-8	.380	.366	1.272	2.649	10.556	18.980	236.763	0	0	0	0	0	.179	728.781
-10	.380	.360	1.272	1.471	9.992	18.980	195.293	0	0	0	0	3.951	.208	768.021
-12	.380	.360	1.272	.885	9.712	18.980	169.171	0	0	0	0	5.916	.208	793.044
-14	.380	.360	1.272	.622	9.586	18.980	151.078	0	0	0	0	6.798	.208	810.644
-16	.380	.360	1.272	.456	9.506	18.980	137.264	0	0	0	0	7.357	.208	824.145
-18	.380	.360	1.272	.338	9.450	18.980	125.778	0	0	0	0	7.750	.208	835.412
-20	.380	.358	1.261	.266	9.415	18.949	115.560	0	0	0	.104	7.994	.217	845.425
-25	.380	.345	1.224	.136	4.814	11.841	62.421	0	0	18.651	.471	8.432	.282	890.934
-30	.380	.340	1.178	.033	1.181	6.177	30.217	0	0	33.383	.927	8.775	.314	917.031
-35	.38	.340	1.123	0	0.280	4.963	22.890	0	.010	36.191	1.468	8.883	.314	922.903
-40	.332	.340	1.064	0	0.223	4.350	18.966	0	.092	37.252	2.045	8.883	.314	926.075

A.8 Physical Properties of Common Gases

Compiled from American Air Liquid, on-line gas data, http://www.airliquide.com.

Air

Normal Composition of Dry Air
- 78.09% N_2, 20.94% O_2, 0.93% Ar.

Molecular Weight
- Molecular weight: 28.95 g/mol.

Density
- Density of air at 1 atm of pressure is 1.291 kg/m^3 at 0°C or 1.222 kg/m^3 at 15.6°C.

Bulk Modulus
- Bulk modulus of air at 1 atm of pressure is 1.01 bar = 0.101 MPa.

Critical Point
- Critical temperature: −140.5 °C.
- Critical pressure: 37.71 bar.

Gaseous Phase
- Gas density (1.013 bar at boiling point): 3.2 kg/m^3.
- Gas density (1.013 bar and 15°C): 1.202 kg/m^3.
- Compressibility factor (Z) (1.013 bar and 15 °C): 0.9992.
- Specific gravity (air =1) (1.013 bar and 21 °C): 1.
- Specific volume (1.013 bar and 21 °C): 0.833 m^3/kg.
- Heat capacity at constant pressure (C_p) (1.013 bar and 21 °C): 0.029 kJ/(mol K).
- Heat capacity at constant volume (C_v) (1.013 bar and 21 °C): 0.02 kJ/(mol K).
- Ratio of specific heats (Gamma: C_P/C_v) (1.013 bar and 21 °C): 1.4028.
- Viscosity (1 bar and 0°C): 0.0001695 Poise.
- Thermal conductivity (1.013 bar and 0°C): 23.94 mW/(m K).

Miscellaneous
- Solubility in water (1.013 bar and 0°C): 0.0292 vol/vol.

Carbon Dioxide (CO_2)

Molecular Weight
- Molecular weight: 44.01 g/mol.

Solid Phase
- Latent heat of fusion (1013 bar, at triple point): 196.104 kJ/kg.
- Solid density: 1562 kg/m^3.

Liquid Phase

- Liquid density (at −20°C and 19.7 bar): 1032 kg/m^3.
- Liquid/gas equivalent (1.013 bar and 15 °C (per kg of solid)): 845 vol/vol.
- Boiling point (sublimation): −78.5°C.
- Latent heat of vaporization (1.013 bar at boiling point): 571.08 kJ/kg.
- Vapor pressure (at 20 °C): 58.5 bar.

Critical Point

- Critical temperature: 31 °C.
- Critical pressure: 73.825 bar.
- Critical density: 464 kg/m^3.

Triple Point

- Triple point temperature: −56.6 °C.
- Triple point pressure: 5.185 bar.

Gaseous Phase

- Gas density (1.013 bar at sublimation point): 2.814 kg/m^3.
- Gas density (1.013 bar and 15°C): 1.87 kg/m^3.
- Compressibility factor (Z) (1.013 bar and 15 °C): 0.9942.
- Specific gravity (air = 1) (1.013 bar and 21 °C): 1.521.
- Specific volume (1.013 bar and 21 °C): 0.547 m^3/kg.
- Heat capacity at constant pressure (C_p) (1.013 bar and 25 °C): 0.037 kJ/(mol K).
- Heat capacity at constant volume (C_v) (1.013 bar and 25 °C): 0.028 kJ/(mol K).
- Ratio of specific heats (Gamma: C_p/C_v) (1.013 bar and 25 °C): 1.293759.
- Viscosity (1.013 bar and 0°C): 0.0001372 Poise.
- Thermal conductivity (1.013 bar and 0°C): 14.65 mW/(m K).

Miscellaneous

- Solubility in water (1.013 bar and 0°C): 1.7163 vol/vol.
- Concentration in air: 0.03 vol %.

Ethane (C$_2$H$_6$)

Molecular Weight

- Molecular weight: 30.069 g/mol.

Critical Point

- Critical temperature: 32.2 °C.
- Critical pressure: 48.839 bar.

Gaseous Phase

- Gas density (1.013 bar at boiling point): 2.054 kg/m^3.
- Gas density (1.013 bar and 15°C): 1.282 kg/m^3.
- Compressibility factor (Z) (1.013 bar and 15 °C): 0.9912.
- Specific gravity (air = 1) (1.013 bar and 15°C): 1.047.
- Specific volume (1.013 bar and 21 °C): 0.799 m^3/kg.
- Heat capacity at constant pressure (C_p) (1 bar and 25 °C): 0.053 kJ/(mol K).
- Heat capacity at constant volume (C_v) (1 bar and 25 °C): 0.044 kJ/(mol K).
- Ratio of specific heats (Gamma: C_p/C_v) (1 bar and 25 °C): 1.193258.
- Viscosity (1.013 bar and 0°C (32°F)): 0.0000855 Poise.
- Thermal conductivity (1.013 bar and 0°C): 18 mW/(m K).

Miscellaneous

- Solubility in water (1.013 bar and 20 °C): 0.052 vol/vol.
- Autoignition temperature: 515 °C.

Methane (CH$_4$)

Molecular Weight

- Molecular weight: 16.043 g/mol.

Critical Point

- Critical temperature: −82.7°C.
- Critical pressure: 45.96 bar.

Gaseous Phase

- Gas density (1.013 bar at boiling point): 1.819 kg/m^3.
- Gas density (1.013 bar and 15 °C): 0.68 kg/m^3.
- Compressibility factor (Z) (1.013 bar and 15 °C): 0.998.
- Specific gravity (air = 1) (1.013 bar and 21 °C): 0.55.
- Specific volume (1.013 bar and 21 °C): 1.48 m3/kg.
- Heat capacity at constant pressure (C_p) (1 bar and 25 °C): 0.035 kJ/(mol K).
- Heat capacity at constant volume (Cv) (1 bar and 25 °C): 0.027 kJ/(mol K).
- Ratio of specific heats (Gamma: Cp/Cv) (1 bar and 25 °C): 1.305 454.
- Viscosity (1.013 bar and 0°C): 0.0001027 Poise.
- Thermal conductivity (1.013 bar and 0°C): 32.81 mW/(mK).

Miscellaneous

- Solubility in water (1.013 bar and 2°C): 0.054 vol/vol.
- Autoignition temperature: 595 °C.

Helium (He)

Molecular Weight
- Molecular weight: 4.0026 g/mol.

Critical Point
- Critical temperature: −268 °C.
- Critical pressure: 2.275 bar.
- Critical density: 69.64 kg/m^3.

Gaseous Phase
- Gas density (1.013 bar at boiling point): 16.891 kg/m^3.
- Gas density (1.013 bar and 15°C): 0.169 kg/m^3.
- Compressibility factor (Z) (1.013 bar and 15 °C): 1.0005.
- Specific gravity (air = 1) (1.013 bar and 21 °C): 0.138.
- Specific volume (1.013 bar and 21 °C): 6.037 m^3/kg.
- Heat capacity at constant pressure (Cp) (1 bar and 25 °C): 0.02 kJ/(mol K).
- Heat capacity at constant volume (Cv) (1 bar and 25 °C): 0.012 kJ/(mol K).
- Ratio of specific heats (Gamma: Cp/Cv) (1 bar and 25 °C): 1.664.
- Viscosity (1.013 bar and 0°C): 0.0001863 Poise.
- Thermal conductivity (1.013 bar and 0°C): 142.64 mW/(mK).

Miscellaneous
- Solubility in water (20°C and 1 bar): 0.0089 vol/vol.

Hydrogen (H$_2$)

Molecular Weight
- Molecular weight: 2.016 g/mol.

Critical Point
- Critical temperature: −240 °C.
- Critical pressure: 12.98 bar.
- Critical density: 30.09 kg/m^3.

Triple Point
- Triple point temperature: −259.3 °C.
- Triple point pressure: 0.072 bar.

Gaseous Phase
- Gas density (1.013 bar at boiling point): 1.312 kg/m^3.
- Gas density (1.013 bar and 15 °C): 0.085 kg/m^3.
- Compressibility factor (Z) (1.013 bar and 15 °C): 1.001.
- Specific gravity (air = 1) (1.013 bar and 21 °C): 0.0696.

- Specific volume (1.013 bar and 21 °C): 11.986 m^3/kg.
- Heat capacity at constant pressure (Cp) (1 bar and 25 °C): 0.029 kJ/(mol K).
- Heat capacity at constant volume (Cv) (1 bar and 25 °C): 0.021 kJ/(mol K).
- Ratio of specific heats (Gamma: Cp/Cv) (1 bar and 25 °C): 1.384 259.
- Viscosity (1.013 bar and 15 °C): 0.0000865 Poise.
- Thermal conductivity (1.013 bar and 0°C): 168.35 mW/(m K).

Miscellaneous
- Solubility in water (1.013 bar and 0°C): 0.0214 vol/vol.
- Concentration in air: 0.0000005 vol/vol.
- Autoignition temperature: 560 °C.

Nitrogen (N$_2$)

Molecular Weight
- Molecular weight: 28.0134 g/mol.

Critical Point
- Critical temperature: −147 °C.
- Critical pressure: 33.999 bar.
- Critical density: 314.03 kg/m^3.

Triple Point
- Triple point temperature: −210.1 °C.
- Triple point pressure: 0.1253 bar.

Gaseous Phase
- Gas density (1.013 bar at boiling point): 4.614 kg/m^3.
- Gas density (1.013 bar and 15 °C): 1.185 kg/m^3.
- Compressibility factor (Z) (1.013 bar and 15 °C): 0.9997.
- Specific gravity (air = 1) (1.013 bar and 21 °C): 0.967.
- Specific volume (1.013 bar and 21 °C): 0.862 m^3/kg.
- Heat capacity at constant pressure (Cp) (1.013 bar and 25 °C): 0.029 kJ/(mol K).
- Heat capacity at constant volume (Cv) (1.013 bar and 25 °C): 0.02 kJ/(mol K).
- Ratio of specific heats (Gamma: Cp/Cv) (1.013 bar and 25 °C): 1.403846.
- Viscosity (1.013 bar and 0°C): 0.0001657 Poise.
- Thermal conductivity (1.013 bar and 0°C): 24 mW/(mK).

Miscellaneous
- Solubility in water (1.013 bar and 0°C): 0.0234 vol/vol.
- Concentration in air: 0.7808 vol/vol.

Oxygen (O$_2$)

Molecular Weight
• Molecular weight: 31.9988 g/mol.

Critical Point
• Critical temperature: −118.6 °C.
• Critical pressure: 50.43 bar.
• Critical density: 436.1 kg/m^3.

Triple Point
• Triple point temperature: −218.8 °C.
• Triple point pressure: 0.00152 bar.

Gaseous Phase
• Gas density (1.013 bar at boiling point): 4.475 kg/m3.
• Gas density (1.013 bar and 15°C): 1.354 kg/m3.
• Compressibility factor (Z) (1.013 bar and 15 °C): 0.9994.
• Specific gravity (air = 1) (1.013 bar and 21 °C): 1.105.
• Specific volume (1.013 bar and 21 °C): 0.755 m3/kg.
• Heat capacity at constant pressure (Cp) (1 bar and 25 °C): 0.029 kJ/(mol K).
• Heat capacity at constant volume (Cv) (1 bar and 25 °C): 0.021 kJ/(mol K).
• Ratio of specific heats (Gamma: Cp/Cv) (1 bar and 25 °C): 1.393365.
• Viscosity (1.013 bar and 0°C): 0.0001909 Poise.
• Thermal conductivity (1.013 bar and 0°C): 24.24 mW/(mK).

Miscellaneous
• Solubility in water (1.013 bar and 0°C): 0.0489 vol/vol.
• Concentration in air: 0.2094 vol/vol.

A.9   Velocity, Moduli, and Density of Carbon Dioxide

Tables A.9.1–A.9.3 summarize velocities, densities, and moduli for CO_2 as functions of temperature and pressure. Z. Wang (Wang, 2000a) provided the data. Bold numerals are actual measurements, while others are interpolations.

Table A.9.1 *Velocity for CO_2 as a function of temperature and pressure*

Velocities (m/s)

P (psi)						T (°C)							P (MPa)
	17	27	37	47	57	67	77	87	97	107	117	127	
14.5	**264**	**268**	**272**	**276**	**280**	**284**	**288**	**291**	**295**	**299**	**302**	**306**	0.10
145	**256**	**261**	**266**	**270**	**275**	**279**	**284**	**288**	**292**	**296**	**300**	**304**	1.00
580	**220**	**231**	**241**	**251**	259	**264**	**271**	**276**	**281**	**287**	**292**	296	4.00
1015	379	**185**	**209**	**219**	237	**252**	**263**	**270**	**276**	**282**	**285**	290	7.00
1450	**483**	**411**	**142**	**183**	212	**238**	**249**	**257**	**265**	**272**	**278**	286	10.01
2030	554	**496**	366	269	247	248	**255**	260	269	278	285	**291**	14.01
2900	637	**586**	502	441	382	358	**353**	345	337	328	320	**314**	20.01
3625	688	**640**	591	538	488	460	**431**	413	397	382	366	**352**	25.01
4350	737	**687**	641	596	552	521	**494**	470	450	429	412	**397**	30.02
5800	814	**766**	714	676	639	615	**592**	564	545	526	504	**487**	40.02

Table A.9.2 *Density for CO$_2$ as a function of temperature and pressure*

Densities (kg/m^3)

P (psi)	17	27	37	47	57	67	77	87	97	107	117	127	P (MPa)
						T (°C)							
14.5	1.86	1.80	1.74	1.68	1.63	1.58	1.54	1.49	1.45	1.41	1.38	1.34	0.10
145	19.63	18.84	18.13	17.48	16.88	16.32	15.80	15.31	14.86	14.44	14.04	13.66	1.00
580	129.00	93.95	87.09	81.69	77.24	73.45	70.13	67.19	64.54	62.15	60.00	57.97	4.00
1015	830.00	680.00	200.72	182.83	163.62	149.60	139.28	130.94	123.91	117.86	112.75	107.94	7.00
1450	917.00	805.00	683.00	449.73	327.61	267.15	235.50	213.81	197.32	184.22	174.09	164.43	10.01
2030	930.00	860.00	780.00	690.00	570.00	480.00	390.00	335.00	300.00	275.00	252.00	235.00	14.01
2900	960.00	910.00	860.00	800.00	740.00	680.00	620.00	560.00	510.00	470.00	420.00	390.00	20.01
3625	990.00	950.00	900.00	850.00	790.00	750.00	700.00	640.00	590.00	550.00	510.00	480.00	25.01
4350	1008.00	970.00	930.00	890.00	850.00	810.00	770.00	720.00	680.00	630.00	600.00	570.00	30.02
5800	1040.00	1000.00	970.00	940.00	900.00	870.00	840.00	800.00	770.00	730.00	700.00	670.00	40.02

Table A.9.3 *Bulk modulus for CO$_2$ as a function of temperature and pressure*

Bulk moduli (GPa) P (psi)	T (°C) 17	27	37	47	57	67	77	87	97	107	117	127	P (MPa)
14.5	0.000 13	0.00013	0.000 13	0.00013	0.00013	0.00013	0.00013	0.00013	0.00013	0.000 13	0.00013	0.000 13	0.10
145	0.001 29	0.001 28	0.001 28	0.001 28	0.001 28	0.001 27	0.001 27	0.001 27	0.001 26	0.001 26	0.001 26	0.001 26	1.00
580	0.006 24	0.005 01	0.005 04	0.005 13	0.005 18	0.005 14	0.005 17	0.005 10	0.005 10	0.00511	0.00511	0.005 07	4.00
1015	0.119 22	0.023 27	0.008 73	0.008 79	0.009 23	0.009 51	0.009 65	0.009 55	0.00941	0.009 37	0.009 16	0.009 10	7.00
1450	0.213 93	0.135 98	0.015 99	0.012 82	0.01469	0.015 20	0.01460	0.01412	0.013 86	0.013 63	0.013 46	0.013 41	10.01
2030	0.28543	0.21157	0.10449	0.049 93	0.03478	0.029 52	0.025 36	0.02265	0.02171	0.021 25	0.02047	0.019 90	14.01
2900	0.389 54	0.31249	0.216 72	0.155 58	0.107 98	0.087 15	0.077 26	0.066 65	0.057 92	0.05056	0.043 01	0.03845	20.01
3625	0.468 61	0.389 12	0.314 35	0.246 03	0.18813	0.158 70	0.13003	0.10916	0.09299	0.08026	0.068 32	0.05947	25.01
4350	0.547 51	0.457 81	0.38212	0.31614	0.259 00	0.219 87	0.18791	0.159 05	0.137 70	0.11595	0.10185	0.089 84	30.02
5800	0.689 10	0.586 76	0.49450	0.429 56	0.36749	0.329 06	0.294 39	0.25448	0.228 71	0.201 97	0.177 81	0.158 90	40.02

A.10   Standard Temperature and Pressure

There is a wide variation in definition of standard temperature and pressure (STP). Table A.10.1 lists some in current use (compiled from Wikipedia: http://en.wikipedia.org/wiki/ Standard_conditions_for_temperature_and_pressure).

Table A.10.1 *Variable definitions of standard temperature and pressure*

Temperature (°C)	Absolute pressure (kPa)	Publishing or establishing entity
0	100.000	International Union of Pure and Applied Chemistry
0	101.325	National Institute of Standards and Technology International Organization for Standardization
15	101.325	ICAO's International Standard Atmosphere International Organization for Standardization European Environment Agency Electricity and Gas Inspection Act of Canada
20	101.325	U.S. Environmental Protection Agency National Institute of Standards and Technology
25	101.325	U.S. Environmental Protection Agency
25	100.000	Standard Ambient Pressure and Temperature
20	100.000	Compressed Air and Gas Institute
15	100.000	Society of Petroleum Engineers

References

Achenbach, J.D., 1984. *Wave Propagation in Elastic Solids*. Amsterdam: Elsevier Science Publication.

Agnolin, I. and Roux, J.-N., 2007. Internal states of model isotropic granular packings. I. Assembling process, geometry, and contact networks. *Physical Review E*, **76**, 061302.

Agnolin, I., Roux, J.-N., Massaad, P., Jia, X. and Mills, P., 2005. Sound wave velocities in dry and lubricated granular packings: Numerical simulations and experiments. Powders & Grains, Stuttgart, Germany, A. A. Balkema, Leiden, 313–317.

Ahmadi, M.M., Mohammadi, S. and Nemati Hayati, A., 2011. Analytical derivation of tortuosity and permeability of monosized spheres: A volume averaging approach. *Physical Review E*, **83**, 026312, 1–8.

Ahmadov, R. 2011. *Microtextural elastic and transport properties of source rocks*, PhD dissertation, Stanford University.

Akaogi, M., Navrotsky, A., Yaki, T. and Akimoto, S., 1987. Pyroxene-garnet transformation: Thermochemistry and elasticity of garnet solid solutions and application to mantle models. In *High-Pressure Research in Mineral Physics*, Geophysical Monograph Series vol. 39., ed. M. H. Manghnani and Y. Syono. Washington, DC: AGU, pp. 251–260.

Aki, K. and Richards, P.G., 1980. *Quantitative Seismology: Theory and Methods*. San Francisco, CA: W. H. Freeman and Co.

Aleksandrov, K.S. and Ryzhova, T.V., 1961a. Elastic properties of rock-forming minerals II. Layered silicates. *Bulletin of the Academy of Sciences of USSR, Geophysics Series*, English translation no. **12**, 1165–1168.

Alexandrov, K.S. and Ryzhova, T.V., 1961b. The elastic properties of crystals. *Soviet Physics Crystallography*, **6**, 228–252.

Alexandrov, K.S., Ryzhova, T.V. and Belikov, B.P., 1964. The elastic properties of pyroxenes. *Soviet Physics Crystallography*, **8**, 589–591.

Alfred, D. and Vernik, L., 2013. A new petrophysical model for organic shales. *Petrophysics*, **54**, 240–247.

Aliyeva, S., 2017. *The effect of induced stress heterogeneity and pore space on solid substitution*, PhD dissertation, Stanford University.

Alkalifah, T. and Tsvankin, I., 1995. Velocity analysis for transversely isotropic media. *Geophysics*, **60**, 1550–1566.

Alsinan, S., 2016. Effect of artificially induced maturation on the elastic properties of kerogen. Annual Meeting Society of Exploration Geophysicists.

Anderson, A.T., Davis, A.M. and Lu, F., 2000. Evolution of Bishop Tuff rhyolitic magma based on melt and magnetite inclusions and zoned phenocrysts. *Journal of Petrology*, **41**, 449–473, doi:10.1093/petrology/41.3.449.

Anderson, D.L., 1988. Temperature and pressure derivatives of elastic constants with application to the mantle. *Journal of Geophysical Research*, **93**, 4688–4700.

Anderson, O.L. and Liebermann, R.C., 1966. Sound velocities in rocks and minerals. *VESIAC State-of-the-Art Report No. 7885-4-x*. University of Michigan.

Angel, Y.C. and Achenbach, J.D., 1985. Reflection and transmission of elastic waves by a periodic array of cracks. *Journal of Applied Mechanics*, **52**, 33–11.

Aramaki, S., 1984. Formation of the Aira caldera, southern Kyushu, approx. 22,000 years ago. *Journal of Geophysical Research*, **89**, 8485–8501, doi:10.1029/JB089iB10p08485.

Archie, G.E., 1942. The electrical resistivity log as an aid in determining some reservoir characteristics. *Transactions American Institute of Mechanical Engineers*, **146**, 54–62.

Arps, J.J., 1953. The effect of temperature on the density and electrical resistivity of sodium chloride solutions. *Transactions American Institute of Mining (AIME)*, **198**, 327–330.

Arts, R.J., Helbig, K. and Rasolofosaon, P.N.J., 1991. General anisotropic elastic tensors in rocks: Approximation, invariants and particular directions. In *Expanded Abstracts, 61st Annual International Meeting, Society of Exploration Geophysicists*, **ST2.4**, pp. 1534–1537. Tulsa, OK: Society of Exploration Geophysicists.

Assouline, S. and Rouault, Y., 1997. The relationships between particle and pore size distributions in multicomponent sphere packs: Application to the water retention curve. *Colloids and Surfaces A*, **127**, 201–210.

Assur, A., 1958. Composition of sea ice and its tensile strength. In *Arctic Sea Ice*, National Academy of Sciences –National Research Council, Publication 598, 106–138.

Auld, B.A., 1990. *Acoustic Fields and Waves in Solids*, 2 vols. Malabar, FL: Robert E. Krieger Publication Co.

Avellaneda, M. and Milton, G.W., 1989. Optimal bounds on the effective bulk modulus of polycrystals. *SIAM Journal of Applied Mathematics*, **49**, 824–837.

Avseth, P., 2000. *Combining Rock Physics and Sedimentology for Seismic Reservoir Characterization of North Sea Turbidite Systems*. PhD dissertation, Stanford University.

Bachrach, R. and Avseth, P., 2008. Rock physics modeling of unconsolidated sands: Accounting for nonuniform contacts and heterogeneous stress fields in the effective media approximation with applications to hydrocarbon exploration. *Geophysics*, **73**, E197–E209.

Backus, G.E., 1962. Long-wave elastic anisotropy produced by horizontal layering. *Journal of Geophysical Research*, **67**, 4427–4440.

Bacon, C.R. and Druitt, T.H., 1988. Compositional evolution of the zoned calcalkaline magma chamber of Mount Mazama, Crater Lake, Oregon. *Contributions to Mineralogy and Petrology*, **98**, 224–256, doi:10.1007/BF00402114.

Bacon, C.R., Newman, S. and Stolper, E., 1992. Water, CO2, Cl, and F in melt inclusions in phenocrysts from three Holocene explosive eruptions, Crater Lake, Oregon. *American Mineralogist*, **77**, 1021–1030.

Bahrami, M., Yovanovich, M.M. and Culham, J.R., 2005. Pressure drop of fully-developed, laminar flow in microchannels of arbitrary cross-section. Proceedings of ICMM 2005 3rd International Conference on Microchannels and Minichannels, June 13–15, Toronto, Ontario, Canada.

Bailey, R., Dalrymple, G. and Lanphere, M., 1976. Volcanism, structure, and geochronology of Long Valley Caldera, Mono County, California. *Journal of Geophysical Research*, **81**, 725–744, doi:10.1029/JB081i005p00725.

Bakhorji, A.M., 2010. *Laboratory measurements of static and dynamic elastic properties in carbonate*, PhD dissertation, University of Alberta.

Bakulin, A., 2003. Intrinsic and layer-induced vertical transverse isotropy. *Geophysics*, **68**, 1708–1713.

Bakulin, A. and Grechka, V., 2003. Effective anisotropy of layered media. *Geophysics*, **68**, 1817–1821.

Bakulin, A., Grechka, V. and Tsvankin, I., 2000. Estimation of fracture parameters from reflection seismic data – Part I: HTI model due to a single fracture set. *Geophysics*, **65**, 1788–1802.

Bakulin, V. and Bakulin, A., 1999. Acoustopolarizational method of measuring stress in rock mass and determination of Murnaghan constants. In *69th Annual International Meeting, SEG, Expanded Abstracts*, pp. 1971–74.

Balberg, I., 1985. "Universal" percolation-threshold limits in the continuum. *Physical Review B*, **31**, 4053–4055.

Balberg, I., Binenbaum, N. and Wagner, N., 1984. Percolation thresholds in the three-dimensional sticks system. *Physical Review Letters*, **52**, 1465–1468.

Balluffi, R.W., 2012. *Introduction to Elasticity Theory for Crystal Defects*. Cambridge: Cambridge University Press.

Bandyopadhyay, K., 2009. *Seismic anisotropy: Geological causes and its implications to reservoir geophysics*, PhD dissertation, Stanford University.

Banik, N.C., 1987. An effective anisotropy parameter in transversely isotropic media. *Geophysics*, **52**, 1654.

Banik, N.C., Lerche, I. and Shuey, R.T., 1985. Stratigraphic filtering, Part I: Derivation of the O'Doherty-Anstey formula. *Geophysics*, **50**, 2768–2774.

Bansal, N.P. and Doremus, R.H., 1986. *Handbook of Glass Properties*. Academic Press, London.

Bardis, S.C., Lumsden, A.C. and Barton, N.L., 1983. Fundamentals of rock joint deformation. *International Journal of Rock Mechanics*, **20**, 249–268.

Bass, J.D., 1984a. Elasticity of single-crystal orthoferrosilite. *Journal of Geophysical Research*, **89**, 4359–4371.

Bass, J.D., 1984b, Elasticity of single-crystal $SmAlO_3$, $GdAlO_3$ and $ScAlO_3$ perovskites. *Physics of the Earth and Planetary Interiors*, **36**, 145–156.

Bass, J.D., 1986. Elasticity of uvarovite and andradite garnets. *Journal of Geophysical Research*, **91**, 7505–7516.

Bass, J.D., 1995. Elasticity of minerals, glasses, and melts. In *Mineral Physics and Crystallography: A Handbook of Physical Constants*, AGU Reference Shelf 2, ed. T. J. Ahrens, 45–63.

Bass, J.D. and Anderson, D.L., 1984. Composition of the upper mantle: Geophysical tests of two petrological models. *Geophysical Research Letters*, **11**, 237–240.

Bass, R., Rossberg, G. and Ziegler, G., 1957. Die elastischen Konstanten des Eises. *Zeitschrit für Physik, Bd.*, **149**, Ht. 2, 199–203.

Batzle, M. and Wang, Z., 1992. Seismic properties of pore fluids. *Geophysics*, **57**, 1396–1408.

Baule, A. and Makse, H.A., 2014. Fundamental challenges in packing problems: From spherical to non-spherical particles. *Soft Matter*, **10**(25), 4423–4429.

Baule, A., Mari, R., Bo, L., Portal, L. and Makse, H.A., 2013. Mean-field theory of random close packings of axisymmetric particles. *Nature Communications*, **4**, 2194.

Bear, J., 1972. *Dynamics of Fluids in Porous Media*. Mineola, NY: Dover Publications, Inc.

Beggs, H.D. and Robinson, J.R., 1975. Estimating the viscosity of crude oil systems. *Journal of Petroleum Technology*, **27**, 1140–1141.

Behar, F., Vandenbroucke, M., Tang, Y., Marquid, F. and Espitalie, J., 1997. Thermal cracking of kerogen in open and closed systems: Determination of kinetic parameters and stoichiometric coefficients for oil and gas generation. *Organic Geochemistry*, **26**, 321–339.

Belikov, B.P., Alexandrov, T.W. and Rysova, T.W., 1970. *Elastic Properties of Rock Minerals and Rocks*. Moscow: Nauka.

Beltzer, A.I., 1988. *Acoustics of Solids*. Berlin: Springer-Verlag.

Ben-Menahem, A. and Singh, S., 1981. *Seismic Waves and Sources*. New York: Springer-Verlag.

Bennett, H.F., 1968. An investigation into velocity anisotropy through measurements of ultrasonic wave velocities in snow and ice cores from Greenland and Antarctica, PhD thesis, University of Wisconsin-Madison.

Bennison, T., 1998. Prediction of Heavy Oil Viscosity. IBC Heavy Oil Field Development Conference, London.

Benveniste, Y., 1987. A new approach to the application of Mori-Tanaka's theory in composite materials. *Mechanics of Materials*, **6**, 147–157.

Beran, M.J., 1968. *Statistical Continuum Theories*. New York: Wiley Interscience.

Beran, M.J., 1981. Field fluctuations in a random medium. In *Continuum Models of Discrete Systems 4*, ed. O. Brulin and R.K.T. Hsieh. North-Holland Publishing Co.

Beran, M.J. and Molyneux, J., 1966. Use of classical variational principles to determine bounds for the effective bulk modulus in heterogeneous media. *Quarterly of Applied Mathematics*, **24**, 107–118.

Berge, P.A., Berryman, J.G. and Bonner, B.P., 1993. Influence of microstructure on rock elastic properties. *Geophysical Research Letters*, **20**, 2619–2622.

Berge, P.A., Fryer, G.J. and Wilkens, R.H., 1992. Velocity-porosity relationships in the upper oceanic crust: Theoretical considerations. *Journal of Geophysical Research*, **97**, 15239–15254.

Bergman, D.J., 1978. The dielectric constant of a composite material in classical physics. *Physics Reports*, **43**, 377–407.

Bergman, D., 1978. The dielectric constant of a composite material – a problem in classical physics, Physics Letters C: Physics, **43**, 377–407.

Bergmann, L., 1954. *Der Ultraschall und seine Anwendung in Wissenschaft und Technik*. Zurich: S. Hirzel.

Berlincourt, D., Jaffe, H. and Shiozawa, L.R., 1963. Electroelastic properties of the sulfides, selenides, and tellurides of Zn and Cd. *Physical Review*, **129**, 1009–1017.

Bernal, J.D. and Mason, J., 1960. Coordination of randomly packed spheres. *Nature*, **188**, 910–911.

Bernstein, B.T., 1963. Elastic constants of synthetic sapphire at 27 degrees Celsius. *Journal of Applied Physics*, **34**, 169–172.

Berryman, J.G., 1980a. Confirmation of Biot's theory. *Applied Physics Letters*, **37**, 382–384.

Berryman, J.G., 1980b. Long-wavelength propagation in composite elastic media. *Journal of the Acoustical Society of America*, **68**, 1809–1831.

Berryman, J.G., 1981. Elastic wave propagation in fluid-saturated porous media. *Journal of the Acoustical Society of America*, **69**, 416–124.

Berryman, J.G., 1983. Dispersion of extensional waves in fluid-saturated porous cylinders at ultrasonic frequencies. *Journal of the Acoustical Society of America*, **74**, 1805–1812.

Berryman, J.G., 1988. Interpolating and integrating three-point correlation functions on a lattice. *Journal of Computational Physics*, **75**, 86–102.

Berryman, J.G., 1992a. Effective stress for transport properties of inhomogeneous porous rock. *Journal of Geophysical Research*, **97**, 17409–17424.

Berryman, J.G., 1992b. Single-scattering approximations for coefficients in Biot's equations of poroelasticity. *Journal of the Acoustical Society of America*, **91**, 551–571.

Berryman, J.G., 1993. Effective stress rules for pore-fluid transport in rocks containing two minerals. *International Journal of Rock Mechanics*, **30**, 1165–1168.

Berryman, J.G., 1995. Mixture theories for rock properties. In *Rock Physics and Phase Relations: A Handbook of Physical Constants*, ed. T.J. Ahrens. Washington, DC: American Geophysical Union, pp. 205–228.

Berryman, J.G., 2004. Measures of microstructure to improve estimates and bounds on elastic constants and transport coefficients in heterogeneous media, Lawrence Livermore National Lab report UCRL-JRNL-207118.

Berryman, J.G., 2005. Bounds and estimates for transport coefficients of random and porous media with high contrasts. *Journal of Applied Physics*, **97**, 063504_1–063504_11.

Berryman, J.G., 2008. Exact seismic velocities for transversely isotropic media and extended Thomsen formulas for stronger anisotropies. *Geophysics*, **73**, D1–D10.

Berryman, J.G., 2011. Bounds and self-consistent estimates for elastic constants of polycrystals composed of orthorhombics or crystals with higher symmetries. *Physical Review E,* **83,** 046130–1–046130–11.

Berryman, J.G. and Berge, P.A., 1996. Critique of two explicit schemes for estimating elastic properties of composites. *Mechanics of Materials,* **22,** 149–164.

Berryman, J.G. and Blair, S.C., 1986. Use of digital image analysis to estimate fluid permeability of porous materials: Application of two-point correlation functions. *Journal of Applied Physics,* **60,** 1930–1938.

Beskok, A. and Karniadakis, G.E., 1999. A model for flows in channels, pipes, and ducts at micro and nano scales. *Microscale Thermophysical Engineering,* **3,** 43–77.

Berryman, J.G. and Milton, G.W., 1988. Microgeometry of random composites and porous media. *Journal of Physics D,* **21,** 87–94.

Berryman, J.G. and Milton, G.W., 1991. Exact results for generalized Gassmann's equation in composite porous media with two constituents. *Geophysics,* **56,** 1950–1960.

Berryman, J.G., Grechka, V.Y. and Berge, P., 1999. Analysis of Thomsen parameters for finely layered VTI media. *Geophysical Prospecting,* **47,** 959–978.

Berryman, J.G., Pride, S.R. and Wang, H.F., 1992. A differential scheme for elastic properties of rocks with dry or saturated cracks. In *Proceedings 15th ASCE Engineering Mechanics Conference.*

Bhalla, A.S., Cook, W.R., Hearmon, R.F.S., *et al.,* 1984. Elastic, piezoelectric, pyroelectric, piezo-optic, electrooptic constants, and nonlinear dielectric susceptibilities of crystals. In *Landolt–Bornstein: Numerical Data and Functional Relationships in Science and Technology. Group III: Crystal and Solid State Physics,* vol. 18 (supplement to vol.), ed. K.-H. Hellwege and A.M. Hellwege. Berlin: Springer-Verlag.

Bhimasenacher, J., 1945. Elastic constants of calcite and sodium nitrate. *Proceedings of the Indian Academy of Sciences A,* **22,** 199–207.

Bigelow, E., 1992. *Introduction to Wireline Log Analysis.* Houston, TX: Western Atlas International, Inc.

Bilodeaux, B., 1997. *Shaley Sand Evaluation,* course notes. Stanford University.

Bina, C.R. and Wood, B.J., 1987. Olivine-Spinel transitions: Experimental and thermodynamic constraints and implications for the nature of the 400-km seismic discontinuity. *Journal of Geophysical Research,* **92,** 4853–4866.

Biot, M.A., 1956. Theory of propagation of elastic waves in a fluid saturated porous solid. I. Low frequency range and II. Higher-frequency range. *Journal of the Acoustical Society of America,* **28,** 168–191.

Biot, M.A., 1962. Mechanics of deformation and acoustic propagation in porous media. *Journal of Applied Physics,* **33,** 1482–1498.

Biot, M.A. and Willis, D.G., 1957. The elastic coefficients of the theory of consolidation. *Journal of Applied Mechanics,* **24,** 594–601.

Birch, F., 1960a. Elastic constants of rutile – a correction to a paper by R. K. Verma, "Elasticity of some high-density crystals." *Journal of Geophysical Research,* **65,** 3855–3856.

Birch, F., 1960b. The velocity of compressional waves in rocks to 10 kilobars. *Journal of Geophysical Research,* **65,** 1083–1102.

Birch, F., 1961. The velocity of compressional waves in rocks to 10 kilobars, Part 2. *Journal of Geophysical Research,* **66,** 2199–2224.

Birch, F., 1966. Compressibility; elastic constants. In *Handbook of Physical Constants,* ed. S.P. Clark, Geological Society of America, Memoir, vol. **97,** pp. 97–174.

Bishop, C.M., 2006. *Pattern Recognition and Machine Learning.* New York: Springer-Verlag.

Blair, D.P., 1990. A direct comparison between vibrational resonance and pulse transmission data for assessment of seismic attenuation in rock. *Geophysics,* **55,** 51–60.

Blakslee, O.L., Proctor, D.G., Seldin, E.J., Sperce, G.B. and Werg, T., 1970. Elastic constants of compression-annealed pyrolitic graphite. *Journal of Applied Physics*, **41**, 3373–3382.

Blangy, J.P., 1992. *Integrated seismic lithologic interpretation: The petrophysical basis*, PhD dissertation, Stanford University.

Bobeth, M. and Diener, G., 1986. Field fluctuations in multicomponent mixtures. *Journal of the Mechanics and Physics of Solids*, **34**, 1–17.

Boggs, S., 2001. *Principles of Sedimentology and Stratigraphy*. Upper Saddle River, NJ: Prentice-Hall.

Bogorodskiy, V.V., 1964. Uprugiye moduli kristalla l'da [Elastic moduli of ice crystals]. *Akusticheskiy Zhurnal*, **10**, 152–155. [Translation in Soviet Physics-Acoustics, 10, 124–126.]

Bordenave, M.L., Espitalié, J., Leplat, P., Oudin, J.L. and Vandenbroucke, M., 1993. Screening techniques for source rock evaluation. In *Applied Petroleum Geochemistry*, ed. M.L. Bordenave. Paris: Editions Technip, pp. 217–278.

Born, M. and Wolf, E., 1980. *Principles of Optics*, 6th ed. Oxford: Pergamon Press.

Bortfeld, R., 1961. Approximation to the reflection and transmission coefficients of plane longitudinal and transverse waves. *Geophysical Prospecting*, **9**, 485–503.

Boucher, S., 1974. On the effective moduli of isotropic two-phase elastic composites. *Journal of Composite Materials*, **8**, 82–89.

Bourbié, T., Coussy, O. and Zinszner, B., 1987. *Acoustics of Porous Media*. Houston, TX: Gulf Publishing Co.

Bowers, G.L., 1995. Pore pressure estimation from velocity data: Accounting for pore pressure mechanisms besides undercompaction. *SPE Drilling and Completion* (June), 89–95.

Boyse, W.E., 1986. *Wave propagation and inversion in slightly inhomogeneous media*, PhD dissertation, Stanford University.

Bracewell, R., 1965. *The Fourier Transform and Its Application*. New York: McGraw-Hill Book Co.

Bradford, I.D.R., Fuller, J., Thompson, P.J. and Walsgrove, T.R., 1998. *Benefits of assessing the solids production risk in a North Sea reservoir using elastoplastic modeling: SPE/ISRM 47360*. Papers presented at the *SPE/ISRM Eurock '98*, Trondheim, Norway, 8–10 July, pp. 261–269.

Bradley, W.B., 1979. Failure of inclined boreholes. *J. Energy Resources Tech, Trans., ASME*, **101**, 232–239.

Brandt, H., 1955. A study of the speed of sound in porous granular media. *Journal of Applied Mechanics*, **22**, 479–486.

Bratli, R.K., Horsrud, P. and Risnes, R., 1983. Rock mechanics applied to the region near a wellbore. *Proceedings 5th International Congress on Rock Mechanics*, F1–F17. Melbourne: International Society for Rock Mechanics.

Brethauer, G.E., 1974. Stress around pressurized spherical cavities in triaxial stress fields. *International Journal of Rock Mechanics and Mining Sciences & Geomechanics Abstracts*, **11**, 91–96.

Brevik, I., 1995. Chalk data, presented at workshop on effective media, Karlsruhe.

Brie, A., Pampuri, F., Marsala, A.F. and Meazza, O., 1995. Shear sonic interpretation in gas-bearing sands. *SPE*, **30595**, 701–710.

Brisco, B., Pultz, T.J., Brown, R.J., *et al.*, 1992. Soil moisture measurement using portable dielectric probes and time domain reflectometry. *Water Resources Research*, **28**, 1339–1346.

Bristow, J.R., 1960. Microcracks, and the static and dynamic elastic constants of annealed heavily coldworked metals. *Brit. Journal of Applied Physics*, **11**, 81–85.

Brito Dos Santos, W.L., Ulrych, T.J. and De Lima, O.A.L., 1988. A new approach for deriving pseudovelocity logs from resistivity logs. *Geophysical Prospecting*, **36**, 83–91.

Brocher, T., 2005. Relations between elastic wavespeeds and density in the Earth's crust. *Bulletin of the Seismological Society of America.*, **95**, 6, 2081–2092.

Brockamp, B. and Querfurth, H., 1965. Untersuchungen über die Elastizitätskonstanten von See- und Kunsteis, Polarforschung, Bd. 5. *Jahrg.*, **34**, Ht. 1–2. 253–262.

Brouwers, H.J.H., 2006. Particle-size distribution and packing fraction of geometric random packings. *Physical Review E*, **74**, 031309.

Brouwers, H.J.H., 2010. Viscosity of a concentrated suspension of rigid monosized particles. *Physical Review E*, **81**, 051402–1–051402–11.

Browaeys, J.T. and Chevrot, S., 2004. Decomposition of the elastic tensor and geophysical applications. *Geophysical Journal International*, **159**, 667–678.

Brown, G.G., 1966. *Unit Operations*. New York: J. Wiley.

Brown, R. and Korringa, J., 1975. On the dependence of the elastic properties of a porous rock on the compressibility of the pore fluid. *Geophysics*, **40**, 608–616.

Brown, R.J.S., 1980. Connection between formation factor for electrical resistivity and fluid-solid coupling factor in Biot's equations for acoustic waves in fluid-filled porous media. *Geophysics*, **45**, 1269–1275.

Brown, S.R. 1989. Transport of fluid and electric current through a single fracture. *Journal of Geophysical Research*, **89**, 9429–9438.

Brown, S.R. and Scholz, C.H., 1986. Closure of random elastic surfaces: I. Contact. *Journal of Geophysical Research*, **90**, 5531–5545.

Bruggeman, D.A.G., 1935. Berechnung verschiedener physikalischer Konstanten von heterogenen Substanzen. *Annalen der Physik, Leipzig*, **24**, 636–679.

Budiansky, B., 1965. On the elastic moduli of some heterogeneous materials. *Journal of the Mechanics and Physics of Solids*, **13**, 223–227.

Budiansky, B. and O'Connell, R.J., 1976. Elastic moduli of a cracked solid. *Int. J. Solids Structures*, **12**, 81–97.

Budiansky, L.V., 1993. Bounds on the effective moduli of composite materials, School on Homogenization, ICTP, Trieste, September 6–17.

Burnham, A.K., Peters, K.E. and Schenk, O., 2016. Evolution of vitrinite reflectance models, AAPG 2016 Annual Convention and Exhibition, Search & Discovery article 41982.

Butkovich, T.R., 1955. Crushing strength of lake ice, U.S. Army Cold Regions Research Engineering Laboratory, Research Report 15.

Cadoret, T., 1993. *Effet de la Saturation Eau/Gaz sur les Proprietes Acoustiques des Roches*, PhD dissertation, University of Paris, VII.

Carcione, J.M., Ursin, B. and Nordskag, J.I., 2007. Cross-property relations between electrical conductivity and the seismic velocity of rocks. *Geophysics*, **72**, E193–E204.

Carman, P.C., 1937. Flow through granular beds. *Transactions Institute of Chemical Engineering*, **15**, 150–166.

Carman, P.C., 1956. *Flow of Gases Porous Media*. Butterworth.

Carman, P.C., 1961. *L'ecoulement des Gaz a Travers les Milieux Poreux*. Paris: Bibliotheeque des Sciences et Techniques Nucleaires, Presses Universitaires de France.

Carmichael, R.S., 1989. *Practical Handbook of Physical Properties of Rocks and Minerals*. Boca Raton, FL: CRC Press.

Caro, M.A., 2014. Extended scheme for the projection of material tensors of arbitrary symmetry onto a higher symmetry tensor, arXiv preprint arXiv:1408.1219.

Castagna, J.P., 1993. AVO analysis-tutorial and review. In *Offset Dependent Reflectivity – Theory and Practice of AVO Analysis, Investigations in Geophysics*, No. 8, ed. J.P. Castagna and M. Backus. Tulsa, OK: Society of Exploration Geophysicists, pp. 3–36.

Castagna, J.P., Batzle, M.L. and Eastwood, R.L., 1985. Relationships between compressional-wave and shear wave velocities in clastic silicate rocks. *Geophysics*, **50**, 571–581.

Castagna, J.P., Batzle, M.L. and Kan, T.K., 1993. Rock physics – The link between rock properties and AVO response. In *Offset-Dependent Reflectivity – Theory and Practice of AVO Analysis*,

Investigations in Geophysics, No. 8, ed. J.P. Castagna and M. Backus. Tulsa, OK: Society of Exploration Geophysicists, pp. 135–171.

Chang, C., Zoback, M.D. and Khaksar, A., 2004. Rock strength and physical property measurements in sedimentary rocks. *SRB Annual Report*, vol. **96**, paper G4.

Chapman, M., 2003. Frequency-dependent anisotropy due to meso-scale fractures in the presence of equant porosity. *Geophysical Prospecting*, **51**, 369–379.

Chapman, M., Liu, E. and Li, X-Y., 2006. The influence of fluid-sensitive dispersion and attenuation on AVO analysis. *Geophysical Journal International*, **167**, 89–105.

Chapman, M., Zatsepin, S.V. and Crampin, S., 2002. Derivation of a microstructural poroelasticity model. *Geophysical Journal International*, **151**, 427–451.

Chelam, E.V., 1961. *Thesis*. Bangalore: Indian Institute of Science.

Chen, D.-L., Yang, P.-F. and Lai, Y.-S., 2013. On non-monotonicity of linear viscoelastic functions. *Mechanics and Physics of Solids*, 20, doi:10.1177/1081286513509100.

Chen, W., 1995. AVO in azimuthally anisotropic media: Fracture detection using P-wave data and a seismic study of naturally fractured tight gas reservoirs, PhD dissertation, Stanford University.

Chen, W.T., 1968. Axisymmetric stress field around spheroidal inclusions and cavities in transversely isotropic material. *Journal of Applied Mechanics*, 35, 770–773.

Cheng, C.H., 1978. *Seismic velocities in porous rocks: Direct and inverse problems*, ScD thesis, MIT, Cambridge, Massachusetts.

Cheng, C.H., 1993. Crack models for a transversely anisotropic medium. *Journal of Geophysical Research*, **98**, 675–684.

Choy, M.M., Cook, W.R., Hearmon, R.F.S., *et al.*, 1979. Elastic, piezoelectric, pyroelectric, piezo-optic, electrooptic constants, and nonlinear dielectric susceptibilities of crystals. In *Landolt—Bo'rnstein: Numerical Data and Functional Relationships in Science and Technology. Group III: Crystal and Solid State Physics*, vol. 11 (revised and extended edition of vols. and III/2), ed. K.-H. Kellwege and A.M. Hellwege. Berlin: Springer-Verlag.

Christensen, N.I., 1965. Compressional wave velocities in metamorphic rocks at pressures to 10 kb. *Journal of Geophysical Research*, **70**, 6147–6164.

Christensen, N.I., 1968. Compressional wave velocities in basic rocks. *Pacific Science*, **22**, 41–44.

Christensen, N.I., 1972. Elastic properties of polycrystalline magnesium, iron, and manganese carbonates to 10 kilobars. *Journal of Geophysical Research*, **77**, 369–372.

Christensen, N.I. and Mooney, W.D., 1995. Seismic velocity structure and composition of the continental crust: A global view. *Journal of Geophysical Research*, **100**, 9761–9788.

Christensen, N.I. and Wang, H.F., 1985. The influence of pore pressure and confining pressure on dynamic elastic properties of Berea Sandstone. *Geophysics*, **50**, 207–213.

Christensen, N.I., Wilkens, R.H., Blair, S.C. and Carlson, R.L., 1980. Seismic velocities, densities, and elastic constants of volcanic breccias and basalt from Deep Sea Drilling Project Leg 59. In *1980, Initial Reports of the Deep Sea Drilling Project*, vol. LIX, ed. L. Kroenke, R. Scott, *et al.* Washington: U.S. Government Printing Office, 515–517.

Christensen, R.M., 1982. *Theory of Viscoelasticity*. Academic Press.

Christensen, R.M., 1991. *Mechanics of Composite Materials*. Malabar, FL: Robert E. Krieger Publication Co.

Christensen, R.M., 2005. *Mechanics of Composite Materials*. New York: Dover Publications.

Cioni, R., Civetta, L., Marianelli, P., Metrich, N., Santacroce, R. and Sbrana, A.A., 1995. Compositional layering and syn-eruptive mixing of a periodically refilled shallow magma chamber: The AD 79 Plinian eruption of Vesuvius. *Journal of Petrology*, **36**, 739–776, doi:10.1093/petrology/36.3.739.

Civan, F., 2010. Effective correlation of apparent gas permeability in tight porous media. *Transport in Porous Media*, **82**, 375–384.

Ciz, R. and Shapiro, S., 2007. Generalization of Gassmann equations for porous media saturated with a solid material. *Geophysics*, **72**, A75–A79.

Ciz, R., Siggins, A.F., Gurevich, B. and Dvorkin, J., 2008. Influence of microheterogeneity on effectives properties of rocks. *Geophysics*, **73**, E7–E14.

Claerbout, J.F., 1985. *Fundamentals of Geophysical Data Processing*. Palo Alto, CA: Blackwell Scientific Publications.

Claerbout, J.F., 1992. *Earth Sounding Analysis: Processing versus Inversion*. Boston, MA: Blackwell Scientific Publications.

Clark, V.A., Tittmann, B.R. and Spencer, T.W., 1980. Effect of volatiles on attenuation (Q-1) and velocity in sedimentary rocks. *Journal of Geophysical Research*, **85**, 5190.

Clavier, C., Coates, G. and Dumanoir, J., 1984. Theoretical and experimental bases for the dual-water model for interpretation of shaley sands. *Society of Petroleum Engineers Journal*, **24**, 153–168.

Cleary, M.P., Chen, I.-W. and Lee, S.-M., 1980. Self-consistent techniques for heterogeneous media. *American Society of Civil Engineers Journal of Engineering Mechanics*, **106**, 861–887.

Clennell, M.B., 1997. Tortuosity: A guide through the maze. In *Developments in Petrophysics*, ed.M. A. Lovell and P.K. Harvey. London: Geological Society of London, pp. 299–344.

Coates, R., Pierce, G. and Fung, H., 2005. Impact of methane loss on Bitument Viscosity-Joint Industry Project, Alberta Research Council.

Cole, K.S. and Cole, R.H., 1941. Dispersion and absorption in dielectrics I. Alternating current characteristics. *Journal of Chemical Physics*, **9**, 341–351.

Connolly, P., 1998. Calibration and inversion of non-zero offset seismic. *Soc. Expl. Geophysics*, 68th Annual Meeting, Expanded Abstracts. Tulsa, OK: Society of Exploration Geophysicists.

Connolly, P., 1999. Elastic impedance. *The Leading Edge*, **18**, 438–152.

Consiglio, R., Baker, D.R., Paul, G. and Stanley, H., 2003. Continuum percolation thresholds for mixtures of spheres of different sizes. *Physica A*, **319**, 49–55.

Coombs, M. and Gardner, J.E., 2001. Shallow-storage conditions for the rhyolite of the 1912 eruption at Novarupta, Alaska. *Geology*, **29**, 775–778, doi:10.1130/0091-7613(2001)029<0775: SSCFTR>2.0.CO;2.

Corson, P.B., 1974. Correlation functions for predicting properties of heterogeneous materials. *Journal of Applied Physics*, **45**, 3159–3179.

Cottrell, E., Gardner, J.E. and Rutherford, M.J., 1999. Petrologic and experimental evidence for the movement and heating of the pre-eruptive Minoan rhyodacite. *Contributions to Mineralogy and Petrology*, **135**, 315–331, doi:10.1007/ s004100050514.

Cowin, S.C., Yang, G. and Mehrabadi, M.M., 1999. Bounds on the effective anisotropic elastic constants. *Journal of Elasticity*, **57**, 1–24.

Crain, E.R. and Holgate, P., 2014. A 12-step program to reduce uncertainty in kerogen-rock reservoirs, Presented at the GeoConvention: Focus Conference, Calgary, May12–16.

Cruts, H.M.A., Groenenboom, J., Duijndam, A.J.W. and Fokkema, J.T., 1995. Experimental verification of stress-induced anisotropy. *Expanded Abstracts, Society of Exploration Geophysicists*, 65th Annual International Meeting, pp. 894–897.

Cumberland, D.J. and Crawford, R.J., 1987. The packing of particles. In *Handbook of Powder Technology*, vol. 6. New York: Elsevier.

Curie, P., 1894. On symmetry in physical phenomena, symmetry of an electric field and of a magnetic field. J*ournal de Physique*, **3**, 401.

Curtis, M.E., Ambrose, R.J., Sondergeld, C.H. and Rai, C.S. 2011. Investigation of the relationship between organic porosity and thermal maturity in the Marcellus Shale, North American Unconventional Gas Conference and Exhibition, SPE, Paper 144370.

Dagan, G., 1993. Higher-order correction for effective permeability of heterogeneous isotropic formations of lognormal conductivity distribution. *Transport in Porous Media*, **12**, 279–290.

Daly, A.R. and Edman, J.D., 1987. Loss of organic carbon from source rocks during thermal maturation, AAPG Bulletin, 7.

Dandekar, D.P., 1968. Pressure dependence of the elastic constants of calcite. *Physical Review*, **172**, 873.

Dantl, G., 1968. Die elastischen Moduln von Eis-Einkristallen. *Physik der kondensierten Materie*, Bd. **7**, Ht. 5, 390–97.

Dantl, G., 1969. Elastic moduli of ice. In *Physics of Ice: Proceedings of the International Symposium on Physics of Ice*, September 9–14, 1968, ed. N. Riehl, B. Bullemer, H. Engelhardt. Munich: Plenum Press, pp. 223–30.

Darcy, H., 1856. *Les Fontaines Publiques de la Ville de Dijon*. Paris: Dalmont.

Das, V., Mukerji, T. and Mavko, G., 2019. Scale effects of velocity dispersion and attenuation in layered viscoelastic medium. *Geophysics*, **84**, T147–T166.

Debye, P., 1945. *Polar Molecules*. Mineola, NY: Dover.

De Ghetto, G., Paone, F. and Villa, M., 1995. Pressure-volume-temperature correlations for heavy and extra heavy oils: International heavy oil symposium. *SPE*, **30316**, 647–662.

Dejtrakulwong, P., 2012. *Rock physics and seismic signatures of sub-resolution sand-shale system*, PhD dissertation, Stanford University.

Dejtrakulwong, P. and Mavko, G., 2016. Fluid substitution for thinly interbedded sand-shale sequences using the mesh method. *Geophysics*, **81**, D599–D609.

de la Cruz, V. and Spanos, T.J.T., 1985. Seismic wave propagation in a porous medium. *Geophysics*, **50**, 1556–1565.

Delameter, W.R., Hermann, G. and Barnett, D.M., 1974. Weakening of an elastic solid by a rectangular array of cracks. *Journal of Applied Mechanics*, **42**, 74–80.

Dellinger, J., 2005. Computing the optimal transversely isotropic approximation of a general elastic tensor. *Geophysics*, **70**, I1–I10.

Dellinger, J., Vasicek, D. and Sondergeld, C., 1998. Kelvin notation for stabilizing elastic-constant inversion. *Revue de l'Institut Français du Petrol*, **53**, 709–719.

Dellinger, J. and Vernik, L., 1992. Do core sample measurements record group or phase velocity? Society of Exploration Geophysicists, 62nd Annual International Meeting, Expanded Abstracts, pp. 662–665.

Demaison, G.J., Holck, A.J.J., Jones, R.W. and Moore, G.T., 1983. Predictive Source bed stratigraphy; a guide to regional petroleum occurrence. Proceedings of the 11th World Petroleum Congress, London, **2**, 17–29.

Denton, W.H., 1957. The packing and flow of spheres. *AERE Report E/R*, 1095.

Desbrandes, R., 1985. *Encyclopedia of Well Logging*. Houston, TX: Gulf Publishing Company.

Dewey, J.M., 1947. The elastic constants of materials loaded with non-rigid fillers. *Journal of Applied Physics*, **18**, 578–581.

Dias, R., Teixeira, J.A., Mota, M. and Yelshin, A., 2006. Tortuosity variation in a low density binary particulate bed. *Separation and Purification Technology*, **51**, 180–184.

di Carlo, I., Pichavant, M., Rotolo, S.G. and Scaillet, B., 2006. Experimental crystallization of a high-K arc basalt: The golden pumice, Stromboli volcano (Italy). *Journal of Petrology*, **47**, 1317–1343, doi:10.1093/petrology/egl011.

Diez, A. and Eisen, O., 2015. Seismic propagation in anisotropic ice – Part 1: Elasticity tensor and derived quantities from ice-core properties. *The Cryosphere*, **9**, 367–384.

Diez, A., Eisen, O., Hofstede, C., Lambrecht, A., Mayer, C., Miller, H., Steinhage, D., Binder, T. and Weikusat, I., 2015. Seismic propagation in anisotropic ice – Part 2: Effects of crystal anisotropy in geophysical data. *The Cryosphere*, **9**, 385–398.

Digby, P.J., 1981. The effective elastic moduli of porous granular rocks. *Journal of Applied Mechanics*, **48**, 803–808.

Dingwell, D.B. and Webb, S.L., 1990. Structural relaxation in silicate melts. *European Journal of Mineralogy*, **2**, 427–444.

Dirksen, O., Humphreys, M.C.S., Pletchov, P., Melnik, O., Demyanchuk, Y., Sparks, R.S.J. and Mahony, S., 2006. The 2001–2004 dome-forming eruption of Shiveluch volcano, Kamchatka: Observation, petrological investigation and numerical modelling. *Journal of Volcanology and Geothermal Research*, **155**, 201–226, doi:10.1016/j.jvolgeores.2006.03.029.

Dix, C.H., 1955. Seismic velocities from surface measurements. *Geophysics*, **20**, 68–86.

Domenico, S.N., 1976. Effect of brine-gas mixture on velocity in an unconsolidated sand reservoir. *Geophysics*, **41**, 882–894.

Donev, A., Cisse, I., Sachs, D., Variano, E., Stillinger, F., Connelly, R., Torquato, S. and Chaikin, P., 2004. Improving the density of jammed disordered packings using ellipsoids. *Science*, **303**, 990–993.

Doraiswami, M.S., 1947. Elastic constants of magnetite, pyrite, and chromite. *Proceedings of the Indian Academy of Sciences, A*, **25**, 414-H6.

Duda, A., Koza, Z. and Matyka, M., 2011. Hydraulic tortuosity in arbitrary porous media flow. *Physical Review E*, **84**, 036319.1–8.

Duda, R.O., Hart, P.E. and Stork, D.G., 2000. *Pattern Classification*. New York: John Wiley & Sons.

Duffaut, K., Alsos, T., Landr0, M., Rogn0, H. and Al-Najjar, N.F., 2000. Shear-wave elastic impedance. *Leading Edge*, **19**, 1222–1229.

Duffy, T.S. and Anderson, D.L., 1989. Seismic velocities in mantle minerals and the mineralogy of the upper mantle. *Journal of Geophysical Research*, **94**, 1895–1912.

Duffy, T.S. and Vaughan, M.T., 1988. Elasticity of enstatite and its relationship to crystal structure. *Journal of Geophysical Research*, **93**, 383–391.

Dullien, F.A.L., 1979. *Porous Media: Fluid Transport and Pore Structure*. Academic Press.

Dullien, F.A.L., 1991. One and two phase flow in porous media and pore structure. In *Physics of Granular Media*, ed. D. Bideau and J. Dodds. New York: Science Publishers Inc., pp. 173–214.

Dullien, F.A.L., 1992. *Porous Media: Fluid Transport and Pore Structure*. San Diego, CA: Academic Press.

Dunbar, N.W., Hervig, R.L. and Kyle, P.R., 1989a. Determination of pre-eruptive H2O, F and Cl contents of silicic magmas using melt inclusions: Examples from Taupo volcanic center, New Zealand. *Bulletin Volcanology*, **51**, 177–184, doi:10.1007/BF01067954.

Dunbar, N.W., Kyle, P.R. and Wilson, C.J.N., 1989b. Evidence for limited zonation in silicic magma systems, Taupo volcanic zone, New Zealand. *Geology*, **17**, 234–236, doi:10.1130/0091-7613 (1989) 017<0234:EFLZIS>2.3.CO;2.

Du Plessis, J.P. and Masliyah, J.H., 1991. Flow through isotropic granular porous media. *Transport in Porous Media*, **6**, 207–221.

Dusseault, 2006. Mechanics of Heavy Oil, short course, US Society of Rock Mechanics.

Dutta, N.C. and Ode, H., 1979. Attenuation and dispersion of compressional waves in fluid-filled porous rocks with partial gas saturation (White model) – Part 1: Biot theory, Part II: Results. *Geophysics*, **44**, 1777–1805.

Dutta, N.C. and Seriff, A.J., 1979. On White's model of attenuation in rocks with partial gas saturation. *Geophysics*, **44**, 1806–1812.

Dvorkin, J.P. and Mavko, G., 2006. Modeling attenuation in reservoir and nonreservoir rock. *Leading Edge*, **25**, 194–197.

Dvorkin, J., Mavko, G. and Gurevich, B., 2007. Fluid substitution in shaley sediment using effective porosity. *Geophysics*, **72**, O1–O8.

Dvorkin, J., Mavko, G. and Nur, A., 1995. Squirt flow in fully saturated rocks. *Geophysics*, **60**, 97–107.

Dvorkin, J., Nolen-Hoeksema, R. and Nur, A., 1994. The squirt-flow mechanism: Macroscopic description. *Geophysics*, **59**, 428–438.

Dvorkin, J. and Nur, A., 1993. Dynamic poroelasticity: A unified model with the squirt and the Biot mechanisms. *Geophysics*, **58**, 524–533.

Dvorkin, J. and Nur, A., 1996. Elasticity of high-porosity sandstones: Theory for two North Sea datasets. *Geophysics*, **61**, 1363–1370.

Eastwood, R.L. and Castagna, J.P., 1986. Interpretation of *VP/VS* ratios from sonic logs. In *Shear Wave Exploration, Geophysical Developments*, no. 1, ed. S.H. Danbom and S.N. Domenico. Tulsa, OK: Society of Exploration Geophysicists.

Eaton, B.A., 1975. The equation for geopressure prediction from well logs. *Paper SPE 5544*. Houston, TX: Society of Petroleum Engineers.

Eberhart-Phillips, D.M., 1989. *Investigation of crustal structure and active tectonic processes in the coast ranges, Central California*, PhD dissertation, Stanford University.

Efron, B. and Tibshirani, R.J., 1993. *An Introduction to the Bootstrap*. New York: Chapman and Hall.

Eimer, C., 1967. Stresses in multi-phase media. *Archives of Mechanics*, **19**, 521.

Eimer, C., 1968. The boundary effect in elastic multiphase bodies. *Arch. Mech. Stos.*, **20**, 87.

Einspruch, N.G. and Manning, R.J., 1963. Elastic constants of compound semi-conductors ZnS, PbTe, GaSb. *Journal of the Acoustical Society of America*, **35**, 215–216.

Einstein, A., 1906. Eine neue Bestimmung der MolekuÈldimensionen, Annal. *Physik*, **19**, 289–306. English translation in Investigation on the Theory of Brownian Motion, Dover, New York, 1956.

Eischen, J.W. and Torquato, S., 1993. Determining elastic behavior of composites by the boundary element method. *Journal of Applied Physics*, **74**, 159–170.

Eissa, E.A. and Kazi, A., 1988. Relation between static and dynamic Young's moduli of rocks. *International Journal of Rock Mechanics*, **25**, 479–482.

Ellis, D., Howard, J., Flaum, C., *et al.*, 1988. Mineral logging parameters: Nuclear and Acoustical Technology Review, **36**(1), 38–55.

Elmore, W.C. and Heald, M.A., 1985. *Physics of Waves*. Mineola, NY: Dover Publications, Inc.

Elrod, H.G., 1979. A general theory for laminar lubrication with Reynolds roughness. *Journal of Lubrication Technology*, **101**, 8–14.

Endres, A.L. and Knight, R., 1992. A theoretical treatment of the effect of microscopic fluid distribution on the dielectric properties of partially saturated rocks. *Geophysical Prospecting*, **40**, 307–324.

Epstein, P.S., 1941. On the absorption of sound waves in suspensions and emulsions. In *Theodore Von Karmen Anniversary Volume*, pp. 162–188.

Epstein, P.S. and Carhart, R.R., 1953. The absorption of sound in suspensions and emulsions: I. Water fog in air. *Journal of the Acoustical Society of America*, **25**, 553–565.

Eshelby, J.D., 1957. The determination of the elastic field of an ellipsoidal inclusion, and related problems. *Proceedings of the Royal Society, London A*, **241**, 376–396.

Fabricius, I.L., 2003. How burial diagenesis of chalk sediments controls sonic velocity and porosity. *AAPG Bulletin*, **87**, 1755–1778.

Fabrikant, V.I., 1989. *Applications of Potential Theory in Mechanics. Selection of New Results*, Kluwer Academic.

Farris, R.J., 1968. Prediction of the viscosity of multimodal suspensions from unimodal viscosity data. *Transactions of the Society of Rheology*, **12**, 281.

Faure, S., Lefebvre-Lepot, A. and Semin, B., 2009. Dynamic numerical investigation of random packing for spherical and nonconvex particles. *ESAIM: Proceedings*, **28**, 13–32.

Faust, L.Y., 1953. A velocity function including lithologic variation. *Geophysics*, **18**, 271–288.

Fedorov, F.I., 1968. *Theory of Elastic Waves in Crystals*. Plenum.

Finney, J.L., 1970. Random packings and the structure of simple liquids – I. The geometry of random close packing. *Proceedings of the Royal Society London A*, **319**, 479–493.

Firouzi, M. and Hashemabadi, S.H., 2009. Analytic solution for Newtonian laminar flow through the concave and convex ducts. *Journal of Fluids Engineering*, **131**, 094501–1–094501–6.

Fjaer, E., Holt, R.M., Horsrud, P., Raaen, A.M. and Risnes, R., 2008. *Petroleum Related Rock Mechanics*. Amsterdam: Elsevier.

Florez, J.-M., 2005. Integrating geology, rock physics, and seismology for reservoir quality prediction, PhD dissertation, Stanford University.

Focke, J.W. and Munn, D., 1987. Cementation exponents (**m**) in Middle Eastern carbonate reservoirs. *Society of Petroleum Engineers Formation Evaluation*, **2**, 155–167.

Folk, R.L. and Ward, W., 1957. Brazos river bar: A study in the significance of grain-size parameters. *Journal of Sedimentary Petrology*, **27**, 3–26.

Formation Evaluation Data Handbook, 1982. Fort Worth, TX: Gearhart Industries, Inc.

Frazer, L.N., 1994. A pulse in a binary sediment. *Geophysical Journal International*, **118**, 75–93.

François, D., Pineau, A. and Zaoui, A., 2013. Mechanical behaviors of materials, volume II: Fracture mechanics and damage. In *Solid Mechanics and Its Applications*, ed. G.M.L. Gladwell. New York: Springer Publishing.

Frankel, J., Rich, F.J. and Homan, C.G., 1976. Acoustic velocities in polycrystalline NaCl at 300 K measured at static pressures from 25 to 27 kbar. *Journal of Geophysical Research*, **81**, 6357–6363.

Frankel, N.A. and Acrivos, A., 1970. The constitutive equation for a dilute emulsion. *Journal of Fluid Mechanics*, **44**, 65–78.

Freyburg, E., 1972. Der Untere und mittlere Buntsandstein SW-Thuringen in seinen gesteinstechnicschen Eigenschaften. *Berichte der Deutsche. Gesellschaft Geologische Wissenschaften A*, **176**, 911–919.

Fukunaga, K., 1990. *Introduction to Statistical Pattern Recognition*. Boston, MA: Academic Press.

Galford, J., Quirein, J., Westcott, D. and Witkowsky, J., 2013. SPWLA 54th Annual Logging Symposium, June 22–26, 2013.

Gammon, P.H., Kiefte, H. and Clouter, M.J., 1980. Elastic constants of ice by Brillouin spectroscopy. *Journal of Glaciol*ogy, **25**, 159–167.

Gammon, P.H., Kiefte, H., Clouter, M.J. and Denner, W.W., 1983. *Journal of Glaciology*, **29**, 433–460.

Gangi, A.F. and Carlson, R.L., 1996. An asperity-deformation model for effective pressure. *Tectonophysics*, **256**, 241–251.

Garboczi, E.J., Snyder, K.A., Douglas, J.F. and Thorpe, M.F., 1995. Geometrical percolation threshold of overlapping ellipsoids. *Physical Review E*, **52**, 819–828.

Garcia, M.O. and Wolfe, E.W., 1988. Petrology of the erupted lava, in The Puu Oo Eruption of Kilauea Volcano, Hawaii: Episodes 1 Through 20, January 3, 1983, Through June 8, 1984, U.S. Geol. Survey Professional Paper 1463, 127–143.

García, X., Araujo, M. and Medina, E., 2004. P-wave velocity–porosity relations and homogeneity lengths in a realistic deposition model of sedimentary rock. *Waves Random Media*, **14**, 129–142.

Garcia, X. and Medina, E.A., 2006. Hysteresis effects studied by numerical simulations: Cyclic loading-unloading of a realistic sand model. *Geophysics*, **71**, F13–F20.

Gardner, G.H.F., 1962. Extensional waves in fluid-saturated porous cylinders. *Journal of the Acoustical Society of America*, **34**, 36–40.

Gardner, G.H.F., Gardner, L.W. and Gregory, A.R., 1974. Formation velocity and density – the diagnostic basics for stratigraphic traps. *Geophysics*, **39**, 770–780.

Gardner, J.E., Rutherford, M., Carey, S. and Sigurdsson, H., 1995. Experimental constraints on pre-eruptive water contents and changing magma storage prior to explosive eruptions of Mount St. Helens volcano. *Bulletin of Volcanology*, **57**, 1–17, doi:10.1007/BF00298703.

Gassmann, F., 1951. Über die Elastizität poröser Medien. *Vierteljahrsschrift der Naturforschenden Gesellschaft in Zurich*, **96**, 1–23.

Geertsma, J., 1961. Velocity-log interpretation: The effect of rock bulk compressibility. *Society of Petroleum Engineers Journal*, **1**, 235–248.

Geertsma, J. and Smit, D.C., 1961. Some aspects of elastic wave propagation in fluid-saturated porous solids. *Geophysics*, **26**, 169–181.

Gelinsky, S. and Shapiro, S., 1997a. Poroelastic Backus averaging for anisotropic layered fluid- and gas-saturated sediments. *Geophysics*, **62**, 1867–1878.

Gelinsky, S. and Shapiro, S., 1997b. Dynamic equivalent medium model for thickly layered saturated sediments. *Geophysical Journal International*, **128**, F1–F4.

Gelinsky, S., Shapiro, S., Muller, T. and Gurevich, B., 1998. Dynamic poroelasticity of thinly layered structures. *International Journal of Solids and Structures*, **35**, 4739–4751.

Ghanbarian, B., Hunt, A.G., Ewing, R.P. and Sahimi, M., 2013. *Soil Science Society of America Journal*, **77**, 1461–1477.

Ghiorso, M.S. and Sack, R.O., 1995. Chemical mass transfer in magmatic processes IV. A revised and internally consistent thermodynamic model for the interpolation and extrapolation of liquid-solid equilibria in magmatic systems at elevated temperatures and pressures. *Contributions to Mineralogy and Petrology*, **119**, 197–212, doi:10.1007/BF00307281.

Gibiansky, L.V., 1993. Bounds on the Effective Moduli of Composite Materials, School on Homogenization, ICTP, Trieste, September 6–17.

Gibiansky, L.V. and Sigmund, O., 2000. Multi phase composites with extremal bulk modulus. *Journal of the Mechanics and Physics of Solids*, **48**, 461–498.

Gibiansky, L.V. and Torquato, S., 1996. Bounds on the effective moduli of cracked materials. *Journal of the Mechanics and Physics of Solids*, **44**, 233–242.

Gibiansky, L. and Torquato, S., 1998. Rigorous connection between physical properties of porous rocks. *Journal of Geophysical Research*, **103**, 23911–23923.

Gibson, I.L. 1970. A pantelleritic welded ash-flow tuff from the Ethiopian Rift Valley. *Contributions to Mineralogy and Petrology*, **28**, 89–111, doi:10.1007/ BF00404992.

Gibson, R.L. and Toksoz, M.N., 1990. Permeability estimation from velocity anisotropy in fractured rock. *Journal of Geophysical Research*, **95**, 15643–15656.

Giordano, D., Russell, J.K. and Dingwell, D.B., 2008. Viscosity of magmatic liquids: A model. *Earth and Planetary Science Letters*, **271**, 123–134.

Glover, P.W.J., 2010. A generalized Archie's law for n phases. *Geophysics*, **75**, E247–E265.

Glover, P.W.J., Hole, M.J. and Pous, J., 2000. A modified Archie's law for two conducting phases. *Earth and Planetary Sciences Letters*, **180**, 369–383.

Goddard, J.D., 1990. Nonlinear elasticity and pressure-dependent wave speeds in granular media. *Proceedings of the Royal Society A*, **430**, 105–131.

Godfrey, J.J., Beaudoin, B.C. and Klemperer, S.L., 1997. Ophiolitic basement to the Great Valley forearc basin, California, from seismic and gravity data: Implications for crustal growth at the North American continental margin. *Geological Society of America Bulletin*, **109**, 1536–1562.

Goepfert, K. and Gardner, J.E., 2010. Influence of pre-eruptive storage conditions and volatile contents on explosive Plinian style eruptions of basic magma. *Bulletin of Volcanology*, **72**, 511–521, doi:10.1007/s00445-010-0343-1.

Gold, N., Shapiro, S., Bojinski, S. and Muller, T.M., 2000. An approach to upscaling for seismic waves in statistically isotropic heterogeneous elastic media. *Geophysics*, **65**, 1837–1850.

Golubev, A.A. and Rabinovich, G.Y., 1976. Resultaty primeneia appartury akusticeskogo karotasa dlja predeleina proconstych svoistv gornych porod na mestorosdeniaach tverdych isjopaemych. *Prikladnaja Geofizika Moskva*, **73**, 109–116.

González, E.F., 2006. Physical and quantitative interpretation of seismic attributes for rocks and fluids identification, PhD dissertation, Stanford University.

Goodman, R.E. 1976. *Methods of Geological Engineering in Discontinuous Rocks*. New York: West Publishing.

Gorjainov, N.N. and Ljachowickij, F.M., 1979. *Seismic Methods in Engineering Geology*. Moscow: Nedra.

Graham, E.K., Jr. and Barsch, G.R., 1969. Elastic constants of single-crystal forsterite as a function of temperature and pressure. *Journal of Geophysical Research*, **74**, 5949–5960.

Graham, E.K. and Kim, H.S., 1986. The bulk modulus of wustite (FeO) (abstract). *EOS Transactions AGU*, **67**, 1240.

Graham, E.K., Sopkin, S.M. and Resley, W.E., 1982. Elastic constants of fayalite, Fe2SiO4 and the olivine solution series (abstract). *EOS Transactions AGU*, **63**, 1090.

Gray, M., 2010. SPE distinguished lecture.

Gray, D., Goodway, B. and Chen, T., 1999. Bridging the gap: AVO to detect changes in fundamental elastic constants. In *Expanded Abstract, SEG International Meeting*. Tulsa, OK: Society of Exploration Geophysicists.

Grechka, V. and Tsvankin, I., 1998. 3-D description of normal moveout in anisotropic inhomogeneous media. *Geophysics*, **63**, 1079–1092.

Greenberg, M.L. and Castagna, J.P., 1992. Shear-wave velocity estimation in porous rocks: Theoretical formulation, preliminary verification and applications. *Geophysical Prospecting*, **40**, 195–209.

Greenhalgh, S.A. and Emerson, D.W., 1986. Elastic properties of coal measure rock from the Sydney Basin, New South Wales. *Exploration Geophysics*, **17**, 157–163.

Greenwood, J.A. and Williamson, J., 1966. Contact of nominally flat surfaces. *Proceedings of the Royal Society, London A*, **295**, 300–319.

Gubernatis, J.E. and Krumhansl, J.A., 1975. Macroscopic engineering properties of polycrystalline materials: Elastic properties. *Journal of Applied Physics*, **46**, 1875.

Guéguen, Y. and Palciauskas, V., 1994. *Introduction to the Physics of Rocks*. Princeton, NJ: Princeton University Press.

Guodong, J., Patzek, T.D. and Silin, D.B., 2004. SPE90084: Direct prediction of flow properties of unconsolidated and consolidated reservoir rocks from image analysis. In SPE Annual Technical Conference and Exhibition, Houston, Texas, USA.

Gurevich, B., 2004. A simple derivation of the effective stress coefficient for seismic velocities in porous rocks. *Geophysics*, **69**, 393–397.

Gurevich, B. and Lopatnikov, S.L., 1995. Velocity and attenuation of elastic waves in finely layered porous rocks. *Geophysical Journal International*, **121**, 933–947.

Gurevich, B., Makarynska, D. and Pervukhina, M., 2009. Ultrasonic moduli for fluid-saturated rocks: Mavko-Jizba relations rederived and generalized. *Geophysics*, **74**, N25–N30.

Guth, E. and Simha, R., 1936. Untersuchungen über die Viskosität von Suspensionen und Lösungen. *Kolloid Zeitschrift*, **74**, 266.

Gutierrez, M.A., Braunsdorf, N.R. and Couzens, B.A., 2006. Calibration and ranking of pore-pressure prediction models. *Leading Edge*, **25**(12), 1516–1523.

Gvirtzman, H. and Roberts, P., 1991. Pore-scale spatial analysis of two immiscible fluids in porous media. *Water Resources Research*, **27**, 1165–1176.

Hacikoylu, P., Dvorkin, J. and Mavko, G., 2006. Resistivity-velocity transforms revisited. *Leading Edge*, **25**, 1006–1009.

Haji-Akbari, A., Engel, M., Keys, A.S., Zheng, X., Petschek, R.G., Palffy-Muhoray, P. and Glotzer, S. G., 2009. Disordered, quasicrystalline and crystalline phases of densely packed tetrahedra. *Nature*, **462**, 773–777.

Halleck, P.M., 1973. The compression and compressibility of grossularite garnet: A comparison of X-ray and ultrasonic methods, PhD thesis, University of Chicago.

Hamilton, G.M., 1983. Explicit equation for the stresses beneath a sliding spherical contact, Proceedings. *Institution of Mechanical Engineers*, **197**, 53–59.

Hammer, J.E., Rutherford, M.J. and Hildreth, W., 2002. Magma storage prior to the 1912 eruption at Novarupta, Alaska. *Contributions to Mineralogy and Petrology*, **144**, 144–162, doi:10.1007/s00410-002-0393-2.

Han, D.-H., 1986. *Effects of porosity and clay content on acoustic properties of sandstones and unconsolidated sediments*, PhD dissertation, Stanford University.

Han, D-H. and Batzle, M., 2000a. Velocity, density, and modulus of hydrocarbon fluids – Data measurement, Expanded Abstracts, SEG Annual Meeting.

Han, D-H. and Batzle, M., 2000b. Velocity, density, and modulus of hydrocarbon fluids – Empirical modeling, Expanded Abstracts, SEG Annual Meeting.

Han, D.-H., Nur, A. and Morgan, D., 1986. Effects of porosity and clay content on wave velocities in sandstones. *Geophysics*, **51**, 2093–2107.

Han, T., 2010. *Joint elastic-electrical properties of reservoir sandstones*, PhD dissertation, University of Southampton.

Hanai, T., 1968. Electrical properties of emulsions. In *Emulsion Science*, ed. P. Sherman. New York: Academic Press, pp. 353–478.

Haniford, V.M. and Weidner, D.J., 1985. Elastic properties of grossular-andradite garnets Ca_3 $(Al,Fe)Si_3O_{12}$ (abstract). *EOS Transactions AGU*, **66**, 1063.

Hanson, M.T., 1992. The elastic field for conical indentation including sliding friction for transverse isotropy. *Applied Mechanics*, **59**, SI23–S130.

Harms, E., Gardner, J.E. and Schmincke, H.-U., 2004. Phase equilibria of the Lower Laacher See Tephra (East Eifel, Germany): Constraints on preeruptive storage conditions of a phonolitic magma reservoir. *Journal of Volcanology and Geothermal Research*, **134**, 125–138, doi:10.1016/j.jvolgeores.2004.01.009.

Hashin, Z., 1965. On elastic behavior of fibre reinforced materials of arbitrary transverse phase geometry. *Journal of the Mechanics and Physics of Solids*, **13**, 119–134.

Hashin, Z., 1983. Analysis of composite materials – A survey. *Journal of Applied Mechanics*, **50**,481–505.

Hashin, Z. and Shtrikman, S., 1962a. A variational approach to the theory of effective magnetic permeability of multiphase materials. *Journal of Applied Physics*, **33**, 3125–3131.

Hashin, Z. and Shtrikman, S., 1962b. A variational approach to the theory of the elastic behaviour of polycrystals. *Journal of Mechanics and Physics of Solids*, **10**, 343–352.

Hashin, Z. and Shtrikman, S., 1962c. On some variational principles in anisotropic and nonhomogeneous elasticity. *Journal of Mechanics and Physics of Solids*, **10**, 335–342.

Hashin, Z. and Shtrikman, S., 1963. A variational approach to the elastic behavior of multiphase materials. *Journal of the Mechanics and Physics of Solids*, **11**, 127–140.

Hastie, T., Tibshirani, R. and Freidman, J., 2001. *The Elements of Statistical Learning: Data Mining, Inference, and Prediction*. New York: Springer-Verlag.

Hay, J. and Herbert, E., 2011. Measuring the complex modulus of polymers by instrumented indentation testing: Experimental Techniques. doi:10.1111/j.1747-1567.2011.00732.x.

Heard, H.C., Abey, A.E., Bonner, B.P. and Duba, A., 1975. Stress-strain behavior of polycrystalline NaCl to 3.2 GPa, Lawrence Livermore Laboratory Rept. UCRL-51743.

Hearmon, R.F.S., 1946. The elastic constants of anistropic materials. *Reviews of Modern Physics*, **18**, 409–140.

Hearmon, R.F.S., 1956. The elastic constants of anistropic materials. *Advances in Physics*, **5**, 323–382.

Hearmon, R.F.S., 1979. The elastic constants of crystals and other anisotropic materials. In *Landolt–Bo'rnstein Tables, III/11*, ed. K.H. Hellwege and A.M. Hellwege. Berlin: Springer-Verlag, pp. 1–244.

Hearmon, R.F.S., 1984. The elastic constants of crystals and other anisotropic materials. In *Landolt–Bo'rnstein Tables, III/18*, ed. K.H. Hellwege and A.M. Hellwege. Berlin: Springer-Verlag, pp. 1–154.

Helbig, K., 1994. *Foundations of Anisotropy for Exploration Seismics*. Tarrytown, NY: Pergamon.

Helbig, K., 1996. Representation and approximation of elastic tensors. In *Seismic Anisotropy*, ed. E. Fjaer, R.M. Holt and J.S. Rathore. Tulsa, OK: Society of Exploration Geophysicists, pp. 36–37.

Helbig, K., 1998. A formalism for the consistent description of non-linear elasticity of anisotropic media. *Revue de l' Institut Français du Pétrole*, **53**, 693–708.

Helbig, K. and Schoenberg, M., 1987. Anomalous polarizations of elastic waves in transversely isotropic media. *Journal of the Acoustical Society of America*, **81**, 1235–1245.

Helgerud, M.B., 2001. *Wave speeds in gas hydrate and sediments containing gas hydrate: A laboratory and modeling study*, PhD dissertation, Stanford University.

Hermance, J.F., 1979. The electrical conductivity of materials containing partial melt, a simple model from Archie's law. *Geophysical Research Letters*, **6**, 613–616.

Hermann, C., 1934. Tensoren und Kristallsymmetrie. *Zeitschrift für Kristallographie, A.*, **89**, 32–48.

Herrick, D.C., 1988. Conductivity models, pore geometry, and conduction mechanisms. *Trans. Soc. Prof. Well Log Analysts, 29th Annual Logging Symposium*, San Antonio, TX. Paper D.

Hertz, H., 1881. Über die Berührung fester elastischer Körper. *Journal für die reine und angewandte Mathematik*, **92**, 156–171.

Hervig, R.L., Dunbar, N., Westrich, H.R. and Kyle, P.R., 1989. Preeruptive water content of rhyolitic magmas as determined by ion microprobe analyses of melt inclusions in phenocrysts. *Journal of Volcanology and Geothermal Research*, **36**, 293–302, doi:10.1016/0377-0273(89)90075-9.

Hicks, W.G. and Berry, J.E., 1956. Application of continuous velocity logs to determination of fluid saturation of reservoir rocks. *Geophysics*, **21**, 739.

Hildreth, W. and Wilson, C.J.N., 2007. Compositional zoning of the bishop tuff. *Journal of Petrology*, **48**, 951–999, doi:10.1093/petrology/egm007.

Hill, R., 1952. The elastic behavior of crystalline aggregate. *Proceedings of the Physical Society, London A*, **65**, 349–354.

Hill, R., 1963. Elastic properties of reinforced solids: Some theoretical principles. *Journal of the Mechanics and Physics of Solids*, **11**, 357–372.

Hill, R., 1964. Theory of mechanical properties of fibre-strengthened materials: I. elastic behaviour. *Journal of the Mechanics and Physics of Solids*, **12**, 199–212.

Hill, R., 1965. A self-consistent mechanics of composite materials. *Journal of the Mechanics and Physics of Solids*, **13**, 213–222.

Hilterman, F., 1989. Is AVO the seismic signature of rock properties? *Expanded Abstracts, Soc. Expl. Geophys, 59th Annual International Meeting*. Tulsa, OK: Society of Exploration Geophysicists, p. 559.

Hinch, E.J., 1977. An averaged-equation approach to particle interactions in a fluid suspension. *Journal of Fluid Mechanics*, **83**, 695–720.

Hinkle, A., 2007. *Relating chemical and physical properties of heavy oils*, MS thesis, Colorado School of Mines.

Hixson, R.S., Winkler, M.A. and Hodgdon, M.L., 1990. Sound speed and thermophysical properties of liquid iron and nickel. *Physical Review B*, **42**, 6485–6491.

Hobbs, M.E., Jhon, M.S. and Eyring, H., 1966. The dielectric constant of liquid water and various forms of ice according to significant structure theory. *Proceedings National Academy of Sciences*, **56**, 31–38.

Hoffmann, R., Xu, X., Batzle, M., *et al.*, 2005. Effective pressure or what is the effect of pressure? *Leading Edge*, December, 1256–1260.

Holtz, F., Sato, H., Lewis, J., Behrens, H. and Nakada, S., 2004. Experimental petrology of the 1991–1995 Unzen dacite, Japan. Part 1: Phase relations, phase composition and pre-eruptive condition. *Journal of Petrology*, **46**, 319–337, doi:10.1093/petrology/egh077.

Hornby, B.E., 1998. Experimental laboratory determination of the dynamic elastic properties of wet, drained shales. *Journal of Geophysical Research*, **103**(B12), 29945–29964, http://dx.doi.org/10.1029/97JB0 2380.

Horsrud, P., 2001. Estimating mechanical properties of shale from empirical correlations. *SPE Drilling Completion*, **16**, 68–73.

Hossain, M.S., Sarica, C., Zhang, Z.-Q. and Rhyne, L, 2005. Assessment and development of heavy-oil correlations, SPE/PS-CIM/CHOA 97907

Hottman, C.E. and Johnson, R.K., 1965. Estimation of formation pressures from log derived shale properties. *Journal of Petroleum Technology*, **17**, 717–722.

Hovem, J.M. and Ingram, G.D., 1979. Viscous attenuation of sound in saturated sand. *Journal of the Acoustical Society of America*, **66**, 1807–1812.

Howells, I.D., 1974. Drag due to the motion of a Newtonian fluid through a sparse random array of small fixed rigid objects. *Journal of Fluid Mechanics*, **64**, 449–475.

Hu, G.K. and Weng, G.J., 2000. Some reflections on the Mori-Tanaka and Ponte Castañeda-Willis methods with randomly oriented ellipsoidal inclusions. *Acta Mechanica*, **140**, 31–40.

Hudson, J.A., 1980. Overall properties of a cracked solid. *Mathematical Proceedings of the Cambridge Philosophical Society*, **88**, 371–384.

Hudson, J.A., 1981. Wave speeds and attenuation of elastic waves in material containing cracks. *Geophysical Journal of the Royal Astronomical Society*, **64**, 133–150.

Hudson, J.A., 1990. Overall elastic properties of isotropic materials with arbitrary distribution of circular cracks. *Geophysical Journal International*, **102**, 465–469.

Hudson, J.A. and Liu, E., 1999. Effective elastic properties of heavily faulted structures. *Geophysics*, **64**, 479–485.

Hudson, J.A., Liu, E. and Crampin, S., 1996. The mechanical properties of materials with interconnected cracks and pores. *Geophysical Journal International*, **124**, 105–112.

Huffman, D.F. and Norwood, M.H., 1960. Specific heat and elastic constants of calcium fluoride at low temperatures. *Physical Review*, **117**, 709–711.

Hughes, D.S. and Jones, H.J., 1950. Variation of elastic moduli of igneous rocks with pressure and temperature. *Bulletin of the Geological Society of America*, **61**, 843–856.

Humbert, P. and Plicque, F., 1972. Propriétés élastiques de carbonate rhomboedriques monocristal-lins: Calcite, magnésite, dolomie. *Comptes Rendus de Académie des Sciences, Paris B*, **275**, 391–394.

Hume, H.R. and Shakoor, A., 1981. Mechanical Properties. In *Physical Properties Data for Rock Salt*, ed. L.H. Gevantman. National Bureau of Standards, 103–203.

Huntington, H.B., 1958. The elastic constants of crystals. In *Solid State Physics*, vol. 7, ed. F. Seitz and D. Turnbull. New York: Academic Press, pp. 213–351.

Irifune, T., 1987. An experimental investigation of the pyroxene-garnet transformation in a pyrolite composition and its bearing on the constitution of the mantle. *Physics of the Earth and Planetary Interiors*, **45**, 324–336.

Isaak, D. and Anderson, O.L., 1987. The high temperature and thermal expansion properties of a grossular garnet (abstract). *EOS Transactions AGU*, **68**, 410.

Itskov, M., 2013. *Tensor Algebra and Tensor Analysis for Engineers*. New York: Springer Publishing.

Izbekov, P., Gardner, J.E. and Eichelberger, J.C., 2004. Comagmatic granophyre and dacite from Karymsky volcanic center, Kamchatka: Experimental constraints for magma storage conditions. *Journal of Volcanology and Geothermal Research*, **131**, 1–18, doi:10.1016/S0377-0273(03)00312-3.

Jackson, J.D., 1975. *Classical Electrodynamics*, 2nd ed. New York: John Wiley and Sons.

Jackson, P.D., Taylor-Smith, D. and Stanford, P.N., 1978. Resistivity-porosity-particle shape relation-ships for marine sands. *Geophysics*, **43**, 1250–1262.

Jaeger, J., Cook, N.G. and Zimmerman, R., 2007. *Fundamentals of Rock Mechanics*, 4th edn. Malden, MA: Blackwell Ltd.

Jaeger, J.C. and Cook, N.G.W., 1969. *Fundamentals of Rock Mechanics*. London: Chapman and Hall Ltd.

Jagadisan, A., Yang, A. and Heidari, Z. 2017. Experimental quantification of the impact of thermal maturity on kerogen density. *Society of Petrophysicists and Well-Log Analysts*, 58, SPWLA-2016-WW.

Jaishankar, A. and McKinley, G., 2012. Power-law rheology in the bulk at the interface: Quasi-properties and fractional constitutive equations. *Proceedings of the Royal Society A. Mathematical, Physical and Engineering Sciences*, **469**, 20120284.

Jakobsen, M., Hudson, J.A. and Johansen, T.A., 2003. T-matrix approach to shale acoustics. *Geophysical Journal International*, **154**, 533–558.

Jakobsen, M., Hudson, J.A., Minshull, T.A. and Singh, S.C., 2000. Elastic properties of hydrate- bearing sediments using effective medium theory. *Journal of Geophysical Research*, **105**, 561–577.

Jakobsen, M., Johansen, T.A. and McCann, C., 2003. The acoustic signature of fluid flow in a complex porous media. *Journal of Applied Geophysics*, **54**, 219–246.

Jaoshvili, A., Esakia, A., Porrati, M. and Chaikin, P.M., 2010. Experiments on the random packing of tetrahedral dice. *Physical Review Letters*, **104**, 185501.

Jardani, A., Revil, A., Boleve, A., Crespy, A., Dupont, J-P., Barrash, W. and Malama, B., 2007. Tomography of the Darcy velocity from self-potential measurements. *Geophysical Research Letters*, **34**, L24403.

Jarvie, D.M. 2012. Shale resource systems for oil and gas: Part 1—Shale-gas resource systems, and Part 2—Shale-oil resource systems. In *Shale Reservoirs—Giant Resources for the 21st century: AAPG Memoir 97*, ed. J.A. Breyer, pp. 69–119. Tulsa, OK: The American Association of Petroleum Geologists.

Jarvie, D.M., Claxton, B., Henk, B. and Breyer, J. 2001. Oil and shale gas from the Barnett shale, Ft. Worth Basin, Texas, AAPG annual convention and exhibition, Denver.

Jarvie, D.M., Hill, R.J., Ruble, T.E. and Pollastro, R.M., 2007. Unconventional shale-gas systems– The Mississippian Barnett Shale of north-central Texas as one model for thermogenic shale-gas assessment. *AAPG Bulletin*, **91**, 475–499.

Javanbakhti, A.R., Lines, L.R. and Gray, D., 2018. Empirical modeling of the saturated shear modulus in oil sands. *Geophysics*, **84**(3), MR129–MR137.

Jeanloz, R. and Thompson, A.B., 1983. Phase transitions and mantle discontinuities. *Reviews of Geophysics*, **21**, 51–74.

Jenkins, G.M. and Watts, D.G., 1968. *Spectral Analysis and Its Applications*. San Francisco, CA: Holden-Day.

Jenkins, J., Johnson, D., La Ragione, L. and Makse, H., 2005. Fluctuations and the effective moduli of an isotropic, random aggregate of identical, frictionless spheres. *Journal of the Mechanics and Physics of Solids*, **53**, 197–225.

Jiao, Y. and Torquato, S., 2011. Maximally random jammed packings of platonic solids: Hyperuniform long-range correlations and isostaticity. *Physical Review E*, **84**, 041309.

Jílek, P., 2002a. *Modeling and inversion of converted-wave reflection coefficients in anisotropic media: A tool for quantitative AVO analysis*, PhD dissertation, Center for Wave Phenomena, Colorado School of Mines.

Jílek, P., 2002b. Converted PS-wave reflection coefficients in weakly anisotropic media. *Pure Appl. Geophysics*, **159**, 1527–1562.

Jizba, D.L., 1991. *Mechanical and acoustical properties of sandstones and shales*, PhD dissertation, Stanford University.

Johansen, T.A., Ruud, B.O. and Jakobsen, M., 2004. Effect of grain scale alignment on seismic anisotropy and reflectivity of shales. *Geophysical Prospecting*, **52**, 133–149.

Johnson, D.L., 1980. Equivalence between fourth sound in liquid He II at low temperatures and the Biot slow wave in consolidated porous media. *Applied Physics Letters*, **37**, 1065–1067.

Johnson, D.L., Koplic, J. and Dashen, R., 1987. Theory of dynamic permeability and tortuosity in fluid-saturated porous media. *Journal of Fluid Mechanics*, **176**, 379–400.

Johnson, D.L. and Plona, T.J., 1982. Acoustic slow waves and the consolidation transition. *Journal of the Acoustical Society of America*, **72**, 556–565.

Johnson, D.L., Plona, T.J., Scala, C., Pasierb, F. and Kojima, H., 1982. Tortuosity and acoustic slow waves. *Physical Review Letters*, **49**, 1840–1844.

Johnson, P.A. and Rasolofosaon, P.N.J., 1996. Nonlinear elasticity and stress-induced anisotropy in rock. *Journal of Geophysical Research*, **100**(B2), 3113–3124.

Johnson, D.L., Schwartz, L.M., Elata, D., *et al.*, 1998. Linear and nonlinear elasticity of granular media: Stress-induced anisotropy of a random sphere pack. *Transactions ASME*, **65**, 380–388.

Johnston, J.E. and Christensen, N.I., 1995. Seismic anisotropy of shales. *Journal of Geophysical Research*, **100**(B4), 5991–6003, http://dx.doi.org/10.1029/ 95JB 00031.

Jona, F. and Scherrer, P., 1952. Die elastischen Konstanten von Eis-Einkristallen. *Helvetica Physica Acta*, **25**, 35–54.

Joshi, S.K. and Mitra, S.S., 1960. Debye characteristic temperature of solids. *Proceedings of the Physical Society, London*, **76**, 295–298.

Juhasz, I., 1981. Normalised Qv – The key to shaley sand evaluation using the Waxman-Smits equation in the absence of core data. *Trans. Soc. Prof. Well Log Analysts, 22nd Annual Logging Symposium*, Paper Z.

Juhasz, I., 1986. Assessment of the distribution of shale, porosity, and hydrocarbon saturation in shaly sands, Tenth European Formation Evaluation Symposium.

Kachanov, M., 1992. Effective elastic properties of cracked solids: Critical review of some basic concepts. *Applied Mechanics Reviews*, **45**, 304–335.

Kachanov, M., Shafiro, B. and Tsukrov, I., 2003. *Handbook of Elasticity Solutions*. Berlin: Springer Science+Business Media.

Kachanov, M., Tsukrov, I. and Shafiro, B., 1994. Effective moduli of solids with cavities of various shapes, Micromechanics of Random Media, ed. M. Ostoja-Starzewski and I. Jasluk. *Applied Mechanics Review*, **47**, no1, part 2. S151–S174.

Kan, T.K. and Swan, H.W., 2001. Geopressure prediction from automatically derived seismic velocities. *Geophysics*, **66**, 1937–1946.

Kandelin, J. and Weidner, D.J., 1988. Elastic properties of hedenbergite. *Journal of Geophysical Research*, **93**, 1063–1072.

Kaneko, K., Kamata, H., Koyaguchi, T., Yoshikawa, M. and Furukawa, K., 2007. Repeated large-scale eruptions from a single compositionally stratified magma chamber: An example from Aso volcano, southwest Japan. *Journal of Volcanology and Geothermal Research*, **167**, 160–180, doi:10.1016/j.jvolgeores.2007.05.002.

Kantor, Y. and Bergman, D.J., 1984. Improved rigorous bounds on the effective elastic moduli of a composite material. *Journal of the Mechanics and Physics of Solids*, **32**, 41–62.

Karniadakis, G., Beskok, A. and Aluru, N., 2005. *Microflows and Nanoflows: Fundamentals and Simulation*. New York: Springer Publishing.

Katahara, K., 2004. Fluid substitution in laminated shaly sands. *SEG Extended Abstracts 74th Annual Meeting*. Tulsa, OK: Society of Exploration Geophysicists.

Katti, D.R., Katti, K.S. and Alstadt, K., 2013. An Insight into Molecular Scale Interactions and In-situ Nanomechanical Properties of Kerogen in Green River Oil Shale. In *Poromechanics V*. Vienna: edited by Christian Hellmich, Bernhard Pichler, Dietmar Adam. American Society of Civil Engineering, pp. 2510–2516.

Katz, D.L., 1942. Prediction of the shrinkage of crude oils. *Drilling and Production Prac*tice, 137.

Kawanabe, Y. and Saito, G., 2002. Volcanic activity of the Satsuma- Iwojima area during the past 6500 years. *Earth Planets Space*, **54**, 295–301.

Keller, J.B., 1964. Stochastic equations and wave propagation in random media. *Proceedings Symposium Applied Mathematics*, **16**, 145–170.

Kelvin, W.T., Lord, 1856. Elements of a mathematical theory of elasticity. *Philosophical Transactions of the Royal Society*, **166**, 481–498.

Kendall, K. and Tabor, D., 1971. An ultrasonic study of the area of contact between stationary and sliding surfaces. *Proceedings of the Royal Society, London A*, **323**, 321–340.

Kennett, B.L.N., 1974. Reflections, rays and reverberations. *Bulletin of the Seismological Society of America*, **64**, 1685–1696.

Kennett, B.L.N., 1983. *Seismic Wave Propagation in Stratified Media*. Cambridge: Cambridge University Press.

Kerimov, A., Mavko, G., Mukerji, T., Dvorkin, J., Al Ibrahim, M., 2018. The influence of convex particles' irregular shape and varying size on porosity, permeability, and elastic bulk modulus of granular porous media: Insights from numerical simulations, JGR Solid Earth, 123, 10563–10582.

King, M.S., 1983. Static and dynamic elastic properties of rocks from the Canadian Shield. *International Journal of Rock Mechanics*, **20**, 237–241.

Kinghorn, R.R.F. and Rahman, M., 1983. Specific gravity as a kerogen type and maturation indicator with special reference to amorphous kerogens. *Journal Petroleum Geology*, **6**, 179–194.

Kitano, T., Kataoka, T. and Shirota, T., 1981. An empirical equation of the relative viscosity of polymer melts filled with various inorganic fillers. *Rheologica Acta*, **20**, 207.

Kjartansson, E., 1979. Constant Q wave propagation and attenuation. *Journal of Geophysical Research*, **84**, 4737–4748.

Klimentos, T. and McCann, C., 1990. Relationships among compressional wave attenuation, porosity, clay content, and permeability in sandstones. *Geophysics*, **55**, 998–1014.

Knight, R. and Dvorkin, J., 1992. Seismic and electrical properties of sandstones at low saturations. *Journal of Geophysical Research*, **97**, 17425–17432.

Knight, R. and Nolen-Hoeksema, R., 1990. A laboratory study of the dependence of elastic wave velocities on pore scale fluid distribution. *Geophysical Research Letters*, **17**, 1529–1532.

Knight, R.J. and Nur, A., 1987. The dielectric constant of sandstones, 60kHz to 4MHz. *Geophysics*, **52**, 644–654.

Knittle, E. and Jeanloz, R., 1987. Synthesis and equation of state of $(Mg,Fe)SiO_3$ perovskite to over 100 gigapascals. *Science*, **235**, 668–670.

Knopoff, L., 1964. *Q. Reviews of Geophysics*, **2**, 625–660.

Knott, C.G., 1899. Reflection and refraction of elastic waves, with seismological applications. *Philosophical Magazine, London*, **48**(64–97), 567–569.

Koesoemadinata, A.P. and McMechan, G.A., 2001. Empirical estimation of viscoelastic seismic parameters from petrophysical properties of sandstone. *Geophysics*, **66**, 1457–1470.

Koesoemadinata, A.P. and McMechan, G.A., 2003. Correlations between seismic parameters, EM parameters, and petrophysical/petrological properties for sandstone and carbonate at low water saturations. *Geophysics*, **68**, 870–883.

Koga, I., Aruga, M. and Yoshinaka, Y., 1958. Theory of plane elastic waves in a piezoelectric crystalline medium and determination of elastic and piezo-electric constants of quartz. *Phyical Review*, **109**, 1467–1473.

Koltermann, C.E. and Gorelick, S.M., 1995. Fractional packing model for hydraulic conductivity derived from sediment mixtures. *Water Resources Research*, **31**, 3283–3297.

Koponen, A., Kataja, M. and Timonen, J., 1997. Permeability and effective porosity of porous media. *Physical Review E*, **56**, 3319–3325.

Korringa, J., 1973. Theory of elastic constants of heterogeneous media. *Journal of Mathematical Physics*, **14**, 509.

Kozeny, J. 1927. Uber kapillare Leitung des Wassers im Bodeu. *Sitzungberichte der Akaddmie der Wissenschaftung in Wein Abteilung IIa*, **136**, 271–301.

Kress, V.C., Williams, Q. and Carmichael, I.S.E., 1988. Ultrasonic investigation of melts in the system Na_{20}-Al_2O_3-SiO_2. *Geochimica et Cosmochimica Acta*, **52**, 283–293.

Krief, M., Garat, J., Stellingwerff, J. and Ventre, J., 1990. A petrophysical interpretation using the velocities of P and S waves (full-waveform sonic). The *Log Analyst*, **31**, November, 355–369.

Krishnamurty, T.S.G., 1963. Fourth-order elastic coefficients in crystals. *Acta Crystallographia*, **16**, 839–840.

Kroner, E., 1967. Elastic moduli of perfectly disordered composite materials. *Journal of the Mechanics and Physics of Solids*, **15**, 319.

Kroner, E., 1977. Bounds for effective elastic moduli of disordered materials. *Journal of the Mechanics and Physics of Solids*, **25**, 137.

Kroner, E., 1986. Statistical modeling. In *Modeling Small Deformations of Polycrystals*, ed. J. Gittus and J. Zarka. New York: Elsevier, pp. 229–291.

Krumbein, W.C., 1934. Size frequency distributions of sediments. *Journal of Sedimentary Petrology*, **4**, 65–67.

Krumbein, W.C., 1938. Size frequency distributions of sediments and the normal phi curve. *Journal of Sedimentary Petrology*, **8**, 84–90.

Kumar, V., Curtis, M.E., Gupta, N., Sondergeld, C.H. and Rai, C.S., 2012. Estimation of elastic properties of organic matter and Woodford shale through nano-indentation measurements, SPE 162778.

Kumazawa, M. and Anderson, O.L., 1969. Elastic moduli, pressure derivatives, and temperature derivative of single-crystal olivine and single-crystal forsterite. *Journal of Geophysical Research*, **74**, 5961–5972.

Kuperman, W.A., 1975. Coherent components of specular reflection and transmission at a randomly rough two-fluid interface. *Journal of the Acoustical Society of America*, **58**, 365–370.

Kuster, G.T. and Toksöz, M.N., 1974. Velocity and attenuation of seismic waves in two-phase media. *Geophysics*, **39**, 587–618.

Lakes, R., 2009. *Viscoelastic Materials*. Cambridge: Cambridge University Press.

Lal, M., 1999. Shale stability: Drilling fluid interaction and shale strength. SPE 54356, presented at *SPE Latin American and Caribbean Petroleum Engineering Conference*, Caracas, Venezuela, 21–23 April. Tulsa, OK: Society of Exploration Geophysicists.

Lamb, H., 1945. *Hydrodynamics*. Mineola, NY: Dover.

Landau, L.D. and Lifschitz, E.D., 1959. *Theory of Elasticity*. Tarrytown, NY: Pergamon.

Landauer, R., 1952. The electrical resistance of binary metallic mixtures. *Journal of Applied Physics*, **23**, 779–784.

Lanfrey, P.-Y., Kuzeljevic, Z.V. and Dudukovic, M.P., 2010. Tortuosity model for fixed beds randomly packed with identical particles. *Chemical Engineering Science*, **65**, 1891–1896. doi:10.1016/j.ces.2009.11.011.

Langlebein. M.P., 1962. Young's modulus for sea ice. *Canadian Journal of Physics*, **40**, 1–8.

Larsen, J.F., 2006. Rhyodacite magma storage conditions prior to the 3430 yBP caldera-forming eruption of Aniakchak volcano, Alaska. *Contributions to Mineralogy and Petrology*, **152**, 523–540, doi:10.1007/s00410-006-0121-4.

Lashkaripour, G.R. and Dusseault, M.B., 1993. A statistical study on shale properties; relationship among principal shale properties. *Proceedings of the Conference on Probabilistic Methods in Geotechnical Engineering*, Canberra, Australia, Rotterdam: Balkema, pp. 195–200.

Lawn, B.R. and Wilshaw, T.R., 1975. *Fracture of Brittle Solids*. Cambridge: Cambridge University Press.

Lazarus, D., 1949. The variation of the adiabatic elastic constants of KCl, NaCl, CuZn, Cu, and Al with pressure to 10000 bars. *Physical Review*, **76**, 545–553.

Levien, L. and Prewitt, C.T., 1981. High pressure structural study of diopside. *American Mineralogist*, **66**, 315–323.

Levin, F.K., 1971. Apparent velocity from dipping interface reflections. *Geophysics*, **36**, 510–516.

Levin, F.K., 1979. Seismic velocities in transversely isotropic media. *Geophysics*, **44**, 918–936.

Li, J.-H. and Yu, B.-M., 2011. Tortuosity of flow paths through a Sierpinski carpet. *Chinese Physics Letters*, **28**, 034701_1–3. doi:10.1088/0256-307X/28/3/034701.

Liebermann, R.C. and Schreiber, E., 1968. Elastic constants of polycrystalline hematite as a function of pressure to 3 kilobars. *Journal of Geophysical Research*, **73**, 6585–6590.

Liu, A., 1996. Stress Intensity Factors, from ASM Handbook, Volume 19: Fatigue and Fracture, ASM International, 980–1000.

Liu, H.P., Anderson, D.L. and Kanamori, H., 1976. Velocity dispersion due to anelasticity: Implications for seismology and mantle composition. *Geophysical Journal of the Royal Astronomical Society*, **47**, 41–58.

Liu, L., 2011. New optimal microstructures and restrictions on the attainable Hashin-Shtrikman bounds for multiphase composite materials. *Philosophical Magazine Letters*, **91**, 7473–7482.

Lo, T.W., Coyner, K.B. and Toksöz, M.N., 1986. Experimental determination of elastic anisotropy of Berea sandstone, Chicopee shale, and Chelmsford granite. *Geophysics*, **51**(1), 164–171, h ttp://dx.doi.org/10.1190/1.1442029.

Lockner, D.A., Walsh, J.B. and Byerlee, J.D., 1977. Changes in velocity and attenuation during deformation of granite. *Journal of Geophysical Research*, **82**, 5374–5378.

Log Interpretation Charts, 1984. Publication SMP-7006. Houston, TX: Schlumberger Ltd.

Lorenz, J., Haas Jr., J.L., Clynne, M.A., Potter II, R.W. and Schafer, C.M., 1981. Geology, mineralogy, and some geophysical and geochemical properties of salt deposits. In *Physical Properties Data for Rock Salt*, ed. L.H. Gevantman. National Bureau of Standards, pp. 1–44.

Loucks, R.G., Reed, R.M., Ruppel, S.C. and Jarvie, D.M., 2009. Morphology, genesis, and distribution of nanometer-scale pores in siliceous mudstones of the Mississippian Barnett Shale. *Journal of Sedimentary Research*, **79**, 848–861.

Lowenstern, J.B., 1993. Evidence for a copper-bearing fluid in magma erupted at the Valley of Ten Thousand Smokes, Alaska. *Contributions to Mineralogy and Petrology*, **114**, 409–421, doi:10.1007/BF01046542.

Lucet, N., 1989. *Vitesse et Attenuation des Ondes Elastiques Soniques et Ultrasoniques dans les Roches sous Pression de Confinement*, PhD dissertation, University of Paris.

Lucet, N. and Zinszner, B., 1992. Effects of heterogeneities and anisotropy on sonic and ultrasonic attenuation in rocks. *Geophysics*, **57**, 1018–1026.

Ludwig, W.J., Nafe, J.E. and Drake, C.L., 1970. Seismic refraction. In *The Sea*, vol. 4, ed. A. E. Maxwell. New York: Wiley-Interscience, pp. 53–84.

Luhr, J.F., 2002. Petrology and geochemistry of the 1991 and 1998–1999 lava flows from Volcán de Colima, México: Implications for the end of the current eruptive cycle. *Journal of Volcanology and Geothermal Research*, **117**, 169–194, doi:10.1016/S0377–0273(02)00243–3.

Macdonald, G.A. and Katsura, T., 1961. Variations in the lava of the 1959 eruption in Kilauea Iki. *Pacific Science Journal*, **15**, 358–369.

Makse, H.A., Gland, N., Johnson, D.L. and Schwartz, L.M., 1999. Why effective medium theory fails in granular materials. *Physical Review Letters*, **83**, 5070–5073.

Makse, H.A., Gland, N., Johnson, D.L. and Schwartz, L.M., 2004. Granular packings: Nonlinear elasticity, sound propagation, and collective relaxation dynamics. *Physical Review E*, **L70**, 061302.1–061302.19.

Makse, H.A., Johnson, D.L. and Schwartz, L.M., 2000. Packing of compressible granular materials. *Physical Review Letters*, **84**, 4160–4163.

Mallick, S., 2001. AVO and elastic impedance. *Leading Edge*, **20**, 1094–1104.

Man, W., Donev, A., Stillinger, F.H., Sullivan, M.T., Russel, W.B., Heeger, D., Inati, S., Torquato, S. and Chaikin, P.M., 2005. Experiments on random packings of ellipsoids. *Physical Review Letters*, **94**, 198001.

Mandeville, C.W., Carey, S. and Sigurdsson, H., 1996. Magma mixing, fractional crystallization and volatile degassing during the 1883 eruption of Krakatau volcano, Indonesia. *Journal of Volcanology and Geothermal Research*, **74**, 243–274, doi:10.1016/S0377-0273(96)00060-1.

Manegold, E. and von Engelhardt, W., 1933. Uber Kapillar-Systeme, XII, Die Berechnung des Stoffgehaltes heterogener Gerutstrukturen. *Kole Zeitschrift*, **63**(2), 149–154.

Manga, M., Castro, J., Cashman, K.V. and Loewenberg, M., 1998. Rheology of bubble-bearing magmas. *Journal of Volcanology and Geothermal Research*, **87**, 15–28.

Manga, M. and Loewenberg, M., 2001. Viscosity of magmas containing highly deformable bubbles. *Journal of Volcanology and Geothermal Research*, **105**, 19–24.

Marianelli, P., Sbrana, A. and Proto, M., 2006. Magma chamber of the Campi Flegrei supervolcano at the time of eruption of the Campanian Ignimbrite. *Geology*, **34**, 937–940, doi:10.1130/G22807A.1.

Marion, D., 1990. *Acoustical, mechanical and transport properties of sediments and granular materials*, PhD dissertation, Stanford University.

Marion, D. and Nur, A., 1991. Pore-filling material and its effect on velocity in rocks. *Geophysics*, **56**, 225–230.

Marion, D., Nur, A., Yin, H. and Han, D., 1992. Compressional velocity and porosity in sand-clay mixtures. *Geophysics*, **57**, 554–563.

Marle, C.M., 1981. *Multiphase Flow in Porous Media*. Houston, TX: Gulf Publishing Company.

Marple, S.L., 1987. *Digital Spectral Analysis with Applications*. Englewood Cliffs, NJ: Prentice-Hall.

Mason, W.P., 1943. Quartz crystal applications. *Bell Syst. Tech. J.*, **22**, 178.

Mason, W.P., 1950. *Piezoelectric Crystals and Their Application to Ultrasonics*. New York: D. Van Nostrand Co., Inc.

Mauret, E. and Renaud, M., 1997. Transport phenomena in multi-particle systems: I. Limits of applicability of capillary model in high voltage beds—Application to fixed beds of fibers and fluidized beds of spheres. *Chemical Engineering Science*, **52**, 1807–1817.

Mavko, G., 1980. Velocity and attenuation in partially molten rocks. *Journal of Geophysical Research*, **85**, 5173–5189.

Mavko, G. and Bandyopadhyay, K., 2009. Approximate fluid substitution in weakly anisotropic VTI rocks. *Geophysics*, **74**, D1–D6.

Mavko, G., Chan, C. and Mukerji, T., 1995. Fluid substitution: Estimating changes in V_P without knowing V_S. *Geophysics*, **60**, 1750–1755.

Mavko, G. and Jizba, D., 1991. Estimating grain-scale fluid effects on velocity dispersion in rocks. *Geophysics*, **56**, 1940–1949.

Mavko, G., Kjartansson, E. and Winkler, K., 1979. Seismic wave attenuation in rocks. *Reviews of Geophysics*, **17**, 1155–1164.

Mavko, G. and Mukerji, T., 1995. Pore space compressibility and Gassmann's relation. *Geophysics*, **60**, 1743–1749.

Mavko, G., Mukerji, T. and Godfrey, N., 1995. Predicting stress-induced velocity anisotropy in rocks. *Geophysics*, **60**, 1081–1087.

Mavko, G. and Nolen-Hoeksema, R., 1994. Estimating seismic velocities in partially saturated rocks. *Geophysics*, **59**, 252–258.

Mavko, G. and Nur, A., 1975. Melt squirt in the asthenosphere. *Journal of Geophysical Research*, **80**, 1444–1448.

Mavko, G. and Nur, A., 1978. The effect of nonelliptical cracks on the compressibility of rocks. *Journal of Geophysical Research*, **83**, 4459–4468.

Mavko, G. and Nur, A., 1997. The effect of a percolation threshold in the Kozeny-Carman relation. *Geophysics*, **62**, 1480–1482.

Mavko, G. and Saxena, N., 2013. Embedded-bound method for estimating the change in bulk modulus under either fluid or solid substitution. *Geophysics*, **78**, L87–L99.

Mazáč, O., Cfslerova, M., Kelly, W.E., Landa, I. and Venhodova, D., 1990. Determination of hydraulic conductivities by surface geoelectrical methods. In *Geotechnical and Environmental Geophysics*, vol. II., ed. S.H. Ward. Tulsa, OK: Society of Exploration Geophysicists, pp. 125–131.

McCain, W.D., 1990. *The Properties of Petroleum Fluids*, 2nd edn., Tulsa, OK: PennWell Publishing Company.

McCann, D.M. and Entwisle, D.C., 1992. Determination of Young's modulus of the rock mass from geophysical well logs. In *Geological Applications of Wireline Logs II*, ed. A. Hurst, C. M. Griffiths and P.F. Worthington. Geological Society Special Publication, vol. 65. London: Geological Society, pp. 317–325.

McCoy, J.J., 1970. On the displacement field in an elastic medium with random variations in material properties. In *Recent Advances in Engineering Science*, vol. 2, ed. A.C. Eringen. New York: Gordon and Breach, pp. 235–254.

McGeary, R.K., 1967. Mechanical packing of spherical particles. *J. Am. Ceram. Soc.*, **44**(10), 513–522.

McNally, G.H., 1987. Estimation of coal measures rock strength using sonic and neutron logs. *Geoexploration*, **24**, 381–395.

McSkimin, H.J., Andreatch, P., Jr. and Thurston, R.N., 1965. Elastic moduli of quartz vs. hydrostatic pressure at 25 and 195.8 degrees Celsius. *Journal of Applied Physics*, **36**, 1632.

McSkimin, H.J. and Bond, W.L., 1972. Elastic moduli of diamond as a function of pressure and temperature. *Journal of Applied Physics*, **43**, 2944–2948.

Meador, R.A. and Cox, P.T., 1975. Dielectric constant logging: A salinity independent estimation of formation water volume. *Society of Petroleum Engineers.*, Paper 5504.

Mehrabadi, M.M. and Cowin, S., 1990. Eigentensors of linear anisotropic elastic materials. *Quarterly Journal of Mechanics Applied Mathematics*, **43**, 15–41.

Mehta, C.H., 1983. Scattering theory of wave propagation in a two-phase medium. *Geophysics*, **48**, 1359–1372.

Meister, R. and Peselnick, L., 1966. Variational method of determining effective moduli of poly-crystals with tetragonal symmetry. *Journal of Applied Physics*, **37**, 4121–4125.

Mese, A. and Dvorkin, J., 2000. Static and dynamic moduli, deformation, and failure in shaley sand. *DOE report*, unpublished.

Miadonye, A., Puttagunta, V.R., Srivastava, R., Huang, S.S. and Dafan, Y., 1997. Generalized oil viscosity model for the effects of temperature, pressure, and gas composition. *The Journal of Canadian Petroleum Technology*, **36**, 50–54.

Middya, T.R. and Basu, A.N., 1986. Self-consistent *T*-matrix solution for the effective elastic properties of noncubic polycrystals. *Journal of Applied Physics*, **59**, 2368–2375.

Middya, T.R., Sarkar, A. and Sengupta, S., 1984. Elastic constants of polycrystalline aggregate. *Acta Physica Polonica A*, **66**, 561–571.

Milholland, P., Manghnani, M.H., Schlanger, S.O. and Sutton, G.H., 1980. Geoacoustic modeling of deep-sea carbonate sediments. *Journal of the Acoustical Society of America*, **68**, 1351–1360.

Militzer, H. and Stoll, R., 1973. Einige Beitrageder geophysics zur primadatenerfassung im Bergbau: Neue Bergbautechnik. *Leipzig*, **3**, 21–25.

Milner, M., McLin, R. and Petriello, J., 2010. Imaging texture and porosity in mudstones and shales: Comparison of secondary and ion milled backscatter SEM methods. Society of Petroleum Engineers Paper 138975, Richardson, Texas, 10 p., doi:10.2118/138975 MS.

Milton, G.W., 1981. Bounds on the electromagnetic, elastic and other properties of two-component composites. *Physical Review Letters*, **46**, 542–545.

Milton, G.W., 1984a. Microgeometries corresponding exactly with effective medium theories. In Physics and Chemistry of Porous Media, AIP Conference Proceedings 107, ed. D.L. Johnson and P.N. Sen. American Institute of Physics, 66.

Milton, G.W., 1984b. Modeling the properties of composites by laminates, talk presented at the Workshop on Homogenization and Effective Moduli of Materials and Media, Institute for Mathematics and Its Applications, Minneapolis.

Milton, G.W., 2002. *The Theory of Composites*. Cambridge: Cambridge University Press.

Milton, G.W. and Phan-Thien, N., 1982. New bounds on effective elastic moduli of two-component materials. *Proceedings Royal Society of London A*, **380**, 305–331.

Mindlin, R.D., 1949. Compliance of elastic bodies in contact. *Journal of Applied Mechanics*, **16**, 259–268.

Miyagi, I. and Yurimoto, H., 1995. Water content of melt inclusion in phenocrysts using secondary ion mass spectrometer. *Bulletin of the Volcanologic Society of Japan*, **40**, 349–355.

Miyake, Y., *et al.*, 2005. On the essential ejecta of the September 2004 eruptions of the Asama volcano, central Japan. *Bulletin of the Volcanologic Society of Japan*, **50**, 315–332.

Moakher, M. and Norris, A.N., 2006. The closest elastic tensor of arbitrary symmetry to an elasticity tensor of lower symmetry. *Journal of Elasticity*, **85**, 215–263.

Modica, C.J. and Lapierre, S.G., 2012. Estimation of kerogen porosity in source rocks as a function of thermal transformation: Example from the Mowry shale in the Powder River Basin. *AAPG Bulletin*, **96**, 87–108.

Morcote, A., Mavko, G. and Prasad, M., 2010. Dynamic elastic properties of coal. *Geophysics*, **75**, E227–E234.

Moos, D., Zoback, M.D. and Bailey, L., 1999. Feasibility study of the stability of openhole multi-laterals, Cook Inlet, Alaska. Presentation at *1999 SPE Mid-Continent Operations Symposium*, Oklahoma City, OK, 28–31 March, SPE 52186.

Mori, T. and Tanaka, K., 1973. Average stress in matrix and average elastic energy of materials with misfitting inclusions. *Acta Metallurgica*, **21**, 571–573.

Morris, P.R., 1969. Averaging fourth-rank tensors with weight functions. *Journal of Applied Physics*, **40**, 447–448.

Mota, M., Teixeira, J.A. and Yelshin, A., 2001. Binary spherical particle mixed beds porosity and permeability relationship measurement. *Transactions of the Filtration Society*, **1**, 101–106.

Mukerji, T., Berryman, J.G., Mavko, G. and Berge, P.A., 1995a. Differential effective medium modeling of rock elastic moduli with critical porosity constraints. *Geophysical Research Letters*, **22**, 555–558.

Mukerji, T., Jørstad, A., Mavko, G. and Granli, J.R., 1998. Near and far offset impedances: Seismic attributes for identifying lithofacies and pore fluids. *Geophysical Research Letters*, **25**, 4557–4560.

Mukerji, T. and Mavko, G., 1994. Pore fluid effects on seismic velocity in anisotropic rocks. *Geophysics*, **59**, 233–244.

Mukerji, T., Mavko, G., Mujica, D. and Lucet, N., 1995. Scale-dependent seismic velocity in heterogeneous media. *Geophysics*, **60**, 1222–1233.

Muller, G., Roth, M. and Korn, M., 1992. Seismic-wave traveltimes in random media. *Geophysical Journal International*, **110**, 29–41.

Muller, T.M. and Gurevich, B., 2005a. A first-order statistical smoothing approximation for the coherent wave field in porous random media. *Journal of the Acoustical Society of America*, **117**, 1796–1805.

Muller, T.M. and Gurevich, B., 2005b. Wave-induced fluid flow in random porous media: Attenuation and dispersion of elastic waves. *Journal of the Acoustical Society of America*, **117**, 2732–2741.

Muller, T.M. and Gurevich, B., 2006. Effective hydraulic conductivity and diffusivity of randomly heterogeneous porous solids with compressible constituents. *Applied Physics Letters*, **88**, 121924.

Muller, T.M., Lambert, G. and Gurevich. B., 2007. Dynamic permeability of porous rocks and its seismic signatures. *Geophysics*, **72**, E149–E158.

Mura, T., 1982. *Micromechanics of Defects in Solids*. The Hague: Kluwer.

Mura, T., 1987. *Micromechanics of Defects in Solids*. 2nd revised edition, The Hague: Kluwer Academic Publishers.

Murai, Y. and Oiwa, H., 2008. Increase of effective viscosity in bubbly liquids from transient deformation. *Fluid Dynamics Research*, **2008**, 565–575.

Murakami, Y., 1987. *Summary of Stress Intensity Factors Handbook: The Society of Materials Science*. Japan: Pergamon Press.

Murata, K.J. and Richter, D.H., 1966. Chemistry of the lavas of the 1959–60 eruption of Kilauea volcano, Hawaii, in The 1959–60 Eruption of Kilauea Volcano, Hawaii, U.S. Geological Survey Professional Paper, 537-A, 1–26.

Murnaghan, F.D., 1951. *Finite Deformation of an Elastic Solid*. New York: John Wiley.

Murphy, W.F., Schwartz, L.M. and Hornby, B., 1991. Interpretation physics of *VP* and *VS* in sedimentary rocks. *Transactions SPWLA 32nd Annual Logging Symposium*, 1–24.

Murphy, W.F., III, 1982. *Effects of microstructure and pore fluids on the acoustic properties of granular sedimentary materials*, PhD dissertation, Stanford University.

Murphy, W.F., III, 1984. Acoustic measures of partial gas saturation in tight sandstones. *Journal of Geophysical Research*, **89**, 11549–11559.

Murphy, W.F., III, Winkler, K.W. and Kleinberg, R.L., 1984. Contact microphysics and viscous relaxation in sandstones. In *Physics and Chemistry of Porous Media*, ed. D.L. Johnson and P. N. Sen. New York: American Institute of Physics, pp. 176–190.

Muzychka, Y.S. and Yovanovich, M.M., 2009. Pressure drop in laminar developing flow in non-circular ducts: A scaling and modeling approach. *Journal of Fluids Engineering*, **131**, 111105–1 –111105–11.

Myer, L., 2000. Fractures as collections of cracks. *International Journal of Rock Mechanics*, **37**, 231–243.

Nakada, S. and Motomura, Y., 1999. Petrology of the 1991–1995 eruption at Unzen: Effusion pulsation and groundmass crystallization. *Journal of Volcanology and Geothermal Research*, **89**, 173–196, doi:10.1016/S0377-0273(98)00131-0.

Nakamura, N. and Nagahama, H., 2000. Curie symmetry principle: Does it constrain the analysis of structural geology? *Forma*, **15**, 87–94.

Neal, C.A., Duggan, T.J., Wolfe, E.W. and Brandt, E.L., 1988. Lava samples, temperatures, and compositions, in The Puu Oo Eruption of Kilauea Volcano, Hawaii: Episodes 1 Through 20, January 3, 1983, Through June 8, 1984, U.S. Geological Survey Professional Paper, 1463, 99–126.

Neumann, F.E., 1885. *Vorlesungen über die Theorie der Elastizität der festen Körper und des Lichtäthers*, ed. O.E. Meyer and B.G. Leipzig. Berlin: Teubner-Verlag.

Nielsen, S.B., Clausen, O.R. and McGregor, E., 2017. Avitrinite reflectance model derived from basin and laboratory data. *Basin Research*, **29**, 515–536.

Nishizawa, O., 1982. Seismic velocity anisotropy in a medium containing oriented cracks–transversely isotropic case. *Journal of the Physics of the Earth*, **30**, 331–347.

Nolet, G., 1987. Seismic wave propagation and seismic tomography. In *Seismic Tomography*, ed. G. Nolet. Dordrecht: D. Reidel Publication Co., pp. 1–23.

Norris, A.N., 1985. A differential scheme for the effective moduli of composites. *Mechanics of Materials*, **4**, 1 16.

Norris, A.N., 1989. An examination of the Mori-Tanaka effective medium approximation for multiphase composites. *Journal of Applied Mechanics*, **56**, 83–88.

Norris, A.N., 2006a. Elastic moduli approximation of higher symmetry for the acoustical properties of an anisotropic material. *Journal of the Acoustical Society of America*, **119**, 2114–2121.

Norris, A.N., 2006b. The isotropic material closest to a given anisotropic material. *Journal of the Mechanics of Materials and Structures*, **1**, 223–238.

Norris, A.N. and Johnson, D.L., 1997. Nonlinear elasticity of granular media. *ASME Journal of Applied Mechanics*, **64**, 39–49.

Norris, A.N., Sheng, P. and Callegari, A.J., 1985. Effective-medium theories for two-phase dielectric media. *Journal of Applied Physics*, **57**, 1990–1996.

Nur, A., 1971. Effects of stress on velocity anisotropy in rocks with cracks. *Journal of Geophysical Research*, **76**, 2022–2034.

Nur, A. and Byerlee, J.D., 1971. An exact effective stress law for elastic deformation of rocks with fluids. *Journal of Geophysical Research*, **76**, 6414–6419.

Nur, A., Marion, D. and Yin, H., 1991. Wave velocities in sediments. In *Shear Waves in Marine Sediments*, ed. J.M. Hovem, M.D. Richardson and R.D. Stoll. Dordrecht: Kluwer Academic Publishers, pp. 131–140.

Nur, A., Mavko, G., Dvorkin, J. and Gal, D., 1995. Critical porosity: The key to relating physical properties to porosity in rocks. In *Proceedings 65th Annual International Meeting, Society of Exploration Geophysicists*, vol. 878. Tulsa, OK: Society of Exploration Geophysicists.

Nur, A. and Simmons, G., 1969a. Stress-induced velocity anisotropy in rocks: An experimental study. *Journal of Geophysical Research*, **74**, 6667.

Nur, A. and Simmons, G., 1969b. The effect of viscosity of a fluid phase on velocity in low-porosity rocks. *Earth and Planetary Sciences Letters*, **7**, 99–108.

Nye, J.F. 1957. *Physical Properties of Crystals*. Oxford: Oxford University Press.

O'Connell, R.J. and Budiansky, B., 1974. Seismic velocities in dry and saturated cracked solids. *Journal of Geophysical Research*, **79**, 4626–4627.

O'Connell, R.J. and Budiansky, B., 1977. Viscoelastic properties of fluid-saturated cracked solids. *Journal of Geophysical Research*, **82**, 5719–5735.

O'Doherty, R.F. and Anstey, N.A., 1971. Reflections on amplitudes. *Geophysical Prospecting*, **19**, 430–458.

Ogston, A.G. and Winzor, D.J., 1975. Treatment of thermodynamic nonideality in equilibrium studies on associating solutes. *The Journal of Physical Chemistry*, **79**, 2496–2500.

Ohno, I., Yamamoto, S. and Anderson, O.L., 1986. Determination of elastic constants of trigonal crystals by the rectangular parallelepiped resonance method. *Journal of Physics and Chemistry of Solids*, **47**, 1103–1108.

Okiongbo, K.R., Aplin, A.C. and Larter, S.R., 2005. Changes in type II kerogen density as a function of maturity: Evidence from the Kimmeridge Clay Formation. *Energy and Fuels*, **19**, 2495–2499.

Okumura, S., Takeuchi, S. and Yamanoi, Y., 2004. Study on mechanism of Vulcanian eruption, in *Report on Grant-in-Aid of Fukada Geological Institute* [in Japanese], Tokyo: Fukada Geol. Inst., pp. 37–48.

Olhoeft, G.R., 1979. Tables of room temperature electrical properties for selected rocks and minerals with dielectric permittivity statistics. *US Geological Survey Open File Report* 79–993.

Ono, K., Soya, T. and Hosono, T., 1982. Geology of the Satsuma-Io-Jima district [in Japanese with English abstract], Quadrangle Ser., scale 1:50000, 80 pp., Geol. Surv. Jpn., Tokyo.

Osborn, J.A., 1945. Demagnetizing factors of the general ellipsoid. *Physical Review*, **67**, 351–357.

Ozkan, H. and Jamieson, J.C., 1978. Pressure dependence of the elastic constants of nonmetamict zircon. *Physics and Chemistry of Minerals*, **2**, 215–224.

Paillet, F.L. and Cheng, C.H., 1991. *Acoustic Waves in Boreholes*. Boca Raton, FL: CRC Press, p. 264.

Pal, R., 2002. Complex shear modulus of concentrated suspensions of solid spherical particles. *Journal of Colloid and Interface Science*, **245**, 171–177.

Pal, R., 2003. Viscous behavior of concentrated emulsions of two immiscible Newtonian fluids with interfacial tension. *Journal of Colloid and Interface Science*, **263**, 296–305.

Palierne, J.F., 1990. Linear rheology of viscoelastic emulsions with interfacial tension. *Rheologica Acta*, **29**, 204–214.

Pallister, J.S., Hoblitt, R.P. and Reyes, A.G., 1992. A basalt trigger for the 1991 eruptions of Pinatubo volcano? *Nature*, **356**, 426–428, doi:10.1038/356426a0.

Pallister, J.S., Hoblitt, R.P., Meeker, G.P., Knight, R.J. and Siems, D.F., 1996. Magma mixing at Mount Pinatubo: Petrographic and chemical evidence from the 1991 deposits. In *Fire and Mud: Eruptions and Lahar of Mt. Pinatubo*, ed. C. Newhall and R. Punonhbayan. Seattle: Univ. of Wash. Press, pp. 687–731.

Papadakis, E.P., 1963. Attenuation of pure elastic modes in NaCl single crystals. *Journal of Applied Physics*, **34**, 1872–1876.

Paris, P.C., Palin-Luc, T., Tada, H. and Saintier, N., 2009. Stresses and crack tip stress intensity factors around spherical and cylindrical voids and inclusions of differing elastic properties and misfit sizes, Proceedings of Crack Paths (CP 2009), Vicenza, Italy, 495–502.

Paterson, M.S. and Weiss, L.E., 1961. Symmetry concepts in the structural analysis of deformed rocks. *Geological Society of America Bulletin*, **72**, 841.

Pech, D., 1984. *Etude de la permiabilité des lits compressibles constitués de copeaux de bois partiellement destructurés*. These de 36me cycle. INP, Grenoble, France

Pedersen, B.K. and Nordal, K., 1999. Petrophysical evaluation of thin beds: A review of the Thomas-Stieber approach. Course 24034 Report, Norwegian University of Science and Technology.

Penny, A.H.A., 1948. A theoretical determination of the elastic constants of ice. *Mathematical Proceedings of the Cambridge Philosophical Society*, **44**, 423–439.

Peselnick, L. and Meister, R., 1965. Variational methods of determining effective moduli of polycrystals: (A) hexagonal symmetry, (B) trigonal symmetry. *Journal of Applied Physics*, **36**, 2879–2884.

Peselnick, L. and Robie, R.A., 1963. Elastic constants of calcite. *Journal of Applied Physics*, **34**, 2494–2495.

Peters, K. and Cassa, M.R., 1994. Applied source rock geochemistry, in AAPG Memoir 60, The Petroleum System – From Source to Trap, ed. L.B. Magoon and W.G. Dow. 93–120.

Pickett, G.R., 1963. Acoustic character logs and their applications in formation evaluation. *Journal of Petroleum Technology*, **15**, 650–667.

Pinkerton, H. and Stevenson, R.J., 1992. Methods of determining the rheologic properties of magmas as sub-liquidus temperatures. *Journal of Volcanology and Geothermal Research*, **53**, 47–66.

Ponte-Castaneda, P. and Willis, J.R., 1995. The effect of spatial distribution on the effective behavior of composite materials and cracked media. *Journal of the Mechanics and Physics of Solids*, **43**, 1919–1951.

Porter, L.B, Ritzi, R.W., Masters, L.J., Dominic, D.F. and Ghanbarian-Alavijeh, B., 2012. The Kozeny-Carman equation with a percolation threshold. *Ground Water*, **50**, 92–99.

Postma, G.W., 1955. Wave propagation in a stratified medium. *Geophysics*, **20**, 780–806.

Pounder, E.R. and Langlebein, M.P., 1964. Arctic sea ice of various ages II Elastic properties. *Journal of Glaciology*, **5**, 99–105.

Poupon, A. and Leveaux, J., 1971. Evaluation of water saturations in shaley formations. *Trans. Soc. Prof. Well Log Analysts, 12th Annual Logging Symposium*, Paper O.

Prager, S., 1969. Improved variational bounds on some bulk properties of a random medium. *Journal of Chemical Physics*, **50**, 4305–4312.

Prasad, M. and Manghnani, M.H., 1997. Effects of pore and differential pressure on compressional wave velocity and quality factor in Berea and Michigan sandstones. *Geophysics*, **62**, 1163–1176.

Prioul, R., Bakulin, A. and Bakulin, V., 2004. Nonlinear rock physics model for estimation of 3D subsurface stress in anisotropic formations: Theory and laboratory verification. *Geophysics*, **69**, 415–125.

Prioul, R. and Lebrat, T., 2004. Calibration of velocity-stress relationships under hydrostatic stress for their use under non-hydrostatic stress conditions. *SEG Expanded Abstracts 74th International Meeting*, October 2004.

Proctor, T.M., Jr., 1966. Low-temperature speed of sound in single-crystal ice. *Journal of the Acoustical Society of America*, **39**, 972–977.

Pšenčík, I. and Martins, J.L., 2001. Weak contrast PP wave displacement R/T coefficients in weakly anisotropic elastic media. *Pure and Applied Geophysics*, **151**, 699–718.

Pšenčík, I. and Vavrycuk, V., 1998. Weak contrast PP wave displacement R/T coefficients in weakly anisotropic elastic media. *Pure and Applied Geophysics*, **151**, 699–718.

Puttagunta, V.R., Singh, B. and Cooper, E., 1988. A generalized viscosity correlation for Alberty heavy oils and bitumens; 4th UNITAR/UNDP Conference on Heavy Crudes and Tar Sands, AOSTRA, Vol 1, 657–673.

Pyrak-Nolte, L.J., Myer, L.R. and Cook, N.G.W., 1990. Transmission of seismic waves across single natural fractures. *Journal of Geophysical Research*, **95**, 8617–8638.

Rafavich, F., Kendal, C.H., St. C. and Todd, T.P., 1984. The relationship between acoustic properties and the petrographic character of carbonate rocks. *Geophysics*, **49**, 1622–1636.

Ransom, R.C., 1984. A contribution towards a better understanding of the modified Archie formation resistivity factor relationship. *The Log Analyst*, **25**, 7–15.

Rasolofosaon, P., 1998. Stress-induced seismic anisotropy revisited. *Revue de l' Institut Français du Pétrole*, **53**, 679–692.

Raymer, L.L., Hunt, E.R. and Gardner, J.S., 1980. An improved sonic transit time-to-porosity transform. *Transactions Society of Professional Well Log Analysts, 21st Annual Logging Symposium*, Paper P.

Renshaw, C.E., 1995. On the relationship between mechanical and hydraulic apertures in rough-walled fractures. *Journal of Geophysical Research*, **100**, 24629–24636.

Resnick, J.R., Lerche, I. and Shuey, R.T., 1986. Reflection, transmission, and the generalized primary wave. *Geophysical Journal of the Royal Astronomical Society*, **87**, 349–377.

Reuss, A., 1929. Berechnung der Fliessgrenzen von Mischkristallen auf Grand der Plastizitatsbedingung fur Einkristalle. *Zeitschrift für Angewandte Mathematik und Mechanik*, **9**, 49–58.

Revil, A., 2012. Spectral induced polarization of shaly sands: Influence of the electrical double layer. *Water Resources Research*, **48**, W02517.

Revil, A. and Florsch, N., 2010. Determination of permeability from spectral induced polarization data in granular media. *Geophysical Journal International*, **181**, 1480–1498.

Revil, A. and Glover, P.W.J., 1997. Theory of ionic surface electrical conduction in porous media. *Physical Review B*, **55**, 1757–1773.

Revil, A. and Glover, P.W.J., 1998. Nature of surface electrical conductivity in natural sands, sandstones, and clays. *Geophysical Research Letters*, **25**, 691–694.

Revil, A. and Jardani, A., 2013. *The Self-Potential Method: Theory and Applications in Environmental Geosciences*. Cambridge: Cambridge University Press.

Revil, A., Woodruff, W.F. and Lu, N., 2011. Constitutive equations for coupled flows in clay materials. *Water Resources Research*, **47**, W05548.

Richardson, J.F., Harker, J.H. and Backhurst, J.R., 2002. *Chemical Engineering, Vol. 2: Particle Technology and Separation Processes*. Oxford: Butterworth-Heinemann.

Richter, D.H. and Murata, K.J., 1966. Petrography of the lavas of the 1959–60 eruption of Kilauea volcano, Hawaii, in The 1959–60 Eruption of Kilauea Volcano, Hawaii, U.S. Geological Survey Professional Paper, 537-D, 1–12.

Ridgway, K. and Tarbuck, K.J., Particulate Mixture Bulk Density. *Chemical Process Engineering*, **49**, 103–105.

Rigden, S.M., Ahrens, T.J. and Stolper, E.M., 1988. Shock compression of molten silicate: Results for a model basaltic composition. *Journal of Geophysical Research*, **93**, 367–382.

Rigden, S.M., Ahrens, T.J. and Stolper, E.M., 1989. High pressure equation of state of molten anorthite and diopside. *Journal of Geophysical Research*, **94**, 9508–9522.

Risnes, R., Bratli, R. and Horsrud, P., 1982. Sand stresses around a borehole. *Society of Petroleum Engineers Journal*, **22**, 883–898.

Rivers, M.L. and Carmichael, S.E., 1987. Ultrasonic studies of silicate melts. *Journal of Geophysical Research*, **92**, 9247–927.

Riznichenko, Y.V., 1949. On seismic quasi-anisotropy. *Izvestiia Akademii Nauk SSSR, Geografie Geofizica*, **13**, 518–544.

Robie, R.A., Bethke, P.M., Toulmin, M.S. and Edwards, J.L., 1966. X-ray crystallographic data, densities, and molar volumes of minerals. In *Handbook of Physical Constants*, ed. S.P. Clark. Boulder, CO: Geological Society of America, pp. 27–74.

Roe, R-J, 1965. Description of crystallite orientation in polycrystalline materials. III General solution to pole figure inversion. *Journal of Applied Physics*, **36**, 2024–2031.

Rojas, M., 2010. *Viscoelastic properties of heavy oils*, PhD dissertation, University of Houston.

Rubinstein, J. and Torquato, S., 1989. Flow in random porous media: Mathematical formulation, variational principles, and rigorous bounds. *Journal of Fluid Mechanics*, **206**, 25–46.

Rosenbaum, J.H., 1974. Synthetic microseismograms: Logging in porous formations. *Geophysics*, **39**, 14–32.

Roth, M., Muller, G. and Sneider, R., 1993. Velocity shift in random media. *Geophysical Journal International*, **115**, 552–563.

Rudnicki, M.D, 2016. Variation of organic matter density with thermal maturity. *AAPG Bulletin*, **100**, 17–22.

Rüger, A., 1995. P-wave reflection coefficients for transversely isotropic media with vertical and horizontal axis of symmetry. *Expanded Abstracts, Society of Exploration Geophysicists, 65th Annual International Meeting*. Tulsa, OK: Society of Exploration Geophysicists, pp. 278–281.

Rüger, A., 1996. Variation of P-wave reflectivity with offset and azimuth in anisotropic media. *Expanded Abstracts, Society of Exploration Geophysicists, 66th Annual International Meeting*. Tulsa, OK: Society of Exploration Geophysicists, pp. 1810–1813.

Rüger, A., 1997. P-wave reflection coefficients for transversely isotropic models with vertical and horizontal axis of symmetry. *Geophysics*, **62**, 713–722.

Rüger, A., 2001. *Reflection Coefficients and Azimuthal AVO Analysis in Anisotropic Media*. Tulsa, OK: Society of Exploration Geophysicists.

Rumpf, H. and Gupte, A.R., 1971. Einfliisse der Porositat und Korngrossenverteilung im Widerstandsgesetz der Porenstromung. *Chemie Ingenieur-Technik*, **43**, 367–375.

Rust, A.C. and Manga, M., 2002. Effects of bubble deformation on the viscosity of dilute suspensions. *Journal of Non-Newtonian Fluid Mechanics*, **104**, 53–63.

Rutherford, M.J. and Devine, J.D., 1996. Preeruption pressure-temperature conditions and volatiles in the 1991 dacite magma of Mount Pinatubo. In *Fire and Mud: Eruptions and Lahars of Mt. Pinatubo*, ed. C. Newhall and R. Punonhbayan. Seattle: Univ. of Wash. Press, pp. 751–766.

Rutherford, M.J., Sigurdsson, II., Carey, S. and Davis, A., 1985. The May 18, 1980, eruption of Mount St. Helens: 1. Melt composition and experimental phase equilibria. *Journal of Geophysical Research*, **90**, 2929–2947, doi:10.1029/ JB090iB04p02929.

Ryzhova, T.V., Aleksandrov, K.S. and Korobkova, V.M., 1966. The elastic properties of rock-forming minerals, V. Additional data on silicates. *Bulletin of the Academy of Sciences of USSR, Earth Physics*, **2**, 111–113.

Rzhevsky, V. and Novick, G., 1971. *The Physics of Rocks*. Moscow: MIR Publishing.

Sahay, P.N., 1996. Elastodynamics of deformable porous media. *Proceedings of the Royal Society, London A*, **452**, 1517–1529.

Sahay, P.N., 2008. On Biot slow S-wave. *Geophysics*, **72**, N19–N33.

Sahay, P.N., Spanos, T.J.T. and de la Cruz, V., 2001. Seismic wave propagation in homogeneous and anisotropic porous media. *Geophysical Journal International*, **145**, 209–222.

Sahimi, M., 1995. *Flow and Transport in Porous Media and Fractured Rock*. Weinheim: VCH Verlagsgesellschaft mbH.

Sain, R., 2010. *Numerical simulation of P-scale heterogeneity and its effects on elastic, electrical, and transport properties*, PhD dissertation, Stanford University.

Sain, R., Mukerji, T. and Mavko, G., 2016. On microscale heterogeneity in granular media and its impact on elastic property estimation. *Geophysics*, **81**, D561–D571.

Saito, G., Kazahaya, K., Shinohara, H., Stimac, J. and Kawanabe, Y., 2001. Variation of volatile concentration in a magma chamber system of Satsuma-Iwojima volcano deduced from melt inclusion analyses. *Journal of Volcanology and Geothermal Research*, **108**, 11–31, doi:10.1016/S0377-0273(00)00276-6.

Saito, G., Stimac, J.A., Kawanabe, Y. and Goff, F., 2002. Mafic-felsic magma interaction at Satsuma-Iwojima volcano, Japan: Evidence from mafic inclusions in rhyolites. *Earth Planets Space*, **54**, 303–325.

Saito, G., Uto, K., Kazahaya, K., Shinohara, H., Kawanabe, Y. and Satoh, H., 2005. Petrological characteristics and volatile content of magma from the 2000 eruption of Miyakejima volcano. *Japan, Bulletin Volcanology*, **67**, 268–280, doi:10.1007/s00445-004-0409-z.

Saomoto, H. and Katagiri, J., 2015. Direct comparison of hydraulic tortousoity and electric tortuosity base on finite element analysis. *Theoretical and Applied Mechanics Letters*, **5**, 177–180.

Sarkar, D., Bakulin, A. and Kranz, R., 2003. Anisotropic inversion of seismic data for stressed media: Theory and a physical modeling study on Berea sandstone. *Geophysics*, **68**, 690–704.

Sato, H., Nakada, S., Fujii, T., Nakamura, M. and Suzuki-Kamata, K., 1999. Groundmass pargasite in the 1991–1995 dacite of Unzen volcano: Phase stability experiments and volcanological implications. *Journal of Volcanology and Geothermal Research*, **89**, 197–212, doi:10.1016/S0377-0273(98)00132-2.

Sawamoto, H., Weidner, D.J., Sasaki, S., Kumazawa, M., 1984. Single crystal elastic properties of the modified spinel (beta) phase of magnesium orthosilicate. *Science*, **224**, 749–751.

Saxena, N. and Mavko, G., 2014a. Exact equations for fluid and solid substitution. *Geophysics*, **79**, L21–L32.

Saxena, N. and Mavko, G., 2014b. Impact of change in pore-fill material on P-wave velocity. *Geophysics*, **79**, D399–D407.

Saxena, N., Mavko, G., Hofmann, R., Dolan, S. and Bryndzia, L.T., 2016. Mineral substitution: Separating the effects of fluids, minerals, and microstructure on P- and S-wave velocities. *Geophysics*, **81**, D63–D76.

Sayers, C.M., 1987. The elastic anisotropy of polycrystalline aggregates of zirconium and its alloys. *Journal of Nuclear Materials*, **144**, 211–213.

Sayers, C.M., 1988a. Inversion of ultrasonic wave velocity measurements to obtain the microcrack orientation distribution function in rocks. *Ultrasonics*, **26**, 73–77.

Sayers, C.M., 1988b. Stress-induced ultrasonic wave velocity anisotropy in fractured rock. *Ultrasonics*, **26**, 311–317.

Sayers, C.M., 1994. The elastic anisotropy of shales. *Journal of Geophysical Research*, **99**, 767–774.

Sayers, C.M., 1995a. Anisotropic velocity analysis. *Geophysical Prospecting*, **43**, 541–568.

Sayers, C.M., 1995b. Simplified anisotropy parameters for transversely isotropic sedimentary rocks. *Geophysics*, **60**, 1933–1935.

Sayers, C.M., 1995c. Stress-dependent seismic anisotropy of shales. *Expanded Abstracts, Society of Exploration Geophysicists, 65th Annual International Meeting*. Tulsa, OK: Society of Exploration Geophysicists, pp. 902–905.

Sayers, C.M., 2004. Seismic anisotropy of shales: What determines the sign of Thomsen delta parameters? *Expanded A&b Abstract, SEG 74th International Exposition*. Tulsa, OK: Society of Exploration Geophysicists.

Sayers, C.M. and Kachanov, M., 1991. A simple technique for finding effective elastic constants of cracked solids for arbitrary crack orientation statistics. *International Journal of Solids and Structures*, **12**, 81–97.

Sayers, C.M. and Kachanov, M., 1995. Microcrack-induced elastic wave anisotropy of brittle rocks. *Journal of Geophysical Research*, **100**, 4149–4156.

Sayers, C.M., Van Munster, J.G. and King, M.S., 1990. Stress-induced ultrasonic anisotropy in Berea sandstone. *International Journal of Rock Mechanics and Mining Sciences & Geomechics Abstracts*, **27**, 429–436.

Scaillet, B. and Evans, W., 1999. The 15 June 1991 eruption of Mount Pinatubo. I. Phase equilibria and pre-eruption P-T-fO2–fH2O conditions of dacite magma. *Journal of Petrology*, **40**, 381–411, doi:10.1093/petrology/40.3.381.

Scaillet, B., Holtz, F. and Pichavant, M., 1998. Phase equilibrium constraints on the viscosity of silicic magmas: 1. Volcanic-plutonic comparison. *Journal of Geophysical Research*, **103**, 27257–27266, doi:10.1029/98JB02469.

Scaillet, B., Pichavant, M. and Cioni, R., 2008. Upward migration of Vesuvius magma chamber over the past 20,000 years. *Nature*, **455**, 216–219, doi:10.1038/nature07232.

Schlichting, H., 1951. *Grenzschicht-Theorie*. Karlsruhe: G. Braun.

Schlumberger, 1989. *Log Interpretation Principles/Applications*. Houston, TX: Schlumberger Wireline & Testing.

Schlumberger, 1991. *Log Interpretation Principles/Applications*. Houston, TX: Schlumberger Wireline & Testing.

Schlumberger, 2009. Log Interpretation Charts. Houston, TX: Schlumberger Wireline & Testing.

Schmincke, H.-U., Park, C. and Harms, E., 1999. Evolution and environmental impacts of the eruption of Laacher See volcano (Germany) 12,900 a BP. *Quarterly International*, **61**, 61–72, doi:10.1016/S1040-6182(99)00017-8.

Schmitt, D.P., 1989. Acoustic multipole logging in transversely isotropic poroelastic formations. *Journal of the Acoustical Society of America*, **86**, 2397–2421.

Schoenberg, M., 1980. Elastic wave behavior across linear slip interfaces. *Journal of the Acoustical Society of America*, **68**, 1516–1521.

Schoenberg, M. and Douma, J., 1988. Elastic wave propagation in media with parallel fractures and aligned cracks. *Geophysical Prospecting*, **36**, 571–590.

Schoenberg, M. and Muir, F., 1989. A calculus for finely layered anisotropic media. *Geophysics*, **54**, 581–589.

Schoenberg, M. and Protazio, J., 1992. "Zoeppritz" rationalized and generalized to anisotropy. *Journal of Seismic Exploration*, **1**, 125–144.

Schön, J.H., 1996. *Physical Properties of Rocks*. Oxford: Elsevier.

Schopper, J.R. 1966. A theoretical investigation on the formation factor/permeability/porosity relationship using a network model. *Geophysical Prospecting*, **14**, 301–341.

Schwab, J.A. and Graham, E.K., 1983. Pressure and temperature dependence of the elastic properties of fayalite (abstract). *EOS Transactions AGU*, **64**, 847.

Schwerdtner, W.M., Tou, J.C.-M. and Hertz, P.B., 1965. Elastic properties of single crystals of anhydrite. *Canadian Journal of Earth Sciences*, **2**, 673–683.

Scott, G.D. and Kilgour, D.M., 1969. The density of random close packing of spheres. *British Journal of Applied Physics (Journal of Physics D)*, **2**(2), 863–866.

Secco, R.A., Manghnani, M.H. and Liu, T.C., 1991. The bulk modulus-attenuation-viscosity systematics of diopside-anorthite melts. *Geophysical Research Letters*, **18**, 93–96.

Segel, L.A., 1987. *Mathematics Applied to Continuum Mechanics*. Mineola, NY: Dover.

Self, S. and King, A.J., 1996. Petrology and sulfur and chlorine emissions of the 1963 eruption of Gunung Agung, Bali, Indonesia. *Bulletin of Volcanology*, **58**, 263–285, doi:10.1007/s004450050139.

Sen, P.N. and Goode, P.A., 1988. Shaley sand conductivities at low and high salinities. *Trans. Soc. Prof. Well Log Analysts, 29th Annual Logging Symposium*, Paper F.

Sen, P.N., Scala, C. and Cohen, M.H., 1981. A self-similar model for sedimentary rocks with application to the dielectric constant of fused glass beads. *Geophysics*, **46**, 781–795.

Seshagiri Rao, T., 1951. Elastic constants of barytes and celestite. *Proceedings of the Indian Academy of Sciences A*, **33**, 251–256.

Shapiro, S.A., 2017. Stress impact on elastic anisotropy of triclinic porous and fractured rocks. *Journal of Geophysical Research Solid Earth*, **122**, 2034–2053.

Shapiro, S.A. and Hubral, P., 1995. Frequency-dependent shear-wave splitting and velocity anisotropy due to elastic multilayering. *Journal of Seismic Exploration*, **4**, 151–168.

Shapiro, S.A. and Hubral, P., 1996. Elastic waves in thinly layered sediments: The equivalent medium and generalized O'Doherty-Anstey formulas. *Geophysics*, **61**, 1282–1300.

Shapiro, S.A. and Hubral, P., 1999. *Elastic Waves in Random Media: Fundamentals of Seismic Stratigraphic Filtering*. Berlin: Springer-Verlag.

Shapiro, S.A., Hubral, P. and Zien, H., 1994. Frequency-dependent anisotropy of scalar waves in a multilayered medium. *Journal of Seismic Exploration.*, **3**, 37–52.

Shapiro, S. and Kaselow, A., 2005. Porosity and elastic anisotropy of rocks under tectonic stress and pore-pressure changes. *Geophysics*, **70**, N27–N38.

Shapiro, S.A. and Zien, H., 1993. The O'Doherty-Anstey formula and localization of seismic waves. *Geophysics*, **58**, 736–740.

Sharma, M.M., Garrouch, A. and Dunlap, H.F., 1991. Effects of wettability, pore geometry, and stress on electrical conduction in fluid-saturated rocks. *The Log Analyst*, **32**, 511–526.

Sharma, M.M. and Tutuncu, A.N., 1994. Grain contact adhesion hysteresis: A mechanism for attenuation of seismic waves. *Geophysical Research Letters*, **21**, 2323–2326.

Sharma, M.M., Tutuncu, A.N. and Podia, A.L., 1994. Grain contact adhesion hysteresis: A mechanism for attenuation of seismic waves in sedimentary granular media. In *Extended Abstracts, Soc. Expl. Geophysics, 64th Annual International Meeting, Los Angeles*. Tulsa, OK: Society of Exploration Geophysicists, pp. 1077–1080.

Sheen, S.-H., Chien, H.-T. and Raptis, A.C., 1996. Ultrasonic methods for measuring liquid viscosity and volume percent of solids, Argonne National Lab, ANL-9714, DOE Contract W-31-109-Eng-38.

Shen, L.C., 1985. Problems in dielectric-constant logging and possible routes to their solution. *The Log Analyst*, **26**, 14–25.

Shepard, J.S., 1993. Using a fractal model to compute the hydraulic conductivity function. *Soil Science Society of America*, **57**, 300–306.

Sheriff, R.E., 1991. *Encyclopedic Dictionary of Exploration Geophysics*, 3rd edn. Tulsa, OK: Society of Exploration Geophysics.

Sherman, M.M., 1986. The calculation of porosity from dielectric constant measurements: A study using laboratory data. *The Log Analyst*, Jan.–Feb., 15–24.

Sherman, M.M., 1988. A model for the frequency dependence of the dielectric permittivity of reservoir rocks. *The Log Analyst*, Sept.–Oct., 358–369.

Shimano, T., Iida, A., Yoshimoto, M., Yasuda, A. and Nakada, S., 2005. Petrological characteristics of the 2004 eruptive deposits of Asama volcano, central Japan. *Bulletin Volcanologic Society of Japan*, **50**, 315–332.

Shuey, R.T., 1985. A simplification of the Zoeppritz equations. *Geophysics*, **50**, 609–614.

Siggins, A.F. and Dewhurst, D.N., 2003. Saturation, pore pressure and effective stress from sandstone acoustic properties. *Geophysical Research Letters*, **30**, 1089. doi:10.1029/2002GL016143.

Simandoux, P., 1963. Dielectric measurements on porous media: Application to the measurement of water saturations: Study of the behavior of argillaceous formations. *Revue de l'Institut Français Pétrole*, **18**, supplementary issue, 193–215.

Simmons, G., 1965. Single crystal elastic constants and calculated aggregate properties. *Journal of the Graduate Research Center, SMU*, **34**, 1–269.

Simmons, G. and Birch, F., 1963. Elastic constants of pyrite. *Journal of Applied Physics*, **34**, 2736–2738.

Sisavath, S., Jing, X. and Zimmerman, R.W., 2001. Creeping flow through a pipe of varying radius. *Physics of Fluids*, **13**, 2762–2772.

Skelt, C., 2004. Fluid substitution in laminated sands. *Leading Edge*, **23**, 485–489.

Skoge, M., Donev, A., Stillinger, F.H. and Torquato, S., 2006. Packing of hyperspheres in high-dimensional Euclidean spaces. *Physical Review E*, **74**, 041127.

Smith, G.C. and Gidlow, P.M., 1987. Weighted stacking for rock property estimation and detection of gas. *Geophysical Prospecting*, **35**, 993–1014.

Smith, W.O., Foote, P.D. and Busand, P.G., 1929. Packing of homogeneous spheres. *Physical Review*, **34**(2), 1271–1274.

Sneddon, I.N., 1948. Boussinesq's problem for a rigid cone. *Proceedings Cambridge Philosophical Society*, **44**, 492–507,

Sneddon, I.N. and Lowergrub, M., 1969. *Crack Problems in the Classical Theory of Elasticity*. New York: John Wiley and Sons.

Soga, N., 1967. Elastic constants of garnet under pressure and temperature. *Journal of Geophysical Research*, **72**, 4227–4234.

Soddy, F., 1936. The hexlet. *Nature*, **137**, 958.

Sondergeld, C.H., Rai, C.S., Margesson, R.W. and Whidden, K.J., 2000. Ultrasonic measurement of anisotropy on the Kimmeridge Shale: 70th Annual International Meeting, SEG, Expanded Abstracts, 1858–1861, http://dx.doi.org/10.1190/1.1815791.

Song, C., Wang, P. and Makse, H.A., 2008. A phase diagram for jammed matter. *Nature*, **453**, 629–632.

Spangenburg, K. and Haussuhl, S., 1957. Die elastischen Konstanten der Alkalihalogenide. *Zeitschrift für Kristallographie*, **109**, 422–137.

Spanos, T.J.T., 2002. *The Thermophysics of Porous Media*. Boca Raton, FL: Chapman & Hall/CRC.

Speight, J.G., 1999. *The Chemistry and Technology of Petroleum*. New York: Marcel Dekker Inc.

Spratt, R.S., Goins, N.R. and Fitch, T.J., 1993. Pseudo-shear – the analysis of AVO. In *Offset Dependent Reflectivity – Theory and Practice of AVO Analysis*, ed. J.P. Castagna and M. Backus, Investigations in Geophysics Series No. 8. Tulsa, OK: Society of Exploration Geophysicists, pp. 37–56.

Standing, M.B., 1962. Oil systems correlations. In *Petroleum Production Handbook*, volume II, ed. T. C. Frick. McGraw-Hill Book Co., part 19.

Standing, M.B., 1981. *Volumetric and Phase Behavior of Oil Field Hydrocarbon Systems*. Richardson, TX: SPE.

Standing, M.B. and Katz, D.L.,1942a. Density of crude oils saturated with natural gas. *Transactions AIME*, **146**, 159.

Standing, M.B. and Katz, D.L.,1942b. Density of natural gases. *Transactions AIME*, **146**, 140.

Stix, J. and Gorton, M.P., 1990. Changes in silicic melt structure between the two Bandelier caldera-forming eruptions, New Mexico, USA: Evidence from zirconium and light rare earth elements. *Journal of Petrology*, **31**, 1261–1283, doi:10.1093/petrology/31.6.1261.

Stix, J. and Layne, G.D., 1996. Gas saturation and evolution of volatile and light lithophile elements in the Bandelier magma chamber between two caldera-forming eruptions. *Journal of Geophysical Research*, **101**, 25181–25196, doi:10.1029/96JB00815.

Stoll, R.D., 1974. Acoustic waves in saturated sediments. In *Physics of Sound in Marine Sediments*, ed. L.D. Hampton. New York: Plenum, pp. 19–39.

Stoll, R.D., 1977. Acoustic waves in ocean sediments. *Geophysics*, **42**, 715–725.

Stoll, R.D., 1989. *Sediment Acoustics*. Berlin: Springer-Verlag, p. 154.

Stoll, R.D. and Bryan, G.M., 1970. Wave attenuation in saturated sediments. *Journal of the Acoustical Society of America*, **47**, 1440–1447.

Stoner, E.C., 1945. The demagnetizing factors for ellipsoids. *Philos. Mag*, **36**, 803–821.

Strandenes, S., 1991. *Rock Physics Analysis of the Brent Group Reservoir in the Oseberg Field*. Stanford Rockphysics and Borehole Geophysics Project, special volume.

Sumino, Y. and Anderson, O.L., 1984. Elastic constants of minerals. In *Handbook of Physical Properties of Rocks*, vol. III, ed. R.S. Carmichael. Boca Raton, FL: CRC Press, pp. 39–137.

Suzuki, I., Anderson, O.L. and Sumino, Y., 1983. Elastic properties of a single-crystal forsterite Mg_2 SiO_4, up to 1200 K. *Physics and Chemistry of Minerals*, **10**, 38–46.

Sumino, Y., Kumazawa, M., Nishizawa, O. and Pluschkell, W., 1980. The elastic constants of single-crystal Fe1_xO, MnO, and CoO, and the elasticity of stochiometric magnesiowustite. *Journal of the Physics of the Earth*, **28**, 475–495.

Suzuki, Y. and Nakada, S., 2007. Remobilization of highly crystalline felsic magma by injection of mafic magma: Constraints from the middle sixth century eruption at Haruna volcano, Honshu, Japan. *Journal of Petrology*, **48**, 1543–1567, doi:10.1093/petrology/egm029.

Sweeney, J.J. and Burnham, A.K., 1990. Evaluation of a simple model of vitrinite reflectance based on chemical kinetics. *AAPG Bulletin*, **74**, 1559–1570.

Sviridov, V., Mayr, S. and Shapiro, S., 2019. Rock elasticity as a function of the uniaxial stress: Laboratory measurements and theoretical modelling of vertical transversely isotropic and orthorhombic shales. *Geophysical Prospecting*, **67**, 867–1881.

Takeuchi, S., 2002. *Petrological study on triggering of eruptions from phenocryst-rich magma chambers*, PhD thesis, 112 pp., Tokyo Institute of Technology.

Takeuchi, S., 2004. Precursory dike propagation control of viscous magma eruptions. *Geology*, **32**, 1001–1004, doi:10.1130/G20792.1.

Takeuchi, S., 2011. Preeruptive magma viscosity: An important measure of magma eruptibility. *Journal of Geophysical Research*, **116**, B10201, 1–19.

Takeuchi, S. and Nakamura, M., 2001. Role of precursory less-viscous mixed magma in the eruption of phenocryst-rich magma: Evidence from the Hokkaido-Komagatake 1929 eruption. *Bulletin of Volcanology*, **63**, 365–376, doi:10.1007/s004450100151.

Tang, X.-M. and Cheng, A., 2004. *Quantitative Borehole Acoustic Methods*. Amsterdam: Elsevier.

Tang, X.-M., Cheng, A. and Toksoz, M.N., 1991. Dynamic permeability and borehole Stoneley waves: A simplified Biot-Rosenbaum model. *Journal of the Acoustical Society of America*, **90**, 1632–1646.

Taylor, G.I., 1932. The viscosity of a fluid containing small drops of another fluid. *Proceedings of the Royal Society of London A*, **138**, 41–48.

Taylor, G.I., 1934. The formation of emulsions in definable fields of flow. *Proceedings Royal Society London A*, **146**, 501–523.

Thomas, D.G., 1965. Transport characteristics of suspension: VIII, A note on the viscosity of Newtonian suspensions of uniform spherical particles. *Journal of Colloid Science*, **20**, 267.

Thomas, E.C. and Stieber, S.J., 1975. The distribution of shale in sandstones and its effect upon porosity. In *Trans. 16th Annual Logging Symposium of the SPWLA*, paper T.

Thomas, E.C. and Stieber, S.J., 1977. Log derived shale distributions in sandstone and its effect upon porosity, water saturation, and permeability. In *Trans. 6th Formation Evaluation Symposium of the Canadian Well Logging Society*.

Thomsen, L., 1986. Weak elastic anisotropy. *Geophysics*, **51**, 1954–1966.

Thomsen, L., 1993. Weak anisotropic reflections. In *Offset Dependent Reflectivity – Theory and Practice of AVO Analysis*, ed. J.P. Castagna and M. Backus, *Investigations in. Geophysics* Series, No. 8. Tulsa, OK: Society of Exploration Geophysicists, pp. 103–111.

Thomson, W., 1878. Mathematical theory of elasticity. *Elasticity, Encyclopedia Britannica*, **7**, 819–825.

Thorpe, M.F. and Jasiuk, I., 1992. New results in the theory of elasticity for two-dimensional composites. *Proceedings Mathematical and Physical Sciences*, **438**, 531–544.

Timoshenko, S.P. and Goodier, J.N., 1934. *Theory of Elasticity*. New York: McGraw-Hill.

Tinder, R.F., 2008. *Tensor Properties of Solids*. Morgan & Claypool.

Tomiya, A., Takahashi, E., Furukawa, N. and Suzuki, T., 2010. Depth and evolution of a silicic magma chamber: Melting experiments on a low-K rhyolite from Usu volcano, Japan, *Journal of Petrology*, **51**, 1333–1354, doi:10.1093/petrology/egq021.

Tissot, B.P., Durand, B., Espitalié, J. and Combaz, A., 1974. Influence of nature and diagenesis of organic matter on formation of petroleum. *AAPG Bulletin*, **58**(3), 499–506.

Tissot, L. and Welte, D.H., 1984. *Petroleum Formation and Occurrence*. Berlin: Springer-Verlag.

Tixier, M.P. and Alger, R.P., 1967. Log evaluation of non-metallic mineral deposits. *Trans. SPWLA 8th Ann. Logging Symp*, Denver, June 11–14, Paper R.

Todd, T. and Simmons, G., 1972. Effect of pore pressure on the velocity of compressional waves in low porosity rocks. *Journal of Geophysical Research*, **77**, 3731–3743.

Topp, G.C., Davis, J.L. and Annan, A.P., 1980. Electromagnetic determination of soil water content: Measurements in coaxial transmission lines. *Water Resources Research*, **16**, 574–582.

Torquato, S., 2002. *Random Heterogeneous Materials: Microstructure and Macroscopic Properties*. New York: Springer Publishing.

Torquato, S. and Rubinstein, J., 1989. Diffusion-controlled reactions. II. further bounds on the rate constant. *Journal of Chemical Physics*, **90**, 1644–1647.

Torquato, S. and Pham, D.C., 2004. Optimal bounds on the trapping constant and permeability of porous media. *Physical Review Letters*, **92**, 255505_1–4.

Tosaya, C.A., 1982. *Acoustical properties of clay-bearing rocks*, PhD dissertation, Stanford University.

Tosaya, C.A. and Nur, A., 1982. Effects of diagenesis and clays on compressional velocities in rocks. *Geophysical Research Letters*, **9**, 5–8.

Truesdell, C., 1965. *Problems of Nonlinear Elasticity*. New York: Gordon and Breach.

Tschoegl, N.W., 1989. *Phenomenological Theory of Linear Viscoelastic Behavior: An Introduction*. Berlin: Springer-Verlag.

Tsukui, M. and Aramaki, S., 1990. The magma reservoir of the Aira proclastic eruption—A remarkably homogeneous high-silica rhyolite magma reservoir. *Bulletin Volcanologic Society of Japan*, **35**, 231–248.

Tsvankin, I., 1997. Anisotropic parameters and P-wave velocity for orthorhombic media. *Geophysics*, **62**, 1292–1309.

Tsvankin, I., 2001. *Seismic Signatures and Analysis of Reflection Data in Anisotropic Media*. New York: Pergamon.

Tucker, M.E., 2001. *Sedimentary Petrology*, 3rd edn. Oxford: Blackwell Science.

Tutuncu, A.N., 1992. *Velocity dispersion and attenuation of acoustic waves in granular sedimentary media*, PhD dissertation, University of Texas, Austin.

Tutuncu, A.N. and Sharma, M.M., 1992. The influence of grain contact stiffness and frame moduli in sedimentary rocks. *Geophysics*, **57**, 1571–1582.

Ucok, H., Ershaghi, I. and Olhoeft, G.R., 1980. Electrical resistivity of geothermal brines, Journal of Petroleum Technology, April, 717–727.

Ungerer, P., 1993. Modeling of petroleum generation and migration, in Applied Petroleum Geochemistry, Editions Technip, p. 411.

Ungerer, P., Collel, J. and Yiannourakou, M., 2014. Molecular modeling of the volumetric and thermodynamic properties of kerogen: Influence of organic type and maturity, *Energy & Fuels*, **29**, 91–105.

Urmos, J. and Wilkens, R.H., 1993. *In situ* velocities in pelagic carbonates: New insights from ocean drilling program leg 130, Ontong Java. *Journal of Geophysical Research*, **98**(B5), 7903–7920.

Vandenbroucke, M. and Largeau, C., 2007. Kerogen origin, evolution, and structure. *Organic Geochemistry*, **38**, 719–833, doi:10.1016/j.orggeochem.2007.01.001.

Vanorio, T., Scotellaro, C. and Mavko, G. 2008. The effect of chemical and physical processes on the acoustic properties of carbonate rocks. *The Leading Edge*, **27**, 1040–1048, doi:10.1190/1.2967558.

Vaughan, M.T. and Guggenheim, 1986. Elasticity of muscovite and its relationship to crustal structure. *Journal of Geophysical Research*, **91**, 4657–4664.

Vavryčuk, V. and Pšenčik, I., 1998. PP-wave reflection coefficients in weakly anisotropic elastic media. *Geophysics*, **63**(6), 2129–2141.

Vazquez, M. and Beggs, H.D., 1980. Correlations for fluid physical property prediction. *Journal of Petroleum Technology* (June), 968.

Venezky, D.Y. and Rutherford, M.J., 1997. Preeruption conditions and timing of dacite-andesite magma mixing in the 2.2 ka eruption at Mount Rainier. *Journal of Geophysical Research*, **102**, 20069–20086, doi:10.1029/97JB01590.

Venezky, D.Y. and Rutherford, M.J., 1999. Petrology and Fe-Ti oxides reequilibration of the 1991 Mount Unzen mixed magma. *Journal of Volcanology and Geothermal Research*, **89**, 213–230, doi:10.1016/S0377–0273(98)00133–4.

Verma, R.K., 1960. Elasticity of some high-density crystals. *Journal of Geophysical Research*, **65**, 757–766.

Vernik, L., 1993. Microcrack-induced versus intrinsic elastic anisotropy in mature HC-source shales. *Geophysics*, **58**, 1703–1706.

Vernik, L., 1994. Source Rock Quality and Maturation Prediction Using Acoustic Velocity and Anisotropy, SRB Special Report, Stanford University.

Vernik, L., 2016. *Seismic Petrophysics in Quantitative Interpretation*, Society of Exploration Geophysicists. doi.org/10.1190/1.9781560803256.

Vernik, L., Bruno, M. and Bovberg, C., 1993. Empirical relations between compressive strength and porosity of siliciclastic rocks. *International Journal of Rock Mechanics and Mining Sciences & Geomechanics Abstracts*, **30**(7), 677–680.

Vernik, L., Castagna, J. and Omovie, S., 2018. Shear-wave velocity prediction in unconventional shale reservoirs. *Geophysics*, 83, MR35–MR45.

Vernik, L., Fisher, D. and Bahret, S., 2002. Estimation of net-to-gross from P and S impedance in deepwater turbidites. *Leading Edge*, **21**, 380–387.

Vernik, L. and Kachanov, M., 2010. Modeling elastic properties of siliciclastic rocks. *Geophysics*, **75**, E171–182.

Vernik, L. and Landis, C., 1996. Elastic anisotropy of source rocks: Implications for hydrocarbon generation and primary migration. *AAPG Bulletin*, **80**(4) 531–544.

Vernik, L. and Liu, X., 1997. Velocity anisotropy in shales: A petrophysical study. *Geophysics*, **62**, 2, 521–532, http://dx.doi.org/10.1190/1.1444162.

Vinogradov, V. and Milton, G.W., 2005. The total creep of viscoelastic composites under hydrostatic or antiplane loading. *Journal of the Mechanics and Physics of Solids*, **53**, 1248–1279.

Vogel, T.A., Eichelberger, J.C., Younker, L.W., Schuraytz, B.C., Horkowitz, J.P., Stockman, H.W. and Westrich, H.R., 1989. Petrology and emplacement dynamics of intrusive and extrusive rhyolites of Obsidian Dome, Inyo craters volcanic chain, Eastern California. *Journal of Geophysical Research*, **94**, 17937–17956, doi:10.1029/JB094iB12p17937.

Voigt, W., 1890. Bestimmung der Elastizitatskonstanten des brasilianischen Turmalines. *Annalen der Physik und Chemie*, **41**, 712–729.

Voigt, W., 1907. Bestimmung der Elastizitatskonstanten von Eisenglanz. *Annalen der Physik*, **24**, 129–140.

Voronov, F.F. and Grigor'ev, S.B., 1976. Influence of pressures up to 100 kbar on elastic properties of silver, sodium and cesium chlorides, Soviet Physics. *Solid State*, **18**, 325–3288.

Wachtman, J.B., Jr., Tefft, W.E., Lam, D.G., Jr. and Strinchfield, R.P., 1960. Elastic constants of synthetic single crystal corundum at room temperature. *Journal of Research of the National Bureau of Standards*, **64A**, 213–228.

Waddell, H., 1932. Volume, shape and roundness of rock particles. *Journal of Geology*, **40**, 443–151.

Wadhawan, V.K., 1987. The generalized Curie principle, the Hermann theorem, and the symmetry of macroscopic tensor properties of composites. *Materials Research Bulletin*, **22**, 651–660.

Wadsworth, J., 1960. Experimental examination of local processes in packed beds of homogeneous spheres. *Nat. Res. Council of Canada, Mech. Eng. Report MT-41* February.

Wallace, P.J. and Anderson, Jr., A.T., 1998. Effects of eruption and lava drainback on the H2O contents of basaltic magmas at Kilauea volcano. *Bulletin of Volcanology*, **59**, 327–344, doi:10.1007/s004450050195.

Walls, J., Nur, A. and Dvorkin, J., 1991. A slug test method in reservoirs with pressure sensitive permeability. *Proceedings 1991 Coalbed Methane Symposium*, University of Alabama, Tuscaloosa, May 13–16, pp. 97–105.

Walsh, J.B., 1965. The effect of cracks on the compressibility of rock. *Journal of Geophysical Research*, **70**, 381–389.

Walsh, J.B., 1969. A new analysis of attenuation in partially melted rock. *Journal of Geophysical Research*, **74**, 4333.

Walsh, J.B. and Brace, W.F., 1984. The effect of pressure on porosity and the transport properties of rock. *Journal of Geophysical Research*, **89**, 9425–9431.

Walsh, J.B., Brace, W.F. and England, A.W., 1965. Effect of porosity on compressibility of glass. *Journal of the American Ceramics Soc*iety, **48**, 605–608.

Walters, C.C., Freund, H., Keleman, S.R., Peczak, P. and Curry, D.J., 2007. Predicting oil and gas compositional yields via chemical structure–chemical yields modeling(CS–CYM): Part 2 – Application under laboratory and geologic conditions. *Organic Geochemistry*, **38**, 306–322.

Walton, K., 1987. The effective elastic moduli of a random packing of spheres. *Journal of the Mechanics and Physics of Solids*, **35**, 213–226.

Wang, H.F., 2000. *Theory of Linear Poroelasticity*. Princeton, NJ: Princeton University Press.

Wang, Z., 2000a. Dynamic versus static properties of reservoir rocks, in *Seismic and Acoustic Velocities in Reservoir Rocks. SEG Geophysics Reprint Series*, **19**, 531–539.

Wang, Z., 2002. Seismic anisotropy in sedimentary rocks, Parts I and II. *Geophysics*, **67**, 1415–1440.

Wang, Z. and Nur, A., 1992. *Seismic and Acoustic Velocities in Reservoir Rocks, vol. 2, Theoretical and Model Studies, Society of Exploration Geophysicists, Geophysics Reprint Series*. Tulsa, OK: Society of Exploration Geophysicists.

Wang, Z. and Nur, A. (eds.), 2000. *Seismic and Acoustic Velocities in Reservoir Rocks, vol. 3, Recent Developments, Geophysics Reprint Series, no. 19*. Tulsa, OK: Society of Exploration Geophysicists.

Waples, D.W. and Marzi, R.W., 1998. The universality of the relationship between vitrinite reflectance and transformation ratio. *Organic Geochemistry*, **28**, 383–388.

Ward, J.A. 2010. Kerogen density in the Marcellus shale: SPE Unconventional Gas Conference, Pittsburgh, Pennsylvania, February 23–25, 2010, SPE-131767-MS, doi:10.2118/131767-MS, 4 p.

Ward, S.H. and Hohmann, G.W., 1987. Electromagnetic theory for geophysical applications. In *Electromagnetic Methods in Applied Geophysics, vol. I, Theory*, ed. M.N. Nabhigian. Tulsa, OK: Society of Exploration Geophysicists, pp. 131–311.

Warren, W.E., 1983. Determination of in-situ stresses from hydraulically fractured spherical cavities. *Journal of Energy Resources Technology*, **105**, 125–127.

Warshaw, C.M. and Smith, R.L., 1988. Pyroxene and fayalites in the Bandelier Tuff, New Mexico: Temperatures and comparison with other rhyolites. *American Mineralogist*, **73**, 1025–1037.

Watt, J.P., 1986. Hashin-Shtrikman bounds on the effective elastic moduli of trigonal (3, $\bar{3}$) and tegragonal (4, $\bar{4}$, 4m) symmetry. *Journal of Applied Physics*, **60**, 3120–3124.

Watt, J.P., Davies, G.F. and O'Connell, R.J., 1976. The elastic properties of composite materials. *Reviews of Geophysics and Space Physics*, **14**, 541–563.

Watt, J.P. and Peselnick, L., 1980. Clarification of the Hashin-Shtrikman bounds on the effective elastic moduli of polycrystals with hexagonal, trigonal, and tetragonal symmetry. *Journal of Applied Physics*, **51**, 1525–1531.

Waxman, M.H. and Smits, L.J.M., 1968. Electrical conductivities in oil-bearing shaley sands. *Society of Petroleum Engineers Journal*, **8**, 107–122.

Webb, S.L., 1997. *Silicate Melts, Lecture Notes in Earth Sciences*, 67. New York: Springer Publishing.

Webster, J.D., Taylor, R.P. and Bean, C., 1993. Pre-eruptive melt composition and constraints on degassing of a water-rich pantellerite magma, Fantale volcano, Ethiopia. *Contributions to Mineralogy Petrology*, **114**, 53–62, doi:10.1007/BF00307865.

Weger, R.J., 2006. *Quantitative pore/rock type parameters in carbonates and their relationship to velocity deviations*, PhD dissertation, University of Miami.

Weidner, D.J., 1986. Mantle model based on measured physical properties of minerals. In *Chemistry and Physics of Terrestrial Planets*, ed. S.K. Saxena. New York: Springer-Verlag, pp. 251–274.

Weidner, D.J., Bass, J.D., Ringwood, E. and Sinclair, W., 1982. The single-crystal elastic moduli of stishovite. *Journal of Geophysical Research*, **87**, 4740–4746.

Weidner, D.J. and Carleton, H.R., 1977. Elasticity of coesite. *Journal of Geophysical Research*, **82**, 1334–1346.

Weidner, D.J. and Hamaya, N., 1983. Elastic properties of the olivine and spinel polymorphs of Mg2GeO4 and evaluation of elastic analogues. *Physics of the Earth and Planetary Interiors*, **33**, 275–283.

Weidner, D.J. and Ito, E., 1985. Elasticity of $MgSiO_3$ in the ilmenite phase. *Physics of the Earth and Planetary Interiors*, **40**, 65–70.

Weidner, D.J., Sawamoto, H. and Sasaki, S., 1984. Single-crystal elastic properties of the spinel phase of Mg_2SiO_4. *Journal of Geophysical Research*, **89**, 7852–7859.

Weidner, D.J., Yeganeh-Haeri, A. and Ito, E., 1987. Elastic properties of majorite (abstract). *EOS Transactions AGU*, **68**, 410.

Weingarten, J.S. and Perkins, T.K., 1995. Prediction of sand production in wells: Methods and Gulf of Mexico case study. *Journal of Petroleum Technology*, **47**, 596–600.

Weng, G.J., 1984. Some elastic properties of reinforced solids with special reference to isotropic ones containing spherical inclusions. *International Journal of Engineering Science*, **22**, 845–856.

Westrich, H.R. and Gerlach, T.M., 1992. Magmatic gas source for the stratospheric SO2 cloud from the June 15, 1991, eruption of Mount Pinatubo. *Geology*, **20**, 867–870, doi:10.1130/0091-7613(1992)020<0867: MGSFTS>2.3.CO;2.

Whitcombe, D.N., 2002. Elastic impedance normalization. *Geophysics*, **67**, 59–61.

Whitcombe, D.N., Connolly, P.A., Reagan, R.L. and Redshaw, T.C., 2002. Extended elastic impedance for fluid and lithology prediction. *Geophysics*, **67**, 63–67.

White, J.E., 1975. Computed seismic speeds and attenuation in rocks with partial gas saturation. *Geophysics*, 40, 224–232.

White, J.E., 1983. *Underground Sound: Application of Seismic Waves*. New York: Elsevier.

White, J.E., 1986. Biot-Gardner theory of extensional waves in porous rods. *Geophysics*, **51**, 742–745.

Widmaier, M., Shapiro, S.A. and Hubral, P., 1996. AVO correction for a thinly layered reflector overburden. *Geophysics*, **61**, 520–528.

Wiggins, R., Kenny, G.S. and McClure, C.D., 1983. A method for determining and displaying the shear-velocity reflectivities of a geologic formation. European Patent Application 0113944.

Williams, D.M., 1990. The acoustic log hydrocarbon indicator. *Society of Professional Well Log Analysts, 31st Annual Logging Symposium*, Paper W.

Williams, S.N., 1983. Plinian airfall deposits of basaltic composition. *Geology*, **11**, 211–214, doi:10.1130/0091-7613(1983)11<211: PADOBC>2.0.CO;2.

Willis, J.R., 1977. Bounds and self-consistent estimates for the overall properties of anisotropic composites. *Journal of the Mechanics and Physics of Solids*, **25**, 185–202.

Wilson, C.J.N., Blake, S., Charlier, B.L.A. and Sutton, A.N., 2005. The 26.5 ka Oruanui eruption, Taupo volcano, New Zealand: Development, characteristics and evacuation of a large rhyolitic magma body. *Journal of Petrology*, **47**, 35–69, doi:10.1093/petrology/egi066.

Winkler, K., 1986. Estimates of velocity dispersion between seismic and ultrasonic frequencies. *Geophysics*, **51**, 183–189.

Winkler, K.W., 1983. Contact stiffness in granular porous materials: Comparison between theory and experiment. *Geophysical Research Letters*, **10**, 1073–1076.

Winkler, K.W., 1985. Dispersion analysis of velocity and attenuation in Berea sandstone. *Journal of Geophysical Research*, **90**, 6793–6800.

Winkler, K.W. and Murphy, W.F., III, 1995. Acoustic velocity and attenuation in porous rocks. In *Rock Physics and Phase Relations: A Handbook of Physical Constants, AGU Reference Shelf 3*. Washington, DC: American Geophysical Union.

Winkler, K.W. and Nur, A., 1979. Pore fluids and seismic attenuation in rocks. *Geophysical Research Letters*, **6**, 1–4.

Winsauer, W.O., Shearin, H.M., Masson, P.H. and Williams, M., 1952. Resistivity of brine saturated sands in relation to pore geometry. *American Association of Petroleum Geologists Bulletin*, **36**, 253–277.

Woeber, A.F., Katz, S. and Ahrens, T.J., 1963. Elasticity of selected rocks and minerals. *Geophysics*, **28**, 658–663.

Wong, S.W., Kenter, C.J., Schokkenbroek, H., van Regteren, J. and De Bordes, P.F., 1993. Optimising shale drilling in the Northern North Sea; borehole stability considerations. SPE paper 26736, *Offshore Europe Conference*, Aberdeen, 7–10 September.

Wood, A.W., 1955. *A Textbook of Sound*. New York: McMillan Co.

Worthington, P.F., 1985. Evolution of shaley sand concepts in reservoir evaluation. *The Log Analyst*, **26**, 23–40.

Wu, T.T., 1966. The effect of inclusion shape on the elastic moduli of a two-phase material. *International Journal of Solids and Structures*, **2**, 1–8.

Wyllie, M.R.J. 1957. *The Fundamentals of Electric Log Interpretation*. New York: Academic Press.

Wyllie, M.R.J., Gardner, G.H.F. and Gregory, A.R., 1963. Studies of elastic wave attenuation in porous media. *Geophysics*, **27**, 569–589.

Wyllie, M.R.J. and Gregory, A.R., 1953. Formation factors of unconsolidated porous media: Influence of particle shape and effect of cementation. *Transactions American Institute of Mechanical Engineers*, **198**, 103–110.

Wyllie, M.R.J., Gregory, A.R. and Gardner, G.H.F., 1958. An experimental investigation of factors affecting elastic wave velocities in porous media. *Geophysics*, **23**, 459–493.

Wyllie, M.R.J., Gregory, A.R. and Gardner, L.W., 1956. Elastic wave velocities in heterogeneous and porous media. *Geophysics*, **21**, 41–70.

Xia, Y. and Mavko, G., 2015. Contact mechanics of cemented grains: The numerical approach, Annual Report: Stanford Rock Physics and Borehole Geophysics Project.

Xu, S. and White, R.E., 1994. A physical model for shear-wave velocity prediction. In *Expanded Abstracts, 56th Eur. Assoc. Expl. Geoscientists Meet. Tech. Exhib*, Vienna, p. 117.

Xu, S. and White, R.E., 1995. A new velocity model for clay-sand mixtures. *Geophysical Prospecting,***43**, 91–118.

Yale, D.P. and Jameison, W.H., Jr., 1994. Static and dynamic rock mechanical properties in the Hugoton and Panoma fields. Kansas Society of Petroleum Engineers, Paper 27939. *Society of Petroleum Engineers Mid-Continent Gas Symposium*, Amarillo, TX, May.

Yale, D.P., Perez, A. and Raney, R., 2018. Novel pore pressure prediction technique for unconventional reservoirs, Unconventional resources technology conference proceedings, URTeC 2901731.

Yamakawa, N., 1962. Scattering and attenuation of elastic waves. *Geophysical Magazine, Tokyo*, **31**, 63–103.

Yamanoi, Y., Takeuchi, S., Okumura, S., Nakashima, S. and Yokoyama, T., 2008. Color measurements of volcanic ash deposits from three different styles of summit activity at Sakurajima volcano, Japan: Conduit processes recorded in color of volcanic ash. *Journal of Volcanology and Geothermal Research*, **178**, 81–93, doi:10.1016/j.jvolgeores.2007.11.013.

Yan, F., Han, D-h, Yao, Q. and Chen, X-L., 2016. Seismic velocities of halite salt: Anisotropy, heterogeneity, dispersion, temperature, and pressure effects. *Geophysics*, **81**, D293–D301.

Yeganeh-Haeri, A. and Vaughan, M.T., 1984. Single-crystal elastic constants of olivine (abstract). *EOS Transactions AGU*, **65**, 282.

Yeganeh-Haeri, A., Weidner, D.J. and Ito, E., 1989. Single-crystal elastic moduli of magnesium metasilicate perovskite. In *Perovskite: A Structure of Great Interest to Geophysics and Materials Science, Geophysics Monograph Series*, vol. 45. Washington, DC: American Geophysical Union, pp. 13–25.

Yeganeh-Haeri, A., Weidner, D.J. and Parise, J.B., 1992. Elasticity of a-cristobalite: A silicon dioxide with a negative Poisson's ratio. *Science*, **257**, 650–652.

Yi, Y.-B. and Sastry, A.M., 2004. Analystical approximation of the percolation threshold for overlapping ellipsoids of revolution. *Proceedings Royal Society of London A*, **460**, 2353–2380.

Yi, Y.-B., Wang, C.-W. and Sastry, A.M., 2004. Two-dimensional vs. three-dimensional clustering and percolation in fields of overlapping ellipsoids. *Journal of the Electrochemical Society*, **151**, A1292–A1300.

Yin, H., 1992. *Acoustic velocity and attenuation of rocks: Isotropy, intrinsic anisotropy, and stress-induced anisotropy*, PhD dissertation, Stanford University.

Yoneda, A., 1990. Pressure derivatives of elastic constants of single crystal MgO, MgAl$_2$O$_4$. *Journal of the Physics of the Earth*, **38**, 19–55.

Yoon, H.S. and Newnham, R.E., 1969. Elastic properties of fluorapatite. *Amer. Mineralog*, **54**, 1193–1197.

Yoon, H.S. and Newnham, R.E., 1973. The elastic properties of beryl. *Acta Crystallographia*, **A29**, 507–509.

Young, H.D., 1962. *Statistical Treatment of Experimental Data*. New York: McGraw-Hill.

Yu, A.B. and Standish, N., 1987. Porosity calculations of multi-component mixtures of spherical particles. *Powder Technology*, **52**, 233–241.

Yu, G., Vozoff, K. and Durney, W., 1993. The influence of confining pressure and water saturation on dynamic properties of some rare Permian coals. *Geophysics*, **58**, 30–38.

Zaman, E. and Jalali, P., 2010. On hydraulic permeability of random packs of monodisperse spheres: Direct flow simulations versus correlations. *Physica A.*, **389**, 205–214.

Zamora, M. and Poirier, J.P., 1990. Experimental study of acoustic anisotropy and birefringence in dry and saturated Fontainebleau sandstone. *Geophysics*, **55**, 1455–1465.

Zarembo, I.K. and Krasil'nikov, V.A., 1971. Nonlinear phenomena in the propagation of elastic waves in solids. *Soviet Physics Uspekhi*, **13**, 778–797.

Zargari, S., Prasad, M., Mba, K.C. and Mattson, E.D., 2013. Organic maturity, elastic properties and textural characteristics of self resourcing reservoirs. *Geophysics*, **78**(4) D223–D235.

Zeller, R. and Dederichs, P.H., 1973. Elastic constants of polycrystals. *Physica Status Solidi B*, **55**, 831.

Zener, C., 1948. *Elasticity and Anelasticity of Metals*. Chicago, IL: University of Chicago Press.

Zeszotarski, J.C., Chromik, R.C., Vinci, R.P., Messmer, M.C, Michels, R. and Larsen, J.W., 2004. Imaging and mechanical property measurements of kerogen via nanoindentation. *Geochimica Cosmochimica Acta*, **68**, 4113.

Zhang, H.P. and Makse, H.A., 2005. Jamming transition in emulsions and granular materials. *Physical Review E*, **72**, 011301.

Zhang, X. and Knackstedt, M.A., 1995. Direct simulation of electrical and hydraulic tortuosity in porous solids. *Geophysical Research Letters*, **22**, 2333–2336.

Zhao, J., Li, S., Zou, R. and Yu, A., 2012. Dense random packings of spherocylinders. *Soft Matter*, **8**, 1003–1009.

Zhao, Y. and Weidner, D.J., 1993. The single-crystal elastic moduli of neighborite. *Physics and Chemistry of Minerals*, **20**, 419–124.

Ziarani, A.S. and Aguilera, R., 2012. Knudsen's permeability correction factor for tight porous media. *Transport in Porous Media*, **91**, 239–260.

Zimmer, M.A., 2003. *Seismic velocities in unconsolidated sands: Measurements of pressure, sorting, and compaction effects*, PhD dissertation, Stanford University.

Zimmerman, R.W., 1984a. Elastic moduli of a solid with spherical pores: New self-consistent method. *International Journal of Rock Mechanics and Mining Sciences & Geomechanics Abstracts*, **21**, 331–343.

Zimmerman, R.W., 1984b. *The effect of pore structure on the pore and bulk compressibilities of consolidated sandstones*, PhD dissertation, University of California, Berkeley.

Zimmerman, R.W., 1984. The elastic moduli of a solid with spherical pores: New self-consistent method. *International Journal of Rock Mechanics and Mining Sciences & Geomechanics Abstracts*, **21**, 339–343.

Zimmerman, R.W., 1986. Compressibility of two-dimensional cavities of various shapes. *Journal of Applied Mechanics*, **53**, 500–504.

Zimmerman, R.W., 1991a. *Compressibility of Sandstones*. New York: Elsevier.

Zimmerman, R.W., 1991b. Elastic moduli of a solid containing spherical inclusions. *Mechanics of Materials*, **12**, 17–24.

Zoback, M.D., 2007. *Reservoir Geomechanics*. Cambridge: Cambridge University Press.

Zoeppritz, K., 1919. Erdbebenwellen VIIIB, On the reflection and propagation of seismic waves. *Gottinger Nachr.*, **I**, 66–84.

Index

Aki–Richards approximation, *see* AVO
analytic signal, *see* Hilbert transform
anellipticity, 59, 62
anisotropy, elastic
  anellipticity, 59, 62
  Backus average, *see* Backus average
  Berryman's weak anisotropy notation, 58
  bulk modulus, anisotropic, 56
  coordinate transformation, 21–25
  dispersion, seismic, 124
  displacement discontinuity model, 75, 286, 289
  ellipsoidal symmetry, 61
  Eshelby–Cheng model, 243, 267–269
  fluid substitution, anisotropic, 387–390
  fracture, 60, 285–290
  frequency-dependent, 189–197
  group, phase, energy velocities, 123–126
  Hooke's law, 44–63
  Hudson's model, 258–267
  layered media, *see* Backus average
  NMO, 128–133
  Ordered sphere packing, effective moduli, 364–366
  Poisson's ratio, anisotropic, 54, 56
  Sayer's notation for weak anisotropy, 61
  squirt, anisotropic, 423–427
  stiffness tensor, 22, 36, 44
  stress-induced anisotropy, 70–79
  Thomsen's notation, 57–61, 77, 302
  T-matrix inclusion models, 241–275
  Voigt notation, 46–48, 64
  Young's modulus, anisotropic, 54, 56
  see also AVO; effective medium models, elastic;
    elasticity; NMO; granular media; Thomsen's
    weak anisotropy; Tsvankin's parameters
API gravity, 460
Archie's law, 589
  cementation exponent, 589
  saturation exponent, 591
  second law, 590
Arp's equation, 608
attenuation, electromagnetic, 582
attenuation, seismic
  Biot model, 367–373
  BISQ model, 419–421

Chapman *et al.* squirt model, 421
characteristic frequency, 429, 441
comparison among fluid-related mechanisms,
    427–433
Dvorkin–Mavko attenuation model, 433–437
empirical relations, 513
Kennett algorithm, 181
Kennett–Frazer theory, 181
Klimentos and McCann relation, 513
Koesoemadinata and McMechan relation,
    514
layered media, *see* stratigraphic filtering
O'Doherty–Anstey formula, 187
scattering, 201–206
squirt model, 372, 413–421,
    423–427
stratigraphic filtering, 185–189
wave in viscous fluid, 455
see also borehole;
    scattering; viscoelasticity and *Q*
autocorrelation, 2
AVO, 136–144
  Aki–Richards approximation, 140
  Anisotropic, 144–154
  approximate forms, 140–144
  arbitrary anisotropy, 153
  AVOA, 145
  Bortfield approximation, 140
  Gardner's equation, 142, 509
  gradient, 141
  Gray *et al.* approximation, 143
  Hilterman's equation, 142
  normal incidence, 141
  orthorhombic, 151
  Poisson's ratio trace (Hilterman's), 142
  P-to-SV, 143, 144
  Schoenberg–Protázio formulation, 145
  Shuey's approximation, 141
  Smith and Gidlow's formula, 142
  shear wave, 143
  transversely isotropic (VTI), 147
  transversely isotropic (HTI), 149
  Zoeppritz (Knott–Zoeppritz) equations, 138, 145
AVOA, *see* AVO, anisotropic

Backus average, 275–285
  arbitrarily anisotropic layers; 279–285
  fractures, parallel, 285–290
  isotropic layers, 275
  poroelastic layers, 280–285
  uniform shear modulus, 236, 277
  viscoelastic, 280–285
BAM, *see* bounding average method
bar resonance, *see* cylindrical rods
Batzle–Wang relations, 457–473
Bayes decision theory, 21
Berryman self-consistent model, *see* self-consistent
  approximation
binary mixture of spheres, *see* granular media
binomial coefficients, 15
binomial distribution, 15
Biot coefficient, *see* Biot–Willis coefficient
Biot model, 367–373
  Geertsma–Smit approximation, 372
  partial saturation, 438–443
  slow P-wave, 368
  slow S-wave, 368, 370
  tortuosity parameter, 368, 552–556
  see also Gassmann's equations
Biot–Rosenbaum theory, 215
Biot–White–Gardner effect in resonant bars, 208
Birch's law, 517
BISQ model, 419–421
Bond coordinate transformation, 22, 23
bootstrap, 19
borehole
  Biot–Rosenbaum theory, 215
  Elliptical, 98
  inclined, stress, 98
  permeable formation, 215–219
  pseudo-Rayleigh in borehole, 211
  Stoneley waves, 211
  stress and deformation, 96, 97, 98
  tube wave, 212
  waves in, 211–219
Bortfield's AVO formula, *see* AVO
bound water, 593
bounding average method (BAM), 292, 400, 412
bounds, cross-property, 596–608
  Berryman–Milton bounds, 597
  Gibiansky–Torquato bound, 598
  Milton bounds, 604
bounds, elastic, 220–308
  Berryman–Milton, 230–233
  bound-filling models, 290–295
  bounding average method, 292
  Hashin–Shtrikman, 222–225, 290, 395
  Hashin–Shtrikman–Walpole, 43, 44–63, 222–233,
    290
  isoframe model, 292
  isostrain average, 220
  isostress average, 221
  microgeometric parameters, 230–233

  modified Hashin–Shtrikman bound, 291,
    474–480
  modified Voigt bound, 474–480
  on fluid substitution, 384
  on Thomsen's parameters, 60
  Poisson's ratio, isotropic, 55
  Poisson's ratio, transversely isotropic, 55
  polycrystalline aggregates, 295–304
  realizability of bounds, 224, 226, 245
  Reuss, 220–222, 234, 290, 298–300, 376
  Voigt, 220–222, 290, 298–300
  Voigt–Reuss–Hill average, 235, 292
  y-transform, 229, 294, 397
bounds, electrical, 577
bounds, permeability, 548
bound-filling models, 290–295
bounding average method, 292, 400
Bower's relation, 511
Bragg scattering, 189
Brandt model, 350
Brie mixing equation, 443
brine, *see* fluid properties
brine resistivity, 608
Brinkman's equation, 532
Brocher's empirical relations, 488
Brown and Korringa equation
  anisotropic form, 387–390
  mixed mineralogy, 385–387
bubble point, 461, *see also* fluid properties
bubbles and particles in a viscoelastic background,
  567–570
Buckley–Leverett equation, 526
bulk modulus, *see* elasticity, effective medium models,
  elastic
Burger's vector, 94

capillary forces, 538–542
capillary number, 568
capillary pressure, 539
  Laplace equation, 539
  Young equation, 538
carbon dioxide properties, 669
Castagna's empirical relations, 487, 492, 493, 494
cation exchange capacity, 592
Cauchy's formula, 36
causality, *see* Fourier transform
cavities in elastic solid, 81–96
  Biot coefficient, 68, 83, 381
  circular hole, stress, 96, 97
  crack closing stress, 71, 89
  cylindrical shell, 96
  ellipsoidal, with finite thickness, 95
  elliptical borehole, 98
  Eshelby ellipsoid, 102–109, 243, 245–247
  needle-shaped (prolate spheroid), 89, 108, 247
  nonelliptical, 94, 287
  penny-shaped crack (oblate spheroid), 71, 88, 107,
    108

cavities in elastic solid (cont.)
  Skempton's coefficient, 83
  spherical cavities, 84–88, 106
  spherical shells, 87
  three-dimensional ellipsoidal cavities, 84–90, 95
  two-dimensional cavities, 90, 92
  see also borehole; effective medium theories;
    inclusion models
cementation models, *see* granular media
circular hole, stress, 96, 97
Chapman *et al.* squirt model, 421
Ciz and Shapiro solid substitution, 406
Classification, 20–21
  Bayes decision theory, 21
  discriminant analysis, 21
  Mahalanobis distance, 21
  supervised, 20
  unsupervised, 20
Clausius–Mossotti formula, 578
clay
  Castagna's velocity relations, 487, 494
  Han's velocity relations, 494
  mudrock line, Tosaya's velocity relations, 486
  V_P–V_S relations, 492–508
  see also electrical conductivity
closing stress, *see* crack closing stress
CO_2 properties, 669
coal, V_P–V_S relations, 497
coated sphere assemblage, 224, 245
Coates equation, 557
Coates–Dumanoir equation, 557
coefficient of determination, 13
coefficient of variation, 12
coherent potential approximation, *see* effective
    medium models
Cole–Cole dielectric model, 583
complex dielectric constant, *see* dielectric constant,
    attenuation electromagnetic
complex modulus, *see* viscoelasticity and Q
compliance tensor, *see* elasticity
compressibility, 238–241
  cavities in elastic solid, 81–96
  definitions and pitfalls, 238–241
  dry-rock, 239, 374
  low-frequency saturated, 83
  pore space, 239
  pitfalls, 238–241
  reservoir fluids, 457–473
  see also elasticity; effective medium models, elastic;
    bounds, elastic
compressive strength, *see* unconfined compressive
    strength
condensates, *see* fluid properties
constant Q model
  see also viscoelasticity and Q
contact angle, 539
contact stiffness, *see* Hertz–Mindlin model
conversion tables, 634–638

convolution theorem, *see* Fourier Transform
coordinate rotation, *see* coordinate transformation
coordinate transformation, 21–25
  Bond transformation, 22
  direction cosines, 22
  Euler angles, 22, 23, 24
coordination number, *see* granular media
correlation coefficient, 12–14
covariance, 12, 14
cracks
  Chapman *et al.* squirt model, *see* squirt model
  crack closing stress, 71, 89
  deformation under stress, *see* cavities in elastic solid
  displacement discontinuity model, 75, 286, 289
  Eshelby–Cheng model, 267–269
  fracture compliance, 285–290
  fractures, flow in, 560
  Hudson model, 258–267
  Kuster–Toksöz model, 243, 244, 248
  nonelliptical, 94, 287
  self-consistent approximation, 224, 244, 248,
    251–255, 273
  squirt model, see squirt model
  stress-induced anisotropy, 70–79
  T-matrix inclusion models, 241–275
  see also cavities in elastic solid; effective medium
    models, elastic; scattering
creeping flow, 535
CRIM formula, 586
critical porosity, 474–478, 498
  modified DEM model, 255–258
  modified Hashin–Shtrikman bound, 291, 474–480
  modified Voigt bound, 474–480
  V_P–V_S relations
cross-correlation, 2
cross-property relations, 596–608
  elastic-electrical, 596–600
  electrical-electrical, 600–601
crystallographic third-order constants, 65
curvilinear coordinates, 80
cylindrical bar/rod, 206–211
  flexural waves, 207
  longitudinal (extensional) waves, 121, 207
  Pochhammer equation, 207
  porous, saturated, 208
  resonance, 208
  torsional waves, 206
cylindrical shell, 96
cylinder, solid, 97

Darcy, unit, 525
Darcy's law, 525–561
  Brinkman's equation, 532
  Buckley–Leverett equation for immiscible
    displacement, 526
  compressible gas, 527
  Forchheimer's relation, 532
dead oil, 460, *see also* fluid properties

Debye dielectric model, 583

DEM model, 224, 243, *see also* effective medium models

density-velocity relations, *see* empirical velocity relations

depolarizing factors, dielectric, 580

Descarte's theorem, 315

dielectric permittivity, 577–581, 611

dielectric susceptibility, 582

differential effective medium model, *see* DEM

diffusion, 527–530, 562

one-dimensional, 562

  nonlinear, 562

  radial, 562

  saturated porous sphere, 530

diffusion equation hyperbolic, 563

diffusion equation nonlinear, 562

diffusion scattering, 201–206

diffusivity, hydraulic, 527–530

Digby model, 347

dimensionless numbers, 639

direction cosines, 22

discriminant analysis, 21

dislocation theory, 94

dispersion, *see* velocity dispersion

displacement discontinuity model, *see* cracks

dissipation, *see* attenuation

distributions, 15–18

binomial coefficients, 15

binomial distribution, 15

logistic distribution, 18

lognormal distribution, 17

normal (Gaussian) distribution, 16

Poisson distribution, 15

truncated exponential distribution, 17

uniform distribution, 16

Weibull distribution, 18

Dix equation, *see* NMO

drainage, 541

drained conditions, 82, 238, 374

dry rock, 82, 238, 239, 374

dual water model, 593

Dutta–Odé model, 444–449

Eaton's pore-pressure relation, 511

Eberhart–Phillips (Han) velocity relations, 485

effective electrical properties, 577–610

effective fluid, 440

effective medium models, dielectric

bounds, 577

Clausius–Mossotti formula, 578

coherent potential approximation, 578

Cole–Cole model, 583

Debye model, 583

DEM model, 224, 579

ellipsoidal inclusions, 84–90

Eshelby ellipsoid, 102–109, 243, 245–247

Hanai–Bruggeman model, 579

Hashin–Shtrikman bounds, 577

layered media, 581

Lorentz–Lorenz equation, 578

Maxwell–Garnett equation, 578

self-consistent approximation, 578

effective medium models, elastic, 220–308

BISQ model, 419–421

bound-filling models, 290–295

bounding average method, 292

Brandt model, 350

cavities in elastic solid, 81–96

cementation models, *see* granular media

Chapman *et al.* squirt model, *see* squirt model

coherent potential approximation, 244

comparison among fluid-related dispersion mechanisms, 427–433

compressibility and pitfalls, 238–241

contact stiffness, grain, *see* Hertz–Mindlin model

coordination number, 312–315

DEM model, 224, 244, 248, 255–258

Digby model, 347

Eshelby–Cheng model, 267–269

Granular dynamics simulations, 313, 363

granular media, 309–366

Hashin–Shtrikman bounds, *see* bounds, elastic

Hertz (Mindlin) model, 287, 337–346

Hill average, 228

Hudson model, 258–267

isoframe model, 292

Jenkins *et al.* model, 349

Johnson *et al.* model, 350

Kuster–Toksöz model, 243, 244, 245, 248

Maxwell approximation, 245

Mori–Tanaka model, 225, 243, 244, 245, 248–249

needle-shaped (prolate) inclusion or cavity, 89, 241–275

O'Connell–Budiansky model, 172, 251

ordered sphere packing, effective moduli, 364–366

penny-shaped (oblate) inclusion or cavity, 88, 107, 108, 241–275

polycrystalline aggregates, 295–304

self-consistent approximation, 224, 244, 248, 251–255, 273

squirt, anisotropic, 413–416, 423–427

squirt model, *see* squirt model

stiff sand model, 354–356, 361

stress-induced anisotropy, 70–79

suspension, 234

T-matrix inclusion models, 241–275

uniform shear modulus (Hill), 228, 236, 277

Walpole bounds, 224, 229–233

Walton model, 346

Wood's formula, 234

Xu–White model, 506

see also bounds, elastic

effective stress, pressure, 66–70

Biot–Willis coefficient, 68, 83, 381

effective stress, pressure (cont.)
  effective stress coefficient, 67
  effective stress law, 66–70
eigentensors and eigenvalues, 35–36, 39, 53
eigenvectors, 36
Einstein's viscosity relation, 568
elastic compliance, *see* elasticity
elastic impedance, 154–160
  extended elastic impedance, 156
  isotropic, 157
  orthorhombic, 159
  pseudo-impedance attribute, 154
  transversely isotropic (VTI), 158
elastic moduli, *see* elasticity
elasticity, 37–63
  2-D elasticity, 40–43
  anellipticity, 59, 62
  anisotropic, 44–63, *see also* anisotropy
  area moduli, 41
  Biot coefficient, 381
  bulk modulus, 37, 56
  cavities in elastic solid, 81–96
  compliance tensor, 22, 44
  compressibility, 37, 82, 83
  cubic, 47, 304
  effective stress properties, 66–70
  eigentensors and eigenvalues, 35–36, 39
  Hooke's law, 37–63
  Isotropic, 37, 44, 47, 302–304
  Kelvin notation, 22, 51–53, 242
  Lame's constant; 37, 38
  moduli, elastic, 37–63
  monoclinic, 44, 48
  nonlinear elasticity, 64
  orthorhombic, 48, 298–300
  plane strain, 40–43, 97
  plane stress, 40, 97
  P-wave modulus, 38
  Poisson's ratio, isotropic, 38, 42
  Poisson's ratio anisotropic, 54, 56
  polycrystalline aggregates, 295–304
  shear modulus, 38
  Skempton coefficient, 83
  static and dynamic moduli, 112–115
  stiffness tensor, 22, 36, 44
  strain energy, 54, 64
  strain tensor, 36, 37, 39, 225, 306
  stress, deviatoric part, 39
  stress, hydrostatic part, 39
  stress tensor, 36, 37, 39, 225
  stress-induced anisotropy, 70–79
  third-order elasticity, 64, 76
  Thomsen parameters, 57–61, 77, 302
  transversely isotropic (hexagonal), 48, 57–61, 275, 300–301
  triclinic, 44, 302
  Tsvankin's parameters for orthorhombic media, 62, 77
  Voigt notation, 22, 45–63, 64

  von Mises stress, 40
  Young's modulus, isotropic, 38, 221
  Young's modulus, anisotropic, 54, 56
  see also AVO; bounds, elastic; effective medium models, elastic; phase velocity
electrical conductivity, 588–596
  Archie's law, 589
  dual water model, 593
  Indonesia formula, 591
  Sen and Goode model, 594
  shaly sands, 591
  Simandoux equation, 591
  Waxman–Smits model, 592
  Waxman–Smits–Juhász model, 592
electromagnetic waves, 582
  reflection coefficients, 584
  transmission coefficients, 584
electromagnetism, units, 637
ellipsoidal cavities, 84–90, 102–109
  see also cavities in elastic solid; effective medium models, elastic
embedded bounds, *see* solid substitution
empirical electrical property relations, 585–588
  Archie's law, 589
  CRIM formula, 586
  cross-property relations, 596–608
  Faust's relation, 601
  Knight and Nur relation, 587
  Koesoemadinata–McMechan relations, 588
  Lichtnecker–Rother formula, 585
  Odelevskii formula, 586
  Olhoeft's relation, 587
  Topp's relation, 586
empirical velocity relations, 474–524
  Birch's law, 517
  Brocher's compilation, 488
  Castagna's velocity relations, 487, 492, 493, 494
  coal, 497
  critical porosity model, 257, 474–478, 498
  cross-property relations, 596–608
  Eberhart–Phillips (Han) relations, 485
  Faust's relation, 601
  Gardner's velocity-density relation, 509
  Geertsma's modulus-porosity relation, 478
  Greenberg–Castagna relations, 502–504
  Han's velocity relations, 494
  interpolators, soft and stiff, 292, 334, 359, 361, 409, 477
  Krief's relations, 500–502
  modified bounds, 474–480
  mudrock line, 487, 494
  Pickett's relation
  Raymer–Hunt–Gardner velocity relations, 474
  static and dynamic moduli, 112–115
  Tosaya's velocity relations, 486
  velocity-density relations, 508
  velocity-strength relations, 515–517
  Vernik's relations, 504

Vernik–Kachanov clastics models, 480–482
V_P-density, 488, 508
V_P-Poisson's ratio, 491
V_P–V_S relations, 488, 489, 492–508
Williams's relation, 505
Wyllie's time average, 478
energy spectrum, *see* spectrum
energy, strain, 54, 64
energy velocity, 124
envelope, *see* Hilbert transform
equations of state fluids, *see* fluid properties
Eshelby–Cheng model, 267–269
Eshelby ellipsoid, 102–109, 243, 245–247
Eshelby tensor, 248
Euler angles, 24
exponential distribution, 17

fast path effect, 198
    Faust's relation, 601
    Formation factor, 588
FFT, *see* Fourier transform
field optimality condition, 225
filtration, *see* diffusion
flexural waves bar, 207
Floquet solution, 188
fluid properties, 457–473
    apparent liquid density, 463
    Batzle–Wang relations for physical properties,
        457–473
    bubble point, 461
    condensates, 469–473
    critical pressure, 469
    critical temperature, 469
    heavy oil, 564–567
    nonwetting, 539
    formation volume factor, 462
    principle of corresponding states, 469
    pseudo critical properties, 469
    pseudo density, 463
    reduced pressure, 469
    reduced temperature, 469
    viscosity, 468, 525, 564–567
    wetting, 539
    Z-factor, gas, 470, 663–669
fluid flow, 525–574
    Brinkman's equation, 532
    creeping flow, 535
    Darcy's law, 525–561
    dynamic permeability, 217, 449–455
    diffusion equation nonlinear, 562
    diffusivity, 527–530
    drainage, 541
    Forchheimer's relation, 532
    fractured formation, 560
    imbibition, 541
    Kozeny–Carman relation, 542–557
    Knudsen flow, 531
    laminar flow, 533–538

multiphase flow
    Navier–Stokes equation, 533
    parallel walls, flow between, 533
    past a sphere, 534
    permeability bounds, 548
    pipes and ducts, 533–536
    relative permeability, 526
    slip flow, 531
    Stokes law, 533
    transition flow; 531
    viscous flow, 531, 533–538
fluid substitution
    anisotropic, 387–390
    Batzle–Wang, *see* fluid properties
    Biot poroelastic theory, 367–373, 381
    Biot–Gassmann fluid substitution, 373, 387–390
    bounding average method (BAM), 292, 400, 412
    bounds on, 384
    Brown and Korringa equations, 385–390
    Dvorkin's model for clay-bearing sandstone, 382
    Gassmann's fluid substitution, 373, 374–384,
        387–390
    partial saturations, 438–443
    P-wave only, 381
    thinly laminated sands, 407–412
    see also solid substitution
fluids, dispersion and attenuation of seismic waves,
    367–473
    Biot theory, 367–373
    from poroelastic heterogeneity, 449–455
    see also squirt model
Folk and Ward formula, *see* granular media
Forchheimer's relation, 532
formation factor, 588
Fourier transform, 1–6
    causality, 9
    convolution theorem, 2
    energy spectrum, 3
    impulse response, 9
    minimum phase, 3
    sampling theorem, 4
    transfer function, 9
    transform pairs, 6
    transform theorems, 6
    zero phase, 3
fractures, *see* cracks
Fresnel's equations, electromagnetic, 584

Gardner's equation, 509
gases, properties, 457–473, 663–669
gas-water ratio, 458
Gassmann equations, 374–384
    anisotropic form, 387–390
    Batzle–Wang relations, *see* fluid properties
    Biot coefficient form, 381
    bounding average method, 292
    compressibility form, 376
    dry rock caveat, 374

Gassmann equations (cont.)
  Dvorkin's model for clay-bearing sandstone, 382
  generalized form for composite porous media, 390
  linear form, 377
  P-wave modulus form, 378, 381
  partial saturations, 438–443
  pore stiffness interpretation, 84, 378
  Reuss average form, 376
  solid pore-filling material, *see* solid substitution
  thinly laminated sands, 407–412
  velocity form, 378
  V_P-only form (no V_S), 378
  see also Brown and Korringa equations
Gaussian distribution, 16
Geertsma's modulus-porosity relations, 478
Geertsma–Smit approximation to the Biot model, 372
granular media, 309–366
  binary mixtures, 316–335
  Brandt model, 350
  cementation models, 354–356, 361
  coefficient of variation (sorting), 337
  coordination number, 312–315
  Digby model, 347
  effective moduli, 337–364
  see also effective medium models, elastic energy of
      deposition
  ellipsoids, packing of, 325–328
  Folk and Ward formula, 337
  Hertz (Mindlin) model, 287, 337–346
  ideal mixtures, 316–317
  Jenkins *et al.* model, 349
  Johnson *et al.* model, 350
  non-ideal mixing, 317–321
  non-spherical particles, 321
  ordered sphere packing, effective moduli,
      364–366
  spheres: packing and sorting of, 309–337
  spheres, packing of identical, 309–316, 337–364
  porosity, 309
  particle shape, 321
  particle size, 335
  percolation, 325–328
  phi-scale of particle size, 336
  psi-scale of particle size, 336
  sorting, 335, 336–337
  sphericity, 322
  standard deviation (sorting), 336, 337
  stiff sand model, 354–356, 361
  Thomas-Stieber model for sand-shale systems,
      329–335
  Yin–Marion models for sand-shale systems,
      329–335
  Udden–Wentworth scale, 335
  uncemented (soft) sand model, 359–360
  Walton model, 346
Gravity, gas, 461
gravity, API, 460
Greenberg–Castagna V_P–V_S relations, 502–504

group velocity, 124
  see also velocity

Han's velocity relations, 494
Hartley transform, 6
Hashin–Shtrikman bounds, *see* bounds, elastic,
      bounds, electrical
Hertz (Mindlin) model, 287, 337–346
hexagonal symmetry, *see* elasticity, transverse isotropy
Hilbert transform, 7–9
  analytic signal, 7–9
  impulse response, 9
  instantaneous envelope, 8
  instantaneous frequency, 8
  instantaneous phase, 8
Hill relations
  uniform shear modulus, 228, 236, 277
  Voigt–Reuss–Hill average, 235, 292
Hilterman's AVO formulas, 142
Hooke's law, *see* elasticity
Hudson's crack model, 258–267
  heavily faulted structures, 264
  scattering attenuation, 201–206
HTI, *see* AVO, elasticity, phase velocity, reflectivity,
      transverse isotropy
hydrate, methane, 659
hydraulic aperture, fractures, 560
hydraulic conductivity, 525
hyperelastic, *see* nonlinear elasticity

ice, 659–661
imbibition, 541
impedance, seismic, 126–136
  see also elastic impedance; AVO
impulse response, see Fourier transform
inclusion models, dielectric, 577–610
  Clausius–Mossotti formula, 578
  coherent potential approximation, 578
  DEM model, 578
  Hanai–Bruggeman model, 578
  depolarizing factors, 579
  ellipsoidal inclusions, 579
  Lorentz–Lorenz equation, 578
  Maxwell–Garnett equation, 578
  self-consistent approximation, 224
inclusion models, elastic, 241–275
  DEM model, 224–248, 255–258, 272–273
  Eshelby ellipsoid, 102–109, 243
  Eshelby–Cheng model, 267–269
  Hudson's model, 258–267
  Kuster–Toksöz model, 243, 244, 245, 248, 249–251
  self-consistent approximation, 224, 244, 248,
      251–255, 273
  T-matrix inclusion models, 241–275
  Xu–White model, 506
  Indonesia formula, 591
instantaneous envelope, *see* Hilbert transform
instantaneous frequency, *see* Hilbert transform

instantaneous phase, *see* Hilbert transform
interfacial tension, *see* surface tension
invariants, tensor, *see* tensor properties and operations
invariant imbedding, *see* Kennett algorithm
irreducible water saturation, 542, 557–560
   Coates equation, 557
   Coates–Dumanoir equation, 557
   permeability relations, 557–560
   Timur equation, 557
   Tixier equation, 557
isoframe model, 292
isostrain average, *see* Voigt average
isostress average, *see* effective fluid, Reuss average,
   Wood's formula

Jenkins *et al.* model, 349
Johnson *et al.* model, 350

Kelvin notation, 22, 51–53, 242
Kennett–Frazer theory, 181
kerogen properties, 519–524
Klinkenberg effect, 530
Knight and Nur relation, 587
Knott–Zoeppritz equations, *see* AVO
Knudsen number, 530
Koesoemadinata–McMechan relations, 514,
   588
Kozeny–Carman relation, 542–557
   apparent radius, 544
   circular pipe, 544
   ellipsoidal particle pack, 546
   elliptical pipe, 533–536
   percolation, 549
   pore throat effect,
   sphere packs, 546, 547
   square pipe, 544
   triangular pipe, 544
Kramers–Kronig relations, 179–181
Krief's relations, 500–502
Kuster–Toksöz model, 243, 244, 245, 248

Lame's constant, *see* elasticity
Landau's third-order constants, 65
Laplace equation for capillary pressure, 539
Laplace transform, 1, 10
layered media
   anisotropy, *see* Backus average, Hudson's model
   Bragg scattering, 189
   dielectric constant, 581
   Floquet solution, 188
   Kennett–Frazer theory, 181, 185
   O'Doherty–Anstey formula, 187
   periodic, 188
   poroelastic, 195–197
   scattering attenuation, 188, 201–206
   stratigraphic filtering, 185–189
   synthetic seismograms, 185–189 *see also*
     Kennet–Frazer theory

uniform shear modulus, 277
waves in elastic layered medium, 181–185, 189–197
waves in viscoelastic layered medium, 181–185,
   189–197
   see also Backus average; stratigraphic filtering;
     velocity dispersion
Lichtnecker–Rother formula, 585
linear regression, 12–14
live oil, 461–469
local flow model, *see* squirt model
logarithmic decrement, *see* viscoelasticity and Q
logistic distribution, 18
lognormal distribution, 17
longitudinal waves, *see* cylindrical rods
Lorentz–Lorenz formula, 578

magma, melt and igneous rocks, 571–574, 653
Mahalanobis distance, 21
Maxwell approximation, *see* effective medium models
Maxwell solid, *see* viscoelasticity and Q
Maxwell–Garnett equation, 578
mean, 12
mean deviation, 12
median, 12
metamorphic rocks, 658
methane hydrate, 659, *see also* ice
microgeometric parameters, *see* bounds, elastic
Mie scattering, 201–206
Mindlin model, 287, 337–346
mineral property tables, 641–652
mineral substitution, *see* solid substitution
minimum phase, *see* Fourier transform
modified Hashin-Shtrikman, *see* bounds, elastic;
   effective medium models, elastic; empirical
   relations
modified Voigt, *see* bounds, elastic;
   effective medium models, elastic;
   empirical relations
moduli, *see* elasticity
modulus defect, 428
   see also viscoelasticity and Q
Mohr's circles, 109–111
monoclinic symmetry, *see* elasticity
Monte Carlo simulations, 18
Mori-Tanaka model, 225, 243, 244, 248–249
mudrock line, 487, 494
Murnaghan third-order constants, 65

Navier-Stokes equations, 533
nearly constant Q model, *see* viscoelasticity and Q
Newtonian flow, *see* fluid flow
NMO (normal moveout), 126–133
   anisotropic, 59, 128–133
   Dix equation, 128
   elliptical anisotropy, 130
   interval velocity, 128
   isotropic, 127
   layered anisotropic earth, 133

NMO (normal moveout) (cont.)
 orthorhombic, 132
 tilted transversely isotropic, 131
 transversely isotropic, 129
 velocity, NMO, 127
 velocity, RMS, 128
normal distribution, 17
Nyquist frequency, 3, 4

O'Connell–Budiansky model, 172, 251
Odelevskii formula, 586
O'Doherty–Anstey formula, 187
oil properties, 460–469
 Batzle–Wang relations, 457–473
 bubble point, 461
 dead oil, 460
 live oil, 461–469
 see also fluid properties
Olhoeft's relation, 587
optical potential approximation, 272
orientation distribution function (ODF), 296
orthorhombic symmetry, *see* AVO, elasticity, phase
  velocity, reflectivity

packing, *see* granular media
partial saturations, 438–443
 Brie mixing equation, 443
 effective fluid, 440
 patchy saturation, 237, 433, 434, 442, 443
 relaxation scale, 441
 relaxation time, 429
 Reuss average mixing, 441
 Voigt average mixing, 443
 White–Dutta–Odé model, 444–449
 see also fluid properties, fluid substitution;
  Gassmann's equations; Biot model
particles and bubbles in a viscoelastic background,
  567–570
partition theorem, 306
pdf, *see distributions*
pendular ring, 540
penny-shaped crack, *see* cavities in elastic solid,
  cracks, effective medium models elastic
percolation porosity for flow, 549
permeability
 Coates equation, 557
 Coates–Dumanoir equation, 557
 dynamic, 217, 449–455
 fractured formations, 560
 gas, 530
 irreducible water saturation relations, 542,
  557–560
 Klinkenberg correction, 530
 Knudsen flow, 530
 Kozeny–Carman relation, 542–557
 relative permeability, 526
 Timur equation, 557
 Tixier equation, 557

phase, 3
 minimum phase, 3
 zero phase, 3
phase velocity, 49–51, 123
 anisotropic, 49–51
 electromagnetic, 123, 582
 isotropic, 49
 orthorhombic, 50
 Rayleigh wave, 122, 207, 211
 transversely isotropic, 49, 57–61
 see also viscoelasticity and Q
phi-scale, *see* granular media
physical constants, tables, 638
Pickett's relation, 492, 493, 499
pipe, flow in, 533–538, *see also* Kozeny–Carman
  relation
 circular pipe, 533–536
 Darcy friction factor, 536
 elliptical pipe, 535–538
 Fanning friction factor, 535–538
 polygonal, 535–538
 rectangular pipe, 534
 sinusoidally varying radius, 535
 triangular pipe, 534
plane stress and strain, *see* elasticity
Pochhammer equation, 207
Poisson distribution, 15
Poisson's ratio, *see* AVO, effective medium models,
  elastic, elasticity
population variance, *see* variance
pore stiffness, 84, 89, 90, 239, 378
 compressibility, *see* compressibility, cavities in
  elastic solid
pore pressure, empirical relations
 Bower's relation, 511
 Eaton's relation, 511
 Kan and Swan, 512
pore pressure, stress-induced, *see* Skempton's
  coefficient
poroelasticity, *see* Biot model
porosity, *see* empirical velocity relations,
  effective medium models,
  elastic
probability, *see* statistics and probability
principal directions, *see* eigenvectors
principal stress, 36, 40
probability distributions, *see* distributions
propagator matrix, 181, 585
psi-scale
 see also granular media
P-wave modulus, *see* effective medium models,
  elastic, elasticity

Q, *see also* viscoelasticity and Q
quality factor, *see also* viscoelasticity and Q

ray parameter, 138
Rayleigh scattering, 201, 205, 262

Rayleigh wave velocity, 122, 207
  pseudo-Rayleigh in borehole, 211
Raymer form, V_s relation, 507
Raymer–Hunt–Gardner velocity relations, 482
reflectivity, electromagnetic, 584
reflectivity, seismic, 126–136, 137
  see also AVO
regression, 12–14
representative elementary volume (REV), 304–306
residual oil, 542
resistivity, electrical, see electrical conductivity
resonant bar, see cylindrical rod
Reuss average, see bounds, elastic, Reuss
Reynolds number, 535
RMS velocity, see NMO
rock property tables and plots, 613–653

sample of laboratory data, 613–625
sample of wireline data, 627–633
  see also mineral properties tables
salt, 648, see also mineral properties tables
sample covariance, see covariance
sample mean, see mean
sample variance, see variance
sampling theorem, see Fourier transform
saturation
  drained conditions, 82, 238, 374
  low-frequency compressibility, 83
  undrained conditions, 82, 238
  see also partial saturations
scattering
  attenuation, 201–206
  Bragg scattering, 189
  Brillouin, 660
  diffusion scattering, 201, 202
  Hudson's model, 241–275
  Kuster–Toksöz model, 241–275
  Mie scattering, 201–206
Rayleigh scattering, 201, 205, 262
  self-consistent approximation, 224, 241–275
  spherical inclusions, 106, 241–275
  stochastic (Mie) scattering, 201–206
  T-matrix inclusion models, 241–275
  see also layered media
second-order elasticity, 64
seismic impedance, see impedance, seismic
seismic velocity, see velocity
self-consistent approximation, see effective medium
  models
Sen and Goode model, 594
shaly sands
  Castagna's velocity relations, 487, 492, 493, 494
  Eberhart–Phillips pressure relation, 485
  electrical conductivity, 591
  Han's velocity relations, 485, 494
  mudrock line, 487, 494
  Tosaya's velocity relations, 486
  V_P–V_S relations, 492–508

shear modulus, see effective medium models, elastic,
  elasticity
shear modulus, uniform, 228, 277
Shuey's approximation, 141
signal velocity; 126
Simandoux equation, 591
Skempton's coefficient, 83
  see also cavities in elastic solid
skin depth, electromagnetic, 582
slow P-wave, see Biot model
slow S-wave, 368, 370
  see also, Biot model
slowness, 123
  see also velocity; stratigraphic slowness
Smith and Gidlow's AVO formula
  see also AVO
Snell's law, 138, 153
Soddy circles, 315
solid substitution, 393–407
  bounding average method (BAM), 400,
    412
  Ciz and Shapiro approximation, 406
  cross-bounds, 596–608
  embedded bounds, 395
  mineral substitution, 404
  P-wave only bounds, 403
  y-transform, 397
sorting, see granular media
specific surface area
spectral ratio, see also viscoelasticity and Q
spectrum, 3, 5, 7–9, 171, 173, 187, 450
sphere packs, see granular media
spherical coordinates, 80
spherical inclusions, see cavities in elastic solid,
  effective medium models, elastic scattering
squirt model, 372
  anisotropic, 423–427
  Chapman et al. squirt model, 421
  extended to all frequencies, 416–419
  Gurevich et al. extension, 415
  Mavko–Jizba model, 413–419
standard deviation, 12
standard linear solid, 370
  see also viscoelasticity and Q
standard temperature and pressure (STP), 672
statistical classification, 20–21
statistics and probability, 12–21
  autocovariance, 15
  Bayes, 20, 21
  bootstrap, 19
  classification, 20–21
  coefficient of determination, 13
  coefficient of variation, 12
  correlation coefficient, 12, 13
  covariance, 12, 14
  distributions, 15–18
  expectation operator, 15
  mean, 12

statistics and probability (cont.)
    mean deviation, 12
    median, 12
    Monte Carlo, 18
    regression, 12–14
    standard deviation, 12
    variance, 12, 13
    variogram, 14, 15
stiff sand model, 354–356, 361
stiffness tensor, *see* elasticity
stochastic (Mie) scattering, 201–206
Stoneley waves in borehole, 211
strain concentration tensor, 103, 242, 248, 306
strain variance, 306
strain, uniform, 225
stratigraphic filtering, 185–189
    Floquet solution, 188
    O'Doherty–Anstey formula, 187
stratigraphic slowness, 185–189
strength, *see* unconfined compressive strength,
    empirical relations
stress-induced anisotropy, 70–79
stress intensity factors, 115–120
surface tension, 538
    capillary number, 568
    contact angle, 539
    Young's relation, 538
suspensions, see bounds, elastic; bounds, Reuss;
    Wood's formula
Sylvester's formula, 26
symmetries,
    cubic, 27, 28, 31, 47, 304
    Curie theorem, 33
    Hermann theorem, 34
    isotropic, 29, 32, 44, 47, 302–304
    monoclinic, 27, 28, 29
    Neumann's principle, 34
    orthorhombic, 28, 30, 56, 70, 298–300
    tetragonal, 27, 30, 303
    triclinic, 27, 44, 302
    trigonal, 27, 30, 303
    transversely isotropic (hexagonal), 27, 28, 31, 275,
    300–301
synthetic seismograms, *see* Kennett algorithm; layered
    media

tensor properties and operations, 25–36
    closest tensor of greater symmetry, 26–33
    see also coordinate transformation
    functions of tensors, 26
    invariants, 34–35, 40
    projection tensors, 32, 242
    Sylvester's formula, 26
    tensor contraction, 25
tensor transformation,
    see coordinate transformation
thinly layered (laminated) sand-shale systems, 331
    see also, fluid substitution

third-order elasticity, 64, 76
Thomas–Stieber model for sand-shale systems,
    329–335
Thomsen's weak anisotropy, 57–61, 77, 302
time-average equation, 478
time-domain reflectometry (TDR), 586
Timur equation, 557
Tixier equation, 557
T-matrix inclusion models, 269–275
    coherent potential approximation, 273
    optical potential approximation, 272
    self-consistent approximation, 224, 251–255, 273
Topp's relation, 586
torsional waves, *see* cylindrical rods
tortuosity, 368, 543, 552–556
Tosaya's velocity relations, 486
transfer function, see Fourier transform
transmissivity
    electromagnetic, 584
    seismic, 126–136
    see also AVO
transverse isotropy, *see* anisotropy, AVO, elasticity,
    NMO, phase velocity, reflectivity, symmetries
truncated exponential distribution, 17
Tsvankin's parameters, 62, 77

Udden–Wentworth scale, see granular media
uncemented (soft) sand model, 359–360
unconfined compressive strength (UCS) empirical
    relations, 515–517
undrained conditions, 82, 238
uniform distribution, 16

variance, 12, 13
variogram, 14, 15
velocity, dispersion, 124
    anomalous, normal, 124
    Biot model, 367–373
    BISQ model, 419–421
    Bragg scattering, 189
    Chapman *et al.* squirt model, 421
    characteristic frequency, 429, 441
    comparison among fluid-related mechanisms,
    427–433
    cylindrical rods, 206–211
    Dvorkin–Mavko attenuation model, 433–437
    electromagnetic, 582
    fast path effect, 198
    features common among models, 427–433
    Floquet solution, 188
    group, phase, energy velocities, 123–126
    heterogeneous media, 197–201
    layered poroelastic, 280–285
    O'Doherty–Anstey formula, 187
    poroelastic heterogeneity, 449–455
    resonant bar, *see* cylindrical bar
    saturation-related, 443
    scattering attenuation, 201–206

squirt model, 372, 413–421, 423–433
stratigraphic filtering, 185–189
stratigraphic slowness, 185–189
three-dimensional heterogeneous media, 197–201
waves in pure viscous fluid, 455
see also layered media; viscoelasticity and Q
velocity, electromagnetic
velocity, seismic, 121–219
  cylindrical rods, 206–211
  Biot, 367–373
  group, phase, energy velocities, 123–126
  Rayleigh wave, 122, 207, 211
  RMS, *see* NMO
  signal velocity, 126
  see also squirt model
  stratigraphic filtering, 185–189
  viscous fluid, in, 455
  see also anisotropy; empirical velocity relations; velocity dispersion
velocity-density relations, 508
velocity-porosity relation, *see* empirical velocity relations
Verniks V_P–V_S relations, 504
viscoelasticity and Q, 160–179
  attenuation, 166, *see also* quality factor
  Backus average, 275–285
  characteristic frequency, 166, 429, 441
  Cole–Cole model, 173
  common models, 165–174
  comparison among fluid-related mechanisms, 427–433
  constant Q model, 168
  creep function, 164
  dispersion, 174–179
  Dvorkin–Mavko attenuation model, 433–437
  fractional Maxwell model, 173
  group, phase, energy velocities, 123–126
  Havriliak–Negami model, 173
  Kramers–Kronig relations, 179–181
  logarithmic decrement, 162
  loss modulus, 165
  Maxwell solid, 161, 169, 177
  melt and magma, 571–574
  modulus, complex, 161
  modulus defect, 167, 428
  nearly constant Q model, 167
  Newtonian viscous fluid, 171, 176, 178
  particles and bubbles in viscoelastic background, 567–570
  power law creep, 173
  quality factor (Q), 162
  relaxation function, 163
  relaxation time, 171
  saturation-related, 443
  spectral ratio, 163

standard linear solid, 161, 165–167, 177, 370
storage modulus, 165
time domain, relaxation and creep, 163
viscosity, 165, 567
Voigt solid, 161, 171
wave number, complex, 174
wave velocity, 174–179
see also Kramers–Kronig relations; velocity dispersion
viscosity of reservoir fluids,
  see fluid properties, viscoelasticity and Q
viscous flow, 531
  circular pipe, 533–536
  elliptical pipe, 535–538
  parallel plates, 533
  rectangular pipe, 534
  triangular pipe, 534
viscous fluid
  see fluid properties, viscoelasticity and Q
  wave velocity in, 455
vitrinite reflectance, 520, 522, 623, 624
Voigt average, *see* bounds, elastic, Voigt
Voigt notation, 22, 45–63, 64
Voigt–Reuss–Hill average, 235, 292
Voigt solid, *see* viscoelasticity and Q
von Mises stress, 40
V_P–V_S relations, 488, 492–508
VTI, *see* AVO, elasticity, NMO, phase velocity, reflectivity, transverse isotropy

Walpole bounds, *see* bounds, elastic
Walton model, 346
water, *see* fluid properties
sea water, 659–662
wavelet, *see* Fourier transform
Waxman–Smits equations, 592
Waxman–Smits–Juhász model, 592
Weibull distribution, 18
White–Dutta–Odé model, 444–449
Williams V_P–V_S relation, 505
Wood's formula, 234
Wyllie's time-average equation, 478

Xu–White model, 506

y-transform, 229, 294, 397
Yin–Marion–Dvorkin–Gutierrez–Avseth model for sand-shale systems, 329–335
Young's modulus, *see* elasticity
Young's relation, 538

zero phase, *see* Fourier transform
Zimmer's unconsolidated sand data
Zoeppritz equations, *see* AVO